Lecture Notes in Computer Science 6741

Commenced Publication in 1973
Founding and Former Series Editors:
Gerhard Goos, Juris Hartmanis, and Jan van Leeuwen

Haralambos Mouratidis   Colette Rolland (Eds.)

Advanced Information Systems Engineering

23rd International Conference, CAiSE 2011
London, UK, June 20-24, 2011
Proceedings

 Springer

Volume Editors

Haralambos Mouratidis
University of East London
School of Computing, IT and Engineering
Docklands Campus, 4/6 University Way, E16 2RD London, UK
E-mail: H.Mouratidis@uel.ac.uk

Colette Rolland
Université Paris1 Panthéon Sorbonne
CRI
90 Rue de Tolbiac, 75013 Paris, France
E-mail: rolland@univ-paris1.fr

ISSN 0302-9743   e-ISSN 1611-3349
ISBN 978-3-642-21639-8   e-ISBN 978-3-642-21640-4
DOI 10.1007/978-3-642-21640-4
Springer Heidelberg Dordrecht London New York

Library of Congress Control Number: 2011928907

CR Subject Classification (1998): H.4, H.3, D.2, C.2, J.1, I.2

LNCS Sublibrary: SL 3 – Information Systems and Application, incl. Internet/Web and HCI

Typesetting: Camera-ready by author, data conversion by Scientific Publishing Services, Chennai, India

Printed on acid-free paper

Springer is part of Springer Science+Business Media (www.springer.com)

Preface

A warm welcome to the proceedings of the 23rd International Conference on Advanced Information Systems Engineering (CAiSE 2011)! The CAiSE series of conferences started in 1989 with the objective to provide a forum for the exchange of experience, research results, ideas and prototypes in the field of information systems engineering. Twenty-two years later, CAiSE has established itself as a leading venue in the information systems area for presenting and exchanging results of emerging methods and technologies that facilitate innovation and create business opportunities.

CAiSE 2011, held in London during June 20–24, 2011 continued this tradition. The theme of CAiSE 2011 was "Information Systems Olympics: Information Systems in a Diverse World." This year's CAiSE conference theme was linked to the coming London Olympic and Paralympic Games 2012, two international multi-sport events that bring together athletes from all continents to celebrate sporting excellence but also human diversity. Diversity is an important concept for modern information systems. Information systems are diverse by nature ranging from basic systems to complex ones and from small to large. The process of constructing such systems is also diverse ranging from ad-hoc methods to structured and formal methods. Diversity is also present among information systems developers, from novice to experienced. Moreover, the wide acceptance of information systems and their usage in almost every aspect of human life has also introduced diversity among users. Users are both novice and experienced and they demonstrate differences related to race, ethnicity, gender, socio-economic status, age, physical abilities, religious beliefs, and so on. It is therefore the responsibility of the information systems engineering community to engineer information systems that operate in such a diverse world.

CAiSE 2011 received 320 submissions, the largest number ever received in the CAiSE conference series. Most of the submissions came from Germany, Spain, Italy, France and China. Following an extensive review process, which included a Program Committee/Program Board meeting during February 13–14, 2011 in London, 42 submissions were accepted as full papers and 5 as short papers. Accepted papers addressed a large variety of issues related to the conference and were organized into ten themes: Requirements, Adaptation and Evolution, Model Transformation, Conceptual Design, Domain-Specific Languages, Case Studies and Experiences, Mining and Matching, Service and Management, Validation and Quality, Business Process Modeling. The program of the conference was also supplemented by a number of tutorials, 11 workshops, a Doctoral Consortium, the CAiSE Forum, and two working conferences. Two keynote speeches were delivered as part of the conference program. Anthony Finkelstein talked about "Open Challenges at the Boundaries Software Engineering and Information Systems," while Dimitrios Beis talked about "Information Systems for the

Olympics Games." Moreover, a panel discussed issues related to "Green and Sustainable Information Systems."

The organization and successful running of a large conference such as CAiSE would not be possible without the valuable help and time of a large number of people. As editors of this volume, we would like to express our gratitude to the Program Committee members, additional reviewers and the Program Board members for their valuable support in selecting the papers for the scientific program of the conference; to the authors of the papers for sending their work to CAiSE; to the presenters of the papers; and to the participants of the conference for their contribution. We also thank our sponsors and the General Chair and Chairs of the various CAiSE 2011 committees for their assistance in creating an exciting scientific program. We would also like to thank the local Organizing Committee at the University of East London for their hospitality and the organization of the social events of the conference.

March 2011

Colette Rolland
Haralambos Mouratidis

Organization

Advisory Committee

Arne Sølvberg	Norwegian University of Science and Technology, Norway
Janis Bubenko Jr.	Royal Institute of Technology, Sweden
Colette Rolland	Université Paris 1 Panthéon Sorbonne, France

General Chair

Pericles Loucopoulos	Loughborough University, UK

Program Chairs

Haralambos Mouratidis	University of East London, UK
Colette Rolland	Université Paris 1 Panthéon Sorbonne, France

Local Arrangements Chairs

Elias Pimenidis	University of East London, UK
Miltos Petridis	University of Greenwich, UK

Workshop and Tutorial Chairs

Oscar Pastor	Valencia University of Technology, Spain
Camille Salinesi	Université Paris 1 Panthéon Sorbonne, France

Forum Chair

Selmin Nurcan	Université Paris 1 Panthéon Sorbonne, France

Panel Chair

Barbara Pernici	Politecnico di Milano, Italy

Doctoral Consortium Chairs

Michel Léonard	Université de Genève, Switzerland
Bernhard Thalheim	Christian Albrechts University Kiel, Germany
Cornelia Boldyreff	University of East London, UK

Publication Chairs

Jolita Ralyté   Université de Genève, Switzerland
David Preston   University of East London, UK

Publicity Chairs

Rebecca Deneckere   Université Paris 1 Panthéon Sorbonne, France
Jaelson Castro   Universidade Federal de Pernambuco, Brazil
Leszek Maciaszek   Macquarie University, Australia
Kecheng Liu   University of Reading, UK
Keng Siau   University of Nebraska-Lincoln, USA

Finance Chair

Mohammad Dastbaz   University of East London, UK

Webmasters

Michalis Pavlidis   University of East London, UK
Sambhu Singh   University of East London, UK

Program Committee Board

Marco Bajec, Slovenia   Barbara Pernici, Italy
Nacer Boudjilida, France   Klaus Pohl, Germany
Eric Dubois, Luxembourg   Jolita Ralyté, Switzerland
Xavier Franch, Spain   Camille Salinesi, France
Marina Jirotka, UK   Janis Stirna, Sweden
Moira Norrie, Switzerland   Roel Wieringa, The Netherlands

Program Committee

Wil van der Aalst, The Netherlands   Boalem Benatallah, Australia
Peggy Aravantinou, Greece   Giuseppe Berio, France
Pär Ågerfalk, Sweden   Mokrane Bouzeghoub, France
Hans Akkermans, The Netherlands   Silvana Castano, Italy
Antonia Albani, The Netherlands   Jaelson Castro, Brazil
Daniel Amyot, Canada   Corine Cauvet, France
Paris Avgeriou, The Netherlands   Donna Champion, UK
Luciano Baresi, Italy   Vasilis Chrisikopoulos, Greece
Ahmad Barfourosh, Iran   Ioanna Constantiou, Denmark
Zohra Bellahsene, France   Panos Constantopoulos, Greece

Additional Referees

Alberto Abelló
David Aguilera-Moncusi
Saeed Ahmadi-Behnam
Naved Ahmed
Reza Akbarinia
Fernanda Alencar
Raian Ali
Christos Anagnostopoulos
Birger Andersson
Vasilios Andrikopoulos
Ion Androutsopoulos
Luca Ardito
George Athanasopoulos
Ahmed Awad
Daniele Barone
Saeed Ahmadi Behnam
Maria Bergholtz
Maxime Bernaert
Devis Bianchini
Riccardo Bonazzi
Boris Brandherm
Glenn J. Browne
Stephan Buchwald
Andrea Capiluppi
Amit Chopra
Remi Coletta
Ajantha Dahanayake
Fabiano Dalpiaz
Rébecca Deneckère
Olfa Djebbi
Vicky Dritsou
Fabien Duchateau
Rami Eid-Sabbagh
Golnaz Elahi
Amal Elgammal
Thibault Estier
Alfio Ferrara
Kunihiko Fujita
Matthias Galster
Dimitris Gavrilis
Andrew Gemino
Sepideh Ghanavati
Emmanuel Giakoumakis

Bas van Gils
Daniela Grigori
Irit Hadar
Stijn Hoppenbrouwers
Ela Hunt
Shareeful Islam
Lei Jiang
Rim Kaabi
Diana Kalibatiene
Christos Kalloniatis
Maya Kaner
Haki Kazem
Takashi Kitamura
David Knuplesch
Spyros Kokolakis
Jens Kolb
Takafumi Komoto
Panos Kourouthanassis
Eleni Koutrouli
Vera Kuenzle
Ales Kumer
Matthias Kunze
Andreas Lanz
Alexei Lapouchnian
Dejan Lavbic
Evaldas Lebedys
Francesco Lelli
Zhan Liu
Mathias Lohrmann
Linh Thao Ly
Alexander Lübbe
Manolis Maragoudakis
Michele Melchiori
Slim Mesfar
Marco Mesiti
Alexandre Métrailler
Wolfgang Molnar
Geert Monsieur
Stefano Montanelli
Gunter Mussbacher
Wanda Opprecht
Sami Ouali
Michael Pantazoglou

Table of Contents

Session 3: Model Transformation 1

Session 4: Conceptual Design 1

Session 5: Conceptual Design 2

Session 6: Domain Specific Languages

Session 7: Case Studies and Experiences

Session 8: Model Transformation 2

Session 9: Mining and Matching

Session 10: Business Process Modelling

Session 11: Validation and Quality

Session 12: Service and Management 1

Session 13: Service and Management 2

Session 14

Ten Open Challenges at the Boundaries of Software Engineering and Information Systems

Anthony Finkelstein

Department of Computer Science, University College London, UK

Abstract. In this talk, intended to provoke discussion, I will suggest ten important open challenges at boundaries where Software Engineering & Information Systems meet. I will focus on challenges are both intellectually demanding and of industrial importance. I will suggest some approaches to meeting these challenges and will lay stress upon the interdisciplinary opportunities they give rise to.

H. Mouratidis and C. Rolland (Eds.): CAiSE 2011, LNCS 6741, p. 1, 2011.
© Springer-Verlag Berlin Heidelberg 2011

Total Integration: The Case of Information Systems for Olympic Games

Dimitrios A. Beis

Global Event Experts

Abstract. The Olympic Games, the most demanding Sports Mega Event, have evolved through the years to complexity levels unparallel in any man designed process. Information Systems play the key role both as an enabler and critical performer for an Olympic Event if not surpassing at least in equal terms with the stars of the Games, the Athletes. Just data registering of records and performance of sports and athletes is not a viable option today for the Olympic Games. The total integration of all functions that make the 15 dream days of the Olympics is the target. Olympic family services, spectator services, cultural integration, media and broadcasting services, world participation and host city performance in transportation, crowd regulation and security must be supported in a first time easy to use, reliable and sub second response performance with 100% availability of all systems. The IS implemented in support of the last three Olympiads in Sydney, Athens and Beijing will be presented, critical issues of success factors analysed and expected developments projected.

H. Mouratidis and C. Rolland (Eds.): CAiSE 2011, LNCS 6741, p. 2, 2011.
© Springer-Verlag Berlin Heidelberg 2011

Requirements Management with Semantic Technology: An Empirical Study on Automated Requirements Categorization and Conflict Analysis

Thomas Moser, Dietmar Winkler, Matthias Heindl, and Stefan Biffl

Christian Doppler Laboratory
Software Engineering Integration for Flexible Automation Systems
Institute of Software Technology and Interactive Systems
Vienna University of Technology, Vienna, Austria
{firstname.lastname}@tuwien.ac.at

Abstract. Requirements managers aim at keeping the set of requirements consistent and up to date throughout the project by conducting the following tasks: requirements categorization, requirements conflict analysis, and requirements tracing. However, the manual conduct of these tasks takes significant effort and is error-prone. In this paper we propose to use semantic technology as foundation for automating the requirements management tasks and introduce the ontology-based reporting approach OntRep. We evaluate the effectiveness and effort the OntRep approach based on a real-world industrial empirical study with professional Austrian IT project managers. Major results were that OntRep provides reasonable capabilities for the automated categorization of requirements, was when compared to a manual approach considerably more effective to identify conflicts, and produced less false positives with similar effort.

Keywords: Requirements categorization, requirements conflict analysis, consistency checking, requirements tracing, case study, empirical evaluation.

1 Introduction

A major goal of requirements engineering is to achieve a common understanding on the set of requirements between all project stakeholders. Modern IT projects are complex due to the high number and complexity of requirements, and geographically distributed project stakeholders with different backgrounds and terminologies. Therefore, adequate requirements management (ReqM) tools are a major contribution to address these challenges. Current ReqM tools typically work with a common requirements database, which can be accessed by all stakeholders to retrieve information on requirements content, state, and interdependencies.

ReqM tools help project managers and requirements engineers to keep the overview on large amounts of requirements by supporting: (a) *Requirements categorization* by clustering requirements into user-defined subsets to help users find relevant requirements more quickly, e.g., by sorting and filtering attribute values; (b) *Requirements conflict analysis* (or consistency checking) by analyzing requirements from different

H. Mouratidis and C. Rolland (Eds.): CAiSE 2011, LNCS 6741, pp. 3–17, 2011.
© Springer-Verlag Berlin Heidelberg 2011

stakeholders for symptoms of inconsistency, e.g., contradicting requirements; and (c) *Requirements tracing* by identifying dependencies between requirements and artifacts to support analyses for change impact and requirements coverage. Unfortunately, ReqM suffers from the following challenges and limitations:

- Incompleteness [7] of requirements categorization and conflict identification, in particular, when performed manually.

- High human effort for requirements categorization, conflict analysis and tracing, especially with a large number of requirements [7].

- Insufficient completeness [6] for conflict analysis and tracing with automated approaches.

- Tracing on syntactic rather than on concept level: requirements are often traced on the syntactic level by explicitly linking requirements to each other. However, requirements engineers actually want to trace concepts, i.e., link requirements based on their meaning, which can be achieved only partially by information retrieval approaches like "keyword matching" [12] [13].

The use of semantic technologies seems promising to address these challenges: Ontologies provide the means for describing the concepts of a domain and the relationships between these concepts in a way that allows automated reasoning [18]. Automated reasoning can support tasks for requirements categorization, requirements conflict analysis, and requirements tracing.

In this paper, we propose OntRep, an automated ontology-based reporting approach for requirements categorization, conflict analysis and tracing based on ontologies and semantic reasoning mechanisms. The main criteria for the evaluation are: correctness and completeness of identified requirements conflicts, effort to develop a project or domain ontology. OntRep aims at lowering the effort for requirements management, while keeping high requirements consistency.

The OntRep approach automatically categorizes requirements into a given set of categories using ontology classes modeled in Protégé and mapping the terms used in the requirements to these classes. Further, OntRep analyzes the content of the requirements and identifies conflicts between requirements. Therefore, conflict analysis is not only based on traditional keyword-matching-approaches, but can also work when different terminologies are used for requirements formulation.

We empirically evaluate OntRep with a real-life project at Siemens Austria, where six project managers in two teams (a) categorized the requirements of the case study project into a set of categories and (b) inspected the given project requirements to identify conflicts between requirements. A requirements engineering expert provided control data for all tasks. Then, we performed the same tasks with OntRep to compare the effort necessary and the quality of results.

The remainder of the paper is organized as follows: Section 2 summarizes related work on requirements categorization, conflict analysis, tracing, and natural language processing technologies; Section 3 introduces the OntRep approach and motivates research issues. Section 4 outlines the case study and Section 5 presents results. Section discusses the results, concludes and suggests further work.

2 Related Work

This section presents related work on natural language processing technologies as foundation for automating the ReqM tasks requirements categorization, conflict analysis, and requirements tracing approaches.

2.1 Requirement Conflicts Detection and Requirements Tracing

Requirements conflict with each other if they make contradicting statements about common software attributes [7]. Requirements authors may use different terminologies for specifying requirements, although the terms used can be derived from the same common concepts.

- In principle there are the following main strategies to identify and eliminate requirements conflicts: Negotiation methods, where stakeholders manually (or with tool support) categorize, discuss, and analyze requirements for conflicts, such as the win-win requirements negotiation approach [1] or its tool-supported variant easy-win-win [2],

- Automation approaches for conflict analysis ([4][6][12]) that use tools to analyze requirements consistency in order to reduce human effort.

"Given that there may be up to n^2 conflicts among n requirements, the number of potential conflicts, could be enormous, burdening the engineer with the time-intensive and error-prone task of identifying the true conflicts" [7]. Several approaches address the issue of automated requirements conflict identification:

The *Trace Analyzer* by Egyed and Grünbacher [7] analyzes the footprint of test cases to generate trace dependencies. If two requirements affect the same part of a system, then their test runs execute overlapping lines of code. Trace dependencies and potential conflicts can be identified among requirements, if their test scenarios execute the same lines of code. However, the Trace Analyzer needs executable code to identify requirements conflicts, which is often not available in early project phases, when conflict analysis is a major goal.

Heitmeyer et al. [11] describe a formal analysis technique, called *consistency checking*, for the automated detection of errors, such as type errors, non-determinism, missing cases, and circular definitions, in requirements specifications. The approach only considers syntactical consistency and does not address semantic conflicts.

Automated requirements tracing approaches are also relevant for requirements conflict analysis: requirements tracing deals with identifying interdependencies between requirements [10] and conflicts between two requirements can be seen as a particular type of interdependency, i.e., tracing is a precondition for conflict analysis. There are reports on several trace automation approaches, such as Egyed's scenario-driven approach to traceability [6], Jackson's key-phrase-based traceability scheme [12]. Further, there are the heterogeneous traceability approach Cleland-Huang et al. [4], and approaches by Pinheiro et al. [20], Leuser [14], and McMillan et al. [16]. These approaches use different techniques to identify requirements interdependencies. Some of them require executable code, so they cannot be used for the identification of interdependencies in early project phases when there is no sufficient code base.

Within these trace automation approaches, information retrieval approaches, such as the RETH approach [13], seem of particular interest as they use keyword-matching techniques to identify requirements interdependencies. However, these techniques do not allow identifying conflicts or other interdependencies between requirements, if they use different terms for similar concepts. In practice these approaches are less effective, because they cannot identify the full set of interdependencies between requirements.

The extended Bakkus-Naur-Form (EBNF) [21] (see Fig. 2) is a general formal language description approach, which is used in the field of requirements analysis to improve the understandability of requirements for humans and machines. EBNF requirements templates contain mandatory and optional elements, e.g. conditions, obligations, actors, process verbs, which are the basis for clear requirements statements.

2.2 Natural Language Processing

Natural language processing (NLP) techniques are useful to parse and extract structure and content of requirements given in natural language for transformation into the structure of an ontology. NLP generally refers to a range of theoretically motivated and computational techniques for analyzing and representing naturally occurring texts [3]. The core purpose of NLP techniques is to achieve human-like language processing for a range of tasks or applications [15].

The core NLP models used in this research are part-of-speech (POS) tagging and sentence parsers [3]. POS tagging involves marking up the words in a text as corresponding to a particular part of speech, based on both its definition, as well as its context. In addition, sentence parsers transform text into a data structure (also called parse tree), which provides insight into the grammatical structure and implied hierarchy of the input text [3]. *Standford parser/tagger*[1] and *OpenNLP*[2] are the core set of NLP tools used in this research.

Another tool that can be used is *WordNet*, a large lexical database of English [17]. Nouns, verbs, adjectives and adverbs are grouped into sets of cognitive synonyms (synsets), each expressing a distinct concept. Synsets are interlinked by means of conceptual-semantic and lexical relations. WordNet is a useful building block for requirements analysis (see below).

These NLP technologies can be used for our purpose, namely to improve the effectiveness of requirements management activities like categorization, conflict analysis, and tracing.

3   Ontology-Based Reporting

Due to the limitations of requirements analysis approaches that address only links between requirements based on syntactic equality, we explore in this work an approach based on semantic equality, i.e., OntRep links similar concepts, if they share the same meaning even if their syntactic representations are different. As ontologies are versatile for representing knowledge on requirements and for deriving new links

---

[1] http://nlp.stanford.edu/software/lex-parser.shtml
[2] http://opennlp.sourceforge.net

between requirements, we introduce an ontology-based approach for reporting analysis results on a set of requirements, so-called ontology-based reporting.

The goal of the ontology-based reporting approach *OntRep* is making ReqM tasks like requirements categorization, conflict analysis and requirements tracing more efficient based on the automation of selected steps in these tasks. The following subsections provide an overview on the approach and motivate research issues.

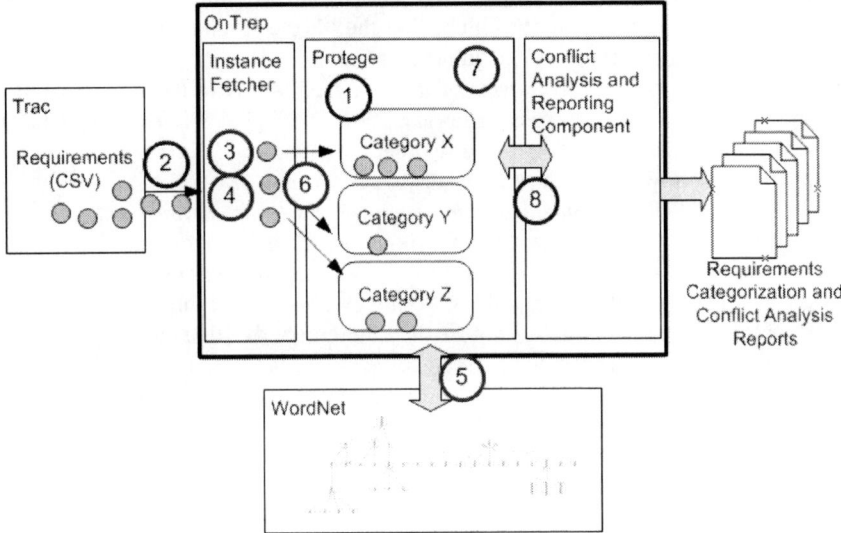

Fig. 1. Components and numbered steps of OntRep

We developed a prototype tool for the *OntRep* approach a plug-in to *Trac*[3], an open source collaboration platform consisting of a Wiki, ticket management system, and subversion integration, which can be extended by Python plug-ins.

Fig. 1 illustrates the *OntRep* tool (together with *Protégé*) consisting of two main components: 1) *Instance fetcher* takes input data, e.g., requirement tickets from *Trac*, analyzes their contents and assigns them requirements categories (classes) defined in the ontology; 2) *Reporting component* reasons on the input data and generates a requirements conflict report based on the analyzed requirements.

3.1 Semantic Requirements Categorization

In a first phase natural language texts have to be linked to semantic categories as preparation for further analysis and reporting. The following steps automate requirements categorization with OntRep (see numbered circles in Fig. 1):

1) *Define the requirement categories* in Protégé, e.g., categories X, Y, Z. Each category is defined as an ontology class in Protégé. It is important to define project-relevant "semantic" categories and not formal ones in order to enable the automated

---

[3] http://trac.edgewall.org/

categorization, e.g., "Security". Typically, these categories can be defined based on a project glossary that contains important project-specific terms.

2) *Provide input data* to be categorized: Requirements are represented as tickets in *Trac*. For our research prototype we export these requirements (the small grey circles in Fig. 1) via CSV from *Trac* and import them into the instance fetcher.

3) *Remove irrelevant stop-words*, like "and", "any", "but", which cannot be used for categorization. This step is performed automatically using a standard stop-word list[4].

4) *Bring all remaining words into their root form*: this process is called "stemming" based on a well-known algorithm, like the "Porter Stemmer" algorithm [19]. An example is to stem "jumping" to "jump".

5) *Get all synonyms and hyponyms* of the analyzed words in the requirements by using the natural language processing library "WordNet" [17]. For example, "house" is a synonym for "building", "dog" is a hyponym of "animal". Further check all relevant substrings of a word like "net" as a substring of "network".

6) *Heuristic-based assignment of each requirement to the defined categories* depending on the number of hits for 1) synonyms, 2) hyponyms and 3) substring matches. The heuristic checks if the hits for synonym, hyponym and substring matches meet the given threshold values. So the number of met thresholds is between 0 and 3. If this number is equal of higher than the number of thresholds that must be met, the word will be related to that category, otherwise not. If several categories reach these thresholds, the requirement will be categorized into all of these categories (multi-dimensional categorization is allowed)

7) *Save the element as an individual of the ontology class*, if it is not already in the class. This can only be checked if one or more of the elements attributes have been declared as primary keys (uniquely identifying the element). If the element has already been saved in another class as well (which could be the case), declare that the new element is the same as the already existing one with the *"owl:sameAs"* property.

8) *Semantic requirements conflict analysis:* If the requirements are formally described using a specified grammar (e.g., EBNF), the information contained in the textual requirement descriptions can be semantically analyzed in order to identify possible inconsistencies and/or conflicts, see subsection 3.2.

3.2 Semantic Conflict Analysis

In the second phase, analysis and reporting approaches build on the mapping of requirements to semantic categories. For formally specified requirement semantics, in our case following an EBNF template (see Fig. 2), semantic analysis can identify inconsistencies and conflicts using a set of assertions that should hold true for all available facts. These assertions are based on the available requirements, while the available facts are based on the environment and properties of the target system.

Fig. 3 depicts examples for conflicts between requirements (the CRR conflict type explained in section 4), e.g. the third conflict contains an inconsistency between requirements nr. 16 and nr.13: The "thing to be processed" part of requirement nr. 16 contains a value of 30 for "number of index updates", whereas requirement nr. 13 contains the value 20 for "number of index updates, which finally is a requirements conflict.

---

[4] http://www.textfixer.com/resources/common-english-words.txt

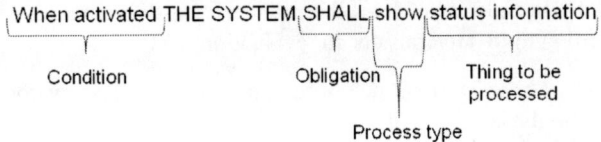

Fig. 2. EBNF requirements structure (sentence level) [21]

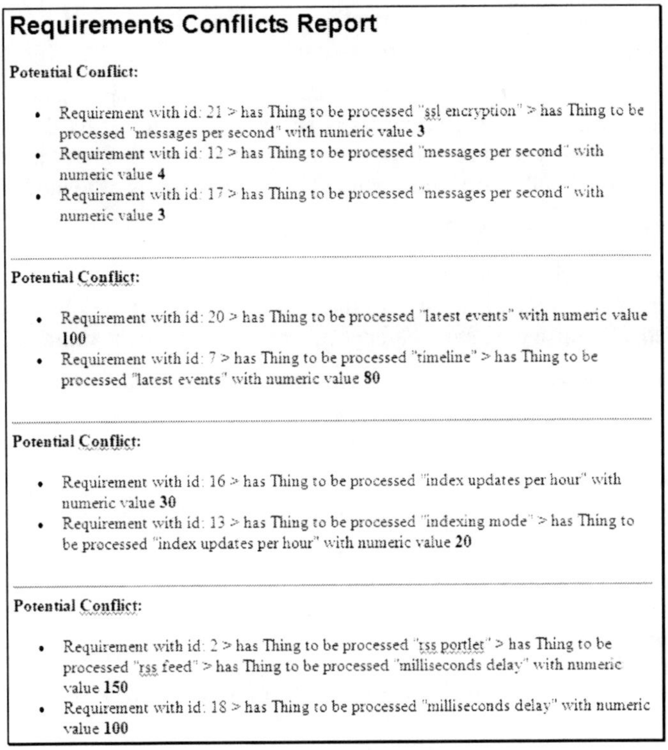

Requirements Conflicts Report

Potential Conflict:

- Requirement with id: 21 > has Thing to be processed "ssl encryption" > has Thing to be processed "messages per second" with numeric value **3**
- Requirement with id: 12 > has Thing to be processed "messages per second" with numeric value **4**
- Requirement with id: 17 > has Thing to be processed "messages per second" with numeric value **3**

Potential Conflict:

- Requirement with id: 20 > has Thing to be processed "latest events" with numeric value **100**
- Requirement with id: 7 > has Thing to be processed "timeline" > has Thing to be processed "latest events" with numeric value **80**

Potential Conflict:

- Requirement with id: 16 > has Thing to be processed "index updates per hour" with numeric value **30**
- Requirement with id: 13 > has Thing to be processed "indexing mode" > has Thing to be processed "index updates per hour" with numeric value **20**

Potential Conflict:

- Requirement with id: 2 > has Thing to be processed "rss portlet" > has Thing to be processed "rss feed" > has Thing to be processed "milliseconds delay" with numeric value **150**
- Requirement with id: 18 > has Thing to be processed "milliseconds delay" with numeric value **100**

Fig. 3. Example report on requirements conflicts

3.3 Research Issues

The underlying idea of this research is to use advanced semantic technologies, like ontologies and reasoning mechanisms, to increase the effectiveness and efficiency of ReqM activities. In a large software project, tasks like requirements categorization, conflict analysis, and tracing would need human effort and duration that often prohibits their use in practice. Therefore, software projects often end up with (a) unstructured requirements and (b) conflicts that get discovered late and expensively.

In this context, the main research question of this paper is: To what extent can a semantic-technology-based approach, like OntRep, increase the effectiveness and efficiency of requirements categorization and conflict analysis compared to a traditional manual approach? In order to address the research question we derived the following variables (according to [8]) to consider for evaluation:

- The *number of requirements* determines the effort necessary for categorization and conflict analysis.

- *Number of requirement categories* used to categorize the requirements. Further, the *total number of true requirements conflicts* existing in a list of requirements, which can be identified by various approaches for conflict detection. This is a baseline measurement for the effectiveness of an approach, i.e., a perfect approach would find 100% of the true requirements conflicts.

- *Approach for categorization/conflict analysis*: e.g., for automation approaches the formal structure of requirements is an important factor. As described above, we use the EBNF template for specifying requirements. Using plain text or other formats probably affects the correctness and completeness of identified conflicts.

  Dependent variables that we want to study by the evaluation are:

- *Number of conflicts identified*. This number consists of two measures: recall (number of correctly identified conflicts), for measuring the effectiveness of an approach, and false positives (number of wrongly identified conflicts).

- *True conflicts that have not been identified* (false negatives): subtracting the *number of correctly identified conflicts* from the *total number of true requirements conflicts* tells us about the recall of conflict identification.

- *Plausibility of requirements classification*: regarding categorization two kinds of error can occur: (1) requirements have been assigned to a wrong category, and (2) requirements have not been assigned to a category they actually belong to. In order to measure these parameters, we take the manual categorization results of a requirements engineering expert as reference. In addition, we count the *number of requirements in each category*.

Besides these parameters we also record the effort for requirements categorization and conflict analysis. This includes *preparation effort* (e.g., creating the ontology that is used for categorization and conflict analysis), *categorization effort*, and *conflict analysis effort*. The case study is described in detail in the following section.

4 Case Study Description

The following subsections describe the characteristics of the pilot study design.

Study Subject. The case study project "Technoweb 2.0" is an IT development project with the goal to design and implement a web application that serves as a platform for communication and networking between technology experts within Siemens. This platform builds on the Java technology *Liferay*, where *portlets* act as components of the web application. The project is performed in an agile way using the software development process "SCRUM" and the configuration & project management platform *Trac*. In Trac all requirements, tasks and bugs are stored as tickets.

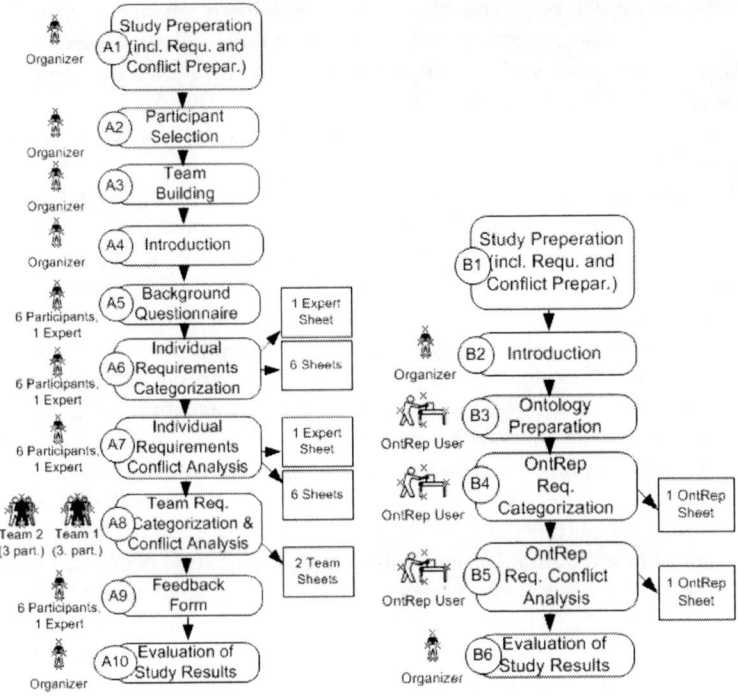

Fig. 4. Evaluation Setting: Manual (left), OntRep (right)

Study Design, Material, Participants, and Process. We applied the standard practices of empirical software engineering research according to Freimut et al. [8] and Wohlin et al. [22]. Fig. 4 illustrates the steps of the empirical study.

A1, B1) *Study preparation:* This step dealt with the creation/preparation of all *artifacts* necessary for the evaluation: 23 requirements in EBNF syntax, 8 categories as input for the requirements categorization step, deployment of 22 seeded conflicts (based on typical requirements conflicts found in practice at Siemens Austria). Further, we used questionnaires to capture the individual background experience of the participants and a feedback questionnaire to capture, whether the participants found the approach useful and usable.

A2, A3) *Participant selection and team building*: There were 6 participants who performed the manual categorization and conflict analysis tasks. They had similar experience on project management (3 to 5 years) and on requirements management but advanced general software engineering know-how. In addition, we had one expert with deeper know-how and experience, especially in ReqM. Finally, there was one OntRep tool user. He was well familiar with OntRep and had similar experience as the other participants. As described below, the evaluation consisted of individual work and team work. For the latter, the teams were assigned randomly.

A4, B2) *Introduction (Guidelines and Data Collection):* Before execution of tasks the study organizer introduced the participants to the project and the manual requirements categorization and conflict detection tasks. Further, the participants were guided step-by-step through the requirements classification and conflict detection process. The participants were sitting in one room without talking to each other. In the team phase, the two teams (3 participants each) worked in separated rooms. The expert, as well as the OntRep tool user, also worked separately.

A5) *Background questionnaire:* before they started with the actual ReqM tasks, the participants filled in the questionnaire.

A6) *Individual requirements categorization:* Then the participants read through the 23 given requirements. The participants individually categorized the requirements into one or more of the given 8 categories. Each participant conducts the requirements categorization individually. In addition, one Requirements Engineering expert also does the categorization. The time needed by each participant is captured.

A7) *Individual requirements conflict analysis:* In addition to the 23 requirements that had been categorized before, further elements were displayed (as rows below the other requirements), namely: 11 constraints (technical and business), and 4 formal documentation rules (documentation guidelines).

The participants again read through the task description and then had to identify conflicts and enter them into the sheet. A conflict can have one of the following types: conflict between requirements (CRR), conflict of a requirement with a constraint (CRC), conflict of a requirement with a formal guideline, i.e., ill-formed requirement (CRG). In total, the case study data contained: 5 conflicts of type CRR, 7 of type CRC, and 10 of type CRG. After the evaluation of the manual approach, we again have the 6 individual results. Again, one Requirements Engineering expert also conducted the conflict analysis. The effort needed by each participant was captured.

A8) *Team requirements categorization & conflict analysis:* Afterwards, the participants harmonized their individual results within 2 randomly assigned groups. Effort was captured for this task. The results are 2 team sheets.

A9) *Feedback forms:* filled in at the end by the participants.

A10, B6) *Evaluation of study results:* The manually created results of the expert and the teams were then compared with the result generated by OntRep.

The process for the automated approach is:

B3) *Ontology preparation:* A tool expert created one ontology class in OntRep (Protégé) for each category and then imported the given requirements from Trac as CSV into OntRep.

B4) *OntRep requirements categorization:* The tool then executed the categorization and generated a final result. We captured the effort to create the ontology classes and to generate the final report.

B5) *OntRep requirements conflict analysis:* Then, we again provided the requirements as CSV-input to OntRep. Further, the tool expert had to model the constraints as facts and the formal guidelines as rules in the ontology. We captured the effort for this. Then, the tool executed the conflict identification and generated a final report.

Data Capturing Analysis and Statistical Evaluation. Finally, we analyzed and evaluated the following results: (a) 6 spreadsheets for requirements categorization and 6 spreadsheets for conflict analysis from each of the 6 individual participants, (b) 1 categorization spreadsheet and 1 conflict analysis spreadsheet from a requirement engineering expert, and finally (c) 1 categorization spreadsheet and 1 conflict analysis spreadsheet created with the OntRep approach. The results were evaluated with descriptive statistics in Excel and R and are described in the following section.

5 Results

The following subsections describe the results of the pilot study regarding requirements categorization and conflict analysis.

5.1 Requirements Categorization

In order to evaluate the requirements categorization task, we took the categorization result of the requirements engineering expert as reference solution for comparing the results of the manual and automated approaches.

Table 1. Results of manual and automated req. categorization

	Individuals avg./std.dev.	Groups avg./std.dev.	OntRep
	MANUAL		AUTOMATED
1. Overfulfilled	9.5/3.9	**12.5/3,5**	6.0
2. Correct	5.7/2.7	6.0/2.8	**8.0**
Sum 1. & 2.	15.2/2.3	**18,5/0.7**	14.0
3. Partly correct	2.0/1.1	0.0/0.0	2.0
4. False	5.8/1.7	**4.5/0.7**	7.0

Table 1 summarizes the results of the manual and automated requirements categorization approaches: the rows in the table contain the quality levels of the categorization: "overfulfilled" means that a requirement was categorized into all correct categories but also into one or more additional ones, "correct" means that a requirement was categorized in the right categories. "Partially correct" means that a requirement was categorized in some but not all of the correct categories, "false" means that a requirement was categorized into wrong categories but not the right ones. The group results are better than the individual results: the number of false categorizations is reduced, and the number of correct and overfulfilled categorization is increased. Overfulfillment is not a problem, because all requirements are categorized into the right categories, and into some more categories, but this is just additional information which is allowed.

Categorization with OntRep was more accurate, i.e., 8 requirements (more than with the manual approach) have been categorized into the right categories without categorizing them in additional categories. On the other hand, comparing the sum of correctly categorized requirements (overfulfilled + correct) shows the lowest value for automation. Further, the number of false categorizations is also the highest. This is due to the fact, 4 requirements were not categorized at all. The reason therefore is that the terms used in these requirements could not be mapped to the categories, neither through substrings, synonyms or hyponyms.

The average effort for manual categorization was around 15 min per person. The group work took ca. 12 min in addition, resulting in an additional group effort of 36 person minutes. With OntRep the following preparations were necessary to enable the automated categorization: conversion of requirements into EBNF form (30 min.), preparation of ontology classes and user-defined synonyms (14 min.). After this, the run time for categorization was ca. 2 minutes. If the requirements exist in EBNF form, which is the case for some larger projects at Siemens Austria, the effort is similar to the manual average effort of manual categorization, but much more scalable.

5.2 Requirements Conflict Analysis

We analyzed conflicts of the three types described above, because conflicts of this type can be modeled in OntRep by means of facts and rules. The OntRep results for these conflicts are complete: all 22 conflicts of the defined conflict classes in the given data were identified, because OntRep works reliably, when the following pre-requisite are met: requirements exist in EBNF as input via CSV, modeling of glossary terms in ontology (10 min.), modeling facts, constraints and rules in the ontology (46 min.). Therefore, the total OntRep preparation time is 100 min. for the given case. The overall report generation took 4 minutes.

Table 2. Results of conflict detection capability analysis

	# correctly identified conflicts (avg./ std.dev.)	Avg. % of true conflicts found (avg./ std.dev.)	# conflicts found (avg./ std.dev.)	False positives in % (avg./ std.dev.) of # conflicts found
Individuals	7.0/3.9	31.8/18	17.0/6.8	58.8/22.0
Groups	10.5/0.7	47.7/3.0	21.5/0.7	51.2/4.9
Expert	15.0	68.2	17.0	11.8
OntRep	**22.0**	**100.0**	**22.0**	**0.0**

In comparison, the manual conflict analysis approach resulted in a lower completeness (see Table 2): the individual participants identified only 31.8% of existing conflicts on average. The harmonization of results within the groups brought an improvement to 47.7%, which means that approximately 3 additional conflicts have been identified by merging of the individual results into one group result. Also the number of false positives was slightly reduced by 1. The correctness of the manual approach was also lower than with the OntRep approach: 58.8% of identified conflicts were false positives. This percentage could only slightly be reduced by the group harmonization. i.e., ca. 1-2 false positives were been eliminated during team work.

In addition to comparing the individual and group results with OntRep results, but also had one expert performing the conflict analysis. Compared to the other participants and the team results, he provided the best results, i.e., the highest number of correctly identified traces, and the lowest percentage of false positives. Regarding effort, the expert was also the best with the manual approach: He needed 45 min. for conflict analysis, whereas the other participants needed 97 min. on average. In addition the group phase took 37 min., resulting in an additional group effort 111 person minutes.

5.3 Threats to Validity

We addressed threats internal validity [10] of the study by two measures: a) intensive reviews of the study concept and materials, and b) a test run of the study conducted by a test person in order to make sure that the guidelines, explanations, and task descriptions are understandable for the participants and to estimate the required effort/time frame. Regarding external validity [30], we performed this initial case study in a professional context at a software development company. The participants had medium requirements management know-how and advanced software engineering know-how. In addition, we had a requirements engineering expert as experimental "control group". Nevertheless, the small number of participants might limit the generalization of results. Therefore, we suggest replicating the study in a larger context.

Further, the requirements in this case study were formulated using the EBNF syntax, which is a major condition for OntRep to analyze the requirements. We did not yet analyze the quality of results with a set of requirements, which is not or only partially formulated in EBNF. Further studies are needed to evaluate this.

6 Discussion and Conclusion

Software and systems engineering projects are complex due to the increasing number and complexity of requirements, and the project participants with different domain backgrounds and terminologies. To keep the overview on requirements, project managers conduct requirements categorization, conflict analysis, and tracing. However, the manual conduct of these tasks takes significant effort and is error-prone.

In this paper we proposed semantic technology as foundation for automating the requirements management tasks and introduced the automated ontology-based reporting approach OntRep based on a project ontology and a reasoning mechanism. We used requirements formulated in EBNF as input to the proposed OntRep approach, which supports automated requirements *categorization* and requirements *consistency checking*. We evaluated the effectiveness and effort the OntRep approach based on a real-world industrial case study with 6 project managers in 2 teams. The study focused on requirements categorization and requirements conflict analysis. During the evaluation the study participants a) categorized the requirements of the case study project into a set of categories and b) inspected the given project requirements to identify conflicts between requirements. In addition a requirements expert and an OntRep user performed the same tasks to enable comparing the quality of results and the effort for all activities.

The case study results suggest that OntRep can be an attractive alternative for requirements categorization in typical software development projects, because it provides slightly lower effectiveness with similar effort compared to manual approaches, but much more scalable. OntRep's performance can be increased by adding additional, synonyms or hyponyms to the ontology (which has to be done manually at the moment), so that all used terms in requirements can be mapped to categories. Regarding conflict analysis, OntRep found all conflicts in the requirements during the empirical study, while manual conflict analysis identified only 50 to 60% of the conflicts and produced more false positives with similar effort. OntRep analyzes three types of conflicts at the moment: conflicts between requirements, conflicts between requirements and some constraints, or conflicts of requirements with some formal guidelines. The OntRep automation approach seems beneficial for project managers who want to manage their requirements with less effort, but in the same turn keep the requirements consistency high. Using the OntRep approach, organizations in software development projects could benefit from reduced manual effort for categorization and conflict analysis, and reduced communication and clarification effort through semi-automated semantic conflict analysis support.

Further work will focus on the replication of this pilot study in a larger context, i.e., with more participants to improve the external validity of results. In addition, we want to increase the number of requirements to be categorized and analyzed for conflicts in order to analyze the correctness, completeness, and especially the effort for larger sets of requirements. We assume that especially the efficiency of OntRep will improve with the number of requirements when compared to a manual approach. Another aspect is to adapt OntRep for application to a set of requirements.

Acknowledgments. We want to thank Alexander Wagner for the prototype implementation of the OntRep concepts and his support during the pilot study. This work has been supported by the Christian Doppler Forschungsgesellschaft and the BMWFJ, Austria.

References

1. Boehm, B., In, H.: Identifying Quality-Requirement Conflicts. IEEE Software (1996)
2. Briggs, R.O., Grünbacher, P.: EasyWinWin: Managing Complexity in Requirements Negotiation with GSS. In: Proceedings of the 35th Hawaii International Conference on System Sciences (2002)
3. Choi, F.Y.Y.: Advances in domain independent linear text segmentation. In: Proceedings of the 1st North American Chapter of the Association for Computational Linguistics Conference. Morgan Kaufmann Publishers Inc., Seattle (2000)
4. Cleland-Huang, J., Zemont, G., Kukasik, W.: A Heterogeneous Solution for Improving the Return on Investment of Requirements Traceability. In: 12th IEEE Int. Conf. on Requirements Engineering (2004)
5. Cruz, I.R., Huiyong, X., Feihong, H.: An ontology-based framework for XML semantic integration. In: International Database Engineering and Applications Symposium (IDEAS 2004), pp. 217–226. IEEE, Los Alamitos (2004)
6. Egyed, A.: A Scenario-Driven Approach to Traceability. In: Proceedings of the 23rd International Conference on Software Engineering (ICSE), Toronto, Canada, pp. 123–132 (2001)
7. Egyed, A., Grünbacher, P.: Identifying Requirements Conflicts and Cooperation: How Quality Attributes and Automated Traceability Can Help. IEEE Software (2004)
8. Freimut, B., Punter, T., Biffl, S., Ciolkowski, M.: State-of-the-Art in Empirical Studies, Report: ViSEK/007/E, Fraunhofer Inst. of Experimental Software Engineering (2002)

9. Gangemi, A., Guarino, N., Masolo, C., Oltramari, A.: Sweetening WordNet with DOLCE. AI Magazine 24(4), 13–24 (2003)
10. Gotel, O., Finkelstein, A.C.W.: An analysis of the requirements traceability problem. In: 1st International Conference on Requirements Engineering, pp. 94–101 (1994)
11. Heitmeyer, C.L., Jeffords, R.D., Labaw, B.G.: Automated consistency checking of requirements specifications. In: 2nd International Symposium on Requirements Engineering (RE 1995), York, England (1995)
12. Jackson, J.: A Keyphrase Based Traceability Scheme. IEEE Colloquium on Tools and Techniques for Maintaining Traceability During Design, 2-1-2/4 (1991)
13. Kaindl, H.: The Missing Link in Requirements Engineering. ACM SigSoft Software Engineering Notes 18(2), 30–39 (1993)
14. Leuser, J.: Challenges for semi-automatic trace recovery in the automotive domain. In: Proceedings of the ICSE Workshop on Traceability in Emerging Forms of Software Engineering, TEFSE (2009)
15. Liddy, E.D.: Natural Language Processing, 2nd edn. Encyclopedia of Library and Information Science. Marcel Decker, Inc., NY (2001)
16. McMillan, C., Poshyvanyk, D., Revelle, M.: Combining textual and structural analysis of software artifacts for traceability link recovery. In: Proceedings of the ICSE Workshop on Traceability in Emerging Forms of Software Engineering, TEFSE (2009)
17. Miller, G.A.: WordNet: A Lexical Database for English. Communications of the ACM 38(11), 39–41 (1995)
18. Pedrinaci, C., Domingue, J., Alves de Medeiros, A.K.: A core ontology for business process analysis. In: Bechhofer, S., Hauswirth, M., Hoffmann, J., Koubarakis, M. (eds.) ESWC 2008. LNCS, vol. 5021, pp. 49–64. Springer, Heidelberg (2008)
19. van Rijsbergen, C.J., Robertson, S.E., Porter, M.F.: New models in probabilistic information retrieval, British Library Research and Development Report, no. 5587 (1980)
20. Pinheiro, F.A.C., Goguen, J.A.: An Object-Oriented Tool for Tracing Requirements. IEEE Software 13(2), 52–64 (1996)
21. Rupp, C.: Requirements Engineering und –Management. Hanser (2002)
22. Wohlin, C., Höst, M., et al.: Controlled Experiments in Software Engineering. Journal for Information and Software Technology, 921–924 (2001)

S³C: Using Service Discovery to Support Requirements Elicitation in the ERP Domain

Markus Nöbauer[1], Norbert Seyff[2], Neil Maiden[3], and Konstantinos Zachos[3]

[1] InsideAX GmbH, Lunzerstraße 64, 4031 Linz, Austria
markus.noebauer@insideAx.at
[2] University of Zurich, Requirements Engineering Research Group, Zurich, Switzerland
seyff@ifi.uzh.ch
[3] City University London, Centre for HCI Design, London EC1V 0HB, UK
n.a.m.maiden@city.ac.uk, kzachos@soi.city.ac.uk

Abstract. Requirements Elicitation and Fit-Gap Analysis are amongst the most time and effort-consuming tasks in an ERP project. There is a potentially high rate of reuse in ERP projects as solutions are mainly based on standard software components and services. However, the consultants' ability to identify relevant components for reuse is affected by the increasing number of services available to them. The work described in this experience paper focuses on providing support for consultants to identify existing solutions informing system design. We report the development of a tool-supported approach called S³C, based on Microsoft Sure Step methodology and SeCSE open source service discovery tools. The S³C approach is tailored to the needs of SME companies in the ERP domain and overcomes limitations of Sure Step. The initial application and evaluation of the S³C approach also allows presenting lessons learned.

Keywords: Information systems, requirements elicitation, service discovery, ERP.

1 Introduction

Enterprise Resource Planning (ERP) systems are software systems that support business operations. They were first introduced in material management but nowadays ERP systems support a broad range of business activities [1]. Woods [2] highlights that the service-oriented paradigm has changed the nature of ERP systems. Novel ERP systems are based on software services and software vendors provide various tool support and frameworks to integrate ERP systems in broader service-oriented systems [3].

The rising number of services makes the task of the consultants, who align an ERP system with a customer's requirements, increasingly difficult: the consultants need to know all these services, their functionality and quality of service in order to do their job properly. We observed this problem in two companies that provide ERP

H. Mouratidis and C. Rolland (Eds.): CAiSE 2011, LNCS 6741, pp. 18–32, 2011.

solutions based on Microsoft Dynamics AX for their customers. Dynamics AX is a business management solution that combines ERP functionality and additional domain specific modules [4]. The consultants at both companies apply Microsoft's Sure Step methodology [5], which provides guidance and support for gathering customers' requirements and identifying which requirements can be satisfied using standard ERP system functionality (Fit-Gap Analysis). However, the increasing number of services both provided by Dynamics AX [6] and developed by partner companies raises challenges for the application of Sure Step. The consultants find it increasingly difficult to keep track of the available services. This limits their ability to identify adequate services and negatively affects later system design and development.

In our current research, we are addressing the challenges resulting from the introduction of software services in the ERP domain. The research focuses on system design activities within ERP projects based on Sure Step. We explore the ways consultants can be supported in handling the increasing number of services within ERP systems. This paper presents the Semantic Service Search & Composition (S³C) approach, a tool-based solution supporting consultants in identifying requirements and relevant services to inform the Fit-Gap Analysis within ERP projects. We further discuss the experience gained from two studies where the S³C Solution Explorer was applied by ERP consultants at two different companies.

The remainder of the paper is organized as follows. Section 2 presents the Sure Step analysis phase and highlights the limitations of this approach and the challenges raised by service-centric system development. Section 3 presents the research goal and research objectives in more detail. In that section we then present the needs of ERP consultants. We further discuss the SeCSE Requirements Process which informed our solution. Section 3 then presents the S³C approach which adapts Sure Step towards service-centric development. In the last part of Section 3 we present novel tools which were developed to support the S³C approach as well as integrated service discovery components from SeCSE. In Section 4 we present and discuss the results of the initial evaluation of the S³C Solution Explorer. Section 5 presents lessons learned and Section 6 concludes and presents further work.

2   Requirements Elicitation Based on Sure Step and Its Limitations

Sure Step defines roles such as that of the application consultant who is responsible for gathering and specifying requirements and for conducting a Fit-Gap Analysis. Figure 1 gives a detailed overview of Sure Step which divides an ERP installation project into six phases: Diagnostic, Analysis, Design, Development, Deployment and Operation. Each phase consists of mandatory and optional activities. In addition, there exist Cross Phase Processes that span across multiple phases. There are four activities relevant to this research which focuses on requirements elicitation and system design. These activities are usually conducted in sequential order. The following paragraphs give a more detailed description:

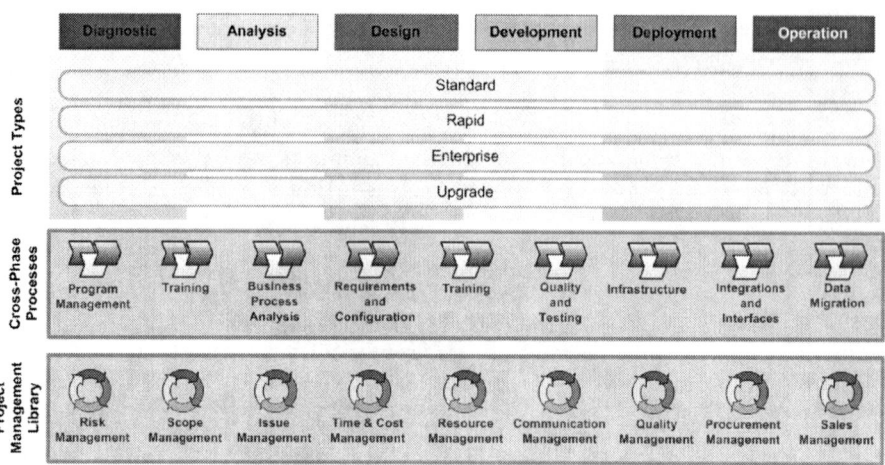

Fig. 1. Microsoft Dynamics Sure Step Phase Model

Conduct Business Process Analysis. A detailed business process analysis is conducted in a workshop, therefore consultants compare the customers' current business process with standard ERP processes. The result is a To-Be process model (see Figure 2) which describes business process steps using natural language text. For example, an item arrival process could include events such as: *A vendor delivers goods from different purchase orders.*

Gather Business Requirements. After defining the To-Be process model consultants and customer start to gather and document requirements following a predefined Word-template. Consultants document upcoming requirements using the Functional and Non-Functional Requirements Document (see Figure 2). An example requirement descriptions supporting the item arrival process could be: *The system should inform the warehouse worker if the delivered quantity is higher than the ordered quantity.*

Conduct Fit-Gap Analysis. After the workshop consultants conduct a Fit-Gap Analysis. For each requirement consultants try to identify an available solution. If the consultants are able to identify a software solution they document it as fit, otherwise it is a gap. In a next step the consultants and the customer jointly investigate ways to resolve the gaps. There are three ways to deal with gaps: (i) Adding a 3rd party solution; (ii) Changing the business process; (iii) Customizing the standard application to fit the requirements. Fits and gap resolutions are documented in the Fit-Gap Worksheet (see Figure 2).

Derive Functional Design Documents. After the Fit-Gap Analysis consultants create a Functional Design Document for Configuration which describes how selected standard software services need to be configured to fulfill customer needs. Furthermore, developers and consultants create a Functional Design Document for Customization. This document describes the planned development work needed to provide solutions for the gaps.

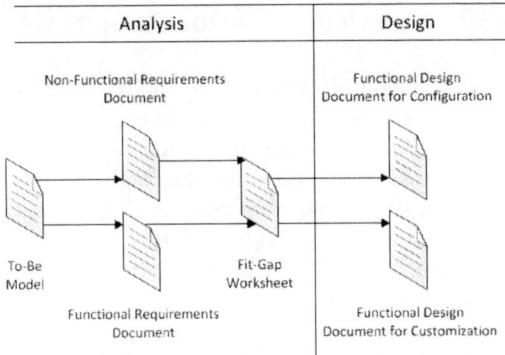

Fig. 2. Information Flow from Analysis to Design Phase in Sure Step

Sure Step was not designed to support the development of service-oriented systems. It is a document-oriented process which separates requirements elicitation workshops, investigation in gap resolutions and solution design. With a growing number of services, it is getting harder for consultants to identify suitable services. These circumstances lead to the following limitations of Sure Step in practice:

Services are not identified. Although existing services could provide a solution and fulfill a customer's request, a consultant might not be able to identify a suitable service solution due to the high number of services available. As a result, new services are developed instead of reusing existing services.

Inadequate services are selected. While the number of available services is increasing, the functionality provided by existing services changes. Consultants therefore cannot keep an overview on all existing services and the provided functionality of their current and depreciated versions. As a result, consultants select services they are familiar with although other services would provide more accurate functionality.

Late service discovery. In a later design phase, developers might identify services more accurate than those specified in the Fit-Gap Worksheet. This causes renegotiation of customer requirements. However, in most cases it is too late to change the planned solution as development has already started.

Inaccurate estimates. Time and cost estimations are based on the Fit-Gap Worksheet. Due to the inadequate selection of services, the Fit-Gap Worksheet may be incorrect. This results in inaccurate time and cost estimates.

Insufficient information for customers. Consultants do not have a deep knowledge of particular service functionality. Therefore, in Sure Step, consultants and customers usually do not discuss the functionality provided by selected services. However, consultants have found that such discussions provide an important input for customers; they even help triggering new customer requirements.

Time consuming approach. After defining the To-Be business process and gathering customers' requirements within an initial workshop, consultants have to conduct a Fit-Gap Analysis. As this task is very time consuming it is done after the workshop. However, consultants need to approach customers again to agree on gap resolutions.

The discussed limitations delay the completion of projects. Furthermore, these limitations lead to an unnecessary increase of costs to realize ERP systems.

3 Semantic Service Search and Composition (S³C)

In the Semantic Service Search & Composition (S³C) project, we are exploring the challenges of introducing services in the ERP domain. We are focusing on SMEs developing ERP systems by following the Microsoft Sure Step methodology. The goal is to develop possible solutions to overcome identified problems and to apply these solutions in practice. We want our research to supports consultants in identifying requirements and relevant services to inform Fit-Gap Analysis within ERP projects. We used action research [7] to conducting this work and aimed at meeting the following research objectives:

RO 1: Identify the needs of consultants regarding requirements elicitation in ERP projects based on Sure Step.

RO 2: Adapt and extend relevant service discovery approaches to support ERP projects based on Sure Step.

RO 3: Evaluate the benefits and limitations of the developed tool-supported approach.

The first research objective focuses on identifying the consultants' needs in order to overcome limitations of the existing approach. The second research objective investigates how to extend Sure Step. This task includes the identification of relevant research whose results could inform a possible tool-supported solution for requirements elicitation in ERP projects based on Sure Step. The third research objective focuses on investigating the usability and utility of the tool-supported approach. Our aim is to investigate if the envisioned tool-supported approach does support the daily work of consultants in ERP projects based on Sure Step.

3.1 Identifying the Needs of Business Consultants

In a first step we identified and analyzed the needs of consultants. We therefore interviewed 4 employees at Terna[1], an Austrian ERP partner following Sure Step. We discussed the identified limitations of Sure Step and asked them about their needs regarding a novel tool-supported approach which would overcome existing limitations. The following paragraphs describe the consultants' key requirements.

Integrated description of business processes and requirements. Sure Step uses Word templates to specify business processes and requirements. This results in the creation of several different documents and the distribution of information across several documents. Handling documents is therefore time consuming and often results in inconsistencies. The new tool-based approach should therefore provide an integrated solution which allows the structured and integrated specification of business process information and requirements.

Linking requirements and use cases. As discussed, Sure Step provides Word templates for documenting business processes and requirements. This means that there is no support for linking a requirement to a particular use case – a feature that should be supported by the new approach.

---

[1] Terna is an Austrian ERP company owned by Allgeier Holding AG. They are Microsoft Dynamics AX partner, Lawson M3 distributor and maintain AMS4U based on AS400.

Identifying existing solutions. Sure Step does not support the automatic identification of existing solutions. The new approach should be able to identify services based on use cases and linked requirements. More specifically, a novel integrated approach should propose a list of candidate services and provide service descriptions.

Narrow down service discovery results. In our discussions consultants empathized that the time to discuss solutions with customers is limited and that they would need a feature supporting them by highlighting the most promising solution(s). A novel approach should highlight the fact that this service only has a low priority and that it is unlikely that it can contribute to the envisioned solution.

Provide details on services of interest. Details of a service of interest need to be immediately accessible for further discussion. The new approach should allow consultants to access important information related to a selected service.

3.2 Identifying Research Informing S³C

An analysis of existing work shows that similarity analysis is a well-described issue in ERP literature [8, 9]. However, we could not identify relevant work focusing on service discovery mechanisms for similarity matching. Also in requirements research, little has been reported on service discovery. Schmid et al. [10] discuss a requirements-led process enabling runtime service discovery but do not report on tool support. Elsewhere, Esmaeilsabzali et al. [11] present new models for requirements-based service discovery that assume formal expression of system operations. Zachos et al. [12] have researched new tools and techniques to form service queries from incomplete requirements specifications as part of the EU-funded SeCSE Integrated Project [13]. We considered the work by Zachos et al. [14] to be most promising for extending Sure Step.

The SeCSE requirements process is depicted in Figure 3. Service queries are extracted from a service request constructed from a requirements specification and then fired at service registries. The retrieved service descriptions enable consultants and customers to select the most appropriate service(s). The main innovation is to expand service queries to handle requirements expressed in natural language. As such, SeCSE appears to be the first approach to integrate requirements and service discovery methods and tools. Therefore it was chosen to be the base for the S³C project.

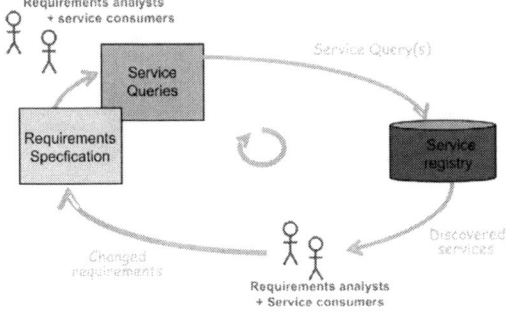

Fig. 3. SeCSE's Requirements Process

3.3 The S³C Approach

The S³C approach is based on Microsoft Sure Step and SeCSE to provide advanced support for consultants. It extends Sure Step by integrating SeCSE's service discovery mechanisms. However, a main contribution of the S³C project are novel tools which provide support for consultants (see Section 3.4). The S³C approach integrates Business Process Analysis, Gathering Business Requirements and Fit-Gap Analysis within the same workshop. Figure 4 shows the modified information flow with integrated service discovery.

Conduct Business Process Analysis. As in Sure Step a detailed business process analysis is conducted in a workshop. Consultants compare the customer's current business process with standard ERP processes and document results with the help of a To-Be process model (see Figure 4).

Gather Business Requirements. In the same workshop consultants and customers identify and discuss requirements on how the planned ERP system will support the To-Be process. These requirements are linked to the business process and are specified in the Requirements Document (see Figure 4).

Identify and Discuss Relevant Solutions (Fits). The To-Be business process description and gathered requirements are used as input to identify existing services which potentially can fulfill the customers' needs. For an item arrival use case the system will suggest a vendor service, described as *"Enables external systems to read, create, update and delete vendors."* Consultants and customers walk through and discuss the listed solutions. This can trigger new requirements which can then cause modifications in the list of relevant solutions. This iterative approach strengthens the interaction between consultants and customers and allows the customer to participate in the solution design. Selected solutions are documented as fits in the Fit-Gap Worksheet (see Figure 4).

Discuss Gap Resolutions. In a next step the consultants and the customer investigate gap resolutions which are documented in the Fit-Gap Worksheet.

Derive Functional Design Documents. As in Sure Step the Fit-Gap Worksheet is used to create the Functional Design Document for Configuration and Functional Design Document for Customization (see Figure 4).

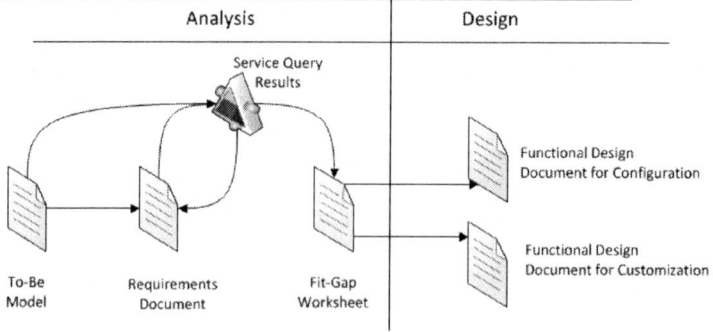

Fig. 4. Information Flow in Sure Step with Service Discovery (S³C)

3.4 The S³C Tool Environment

We developed adequate tool support based on the conceptual solution – as outlined in the previous section. The S³C Tool Environment consists of a variety of applications, services, prototypes and databases. The main technical contribution are novel S³C tool which were built on top of selected SeCSE components. Figure 5 shows the S³C Tool Environment including newly developed tools as well as original SeCSE components.

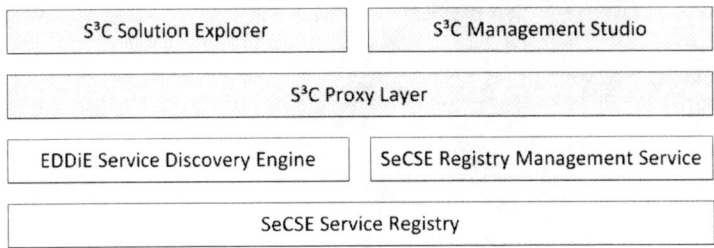

Fig. 5. S³C Tool Environment

The *S³C Solution Explorer* is an application for onsite consultants; it enables them to document the To-Be business model in the form of use cases and to link and document requirements. This information can be used as input to perform service discovery requests. To do so, the use case description and the linked requirements are compiled into an XML query document. The query document is used as input for the EDDiE service discovery engine. The query result is a XML document including candidate services, in order of relevance (calculated by the EDDiE service discovery engine). With this information, the S³C Solution Explorer presents a ranked list of relevant services also highlighting the matching probability for each service. The descriptions of these candidate services provide input for further discussions with customers and support solution selection. Figure 6 shows the S³C Solution Explorer representing the Item Arrival Use Case.

The *S³C Management Studio* is used by system administrations to access the service registry to keep the stored information up to date. Managing the service registry includes providing and updating information about services, such as the service provider, a description and other meta-data. Accurate information about services is vital for service discovery requests as it enables the system to identify accurate candidate services.

The *S³C Proxy Layer* was introduced to overcome the heterogonous nature of the SeCSE platform. Due to the involvement of different research partners in SeCSE the provided solution is a mix of different platforms and technologies. The S³C Proxy Layer is an abstraction layer that provides an interface for performing service discovery and manipulating services in the service registry. The S³C Proxy is built using Microsoft .NET technology and translates either in Java or XML to the SeCSE components. All S³C tools are built upon the S³C Proxy Layer and have no contact with the underlying SeCSE components.

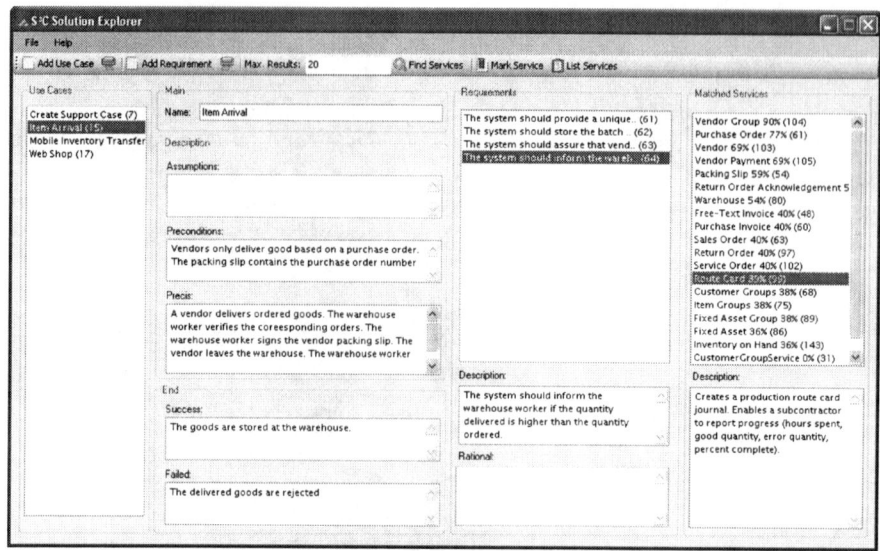

Fig. 6. The S³C Solution Explorer Application

The *EDDiE Service Discovery Engine* uses the information gathered with the S³C Solution Explorer as input to retrieve services. EDDiE implements advanced term disambiguation and query expansion algorithms to add different terms with similar meanings to the query using the WordNet online lexicon [15, 16], thus increasing the number of web services retrieved from the registries. Furthermore, EDDiE provides capabilities to calculate the relevance of a discovered service [14].

The *SeCSE Management Service* provides an application interface to the SeCSE Registry via a web service. It is used to manage providers, services and meta-data such as description, quality of service, commercial information and signatures.

The *SeCSE Service Registry* is a XML database that holds all the information about services and their descriptions. While SeCSE tools used a federated online database, the S³C solution is based on a locally installed database.

4   Initial Evaluation of the S³C Solution Explorer

Our evaluation strategy focused on investigating the utility of the S³C Tool Environment. We conducted two studies at different ERP partners who both follow Sure Step. We first performed an initial utility and usability study at Terna to figure out whether the S³C Solution Explorer application fulfills the key requirements of consultants. The second evaluation was conducted at InsideAx[2] to investigate the utility of the S³C Solution Explorer in more detail. This study focused on comparing the time needed to identify relevant services with the help of the S³C Solution Explorer to Sure Step and investigated the correctness and completeness of the solution.

---

[2] InsideAx is an Austrian ERP company focusing on Microsoft Dynamics AX Solutions.

4.1 S³C Solution Explorer Initial Utility and Usability Evaluation at Terna

Three consultants participated in the initial utility evaluation of the S³C Solution Explorer at Terna. Each has a master degree in business informatics and more than three years of experience in ERP system customization. None of the consultants is an author of this paper, and none had had contact with the S³C Solution Explorer prior to the evaluation. However, two of the consultants also participated in the first interview where requirements for S³C approach were identified (see Section 3.1).

Each evaluation was structured in 3 parts – briefing, evaluation and debriefing. During the briefing each participant was informed of the study's purpose and the task they were intended to perform. The briefing also included a short introduction to the S³C Solution Explorer. For the study we prepared a typical ERP use case describing how a warehouse worker handles the receipt of goods – the item arrival use case. The example use case also included four requirements which were gathered from customers in previous projects (e.g. *The system should provide a unique identification number for each vendor*). During the evaluation each consultant used the S³C Solution Explorer to enter the use case and the related requirements. Furthermore, they were asked to use the Solution Explorer to identify relevant services for that use case. One of the authors observed each consultant who spoke loudly throughout the process. The utility of the S³C Solution Explorer concerning the previously discovered consultants' requirements (see Section 3.1) was discussed during the debriefing. The debriefing was then used to discuss usability problems of the S³C Solution Explorer. All 3 consultants completed the evaluation and debriefing, which lasted on average 20 and 25 minutes respectively.

Observations and qualitative feedback were encouraging but they also identified some limitations. The study revealed that the S³C Solution Explorer does fulfill most of the consultants' requirements (see Section 3.1):

- The consultants were able to document use cases and requirements.
- The consultants said that the tool provides a clear structure in linking use cases and requirements.
- The consultants were able to use the tool to run service queries.
- The consultants said that the tool suggested relevant service solutions.
- The consultants said that a ranked list of relevant services and the provided matching probabilities supported them in narrowing down the discovered results.
- The consultants said that the current service description does provide key information which (in most cases) allows them to decide whether a service is relevant. However, they requested more detailed service descriptions to improve the decision process. For example, consultants requested information about the usage of the service in previous projects.

In the debriefing meetings, the consultants also highlighted the fact that they would be willing to use a tool such as the S³C Solution Explorer for their daily work with customers. However, the interview also revealed several usability issues. They pointed out that the first prototype does present too much information at a time and that it is therefore hard to keep an overview. They argued that service discovery queries take too much time (on the average 3 seconds) and that they would prefer quicker responses in order to strengthen seamless discussions with customers. They requested that the list of relevant services should be automatically updated in time when entering new requirements.

However, results from the debriefing sessions indicated that most usability deficiencies were not critical to the main tasks and that the current S^3C Solution Explorer prototype does fulfill the consultants' key requirements. With these results, we decided to use the S^3C Solution Explorer without further development in more naturalistic studies in order to explore its utility in more detail.

4.2 S^3C Solution Explorer Utility Evaluation at InsideAx

This study investigated the effect of the S^3C Solution Explorer on the discovery of relevant services. We explored whether consultants need less time to identify relevant services with the help of the S^3C Solution Explorer compared to non-tool supported service identification. Furthermore, we investigated the correctness and completeness of these solutions. Three consultants from InsideAx participated in this S^3C Solution Explorer evaluation. One has a master degree in business informatics and more than three years of experience in ERP system customization (Consultant 1). The second consultant had more than three years of experience in ERP system customization while the 3rd consultant only had one year experience. None of the consultants is an author of this paper, and none had had contact with the S^3C Solution Explorer prior to the evaluation.

As for the first study, each evaluation was structured in briefing, evaluation and debriefing. For the evaluation we prepared an application example which included three use cases and typical use case requirements derived from previous projects. For each of the use case we reviewed existing implementations and ERP system documentations and defined a list of relevant services out of more than 70 available services. This standard solution was used as a basis for evaluating consultants' solutions. The application example consisted of following use cases:

- *Item Arrival* describing the handling of delivered goods from vendors. Five relevant services were identified as standard solution for this use case.
- *Transfer production goods to warehouse* representing events relevant for inventory management. Again, five relevant services were identified as standard solution for this use case.
- *Buy Item in a Web Shop* discussing sale related events. Six services were considered to be the standard solution for this use case.

In the briefing session, the consultants were told about the purpose of the study. We further discussed the application example with the consultants. Each of them was asked to estimate the time he would need for the identification of relevant services without tool support (see Table 1). During the evaluation each consultant used the S^3C Solution Explorer to identify relevant services for the three use cases and to document the results. The debriefing session was used to discuss their solution and to identify utility issues regarding the S^3C Solution Explorer. We compared particularly the time required for the identification of relevant services to the estimates. Furthermore, the developed solutions were compared to the standard solution in terms of correctness and completeness (see Figure 7).

$$Correctness = \frac{|\text{ Identified Services } \cap \text{ Relevant Services }|}{|\text{ Identified Services }|}$$

$$Completeness = \frac{|\text{ Identified Services } \cap \text{ Relevant Services }|}{|\text{ Relevant Services }|}$$

Fig. 7. Correctness and Completeness metrics

The results of the evaluation are shown in Table 1. The consultants' estimates for performing the service identification without tool support ranged from 2:30 hours to 4:20 hours. The experiment revealed that using the S³C Solution Explorer consultants needed between 45 minutes and 1:20 hours to actually conduct the task. Consultants were able to indentify correct services – apart from one exception; all identified services were included in the standard solution. Services completeness ranged from 50% to 80%. On the average consultants identified 62% of the relevant services.

Table 1. InsideAx Evaluation Results

	Time Estimation (no tool support)	Actual Time (with tool support)	Correctness			Completeness		
			Item Arrival Use Case	Invent Transfer Use Case	Web shop Use Case	Item Arrival Use Case	Invent Transfer Use Case	Web shop Use Case
Consultant 1	02:30	01:05	100%	100%	100%	60%	80%	50%
Consultant 2	04:00	00:45	100%	100%	100%	60%	60%	66%
Consultant 3	04:20	01:20	75%	100%	100%	60%	60%	66%

The results reveal that the S³C Solution Explorer enables consultants to discover relevant services in significantly less time than their estimates for service identification without tool support. In debriefing meetings consultants said that service identification with the help of S³C Solution Explorer was faster because services were presented using a ranked list with matching probabilities.

The two experienced consultants had a 100% correctness rate. In the debriefing meeting they argued that having service descriptions and requirements presented next to each other did support their matching process and did speed up the actual decision. However, they also mentioned that their experience did support them in taking decisions. The two experienced consultants argued that their knowledge about ERP systems did in particular support them in taking the right decision when the provided service description was limited (e.g. the Inventory Transaction Service was described as "Describes the inventory transactions document"). The less experienced consultant (Consultant 3) argued that he needed more time to make decisions. He said that in most of the cases the provided service description did allow him to make a correct decision. However, he also mentioned that more information on the services would have been helpful and would have made him more confident in his decisions.

The consultants did indentify key services relevant for the use case. However, other relevant services were not identified in this first iteration. Discussions with the consultants revealed that they did focus on correctness rather than completeness. They explained that they only selected services where they felt certain that they were relevant. They also explained that they did consider other services to be relevant. However, in most cases they preferred not to select them because of doubts caused by limited descriptions.

As for the initial evaluation all consultants' said that a tool such as the S^3C Solution Explorer can support their daily work. However, they also mentioned utility and usability issues.

4.3 Threats to Validity

The validity of the reported results was subject to the following possible threats. The limited number of experiments and the limited number of participants within the experiments does not allow drawing any statistically relevant conclusions. No comparative evaluation has been undertaken. Because of time and resource constraints, we decided not to have a control group to conduct the experiment without tool support. Although time estimations made by consultants do reflect experiences from practice, not having a control group means that these results need to be interpreted with care. This issue also applies to the presented results on service selection correctness and completeness. However, the uniqueness of the presented S^3C approach and the paucity of data on service discovery tools within the ERP domain means that the results presented provide a valuable input for further evaluations.

5 Lessons Learned

In the following we do present lessons learned that highlight benefits and weaknesses and discuss interesting facts regarding the conducted research:

Lesson 1: Adequate service descriptions are essential to decide about the relevance of a service. Our studies revealed that, in some cases, the provided service descriptions are too short to make a clear decision. In such cases, it was easier for experienced analysts compared with inexperienced to make a decision. We do see great potential in providing more detailed service descriptions. This might have positive effects on completeness and might further enable analysts to make decisions faster.

Lesson 2: Using the S^3C solution did allow substantial time savings, compared to estimates for non-tool supported Sure Step. One of the consultants was able to perform the analysis in about 20% of the estimated time. Even the analyst who had the lowest estimate was able to finish the task in less than half of the estimated time.

Lesson 3: Discussions with ERP experts of the two companies revealed that time estimates are normally of high quality. However, there are no records showing the correctness and completeness regarding the output of the analysis phase. The lack of such a baseline did not allow a more detailed comparison of our study results. However, ERP domain experts did consider the quality of the output of the second evaluation as good.

Lesson 4: Companies argue that the output of the analysis strongly depends on the experience of the consultant. Our experiment indicates that the S³C Solution Explorer particularly supports inexperienced consultants to come up with good quality results (which were similar to results from experienced consultants).

Lesson 5: Current usability and utility issues still limit the usage of the S³C Solution Explorer in real-world projects. However, the presented studies revealed its potential. Consultants were able to work with the S³C Solution Explorer without training and all confirmed that they want to use an improved version of the S³C Solution Explorer in customer workshops.

6 Conclusions and Further Work

In this paper, we present the tool-supported S³C approach which was built to overcome limitations of Microsoft Dynamics Sure Step, an existing approach used for requirements elicitation and Fit-Gap Analysis within ERP projects. In a first step, we discussed the limitations of Sure Step which result from the increasing use of software services in the ERP domain. The high number of services hinders consultants in discovering and selecting appropriate service-based solutions. As a result, projects take longer which leads to increased costs.

To figure out how to overcome these limitations (RO 1), we asked consultants in an ERP company about their needs on a tool-supported approach which enables them to identify relevant service solutions. Considering the identified needs and we identified SeCSE as a base for our research and realized the tool-supported S³C approach (RO 2). It is based on Sure Step but provides sophisticated service discovery mechanisms and novel tool support for consultants. The S³C Solution Explorer enables consultants to identify relevant solutions while they discuss upcoming requirements with customers. The evaluation of the benefits and limitations of the S³C solution (RO 3) suggests that the tool-supported S³C approach fulfils the consultants' needs and has the potential to support them in their daily work.

The development and evaluation of the S³C solution is the main contribution of our work. Although our research is focusing on ERP projects based on Sure Step, we assume that companies in the ERP domain that follow other approaches face similar challenges. We envision that the tool-supported S³C approach will stimulate further research in the field and support other ERP companies in overcoming the issues raised by introducing services in the ERP domain.

Our future research will focus on case studies to investigate the benefits and limitations of the S³C approach in real-world projects. We plan to evaluate whether the usage of the S³C Solution Explorer leads to more and more complete requirements. Informed by the results of such studies, we plan to provide an improved version of the S³C tools and more guidance and support for consultants on how to apply the S³C approach.

Acknowledgements

The research conducted was in part funded by the Austrian Research Promotion Agency (FFG, Project Nr.: 821614).

References

1. Daneva, M., Wieringa, R.: Requirements Engineering for Cross-organizational ERP Implementation: Undocumented Assumptions and Potential Mismatches (2005)
2. Woods, J.: Business Managers Need to Care About SOA in ERP. Gartner Research (2008)
3. Microsoft Corporation: Enabling "Real World SOA" through the Microsoft Platform (2006)
4. Microsoft Corporation: Microsoft Dynamics AX 2009: Designed to Enhance Productivity (2010), http://download.microsoft.com/download/E/2/2/E228B46E-F0E2-4C38-8F02-A21B7B544B39/AX_User_Productivitywp.xps
5. Microsoft Corporation: Microsoft Dynamics Sure Step Methodology, Platform 2.6.4.0 (2009)
6. Microsoft Corporation: Standard Axd Documents, http://msdn.microsoft.com/en-us/library/aa859008.aspx
7. Davison, R.M., Martinsons, M.G., Kock, N.: Principles of canonical action research. Information Systems Journal (2004)
8. Rolland, C., Prakash, N.: Bridging the gap between organizational needs and ERP functionality. Requirements Eng. 41, 180–193 (2000)
9. Salinesi, C., Bouzid, M., Elfassy, E.: An Experience of Reuse Based Requirements Engineering in ERP Implementation Projects. In: 11th IEEE International Enterprise Distributed Object Computing Conference
10. Schmid, K., Eisenbarth, M., Grund, M.: From Requirements Engineering to Knowledge Engineering: Challenges in Adaptive Systems. In: Proceedings SOCCER Workshop, RE 2005 Conference, Paris (2005)
11. Esmaeilsabzali, S., Day, N., Mavadatt, F.: Specifying Search Queries for Web Service Discovery. In: Proceedings SOCCER (Service-Oriented Computing: Consequences for Engineering Requirements) Workshop, RE 2005 Conference, Paris (2005)
12. Zachos, K., Maiden, N.A.M., Zhu, X., Jones, S.V.: Discovering Web Services to Specify More Complete System Requirements. In: Krogstie, J., Opdahl, A.L., Sindre, G. (eds.) CAiSE 2007 and WES 2007. LNCS, vol. 4495, pp. 142–157. Springer, Heidelberg (2007)
13. SeCSE Home (2010), http://www.secse-project.eu/
14. Zachos, K.: Using Discovered Services to Create Requirements for Service-centric System, Centre of Human-Computer Interaction Design, City University London (2008)
15. WordNet (2010), http://wordnet.princeton.edu/
16. Voorhees, E.: Using wordnet to disambiguate word senses for text retrieval. In: Proceedings of the 16th ACM-SIGIR Conference, pp. 171–180 (1993)

Requirements Engineering for Self-Adaptive Systems: Core Ontology and Problem Statement

Nauman A. Qureshi[1], Ivan J. Jureta[2], and Anna Perini[1]

[1] Fondazione Bruno Kessler - IRST, Software Engineering Research Group
Via Sommarive, 18, 38050 Trento, Italy
{qureshi,perini}@fbk.eu
[2] FNRS & Louvain School of Management, University of Namur, Belgium
ivan.jureta@fundp.ac.be

Abstract. The vision for self-adaptive systems (SAS) is that they should continuously adapt their behavior at runtime in response to changing user's requirements, operating contexts, and resource availability. Realizing this vision requires that we understand precisely how the various steps in the engineering of SAS depart from the established body of knowledge in information systems engineering. We focus in this paper on the requirements engineering for SAS. We argue that SAS need to have an internal representation of the requirements problem that they are solving for their users. We formally define a minimal set of concepts and relations needed to formulate the requirements problem, its solutions, the changes in its formulation that arise from changes in the operating context, requirements, and resource availability. We thereby precisely define the *runtime requirements adaptation problem* that a SAS should be engineered to solve.

Keywords: Requirements Engineering, Runtime, Adaptation Problem, Self-Adaptive Systems.

1 Introduction

Contemporary software systems, such as service-based mobile applications that are increasingly immersed in users' everyday life, must continuously adapt their behavior to changes in users' requirements, operating conditions, and resource availability [5]. For instance, a music download application may need to behave differently when the user's device is connected through the mobile phone to the Internet, than when it is connected via Wi-Fi, when the device is plugged to a docking station rather than on battery power, when the user's preferred music delivery service is not available, and another needs to be selected, and so on. Such software has to cope with such problems as incomplete specifications of its operating conditions, unanticipated events, variable quality of service from third-party services.

The vision for self-adaptive systems (SAS) is that they should continuously adapt their behavior at runtime in response to changing user's requirements, operating contexts, and resource availability. Realizing this vision requires that we understand precisely how the various steps in the engineering of SAS depart from the established body of knowledge in information systems engineering. Research agendas for SAS

H. Mouratidis and C. Rolland (Eds.): CAiSE 2011, LNCS 6741, pp. 33–47, 2011.

have been proposed in various communities. For instance, the Software Engineering for Self-Adaptive Systems (SEAMS) community focuses on issues pertaining to system architectures[1], while the Requirements Engineering community has proposed methods for analyzing requirements for self-adaptivity and suggested that requirements should become artifacts used, processed, and changed at runtime [2,11,18,21]. This led to proposals for various methods to engineer requirements for SAS. However, there has been limited consensus among the research communities on two issues: (i) what main concepts and relations are needed to define the requirements for SAS, and (ii) how does the requirements problem (i.e., the problem to solve during requirements engineering) differ for SAS, compared to systems that are not self-adaptive.

Taking the perspective of requirements engineering (RE) for SAS, we envision SAS to have an internal representation of their user's requirements, and of their operational environment, by being equipped with automated reasoning capabilities for monitoring, analyzing changes that occur dynamically at runtime and finding solutions (i.e. set of tasks that can satisfy the requirements using an available services or otherwise) to meet them, thus ensuring the consistency with the intended system's requirements.

The aim of this paper is to identify concepts and relations, which are necessary to deal with while eliciting and analyzing requirements for SAS and are important to take adaptation decision at runtime by the system itself. This leads us to formulate the *runtime requirements adaptation problem* that a SAS should be engineered to solve.

Section 2 presents the conceptual tools used in the rest of the paper. We introduce the runtime requirements adaptation problem in Section 3. In Section 4, we present how adaptation problem is connected to RE by extending the core ontology for RE and proposing concepts and relationships needed to determine the runtime requirements adaptation problem using examples related to travel planning software. Related work and discussion are presented in Section 5 and 6. The paper ends with conclusions and a summary of directions for future work.

2   Preliminaries

2.1   General Requirements Problem

The overall aim of RE is to identify the purpose of the system-to-be and to describe as precisely and completely as possible the properties and behaviors that the system-to-be should exhibit in order to satisfy that purpose. This is also a rough statement of the requirements problem that should be solved when engineering requirements.

Zave & Jackson [24] formalized the requirements problem as finding a specification (S) in order to satisfy requirements (R) and not violate domain assumptions (K), to ensure that $K, S \vdash R$. This formulation highlights the importance of the specification to be consistent with domain assumptions, and that requirements should be derivable from K and S. It was subsequently argued that there is more to the requirements problem than this formulation states [7]. Namely, a new core ontology for requirements (CORE) was suggested along with a new formulation of the requirements problem to recognize that in addition to goals and tasks, different stakeholders have different preferences

---

[1] http://www.hpi.uni-potsdam.de/giese/public/selfadapt/front-page

over requirements, that they are interested in choosing among candidate solutions to the requirements problem, that potentially many candidate solutions exist (as in the case of service-/agent-oriented systems, where different services/agents may compete in offering the same functions), and that requirements are not fixed, but change with new information from the stakeholders or the operational environment. In absence of preferences, it is (i) not clear how candidate solutions to the requirements problem can be compared, (ii) what criteria (should) serve for comparison, and (iii) how these criteria are represented in requirements models.

Techne [8], an abstract requirements modeling language was recently introduced as a starting point for the development of new requirements modeling languages that can be used to represent information and perform reasoning needed to solve the requirements problem. Techne is abstract in that it assumes no particular visual syntax (e.g., diagrammatic notation such as present in Tropos [14]), and it includes only the minimum concepts and relations needed to formalize the requirements problem and the properties of its solutions. Techne is a convenient formalism for the formulation of the runtime requirements adaptation problem, as it is adapted to the concepts, such as goal, task, domain assumption, and relations (e.g. Consequence, Preference, Is-mandatory, Is-optional) that remain relevant for the RE of SAS. To keep the discussion simple in this paper, we assume that requirements and other information is available in propositional form, so that every proposition is nothing but a natural language sentence. We overview below the requirements problem formulation using parts of Techne that we need in the rest of the paper. In [15], we provide definitions of the consequence relation $\vdash_\gamma$ used in the definition of the candidate solution below.

Definition 1. *Requirements problem:* **Given** *the elicited or otherwise acquired: domain assumptions (in the set* K*),tasks in* T*,goals in* G*, quality constraints in* Q*, softgoals in* S*, and preference, is-mandatory and is-optional relations in* A*,* **find** *all* **candidate solutions** *to the requirements problem and compare them using preference and is-optional relations from* A *to identify the most desirable candidate solution.*

Definition 2. **Candidate solution:** *A set of tasks* T^* *and a set of domain assumptions* K^* *are a* **candidate solution** *to the requirements problem if and only if:*

1. K^* *and* T^* *are not inconsistent,*
2. $K^*, T^* \vdash_\gamma G^*, Q^*$*, where* $G^* \subseteq G$ *and* $Q^* \subseteq Q$*,*
3. G^* *and* Q^* *include, respectively, all mandatory goals and quality constraints, and*
4. *all mandatory softgoals are approximated by the consequences of* $K^* \cup T^*$*, so that* $K^*, T^* \vdash_\gamma S^M$*, where* S^M *is the set of mandatory softgoals.*

We start below from the CORE ontology and problem formulation in Techne, and add concepts specific to the RE for SAS, which leads us to an ontology for requirements in SAS and the formulation of the *requirements problem in context* for SAS. We subsequently show how to formulate the *runtime requirements adaptation problem* as a dynamic RE problem of changing (e.g. switching, re-configuring, optimizing) the SAS from one requirements problem to another requirements problem, whereby the changing is due to change in requirements, context conditions, and/or resource availability.

2.2   Adaptive Requirements and Continuous Adaptive RE (CARE) Framework

To support the analysis at design-time, we proposed *adaptive (functional or non-functional) requirements*. They have some degree of flexibility in their achievement conditions, which in turn requires the monitoring of the specification, while taking into account the changes in the operating context, evaluation criteria and alternative software behaviors [16].

More recently, we have proposed a Continuous Adaptive Requirements Engineering (CARE) framework [17,18,19] that views adaptive requirements as runtime artifact that reflect the system's monitoring and adaptation capabilities. RE is performed at runtime by updating initial requirements with new ones, or by removing requirements.

We proposed a classification of adaptation types in the CARE framework, that can be performed by the system itself or by involving the user (both the end-user or the designer) [19]. Mainly, type 1 and 2 adaptations are performed by the system itself, type 1 corresponds to system exploiting existing available solutions when needed, where type 2 is related to monitored information i.e. exploiting it to evaluate changes and select alternative solution. Type 3 and 4 adaptations involves users. In type 3 end-users may express new requirements or change existing ones at any given time, by giving input information correspondingly system analyzes it by finding solutions (adapts) for new/refined needs of the end-users. In type 4, requirements for which there are no possible solutions available analyst/designers are involved for offline evolution of the system.

3   Runtime Requirements Adaptation Problem

Various definitions of SAS have been offered in the literature. We remain aligned with the usual conception, namely, that a SAS is a software system that can alter its behavior in response to the changes that occur dynamically in its operating environment. The operating environment can include anything that is observable by the software itself including operational setting in a context, end-user's input, profile and resources.

SAS must be "aware" at runtime of the changes in requirements, its operating context, and in the availability of resources. SAS at runtime need the ability to sense changes. We interpret this as a sort of RE that we call "RE@runtime", where SAS plays – to the feasible extent – the role of an analyst. It has a representation of requirements, of the conditions in its operating contexts (acquired through sensors, for instance), and of the resources it uses. It can add, remove, or otherwise change these, depending on the changes detected through interfaces with the users, environment, and resources. New information thereby acquired can affect the "requirements problem" leading the SAS to query the user for new requirements, or otherwise adapt following the adaptation types [19]. E.g., (a) select a predefined available behavior or look for an alternative behavior the SAS has been designed and implemented for, by exercising its internal monitor-eval-adapt loop (i.e., adaptation types 1 and 2); (b) compose a new solution by exploiting knowledge on available services or explicitly acquired through user's input or change in context (i.e., adaptation type 3); (c) if no solution can be found for the new information, SAS must inform the user for further instructions (adaptation type 4).

We see runtime adaptation as a *dynamic RE problem*, where changes in requirements, context, and resources lead to a new requirements problem, and this in turn requires the resolution of that new problem. The dynamic aspect is that move from one requirements problem formulation to another due to the changes that the SAS can detect. The problem of reformulating the requirements problem when changes occur, and then solving the changed problem is what we call the *Runtime requirements adaptation problem*. To resolve this problem, we argue that it is necessary to know what kinds of information and which relations are crucial for the system to capture. We therefore make explicit the dynamic parts in the requirements problem formulation based on the CORE ontology.

4   RE for SAS and Its Core Ontology

Building upon the above considerations, we argue that concepts such as user's context and resources must be considered as first class citizens on top of the existing CORE ontology to engineer requirements for SAS. We add two new concepts, Context and Resource on top of the CORE ontology to accommodate the changes that might occur at runtime, which not only demands adaptation (i.e. dynamically changing from one requirements problem to another) but also requires an update to the specification (i.e. refinement of requirements).

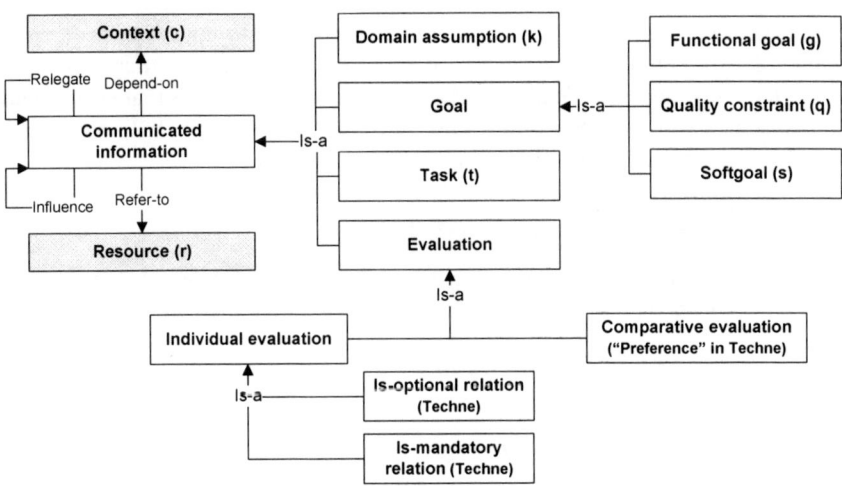

Fig. 1. Revised ontology with new concepts (Context and Resource) and relations (Relegation and Influence) related to the concepts of the CORE ontology for RE (for details, see [7])

The revised ontology of the concepts is proposed taking into account the concepts proposed in Techne [8] to support the definition of *runtime requirements adaptation problem*. In Fig. 1, concepts (Context and Resource) and relations (Relegation and Influence) are presented, relating them to the concepts of the CORE ontology. Below, we start by introducing the concept of a requirements database.

Definition 3. *Requirements database:* *A requirements database, denoted Δ is the set of all information elicited or otherwise acquired during the* RE *of a system-to-be.*

Remark 1. Since Δ should include all information elicited or otherwise acquired in RE, it should include all instances of domain assumptions (i.e. invariants in the application domain), goals, softgoals, quality constraints, and tasks that we elicited, found through refinement or otherwise identified during RE. One can view Δ as a repository of information that is usually found in what is informally referred to as a "requirements model". The notation used in the definition of the requirements problem, note that $\Delta = \mathbf{K} \cup \mathbf{T} \cup \mathbf{G} \cup \mathbf{Q} \cup \mathbf{S}$.   ■

Remark 2. Below, we will use the term *requirement* to abbreviate "member of the requirements database Δ". I.e., we will call every member of Δ a requirement.   ■

To get to the definition of the *runtime requirements adaptation problem*, we start introducing the Context concept. In [15], we discussed how this definition relates to existing conceptions and use of Context in the AI and RE literatures (e.g., [12,6,20,1]).

Definition 4. *Context:* *An instance C of the Context concept is a set of information that is presupposed by the stakeholders to hold when they communicate particular requirements. We say that every requirement depends on one or more contexts to say that the requirement would not be retracted by the stakeholders in every one of these contexts.*

Firstly, we need a language to write this information that is presupposed, and is thereby in the set of information that we call a particular context. We develop that language below. Secondly, the dependence of a requirement on a context means that every requirement is specific to one or more contexts, and thus, requirements need to be annotated by contexts, which begs additional questions on how the engineer comes to determine contexts. At this point, we revise the Techne language to allow information that is included in contexts. This results in adding one more sort.

Definition 5. *Techne for SAS:* *The language $\mathcal{L}_{SAS}$ is a finite set of expressions, in which every expression $\phi \in \mathcal{L}_{SAS}$ satisfies the following* BNF *specification[2]:*

$$x ::= \mathbf{k}(p) \mid \mathbf{g}(p) \mid \mathbf{q}(p) \mid \mathbf{s}(p) \mid \mathbf{t}(p) \tag{4.1}$$

$$q ::= \mathbf{c}(p) \tag{4.2}$$

$$w ::= x \mid q \tag{4.3}$$

$$y ::= \bigwedge_{i=1}^{n} w_i \rightarrow w \mid \bigwedge_{i=1}^{n} w_i \rightarrow \bot \tag{4.4}$$

$$\phi ::= w \mid \mathbf{k}(y) \mid \mathbf{c}(y) \tag{4.5}$$

Remark 3. We used (indexed/primed p, q, r) as an arbitrary atomic statement, every ϕ an arbitrary complex statement, and every x an arbitrary label to represent Techne labeled propositions i.e. domain assumption ($\mathbf{k}(p)$), a goal ($\mathbf{g}(p)$), etc., to distinguish from these basic labeled propositions the context propositions (i.e., propositions about

---

[2] In BNF: "::=" reads "defines"; "|" reads "or".

context), $\mathbf{c}(p)$ is added separately in the BNF specification, via q, and every w can either be x or q. Every y represents a complex statement as a formula with conjunction and implication such that y can be either w or $\bot$, where w is some requirement in a context propositions and $\bot$ refers to logical inconsistency. We can then rewrite ϕ as a complex statement consists of either w or $\mathbf{k}(y)$ or $\mathbf{c}(y)$. ■

Definition 6. Consequence relation of Techne in context: *Let* $\Pi \subseteq \mathcal{L}_{SAS}$, $\phi \in \mathcal{L}_{SAS}$, *and* $z \in \{\phi, \bot\}$, *then:*

1. $\Pi \vdash_{\mathcal{C}T} \phi$ *if* $\phi \in \Pi$, *or*
2. $\Pi \vdash_{\mathcal{C}T} z$ *if* $\forall 1 \leq i \leq n$, $\Pi \vdash_{\mathcal{C}T} \phi_i$ *and* $\mathbf{k}(\bigwedge_{i=1}^{n} \phi_i \rightarrow z) \in \Pi$.

The consequence relation $\vdash_{\mathcal{C}T}$ is sound w.r.t. standard entailment in propositional logic. It deduces only positive statement by being paraconsistent, thus all admissible candidate solutions are found via paraconsistent and non-monotonic reasoning. Reasoning is paraconsistent because an inconsistent Δ or C should not allow us to conclude the satisfaction of all requirements therein; it is non-monotonic in that prior conclusions drawn from a Δ or a C may be retracted after new requirements are introduced.

We also need a function that tells us which contexts a requirement applies to.

Definition 7. Contextualization function: *Let* C *be the set of all contexts.* $\mathcal{C} : \wp(\mathcal{L}_{SAS})$ $\longrightarrow \wp(C)$ *(where $\wp$ returns the powerset) is called the contextualization function that for a given set of formulas returns the set of contexts to which these formulas apply to. By "apply to", we mean that* $C \in \mathcal{C}(\phi)$ *iff the following conditions are satisfied:*

1. $C, \phi \nvdash_{\mathcal{C}T} \bot$, *i.e.*, ϕ *is not inconsistent with context* C,
2. C *is such that* $\exists X \subseteq \Delta$ *such that* $C, X \vdash_{\mathcal{C}T} \phi$, *i.e., the context* C *together with some requirements* X *from* Δ *lets us deduce* ϕ.

Several remarks are in order. Firstly, with $\mathcal{L}_{SAS}$, we now have a new sort for expressions that are members of a set that defines a context. Recall that we defined an instance C of Context as a set of information, so that now $\mathcal{L}_{SAS}$ tells us that one member of that set can either be a proposition p, denoted $\mathbf{c}(p)$, or can be a formula with implication, denoted $\mathbf{c}(y)$ in the BNF specification. E.g., if the engineer assumes that the stakeholders wants that her goal $\mathbf{g}(p)$ for "arrive at destination" be satisfied both in the context C_1 in which the context proposition "$\mathbf{c}(q)$: flight is on time" holds (i.e., $\mathbf{c}(q) \in C_1$), and in the context C_2 in which the context proposition "$\mathbf{c}(r)$: flight is delayed but not more than 5 hours" holds (i.e., $\mathbf{c}(r) \in C_2$), then $C_1 \in \mathcal{C}(\mathbf{g}(p))$ and $C_1 \in \mathcal{C}(\mathbf{g}(p))$.

Secondly, observe that the BNF specification lets us write formulas in which we combine context propositions and requirements, e.g.: $\mathbf{k}(p) \wedge \mathbf{c}(q) \rightarrow \bot$, which the requirements engineer can use to state that the domain assumption $\mathbf{k}(p)$ that was communicated by the stakeholder does not hold in contexts in which the context proposition $\mathbf{c}(q)$ holds.

Since we can combine context formulas and requirements, we can state very useful relations, such as that some requirements conflict with some contexts, by saying that these requirements are inconsistent with some of the context formulas in these contexts. As an aside, rules that connect requirements and context formulas need not be specified in a definite way by the requirements engineer. It is also possible to learn them by asking feedback to the user i.e. a SAS at runtime can perform this task. For example, if the system asks the user a question of the form: *Your flight is delayed by 5 hours or more. Do you wish to rebook a flight for the next day?* This question can be reformulated as a question on which of these two formulas to add to the current context of the user (i.e. the context in which we asked the user that question):

$$\mathbf{c}(\mathbf{c}(p) \wedge \mathbf{g}(q_1) \rightarrow \bot) \tag{4.6}$$

$$\mathbf{c}(\mathbf{c}(p) \wedge \mathbf{g}(q_2) \rightarrow \bot) \tag{4.7}$$

where $\mathbf{c}(p)$ is for "flight delayed by more than 5 hours", $\mathbf{g}(q_1)$ is for the goal "keep the booked flight", and $\mathbf{g}(q_2)$ is for the goal "rebook the same flight for the next day". If the user answers "yes", then add formula $\mathbf{c}(\mathbf{c}(p) \wedge \mathbf{g}(q_1)) \rightarrow \bot$ to the context in which we asked the user that question; if the user answers "no", then we add $\mathbf{c}(\mathbf{c}(p) \wedge \mathbf{g}(q_2)) \rightarrow \bot$ to the current context.

We now add the Resource concept. Since the formal language that we use in this paper is propositional, we will keep the resource concept out of it.

Definition 8. *Resource: An instance R of the Resource concept is an entity referred to by one or more instances of **Communicated information**.*

In order to introduce resources in the definition of the runtime requirements adaptation problem, we need a function that tells us which resources are referred to by a task, domain assumption, or a context proposition, as these resources will need to be available and used in some way in order to ensure that the relevant domain assumptions and context propositions hold, and that the tasks can be executed.

Definition 9. *Resource selector function: Let C be the set of all contexts. Given a set of tasks, domain assumptions, and/or context propositions, the resource selector function returns the identifiers of resources necessary for the domain assumptions and/or context propositions to hold, and/or tasks to be executed:*

$$\mathcal{R} : \wp(T \cup K \cup \bigcup C) \longrightarrow \wp(R) \tag{4.8}$$

The domain of $\mathcal{R}$ are domain assumptions, context propositions, and tasks. The reason that goals, softgoals, and quality constraints are absent is that the resources will be mobilized to realize a candidate solution to the requirements problem, and the candidate solution includes only domain assumptions and tasks. Since these domain assumptions and tasks are contextualized, we need to ensure the availability of resources that are needed in the context on which these domain assumptions and tasks depend on. Note also that we have $\bigcup C$ because C is a set of sets, so that we need to get the union of all of the sets in C.

We can now formulate the runtime requirements adaptation problem for SAS.

Definition 10. *Runtime requirements adaptation problem: Given a candidate solution $CS(C_1)$ in the context $C_1 \in C$ to the requirements problem $RP(C_1)$ in context $C_1 \in C$, and a change from context C_1 to $C_2 \neq C_1$, **find** the requirements problem $RP(C_2)$ in context $C_2 \in C$ and choose among candidate solutions to $RP(C_2)$ a solution $CS(C_2)$ in the context C_2 to the requirements problem $RP(C_2)$ in the context $C_2 \in C$.*

The definition of the runtime requirements adaptation problem reflects the intuition that by changing the context, the requirements problem may change – as requirements can change – and from there, a new solution needs to be found to the requirements problem in the new context.

We now reformulate the requirements problem so as to highlight the role of context in it, as well as of the resources.

Definition 11. *Requirements problem RP(C) in context C: Given the elicited or otherwise acquired: domain assumptions in the set **K**, tasks in **T**, goals in **G**, quality constraints in **Q**, softgoals in **S**, preference, is-mandatory and is-optional relations in **A**, a context C on which $K \cup T \cup G \cup Q \cup S$ and **A** depend on, **find** all **candidate solutions in context** C to RP(C) and compare them using preference and is-optional relations from **A** to identify the most desirable candidate solution.*

Definition 12. *Candidate solution CS **in the context** C: A set of tasks T^* and a set of domain assumptions K^* are a **candidate solution in the context** C to the requirements problem RP(C) in context C if and only if:*

1. *K^* and T^* are not inconsistent,*
2. *$C, K^*, T^* \vdash_{\partial \tau} G^*, Q^*$, where $G^* \subseteq G$ and $Q^* \subseteq Q$,*
3. *G^* and Q^* include, respectively, all mandatory goals and quality constraints,*
4. *all mandatory softgoals are approximated by the consequences of $C, K^* \cup T^*$, so that $K^*, T^* \vdash_{\partial \tau} S^M$, where S^M is the set of mandatory softgoals, and*
5. *resources $\mathcal{R}(C \cup K^* \cup T^*)$ needed to realize this candidate solution are available.*

As discussed earlier, we view runtime requirements adaptation problem as a dynamic RE problem. To support the analysis, we add two relations in the CORE ontology. We now define the *relegation relation* via the inference and preference relations in Techne.

Definition 13. *Relegation relation: A relegation relation is an $(n+1)$-ary relation that stands between a requirement $\phi \in \Delta$ and n other sets of requirements $\Pi_1, \Pi_2, \ldots, \Pi_n \subseteq \Delta$ if and only if there is an inference relation from every Π_i to ϕ and there is a binary relation: $\succ_\phi \subseteq \{\Pi_i \mid 1 \leq i \leq n\} \times \{\Pi_i \mid 1 \leq i \leq n\}$ whereby $\Pi_i \succ_\phi \Pi_j$ if it is strictly more desirable to satisfy ϕ by ensuring that Π_i holds, than to satisfy ϕ by ensuring that Π_j holds.*

The inference relations required by a relegation relations indicate that a relegation relation can only be defined for requirements that we know how to satisfy in different ways. For example, if we have a goal $\mathbf{g}(p)$, and we have two ways to satisfy that goal, e.g.:

$$\Pi_1 = \{\mathbf{t}(q_1), \mathbf{b}(\mathbf{t}(q_1) \to \mathbf{g}(p))\} \qquad (4.9)$$

$$\Pi_2 = \{\mathbf{t}(q_2), \mathbf{b}(\mathbf{t}(q_2) \to \mathbf{g}(p))\} \qquad (4.10)$$

then we have satisfied the first condition from the definition of the relegation relation, since $\Pi_1 \vdash_{\partial \tau} \mathbf{g}(p)$ and $\Pi_2 \vdash_{\partial \tau} \mathbf{g}(p)$.

The second condition in the definition of the relegation relation says that we need to define a preference relation $\succ_{\mathbf{g}(p)}$ between different ways of satisfying $\mathbf{g}(p)$. Observe that we define $\succ_{\mathbf{g}(p)}$ between *sets* of information, not pieces of information. The Techne preference relation defines preference between individual pieces of information, so we can use preference relations between members of Π_1 and Π_2 to define $\succ_{\mathbf{g}(p)}$.

Suppose that $\mathbf{t}(q_1) \succ \mathbf{t}(q_2)$, i.e., that we prefer to execute task $\mathbf{t}(q_1)$ to executing the task $\mathbf{t}(q_2)$. We can define $\succ_{\mathbf{g}(p)}$ as a function of the Techne preference relation, i.e., $\succ_{\mathbf{g}(p)} \overset{def}{=} f(\succ)$, that is, from the information that the preference relation already includes. Namely, in this example it is appropriate to say that, if $\mathbf{t}(q_1) \succ \mathbf{t}(q_2)$, then $\Pi_1 \succ_{\mathbf{g}(p)} \Pi_2$. Since we have only Π_1 and Π_2, it is enough to know that $\Pi_1 \succ_{\mathbf{g}(p)} \Pi_2$ to know everything we need to define the relegation relation.

Namely, the relegation relation $(\mathbf{g}(p), \Pi_1, \Pi_2, \succ_{\mathbf{g}(p)})$ tells us that, if we cannot satisfy $\mathbf{g}(p)$ through Π_1 then we will relegate to Π_2, i.e., satisfy $\mathbf{g}(p)$ through Π_2.

Finally, we define the *influence relation*. Note that it is simple here, since we have no numerical values, so we cannot speak about influence as correlation. We can only say that some information influences some other information if the absence of the former makes it impossible for us to satisfy the latter.

Definition 14. *Influence relation:* *An influence relation is a binary relation from* $\psi \in \mathcal{L}_{SAS}$ *to* $\phi \in \mathcal{L}_{SAS}$, *iff either:*

1. $\exists \Pi \subseteq \Delta \cup C$ *s.t.* $\Pi \vdash_{\partial \tau} \phi$ *and* $\Pi \setminus \psi \nvdash_{\partial \tau} \phi$, *or*
2. $\forall \Pi \subseteq \Delta \cup C$ *s.t.* $\Pi \vdash_{\partial \tau} \phi$ *and* $\Pi \setminus \psi \nvdash_{\partial \tau} \phi$.

In the first case above, we say that ψ *weakly influences* ϕ, *denoted* $\psi \overset{wi}{\longrightarrow} \phi$. *In the second case above, we say that* ψ *strongly influences* ϕ, *denoted* $\psi \overset{si}{\longrightarrow} \phi$.

Remark 4. If $\psi \overset{si}{\longrightarrow} \phi$, then we have no way to satisfy ϕ if ψ is not satisfied. If $\psi \overset{wi}{\longrightarrow} \phi$, then some ways of satisfying ϕ cannot be used to do so if ψ is not satisfied.   ∎

4.1 Runtime Requirements Adaptation Problem Illustration

We now revisit the above definitions and use scenarios from travel exemplar case study to illustrate how SAS, instantiating CARE and running on user's mobile phone, resolves the "runtime requirements adaptation problem" at runtime.

For example, user arrives at the airport to avail her flight from Italy to Canada via Paris for a business meeting. While at the airport after the boarding, user want to connect to the Internet using her mobile phone to check emails and flight details before checking in for the plane. Moreover, user wants to be informed about any flight delay.

Taking the above example, we now present SAS adaptation sequence at runtime in case of change in context C along the time $T = t^1, \dots t^n$ as shown in the Fig.2. Let CS be a set of candidate solution, thereby determining the runtime requirements adaptation problem as a combination of instances of the tasks $\mathbf{T}^*$) and domain assumptions $\mathbf{K}^*$ such that $\mathbf{G}^*, \mathbf{Q}^*$ and $\mathbf{S^M}$ are satisfied. In case of changes in the context $C = C_1, \dots C_n$ overtime for which CS needs to be re-evaluated by the system and R is required to be used or identified in a given context C to realize CS. By re-evaluation we mean that system at runtime exploits its monitored information, evaluate all the possible alternative CS or search for new ones (i.e. exploiting available services) that can satisfy the runtime adaptation problem in response to changes in the C therefore adapting to the

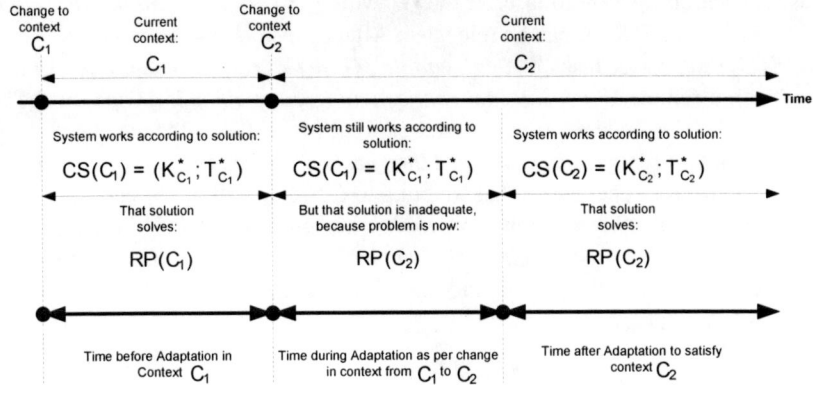

Fig. 2. System Adaptation Sequence in Time

candidate solution CS. At this SAS may perform at sub-optimal level and can exploit automated reasoning techniques such as planning or decision making techniques such as analytic hierarchy process (AHP)[3].

Before Adaptation: Assume that at time t^1, the user's goals $\mathbf{G}^*$ are *to connect to the Internet for checking details of itinerary* and *inform about the flight delays* with the quality constraint $\mathbf{Q}$ *to have the Internet connectivity not less than 256Kbps*. To satisfy these requirements, SAS is running according to its candidate solution CS i.e., using the set of tasks $\mathbf{T}^*$, e.g. *search for available connection, enable Wi-Fi, get itinerary details* and *show flight itinerary*, in the current context i.e. C_1 is *at the airport*, and the domain assumption K_{t1}^* *Internet must be available at the airport*, is not violated. This implies that, $CS(C_1) = (K_{C_1}^*, T_{C_1}^*)$ and $CS(C_1)$ satisfies the runtime requirements adaptation problem i.e. $RP(C_1)$. We can rewrite this as:

$$C_1, \mathbf{K}_{C_1}^*, \mathbf{T}_{C_1}^* \vdash_{\!\mathcal{T}} \mathbf{G}_{C_1}^*, \mathbf{Q}_{C_1}^*$$

where $\mathcal{R}(C_1 \cup \mathbf{K}_{C_1}^* \cup \mathbf{T}_{C_1}^*)$ identifies the set of resources $\mathbf{R}$ available, e.g. (Airport Wi-Fi hotspot, Mobile Phone of the user) to perform $\mathbf{T}_{C_1}^*$.

During Adaptation: SAS while executing $CS(C_1)$, observes a change in context i.e. *the airport Wi-Fi connection becomes unavailable* at time t^2. Due to this change in context from C_1 to C_2, the existing candidate solution $CS(C_1)$ might be valid but is not adequate to satisfy the current context C_2. As a consequence, the SAS needs to re-evaluate its candidate solutions CS by searching in its solution base i.e. Δ or looking for solutions that can be realized through available services. For instance, a new candidate solution $CS(C_2)$ could be, e.g. *connect to the Internet using data services either 3G or Edge on mobile phone R*; or *recommending user to move to the area where the signal strength is stronger*; or *avail the Internet on the free booths*. At this stage, SAS

---

[3] Discussing such techniques are out the scope of this paper. We present three scenarios to illustrate how the SAS can adapt at runtime by resolving the runtime requirements adaptation problem.

may use relegation relation to infer, if the $\mathbf{G}^*$ with a $\mathbf{Q}$ is *to have the Internet connectivity not less than 256Kbps* can be relegated. After re-evaluating the possibilities, SAS finds $CS(C_2)$ i.e. set of tasks $\mathbf{T}^*$, e.g. *enable 3G or Edge service* and *connect to the Internet* with a refined $\mathbf{Q}$ i.e. *Internet connectivity greater than 256Kbps* for the user. At this stage the influence relation is also used to ascertain the influence of $CS(C_2)$ on user's goals and preference, e.g. *Hi-speed Internet is preferred than no Internet connection.* SAS can derive conclusions that adapting to $CS(C_2)$ will not affect $K_{C_2}^*$ i.e. *Any flight information must be communicated to the customer* and goal $\mathbf{G}^*$ i.e. *to connect to the Internet to view itinerary* and *inform about the flight delays* will be satisfied. Therefore, $CS(C_2) = (K_{C_2}^*, T_{C_2}^*)$; satisfying the runtime requirements adaptation problem i.e. $RP(C_2)$. We can rewrite this as:

$$C_2, \mathbf{K}_{C_2}^*, \mathbf{T}_{C_2}^* \vdash_\tau \mathbf{G}_{C_2}^*, \mathbf{Q}_{C_2}^*$$

where $\mathcal{R}(C_2 \cup \mathbf{K}_{C_2}^* \cup \mathbf{T}_{C_2}^*)$ identifies the set of resources $\mathbf{R}$ available, e.g. (Access 3G or Edge data services, Mobile Phone of the user) to perform $\mathbf{T}_{C_2}^*$.

After Adaptation: At time t^3, SAS adapts to the candidate solution $CS(C_2)$ taking into account the context C_2 and available resources R i.e. Access 3G or Edge data services, Mobile Phone of the user, thus not violating the $K_{C_2}^*$. Adaptation is performed dynamically at runtime by changing (e.g. switching, re-configuring, optimizing) SAS from one requirements problem to another i.e. $RP(C_1)$ to $RP(C_2)$, in response to changes in the context, user's needs or resource variability.

5   Related Work

Requirements engineering is carried out at the outset of the whole development process, but in the context of SAS, RE activities are also needed at runtime thus enabling a seamless adaptation. Research on SAS has recently received attention from variety of research communities mainly targeting the software engineering of SAS from requirements, design and implementation perspectives. Focusing on requirements engineering for SAS, research agenda from SEAMS [5] and RE community has identified key challenges that must be addressed while developing such systems.

For instance, in [23], a declarative language (RELAX) is proposed to capture uncertainty, using temporal operators (such as eventually, until) by formalizing the semantics in Fuzzy Branching Temporal logic, to specify requirements for SAS. Similarly, adopting goal-oriented analysis for adaptive systems, mitigation strategies are proposed to accommodate uncertainty and failures by using obstacle analysis in [4]. Requirements reflection is another aspect, where ideas from computational reflection has been borrowed to provide SAS the capability to be aware of its own requirements [21]. Similarly, online goal refinement [9] is of prime importance considering the underline architecture of the intended SAS. Taking the engineering perspective, making the role of feedback loops more explicit in the design of SAS has been recognized as a key requirement for developing SAS in [3].

In our previous work, we proposed similar ideas to engineer adaptive requirements using goal models and ontologies by making explicit the requirements for feedback loop (i.e. monitoring, evaluation criteria, and adaptation alternative) more explicit in [16]. We extend this work in [17,18,19] by proposing a Continuous Adaptive RE (CARE) framework for continuous online refinement of requirements at runtime by the system itself involving the end-user. We proposed an architecture of an application that instantiate CARE. We proposed a classification of adaptation at runtime by exploiting incremental reasoning over adaptive requirements represented as runtime artifact. Similar ideas has been proposed treating goals as fuzzy goals formalized using fuzzy logic representing strategies for adaptation and operationalizing them as BPEL processes in [2]. Another variation of this idea has been advocated in [22] as "Awareness Requirements", as a way to express constraints on requirements as meta requirements to deal with uncertainty while developing SAS.

In goal-oriented modeling, *Tropos* has been extended to capture the contextual variability (mainly location) [1] by leveraging the concept of variation points [10] exploiting the decomposition rules in a goal tree. Mainly, it helps in linking the alternative in the goal model to the corresponding context (location) that helps in monitoring facts and reasoning for adaptation in case of change in the context (location). Extended design abstractions, including environment models, explicit goal types, and conditions for building adaptive agents have been proposed as an extension of *Tropos*, in Tropos4AS [13].

6   Discussion

It is worth to further discuss assumptions underlying the suggested problem formulation, its generality as well as its practical impact. The problem formulation suggested in this paper makes no assumptions and imposes no constraints on how the information that is used and acquired. We thereby recognize that not all information can be collected during requirements engineering, or at design time, but that this will depend on the technologies used to implement the system. For example, the information about the context, the formulas in C may – if the implementation technology allows – be obtained by recognizing patterns in the data that arrives through sensors, then matching patterns of data to templates of proposition or implications. We stayed in the propositional case, since this was enough to define the main concepts and relations, and subsequently use them to formulate the runtime requirements adaptation problem. The actual system will operate using perhaps more elaborate, first-order formalisms to represent information, so as to make that information useful for planning algorithms applied to identify candidate solutions. However, regardless of the formalism used, the system still needs to be designed to ensure the general conditions and relations that the problem formulation states: e.g., that the system needs an internal representation of information pertaining to contexts, domain assumptions, tasks, goals, and so on, that goals and quality constraints are satisfied through consistent combinations of C, $\mathbf{K}$ and $\mathbf{T}$, among others.

Concerning generality of the proposed problem formulation and practical implication, our aim in this work was first to understand the general problem, and then focus on developing particular requirements modeling languages to handle it. In this regards we believe that recently proposed frameworks for engineering requirements for SAS can be reconnected to our problem formulation. Consider, for example RELAX [23], which proposes a formalism for the specification of requirements and a particular way to relax them at runtime: if a requirement cannot be satisfied to the desired extent, then alternative requirements can be specified in RELAX, stating thereby how achievement conditions for the original requirement can be relaxed. This mechanism can be considered a particular way to implement the Relegate relation, that is the RELAX mechanism obtains a straightforward interpretation in the language we used here. There can be other ways to handle uncertainty and relaxation of requirements, and our aim in this paper was to remain independent of particular approaches.

7   Conclusions and Future Work

We argued in this paper for a general formulation of the runtime requirements adaptation problem, using recent work on the revised general requirements problem and its core ontology. Taking into account our work on continuous adaptive RE in CARE, and the types of adaptation defined in CARE, in this paper, we proposed to make explicit the dynamic parts in requirements representation and formulated the runtime requirements adaptation problem. In particular, two key concepts help managing runtime requirements changes, namely the concept of *context* and *resource*, while the *relegation* and *influence* relations capture changes at runtime. The proposed *runtime requirements adaptation problem* is envisioned as dynamic RE problem for adaptive systems i.e. reformulating the requirements problem when changes occur that a SAS can detect, and then solving the changed problem at runtime.

Ongoing work aims at exploiting these formal definitions of concepts and relations into a more concrete modeling and analysis language. The concept of requirements database Δ introduced in this paper provides premise to define operations (e.g. adding, removing, substituting requirements) that a SAS may perform over Δ to update its own specification at runtime thus help realizing continuous adaptive RE (see CARE [17,18]). Moreover, the application at runtime of automated reasoning (such as in AI planning) and decision-making techniques (e.g., Analytic Hierarchy Process) may be relevant for the engineering and running of SAS. They require further investigation.

References

1. Ali, R., Dalpiaz, F., Giorgini, P.: Location-based software modeling and analysis: Tropos-based approach. In: Li, Q., Spaccapietra, S., Yu, E., Olivé, A. (eds.) ER 2008. LNCS, vol. 5231, pp. 169–182. Springer, Heidelberg (2008)
2. Baresi, L., Pasquale, L., Spoletini, P.: Fuzzy goals for requirements-driven adaptation. In: 18th IEEE Int. Requirements Eng. Conf., Sydney, Australia, pp. 125–134 (2010)
3. Brun, Y., Serugendo, G.D.M., Gacek, C., Giese, H.M., Kienle, H., Litoiu, M., Müller, H.A., Pezzè, M., Shaw, M.: Engineering self-adaptive systems through feedback loops. Software Engineering for Self-Adaptive Systems 5525, 48–70 (2009)
4. Cheng, B.H.C., Sawyer, P., Bencomo, N., Whittle, J.: A goal-based modeling approach to develop requirements of an adaptive system with environmental uncertainty. In: Schürr, A., Selic, B. (eds.) MODELS 2009. LNCS, vol. 5795, pp. 468–483. Springer, Heidelberg (2009)
5. Cheng, B.H., de Lemos, R., Giese, H., Inverardi, P., Magee, J.: Software Engineering for Self-Adaptive Systems: A Research Roadmap. In: Cheng, B.H.C., de Lemos, R., Giese, H., Inverardi, P., Magee, J. (eds.) Software Engineering for Self-Adaptive Systems. LNCS, vol. 5525, pp. 1–26. Springer, Heidelberg (2009)

6. Dey, A.K.: Understanding and using context. Personal Ubiquitous Comput. 5(1), 4–7 (2001)
7. Jureta, I.J., Mylopoulos, J., Faulkner, S.: Revisiting the core ontology and problem in requirements engineering. In: 16th IEEE Int. Requirements Eng. Conf., pp. 71–80 (2008)
8. Jureta, I.J., Borgida, A., Ernst, N.A., Mylopoulos, J.: Techne: Towards a new generation of requirements modeling languages with goals, preferences, and inconsistency handling. In: 18th IEEE Int. Requirements Eng. Conf., Sydney, Australia, pp. 115–124 (2010)
9. Kramer, J., Magee, J.: Self-managed systems: an architectural challenge. In: Future of Software Engineering, 2007. FOSE 2007, pp. 259–268 (May 2007)
10. Liaskos, S., Lapouchnian, A., Yu, Y., Yu, E., Mylopoulos, J.: On goal-based variability acquisition and analysis. In: 14th IEEE Int. Requirements Eng. Conf., pp. 79–88 (2006)
11. Liaskos, S., McIlraith, S.A., Mylopoulos, J.: Integrating preferences into goal models for requirements engineering. In: 18th IEEE Int. Requirements Eng. Conf., Sydney, Australia, pp. 135–144 (2010)
12. McCarthy, J.: Notes on formalizing context. In: Proceedings of the 13th International Joint Conference on Artifical Intelligence, vol. 1, pp. 555–560. Morgan Kaufmann Publishers Inc., San Francisco (1993)
13. Morandini, M., Penserini, L., Perini, A.: Towards goal-oriented development of self-adaptive systems. In: ICSE Workshop on Software Engineering for Adaptive and Self-Managing Systems (SEAMS 2008), pp. 9–16. ACM, New York (2008)
14. Penserini, L., Perini, A., Susi, A., Mylopoulos, J.: High variability design for software agents: Extending Tropos. TAAS 2(4) (2007)
15. Qureshi, N.A., Jureta, I., Perini, A.: On runtime requirements adaptation problem for self-adaptive systems, SE Research Group Technical Report (TR-FBK-SE-2010-1), FBK, Trento, Italy (2010), http://se.fbk.eu/files/TR-FBK-SE-2010-1.pdf
16. Qureshi, N.A., Perini, A.: Engineering adaptive requirements. In: ICSE Workshop on Software Engineering for Adaptive and Self-Managing Systems (SEAMS 2009), pp. 126–131. IEEE Computer Society, Washington, DC, USA (2009)
17. Qureshi, N.A., Perini, A.: Continuous adaptive requirements engineering: An architecture for self-adaptive service-based applications. In: First Int. Workshop on Requirements@Run.Time (RE@RunTime), Sydney, Australia, pp. 17–24 (2010)
18. Qureshi, N.A., Perini, A.: Requirements engineering for adaptive service based applications. In: 18th IEEE Int. Requirements Eng. Conf., Sydney, Australia, pp. 108–111 (2010)
19. Qureshi, N.A., Perini, A., Ernst, N.A., Mylopoulos, J.: Towards a continuous requirements engineering framework for self-adaptive systems. In: First Int. Workshop on Requirements@Run.Time (RE@RunTime), held at (RE 2010), Sydney, Australia, pp. 9–16 (2010)
20. M., Salifu, Yu, Y., Nuseibeh, B.: Specifying monitoring and switching problems in context. In: 15th IEEE Int. Requirements Eng. Conf., pp. 211–220 (2007)
21. Sawyer, P., Bencomo, N., Whittle, J., Letier, E., Finkelstein, A.: Requirements-aware systems a research agenda for re for self-adaptive systems. In: 18th IEEE Int. Requirements Eng. Conf., Sydney, Australia, pp. 95–103 (2010)
22. Souza, V.E.S., Lapouchnian, A., Robinson, W.N., Mylopoulos, J.: Awareness requirements for adaptive systems, technical Report DISI-10-049, DISI, Universit'a di Trento, Italy (2010)
23. Whittle, J., Sawyer, P., Bencomo, N., Cheng, B.H., Bruel, J.-M.: RELAX: Incorporating Uncertainty into the Specification of Self-Adaptive Systems. In: 17th IEEE Int. Requirements Eng. Conf., Atlanta, pp. 79–88 (2009)
24. Zave, P., Jackson, M.: Four dark corners of requirements engineering. ACM Trans. Softw. Eng. Methodol. 6(1), 1–30 (1997)

A Fuzzy Service Adaptation Based on QoS Satisfaction

Barbara Pernici and Seyed Hossein Siadat

Politecnico di Milano
Dipartimento di Elettronica e Informazione
Piazza Leonardo da Vinci 32, 20133 Milano, Italy
{pernici,siadat}@elet.polimi.it

Abstract. Quality of Service (QoS) once defined in a contract between two parties may change during the life-cycle of Service-Based Applications (SBAs). Changes could be due to system failures or evolution of quality requirements from the involved parties. Therefore, Web Services need to be able to adapt dynamically to respond to such changes. Adaptation and evolution of services are playing an important task in this domain. An essential issue to be addressed is how to efficiently select an adaptation while, there exists different strategies. We propose a fuzzy service adaptation approach that works based on the degree of QoS satisfaction. In particular, we define fuzzy parameters for the QoS property descriptions of Web Services. This way, partial satisfaction of parameters is allowed through measuring imprecise requirements. The QoS satisfaction degree is measured using membership functions provided for each parameter. Experimental results show the effectiveness of the fuzzy approach using the satisfaction degree in selecting the best adaptation strategy.

Keywords: QoS, service adaptation and evolution, fuzzy logic.

1 Introduction

In Service-Based Applications (SBAs), Quality of Service (QoS) parameters may change during the life cycle of the application. Web service adaptation is an important phase to deal with such changes. Handling changes in a demanding and adaptive environment is a vital task. One main issue lies in QoS property descriptions of Web Services. This involves specifying service requirements in a formal way, monitoring and dynamically adapting and evolving the services with respect to the QoS changes. Static adaptation is impractical due to the changing environment and high cost of maintenance and development. Specifying all possible alternative behaviour for adaptation at design time is impossible. Therefore, a declarative approach is required at run-time to support adaptation decisions.

In order to perform run-time decisions for adaptation in a volatile environment, one issue is to consider the imprecise evaluation of QoS properties. Existing approaches do not allow partial satisfaction of parameters. It is required that

H. Mouratidis and C. Rolland (Eds.): CAiSE 2011, LNCS 6741, pp. 48–61, 2011.

services should be able to tolerate a range of violation in their quality description. However, handling this toleration need to be done with special care. An important issue to address in SBAs is to what extent the QoS parameters of a Web Service are satisfiable. The answer to this issue could be a basic for making adaptation decisions. However, this issue has not been addressed adequately in the literatures. Evaluating the extent of parameter satisfaction is necessary to help the selection of best adaptation strategy.

As an initial step to this, in [3] we provided conditions under which QoS changes are acceptable. We used a temporal logic namely Allen's Interval Algebra (AIA) [2] to formally specify the non-functional properties of web services. We then used the AIA to reason about changes of quality parameters and their evolution. In this paper, we extend [3] and propose a fuzzy approach to support service adaptation and evolution. We define *fuzzy parameters* for QoS property description of Web Services. Fuzzy parameters could be considered as fuzzy sets and measured based on their value of membership. Satisfaction degree of fuzzy parameters is measured according to their actual distance of the agreed quality ranges in the contract. The goal of this paper is to provide flexibility for service specification by applying fuzzy parameters. Using a fuzzy approach allows us to deal with reasoning on the quality violations that is approximate rather than accurate. At the end, we propose different categories of adaptation that perform based on the satisfaction degree. Experimental results show the effectiveness of using the fuzzy approach over the non-fuzzy one in making decisions for adaptation.

The remainder of the paper is structured as follows. Section 2 describes the major related work. In Section 3 we present a definition for QoS property description of services through introducing fuzzy parameters. In Section 4 we specify satisfaction functions for each parameter to measure to what extend the QoS is achieved with respect to the existing contract. We explain the decision making mechanism in Section 5 that works based on the satisfaction degree. Section 6 provides experimental result using a simulator and evaluates the effectiveness of the proposed approach. Section 7 concludes the paper and discusses our future work.

2   Related Work

Deviation of quality ranges from the existing contract may produce a system failure and bring dissatisfaction for customers. To this end, the evolution and adaptation of web services are becoming two important issues in reacting to the various changes in order to provide the agreed QoS stated in the contract. Recently, many adaptation strategies and methods have been proposed in the literature. However, most of the work in service adaptation concentrates on the technical issues and definition of mechanisms for adaptation rather than considering QoS perspective. A list of adaptation strategies for repair processes in SBAs is provided in [6] and [1]. For example, [7] proposed a service replacement approach for adaptive Web Service composition and execution, while Canfora et al. [5] presented a re-composition approach dealing QoS replanning issues at run time using late binding technique. However, none of these works consider the consequences and potential overheads of adaptation. To this end, for example, an environment for compensation of Web Service transactions is proposed in [25]. In order to consider the overall value of a change, [15] presented an approach called *value of changed information* (VOC). Furthermore, an adaptation mechanism is proposed based on VOC in [8]. However, these works have the limitation that they do not take into account the satisfaction level of services. Making adaptation decisions and evaluating them is therefore complicated and has consequences that are often neglected. Some qualitative and quantitative techniques has been proposed, however evaluating impacts of adaptation still remains as an open challenge.

One core issue to address is the definition of a flexible description for Web Services. Formulation of service specifications/requirements has been studied in the literature. In autonomic systems and in particular web services, reasoning about such specification is a hard job due to the changing environment that affects service requirements. Although a lot of research has been conducted for functional Web Service description, only a few efforts have been done with respect to non-functional properties description of Web Service. Among syntactic and semantic WS description we refer to the work done in [30], [21] and [16] which they also provided algorithms for service selection based on such description. A major limitation of those papers and other similar ones is due to not considering the partial satisfaction of the QoS attributes. With this regards, [23] provided a semantic Policy Centered Meta-model (PCM) approach for non-functional property description of web services. A number of operators (e.g. greaterEqual, atLeast) for numeric values are defined in the model for determining tradeoffs between various requests. Therefore, the approach can support the selection of Web Services that partially satisfy user constraints. In [20] and [19], the authors extend the approach proposed in [23] by proving a solution for Web Service evaluation based on constraint satisfaction problem. The approach uses utility function to present the level of preferences for each value ranges defined in the service description. However, it does not take care of adaptation issues and controlling values at run-time. In [22], the authors discuss about fixable and non-fixable properties to deal with bounded uncertainty issue. Constraint programming is used as a solution, however, there is no evaluation of the work.

It is required to provide a framework to evaluate alternatives and quantify their impact for making decision decisions. Each alternative has different degree of satisfaction and their impact has to be evaluated in order to select the best adaptation strategy. A quantitative approach applying a probabilistic modeling is used for partial goal satisfaction in [18]. Dealing with the uncertainty issue is one major problem in order to formulate and manage service specification. Thus, recently researchers are investigating to incorporate this uncertainty into the service specification. In [14], the author provides support for reasoning about uncertainty. A goal-base approach for requirement modeling in adaptive systems is proposed in [9] which uncertainty of the environment is taken into account.

Furthermore, a language named RELAX is developed for specifying requirements in adaptive systems [26,27] in which certain requirements could be temporarily relaxed in favor of others. In general, different temporal logics have been used for formal specification of requirement. Linear Temporal Logic (LTL) has been used in [4,11] to formally specify requirements in a goal oriented approach. In particular, LTL is extended in [31] and named A-LTL to support adaptive program semantics by introducing an adaptation operator. [3] uses Allen's interval algebra for the formal specification of service requirement. Those approaches have limitations such that they are unable to consider environmental uncertainty and behave in a binary satisfaction manner.

Fuzzy approach [29] is an alternative to concur such limitations of aforementioned approaches. However, the fuzzy approach may not be the only alternative to deal with uncertainty. Different mathematical and frameworks are presented in the literature to address the uncertainty issue and partial satisfaction of the requirements. For example, making decisions about non-functional properties using Bayesian networks is proposed in [13] while [17] used a probabilistic method for this purpose. Applying fuzzy logic to incorporate uncertainty and making decisions has been proposed in other domains such as management, economy and many aspects of computer science, however, to the best of our knowledge there is very little of such application in adaptation of web services. As of such, [10] proposed a fuzzy approach for assigning fitness degrees to service policies in a context-aware mobile computing middleware. A trade-off analysis using fuzzy approach for addressing conflicts using imprecise requirements in proposed in [28]. With respect to partial satisfaction of requirements, [12] provided a web service selection approach using imprecise QoS constraints.

There are several different approaches towards adaptation of web services. This diversity yields from a missing consensus on the required decision making to automatically perform web service adaptation. Therefore, in this paper we propose a fuzzy adaptation approach as a possible way in providing a foundation of such a consensus which is based on the satisfaction degree of QoS parameters.

3   Fuzzy Parameters for QoS Property Description

This section is devoted to present a formal definition of quality *parameters* in a service description and is concerned with QoS property descriptions of Web Services. The formal specification we propose has been inspired and is an extension of our previous work [3]. We extended the work by defining *fuzzy parameters* for such service description. Fuzzy parameters could be considered as fuzzy sets and measured based on their value of membership. Satisfaction degree of fuzzy parameters is measured according to their actual distance of the agreed quality ranges in the contract. Having introduced the fuzzy parameters it is possible to understand to what extent the quality parameters are violated/satisfied. This way, partial satisfaction of parameters is allowed through measuring imprecise requirements.

We define set $\mathcal{D}$ to contain the quality dimensions (such as availability, execution time, price or throughput) identified and agreed by the service provider and consumer. Each quality dimension has a domain and range; e.g., availability is a probability usually expressed as a percentage in the range 0-100% and execution time is in the domain of real numbers in the range $0.. + \infty$. A quality dimension d can be considered *monotonic* (denoted by d^+) or *antitonic* (d^-); monotonicity indicates that values closer to the upper bound of the range are considered better, whilst with antitonic dimensions values closer to the lower bound are considered better. A *parameter m* associates a quality dimension to a value range [3].

If a parameter is non-fuzzy (strict) its satisfaction degree will be evaluated in a binary manner (Yes or No). In contrast, fuzzy parameters (relaxed) will be evaluated in a fuzzy manner which shows different degree of satisfaction ($x \in [0, 1]$). Note that we also provide value ranges for both parameters regardless of being fuzzy or non-fuzzy. The satisfaction degree of fuzzy parameters will be measured using *membership functions* provided for each parameter. In the following we provide the extended definition of a *parameter* based on the definition introduced in [3]. In particular, we define a type t for a parameter that can be either strict or fuzzy.

Definition 1 (Parameter). *We define a Parameter $m \in \mathcal{M}$ as a tuple $m := (d, v, t), d \in \mathcal{D}, v \in \mathcal{V}, t \in \{s, f\}$. where $\mathcal{D}$ is the set of quality dimensions, $\mathcal{V}$ is the set of ranges for all quality dimensions $\mathcal{D}$, s represent a strict parameter and f represent a fuzzy parameter.*

QoS once defined in a contract between two parties may change during a service life-cycle. Changes could be due to system failures or evolution of quality requirements from the involved parties. Therefore, Web Services need to be able to adapt dynamically to respond to such changes. Adaptation and evolution of services are playing an important task in this domain. However, adaptation of web services needs to be performed in an appropriate manner to accommodate QoS changes/violations by choosing the best adaptation strategy. Defining service description with the proposed fuzzy parameters provides a flexible situation in dealing with adaptation decisions. We discuss how it can facilitate the adaptation of web services through an example. According to the new definition of parameters, we consider availability and response time as fuzzy parameters. Let us assume an example of a contract with initial value ranges of availability between 80% to 90% and response time between 2 to 5 seconds. We use this example throughout the paper.

We provided situations in which new QoS ranges could be still acceptable for both parties according to the existing contract [3]. We introduced a compatibility mechanism that used parameter subtyping relation and Allen's Interval Algebra [2] for comparing value ranges and their evolution. The provider and requestor are compatible with each other according to the existing contract if the QoS changes are in one of the acceptable situations. If the compatibility is not provided, however it does not give any information about the degree of satisfaction/dissatisfaction of the offered service. For example if the new range

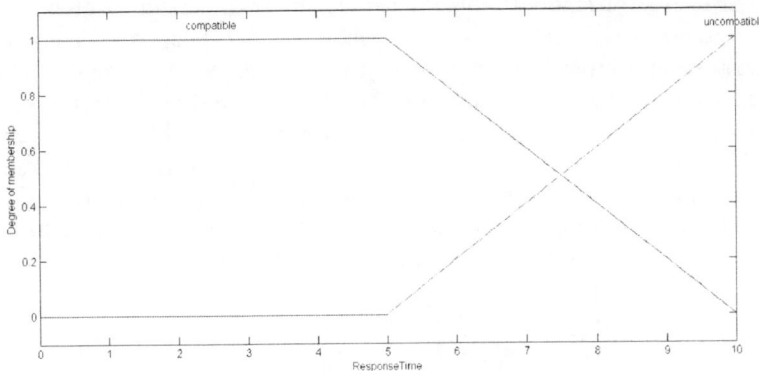

Fig. 1. Membership functions for response time

of availability is less than 80%, this is not considered as an acceptable situation and it is considered as a violation. In such cases, we would also like to understand to what extent the quality parameter and the aggregated service quality are satisfactory. An availability of 75% might still be acceptable if we consider the partial satisfaction of quality ranges.

4   Specifying Satisfaction Function

Having defined the fuzzy parameters we are able to apply the fuzzy logic. As for the first step we need to know the right amount of quality satisfaction. Previously in [3], we provided a compatibility mechanism to understand under which conditions the changes are acceptable. The approach suffers from the limitation that changes are considered either compatible or incompatible with the contract. This means, quality changes are calculated in a binary approach which it does not take into account clearly the relation of quality parameters with their satisfaction. To say it in other way, the QoS parameters are measured in a precise manner and their partial satisfaction is not taken into account. In the following we provide mechanisms to allow partial satisfaction of quality parameters imprecisely using fuzzy sets.

The main point of using fuzzy logic is to find a relation and to map our input space to the output space. The inputs here are namely service availability and response time and the output is the overall satisfaction degree of them. For each QoS parameter in the service description we provide a membership function that represent the level of satisfaction of each parameter. The membership functions map the value of each parameter to a membership value between 0 and 1. We use a piece-wise linear function, named *trapezoidal* membership function, for this purpose. Membership functions for ResponseTime and availability are shown in Figures 1 and 2. Both membership functions have two linguistic states namely *compatible* and *incompatible* and they are identified according to the initial value ranges of the contract. Figure 1 shows that the response time of 0 to 5 has the maximum degree of compatibility; however, the membership degree decreases while the response time increases. It also shows that response time has the minimum degree of incompatibility between 0 to 5 seconds; however, the membership degree increases while the response time increases. Note that the response time is set to be 2 to 5 seconds in the contract; however, the range between 0 to 2 is also acceptable with the same membership value as the initial range in the contract has [3]. With the same reasoning, the availability of 80 to 100 has the maximum degree of compatibility illustrated in Figure 2. However, the membership degree of compatibility decreases while the availability decreases.

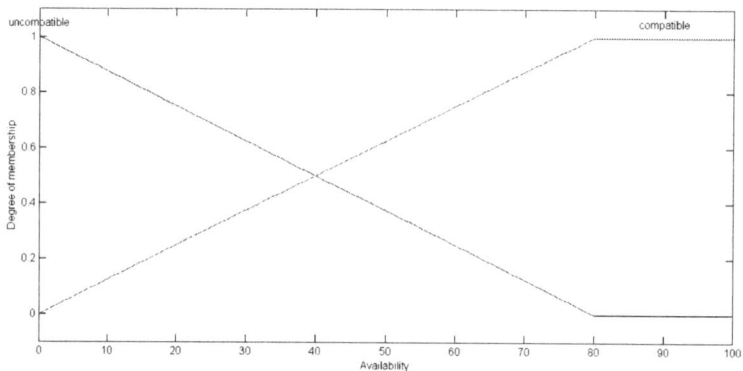

Fig. 2. Membership functions for ResponseTime and Availability

Having defined the membership functions, the mapping between the input and output space will be done by defining a list of *if-then statements* called *rules*. We have already defined what do we mean being compatible and incompatible for the quality parameters and specified their ranges using membership functions. Since we are relaxing the antecedent using a fuzzy statement, it is also required to represent the membership degree of the output (i.e. here satisfaction). Therefore, the satisfaction degree is also represented as fuzzy sets: satisfaction is low, satisfaction is average and satisfaction is high.

We define three if-then rules as below. it represents the *antecedent* and *consequent* of the rule. All the rules are applied in parallel and their order in unimportant. We define the fuzzy union/disjunction (OR) and the fuzzy conjunction/intersection (AND) using *max* and *min* functions respectively. Therefore $A\ AND\ B$ is represented as $min(A, B)$ and $A\ OR\ B$ is represented as $max(A, B)$.

1. *If* (*ResponseTime is compatible*) *and* (*Availability is Compatible*) *then* (*Satisfaction is high*).

2. *If* (*ResponseTime is incompatible*) *or* (*Availability is incompatible*) *then* (*Satisfaction is average*).

3. *If* (*ResponseTime is incompatible*) *and* (*Availability is incompatible*) *then* (*Satisfaction is low*).

5   Decision Making for Adaptation and Evolution

We use the satisfaction degree calculated using the fuzzy inference system for the adaptation and evolution decision making. The decision making mechanism works based on the algorithm we provided in [24]. The algorithm evaluates the evolution of the service and decides which adaptation strategy to take with respect to the predefined threshold degree for QoS satisfaction. The two main decisions are the *internal renegotiation* in which the changes are compatible with the service description in the contract and *service replacement* in which the changes are incompatible with the existing contract. The former case deals with the internal contract modification with the same provider and requester while the earlier case requires the selection of a new service and establishment of a new contract which can result in a huge loss of time and money.

Having provided such a decision making mechanism allows us to offer a flexible adaptation mechanism. This is done by identifying threshold to what constitutes *compatible* and *incompatible*. Using satisfaction degree allows us to define the criticality of a change/violation. Therefore, we are able to understand whether a violation is critical and it results in a service replacement or the violation is still acceptable. This way, a slight change from the quality ranges defined in the contract will not trigger the adaptation. Table 1 shows the result of checking for compatibility for a possible set of changes. The comparison is between our fuzzy approach and a traditional non-fuzzy one that works based on the precise evaluation of the quality ranges in the contract.

Table 1. Comparing the adaptation decisions using fuzzy and non-fuzzy approach

Change	Replacement? (Non-fuzzy/Fuzzy)	Change	Replacement? (Non-fuzzy/Fuzzy)
$S_1 = (6, .90)$	Yes/No	$S_2 = (7, .75)$	Yes/Yes
$S_3 = (5, .85)$	No/No	$S_4 = (3, .70)$	Yes/No
$S_5 = (2, .85)$	No/No	$S_6 = (2, .78)$	Yes/No
$S_7 = (2, .60)$	Yes/Yes	$S_8 = (3, .90)$	No/No
$S_9 = (7, .95)$	Yes/Yes	$S_{10} = (6, .50)$	Yes/Yes

In the fuzzy approach the replacement is based on the satisfaction degree. However, in the non-fuzzy approach a service replacement is necessary if any parameters are violated from the initial range, albeit minor deviation. For example in $S_1 = (6, .90)$, changing the response-time to 6s will not result a service replacement applying the fuzzy approach since it has the satisfaction degree of almost 83%. While applying a non-fuzzy approach, it is considered a violation because it does not respect the initial response-time range $(2, 5)$ in the contract. However, if a change results in a low satisfaction degree, service replacement is necessary in both approaches as in the case $S_{10} = (6, .50)$ which the satisfaction degree is around 62%.

6 Experiments and Implementation

Having defined the membership functions and rules in the previous sections, we have built and simulated a fuzzy inference system to interpret rules. The process has different steps including: fuzzification of input quality parameters, applying fuzzy operators to the antecedent, implication from the antecedent to the consequent, aggregation of the results for each rule, and defuzzification. A view of the simulator including the previous steps is illustrated in Figure 3 in which a complete fuzzy inference system is represented.

The first step is to apply the membership functions to map each QoS parameters to the appropriate fuzzy set (between 0 and 1). We used two inputs of Availability (interval between 0 to 100) and Response-time (interval between 0 to 10). The inputs are mapped to fuzzy linguistic sets: availability is compatible, availability is incompatible, response-time is compatible, and response-time is incompatible. Figures 1 and 2 show to what extent the availability and response-time are compatible. The next step is to give the result of the fuzzified input parameters to the fuzzy operators. According to the rules, AND and OR operators are applicable. This will give us a degree of support for each rule. Next is applying the implication method that uses the degree of support to calculate the output fuzzy set. We used a *minimum* method to truncate the output fuzzy set for all the rules separately. However, we apply all the rules in parallel and we do not define any priority and weight for them.

At the end of the implication, we apply an aggregation method to combine all the rules. This way, the outputs of each rule represented in fuzzy sets are combined into a single fuzzy set. A maximum method is used for the aggregation. The last step is to defuzzify the fuzzy set resulted after the aggregation step. We applied a *centroid* method to calculate the defuzzification process. The method returns the center of the area under the curve. Figure 3 shows that the

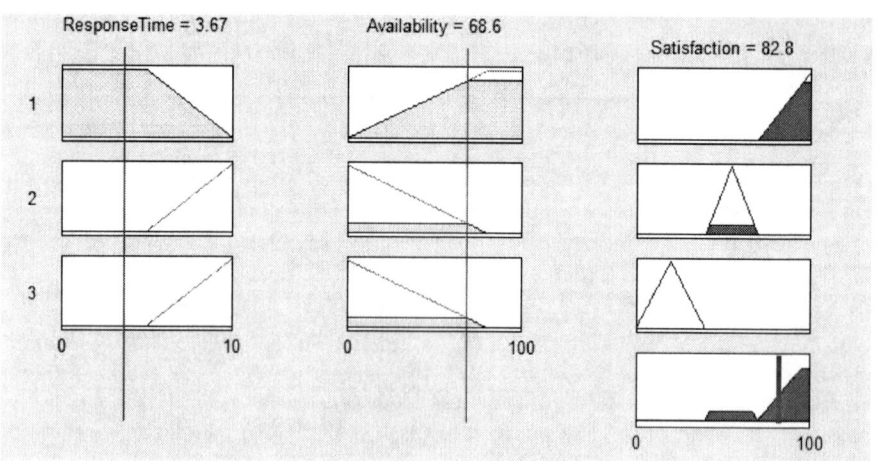

Fig. 3. A view of the simulator for fuzzy inference system

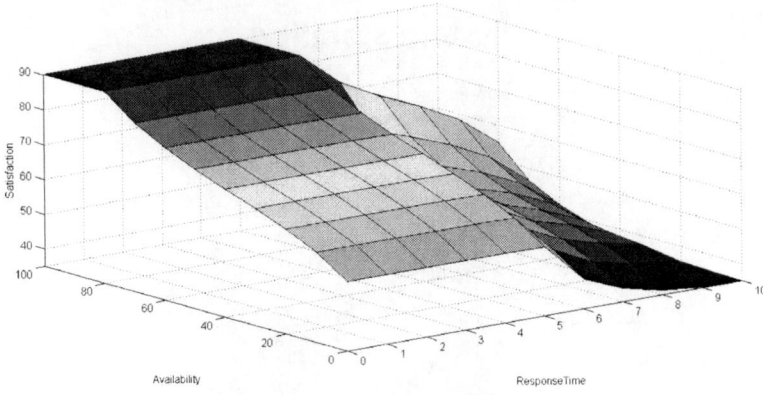

Fig. 4. The output of satisfaction degree according to ResponseTime and Availability membership function

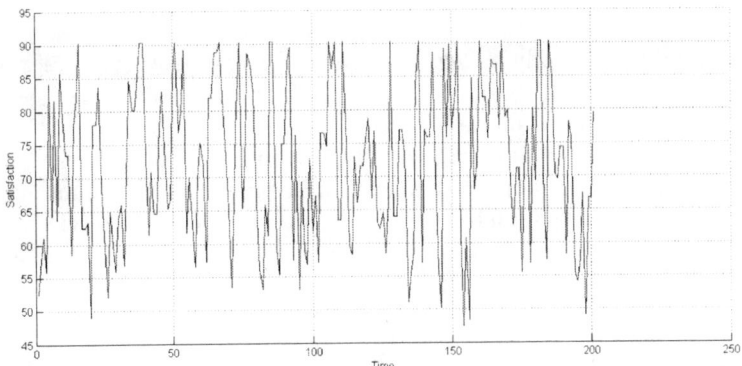

Fig. 5. Satisfaction degree

response-time of 3.67 seconds and availability of 68.6% result a satisfaction degree of 82.8. Figure 4 shows a surface map for the system and the dependency of the satisfaction degree on the response-time and availability.

We evaluate the effectiveness of the fuzzy approach with a non-fuzzy approach with respect to the stability of the system in terms of number of times a service needs to be replaced. The fuzzy approach performs the adaptation based on the QoS satisfaction. Only if the result of the satisfaction is lower than a threshold a service replacement occurs. While in the non-fuzzy approach, the replacement decision is done based on the precise evaluation of the QoS value ranges. We have conducted our experiment 200 times, each time providing random data for the input parameters. Figure 5 illustrates the output (satisfaction degree) of the experiment. The satisfaction threshold was set to 70%.

Figure 6 represents the stability of the fuzzy and non-fuzzy systems. As it is shown, the number of service replacement in a non-fuzzy approach is much

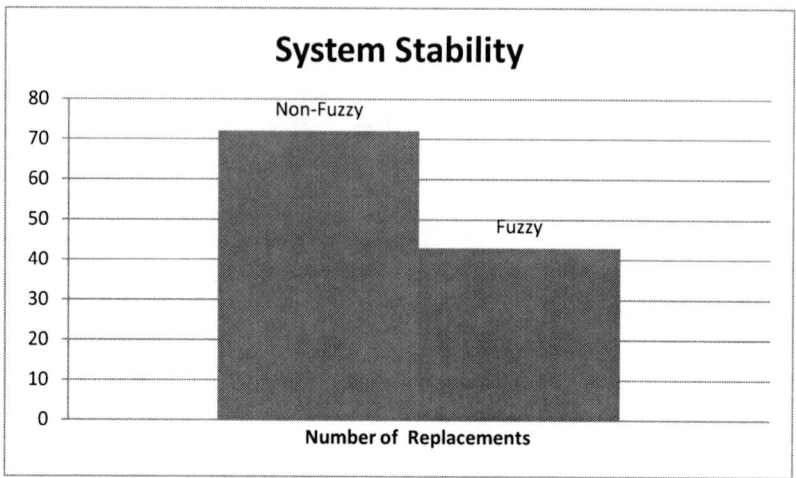

Fig. 6. System stability of using fuzzy and non-fuzzy approach

higher than when we apply a fuzzy approach. This actually is a direct proof of our approach. Using fuzzy parameters we allow partial satisfaction of the parameters. Therefore, the decision making for adaptation is not based on the precise evaluation of the quality ranges and it is rather imprecise and allows the parameters to be relaxed. The non-fuzzy approach involved the maximum number of service replacement which includes more queries for the service selection. This can results in a huge loss of time and money. The cost of establishing a new contract is also considerable.

7   Conclusions and Future Work

In this paper, we used fuzzy parameters for the QoS property descriptions of Web Services and a fuzzy approach is taken in order to select adaptation strategy. However, interpreting and presenting adaptation decisions based on fuzzy logic is still a hot research area that requires to be investigated more in the research community of software and service engineering.

In particular, we used linear *trapezoidal* membership function for the sake of simplicity. Currently, we are conducting more experiment to investigate the usage of *Gaussian* distribution function and *Sigmoid* curve that have the advantages of being smooth and non-zero all the time.

As for the future work, we aim to continue exploring the use of fuzzy parameters for the QoS matching. Applying more sophisticated functions using AI to Map the satisfaction degree to the appropriate adaptation decision might be worth exploring. However, there are still challenges that need to be addressed. For example, to what extent a parameter could be relaxed yet consider no violation? We also plan to incorporate more QoS parameters for calculating the overall satisfaction degree that influence the process of decision making. This requires the definition of more complex rules that represent the relation and dependencies between parameters.

Last but not least, applying an appropriate decision making requires an analytical evaluation based on a cost model. We would like to know under which circumstances the proposed approach is beneficial considering both QoS and business value criteria.

Acknowledgements

The research leading to these results has received funding from the European Community's Seventh Framework Programme FP7/2007-2013 under grant agreement 215483 (S-Cube).

References

1. Di Nitto, E., Kazhamiakin, R., Mazza, V., Bucchiarone, A., Cappiello, C., Pistore, M.: Design for adaptation of service-based applications: Main issues and requirements. In: The Fifth International Workshop on Engineering Service-Oriented Applications: Supporting Software Service Development Lifecycles, WESOA (2009)
2. Allen, J.F.: Maintaining Knowledge about Temporal Intervals. Communications of the ACM 26(11), 832–843 (1983)
3. Andrikopoulos, V., Fugini, M., Papazoglou, M.P., Parkin, M., Pernici, B., Siadat, S.H.: Qos contract formation and evolution. In: EC-Web, pp. 119–130 (2010)
4. Brown, G., Cheng, B.H.C., Goldsby, H., Zhang, J.: Goal-oriented specification of adaptation requirements engineering in adaptive systems. In: SEAMS 2006: Proceedings of the 2006 International Workshop on Self-Adaptation and Self-Managing Systems, pp. 23–29. ACM, New York (2006)
5. Canfora, G., Di Penta, M., Esposito, R., Villani, M.L.: Qos-aware replanning of composite web services. In: ICWS, pp. 121–129 (2005)
6. Cappiello, C., Pernici, B.: Quality-aware design of repairable processes. In: The 13th International Conference on Information Quality (ICIQ 2008), pp. 382–396 (2008)
7. Chafle, G., Dasgupta, K., Kumar, A., Mittal, S., Srivastava, B.: Adaptation in web service composition and execution. In: ICWS, pp. 549–557 (2006)
8. Chafle, G., Doshi, P., Harney, J., Mittal, S., Srivastava, B.: Improved adaptation of web service compositions using value of changed information. In: ICWS, pp. 784–791 (2007)
9. Cheng, B.H., Sawyer, P., Bencomo, N., Whittle, J.: A Goal-Based Modeling Approach to Develop Requirements of an Adaptive System with Environmental Uncertainty. In: Schürr, A., Selic, B. (eds.) MODELS 2009. LNCS, vol. 5795, pp. 468–483. Springer, Heidelberg (2009)
10. Cheung, R., Cao, J., Yao, G., Chan, A.T.S.: A fuzzy-based service adaptation middleware for context-aware computing. In: Sha, E., Han, S.-K., Xu, C.-Z., Kim, M.-H., Yang, L.T., Xiao, B. (eds.) EUC 2006. LNCS, vol. 4096, pp. 580–590. Springer, Heidelberg (2006)
11. Dardenne, A., van Lamsweerde, A., Fickas, S.: Goal-directed requirements acquisition. Sci. Comput. Program 20, 3–50 (1993)

60   B. Pernici and S.H. Siadat

12. Cock, M.D., Chung, S., Hafeez, O.: Selection of web services with imprecise QoS constraints. In: Proceedings of the IEEE/WIC/ACM International Conference on Web Intelligence, WI 2007, pp. 535–541. IEEE Computer Society, Washington, DC, USA (2007)

13. Fenton, N., Neil, M.: Making decisions: using bayesian nets and mcda. Knowledge-Based Systems 14(7), 307–325 (2001)

14. Halpern, J.Y.: Reasoning about Uncertainty. MIT Press, Cambridge (2003)

15. Harney, J., Doshi, P.: Adaptive web processes using value of changed information. In: International Conference on Service-Oriented Computing (ICSOC), pp. 179–190 (2006)

16. Kritikos, K., Plexousakis, D.: Semantic QoS-based web service discovery algorithms. In: ECOWS, pp. 181–190 (2007)

17. Kwiatkowska, M., Norman, G., Parker, D.: Probabilistic symbolic model checking with PRISM: A hybrid approach. In: Katoen, J.-P., Stevens, P. (eds.) TACAS 2002. LNCS, vol. 2280, pp. 52–66. Springer, Heidelberg (2002)

18. Letier, E., van Lamsweerde, A.: Reasoning about partial goal satisfaction for requirements and design engineering. SIGSOFT Softw. Eng. Notes 29, 53–62 (2004)

19. Li, P., Comerio, M., Maurino, A., De Paoli, F.: Advanced non-functional property evaluation of web services. In: Proceedings of the 2009 Seventh IEEE European Conference on Web Services, ECOWS 2009, pp. 27–36. IEEE Computer Society, Washington, DC, USA (2009)

20. Li, P., Comerio, M., Maurino, A., De Paoli, F.: An approach to non-functional property evaluation of web services. In: Proceedings of the 2009 IEEE International Conference on Web Services, ICWS 2009, pp. 1004–1005. IEEE Computer Society, Washington, DC, USA (2009)

21. Martín-Díaz, O., Cortés, A.R., Benavides, D., Durán, A., Toro, M.: A quality-aware approach to web services procurement. In: Benatallah, B., Shan, M.-C. (eds.) TES 2003. LNCS, vol. 2819, pp. 42–53. Springer, Heidelberg (2003)

22. Martín-Díaz, O., Cortés, A.R., García, J.M., Toro, M.: Dealing with fixable and non-fixable properties in service matchmaking. In: Dan, A., Gittler, F., Toumani, F. (eds.) ICSOC/ServiceWave 2009. LNCS, vol. 6275, pp. 228–237. Springer, Heidelberg (2010)

23. De Paoli, F., Palmonari, M., Comerio, M., Maurino, A.: A meta-model for non-functional property descriptions of web services. In: Proceedings of the 2008 IEEE International Conference on Web Services, pp. 393–400. IEEE Computer Society, Washington, DC, USA (2008)

24. Pernici, B., Siadat, S.H.: Adaptation of web services based on QoS satisfaction. In: WESOA 2010: Proceedings of the 6th International Workshop on Engineering Service-Oriented Applications. Springer, Heidelberg (2010)

25. Schäfer, M., Dolog, P., Nejdl, W.: An environment for flexible advanced compensations of web service transactions. ACM Trans. Web 2, 14:1–14:36 (2008)

26. Whittle, J., Sawyer, P., Bencomo, N., Cheng, B.H.C., Bruel, J.-M.: Relax: Incorporating uncertainty into the specification of self-adaptive systems. In: RE, pp. 79–88 (2009)

27. Whittle, J., Sawyer, P., Bencomo, N., Cheng, B.H.C., Bruel, J.-M.: Relax: a language to address uncertainty in self-adaptive systems requirement. Requir. Eng. 15(2), 177–196 (2010)

28. Yen, J., Tiao, W.A.: A systematic tradeoff analysis for conflicting imprecise require-
    ments. In: Proceedings of the 3rd IEEE International Symposium on Requirements
    Engineering, RE 1997, pp. 87–96. IEEE Computer Society, Washington, DC, USA
    (1997)
29. Zadeh, L.A.: Fuzzy sets. Information and Control 8, 338–353 (1965)
30. Zeng, L., Benatallah, B., Ngu, A.H.H., Dumas, M., Kalagnanam, J., Chang,
    H.: Qos-aware middleware for web services composition. IEEE Trans. Softw.
    Eng. 30(5), 311–327 (2004)
31. Zhang, J., Cheng, B.H.C.: Using temporal logic to specify adaptive program se-
    mantics. Journal of Systems and Software 79(10), 1361–1369 (2006)

Dealing with Known Unknowns: Towards a Game-Theoretic Foundation for Software Requirement Evolution*

Le Minh Sang Tran and Fabio Massacci

Università degli Studi di Trento, I-38100 Trento, Italy
{tran,fabio.massacci}@disi.unitn.it

Abstract. Requirement evolution has drawn a lot of attention from the community with a major focus on management and consistency of requirements. Here, we tackle the fundamental, albeit less explored, alternative of modeling the future evolution of requirements.

Our approach is based on the explicit representation of *controllable* evolutions vs *observable* evolutions, which can only be estimated with a certain probability. Since classical interpretations of probability do not suit well the characteristics of software design, we introduce a game-theoretic approach to give an explanation to the semantic behind probabilities. Based on this approach we also introduce quantitative metrics to support the choice among evolution-resilient solutions for the system-to-be.

To illustrate and show the applicability of our work, we present and discuss examples taken from a concrete case study (the security of the SWIM system in Air Traffic Management).

Keywords: software engineering, requirement evolution, observable and controllable rules, game-theoretic.

1 Introduction

> "...There are known unknowns: that is to say, there are things that we now know we don't know..."
>
> — Donald Rumsfeld, United States Secretary of Defense

In the domain of software, evolution refers to a process of continually updating software systems in accordance to changes in their working environments such as business requirements, regulations and standards. While some evolutions are unpredictable, many others can be predicted albeit with some uncertainty (e.g. a new standard does not appear overnight, but is the result of a long process).

The term *software evolution* has been introduced by Lehman in his work on laws of software evolution [17, 18], and was widely adopted since 90s. Recent studies in software evolutions attempt to understand causes, processes, and effects of the phenomenon [2, 14, 16]; or focus on the methods, tools that manage the effects of evolution [19, 25, 28].

---

* This work is supported by the European Commission under project EU-FET-IP-SECURECHANGE.

H. Mouratidis and C. Rolland (Eds.): CAiSE 2011, LNCS 6741, pp. 62–76, 2011.

Requirement evolution has also been the subject of significant research [12, 15, 24, 26, 31]. However, to our understanding, most of these works focus on the issue of management and consistency of requirements. Here, we tackle a more fundamental question of modeling uncertain evolving requirements in terms of evolution rules. Our ultimate goal is to support the decision maker in answering such a question "Given these anticipated evolutions, what is a solution to implement an evolution-resilient system?".

This motivates our research in modeling and reasoning on a requirement model of a system which might evolve sometime in the future. We assume that stakeholders will know the tentative possible evolutions of the system-to-be, but with some uncertainty. For example, the Federal Aviation Authority (FAA) document of the System Wide Information Management (SWIM) for Air Traffic Management (ATM) lists a number of potential alternatives that subject to other high-level decisions (*e.g.*, the existence of an organizational agreement for nation-wide identity management of SWIM users). Such organization-level agreements do not happen overnight (and may shipwreck at any time) and stakeholders with experience and high-level positions have a clear visibility of the likely alternatives, the possible but unlikely solutions, and the politically impossible alternatives.

Our objective is to model the evolution of requirements when it is known to be possible, but it is unknown whether it would happen: the *known unknown*.

1.1   The Contributions of This Paper

We set up a game-theoretic foundation for modeling and reasoning on evolutionary requirement models:

- A way to model requirement evolutions in terms of two kinds of evolution rules: *controllable* and *observable* rules that are applicable to many requirement engineering models (from problem frames to goal models).
- A game-theoretic based explanation for probabilities of an observable evolution.
- Two quantitative metrics to help the designer in deciding optimal things to implement for the system-to-be.

This paper is started by a sketch of a case study (§2). To our purpose, we only focus on requirements of a part of the system-under-study. We distinguish which requirements are compulsory, and which are optional at design time. Based on these, we construct simple evolution scenario to illustrate our approach in subsequent sections, *i.e.* some compulsory requirements become obsoleted, and some optional ones turn to be mandatory.

Then, we discuss how to model requirement evolution (§3) using evolution rules and probabilities of evolution occurrences. We employ the game-theoretic interpretation to account for the semantic of probabilities.

We also introduce two quantitative metrics to support reasoning on rule-based evolutionary requirement models (§4). The reasoning is firstly performed on a simple scenario. Then we show a programmatic way to adapt the technique to a more complex scenario (*e.g.*, large model, multiple evolutions) (§5).

Table 1. High level requirements of ISS-ENT and ISS-BP in SWIM Security Services

ID	Requirement	Opt.
RE1	Manage keys and identities of system entities (human, software, devices,...)	
RE2	Support Single Sign-On (SSO)	•
RE3	Support a robust Identity and Key Management Infrastructure (IKMI) that can be scaled up to large number of applications and users.	•
RE4	Intrusion detection and response	
RB1	Less cross-program dependencies for External Boundary Protection System	
RB2	More robust and scalable common security solution	•
RB3	Simpler operation of External Boundary Protection System	•
RB4	Support overall security assessment	•

The Opt(ional) column determines whether a requirement is compulsory or not at current design time. Due to evolution, optional requirements may turn to be compulsory, and current compulsory ones may no longer be needed in the future.

In addition, we discuss current limits of our work, but not the approach, as well as our plan to address them (§6). Finally, we review related works (§7) and conclude the paper(§8).

2   Case Study

Throughout this work, to give a clearer understanding of the proposed approach we draw examples taken from the design architecture of SWIM [23, 7] in ATM.

SWIM provides a secure, overarching, net-centric data network, and introduces a Service-Oriented Architecture (SOA) paradigm for airspace management. The United States FAA [7] has proposed a logical architecture of SWIM which consists of several function blocks, among which we choose to consider the Security Services block. At high level analysis of Security Services, there are five security areas: *i)* Enterprise Information Security System (ISS-ENT), *ii)* Boundary Protection ISS (ISS-BP), *iii)* SWIM Core ISS, *iv)* National Air Space (NAS) End System ISS, and *v)* Registry control. To avoid a detailed discussion on the architecture of SWIM Secure Services, which are not main topic of this work, while providing enough information for illustrating our work we refine our scope of interest on two areas: ISS-ENT and ISS-BP.

– ISS-ENT includes security requirements that are provided as part of an underlying IT/ISS infrastructure used by systems throughout the NAS.
– ISS-BP includes requirements with regard to control connections and information exchanges between internal NAS and external entities. These requirements refer to both network layer control. (*e.g.,* VPNs, firewalls) and application layer control.

Table 1 lists high level requirements of ISS-ENT and ISS-BP. For convenience, each requirement has a corresponding identifier: two characters for the security area (RE - stands for ISS-ENT requirements, RB - stands for ISS-BP ones),

Table 2. Design elements that support requirements listed in Table 1

ID	Element Description	RE1	RE2	RE3	RE4	RB1	RB2	RB3	RB4
A	Simple IKMI	•							
B1	OpenLDAP based IKMI	•		•					
B2	Active Directory based IKMI	•	•	•					
B3	Oracle Identity Directory based IKMI	•	•	•				•	
C	Ad-hoc SSO	•							
D	Network Intrusion Detection System					•			
E	Common application gateway for External Boundary Protection System							•	•
F	Centralized Policy Decision Point (PDP)							•	
G	Application-based solution for External Boundary Protection System						•		

Each element in this table can support (or fulfill) requirements listed in columns. To prevent useless redundancy, some elements are exclusive to due to functionality overlapping (*e.g.*, A, B1, B2 and B3 are mutual exclusive each other).

and a sequence number. There are compulsory requirements (*i.e.* they are essential at the time the system is designed) and optional ones (*i.e.* they can be ignored at present, but might be critical sometime in the future). Solutions for these requirements are listed in Table 2. Each solution has an IDentifier, a short description and a checklist of requirements that it can fulfill.

3   Modeling Requirement Evolution

In this section, we describe how we model evolution, which essentially affects to any further analysis. We capture evolutions by classifying them into two groups: *controllable* and *observable*. Furthermore, we include in this section the game-theoretic account for probability.

3.1   Evolution on Requirement Model: Controllable and Observable

Stakeholder requirements, mostly in textual format, are their wishes about the system-to-be. Analyzing requirements in such format is difficult and inefficient. Designer thus has to model requirements and design decisions by using various approaches (*e.g.*, model-based, goal-based) and techniques (*e.g.*, DFD, UML).

Generally, a requirement model is a set of elements and relationships, which depend on particular approach. For instance, according to Jackson and Zave [30], model elements are *Requirements, Domain assumptions, Specifications*; in a goal-based model (*e.g.*, i*), elements are goals, actors and so on.

Here we do not investigate any specific requirement model (*e.g.*, goal-based model, UML models), nor go to details about how many kinds of element and relationship a model would have. The choice of a one's favorite model to represent these aspects can be as passionate as the choice of a one's religion or football team, so it is out of scope. Instead, we treat elements at abstract meaning, and only be interested in the satisfaction relationship among elements.

In our work, we define the satisfaction relationship in terms of usefulness. That an element set X is useful to another element set Y depends on the ability to satisfy (or fulfill) Y if X is satisfied. We define a predicate useful(X, Y) which is true (1) if X can satisfy all elements of Y, otherwise false (0). The implementation of useful depends on the specific requirement model. For examples:

- Goal models [20]: useful corresponds to Decomposition and Means-end relationships. The former denotes a goal can be achieved by satisfying its subgoals. The later refers to achieving an (end) goal by doing (means) tasks.
- Problem frames [13]: useful corresponds to requirement references and domain interfaces relationships. Requirements are imposed by machines, which connect to problem world via domain interfaces. Problem world in turn connects to requirements via requirement references.

For evolutionary software systems which may evolve under some circumstances (e.g., changes in requirements due to changes in business agreements, regulations, or domain assumption), their requirement models should be able to express as much as possible the information about known unknowns i.e. potential changes. These potential changes are analyzed by evolution assessment algorithms to contribute to the decision making process, where a designer decides what would be in the next phase of the development process.

Based on the actor who can decide which evolution would happen, we categorize requirement evolutions into two classes:

- *controllable evolution* is under control of designer to meet high level requirements from stakeholder.
- *observable evolution* is not under control of designer, but its occurrence can be estimated with a certain level of confidence.

Controllable evolutions, in other words, are designer's moves to identify different design alternatives to implement a system. The designer then can choose the most "optimal" one based on her experience and some analyses on these alternatives. In this sense, controllable evolution is also known as design choice.

Observable ones, in contrast, correspond to potential evolutions of which realization is outside the control of the designer. They are moves of reality to decide how a requirement model looks like in the future. Therefore, the stakeholder and designer have to forecast the reality's choice with a level of uncertainty. The responses are then incorporated into the model.

We capture the evolution in terms of evolution rule. We have *controllable rule* and *observable rule* corresponding to controllable and observable evolution.

Definition 1. *A controllable rule r_c is a set of tuples $\langle RM, RM_i \rangle$ that consists of an original model RM and its possible design alternative RM_i.*

$$r_c = \bigcup_i^n \left\{ RM \xrightarrow{*} RM_i \right\}$$

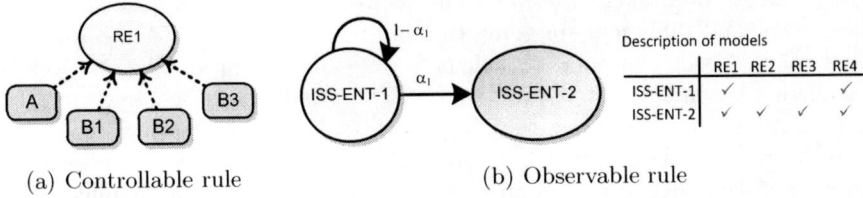

(a) Controllable rule   (b) Observable rule

Fig. 1. Example of controllable rule (a), and observable rule (b)

Definition 2. *An observable rule r_o is a set of triples $\langle RM, p_i, RM_i \rangle$ that consists of an original model RM and its potential evolution RM_i. The probability that RM evolves to RM_i is p_i. All these probabilities should sum up to one.*

$$r_o = \bigcup_{i=1}^{n} \left\{ RM \xrightarrow{p_i} RM_i \right\}$$

Fig. 1 is a graphical representation of evolution rules taken from SWIM case study. Left, Fig. 1(a) describes a controllable rule where a requirement model containing IKMI (RE1) has four design choices: A, B1, B2, and B4 (see Table 1 and Table 2). Right, Fig. 1(b) shows that the initial model ISS-ENT-1 (including RE1 and RE4) can evolve to ISS-ENT-2 (including RE1 to RE4), or remain unchanged with probabilities of α and $1 - \alpha$. These rules are as follows:

$$r_c = \left\{ RE1 \xrightarrow{*} A, RE1 \xrightarrow{*} B1, RE1 \xrightarrow{*} B2, RE1 \xrightarrow{*} B3 \right\}$$
$$r_o = \left\{ \text{ISS-ENT-1} \xrightarrow{\alpha_1} \text{ISS-ENT-2}, \text{ISS-ENT-1} \xrightarrow{1-\alpha_1} \text{ISS-ENT-1} \right\}$$

3.2   Game-Theoretic Account for Probability

Here, we discuss about why and how we employ game-theoretic (or betting interpretation) to account for probabilities in observable rules.

As mentioned, each potential evolution in an observable rule has an associated probability; these probabilities sum up to 1. However, who tells us? And what is the semantic of probability? To answer the first question, we, as system Designers, agree that Stakeholder will tell us possible changes in a period of time. About the second question, we need an interpretation for semantic of probability.

Basically, there are two broad categories of probability interpretation, called "physical" and "evidential" probabilities. Physical probabilities, in which frequentist is a representative, are associated with a random process. Evidential probability, also called Bayesian probability (or subjectivist probability), are considered to be degrees of belief, defined in terms of disposition to gamble at certain odds; no random process is involved in this interpretation.

To account for probability associated with an observable rule, we can use the Bayesian probability as an alternative to the frequentist because we have no event to be repeated, no random variable to be sampled, no issue about measurability (the system that designers are going to build is often unique in some respects). However, we need a method to calculate the value of probability as well as to explain the semantic of the number. Since probability is acquired from the requirement eliciting process involving the stakeholder, we propose using the game-theoretic method in which we treat probability as a price. It seems to be easier for stakeholder to reason on price (or cost) rather than probability.

The game-theoretic approach, discussed by Shafer et al. [27] in Computational Finance, begins with a game of three players, *i.e.* Forecaster, Skeptic, and Reality. Forecaster offers prices for tickets (uncertain payoffs), and Skeptic decides a number of tickets to buy (even a fractional or negative number). Reality then announces real prices for tickets. In this sense, probability of an event E is the initial stake needed to get 1 if E happens, 0 if E does not happen. In other words, the mathematics of probability is done by finding betting strategies.

In this paper, we do not deal with stock market but the design of evolving software, *i.e.* we extend it for software design. We then need to change rules of the game. Our proposed game has three players: *Stakeholder*, *Designer*, and *Reality*. For the sake of brevity we will use "he" for Stakeholder, "she" for Designer and "it" for Reality. The sketch of this game is denoted in protocol 1.

Protocol 1
Game has n round, each round plays on a software C_i
FOR i = 1 to n
      Stakeholder announces p_i
      Designer announces her decision d_i: believe, don't believe
      If Designer believes
            $K_i = K_{i-1} + M_i \times (r_i - p_i)$
      Designer does not believe
            $K_i = K_{i-1} + M_i \times (p_i - r_i)$
      Reality announces r_i

The game is about Stakeholder's desire of having a software C. He asks Designer to implement C, which has a cost of $M\$$. However, she does not have enough money to do this. So she has to borrow money from either Stakeholder or National Bank with the return of interest (ROI) p or r, respectively.

Stakeholder starts the game by announcing p which is his belief about the minimum ROI for investing $M\$$ on C. In other words, he claims that r would be greater than p. If M equals 1, p is the minimum amount of money one can receive for 1\$ of investment. Stakeholder shows his belief on p by a commitment that he is willing to buy C for price $(1 + p)M$ if Designer does not believe him and borrow money from someone else.

If Designer believes Stakeholder, she will borrow M from Stakeholder. Later on, she can sell C to him for $M(1+r)$ and return $M(1+p)$ to him. So, the final amount of money Designer can earn from playing the game is $M(r - p)$.

If Designer does not believe Stakeholder, she will borrow money from National Bank, and has to return $M(1 + r)$. Then, Stakeholder is willing to buy C with $M(1 + p)$. In this case, Designer can earn $M(p - r)$.

Suppose that Designer has an initial capital of K_0. After round i-th of the game, she can accumulate either $K_i = K_{i-1} + M(r - p)$ or $K_i = K_{i-1} + M(p - r)$, depend on whether she believes Stakeholder or not. Designer has a winning strategy if she can select the values under her control (the $M\$$) so that she always keeps her capital never decrease, intuitively, $K_i >= K_{i-1}$ for all rounds.

The law of large numbers here corresponds to say that if unlikely events happen then Designer has a strategy to multiply her capital by a large amount. In other words, if Stakeholder estimates Reality correctly then Designer has a strategy for costs not to run over budget.

4 Making Decision: What Are the Best Things to Implement

One of the main objectives of modeling evolution is to provide a metric (or set of metrics) to indicate how well a system design can adapt to evolution. Together with other assessment metrics, designers have clues to decide what an "optimal" solution for a system-to-be is.

The major concern in assessment evolution is answering the question: "Whether a model element (or set of elements) becomes either useful or useless after evolution?". Since the occurrence of evolution is uncertain, so the usefulness of an element set is evaluated in term of probability. In this sense, this work proposes two metrics to measure the usefulness of element set as follows.

Max Belief. (MaxB): of an element set X is a function that measures the maximum belief supported by Stakeholder such that X is useful to a set of top requirements after evolution happens. This belief of usefulness for a set of model element is inspired from a game in which Stakeholder play a game together with Designer and Reality to decide which elements are going to implementation phase.

Residual Risk. (RRisk): of an element set X is the complement of total belief supported by Stakeholder such that X is useful to set of top requirements after evolution happens. In other words, residual risk of X is the total belief that X is not useful to top requirements with regard to evolution. Importantly, do not confuse this notion of residual risk with the one in risk analysis studies which are different in nature.

Given an evolutionary requirement model $RM = \langle RM, r_o, r_c \rangle$ where $r_o = \bigcup_i \left\{ RM \xrightarrow{p_i} RM_i \right\}$ is an observable rule, and $r_c = \bigcup_{ij} \left\{ RM_i \xrightarrow{*} RM_{ij} \right\}$ is a controllable rule, the calculation of max belief and residual risk is illustrated in Eq. 1, Eq. 2 as follows.

$$MaxB(X) = \max_{RM \xrightarrow{p_i} RM_i \in S} p_i \tag{1}$$

$$RRisk(X) = 1 - \sum_{RM \xrightarrow{p_i} RM_i \in S} p_i \tag{2}$$

where S is set of potential evolutions in which X is useful.

$$S = \left\{ RM \xrightarrow{p_i} RM_i | \exists (RM_i \xrightarrow{*} RM_{ij}) \in r_c \; st.\mathsf{useful}(X, RM_{ij}) \right\}$$

One may argue about the rationale of these two metrics. Because he (or she) can intuitively measure the usefulness of an element set by calculating the *Total Belief* which is exactly the complement of our proposed *Residual Risk*. However, using only Total Belief (or *Residual Risk*) may mislead designers in case of a *long-tail* problem.

The long-tail problem, firstly coined by Anderson [1], describes a larger population rests within the tail of a normal distribution than observed. A long-tail example depicted in Fig. 2 where a requirement model RM might evolve to several potential evolutions with very low probabilities (say, eleven potential evolutions with 5% each), and another extra potential evolution with dominating probability (say, the twelfth one with 45%). Suppose that an element A appears in the first eleven potential evolutions, and an element B appears in the last twelfth potential evolution. Apparently, A is better than B due to A's total belief is 55% which is greater than that of B, say 45%. However, at the end of the day, only one potential evolution becomes effective (*i.e.* chosen by Reality) rather than 'several' potential evolutions are together chosen. If we thus consider every single potential evolution to be chosen, the twelfth one (45%) seems to be the most promising and *Max Belief* makes sense here. Arguing that A is better than B or versa is still highly debatable. Ones might put their support on the long tail [1], and ones do the other way round [5]. Therefore, we introduce both *Residual Risk* and *Max Belief* to avoid any misleading in the decision making process that can be caused when using only Total Belief.

For a better understanding of *Max Belief* and *Residual Risk*, we conclude this section by applying our proposed metrics to the evolution of SWIM Security Services discussed in previous section. In Fig. 3, here we have an initial requirement model RM0(ISS-ENT-1,ISS-BP-1) that will evolve to RM1(ISS-ENT-2,ISS-BP-1), RM2(ISS-ENT-1,ISS-BP-2), and RM3(ISS-ENT-2,ISS-BP-2) with probabilities of 28%, 18% and 42%, respectively. There are 12% that RM0 stays

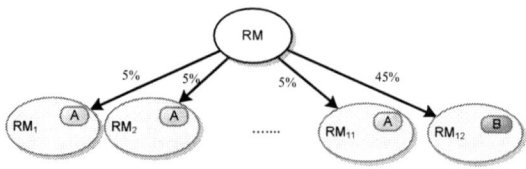

Fig. 2. The long-tail problem

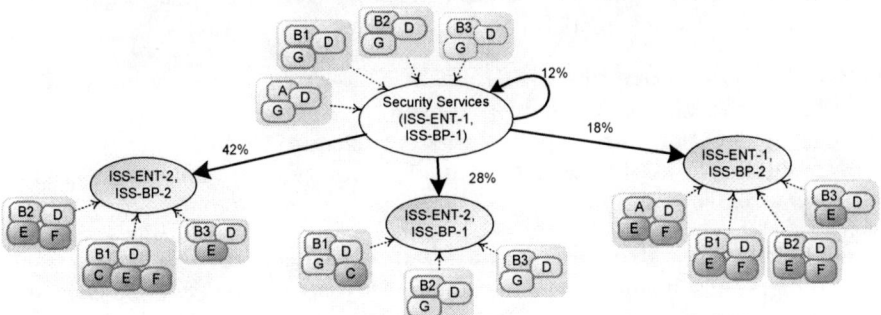

Fig. 3. Evolution of the SWIM Security Service

Table 3. Examples of Max Belief and Residual Risk

Element set	Max Belief	Residual Risk
{A, D,}	n/a	n/a
{A, E, D, G, F}	18%	70%
{B3, D, G}	28%	60%
{B1, D, G, C}	28%	60%
{B3, D, E, G}	42%	0%
{B2, D, E, F, G}	42%	0%

unchanged. Each requirement model is represented as a bubble in which there is a controllable rule with several design alternatives. Each design alternative is an element set represented as a rounded rectangle that contains elements (such as A, D, and G) to support (fulfill) requirements of that requirement model.

Table 3 shows some examples, where the first column displays element sets, and the two next columns show values of max belief and residual risk. Notice that the max belief and residual risk in the first row, where the element set is $\{A, D\}$, are n/a which means that we are unable to find any potential evolution that $\{A, D\}$ can support all top requirements.

In Table 3, $\{B3, D, E, G\}$ and $\{B2, D, E, F, G\}$ seem to be the best choices, since they have a high max belief (42%) and low residual risk (0%). The zero residual risk means these element sets are surely still useful after evolution. If the cost of implementation is the second criteria and assume that each element has equal cost, then $\{B3, D, E, G\}$ seems to be better.

5   Handling Complex Evolution

If a model is too large and complex, instead of dealing with the evolution of the whole model, we can consider evolution in each subpart. If a subpart is still too large and complex, we can recursively divide it into smaller ones, each with its local evolution rule, until we are able to deal with.

We then need to combine these local rules together to produce a global evolution one for the whole model. For simplicity, we assume that:

ASS-1: Independence of evolutions. All observable rules are independent. It means that they do not influent each other. In other words, the probability that an evolution rule is applied does not affect to that of other rules.

ASS-2: Order of evolutions. Controllable evolutions are only considered after observable evolutions.

As discussed, observable rules are analyzed on independent subparts. Prevailing paradigms of software development (*e.g.*, Object-Oriented, Service-Oriented) encourage encapsulation and loosely coupling. Evolutions applying to subparts, therefore, are often independent. Nevertheless, if there are two evolution rules which influent each other, we can combine them into a single one. We assume that dependent evolutions do happen, but not a common case. Hence manual combination of these rules is still doable.

The second assumption is the way we deal with controllable rules. If we apply controllable rules before observable ones, it means we look at design alternatives before observable evolutions happen. This makes the problem more complex since under the effect of evolution, some design alternatives are no longer valid, and some others new are introduced. Here, for simplicity, we look at design alternatives for evolved requirement models that will be stable at the end of their evolution process.

After all local evolutions at subparts are identified, we then combine these rules into a global evolution rule that applies to the whole model. The rationale of this combination is the effort to reuse the notion of Max Belief and Residual Risk (§4) without any extra treatment. In the following we discuss how to combine two independent observable evolution rules.

Given two observable rules:

$$r_{o1} = \bigcup_{i=1}^{n} \left\{ RM1 \xrightarrow{p1_i} RM1_i \right\} \text{ and } r_{o2} = \bigcup_{j=1}^{m} \left\{ RM2 \xrightarrow{p2_j} RM2_j \right\}$$

Let r_o is combined rule from r_{o1} and r_{o2}, we have:

$$r_o = \bigcup_{\substack{1 \le i \le n \\ 1 \le j \le m}} \left\{ RM1 \cup RM2 \xrightarrow{p1_i * p2_j} RM1_i \cup RM2_j \right\}$$

Fig. 4 illustrates an example of combining two observable rules into a single one. In this example, there are two subparts of SWIM Security Service: ISS-ENT and ISS-BP. The left hand side of the figure displays two rules for these parts, and in the right hand side, it is the combined rule.

In general case, we have multiple steps of evolution *i.e.* evolution happens for many times. For the ease of reading, *step 0* will be the first step of evolution, where no evolution is applied. We use RM_i^d to denote the i-*th* model in step d, and $r_{od,i}$ to denote the observable evolution rule that applies to RM_i^d, *i.e.* $r_{od,i}$ takes RM_i^d as its original model.

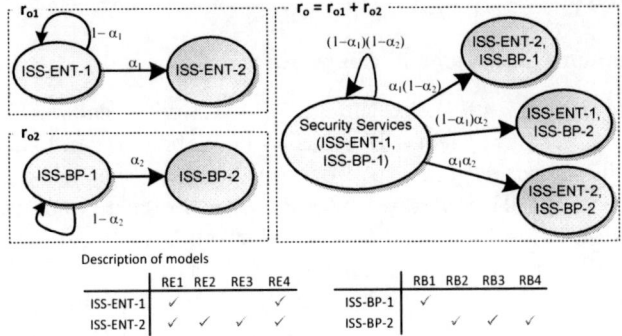

Description of models

	RE1	RE2	RE3	RE4		RB1	RB2	RB3	RB4
ISS-ENT-1	✓			✓	ISS-BP-1	✓			
ISS-ENT-2	✓	✓	✓	✓	ISS-BP-2		✓	✓	✓

Fig. 4. Example of combining two observable evolution rules

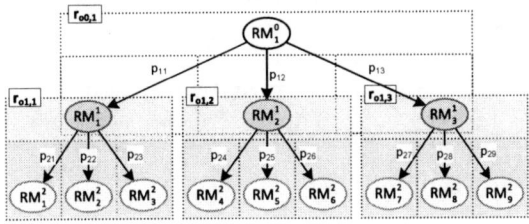

Fig. 5. Multiple steps (phases) evolving requirement model

The multi-step evolution begins with an original model RM_1^0. This model can evolve to one of the potential evolutions RM_i^1. In the second step, each RM_i^1 then also evolves to one of many potential evolutions RM_j^2. The evolution stops after k steps. If we represent a model as a node, and connect a model to its potential evolutions as we have done as aforementioned, then we have a tree-like graph, called *evolution tree* with k-depth.

Fig. 5 illustrates a two-step evolution, in which observable rules are denoted as dotted boxes. The original model lays on top part of a box, and all potential evolutions are in sub boxes laid at the bottom. There are directed edges connecting the original model to potential evolutions. The label on each edge represents the probability such that original model evolves to target model.

In Fig. 5, an initial requirement model RM_1^0 evolves to either RM_1^1, RM_2^1 or RM_3^1. Likewise, RM_i^1 evolves to RM_j^2, where i=1..3 and j=1..9. Here, we have a ternary complete tree of depth 2. Generally, the evolution tree of a k-step consecutive evolution is a complete k-depth, m-ary tree.

We can always collapse a k-step evolution into an equivalent 1-step one in terms of probability by letting the original model evolve directly to the very last models with the probabilities that are multiplication of probabilities of intermediate steps. Therefore, any k-step evolution has an equivalent 1-step evolution. Hence all analyses discussed in §4 are applicable without any modification.

6   Limitation

Obviously there are limitations in this work:

- *Real world applicability.* Even though we work on a real world case study, this work is still pure theory. It needs to be elaborated and then evaluated with the industry. We plan to prove our work in the field of Air Traffic Management (ATM), where we interact with designers and stakeholder of an ATM system, and get their feedback for validation.
- *Obtaining probability.* Since evolution probabilities are obtained from stakeholder, they are individual opinions. To deal with the problem, we shall work on an interaction protocol with stakeholder to minimize inaccuracy, as well as equip an appropriate mathematic foundation (*e.g.,* Dempster and Shafer's theory) for our reasoning.
- *Independence of evolution.* Complex models may require many probabilities that are not independent. This breaks the assumptions discussed in §5. Even though designers can solve this problem by manually combining dependent evolutions, we still need a more systematic way to deal with them.

7   Related Works

A majority of approaches to software evolution has focused on the evolution of architecture and source code level. However, in recent years, changes at the requirement level have been identified as one of the drivers of software evolution [4, 12, 31]. As a way to understand how requirements evolve, research in PROTEUS [24] classifies changing requirements (that of Harker et al [11]) into five types, which are related to the development environment, stakeholder, development processes, requirement understanding and requirement relation. Later, Lam and Loomes [15] present the EVE framework for characterizing changes, but without providing specifics on the problem beyond a meta model.

Several approaches have been proposed to support requirements evolution. Zowgi and Offen [31] work at meta level logic to capture intuitive aspects of managing changes to requirement models. Their approach involves modeling requirement models as theories and reasoning changes by mapping changes between models. However, this approach has a limitation of overhead in encoding requirement models into logic.

Russo et al [26] propose an analysis and revision approach to restructure requirements to detect inconsistency and manage changes. The main idea is to allow evolutionary changes to occur first and then, in the next step, verify their impact on requirement satisfaction. Also based on this idea, Garcez et al [4] aim at preserving goals and requirements during evolution. In the analysis, a specification is checked if it satisfies a given requirement. If it does not, diagnosis information is generated to guide the modification of specification in order to satisfy the requirement. In the revision, the specification is changed according to diagnosis information generated. Similar to Garcez et al, Ghose's [9] framework is based on formal default reasoning and belief revision, aims to address the problem of inconsistencies due to requirement evolution. This approach is supported by automated tools [10]. Also relating to inconsistencies, Fabrinni et al [6] deal with requirement evolution expressed in natural language, which is challenging to capture precisely requirement changes. Their approach employs formal concept analysis to enable a systematic and precise verification of consistency among different stages, hence, controls requirement evolution.

Other notable approaches include Brier et al.'s [3] to capture, analyze, and understand how software systems adapt to changing requirements in an organizational context; Felici et al [8] concern with the nature of requirements evolving in the early phase of systems; Stark et al [29] study the information on how change occurs in the software system and attempt to produce a prediction model of changes; Lormans et al [21] use a formal requirement management system to motivate a more structural approach to requirement evolution.

8   Conclusion

We have discussed a rule-based representation of evolutions on requirement models. We proposed game-theoretic approach to explain the uncertainty of evolutions. We also introduced two notions of max belief and residual risk to reason on evolutionary models, in which the higher max belief and lower residual risk models seem to be more evolution-resilient than others. Together with other analyses (e.g., cost, risk), these values can help designers in making decision.

During the discussion, we provided many examples taken from a real world project, SWIM. These examples not only help to explain better our idea, but also show the promising applicability of our approach.

For future work, we plan to instantiate our approach to a concrete modeling language (e.g., goal-based language) and apply to a more convincing case study. We shall interact with stakeholder and designers, show them our approach and get their feedback to validate the usability of proposed approach.

References

1. Anderson, C.: The long tail. Wired (October 2004)
2. Anton, A., Potts, C.: Functional paleontology: The evolution of user-visible system services. TSE 29(2), 151–166 (2003)
3. Brier, J., Rapanotti, L., Hall, J.: Problem-based analysis of organisational change: a real-world example. In: Proc. of IWAAPF 2006. ACM, New York (2006)
4. d'Avila Garcez, A., Russo, A., Nuseibeh, B., Kramer, J.: Combining abductive reasoning and inductive learning to evolve requirements specifications. IEEE Proceedings - Software 150(1), 25–38 (2003)
5. Elberse, A.: Should you invest in the long tail? Harvard Business Review (2008)
6. Fabbrini, F., Fusani, M., Gnesi, S., Lami, G.: Controlling requirements evolution: a formal concept analysis-based approach. In: ICSEA 2007 (2007)
7. FAA. System wide information management (SWIM) segment 2 technical review. Tech. report (2009)
8. Felici, M.: Observational Models of Requirements Evolution. PhD thesis (2004)

9. Ghose, A.: A formal basis for consistency, evolution and rationale management in requirements engineering. In: ICTAI 1999 (1999)
10. Ghose, A.: Formal tools for managing inconsistency and change in re. In: IWSSD 2000. IEEE Computer Society, Washington, DC, USA (2000)
11. Harker, S., Eason, K., Dobson, J.: The change and evolution of requirements as a challenge to the practice of software engineering. In: RE 2001 (1993)
12. Hassine, J., Rilling, J., Hewitt, J., Dssouli, R.: Change impact analysis for requirement evolution using use case maps. In: IWPSE 2005 (2005)
13. Jackson, M.: Problem Frames: Analysing & Structuring Software Development Problems. Addison-Wesley, Reading (2001)
14. Kemerer, C.F., Slaughter, S.: An empirical approach to studying software evolution. TSE 25(4), 493–509 (1999)
15. Lam, W., Loomes, M.: Requirements evolution in the midst of environmental change: a managed approach. In: CSMR 1998 (1998)
16. LaMantia, M., Cai, Y., MacCormack, A., Rusnak, J.: Analyzing the evolution of large-scale software systems using design structure matrices and design rule theory: Two exploratory cases. In: Proc. of WICSA 2008, pp. 83–92 (2008)
17. Lehman, M.: On understanding laws, evolution and conservation in the large program life cycle. J. of Sys. and Soft. 1(3), 213–221 (1980)
18. Lehman, M.: Programs, life cycles, and laws of software evolution. Proc. IEEE 68(9), 1060–1076 (1980)
19. Lin, L., Prowell, S., Poore, J.: The impact of requirements changes on specifications and state machines. SPE 39(6), 573–610 (2009)
20. Liu, L., Eric Yu, E.: Designing information systems in social context: A goal and scenario modelling approach. Info. Syst. 29, 187–203 (2003)
21. Lormans, M., van Dijk, H., van Deursen, A., Nocker, E., de Zeeuw, A.: Managing evolving requirements in an outsourcing context: an industrial experience report. In: IWPSE 2004, pp. 149–158 (2004)
22. Mens, T., Ramil, J., Godfrey, M.: Analyzing the evolution of large-scale software. J. of Soft. Maintenance and Evolution: Research and Practice 16(6), 363–365 (2004)
23. Program SWIM-SUIT. D1.5.1 Overall SWIM users requirements. Tech. report, 2008.
24. Project PROTEUS. Deliverable 1.3: Meeting the challenge of chainging requirements. Tech. report, Centre for Soft. Reliab., Univ. of Newcastle upon Tyne (1996)
25. Ravichandar, R., Arthur, J., Bohner, S., Tegarden, D.: Improving change tolerance through capabilities-based design: an empirical analysis. J. of Soft. Maintenance and Evolution: Research and Practice 20(2), 135–170 (2008)
26. Russo, A., Nuseibeh, B., Kramer, J.: Restructuring requirements specifications. IEEE Proceedings: Software 146, 44–53 (1999)
27. Shafer, G., Vovk, V., Chychyla, R.: How to base probability theory on perfect-information games. BEATCS 100, 115–148 (2010)
28. Soffer, P.: Scope analysis: identifying the impact of changes in business process models. J. of Soft. Process: Improvement and Practice 10(4), 393–402 (2005)
29. Stark, G., Oman, P., Skillicorn, A., Ameele, A.: An examination of the effects of requirements changes on software maintenance releases. J. of Soft. Maintenance: Research and Practice, 293–309 (1999)
30. Zave, P., Jackson, M.: Four dark corners of req. eng. TSEM 6(1), 1–30 (1997)
31. Zowghi, D., Offen, R.: A logical framework for modeling and reasoning about the evolution of requirements. In: ICRE 1997 (1997)

Goal-Based Behavioral Customization of Information Systems

Sotirios Liaskos[1], Marin Litoiu[1], Marina Daoud Jungblut[1], and John Mylopoulos[2]

[1] School of Information Technology, York University, Toronto, Canada
{liaskos,mlitoiu,djmarina}@yorku.ca
[2] Department of Information Engineering and Computer Science, University of Trento, Italy
jm@disi.unitn.it

Abstract. Customizing software to perfectly fit individual needs is becoming increasingly important in information systems engineering. Users want to be able to customize software behavior through reference to terms familiar to their diverse needs and experience. We present a requirements-driven approach to behavioral customization of software systems. Goal models are constructed to represent alternative behaviors that users can exhibit to achieve their goals. Customization information is then added to restrict the space of possibilities to those that fit specific users, contexts or situations. Meanwhile, elements of the goal model are mapped to units of source code. This way, customization preferences posed at the requirements level are directly translated into system customizations. Our approach, which we apply to an on-line shopping cart system, does not assume adoption of a particular development methodology, platform or variability implementation technique and keeps the reasoning computation overhead from interfering with execution of the configured application.

Keywords: Information Systems Engineering, Goal Modeling, Software Customization, Adaptive Systems.

1 Introduction

Adaptation is emerging as an important mechanism in engineering more flexible and simpler to maintain and manage information systems. To cope with changes in the environment or in user requirements, adaptive systems are able to change their structure and behavior so that they fit to the new conditions [1,2]. An important manifestation of adaptivity is the ability of individual organizations and users to *customize* their software to their unique and changing needs in different situations and contexts.

Consider, for example, an on-line store where users can browse and purchase items. Normally, an anonymous user can browse the products, view their price information and user comments, add them to the cart, log-in and check-out. But different shop-owners may want variations of this process for different users. They may need, for example, to withhold prices, user comments or other product information unless the user has logged in, or only if the user's IP belongs to a certain set of countries. Or they may wish to rearrange the sequence of screens that guide the buyer through the check-out process. Or, finally, they may wish to disable purchasing and allow just browsing, with only some frequent buyers allowed to add comments – with or without logging in first.

H. Mouratidis and C. Rolland (Eds.): CAiSE 2011, LNCS 6741, pp. 77–92, 2011.

The shop-owner should be able to devise, specify and change such rules every time she feels it is necessary and then just observe the system reconfigure appropriately without resorting to expert help. But how easy is this?

Satisfying a great number of behavioral possibilities and switching from one to the other is a challenging problem in information systems engineering. While there is significant research on modeling and implementing variability and adaptation, e.g. in the areas of Software Product-Lines and Adaptive Systems, two aspects of the problem seem to still require more attention. Firstly, the need to easily communicate and actuate the desired customization, using language and terms that reflect the needs and experience of the stakeholders, such us the shop owner of our example. Secondly, the need to allow the stakeholders to construct their customization preferences themselves, instead of selecting from a restricted set of predefined ones, allowing them, thus, to acquire a customization that is better tailored to their individual needs.

To address these issues, in this paper we extend our earlier work on goal variability analysis [3,4] and introduce a goal-driven technique for customizing the behavioral aspect of a software system. A generic goal-decomposition model is constructed to represent a great number of alternative ways by which human agents can use the system to achieve their goals through performance of various tasks. The system-to-be is developed and instrumented in a way that the chunks of code that can enable or prevent performance of such user tasks are clearly located and controlled in the source code. After completion and deployment of the application, to address their specific needs and circumstances, individual stakeholders can refine the goal model by specifying additional constraints to the ways by which human and machine actions are selected and ordered in time. A preference-based AI planner is used to calculate such admissible behaviors and a tree structure representing these behavioral possibilities is constructed. Thanks to having appropriately instrumented the source code, that tree structure can be used as a plug-in which is inserted in the system and enforces the desired system behavior. This way, high-level expressions of desired arrangements of user actions are automatically translated into behavioral configurations of the software system. Amongst the benefits of our approach are both that it brings the customization practice to the requirements level and that it allows leverage of larger number of customization possibilities in a flexible way, without imposing restrictions to the choice of development process, software architecture or platform technology.

The paper is organized as follows. In Section 2 we present the core goal modeling language and the temporal extension that we are using for representing behavioral alternatives. In Section 3 we show how we connect the goal model with the source code, how we express goal-level customization desires and how we translate them into behaviors of the system. We discuss the feasibility of our approach in Section 4. Finally, in Section 5 we discuss related work and conclude in Section 6.

2   Goal Models

Goal models [5,6] are known to be effective in concisely capturing alternative ways by which high-level stakeholder goals can be met. This is possible through the construction of AND/OR goal decomposition graphs. Such a graph can be seen in Figure 1. The

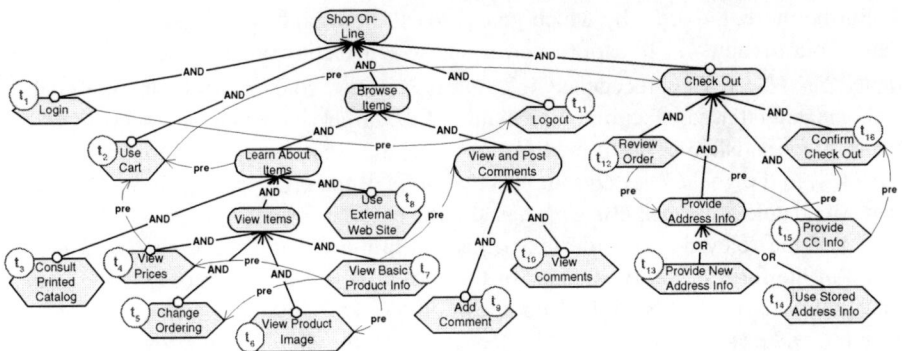

Fig. 1. A goal model

model shows alternative ways by which an on-line store can be used for browsing and purchasing products.

The graph consists of *goals* and *tasks*. Goals – the ovals in the figure – are states of affairs or conditions that one or more actors of interest would like to achieve [6]. Tasks – the hexagonal elements – describe particular low-level activity that the actors perform in order to fulfill their goals. To ease our presentation, next to each task shape a circular annotation containing a literal of the form t_i has been added, which we will use in the rest of the paper to concisely refer to the task. For example, t_7 refers to the task *View Basic Product Info*.

Tasks can be classified into two different categories depending on what the system involvement is during their performance. Thus, *human-agent* tasks are to be performed by the user alone without the support or other involvement of the system under consideration – an external system outside the scope of the analysis may be used though. For example *Consult Printed Catalog* (t_3) belongs to this category because it is performed without involvement of the system. On the other hand, *mixed-agent* tasks are tasks that are performed in collaboration with the system under consideration. Thus *Add Comment* is a mixed-agent task as the user will add the comment and the system will offer the facility to do so. Another example of a mixed-agent task is *View Image*: the system needs to display an image and the user must view it in order for the task to be considered performed. All tasks of Figure 1 are mixed-agent except for t_3 and t_8 which are human-agent tasks.

Goals and tasks are connected with each other using AND- and OR-decomposition links, meaning, respectively, that all (resp. one) of the subgoals of the decomposition need(s) to be satisfied for the parent goal to be considered satisfied. In addition, children of AND-decompositions can be designated as *optional*. This is visually represented through a small circular decoration on top of the optional goal. In the presence of optional goals, the definition of an AND-decomposition is refined to exclude optional sub-goals from the sub-goals that must necessarily be met in order for the parent goal to be satisfied. For example, for the goal *View Items* to be fulfilled, the task *View Basic Product Info* is only mandatory – tasks *View Prices*, *Change Ordering* and *View Product Image* may or may not be chosen to be performed by the user.

Furthermore, the order by which goals are fulfilled and tasks are performed is relevant in our framework. To express constraints in satisfaction ordering we use the *precedence link* ($\xrightarrow{pre}$). The precedence link is drawn from a goal or task to another goal or task, meaning that satisfaction/performance of the target of the link cannot begin unless the origin is satisfied or performed. For example the precedence link from the task *Use Cart* (t_2) to the goal *Check Out* implies that none of the tasks under *Check Out* can be performed unless the task *Use Cart* has already been performed.

Given the relevance of ordering in task fulfillment, solutions of the goal model come in the form or *plans*. A plan for the root goal is a sequence of leaf level tasks that both satisfy the AND/OR decomposition tree and possible precedence links. In plan $[t_1, t_7, t_4, t_2, t_{12}, t_{14}, t_{15}, t_{16}, t_{11}]$ for example, the user logs-in, browses the products with their prices, adds some of them to the cart and then checks out. In plan $[t_1, t_7, t_4, t_9, t_{10}, t_2, t_{12}, t_{14}, t_{15}, t_{16}, t_{11}]$, the user also views and adds comments.

The goal model implies a potentially very large variety of such plans, which are understood as a representation of the variability of *behaviors* that an actor may exhibit in order to achieve their goals. Note that this behavioral variability is to be contrasted with variability of the actual software system, in that the same system variant may be used in a variety of ways by the user. For example, the user of our on-line store may variably choose to use or not to use the *Add Comment* feature, even if that feature is invariably available to them.

3   Enabling Goal-Driven Customization

Let us now see how our framework allows specification of preferred user behaviors and enables subsequent customization of the software system in a way that these preferred behaviors are actually enforced. A schematic of our overall approach can be seen in Figure 2. At design time the system is developed in a way that the code that enables each leaf level task is clearly identified in the source code (frame B in the Figure) and can be disabled or enabled using information appropriately acquired from replaceable customization plug-ins, whose construction takes place after deployment, as described below. After deployment of the application, the users can define behavioral customization constraints at a high-level using structured English (frame C). These constraints are translated into formulae in Linear Temporal Logic (D), which, together with the goal model (A) are provided to a preference-based planner. The latter produces plans of the goal model that best satisfy the given behavioral constraints (E). These plans are finally merged into a structure called *policy tree* (F) which is then plugged into the application so that the latter, thanks to the instrumentation that took place at design time (B), exhibits the behavior that is desired in the original customization constraints. In the rest of this section we describe each of these steps in more detail.

3.1   Connecting Goal Models with Code

To allow interpretation of preferred plans into preferred software customizations, the system is developed in a way such that elements of the source code are associated with tasks of the goal model. In our framework, the nature of this association as well as

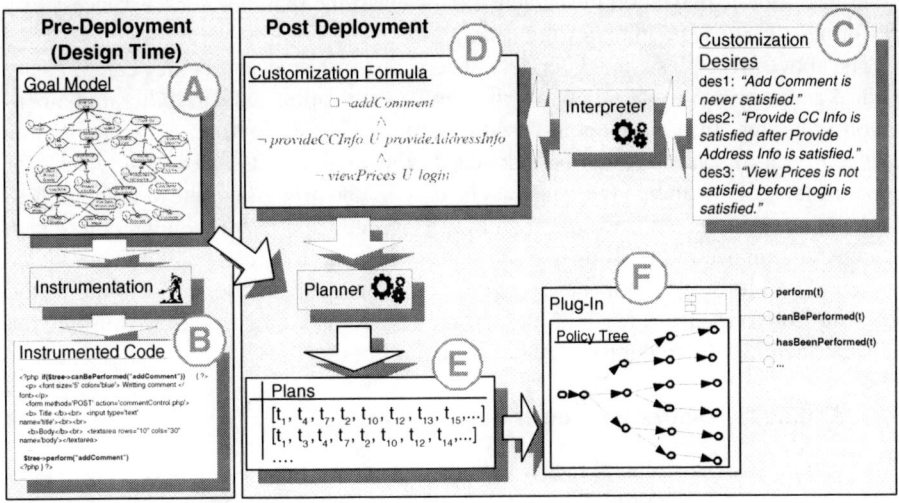

Fig. 2. From Customization Desires to Policy Trees

the way it is established is transparent from a particular implementation technology or architectural approach (e.g. agent-, service- or component-orientation) or particular development process that, for example, goal-oriented development methodologies propose (e.g. [7]). It is also independent of variability implementation and composition techniques (e.g. [8,9,10]) in a sense that any such technique could potentially be chosen and applied. Thus, to establish the association between goal models and code we only identify two general principles, which, if applied during development – in whatever architectural or process context – our framework becomes applicable. These principles refer to *task separation* and *task instrumentation*, explained below.

Task Separation. For every mixed-agent task in the goal model there exists a set of statements which are dedicated to exclusively supporting that task – and, thus, serve no other purpose. Furthermore, it should be possible to prevent these statements from executing, preventing in effect the user from performing the task. There is no requirement that these statements are located in the same part of the implementation and not scattered across components, modules, classes etc. – thus the principle is not a suggestion of task-oriented modularization. We call this code *mapped code (fragment)* to the task. Back in the on-line cart example, the mapped code for task *Login* is the code for drawing the username and password text boxes as well as the "Submit" and "Clear" button on the user screen. This code exists exclusively for allowing the user to perform this task. Not drawing those widgets, through conditioning the mapped code, effectively prevents execution of the task. As we will see, we found that the mapped code is predominantly code that conveniently exists in the view layer of an application.

Task Instrumentation Points. For every mixed-agent task, there is a location in the source code where the state of the system suggests that a task has been performed. In the *Login* example this might be the point in which confirmation that the login credentials are correct is sent back from the database and the application is ready to redirect control elsewhere. In the task *Review Order*, this can be the point where a summary of the order has been displayed on the screen – and we assume that the user has successfully performed the subsequent reviewing task.

The above principles are deliberately general and informal so that they can be easily refined and applied in a variety of architectural, composition and variability implementation scenarios. In a component-based or service-oriented setting, for example, the mapped code of each task can be associated with existing interfaces or services – or adapters thereof – which may or may not be used by the process engine or other orchestration/composition environment. In an aspect-oriented application, on the other hand, modularization need not follow task separation. Instead tasks can be written as advice to be weaved (or not) in appropriate locations in the source code. Later in the paper, drawing from our case study with the on-line cart system, we show how fulfilling the above principles turned out to be a very natural process.

3.2   Adding Customization Constraints

The temporally extended goal model with its precedence links is intended to be an unconstrained and behaviorally rich model of the domain at hand. Indeed, the goal model of Figure 1 describes a large variety of ways by which the user could go about fulfilling the root goal, as long as each of these ways is physically possible and reasonable. However the shop owner may wish to restrict certain possibilities. For example, she may want to disallow the user to view the prices unless he logs in first or prevent the user from viewing and/or adding comments, before logging in or in general. She may even go on to disallow use of the cart, again prior to logging in or even for the entire session. In the last case, this would effectively imply turning the system into a tool for browsing products only.

To express additional constraints on how users can achieve their goals we augment the goal model with the appropriate *customization formulae* (CFs - frame D in Figure 2). CFs are formulae in linear temporal logic (LTL) grounded on elements of the goal model. Different stakeholders in different contexts and situations may wish to augment the goal model with a different set of CFs, restricting thereby the space of possible plans to fit particular requirements. To construct CFs we use 0-argument predicates such as *useCart* or *browseItems* to denote satisfaction of tasks and goals. These predicates become (and stay) true once the task or goal they represent is respectively performed or satisfied. Furthermore, symbols $\square, \diamond$, $\circ$ and U are used to represent the standard temporal operators *always, eventually, next* and *until*, respectively.

Using CFs we can represent interesting temporal constraints that performance of tasks or satisfaction of goals must obey. Back to our on-line shop example, assume that the shop owner would like to disallow certain users from browsing the products without them having logged in first. This could be written as a CF as follows:

$$\neg \; viewBasicProductInfo \; U \; login$$

The above means that, in a use scenario, the task *View Basic Product Info* (t_7) should not be performed (signified by predicate *viewBasicProductInfo* becoming true) before the task *Login* (t_1) is performed for the first time (thus, predicate *login* becoming true). For another class of users there may be a more relaxed constraint:

$$\neg \; viewPrices \; U \; login$$

Universal and existential constraints are also relevant. For example the shop owner may want to disallow users from adding comments, thus:

$$\Box \neg addComment$$

If, in addition to these, she wants to prevent them from viewing prices, logging in and using the cart, this translates into a longer conjunction of universal properties seen in Figure 3. In effect, with the property of the figure the shop owner allows the users to only browse the products, their basic information and their images.

$$(\Box \neg addComment) \land (\Box \neg viewPrices) \land$$
$$(\Box \neg login) \land (\Box \neg useCart)$$

Fig. 3. A Customization Formula

While CFs, as LTL formulae, can in theory be of arbitrary complexity, we found in our experimentation that most CFs that are useful in practical applications are of specific and simple form. Thus simple existence, absence and precedence properties are enough to construct useful customization constraints. Hence, LTL patterns such as the ones introduced by Dwyer et al. [11], can be used to facilitate construction of CFs without reference to temporal operators. In our application, we used patterns in the form of templates in structured language. Thus, CFs can be expressed in forms such as "h_1 is [not] satisfied before/after h_2 is satisfied" to express precedence as well as "h is eventually [not] satisfied" to express existential properties, where h, h_1, h_2 are goals or tasks of the goal model. Examples of customization desire expressions can be seen in frame C of Figure 2. A simple interpreter performs the translation of such customization desires into actual LTL formulae. In this way, construction of simple yet useful CFs is possible by users who are not trained in LTL.

3.3 Identifying Admissible Plans

Adding CFs significantly restricts the space of possible plans by which the root goal can be satisfied. Given a CF, we call the plans of the goal model that satisfy the CF *admissible plans* for the CF. Thus, all $[t_7]$, $[t_7, t_5]$, $[t_7, t_{10}, t_6]$, $[t_8, t_7, t_6, t_5]$ and $[t_3, t_7, t_{10}]$ are examples of admissible plans for the CF of Figure 3. However, plan $[t_1, t_7, t_4, t_9, t_2, t_{12}, t_{14}, t_{15}, t_{16}, t_{11}]$, although it satisfies the goal model and its precedence constraints, it is not admissible because it violates the CF – all its conjuncts actually.

To allow the identification of plans that satisfy a given CF, we are adapting and using a preference-based AI planner, called PPLan [12]. The planner is given as input a goal model, automatically translated to a planning problem specification as well as a CF and returns the set of all admissible plans for the CF (frame E in Figure 2). Unless interrupted, the planner will continue to immediately output plans it finds until there are no more such. Details on how the planner is adapted can be found in [3].

3.4 Constructing and Using the Policy Tree

We saw that the introduction of a CF dramatically decreases the number of plans that are implied by the goal model into a smaller set of admissible ones that also satisfy

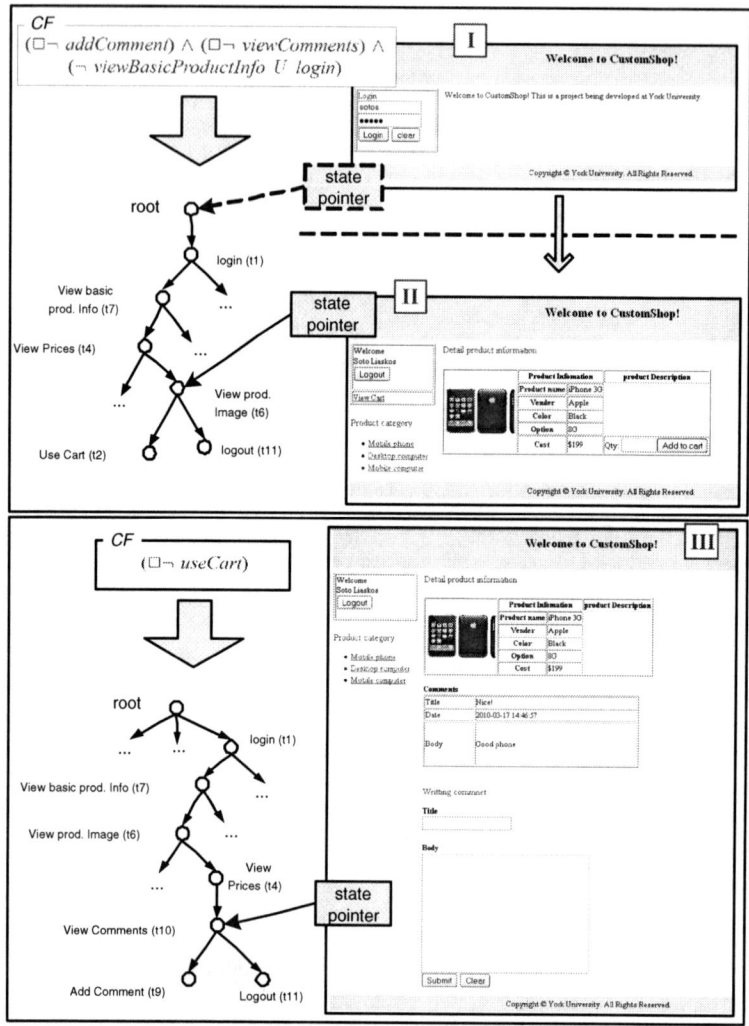

Fig. 4. The effect of Customization Formulae

the CF. The policy tree is simply a concise representation of those admissible plans – with the difference that it includes only the mixed-agent tasks. In particular, each node of the policy tree represents a task in the goal model. Given a set of plans P – where human-agent tasks have been removed – the policy tree is constructed in a way that every sequence of nodes that constitutes a path from the root to a leaf node is a plan in P and vice versa. It follows that every intermediate node in the policy tree represents both a plan prefix – i.e. the first n tasks of a plan – that can be found in P (by looking at the path from the root) and a set of continuation possibilities that yield complete plans of P (by looking at possible paths towards the leafs).

The policy tree is also supplied with a pointer that points to one of the nodes of the tree. We call this the *state pointer*. The role of the state pointer is to maintain information about what tasks have been performed in a given use scenario at run time. Thus, the state pointer pointing to a given node means that the tasks of the plan prefix associated to that node (the *associated prefix*) have already been performed. On the other hand, the tasks that can possibly be performed from that point are restricted to the children of the node currently pointed at, or any of the tasks in the associated prefix – in a sense that these tasks can be repeated.

In Figure 4, for example, on the left side of the bottom frame, part of a policy tree can be seen together with the CF it originated from ($\Box \neg useCart$). Through use of the planner, that CF results in a set of admissible plans, say P. Some of those plans have a prefix $[t_1, t_7, t_6, t_4, t_{10}, \dots]$. Thus, in the resulting policy tree that is depicted, there is a path from the root to the node t_{10} that constructs this prefix. By looking at the children of node t_{10}, we infer that only two expansions of the prefix at hand will yield a longer prefix that also exists in P and therefore is admissible with respect to the CF: t_9 and t_{11}. In practice, this means that if we are to keep satisfying the CF, we should either perform one of those two actions or repeat actions of the existing prefix (but without moving the state pointer).

An algorithm for constructing a policy tree, from a list of admissible plans that the planner returns can be found in our technical report [13]. It is important to note here that a new plan can always be appended to an existing policy tree in linear time and enrich the behavioral possibilities. This allows us to use partial outputs of the planner immediately while gradually enriching the tree as new plans are generated.

3.5   Conditioning and Instrumenting the Source Code

Let us now see how the policy tree can be plugged into the software system to enable a behaviors that comply with the expressed customization desires. Preparation for this needs to actually happen at design time, when the application is developed. Recall that the system is built following the principles of task separation and task instrumentation. This means that, on one hand, each mixed-agent task is associated with a set of statements (the mapped code) whose removal can prevent execution of the task, and on the other hand, for each task there is a well defined location in the code that marks completion of the task. The policy tree is integrated by conditioning access to the mapped code based on the position of the state pointer, and by adding statements in the instrumentation points that advance the position of the state pointer accordingly.

More specifically, the former is implemented through the use of the function *canBePerformed(t)*. The function *canBePerformed(t)* returns true iff task t is one of the children of the node currently pointed at by the state pointer or part of the associated prefix. In other words, the code fragment can be entered only if the new plan prefix that would result from performing the task that maps to that fragment belongs to at least one of the admissible plans. For example the mapped code of the task *Use Cart* involves buttons for adding items to the cart, text fields for specifying quantities, links for viewing the cart content etc. All these will be displayed only if *canBePerformed(useCart)* is true, that is the task *Use Cart* is in one of the children of the state pointer, or it is part of the path from the root to the state pointer. If this is not the case, the mapped code

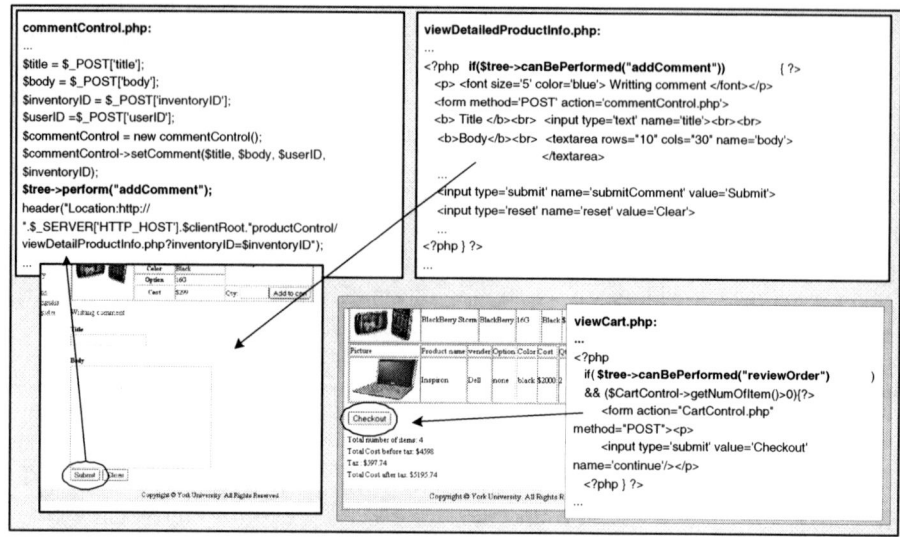

Fig. 5. Conditioning and Instrumenting Code

will not be accessed, preventing rendering of the user interface elements, which in turn prevents performance of the task by the user.

Advancement of the position of the state pointer, on the other hand, is implemented through simple *perform(t)* statements inserted in the instrumentation points, where *t* is the task that was just performed. The effect of the *perform(t)* statement is that the state pointer advances to the child labeled with *t* or stays where it is if *t* is part of the path from the root to the state pointer.

In Figure 5, examples of conditioning and instrumentation are shown for our PHP-based on-line cart system. The upper right frame shows how displaying the widgets for performing the task *Add Comment* is conditional to *canBePerformed(addComment)* being true. Once the user presses the submit button, a different file (commentControl.php) arranges to insert the comment to the database and, among other workings, a call to *perform(addComment)* is made (seen in upper left frame), so that the policy tree advances to the corresponding node. In the lower right frame, how customization conditions are mixed with run-time conditions is illustrated. Thus, the "Checkout" button is visible if "Checkout" is allowed by the current customization policy and the cart is non-empty, which is something irrelevant of policy tree. It is important to notice, therefore, that the policy tree is not used to completely arrange the details of the control flow of the application but to only enforce more abstract customization decisions that have been made at the requirements level. Note also that use of the policy tree is not restricted to the functions discussed above. For example the function *hasBeenPerformed(t)*, which returns true iff task *t* is part of the associated prefix of the node currently pointed, proved in our application to be helpful in handling large numbers of task permutation possibilities.

Note, again, that the injection of conditioning and instrumentation code discussed above is taking place at design time and based on the goal model. It is therefore independent of the actual structure of the policy tree, which, once the system is up and running, varies based on the customization constraints that are in effect each time.

3.6   In Action

Let us now see a complete example of how a system is customized through expression of high-level customization desires. Back to our on-line shop, consider the scenario in which the shop-owner wants to construct CFs for newly identified groups within her customer base. In Figure 4, two different CF scenarios she devised can be seen together with screen-shots showing the effect they have to system behavior. On the scenario on the top frame the CF prevents the users from – among other things – viewing any product information before they login. In effect this means that once the session starts the only user action that is allowed is logging in. Indeed, in the policy tree, login is the only child of the root. This explains the bare-bones screen that is offered to the users (upper screen-shot labeled [I]). Later in the same scenario of the top frame the user has logged in and is browsing products. However, the CF prevents the user from adding any comments. Hence, this facility is absent when viewing detailed product information (screen-shot [II]). Nevertheless, at that stage, making use of the cart or logging out is possible as seen in the policy tree. Thus, the button "Add to cart" is visible next to the product and the button "Logout" on the top left of the screen. The scenario on the lower frame of Figure 4, on the other hand, tailored to e.g. customers from a particular country overseas, prevents use of the cart but does not prevent addition of comments. Thus, at a stage where detailed product information is viewed, the user cannot add the item to the cart as before, but she can post a comment or log-out (screen-shot [III]). This is exactly what the state pointer indicates.

4   Applying Goal-Based Customization

Let us now discuss some of the experiences we acquired from our case study with our on-line cart system. A detailed account on this application can be found in [13].

Code Development and Instrumentation. The on-line cart system we built is a 5 thousand lines-of-code (5KLOC) application in PHP, following a common 3-layer architectural style – i.e. separating view, application logic and storage layers. Two developers, senior undergraduate students at that time, where asked to develop the system following a standard textbook object-oriented approach with the only goal model related restriction that the leaf level tasks of the goal model (which was maintained exclusively by the first author) would be treated as acceptance tests for the end-product and that optional and alternative tasks maintain that status in the implementation. Looking at the result afterwards we found that task separation not only was possible but emerged naturally in the development process. Interestingly, the mapped code would tend to appear at the view layer of the application. Furthermore, subsequent conditioning and instrumentation of the mapped code did not pose difficulties either. Policy trees, on the other hand, are plugged as separate globally visible PHP classes in the application. The use of the methods *canBePerformed(t)* and *performed(t)* to query/manipulate the tree did not pose any obvious perception problems or design issues requiring intense problem solving effort.

Anchoring the Policy Control Process. An issue that triggered further investigation is that of scoping behaviors. In our example, a plan prefix reflects the use of the system by one user at a particular time. The same or a different behavior may unfold from the beginning in a different client system (some other customer trying to buy something), or by the same customer later that day. With the term *anchor* we refer to any type of entity, or group thereof, whose lifetime is bound to a plan prefix. In our example, the anchor is the web session. If, for example, the session expires so does the plan prefix that has been constructed to that point. A new session always means an empty plan prefix (i.e. state pointer points to the root of the policy tree) waiting to be expanded through user actions. In different applications different anchoring entities can be thought. In an application processing business process, e.g. for academic admissions, a student application can be considered as the anchoring entity. Thus, for each new application that arrives a new empty prefix is constructed which is then augmented (through progression of the state pointer) based on tasks that are performed to process that particular application. Interestingly, different anchoring entities can be treated by different policy trees. For example different users of our on-line store (identified through e.g. a cookie mechanism) may experience different behavioral customizations, through assigning a separate policy tree to each of them.

Performance and Tool Considerations. The construction of a policy tree is an off-line activity and can afford longer computation times on separate computing infrastructure. This way, we avoid unpredictably expensive computational steps to intervene in the normal control flow. In our experimentation with several CFs over the bookseller goal model we found that the first hundred of admissible plans can be calculated within a time period ranging between one and 30 minutes. It is important to note that a working customization can be achieved even if a subset (in our case some tens) of all admissible plans is provided, though the resulting policy may prevent behaviors that are otherwise desired. The policy tree can keep being updated as the planner returns new plans. We definitely anticipate improved performance as the field of preference-based planning is fast progressing. For example, an HTN-based planner with preferences has recently been introduced which offers dramatically better performance through utilization of the domain knowledge expressed as task hierarchies ([14]). The principles applied in this paper are applicable to any preference-based planner that can generate sets of plans.

5   Related Work

Our proposal for requirements-driven software customization relates to research on a variety of topics including adaptive systems, product lines and software/service composition.

General goal-driven adaptation has been proposed by several authors. Thus, Zhang et al. [15] use temporal logic to specify adaptive program semantics. Further, work by Brown et al. [16] uses goal models to explicitly specify what should occur during adaptation. Their approach uses goal models to specify the adaptation process; in our approach the adaptation is the indirect result of imposing customization and precedence constraints on goals. Strategy trees have also been used to evaluate alternative reconfigurations of software systems in the context of QoS and structural changes [17]. Our approach differs in that it deals with user goals and behavior adaptation.

Researchers have also proposed different ways to model and bind variability in business processes. Lapouchnian et al. use goal models for analyzing alternative business process configurations [18]. Lu et al. propose the construction of flexible business process templates that lay the basic constraints that must be met [19]. Elsewhere [20,21] variability constructs are added to existing business process notations. In requirements engineering, a constraint language with temporal features has been proposed to analyze families of scenarios [22]. In general, such frameworks do not include an implementation approach, and when they do, this is restricted to specialized frameworks such as workflow engines [21] or e.g. BPEL-based service composition platforms [18].

The extensive literature on software composition, on the other hand (e.g. [9] for a taxonomy), is focusing on specific technologies, frameworks or techniques by which composition can be implemented – e.g. composition of services ([23]), the AHEAD framework and its descendants [24,25] or Aspect Orientation [26], Domain Specific Languages and Generators [27,28]. Use of existing AI planning applications to service composition, in particular, ([29,30] – cf. [10] for a survey), requires certain assumptions such as, for example, availability of cleanly defined services, limited degree of user intervention or the existence of some implementation and execution technique of the desired composition that also alleviates increased reasoning times. Our customization framework attempts to be more generally applicable, has a stronger focus on the implementation aspect without making platform or architectural assumptions and it also focuses on user interactions and therefore families of behaviours (system customizations) rather than single-purpose compositions. At the same time, it focuses on the requirements aspect of the problem, that is how the desired customization result can be communicated through reference to terms related to the experience and the goals of the actual users, rather than technical features of the system.

6   Conclusions

Tailoring the behavior of a software system to the needs of individual stakeholders, contexts and situations as these change over time has emerged as an important need in today's systems development. However, it also poses a challenging engineering and maintenance problem.

The main contribution of our paper is a technique to exactly allow this translation of high-level customization requirements into an appropriately configured system, in a flexible and accessible way. The merits of our approach lie in the following features. Firstly, it offers a direct linkage of software customization with user requirements using goal models and high-level customization desire specifications. This way customization is performed through talking about the user activity and experience rather than features of the system to be. Secondly, our proposal for constructive customization, where users express their exact needs, versus selective, where users select from predefined options, allows for flexibly leveraging a much larger space of customization possibilities, leading to systems that are better tailored to the exact needs of users. Thirdly, the proposed approach implies minimum impact to the implementation process, being transparent to the architectural, modularization, process and platform choices the engineers have made, as long as two simple mapping principles are followed and the ability to maintain and query the policy tree is arranged. Our application in the on-line cart system offered us strong evidence that both the customization practice per se and the engineering and development intervention that enables it are feasible and exhibit the above advantages.

Our proposal opens a variety of possibilities for future research. One of them is an extended empirical investigation on the applicability and generality of our basic implementation principles. Such empirical work also includes evaluating with end-users the extent and manner by which they can construct customization desires of various levels of complexity. Furthermore, application of the technique in a variety of system types would allow better understanding of whether the current form of the policy tree offers the right level of information or whether adding more expressiveness should be attempted. This could include, for example, adaptation of the semantics of satisfaction predicates so that task repetition also becomes subject to CF compliance or addition of run-time instance-level information to the produced policy structure. Such extensions would potentially allow for finer grain customization, but at the significant expense of simplicity, of impact minimality to the design and of maintaining a modest computational cost.

References

1. Oreizy, P., Medvidovic, N., Taylor, R.N.: Architecture-based runtime software evolution. In: Proceedings of the 20th International Conference on Software Engineering (ICSE 1998), Washington, DC, USA, pp. 177–186 (1998)
2. Kramer, J., Magee, J.: Self-managed systems: an architectural challenge. In: Future of Software Engineering (FOSE 2007), Washington, DC, USA, pp. 259–268 (2007)
3. Liaskos, S., McIlraith, S.A., Mylopoulos, J.: Towards augmenting requirements models with preferences. In: Proceedings of the 24th International Conference on Automated Software Engineering (ASE 2009), Auckland, New Zealand, pp. 565–569 (2009)
4. Liaskos, S., McIlraith, S.A., Mylopoulos, J.: Integrating preferences into goal models for requirements engineering. In: Proceedings of the 10th International Requirements Engineering Conference (RE 2010), Sydney, Australia, pp. 135–144 (2010)
5. Dardenne, A., van Lamsweerde, A., Fickas, S.: Goal-directed requirements acquisition. Science of Computer Programming 20(1-2), 3–50 (1993)
6. Yu, E.S.K., Mylopoulos, J.: Understanding "why" in software process modelling, analysis, and design. In: Proceedings of the Sixteenth International Conference on Software Engineering (ICSE 1994), pp. 159–168 (1994)
7. Penserini, L., Perini, A., Susi, A., Mylopoulos, J.: High variability design for software agents: Extending Tropos. ACM Transactions on Autonomous and Adaptive Systems (TAAS) 2(4) (2007)
8. Gacek, C., Anastasopoules, M.: Implementing product line variabilities. SIGSOFT Software Engineering Notes 26(3), 109–117 (2001)

9. McKinley, P.K., Sadjadi, S.M., Kasten, E.P., Cheng, B.H.C.: Composing adaptive software. IEEE Computer 37(7), 56–64 (2004)
10. Rao, J., Su, X.: A survey of automated web service composition methods. In: Cardoso, J., Sheth, A.P. (eds.) SWSWPC 2004. LNCS, vol. 3387, pp. 43–54. Springer, Heidelberg (2005)
11. Dwyer, M.B., Avrunin, G.S., Corbett, J.C.: Patterns in property specifications for finite-state verification. In: Proceedings of the 21st International Conference on Software Engineering (ICSE 1999), Los Alamitos, CA, USA, pp. 411–420 (1999)
12. Bienvenu, M., Fritz, C., McIlraith, S.: Planning with qualitative temporal preferences. In: Proceedings of the 10th International Conference on Principles of Knowledge Representation and Reasoning (KR 2006), Lake District, UK, pp. 134–144 (2006)
13. Liaskos, S., Litoiu, M., Jungblut, M.D., Mylopoulos, J.: Goal-based Behavioral Customization of Information Systems. Technical Report CSE-2010-10, York University (2010)
14. Sohrabi, S., Baier, J.A., McIlraith, S.: HTN planning with preferences. In: Proceedings of the 21st International Joint Conference on Artificial Intelligence (IJCAI 2009), Pasadena, CA, USA, pp. 1790–1797 (2009)
15. Zhang, J., Cheng, B.H.C.: Using temporal logic to specify adaptive program semantics. Journal of Systems and Software (Special Issue on Architecting Dependable Systems) 79(10), 1361–1369 (2006)
16. Brown, G., Cheng, B.H.C., Goldsby, H., Zhang, J.: Goal-oriented specification of adaptation requirements engineering in adaptive systems. In: Proceedings of the 2006 International Workshop on Self-Adaptation and Self-Managing Systems (SEAMS 2006), pp. 23–29. ACM, New York (2006)
17. Simmons, B.: Strategy-trees: A Novel Approach to Policy-Based Management. PhD thesis, University of Western Ontario (February 2010)
18. Lapouchnian, A., Yu, Y., Mylopoulos, J.: Requirements-driven design and configuration management of business processes. In: Alonso, G., Dadam, P., Rosemann, M. (eds.) BPM 2007. LNCS, vol. 4714, pp. 246–261. Springer, Heidelberg (2007)
19. Lu, R., Sadiq, S., Governatori, G.: On managing business processes variants. Data and Knowledge Engineering 68(7), 642–664 (2009)
20. Gottschalk, F., van der Aalst, W.M.P., Jansen-Vullers, M.H., La Rosa, M.: Configurable workflow models. International Journal of Cooperative Information Systems (IJCIS) 17(02), 177–221 (2008)
21. Sadiq, S.W., Orlowska, M.E., Sadiq, W.: Specification and validation of process constraints for flexible workflows. Information Systems 30(5), 349 (2005)
22. Sutcliffe, A.G., Maiden, N.A.M., Minocha, S., Manuel, D.: Supporting scenario-based requirements engineering. IEEE Transactions on Software Engineering 24(12), 1072–1088 (1998)
23. Baresi, L., Pasquale, L.: Live goals for adaptive service compositions. In: Proceedings of the 2010 ICSE Workshop on Software Engineering for Adaptive and Self-Managing Systems (SEAMS 2010), pp. 114–123 (2010)
24. Batory, D., Sarvela, J.N., Rauschmayer, A.: Scaling step-wise refinement. In: Proceedings of the 25th International Conference on Software Engineering (ICSE 2003), Washington, DC, USA, pp. 187–197 (2003)
25. Apel, S., Kastner, C., Lengauer, C.: Featurehouse: Language-independent, automated software composition. In: Proceedings of the 31st International Conference on Software Engineering (ICSE 2009), pp. 221–231 (2009)
26. Kiczales, G., Lamping, J., Mendhekar, A., Maeda, C., Lopes, C., Loingtier, J.-m., Irwin, J.: Aspect-oriented programming. In: Liu, Y., Auletta, V. (eds.) ECOOP 1997. LNCS, vol. 1241, p. 313. Springer, Heidelberg (1997)

27. Cleaveland, J.C.: Building application generators. IEEE Software 5(4), 25–33 (1988)
28. Czarnecki, K., Eisenecker, U.W.: Generative Programming - Methods, Tools, and Applications. Addison-Wesley, Reading (2000)
29. Sohrabi, S., Prokoshyna, N., McIlraith, S.A.: Web service composition via generic procedures and customizing user preferences. In: Cruz, I., Decker, S., Allemang, D., Preist, C., Schwabe, D., Mika, P., Uschold, M., Aroyo, L.M. (eds.) ISWC 2006. LNCS, vol. 4273, pp. 597–611. Springer, Heidelberg (2006)
30. Wu, D., Parsia, B., Sirin, E., Hendler, J., Nau, D.S.: Automating DAML-S web services composition using SHOP2. In: Fensel, D., Sycara, K., Mylopoulos, J. (eds.) ISWC 2003. LNCS, vol. 2870, pp. 195–210. Springer, Heidelberg (2003)

From Requirements to Models: Feedback Generation as a Result of Formalization*

Leonid Kof and Birgit Penzenstadler

Fakultät für Informatik, Technische Universität München,
Boltzmannstr. 3, D-85748, Garching bei München, Germany
{kof,penzenst}@informatik.tu-muenchen.de

Abstract. Natural language is the main presentation means in industrial requirements documents. In addition, communication between the different stakeholders is often insufficient, therefore requirements documents are frequently incomplete and inconsistent. This causes problems during modeling or programming.

The aim of the presented paper is to make deficiencies in behavior specifications apparent in the early project stage. The basic idea is to model the required system behavior and to generate feedback for human analysts, based on the deficiencies of the resulting models. The presented feedback generation was evaluated in an experiment. It was found that it can address genuine problems of requirements documents.

Keywords: Requirements Engineering, Model Extraction, Feedback Generation.

1 Requirements Documents Suffer from Missing Information

At the beginning of a software project, the requirements of different stakeholders are usually gathered in a document. The majority of these documents are written in natural language, as the survey by Mich et al. shows [1]. Diversity of stakeholders and insufficient communication results in imprecise, incomplete, and inconsistent requirements documents, because precision, completeness and consistency are extremely difficult to achieve using mere natural language as the main presentation means.

In software development, the later an error is found, the more expensive its correction. Thus, it is one of the goals of requirements analysis to find and to correct the deficiencies of requirements documents. A practical way to detect errors in requirements documents is to convert informal specifications to system models. In this case, errors in documents would lead to inconsistencies or omissions in models, and, due to the more formal nature of models, inconsistencies and omissions are easier to detect in models than in textual documents.

Although there exist a number of automatic approaches that analyze specifications written in natural language and provide a model, the existing approaches go in one direction only: they transform a textual specification into a formal model. However, in the case that the specification exhibits some deficiencies, they either heuristically compensate these deficiencies or fail silently.

---

* This work was supported by the German Research Council (DFG), Grant BR 887/26-1.

H. Mouratidis and C. Rolland (Eds.): CAiSE 2011, LNCS 6741, pp. 93–107, 2011.

Contribution: The goal of the presented paper is to show, how the deficiencies of the models resulting from the text can be used to generate feedback for human analysts. The feedback can be presented in two forms: (1) in natural language and (2) by special markings on the produced models. The effectiveness of the generated feedback was evaluated in an experiment and it was found that the generated feedback can address genuine problems of requirements specifications that would be overseen by human analysts.

Outline: The remainder of the paper is organized as follows: Section 2 presents our approaches to text-to-model translation, used as the basis for the presented work on feedback generation. Sections 3 and 4 are the technical core of the paper, they present the feedback generation and its evaluation. Finally, Section 5 gives an overview of related work and Section 6 summarizes the paper.

2   From Text to Models: Our Existing Approaches

In our survey of existing modeling techniques [2] it was shown that all existing industrially relevant formalisms are based either on interaction sequences or on finite automata. For this reason, the target model types for the behavior modeling are either finite automata or Message Sequence Charts (MSCs), serving as a representative for interaction-based modeling techniques. The translation from text to MSCs is presented in Section 2.1, and the translation to finite automata in Section 2.2.

2.1   From Scenarios to Message Sequence Charts

Translation of textual scenarios to message sequence charts was presented in [3,4]. For the translation we assume that every message sequence chart (MSC) consists of a set of *actors*, a sequence of *messages* sent and received by these actors, and a sequence of *conditions* (or *assertions*) interleaved with the message sequence. This terminology is illustrated in Figure 1.

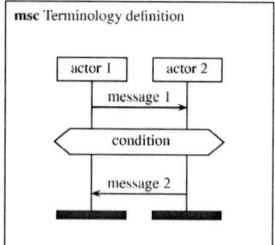

Fig. 1. MSCs, terminology

The basic idea of the scenario-to-MSC translation can be illustrated on the following scenario, taken from the Instrument Cluster Specification [5]:

1. The driver switches on the car (ignition key in position ignition on).
2. The instrument cluster is turned on and stays active.
3. After the trip the driver switches off the ignition.
4. The instrument cluster stays active for 30 seconds and then turns itself off.
5. The driver leaves the car.

A possible manual translation of this scenario to an MSC is shown in Figure 2. There are two challenge/response interactions in this MSC: The instrument cluster replies to the requests of the main controller ("car").

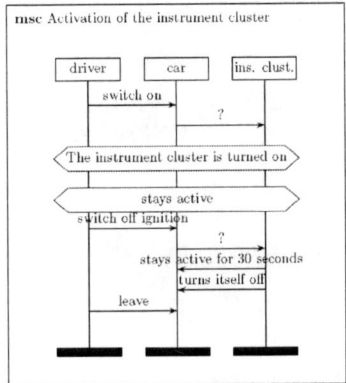

Fig. 2. Possible interpretation of the example scenario

To model challenge/response patterns in MSCs translated from textual scenarios, we organize messages in a stack: If a new message m represents an answer to some previously pushed message m', m' and the messages above it are popped from the stack. Otherwise, the new message m is pushed onto the stack.

We assume that the actors involved in the MSC are provided before the actual text-to-MSC translation. They can be extracted from the requirements document (cf. [4]) or listed manually. The list of actors allows us to decide, which sentences should be translated to messages, and which to assertions: A sentence is translated to a message, if its subject is contained in the set of actors, the sentence is in active voice and contains a grammatical object. All other sentences are translated to assertions (cf. [4]).

For the sentences translated to messages, we assume that the sentence subject is the message sender. For the message receiver there are two possibilities: if the sentence object is contained in the set of actors, the sentence object becomes the message receiver. If the sentence object is not contained in the set of actors, we have to infer the message receiver from the message stack: Let m_{top} be the message on the top of the stack and m_{new} the message under analysis. Then, we assume that m_{new} is the response to m_{top} and, therefore, the receiver of m_{new} is the sender of m_{top}. In a similar way, we can infer missing messages: if the sender of m_{new} is not equal to the receiver of m_{top}, we assume that there is a missing message m' from the receiver of m_{top} to the sender of m_{new}. We put this missing message m' on the stack too. The details of stack management are presented in [3].

2.2   From Automata Descriptions to Automata

Similarly to the text-to-MSC translation, it is possible to translate text pieces describing automata to automata themselves, as presented in [6]. The difference lies in the relation between sentences that we have to model. This can be illustrated on the specification excerpt in Table 1 (from [7]). The header and the first sentence of this excerpt set the context ("normal mode"), and further sentences refer to this context. Thus, instead of a message stack, we have to model the context setting and the usage of the set context.

Table 1. The Steam Boiler, specification excerpt (copied from [7])

Normal mode
1. The normal mode is the standard operating mode in which the program tries to maintain the water level in the steam-boiler between N1 and N2 with all physical units operating correctly.   2. As soon as the water level is below N1 or above N2 the level can be adjusted by the program by switching the pumps on or off.   3. The corresponding decision is taken on the basis of the information which has been received from the physical units.

We model the context by assigning every (sub)sentence to one of the four categories: "state transition", "transition condition", "context setting", or "irrelevant". The assignment of sentence segments to categories takes place in the following steps: (1) splitting of every sentence to segments, (2) assignment of segments to categories on the basis of grammatical information only, and (3) re-assignment of segments to categories, by using context information. Each of these steps is described below.

Sentence splitting: Punctuation symbols, the words "if" and "when" as well as the conjunctions "and" and "or" are used as splitting marks, unless they directly follow an adjective or a number.[1] A splitting example is shown in Table 2.

Table 2. Splitting example

Original sentence
As soon as this signal has been received, the program enters either the mode normal if all the physical units operate correctly or the mode degraded if any physical unit is defective
Splitting
1. As soon as this signal has been received   2. the program enters either the mode normal   3. all the physical units operate correctly   4. the mode degraded   5. any physical unit is defective

---

[1] These heuristics prevent splitting of, e.g., "if the water level lies between N1 and N2, ...".

Assignment of segments to categories on the basis of grammatical information: Identification of the four segment classes is possible on the basis of the Part-of-Speech (POS) tags: A POS tagger decides, for every word, if this word is a noun, verb, adjective, ... The applied tagger has the precision of about 97% [8] which makes it unlikely to become an error source. Furthermore, we assume that the names of the automaton states are extracted from the specification before the actual text-to-automaton translation, cf. [6]. The assignment of the sentence segment to one of the four classes takes place in the following way:

- If the sentence segment does not contain any reference to a state, it is marked as "irrelevant". This holds, for example, for the first segment in Table 2.
- If the sentence segment contains a reference to a state, but first occurrence of the state is not preceded by a verb, this segment is marked as "context setting". For example, in Table 1, the header ("normal mode") and the first sentence set the context for the translation of the following sentences.
- Otherwise, the sentence segment is marked as "state transition".

Here it is important to emphasize that in the first phase no sentence segment is marked as "transition condition".

Re-assignment of segments to categories, by using context information: To take context into account, it is necessary to revise the "context setting"-marks first. Here, the following heuristics is applied: If, for a given sentence, any of its segments is marked as "state transition", then all segments marked as "context setting" are relabeled to "state transition". This compensates for potentially missing verbs in some sentence segments. In the case of the example shown in Table 2, it marks the fourth segment as "state transition" and leaves the other marks unchanged.

When the marking of segments as "state transition" is finished, it is possible to identify transition conditions:

- If a sentence segment is marked as "irrelevant" and directly precedes a segment marked as "state transition", then the former segment is relabeled to "transition condition" (e.g., the first segment of the example in Table 2).
- If a sentence segment is marked as "irrelevant" and directly follows a segment marked as "state transition", then the former segment is relabeled to "transition condition". This allows to treat conditions like "⟨some transition⟩ if ⟨some condition⟩".

When this relabeling process is finished, transitions are created from the sentence segments marked as "state transition". Transition conditions, as well as source and target state of transitions, are inferred from the adjacent "context setting" and "transition condition" segments.

The presented approach is domain-independent and only relies on a special writing style with guidelines for industrial requirements specifications: The complete set of system states is known, sentences describing state transitions contain a reference to the target state, and context setting is stated explicitly (for details, see [6]).

3   Feedback Generation Instead of Inference Assumptions

The translation approaches presented in Section 2 use different inference rules in order
to complete information not explicitly stated in the text. The main idea of feedback generation is to turn off the inference, and, whenever the inference would become necessary, to generate feedback questions addressing the missing information. For MSCs, this
missing information can be the unspecified message receiver or an unspecified intermediate message that is inferred by the means of the stack model. Feedback generation
for MSCs is presented in Section 3.1. For automata, the missing information is always
the source state of a state transition (due to peculiarities of the algorithm presented in
Section 2.2). Feedback generation for automata is presented in Section 3.2.

In general, the generated questions can refer either to the specification text or to the
extracted model. Reference to the model means that, in addition to the actual models,
special markers are generated that pinpoint the problematic model elements. Reference
to the text means that hints in natural language are generated, in the following form:

- **Problematic sentence:** ⟨sentence cited⟩
- **Detected problem:** ⟨problem description⟩

Both for automata and for MSCs, we implemented the generation of both representations.

3.1   Feedback Generation for MSCs

The algorithm for MSC generation presented in Section 2.1 can infer unspecified message receivers and whole missing messages. Correspondingly, the feedback questions
that are generated for MSCs, address (1) unspecified message receivers or (2) messages
that are necessary from the point of view of a continuous message flow, but not explicitly specified in the document.

Every type of the feedback question (addressing the missing receiver or the missing
messages) can refer both to the model and to the specification text. Reference to the text
means that we generate questions or hints in natural language, addressing the detected
problem. Reference to the model means that we use special markers in the generated
MSCs in order to address the detected problems.

Missing message receivers result from active sentences only, as passive sentences
are translated to MSC assertions, cf. [4]. Furthermore, an active sentence needs a grammatical object to be a candidate for an MSC message. Nevertheless, if the sentence
object is not contained in the set of potential MSC actors, inference rules sketched in
Section 2.1 must be applied. Thus, in order to address missing message receivers, we
generate schematic hints referring to the specification text, structured as follows:

⟨sentence cited⟩
Problem: Problematic active sentence: object ⟨sentence object⟩ is not contained in the list of actors. Message receiver cannot be identified.

The inference of missing messages becomes necessary when the sender (basically,
grammatical subject) of the sentence under consideration does not coincide with the
receiver of the last message. In this case we generate coherence questions:

⟨sentence cited⟩
Problem: COHERENCE: Which component / who controls the ⟨sentence subject⟩ to perform this action?

Additionally to the two above question types, based completely on the inference rules presented in Section 2.1, we address a further problem, resulting from passive sentences: whenever we come across a passive sentence, we check if the subject of the sentence coincides with the receiver of the last message. If this is not the case, we generate the following question:

⟨sentence cited⟩
Problem: Problematic sentence: passive, actor unspecified (Who / which component controls the ⟨sentence subject⟩?)

As an example, the set of questions addressing the problems of the scenario on page 95 is presented in Table 3.

Table 3. Questions referring to the specification text, generated for the scenario on page 95

1. The instrument cluster is turned on
   Problem: Problematic sentence: passive, actor unspecified (Who / which component controls the instrument cluster?)
2. The instrument cluster stays active for 30 seconds
   (a) **Problem:** Problematic active sentence: object "seconds" is not contained in the list of actors. Message receiver cannot be identified.
   (b) **Problem:** COHERENCE: Which component / who controls the instrument cluster to perform this action?

For the questions referring to the model, we further develop the idea shown in Figure 2: there, the inferred missing message was represented as a "?"-marked message. To generate the same types of questions as above, we introduce a special actor named "???" and send all the messages where the message receiver is not explicitly specified, to the "???"-actor. Similarly, every missing message is "?"-marked. Missing messages in the sense of Section 2.1 are split into two: a message from the inferred message sender to the "???"-actor, and a message from the "???"-actor to the inferred receiver.

The set of questions generated for the example on page 95 is shown in Figure 3. It is easy to see the correspondence to the questions referring to the specification: Question 1 from Table 3 is represented by two "?"-messages before the first assertions, Question 2a is represented by two messages between the assertions, and Question 2b by the "?"-messages after the second assertion.

3.2   Feedback Generation for Automata

The algorithm for automata generation presented in Section 2.2 infers, if necessary, the source state of the transition from the discourse context. It is easy to turn this inference

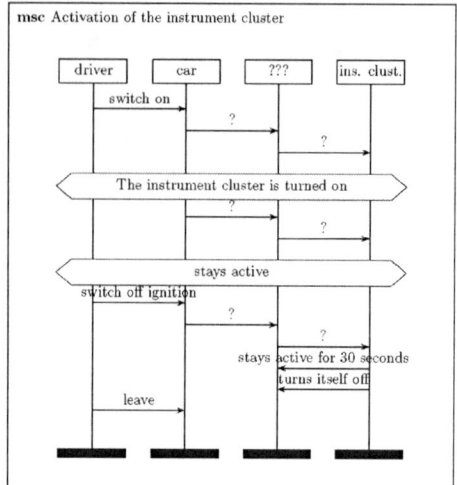

Fig. 3. Questions referring to the model, generated for the scenario on page 95

into feedback generation: whenever the source state of a transition is not explicitly specified, we can generate a corresponding question.

To generate questions referring to the specification text, we create schematic hints structured as follows:

- **Sentence:** ⟨sentence cited⟩
- **Unspecified source state:** transition ⟨transition condition⟩ to state ⟨target state⟩

For the specification excerpt shown in Table 1, this results in the set of questions presented in Table 4.

To generate questions referring to the model, we use the same representation as used in [6] for generated automata. In [6], the generated automata are represented as a three-column table: Every line of the table represents exactly one state transition: it contains the source and the target state, and the transition condition. The generated questions are represented in a similar table, with the only difference that the states that are not explicitly specified in the text are marked as "unknown". For the specification excerpt shown in Table 1, this results in the set of questions presented in Table 5.

4   Evaluation

The presented approach to question generation was evaluated in an experiment. The goal of the experiment was to see, if the generated questions address genuine specification problems that would be overseen by human analysts. The experiment setting is presented in Section 4.1. In the experiment, it was found that the generated questions can address genuine specification problems, not detected by human analysts. The results of the experiment are presented in Section 4.2. It was found, furthermore, that the problems not addressed by the generated questions can be detected by other techniques. The lessons learned in the experiment are presented in Section 4.3.

Table 4. Questions referring to the text, generated for the specification in Table 1

Specification	Generated Question
Sentence: the unit for detection of the level of steam is defective – that is , when v is not equal to zero – the program enters the emergency stop mode.	**Unspecified source state:** transition "the unit for detection of the level of steam is defective – that is , when v is not equal to zero – the program enters" to state "emergency stop mode".
Sentence: the program realizes a failure of the water level detection unit it enters the emergency stop mode.	**Unspecified source state:** transition "the program realizes a failure of the water level detection unit it enters" to state "emergency stop mode".
Sentence: as soon as this signal has been received , the program enters either the mode normal all the physical units operate correctly the mode degraded any physical unit is defective.	**Unspecified source state:** transition "all the physical units operate correctly" to state "mode normal".
Sentence: as soon as this signal has been received , the program enters either the mode normal all the physical units operate correctly the mode degraded any physical unit is defective.	**Unspecified source state:** transition "any physical unit is defective." to state "mode degraded".
Sentence: a transmission failure puts the program into the mode emergency stop.	**Unspecified source state:** transition "a transmission failure puts" to state "mode emergency stop".

4.1 Experiment Setting

In order to evaluate the generated feedback questions, the following evaluation hypothesis was put forward:

H: The automatically generated questions address deficiencies of the specification that would be overseen by human analysts.

In total, 9 PhD candidates/postdocs in computer science participated in the evaluation. The evaluation took place in the following way:

1. First, the experiment subjects were given the chapter of the Steam Boiler Specification [7] that refers to the system behavior (2 pages). Each subject was asked to read

Table 5. Questions referring to the model, generated for the specification in Table 1

Source	Target	Condition
unknown state	emergency stop mode	the unit for detection of the level of steam is defective – that is , when v is not equal to zero – the program enters
unknown state	emergency stop mode	the program realizes a failure of the water level detection unit it enters
unknown state	mode normal	all the physical units operate correctly
unknown state	mode degraded	any physical unit is defective .
unknown state	mode emergency stop	a transmission failure puts

the specification and to mark words, word sequences or sentences that, in his/her opinion, would cause problems if we model the system behavior.

2. Then, every subject was given a set of questions generated as presented in Section 3.2, and asked, for every generated question, to evaluate if the question addresses a genuine problem of the specification.

3. Similarly to step 1, every subject was given 15 (out of 41) scenarios from the Instrument Cluster Specification [5]. The 15 scenarios were selected in such a way that they represent typical system behavior. Thus, the results obtained with the 15 scenarios can be extrapolated to all scenarios provided in the specification.

   Analogously to step 1, every subject was asked to read the specification and to mark words, word sequences or sentences that, in his/her opinion, would cause problems if we model the system behavior.

4. Analogously to step 2, every subject was given a set of generated questions and asked, for every generated question, to evaluate if the question addresses a genuine problem of the specification.

In order to address different representations of the generated questions, the subjects were separated in two groups: One group was given questions referring to the specification text for the Steam Boiler and questions referring to the model for the Instrument Cluster. The other group, inversely, was given questions referring to the model for the Steam Boiler and questions referring to the specification text for the Instrument Cluster. Due to this setting, we avoid two potential threats to validity of the results:

– Every subject was given one set of questions referring to the model and one set of questions referring to the specification. Thus, the differences in subjects' responses cannot be attributed to the differences in the representation of the questions.
– Every subject was given one set of questions about the Steam Boiler Specification, and one set of questions about the Instrument Cluster Specification. This way, we avoid learning effects that would become important if we would have given two sets of questions about the same specification to the same subject.

The evaluation was performed in groups, but the subjects were not allowed to communicate with each other. In total, every subject spent approximately two hours performing the above steps 1-4.

4.2 Experiment Results

The results of the experiment are presented in Table 6. The numbers in the cells represent the number of manually marked problems or generated/confirmed questions. The number of questions generated for the Steam Boiler Specification (automaton-based) coincide for the questions referring to the model and to the text, as the differences in the representations are not too big. For the Instrument Cluster (MSCs), however, differences in representations result in the different number of generated questions (cf. Section 3.1).

Questions generated for the Steam Boiler Specification (automaton-based) were considered by the most subjects as spurious. With one notable exception, most subjects

Table 6. Experiment results. Numbers of found/addressed deficiencies.

	Generated	Subject	Manual	Generated$_{confirmed}$	Manual ∩ Generated	Manual ∩ Generated$_{confirmed}$
Automaton, questions referring to the text	21	1	32	1	1	1
		2	30	1	1	1
		3	13	0	0	0
		4	29	19	0	0
MSCs, questions referring to the model	87	1	38	73	0	0
		2	38	46	5	5
		3	25	43	0	0
		4	16	52	0	0
Automaton, questions referring to the model	21	1	9	1	0	0
		2	4	5	0	0
		3	11	0	0	0
		4	14	2	0	0
		5	12	1	0	0
MSCs, questions referring to the text	50	1	26	31	8	8
		2	18	7	1	1
		3	19	28	3	1
		4	11	27	0	0
		5	14	1	0	0

found that the generated questions do not address genuine problems of the specification. Thus, the experiment hypothesis has to be rejected for the automaton-based specification. For the Instrument Cluster Specification (MSCs), however, the subjects found that many more generated questions address genuine problems of the specification text. The fraction of questions addressing genuine problems (column "Generated$_{confirmed}$" of Table 6) varies from 2% (1 out of 50) to 84% (73 out of 87). It looks like the number of generated questions addressing genuine problems is higher for the questions referring to the model than for the questions referring to the text: For the questions referring to the text, at most 31 out 50 questions (62%) are found to address genuine problems. For the questions referring to the model, though, this rate varies from 49% (43 out of 87) to 84% (73 out of 87). The size of the sample, though, is too small to claim that questions referring to the model better help to detect specification problems.

The most interesting finding of the experiment is presented in the last two columns of Table 6: the manually found problems and the problems addressed by the automatically generated questions are almost always disjoint. The figures in both columns coincide, except for one line. This line results from the following situation: the subject marked two sentences as problematic. The tool marked the same two sentences as problematic too. However, the subject did not confirm the generated questions as addressing genuine problems. The most probable reason is that the subject saw different problems in these sentences than the tool. However, as the subject did not explicitly write down the identified problems, this remains solely our interpretation of the discrepancy.

Two last columns of Table 6, together with the number of generated questions that the subjects found to address genuine specification problems, allow us to claim that the hypothesis is confirmed for the Instrument Cluster Specification, namely, that the automatically generated questions address deficiencies of the specification that would be overseen by human analysts.

4.3 Lessons Learned

The specification problems that were detected by our subjects were highly different from the problems addressed by the automatically generated questions. The problems most often marked by the human analysts were pertinent to unclear phrasing. The automatically generated questions, to the contrary, addressed missing specification pieces. For example, our subjects marked following phrases as problematic:

- **For the Steam Boiler Specification:**
    - "Once all the units which were defective *have been repaired*, the program comes back to normal mode."
      Evaluator's comment: how is it detected that all the units have been repaired?
    - "...when either the *vital units* have a failure..."
      Evaluator's comment: which units count as "vital"?
    - "*As soon as* the water measuring unit is repaired..."
      Evaluator's comment: what does "as soon as" really mean?
- **For the Instrument Cluster Specification:**
    - "...and the instrument cluster *is activated temporarily*."
      Evaluator's comment: by whom is the instrument cluster activated? And what does "temporarily" really mean?
    - "The system displays the *calculated speed*"
      Evaluator's comment: how is the speed calculated?
    - "The input signals from the motor *are sent regularly*"
      Evaluator's comment: to which component are they sent? And what does "regularly" mean?

These phrases are indeed problematic if we want to build a system model, but they represent fuzzy specification, not completely missing facts.

In order to address such fuzzy phrasings, it makes sense to apply the previously developed tool that detects poor phrasings [9]. In its current version, the tool detects poor phrasings listed in the Ambiguity Handbook [10]. Adding further patterns for problematic phrases, however, is a matter of minor extension of the configuration files. An integrated tool, detecting both missing specification pieces (as in the presented paper) and poor phrasings, would provide a reliable means of specification analysis.

5 Related Work

Work related to the presented paper can be subdivided in two areas: work on text-based modeling and work on natural language processing (NLP) in requirements engineering. Both areas are presented below.

5.1 Text-Based Modeling

Saeki at al. [11], Overmyer et al. [12], and Ermagan et al. [13] introduced tools providing modeling approaches. The approach by Saeki et al. allows the user to mark words in the requirements documents, and then to assign the marked word to some noun or verb type. Then, the approach maps nouns to classes (in the sense of object oriented design), and verbs to operations. The approach can handle four predefined verb classes.

Overmyer et al. developed a tool allowing the user to mark words or word sequences and map them to classes, roles, and operations. As opposed to the approach by Saeki et al., they do not assume that the verb must fall into one of the four predefined categories.

Ermagan et al. developed the tool SODA, allowing to link textual use cases to behavior models. However, SODA sticks to manual modeling and does not provide automatic extraction of model elements.

Our approach has the important advantage that it does not assume the textual document to contain sufficient information to generate models. Thus, it can cope with incomplete documents and make the incompleteness apparent.

5.2   Natural Language Processing in Requirements Engineering

There are three areas where natural language processing is applied to requirements engineering: assessment of document quality, identification and classification of application specific concepts, and analysis of system behavior.

Approaches to the assessment of document quality were introduced, for example, by Rupp [14], Fabbrini et al. [15], Kamsties et al. [16], and Chantree et al. [17]. These approaches define writing guidelines and measure document quality by the degree to which the document satisfies the guidelines. These approaches have a different focus from our work: their aim is to detect poor phrasing and to improve it, they do not target system modeling, and do not generate any model-related feedback, as our approach does.

Another class of approaches, as, for example, those by Goldin and Berry [18], Abbott [19], or Sawyer et al. [20] analyzes the requirements documents, extracts application-specific concepts, and provides an initial static model of the application domain. However, these approaches do not generate feedback addressing specification incompleteness either.

The approaches analyzing system behavior translate requirements documents to executable models by analyzing linguistic patterns. Vadera and Meziane [21] propose a procedure to translate certain linguistic patterns into first order logic and then to the specification language VDM, but they do not provide automation for this procedure. Gervasi and Zowghi [22] go further and introduce a restricted language, a subset of English. They automatically translate textual requirements written in this restricted language to first order logic. Similarly, Breaux et al. [23] introduce a restricted language and translate this language to description logic. Avrunin et al. [24] translate natural language to temporal logic. Our work goes further than the above approaches, as we not only translate textual descriptions to models, but also use the resulting models to make the deficiencies of the textual specification apparent.

To summarize, to the best of our knowledge, there is no approach to documents analysis, yet, able not only to translate model descriptions to models themselves, but also to generate feedback concerning the deficiencies of the document.

6   Summary

Even though many formal and semi-formal specification techniques exist, requirements specifications in natural language remain a de-facto standard. Apart from being readable by all stakeholders, specifications in natural language entail a lot of problems with poor phrasings, inconsistencies, and missing information.

The approach presented in our paper analyzes missing information in behavior specifications and automatically generates questions addressing these findings. As evaluated in our experiment, the generated questions can address specification deficiencies that would otherwise be overseen by human analysts. Thus, the presented method should be one of the means to ensure the quality of requirements specifications.

Acknowledgments

We want to thank the participants of the experiment: Mou Dongyue, Florian Hölzl, Maged Khalil, Alexander Krauss, Christian Leuxner, Daniel Méndez Fernández, David Trachtenherz, and Andreas Vogelsang.

References

1. Mich, L., Franch, M., Novi Inverardi, P.: Market research on requirements analysis using linguistic tools. Requirements Engineering 9(1), 40–56 (2004)
2. Kof, L., Schätz, B.: Combining aspects of reactive systems. In: Broy, M., Zamulin, A.V. (eds.) PSI 2003. LNCS, vol. 2890, pp. 344–349. Springer, Heidelberg (2004)
3. Kof, L.: Scenarios: Identifying missing objects and actions by means of computational linguistics. In: 15th IEEE International Requirements Engineering Conference, October 15–19, pp. 121–130. IEEE Computer Society Conference Publishing Services, New Delhi (2007)
4. Kof, L.: From Textual Scenarios to Message Sequence Charts: Inclusion of Condition Generation and Actor Extraction. In: 16th IEEE International Requirements Engineering Conference, September 10-12, pp. 331–332. IEEE Computer Society Conference Publishing Services, Barcelona (2008)
5. Buhr, K., Heumesser, N., Houdek, F., Omasreiter, H., Rothermehl, F., Tavakoli, R., Zink, T.: DaimlerChrysler demonstrator: System specification instrument cluster (2004), http://www.empress-itea.org/deliverables/D5.1_Appendix_B_v1.0_Public_Version.pdf (accessed 16.02.2010)
6. Kof, L.: Translation of Textual Specifications to Automata by Means of Discourse Context Modeling. In: Glinz, M., Heymans, P. (eds.) REFSQ 2009. LNCS, vol. 5512, pp. 197–211. Springer, Heidelberg (2009)
7. Abrial, J.-R., Börger, E., Langmaack, H.: The steam boiler case study: Competition of formal program specification and development methods. In: Abrial, J.-R., Börger, E., Langmaack, H. (eds.) Dagstuhl Seminar 1995. LNCS, vol. 1165. Springer, Heidelberg (1996), citeseer.nj.nec.com/abrial96steam.html
8. Clark, S., Curran, J.R.: Parsing the WSJ using CCG and log-linear models. In: ACL 2004: Proceedings of the 42nd Annual Meeting on Association for Computational Linguistics, p. 103. Association for Computational Linguistics, Morristown (2004)
9. Gleich, B., Creighton, O., Kof, L.: Ambiguity detection: Towards a tool explaining ambiguity sources. In: Wieringa, R., Persson, A. (eds.) REFSQ 2010. LNCS, vol. 6182. Springer, Heidelberg (2010)
10. Berry, D.M., Kamsties, E., Krieger, M.M.: From contract drafting to software specification: Linguistic sources of ambiguity, http://se.uwaterloo.ca/~dberry/handbook/ambiguityHandbook.pdf (accessed 18.11.2004)

11. Saeki, M., Horai, H., Enomoto, H.: Software development process from natural language specification. In: Proceedings of the 11th International Conference on Software Engineering, pp. 64–73. ACM Press, New York (1989)
12. Overmyer, S.P., Lavoie, B., Rambow, O.: Conceptual modeling through linguistic analysis using LIDA. In: ICSE 2001: Proceedings of the 23rd International Conference on Software Engineering, pp. 401–410. IEEE Computer Society, Washington, DC, USA (2001)
13. Ermagan, V., Huang, T.-J., Krüger, I., Meisinger, M., Menarini, M., Moorthy, P.: Towards Tool Support for Service-Oriented Development of Embedded Automotive Systems. In: Proceedings of the Dagstuhl Workshop on Model-Based Development of Embedded Systems (MBEES 2007), Informatik-Bericht 2007-01, Fakultät für Informatik, Technische Universität Braunschweig (2007)
14. Rupp, C.: Requirements-Engineering und -Management. Professionelle, iterative Anforderungsanalyse für die Praxis, 2nd edn. Hanser–Verlag (May 2002) ISBN 3-446-21960-9
15. Fabbrini, F., Fusani, M., Gnesi, S., Lami, G.: The linguistic approach to the natural language requirements quality: benefit of the use of an automatic tool. In: 26th Annual NASA Goddard Software Engineering Workshop, pp. 97–105. IEEE Computer Society, Greenbelt (2001), http://fmt.isti.cnr.it/WEBPAPER/fabbrini_nlrquality.pdf (accessed 08.02.2010)
16. Kamsties, E., Berry, D.M., Paech, B.: Detecting ambiguities in requirements documents using inspections. In: Workshop on Inspections in Software Engineering, Paris, France, pp. 68–80 (2001)
17. Chantree, F., Nuseibeh, B., de Roeck, A., Willis, A.: Identifying nocuous ambiguities in natural language requirements. In: RE 2006: Proceedings of the 14th IEEE International Requirements Engineering Conference (RE 2006), pp. 56–65. IEEE Computer Society, Washington, DC, USA (2006)
18. Goldin, L., Berry, D.M.: AbstFinder, a prototype natural language text abstraction finder for use in requirements elicitation. Automated Software Eng. 4(4), 375–412 (1997)
19. Abbott, R.J.: Program design by informal English descriptions. Communications of the ACM 26(11), 882–894 (1983)
20. Sawyer, P., Rayson, P., Cosh, K.: Shallow knowledge as an aid to deep understanding in early phase requirements engineering. IEEE Trans. Softw. Eng. 31(11), 969–981 (2005)
21. Vadera, S., Meziane, F.: From English to formal specifications. The Computer Journal 37(9), 753–763 (1994)
22. Gervasi, V., Zowghi, D.: Reasoning about inconsistencies in natural language requirements. ACM Trans. Softw. Eng. Methodol. 14(3), 277–330 (2005)
23. Breaux, T.D., Antón, A.I., Doyle, J.: Semantic parameterization: A process for modeling domain descriptions. ACM Trans. Softw. Eng. Methodol. 18(2), 1–27 (2008)
24. Smith, R.L., Avrunin, G.S., Clarke, L.A., Osterweil, L.J.: Propel: an approach supporting property elucidation. In: ICSE 2002: Proceedings of the 24th International Conference on Software Engineering, pp. 11–21. ACM, New York (2002)

A Web Usability Evaluation Process for Model-Driven Web Development

Adrian Fernandez, Silvia Abrahão, and Emilio Insfran

ISSI Research Group, Departamento de Sistemas Informáticos y Computación,
Universitat Politècnica de València, Camino de Vera, s/n, 46022, Valencia, Spain
{afernandez,einsfran,sabrahao}@dsic.upv.es

Abstract. Most of usability evaluation methods for the Web domain have several limitations such as: the concept of usability is only partially supported; usability evaluations are mainly performed when the Web application has been completed; there is a lack of guidelines on how to properly integrate usability into Web development. This paper addresses these issues through the presentation of a Web Usability Evaluation Process (WUEP) for integrating usability evaluations at different stages of model-driven Web development processes. A case study was performed in order to analyze the feasibility of the approach by applying WUEP to evaluate the usability of a real Web application that was developed by using a specific model-driven development process in the industrial domain. The results suggest that WUEP provides good insights into the performance of usability evaluations. The usability problems are corrected at the model and model-transformation levels, thereby improving the usability of the final Web application.

Keywords: Web Usability, Evaluation Process, Web Metrics, Model-Driven Web Development.

1 Introduction

Usability is considered to be one of the most important quality factors for Web applications, along with others such as reliability and security [23]. It is not sufficient to satisfy the functional requirements of a Web application in order to ensure its success. The ease or difficulty experienced by users of these Web applications is largely responsible for determining their success or failure. Usability evaluations and technologies that support the usability design process have therefore become critical in ensuring the success of Web applications [20].

The challenge of developing more usable Web applications has promoted the emergence of a large number of usability evaluation methods. However, most of these methods only consider usability evaluations during the last stages of the Web development process. Works such as Juristo *et al.* [17] claim that usability evaluations should also be performed during the early stages of the Web development process in order to improve user experience and decrease maintenance costs.

This is in line with the results of a systematic mapping study that we performed to investigate which usability evaluation methods have been used to evaluate Web

H. Mouratidis and C. Rolland (Eds.): CAiSE 2011, LNCS 6741, pp. 108–122, 2011.

artifacts and how they were employed [12]. The study suggests several areas for further research, such as the need for usability evaluation methods that can be applied during the early stages of the Web development process, methods that evaluate different usability aspects depending on the underlying definition of the usability concept, the need for evaluation methods that provide explicit feedback or suggestions to improve Web artifacts created during the process, and guidance for Web developers on how the usability evaluation methods can properly be integrated at relevant points of a Web development process.

The majority of Web development processes do not take advantage of the artifacts produced at the requirements and design stages. These intermediate artifacts are principally used to guide developers and to document the Web application. Since the traceability between artifacts and the final Web application is not well-defined, performing usability evaluations of these artifacts can be difficult. This problem is alleviated in Model-Driven Web Development processes (MDWD) where intermediate artifacts (models), which represent different views of a Web application, are used in all the steps of the development process, and the final source code is automatically generated from these models.

Most MDWD processes break up the Web application design into three models: content, navigation and presentation. These dimensions allow proper levels of abstraction to be established [7]. An MDWD process basically transforms models that are independent of technological implementation details (i.e., Platform-Independent Models - PIMs) such as structural models, navigational models or abstract user interface (UI) models into other models that contain specific aspects from a specific technological platform (i.e., Platform-Specific Models - PSMs) such as specific UI models, database schemas. This is done by automatically applying transformation rules. PSMs can be automatically compiled to generate the source code of the final Web application (Code Model - CM). This approach is followed by several methods such as: OO-H [13] or WebML [8]. The evaluation of these models (PIMs, PSMs, and CMs) can provide usability evaluation reports which propose changes that can be directly reflected in the final source code.

In a previous work [11] we followed these ideas to present a Web Usability Model that decomposes the usability concept into sub-characteristics and measurable attributes, which are then associated with Web metrics in order to quantify them. The aim was to evaluate usability attributes in several artifacts obtained from a Web development process that follows an MDWD approach. In this paper, we present an inspection method called Web Usability Evaluation Process (WUEP) that employs our Web Usability Model in order to integrate usability evaluations into several stages of MDWD processes.

This paper is organized as follows. Section 2 discusses related works that report usability evaluation processes based on inspection methods for Web development. Section 3 presents the Web Usability Evaluation Process. Section 4 presents a real case study that has been performed to illustrate the feasibility of WUEP. Finally, Section 5 presents our conclusions and further work.

2   Related Work

Usability evaluation methods can be mainly classified into two groups: empirical methods and inspection methods. Empirical methods are based on capturing and analyzing usage data from real end-users, while inspection methods are performed by expert evaluators or designers and are based on reviewing the usability aspects of Web artifacts, which are commonly user interfaces, with regard to their conformance with a set of guidelines.

Usability inspection methods have emerged as an alternative to empirical methods as a means to identify usability problems since they do not require end-user participation and they can be employed during the early stages of the Web development process [9]. There are several proposals based on inspection methods to deal with Web usability issues, such as the Cognitive Walkthrough for the Web (CWW) [4] and the Web Design Perspectives (WDP) [10]. CWW assesses the ease with which a user can explore a Website by using semantic algorithms. However, this method only supports ease of navigation. WDP extends and adapts the heuristics proposed by Nielsen [21] with the aim of drawing closer to the dimensions that characterize a Web application: content, structure, navigation and presentation. However, this kind of methods tends to present a considerable degree of subjectivity in usability evaluations.

Other works present Web usability inspection methods that are based on applying metrics in order to minimize the subjectivity of the evaluation, such as the WebTango methodology [14] and Web Quality Evaluation Method (WebQEM) [22]. The Web-Tango methodology allows us to obtain quantitative measures, which are based on empirically validated metrics for user interfaces, to build predictive models in order to evaluate other user interfaces. WebQEM performs a quantitative evaluation of the usability aspects proposed in the ISO 9126-1 standard [15], and these quantitative measures are aggregated in order to provide usability indicators.

The aforementioned inspection methods are oriented towards application in the traditional Web development context; they are therefore principally employed in the later stages of Web development processes. As mentioned above, model-driven Web development offers a suitable context for early usability evaluations since it allows models, which are applied in all the stages, to be evaluated. This research line has emerged recently, and only a few works address Web usability issues, such as those of Atterer [3], Abrahão and Insfran [2], and Molina and Toval [19].

Atterer [3] proposed a prototype of a model-based usability validator with which to analyze models that represent enriched Web user interfaces. This approach takes advantage of models that represent the navigation (how the Website is traversed), and the UI of a Web application (abstract properties of the page layout).

Abrahão and Insfran [2] proposed a usability model to evaluate software products that are obtained as a result of model-driven development processes. Although this model is based on the usability sub-characteristics proposed in the ISO 9126 standard [15], it is not specific to Web applications and does not provide specific metrics. The same model was used in [24] with the aim of providing metrics for a set of attributes which would be applicable to the conceptual models that are obtained as a result of a specific MDWD process.

Molina and Toval [19] presented an approach to extend the expressivity of models that represent the navigation of Web applications in order to incorporate usability requirements. It improves the application of metrics and indicators to these models.

Nevertheless, to the best of our knowledge, there is no generic process for integrating usability evaluations into Model-Driven Web development processes.

3 Web Usability Evaluation Process

The Web Usability Evaluation Process (WUEP) has been defined by extending and refining the quality evaluation process that is proposed in the ISO 25000 standard [16]. The aim is to integrate usability evaluations into model-driven Web development processes by employing a Web Usability Model as the principal input artifact. This model decomposes the usability concept into sub-characteristics and measurable attributes, which are then associated with metrics in order to quantify them. These metrics provide a generic definition, which should be operationalized in order to be applicable to artifacts from different abstraction levels (PIMs, PSMs, and, CMs) in different MDWD processes (e.g., OO-H, WebML, UWE).

Figure 1 shows an overview of the main stages of WUEP. Three roles are involved: evaluation designer, evaluator, and Web developer. The *evaluation designer* performs the first three stages: 1) Establishing the requirements of the evaluation, 2) Specification of the evaluation, and 3) Design of the evaluation. The *evaluator* performs the fourth stage: 4) Execution of the evaluation. Finally, the *Web developer* performs the last stage: 5) Analysis of changes. The following sub-sections describe each of the main stages by including the activities into which they are decomposed.

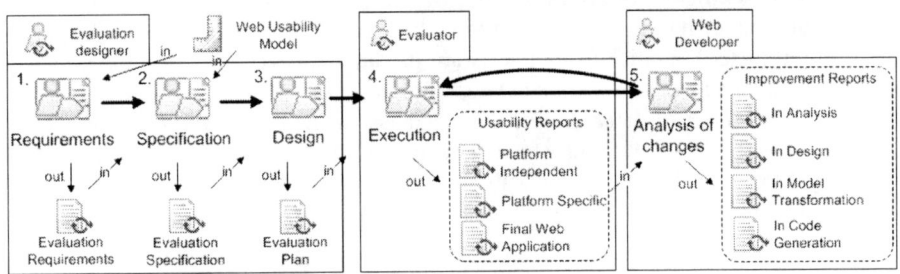

Fig. 1. An overview of the Web Usability Evaluation Process

3.1 Establishment of Evaluation Requirements

The aim of this stage is to establish the requirements of the evaluation to delimit the scope of the evaluation. The activities involved in this stage are described below.

1. Establish the Purpose of Evaluation. This activity determines the aim of the usability evaluation, i.e., whether the evaluation will be performed in a Web application in order to provide feedback during the Web development process, or whether different Web applications, belonging to the same Web application family, are compared in order to establish a ranking among them.

2. Specify Profiles. The different factors that will condition the evaluation are determined. These factors are: the *type of Web application*, since each family of Web applications has different goals that make an impact on the selection of usability attributes, i.e., navigability might be more relevant to Intranets whereas attractiveness might be more relevant to social networks; the *Web development method*, since knowledge about its process and artifacts is needed in order to properly integrate the usability evaluations; and the *context of use*, which takes into account parameters such as type of users, user age, work environment, etc.

3. Select the Web Artifacts to Be Evaluated. The artifacts selected may depend on either the Web development method or the technological platform. The artifacts to be considered might be: *Platform-Independent models* (e.g., content/domain models, navigational models, abstract user interface models) which are obtained as output from the analysis and design stages of an MDWD process; *Platform-Specific models* (e.g., specific user interface models, database schemas) which are obtained as output from the model transformation stage of an MDWD process; and *Code models* (e.g., source code, final user interfaces) which are obtained as output from the code generation stage of an MDWD process.

4. Select Usability Attributes. The Web Usability Model is used as a catalog in order to select which usability attributes will be evaluated. This Web Usability Model is based on the decomposition of the usability characteristic proposed in the ISO 25000 (SQuaRE) [16] quality model. The first version of this model was presented in [11] and has been improved by considering other usability guidelines (e.g., [18]), in order to discover 15 new usability attributes that are relevant to the Web domain. The Web Usability Model currently considers two different views: usability of the software product, and usability in use. In this paper, we focus solely on the usability of the software product, since it can be assessed during the Web development process by inspecting Web artifacts. Usability from a software product perspective is decomposed into seven sub-characteristics: *Appropriateness recognisability*, *Learnability*, *Ease of use*, *Helpfulness*, *Technical accessibility*, *Attractiveness*, and *Compliance*. These are also decomposed into other sub-characteristics and measurable attributes. The Web Usability Model, including all the sub-characteristics attributes and their associated metrics, is available at `http://www.dsic.upv.es/~afernandez/CAiSE11/WebUsabilityModel`.

The outcomes of the above activities represent the *Evaluation Requirements* that will be used as input by the next stage.

3.2 Specification of the Evaluation

The aim of this stage is to specify the evaluation in terms of which metrics are intended to be applied and how the values obtained by these metrics allow usability problems to be detected. The activities involved in this stage are described below.

1. Select the Metrics to Be Applied. The Web Usability Model is used to discover which of the metrics are associated with the usability attributes selected. Metrics allow us to interpret whether or not these attributes contribute to achieving a certain degree of usability in the Web application. The Web metrics that are included in the Web Usability Model were taken from different sources: surveys that contain metrics that had been theoretically and empirically validated (e.g., Calero *et al.* [6]); quality standards (e.g., SQuaRE [16]) and Web guidelines (e.g., W3C [26]). Each metric was studied by considering its parameters (i.e., purpose, interpretation, artifacts to which it can be applied, etc.) in order to provide a generic definition of the metric. The aim of providing a generic definition is to allow metrics to be applied to artifacts of different abstraction levels and from different MDWD processes. Appendix A includes two examples of metrics from the Web Usability Model with their generic definition.

2. Operationalize the Metrics. The calculation formulas of the selected metrics should be operationalized by identifying variables from the generic definition of the metric in the modeling primitives of the selected artifacts, in other words, by establishing a mapping between the generic description of the metric and the concepts that are represented in the artifacts. In the evaluation of models (PIM, PSM, and CM), the calculation of the operationalized formulas may require assistance from an evaluator to determine the values of the variables involved, or it may require a verification tool if these formulas are expressed in variables that can be automatically computed from the input models by query languages such as the Object Constraint Language (OCL).

3. Establish Rating Levels for Metrics. Rating levels are established for ranges of values obtained for each metric by considering their scale type and the guidelines related to each metric whenever possible. These rating levels allow us to discover whether the associated attribute improves the Web application's level of usability, and are also relevant in detecting usability problems that can be classified by their level of severity.

The outcomes of the above activities represent the *Evaluation specification* that will be used as input by the next stage.

3.3 Design of the Evaluation

The aim of this stage is to design how the evaluation will be performed and what information will be collected during the evaluation.

1. Define the Template for Usability Reports. This template is defined in order to present all the data that is related to the usability problems detected. A usability report is commonly a list of usability problems (UP). Each UP can be described by the following fields: *ID*, which refers to a single UP; *description* of the UP; affected attribute from the Web Usability Model; *severity level*, which could be low, medium or critical; *artifact evaluated*, in which metrics have been applied; *source of the problem*, which refers to the artifact that originates the usability problem (e.g., PIMs, PSMs, CMs, and transformation rules); *occurrences*, which refer to the number of appearances of the same UP; and *recommendations* to correct the UP detected (some recommendations might also be automatically provided by interpreting the range values). Other fields that are useful to post-analyze the UP detected can also be added, such as *priority* of the UP; *effort* that is needed to correct the UP; and *changes* that must be performed in order to take the aforementioned fields into consideration.

2. Elaborate an Evaluation Plan. Designing the evaluation plan implies: establishing an evaluation order of artifacts; establishing a number of evaluators; assigning tasks to these evaluators, and considering any restrictions that might conditioned the evaluation. The recommended order is to first evaluate the artifacts that belong to a higher abstraction level (PIMs), since these artifacts drive the development of the final Web application. This allows us to detect usability problems during the early stages of the Web development process. The artifacts that belong to a lower level of abstraction (PSMs and CMs) are then evaluated.

The outcomes of the above activities represent the *Evaluation Plan* that will be used as input by the next stage.

3.4 Execution of the Evaluation

The aim of this stage is to execute the evaluation in accordance with the Evaluation Plan. The evaluator applies the operationalized metrics to the artifacts that have been selected. If the rating levels obtained identify a UP, the elements of the artifact involved that contribute to achieving this metric value are analyzed. This helps us to determine the source of the usability problem thanks to the traceability that exists among the models in an MDWD process.

The outcomes of this stage are: a *platform-independent usability report*, which collect the UPs that are detected during the evaluation of PIMs; a *platform-specific usability report*, which collects the UPs that are detected during the evaluation of PSMs; and a *final Web application usability report*, which collects the UPs that are detected during the evaluation of CMs.

3.5 Analysis of Changes

The aim of this stage is to classify all the UPs detected from each of the usability reports shown above and to analyze the recommendations provided in order to propose changes with which to correct the artifacts. Usability problems whose source is located in PIMs that are related to content and navigation (e.g., UP detected in structural models, navigational models) are collected to create the *improvement report in analysis*. Usability problems whose source is located in PIMs that are related to presentation (e.g., UP detected in abstract user interfaces models) are collected to create the *improvement report in design*. Usability problems whose source is located in PSMs or transformation rules among PIMs and PSMs are collected to create the *improvement report in model transformation*. Finally, usability problems whose source is located in the generation rules among PSMs and CMs are collected to create the *improvement report in code generation*. The last two reports are useful for providing feedback in order to improve the Computer-Aided Web Engineering tool (CAWE) that supports the Web development method and performs the transformations among models, along with the generation rules among models and the final source code.

It is important to note that after applying the changes suggested by the improvement reports, re-evaluations of the artifacts might be necessary.

4   Case Study

An exploratory case study was performed by following the guidelines presented in [25] in order to study the feasibility of applying WUEP in industrial contexts. The stages of which the case study was comprised were: design, preparation, collection of data, and analysis of data, each of which is explained below.

4.1 Design of the Case Study

The case study was designed by considering the five components that are proposed in [25]: purpose of the study, underlying conceptual framework, research questions to be addressed, sampling strategy, and methods employed.

The purpose of the case study is to show the feasibility of applying WUEP to discover usability problems at different levels of abstraction of a MDWD process. The conceptual framework that links the phenomena to be studied is the idea based on integrating usability evaluations into MDWD processes [11]. The research questions that are intended to be addressed are: *a) What type of usability problems can be detected in each phase of a model-driven Web development process and what are their implications for intermediate artifacts (models)?*, and *b) What limitations does the Web Usability Evaluation Process present?*.

The sampling strategy of the case study is based on an embedded single-case design. We contacted a Web development company located in Alicante (Spain) in order to apply WUEP to a real Web application from a project in progress. The company provided us with access to a task management system. The aim was to perform evaluations on the artifacts that define this Web application at different abstraction levels in an MDWD process.

WUEP was applied as follows: two of the authors performed the evaluation designer role in order to design an evaluation plan (in critical activities such as the selection of usability attributes, we required the help of two external Web usability experts); and the evaluator role was performed by five evaluators: the other author and four other independent evaluators with an average of five years' experience in usability evaluations. The technique that was used to obtain feedback about the feasibility of WUEP was the analysis of the usability reports and observation of the evaluators when performing the execution of the evaluation. We have not considered the *analysis of changes* stage since the aim of this case study was to show the feasibility of WUEP in discovering usability problems.

4.2   Preparation of the Case Study

An evaluation plan was defined by following WUEP. With regard to the *establishment of evaluation requirements* stage of WUEP, the purpose of the evaluation was to perform a usability evaluation during the development of the Web application mentioned above. The type of Web application was an Intranet that was developed using the OO-H method [13] supported by the VisualWade tool (www.visualwade.com). The context in which the Web application will be used is a software development company, and there are two kinds of users: project manager and programmers.

OO-H is an MDWD method that provides the semantics and notation for developing Web applications. The platform-independent models (PIMs) that represent the different concerns of a Web application are: a class model, a navigational model, and a presentation model. The class model is UML-based and specifies the content requirements; the navigational model is composed of a set of Navigational Access Diagrams (NADs) that specify the functional requirements in terms of navigational needs and users' actions; and the presentation model is composed of a set of Abstract Presentation Diagrams (APDs), whose initial version is obtained by merging the former models, which are then refined in order to represent the visual properties of the final UI. The platform-specific models (PSMs) are embedded into a model compiler, which automatically obtains the source code (CM) from the Web application by taking all the previous PIMs as input.

The Web artifacts selected to be evaluated were all the platform-independent models and the final UIs (i.e., 1 class model, 4 NADs, 4 APDs and 4 final UIs). PSMs cannot be evaluated since they are embedded in the model compiler. Figure 2(a) shows an excerpt of a NAD and Figure 2(b) shows an excerpt of an APD.

A set of 20 usability attributes were selected from the Web Usability Model through the consensus of the evaluator designers and the Web usability experts. The attributes were selected by considering which of them would be more relevant to the type of Web application and the context in which it is going to be used (e.g., Interconnectivity, Clickability, Permanence of links, etc.). Only 20 attributes were selected in order to avoid extending the duration of the execution stage.

With regard to the *specification of the evaluation* stage of WUEP, metrics associated with the selected attributes were obtained from the Web Usability Model, and then associated with the artifact in which they could be applied. Since metrics can be applied at different abstraction levels, the highest level of application was selected. For example, the metrics presented in Appendix A indicate that: the Compactness metric could be applied to models that specify the navigation (i.e., NADs); and the Discernible links metric could be applied to models that specify the UI (i.e., APDs).

Once the metrics had been associated with the artifacts, metrics were operationalized in order to provide a calculation formula for artifacts from the OO-H method (in this case), and to establish rating levels for them. Appendix B shows an example of operationalization and rating levels for the metrics from Appendix A.

With regard to the *design of the evaluation* stage of WUEP, a template for usability reports was defined by considering the same fields proposed in the previous section. The evaluation plan that was elaborated takes into consideration the following evaluation order of the artifacts: class model; NADs, APDs, and final UIs. Five evaluators were selected to perform the execution stage of WUEP, as mentioned above. This number was selected by following the recommendations of studies which claim that five or more evaluators are needed to perform usability inspections [21]. The evaluators attended a training session of 2 hours at which they were informed of the artifacts from the OO-H method and the tasks to be performed.

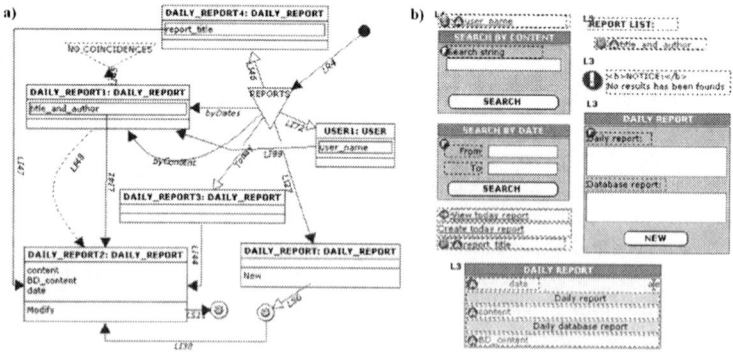

Fig. 2. Excerpts from artifacts (PIMs) of the Web application developed in OO-H

4.3 Collection of the Data

The data for this case study was collected during the *execution of the evaluation* stage of WUEP. As an example, we show values obtained after the evaluators applied certain operationalized metrics to the excerpts of models that are shown in Figure 2.

In the NAD (Fig. 2 (a)), the value of the Compactness metric is: Cp (NAD) = (448-347) / (448-56) = 0.26 since the components of the formula are: n=k=8; Max=448; Min=56; and $\sum i \sum j$ Dij=347. The rating level of this metric ($0.2 < x \leq 0.8$) therefore indicates that there is no usability problem related to the Interconnectivity attribute in this NAD.

In the APD (Fig. 2 (b)), the value of the Discernible links metric is: DL (APD) = 1-0.5(2/5+3/13) = 0.68 since there are 2 links (*title_and_author* and *report_title*) that might be confused with labels, and there are 3 labels (*date, content* and *BD_content*) that might be confused with links as a result of their visual properties. The rating level of this metric ($0.5 < x \leq 0.7$) therefore indicates the existence of a medium usability problem (UP001) related to the Clickability attribute in this APD.

Each evaluator presented the usability problems detected in usability reports by using the defined template. As an example, Table 1 shows an excerpt of the usability report that is presented in the UP001.

Table 1. Usability problem detected: UP001

ID	UP001
Description	*title_and_author* and *report_title* links are confused with labels, whereas *date, content* and *BD_content* labels are confused with links.
Affected attribute	Appropriateness recognisability / Navigability / Clickability
Severity level	Medium (0.68)
Artifact evaluated	Abstract Presentation Diagram (APD no. 3)
Source problem	Abstract Presentation Diagram (APD no. 3)
Occurrences	5 errors in the same APD
Recommendations	Change the visual properties of links and labels to another face/color in order for them to be discernible. The use of typographies, such as blue, and an underlined face for links is recommended.

4.4 Analysis of Data

The usability problems reported in the execution stage were analyzed to address our research questions.

Usability Problems and their Implications for Intermediate Artifacts. The evaluation performed for the case study only considers two different abstraction levels: the platform-independent models: class model, navigational model (NADs), and presentation model (APDs); and the code model: final user interfaces.

With regard to the class model, a mean of 0.6 UPs (std. dev. 0.54) were detected by the evaluators since the expressiveness of its modeling primitives makes it difficult to operationalize metrics. However, UPs related to the minimization of user effort were detected (e.g., *"does not provide default values for elements that are needed for user input"*). These can be solved by correcting the class model itself by adding default values to attributes from classes.

With regard to the navigational model, a mean of 2.2 UPs (std. dev. 0.83) were detected by the evaluators since the navigation steps for carrying out possible actions are considered to be proper by values of metrics such as navigation density, navigation depth, etc. However, UPs related to the lack of return paths were detected (e.g., *"does not provide backs link"* or *"no link to go home"*). These can be solved in the navigational model itself by adding links that connect the information previously obtained.

With regard to the presentation model, a mean of 10.6 UPs (std. dev. 1.51) were detected by the evaluators since the expressiveness of its modeling primitives allows several metrics to be operationalized (e.g., *"error messages are not meaningful"* and *"links and labels are not discernible"*). The former can be solved in the presentation model itself by correcting the label that shows the message, and the latter can be solved in the navigational model since names for links can be provided as properties of them.

With regard to the final user interface, a mean of 12.2 UPs (std. dev. 1.78) were detected by the evaluators. Some examples of the UPs detected were related to the support to operation cancellation (e.g., *"the creation of a new task cannot be canceled"*). This can be solved in the navigational model since links can be added to permit cancelation paths. Other UPs related to the immediate feedback that provide UI controls were also detected (e.g., *"tabs of the main menu do not show the current user state"*). This can be solved in the transformation rules that map the representation of tabs with a specific UI control of the technological platform.

It is important to note that some usability attributes such as the uniformity of the UI position are directly supported by the modeling primitives of the APDs.

Lessons Learned. The case study has been useful in that it has allowed us to learn more about the potentialities and limitations of our proposal and how WUEP can be improved.

WUEP can detect several usability problems from a wide range of types in several artifacts employed during the early stages of an MDWD process. The application of operationalized metrics and traceability among models not only provides a list of usability problems, but also facilitates the provision of recommendations with which to correct them. Metric operationalization also allows WUEP to be applied to different MDWD processes by establishing a mapping between the generic description of the metric and the modeling primitives of artifacts, in addition to other traditional development processes, by operationalizing metrics only in final user interfaces. However, usability reports will only provide feedback to the implementation stage since traceability between artifacts is not well-defined. Moreover, the evaluation process may be a means to discover which usability attributes are directly supported by the modeling primitives or to discover limitations in the expressiveness of these artifacts.

During the execution of the case study, several aspects related to how WUEP is applied were detected to be improved. For example, the application of metrics was detected as being a very tedious task when performed manually, particularly when the metrics provided a complex calculation formula (e.g., Compactness, Depth of the navigation, etc.). In addition, some metrics obtained different values depending on the evaluator since these metrics imply a certain degree of subjectivity (e.g., Discernible links, properly chosen Metaphors, etc.). These issues could be alleviated, at least to some extent, by providing better guidelines in order to minimize the subjectivity in the employment of operationalized metrics and by developing a tool that automates a large part of the evaluation process.

Limitations of the Study. Since the aim of the case study was to show the feasibility of applying WUEP in an MDWD process in industrial contexts, we selected a real Web application that was under development by a company, and we focused principally on the first four stages of WUEP. However, this case study presents some limitations, such as the fact that only one type of Web application was considered since the company imposed certain restrictions. Although the usability attributes to be evaluated were selected through the consensus of the evaluation designers and the two independent Web usability experts, they might not have been very representative. In order to overcome such limitations it is necessary to determine which usability attributes are most relevant for each family of Web applications by considering several opinions from Web usability experts.

5   Conclusions and Further Work

This paper presents the Web Usability Evaluation Process (WUEP) as a usability inspection method that integrates usability evaluations during several stages of Model-Driven Web Development (MDWD) processes. WUEP provides broad support to the concept of usability since its underlying Web Usability Model has been extended and adapted to the Web domain by considering the new ISO 25000 series of standards (SQuaRE), along with several usability guidelines. The explicit definition of the activities and artifacts of WUEP also provides evaluators with more guidance and offers the possibility of automating (at least to some extent) several activities in the evaluation process by means of a process automation tool.

We believe that the inherent features of MDWD processes (e.g., traceability between models by means of model transformations) provide a suitable environment for performing usability evaluations. The integration of WUEP into these environments is thus based on the evaluation of artifacts, particularly intermediate artifacts (models), at several abstraction levels from different MDWD processes. The evaluation of these models (by considering the traceability among them) allows the source of the usability problem to be discovered and facilitates the provision of recommendations to correct these problems during the earlier stages of the Web development process. This signifies that if the usability of an automatically generated user interface can be assessed, the usability of any future user interface produced by MDWD processes could be predicted. In other words, we are referring to a user interface that can be usable by construction [1], at least to some extent. Usability can thus be taken into consideration throughout the entire Web development process. This enables better quality Web applications to be developed, thereby reducing effort at the maintenance stage.

In the future we intend to do the following: to analyze different proposals concerning the inclusion of aggregation mechanisms to merge values from metrics in order to provide scores for usability attributes that will allow different Web applications from the same family to be compared; to determine the most relevant usability attributes for different families of Web applications according to Web domain experts in order to provide pre-defined selections of operationalized metrics; to apply WUEP to different MDWD processes; to evaluate different types of Web applications such as Web applications 2.0 and Rich Internet Applications (RIA) that are developed thorough an MDWD process; to perform controlled experiments in order to assess the effectiveness and ease of use of WUEP by comparing it to other usability evaluation methods; and finally, to develop a tool with the capability of automating a large part of the evaluation process.

Acknowledgments. This research work is funded by the MULTIPLE project (TIN2009-13838) and the FPU program (AP2007-03731) from the Spanish Ministry of Science and Education.

References

1. Abrahão, S., Iborra, E., Vanderdonckt, J.: Usability Evaluation of User Interfaces Generated with a Model-Driven Architecture Tool. In: Maturing Usability: Quality in Software, Interaction and Value, Springer, pp. 3-32 (2007)
2. Abrahão, S., Insfran, E.: Early Usability Evaluation in Model-Driven Architecture Environments. In: 6th IEEE International Conference on Quality Software (QSIC 2006), pp. 287–294. IEEE Computer Society, Beijing (2006)
3. Atterer, R., Schmidt, A.: Adding Usability to Web Engineering Models and Tools. In: Lowe, D.G., Gaedke, M. (eds.) ICWE 2005. LNCS, vol. 3579, pp. 36–41. Springer, Heidelberg (2005)
4. Blackmon, M.H., Polson, P.G., Kitajima, M., Lewis, C.: Cognitive Walkthrough for the Web. In: Proc. of the ACM CHI 2002, USA, pp. 463–470 (2002)
5. Botafogo, R., Rivlin, E., Shneiderman, B.: Structural analysis of hypertexts: Identifying hierarchies and useful metrics. ACM Trans. Inf. Systems 10(2), 142–180 (1992)
6. Calero, C., Ruiz, J., Piattini, M.: Classifying Web metrics using the Web quality model. Online Information Review Journal. Emerald Group. 29(3), 227–248 (2005)
7. Casteleyn, S., Daniel, F., Dolog, P., Matera, M.: Engineering Web Applications. Springer, Heidelberg (2009)
8. Ceri, S., Fraternali, P., Bongio, A.: Web Modeling Language (WebML): A Modeling Language for Designing Web Sites. In: Proc. of the 9th WWW Conf., pp. 137–157 (2000)
9. Cockton, G., Lavery, D., Woolrychn, A.: Inspection-based evaluations. In: Jacko, J.A., Sears, A. (eds.) The Human-Computer Interaction Handbook, 2nd edn., pp. 1171–1190 (2003)
10. Conte, T., Massollar, J., Mendes, E., Travassos, G.H.: Usability Evaluation Based on Web Design Perspectives. In: 1st Int. Symposium on Empirical Software Engineering and Measurement (ESEM 2007), Spain, pp. 146–155 (2007)
11. Fernandez, A., Insfran, E., Abrahão, S.: Integrating a Usability Model into Model-Driven Web Development Processes. In: Vossen, G., Long, D.D.E., Yu, J.X. (eds.) WISE 2009. LNCS, vol. 5802, pp. 497–510. Springer, Heidelberg (2009)
12. Fernandez, A., Insfran, E., Abrahão, S.: Usability Evaluation Methods for the Web: A Systematic Mapping Study. In: Information and Software Technology (2011), doi:10.1016/j.infsof.2011.02.007
13. Gomez, J., Cachero, C., Pastor, O.: Conceptual Modeling of Device-Independent Web Applications. IEEE MultiMedia 8(2), 26–39 (2001)
14. Ivory, M.Y.: An Empirical Foundation for Automated Web Interface Evaluation. PhD Thesis, University of California, Berkeley, Computer Science Division (2001)
15. ISO/IEC 9126-1 Standard, Software Engineering, Product Quality - Part 1: Quality Model (2001)
16. ISO/IEC 25000 series, Software Engineering, Software Product Quality Requirements and Evaluation, SQuaRE (2005)

17. Juristo, N., Moreno, A., Sanchez-Segura, M.I.: Guidelines for eliciting usability function-alities. IEEE Transactions on Software Engineering 33(11), 744–758 (2007)
18. Leavit, M., Shneiderman, B.: Research-Based Web Design & Usability Guidelines. U.S. Government Printing Office (2006),
    http://usability.gov/guidelines/index.html (last access: March 2011)
19. Molina, F., Toval, J.A.: Integrating usability requirements that can be evaluated in design time into Model Driven Engineering of Web Information Systems. Advances in Engineering Software 40(12), 1306–1317 (2009)
20. Neuwirth, C. M., Regli, S. H.: IEEE Internet Computing Special Issue on Usability and the Web, vol. 6(2) (2002)
21. Nielsen, J., Molich, R.: Heuristic evaluation of user interfaces. In: Proc. ACM CHI 1990 Conf., Seattle, WA, pp. 249–256 (1990)
22. Olsina, L., Rossi, G.: Measuring Web Application Quality with WebQEM. IEEE Multimedia 9(4), 20–29 (2002)
23. Offutt, J.: Quality Attributes of Web Software Applications. IEEE Software: Special Issue on Software Engineering of Internet Software, 25–32 (2002)
24. Panach, I., Condori, N., Valverde, F., Aquino, N., Pastor, O.: Understandability measurement in an early usability evaluation for model-driven development: an empirical study. In: Int. Empirical Software Engineering and Measurement (ESEM 2008), pp. 354–356 (2008)
25. Runeson, P., Höst, M.: Guidelines for Conducting and Reporting Case Study Research in Software Engineering. Empirical Software Engineering 14(2) (2009)
26. World Wide Web Consortium W3C: Web Content Accessibility Guidelines (WCAG) 2.0 (2008), http://www.w3.org/TR/WCAG/ (last access: March 2011)

Appendix A: Examples of Metrics with Their Generic Description

Name	Compactness (Cp)
Attribute	Appropriateness recognisability / Navigability / Interconnectivity.
Description	The degree of interconnection among nodes belonging to a hypermedia graph. Its formula is $Cp = (Max - \sum i \sum j \; Dij) / (Max - Min)$, where $Max = (n2 - n)*k$; $Min = (n2 - n)$; n = quantity of nodes in the graph; k = constant superior to the amount of nodes; $\sum i \sum j \; Dij$ = the sum of distances taken from the matrix of converted distances (with factor k); and Dij = the distance between the nodes i and j.
Scale type	Ratio between 0 and 1.
Interpretation	Values closer to 1 signify that each node can easily reach any other node in the graph. However, a fully connected graph may lead to disorientation.
Application level	At the PIM/PSM level, if navigation is modeled as a graph where nodes represent pages and edges are links among pages; and at the CM level by analyzing the link's targets included in the source code for each page.

Name	Discernible links (DL)
Attribute	Appropriateness recognisability / Navigability / Clickability.
Description	The degree to which links are discernible from other textual elements. Its formula is $DL = 1 - 0.5(CL / NL + CT / NT)$, where CL = the number of links that may be confused with textual elements; NL= total number of links; CT = number of textual elements that may be confused with links; and NT = total number of textual elements.
Scale type	Ratio between 0 and 1.
Interpretation	Values closer to 1 indicate that users can differentiate links from textual elements.
Application level	At the PIM/PSM level, if the expressiveness of abstract user interface models allows aesthetic properties to be defined for links such as color or style; and at the CM level by assessing the elements from the source code that define the visual appearance of links in the final user interface (e.g., Cascading Style Sheets).

Appendix B: Examples of Operationalized Metrics for OO-H

Compactness (Cp): In an OO-H navigational model, the compactness is calculated by considering the navigational classes and links of each NAD as a hypermedia graph in which classes are nodes and links are edges. If we consider the scale type and the recommendations suggested by Botafogo *et al.* [5], then the rating levels are established as no UP ($0.2 < x \leq 0.8$), and medium UP ($0 \leq x < 0.2$ and $0.8 < x \leq 1$).

Discernible links (DL): In an OO-H presentation model, the visual properties of textual elements in each APD (i.e., links and labels) are represented as attributes such as font, color, and face. The operationalized metric is:

$$DL \; (APD) = 1 - 0.5(CLi \, / \, NTLi + CLa \, / \, NTLa)$$

where *CLi* signifies the number of links with similar visual properties to most of the labels, *CLa* signifies the number of labels with similar visual properties to most of the links, whereas *NTLi* and *NTLa* signify the total number of links and labels, respectively. If we consider the scale type, then the rating levels are established as no UP ($0.99 < x \leq 1$), low UP ($0.7 < x \leq 0.99$), medium UP ($0.5 < x \leq 0.7$), and critical UP ($0 < x \leq 0.5$).

A Trace Metamodel Proposal Based on the Model Driven Architecture Framework for the Traceability of User Requirements in Data Warehouses

Alejandro Maté and Juan Trujillo

Lucentia Research Group
Department of Software and Computing Systems
University of Alicante
{amate,jtrujillo}@dlsi.ua.es

Abstract. The complexity of the Data Warehouse (DW) development process requires to follow a methodological approach in order to be successful. A widely accepted approach for this development is the hybrid one, in which requirements and data sources must be accommodated to a new DW model. The main problem is that we lose the relationships between requirements, elements in the conceptual models and data sources in the process, since no traceability is explicitly specified. Therefore, this hurts requirements validation capability and increases the complexity of Extraction, Transformation and Load processes. In this paper, we propose the first trace metamodel for DWs and focus on the relationships between requirements and conceptual models. We propose a set of Query/View/Transformation rules to include traceability in DWs in an automatic way, allowing us to trace every requirement to the conceptual model and further increasing user satisfaction.

Keywords: Data Warehouses, traceability, user requirements, MDA.

1 Introduction

Data Warehouses (DW) integrate several heterogeneous data sources in multidimensional structures (i.e. facts and dimensions) in support of the decision-making process [10, 12]. Therefore, the development of the DW is a complex process which must be carefully planned in order to meet user needs. In order to develop the DW, three different approaches, similar to the existing ones in Software Engineering (bottom-up, top-down, and hybrid), were proposed [21, 4].

The first approach follows a bottom-up process and makes use of the information in the data sources while ignoring the user requirements. As the schema is not adapted to the user needs [21] the DW fails to meet the user expectations. The second approach follows a top-down process and focuses on the user requirements while ignoring the data sources. Therefore, it is possible that some of the user needs cannot be satisfied because the necessary data has not been

H. Mouratidis and C. Rolland (Eds.): CAiSE 2011, LNCS 6741, pp. 123–137, 2011.

stored [4]. The third approach (hybrid) makes use of both data sources and user requirements [16]. With this approach, user requirements which cannot be satisfied are noticed in earlier stages. Once the information from both worlds is collected, the incompatibilities have to be solved by acommodating both data sources and requirements in a single model.

However, with the hybrid approach a new problem arises. In the top-down and bottom-up approaches, every element used for the implementation of the DW comes from a single source only (either requirements or data sources), thereby allowing us to trace elements by name matching. Nevertheless, in the hybrid approach, additional effort is required in order to check which parts of the DW match, not only with each requirement, but also with each part of the data sources. Due to our experience, by following the hybrid approach, changes are done almost in every project, since it is very common that user requirements and data sources do not match, thus losing the implicit traceability.

In this process, the relationships between the elements are not recorded and lost, since there is no explicit traceability included in the development process. In turn, this hurts requirements validation [24, 26, 29], making unable to check the current status of each requirement or take decisions about alternative implementations if a given requirement cannot be fulfilled. Although the traceability aspect has been thoroughly studied [1, 5, 11, 22, 2, 8, 9, 23, 26], it has been almost completely overlooked in DW development. To the best of our knowledge, the only references to requirements traceability in DW are those from [15], which only mention implicit traceability by name matching.

In our previous works [14, 15, 16, 17], we defined a hybrid DW development approach in the context of the Model Driven Architecture (MDA) framework [19]. DWs are sensitive to be developed by using MDA, cutting development time and making the process less error prone, since transformations from the top layer to the final implementation are performed in an semi-automatic way. In our approach, requirements are specified in a Computation Independent Model (CIM) by means of a UML profile [15] based on the i* framework [30]. Then, they are automatically derived, reconciled with the data sources in a hybrid model, and refined through a series of layers (Platform Independent Model (PIM) layer and Platform Specific Model (PSM) layer) until the final implementation is achieved, as seen in figure 1.

The automatic derivation is done by means of model to model transformations specified by Query/View/Transformation (QVT) [20] rules. QVT is a language defined by the Object Management Group (OMG) and proposed as a standard to create model to model transformations. However, due to our experience in real-world projects, the lack of traceability does not allow us to adequately validate requirements and incurs in additional costs when requirements change.

In this paper, we complement our previous works with the inclusion of the first traceability metamodel for DWs and an automatic derivation of the correspoding trace models. In this way, by including traceability, we improve the reusability, maintainability and rationale comprehension of the models [24, 29], and we are

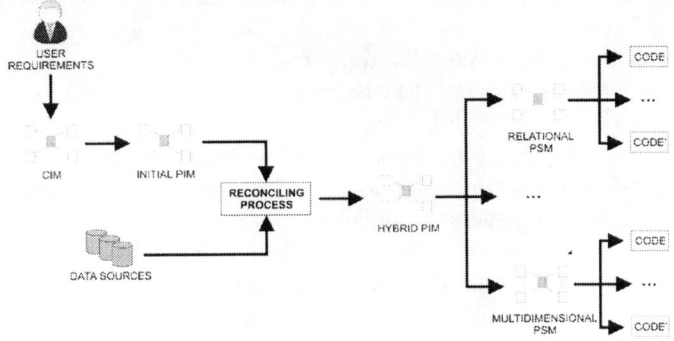

Fig. 1. Layers in our DW approach

able to easily analyze which requirements have been met and which elements from the models will be affected by a change in a given requirement.

The rest of the paper is structured as follows. Section 2 presents related work about traceability. Section 3 introduces our traceability metamodel for DWs and the inclusion of trace models in our approach. Section 4 presents the QVT rules for automatic derivation of traces in the DW context. Section 5 presents an example of application, in order to show the benefits of our proposal and Section 6 outlines the conclusions and sketches the future work to be done in this area.

2   Related Work

In this section, we will discuss the existing traceability research in other fields, its benefits and problems, and we will also discuss its current status in the DW field. Currently, traceability can be studied from two different points of view. The first one is the Requirements Engineering (RE) field, whereas the second one is the Model Driven Development (MDD) field. Although both fields are focused in different aspects of traceability, they also have some common issues.

Most of the work done until now has been in the RE field [1,2,3,8,9,23,26,31]. Some authors [8, 29] consider pre-requirement specification (pre-RS) as a more complex scenario, since it has to deal with artifacts written in natural language and different points of view, and post-requirement specification (post-RS) as a simpler, one since the requirements are already modeled.

The main benefits provided by traceability have been studied in this field [2, 23, 24]. Traceability helps assesing the impact of changes and rationale comprehension, by identifying which parts of the implementation belong to each requirement [2]. It also helps the reusability and maintainability, since the scope of each part of the project is known and defined thanks to the traces. In turn, these benefits help lowering the costs associated with the project [23, 24].

The main drawbacks mentioned about traceability are the non-existence of a standard traceability definition or metamodel, the manual recording of traces, and that traceability itself is seen as a burden until it is necessary later on although its benefits have been tested [24]. This situation creates a problem which makes difficult to succesfully apply traceability.

In order to alleviate the first drawback, a classification of eight categories for traces was presented in [26]. The second and third drawbacks can be solved by automating the trace recording. However, in the RE field, the trace recording is focused on pre-RS traceability, and needs to find traces in documents in natural language. In turn, this generates models that must be supervised, with a high percentage of irrelevant traces that difficult the comprehension and visualization of the trace model, which usually has a huge number of traces already [28].

On the other hand, in the MDD field, the MDA framework is used [1, 5, 11, 22, 28]. The automatic derivation process starts from a CIM layer, where the requirements are specified as models, usually by means of goal-oriented models [7, 13, 15, 18, 21, 25, 32]. In this sense, the traceability research in the MDD field is mainly focused on post-RS, which makes the automation of traces an easier task and less prone to errors, since everything is either a model or an element in a model. However, although more restrictive, the traceability definitions in this field are not standard either. There are mainly two definitions of traceability in the MDD community. The definition we will use in this paper comes from [22]; They define traceability as "[...] the ability to chronologically interrelate uniquely identifiable entities in a way that matters. [...] [It] refers to the capability for tracing artifacts along a set of chained [manual or automated] operations."

In the DW field, as we previously stated, there is no mention of traceability being included in the process, even though there are approaches which would benefit from it. These approaches are based on model transformations through multiple layers, either following MDA [16] or a similar set of layers [27]. Currently, whenever a change to an element is done, the traceability as defined in [22] is lost, since the elements are associated by name matching. Therefore, we lose all the aforementioned benefits of traceability, which can be obtained in the DW field at a low cost, since trace recording can be automated. Moreover, the quality metrics presented in [27] could be provided in a more automated way with traceability support, as opposed to performing the process manually, allowing us to increase the quality of the final implementation.

Due to the peculiarity and idiosyncrasy of data warehouses, we will need to difference between (i) the traces coming from the requirements (for requirements validation and impact change analysis), (ii) the traces coming from the data sources (for querying and derivation of initial Extraction, Transformation and Load processes) and (iii) the traces linking elements in the multidimensional conceptual models, based on their particular relationships [14].

3   A Traceability Approach and a Trace Metamodel for Data Warehouses

As previously stated, if we wish to perform automatic operations with traces, we must be able to identify the meaning of each trace. In order to do this, we

need to elaborate a set of trace types, which define the semantic of the relationships between elements. In this section, we will introduce the trace metamodels
proposed in the MDD field along with our proposed metamodel for DW.

3.1   Model Driven Architecture Metamodels for Traceability

Our traceability approach is based on the trace framework proposed by the
OMG, which is included in the MDA framework [19].

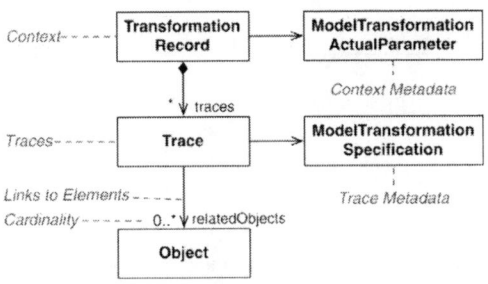

Fig. 2. Metamodel for traceability in MDA

The metamodel, presented in figure 2, is composed by a *transformation record*
which represents the transformation that generated the traces. The transformation record contains the set of traces produced and can have associated metadata
as, for example, the parameters passed to the transformation when it was executed. For each trace recorded, there is a set of model elements which are linked
by the previously-mentioned trace, varying from 0 to N elements. As in the previous case, the trace can have associated metadata as, for example, which was
the rule of the transformation created each trace.

According to this proposal there is a core metamodel for the ATLAS Model
Weaver (AMW) [6], used for linking elements from models. This core metamodel
constitutes the base for the traceability metamodel which we extend.

3.2   Proposed Metamodel

Our proposed metamodel for traceability extends AMW metamodel for traceability, including the necessary semantic types for traceability in DW. The result
can be seen in figure 3.

In this metamodel, a *TraceModel* has a set of models (*wovenModels*) linked
by the trace model. Each of these woven models has the list of references (*Ele-
mentRef*) which identify the elements linked by the traces. The trace model has
also a set of *TraceLinks*, which define the relationships between the elements in
the woven models. Each trace link has a set of *sourceElements*, which were the
source of the automatic derivation, and a set of *targetElements* which were the

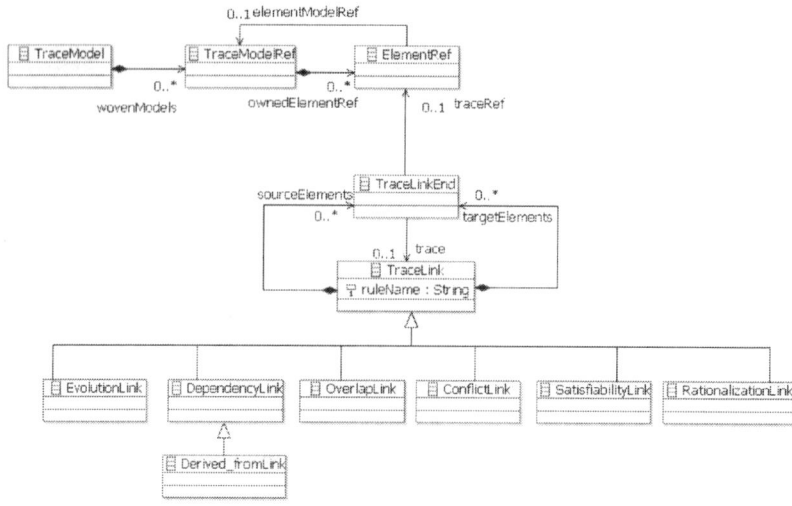

Fig. 3. AMW Metamodel for traceability extended with semantic links for DWs

result of the automatic derivation. A trace link can also have one parent, as well as a set of children trace links. This is an important feature, since it allows us to group traces forming hierarchies, providing different levels of detail in the trace models. Therefore, the trace models allow us to visualize over a hundred traces, which they typically store, in a scalable way. The elements linked by the traces are represented by the *TraceLinkEnds*, which reference the identifiers listed in the woven models.

In order to add semantics to the traces in the metamodel, we extend the *TraceLink* element, aligning the types with the classification made in [26]. We could use a reduced set of links, since in our case *Overlap* and *Conflict* are very similar. However, for the sake of standarization, we include the whole set. Nevertheless, in our case, each trace will only have one semantic type attached (since we do not include roles because they are included at the CIM level). Therefore, the definition of each trace link type is as follows:

- *Satisfiability* and *Dependency* will be used for vertical traceability (between different layers). In the first case, the traces with this type will be those coming from the requirements (in the CIM layer) to the elements in the PIM. In the second case, we will use a specialization of the *Dependency* type, *Derived_from*, in order to specify the traces coming from the data sources to the multidimensional elements at the PIM level.
- *Evolution* links will be included to handle horizontal traceability which takes care of element changes at the same layer (e.g. from PIM to PIM).
- *Overlap* and *Conflict* will be used for solving conflicts where the same element comes both from the requirements and from the data sources in a

different shape. In this case, the designer will decide which derived element is the correct solution to the conflict.

— *Rationalization* links will be included as means of enabling the user to record his own annotations in the trace model about changes or decisions taken.

Once we have defined our trace metamodel, we need to define an approach to create the trace models in an automatic way, which models will be created and what information will they store. In order to include traceability in our approach [15, 16], we will introduce the trace models shown in figure 4. The first step to include traceability and support for automated operations (like requirements and transformation validation, calculation of traceability measures and derivation including source datatypes) in our approach, is to make the relationships (shown in [15, 16]) between the CIM and PIM elements explicit. This is a vertical traceability (source and target models are in a different MDA layer) case in which all the relations are perfectly known, since they are created in an automatic way, so we just need to create the elements which correspond to the traces simultaneously as the transformation is executed and the target model is created. This CIM2PIMTrace model will be storing mainly *Satisfiability* traces.

The second step,is to be able to record the traces between the data sources represented in the PSM and the hybrid PIM. The hybrid PIM is the result of a transformation using as input the first PIM (from now on "initial PIM") and the data sources. The hybrid PIM should be traced both to the initial PIM, in order to be able to trace the original requirements, and to the data sources, in order to keep track of the source tables and attributes. This hybrid PIM can contain conflicts between concepts that come from both the requirements and from the data sources, either defining the same concept differing only in their name (overlap) or totally differing both in name and attributes (conflict). In this sense, this hybrid PIM will have both vertical (between the data sources and the hybrid PIM, DS2PIMTrace) and horizontal (between the initial PIM and the hybrid PIM, PIM2PIMTrace1) traceability models. The vertical trace

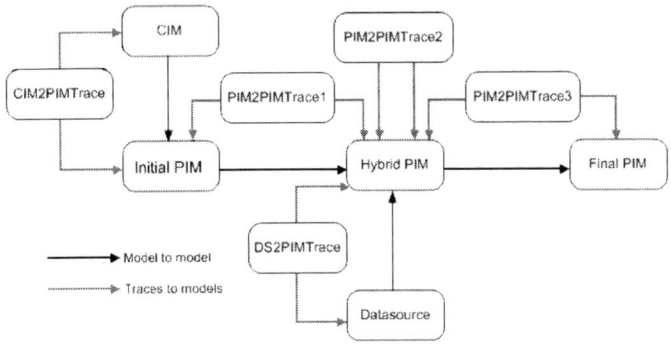

Fig. 4. Inclusion of traceability models in DW development process

model between the PSM and the hybrid PIM will record the *Derived_from* traces, whereas the horizontal trace model between the initial PIM and the hybrid PIM will record the *Evolution* traces. An additional PIM2PIMTrace2 model can be added to record the existing *Overlaps* and *Conflicts*, but these should be manually added, since only the designer knows which elements in the model refer to the same concept. It is important to note that, these kind of traces, are not less important than the automatic ones, since they will act as a bridge to map certain requirements to the data sources.

The last step is deriving the final PIM, which will be used to generate the target DW. This final PIM retains only the elements from the hybrid PIM which will be finally used. In this sense, this PIM is the result of filtering the undesired elements and resolving the conflicts which appeared in the hybrid PIM. Therefore, the traces from the hybrid PIM to the final PIM will show which concepts were the ones chosen as a solution to each existing conflict. The type of these traces will be *Evolution* and will be stored in the PIM2PIMTrace3 model.

Since the development process is performed by succesive deriving, adding, and filtering elements while most elements are not altered, the traces have low volatility. In this sense, developing a reactive framework which automatically updates the corresponding traces whenever a change (update or delete) is made would minimize the maintenance effort.

Once we have defined the trace metamodel and we have shown all the required trace models, we need to formally define the automatic derivation of the traces by means of QVT [20] rules.

4   Automatic Derivation of Traceability Models in Data Warehouses

In this section, we will discuss the necessary transformations to automatically generate the aforementioned traces and store them in trace models, which can be updated over time. Due to paper constraints, we will focus on the traces coming from CIM to PIM, both in this section and in our example.

According to our proposal for developing DWs [15, 16], we use a hybrid approach deriving the elements in an initial PIM model from the requirements by means of QVT rules. QVT rules specify a transformation by checking for a defined pattern in the source model. Once the pattern is found, a QVT rule transforms elements from the source metamodel into the target metamodel. A QVT, which creates the links between the CIM business process element and the PIM fact and fact attribute elements, is shown in figure 5.

On the left hand of the transformation rule is the source metamodel, which in our case, is our i* profile in order to model requirements for DWs. In this QVT rule, we have a business process with its associated rationale, which is modeled by means of strategic, decisional and information goals. These goals model the business logic from which we obtain the information requirements which, in turn are decomposed into measures, which are indicators of business performance, and contexts.

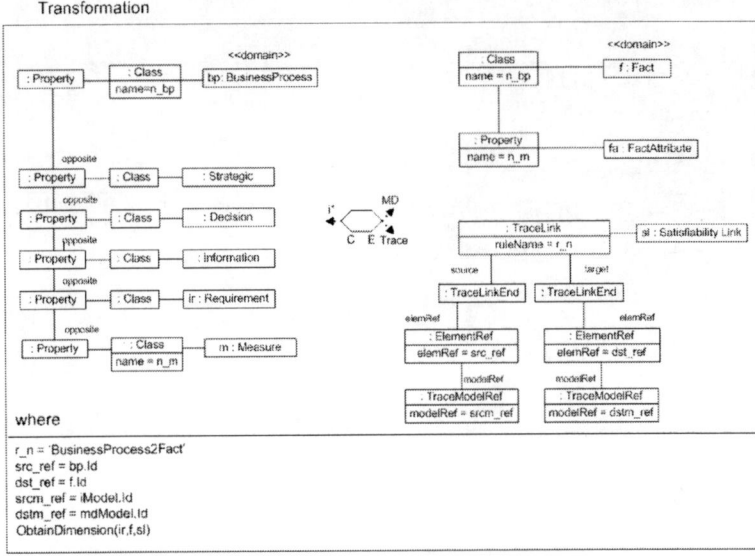

Fig. 5. QVT rule to derive a fact from a business process along with the associated trace

On the right hand of the transformation rule are the target metamodels. On the one hand, we have our multidimensional profile, composed by the fact (focus of the analysis, related to the business process) and the associated fact attribute (indicator of performance, related to the measure). On the other hand we also have the trace metamodel we proposed, composed in this case by the trace link which makes explicit the relationship between the business process and the fact. In an analogous way, the trace link for making explicit the relationship between the measure and the fact attribute would be also created, but due to space constraints it is omitted.

The "C" at the center of the figure means that the source model is checked, whereas the "E" means that the target models are enforced. This means that, each time that the described pattern is found in the source models, the target patterns are enforced (generated) in the resulting target models. The rest of the QVT relationships between models in our approach (without including traceability) can be found in [16]. The corresponding QVT rule for transforming a context into a dimension can be seen in figure 6.

In this figure, a context "c" from the CIM, on the left hand, is transformed into a dimension "d" and its associated base level "b" in the PIM, at the right hand of the figure. The associated trace link to this transformation has a source element reference, which is the context from the CIM, and three (one omitted) trace link ends. Each trace link end references a different element but has the same model reference (since the elements are in the same target model). The

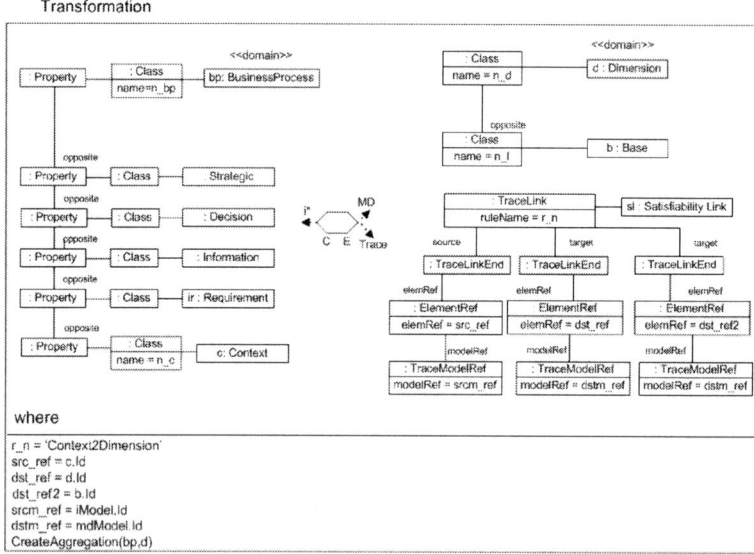

Fig. 6. QVT transformation for deriving a context into a dimension with its associated trace

omitted trace link corresponds to the association between the base level and the dimension. The rest of the relationships between the CIM and PIM elements is summarized in table 1, along with its corresponding rule name.

Table 1. Relationships between CIM elements and PIM elements

Transformation rule name	Source element (CIM)	Target Element (PIM)
BusinessProcess2Fact	BusinessProcess	StarPackage
BusinessProcess2Fact	BusinessProcess	FactPackage
BusinessProcess2Fact	BusinessProcess	Fact
Measure2FactAttribute	Measure	FactAttribute
Context2Dimension	Context	DimensionPackage
Context2Dimension	Context	Dimension
Context2Dimension, Context2Base	Context	Level
Context2Dimension, Context2Base	Context	Association

The contexts can derive either into a dimension package, its corresponding dimension and first level if the context is the base of the dimension, or into a level and an association with the previous level in the dimension hierachy. A graphic example of the generic relationships is shown in figure 7.

In this figure, we can see the previously described elements in the QVT from the CIM at left side. The business process has its associated rationale represented by means of succesive goals, which derive into an information requirement for

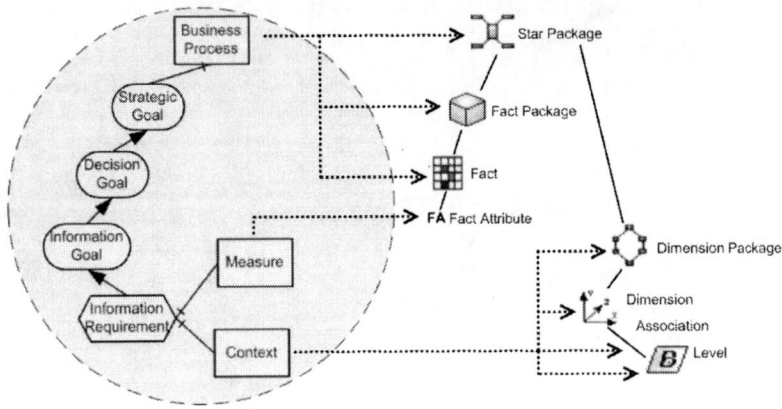

Fig. 7. Generic relationships between CIM elements and PIM Elements

the DW. In turn, this information requirement is decomposed into contexts and measures. At the right hand of the figure, the corresponding multidimensional elements from the PIM appear. As aforementioned, the business process is related to the fact (and the corresponding packages), whereas the first context is related to the dimension package, the dimension, the base level and its association with the dimension. On the other hand, the second context is associated with the second level in the dimension. With this approach, if we wish to check the result of the transformations for debugging, we can check which rule created each element with the information stored in the traces. In addition, these traces allow us to keep track of which elements in the PIM model match with each requirement in the CIM model, making requirements validation easier. Moreover, if any requirement is changed, we know which elements are affected and which rule created them, being able to execute the corresponding rule of the transformation to regenerate the affected part of the PIM.

Although in this section we have presented the generation of traces from a goal-based CIM, the traces could be used to trace any element in a different CIM model from another proposal, as long as it has a unique reference identifier.

5   Example of Application

In this section, we will present an example of application for our traceability proposal, showing how the traces can be navigated to retrieve useful information.

A University wishes to build a DW in order to analyze the factors which influence the performance of the students. This university has a transactional database created for managing the information about professors, degrees, subjects and students, which will serve as data source for the data warehouse. The analysts wish to analyze the *students grades* by *subject*, *professor* who teaches them, and *student*, taking into account how many *hours* they spend studying

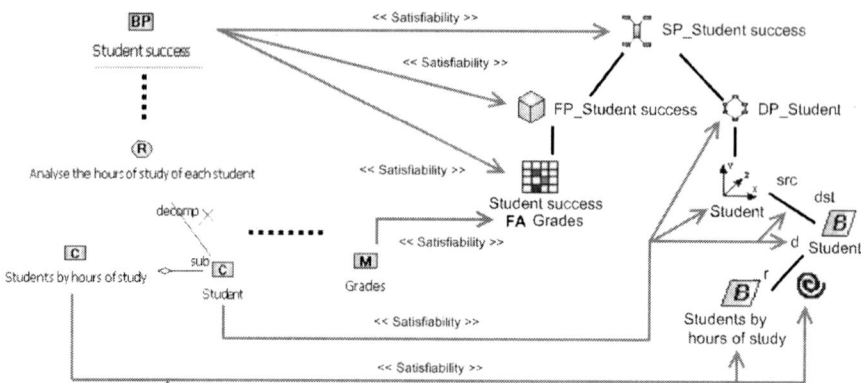

Fig. 8. Requirements modeled by using our i* profile for DW and its corresponding multidimensional model

per week. These requirements are recorded in a CIM which acts as the starting point for the process.

In figure 8, we can see the requirements on the left side, where the business process (focus of the analysis) in this case is the student success. Its corresponding contexts are the students and the students grouped by hours of study per week (shown in the figure), the subjects and the professors (omitted due to space constraints). On the right hand of the figure, we can see the PIM with the StarPackage "SP_student success, the FactPackage "FP_student success and the Fact "Student success", associated with the business process. The measure "student grades" is associated with its corresponding FactAttribute, whereas the "student" context derives into the DimensionPackage "DP_Student", the Dimension "Student", the base level "Student" and its association with the dimension. Lastly, the aggregated context "Students by hours of study" derives into its corresponding level and the association towards the previous level in the dimension (in this case "student" level). Table 2 summarizes the relationships between elements and its corresponding transformation rule.

Table 2. Elements linked between CIM and PIM models by satisfiability links

Satisfiability link rule name	Source Elements (CIM)	Target Elements (PIM)
BusinessProcess2Fact	Student success	SP_Student sucess, FP_Student sucess, Student sucess
Measure2FactAttribute	Grades	Grades
Context2Dimension	Student	DP_Student,Student(Dime) Student(Level),Association
Context2Base	Students by hours of study	Students by hours of study (Level), Association

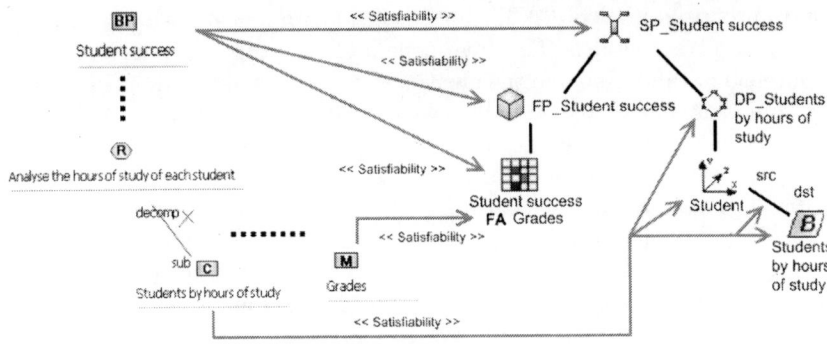

Fig. 9. New version of the CIM and its corresponding PIM obtanied by applying our trace metamodel and the transformations

Once the PIM has been derived, the analysts wish to change the requirements model. They wish to remove the "Student" context from the requirements. The context is removed from the model, and the previously aggregated context "Students by hours of study" now satisfies the information requirement from which "Students" derived in the CIM. With our approach we can track the changes done in the CIM model and identify the affected elements. In this case, the contexts associated with the information requirement in the CIM model will be affected, as well as the dimension "DP_Students" and its corresponding elements in the PIM model. The rule used to generate the PIM elements was Context2Dimension, so we will need to execute this rule again with the new parameters (in this case "Students by hours of study" context) to regenerate the affected part in the derived model. The result is shown in figure 9.

In this model, the "Students by hours of study" dimension replaces the previous "Students" dimension, association and base level while the other elements remain the same.

This example is a simplification from a real-world project of another university. Three different data marts were designed, each data mart CIM averaging forty to fifty decisional goals, deriving into an average of twenty eight to thirty four contexts and measures per data mart. A change in a decisional goal would affect an average of other three to four goals and their derived contexts and measures, with the corresponding changes at the PIM level. Thanks to the traces, the designer knew which elements had to be changed in the final implementation, cutting the development time of the project.

6   Conclusions and Future Work

In this paper, we have proposed the first trace metamodel for DW development based on the MDA framework, in order to include semantic traces. We have shown the necessary trace models to be included in our development process. Furthermore, we have focused on the relationships between the CIM and the PIM and have proposed a set of QVT transformations to automatically generate the corresponding trace models. The great benefit of our proposal is the improvement in requirements validation and the identification of the corresponding elements in the PIM models, being able to easily assess the impact of changes and regenerate the affected parts. This was shown by means of the presented example.

Our plans for the immediate future are developing a new set of QVT transformations to explore the relationships between the PIM and PSM and explore the potential of using the information recorded in the traces in order to support automated analysis. We will also develop a traceability framework in order to make the maintenance of traces as automatic as possible.

Acknowledgments. This work has been partially supported by the MESO-LAP (TIN2010-14860) and SERENIDAD (PEII-11-0327-7035) projects from the Spanish Ministry of Education and the Junta de Comunidades de Castilla La Mancha. Alejandro Maté is funded by the Generalitat Valenciana under an ACIF grant (ACIF/2010/298).

References

1. Aizenbud-Reshef, N., Nolan, B., Rubin, J., Shaham-Gafni, Y.: Model traceability. IBM Systems Journal 45(3), 515–526 (2006)
2. Antoniol, G., Canfora, G., Casazza, G., De Lucia, A., Merlo, E.: Recovering traceability links between code and documentation. IEEE Transactions on Software Engineering 28(10), 970–983 (2002)
3. Arkley, P., Mason, P., Riddle, S.: Position paper: Enabling traceability. In: Proceedings of the 1st International Workshop on Traceability in Emerging Forms of Software Engineering, Edinburgh, Scotland, pp. 61–65 (2002)
4. Ballou, D., Tayi, G.: Enhancing data quality in data warehouse environments. Communications of the ACM 42(1), 73–78 (1999)
5. Barbero, M., Del Fabro, M., Bézivin, J.: Traceability and provenance issues in global model management. In: ECMDA-TW, pp. 47–56 (2007)
6. Del Fabro, M., Bézivin, J., Valduriez, P.: Weaving Models with the Eclipse AMW plugin. In: Eclipse Modeling Symposium, Eclipse Summit Europe 2006, Esslingen, Germany (2006)
7. Franch, X.: Incorporating Modules into the i^* Framework. In: Pernici, B. (ed.) CAiSE 2010. LNCS, vol. 6051, pp. 439–454. Springer, Heidelberg (2010)
8. Gotel, O., Finkelstein, A.: An analysis of the requirements traceability problem. In: ICRE, pp. 94–101. IEEE, Los Alamitos (1994)
9. Gotel, O.C.Z., Morris, S.J.: Macro-level Traceability Via Media Transformations. In: Rolland, C. (ed.) REFSQ 2008. LNCS, vol. 5025, pp. 129–134. Springer, Heidelberg (2008)
10. Inmon, W.H.: Building the data warehouse. Wiley-India, Chichester (2009)
11. Jouault, F.: Loosely coupled traceability for atl. In: ECMDA-TW, Nuremberg, Germany, pp. 29–37 (2005)
12. Kimball, R.: The data warehouse toolkit. Wiley-India, Chichester (2009)
13. Kolp, M., Giorgini, P., Mylopoulos, J.: Organizational Patterns for Early Requirements Analysis. In: Advanced Information Systems Engineering (CAiSE), LNCS, vol. 2681, pp. 617–633. Springer Berlin (2003)

14. Luján-Mora, S., Trujillo, J., Song, I.-Y.: A UML profile for multidimensional modeling in data warehouses. DKE 59(3), 725–769 (2006)
15. Mazón, J.-N., Trujillo, J.: A model driven modernization approach for automatically deriving multidimensional models in data warehouses. In: Parent, C., Schewe, K.-D., Storey, V.C., Thalheim, B. (eds.) ER 2007. LNCS, vol. 4801, pp. 56–71. Springer, Heidelberg (2007)
16. Mazon, J.N., Pardillo, J., Trujillo, J.: An MDA approach for the development of data warehouses. DSS 45(1), 41–58 (2008)
17. Mazón, J.-N., Trujillo, J., Lechtenbörger, J.: Reconciling requirement-driven data warehouses with data sources via multidimensional normal forms. DKE 63(3), 725–751 (2007)
18. Mouratidis, H., Giorgini, P., Manson, G.: Integrating Security and Systems Engineering: Towards the Modelling of Secure Information Systems. In: Eder, J., Missikoff, M. (eds.) CAiSE 2003. LNCS, vol. 2681, pp. 63–78. Springer, Heidelberg (2003)
19. OMG: A Proposal for an MDA Foundation Model (2005)
20. OMG: The Meta-Object Facility 2.0 Query/View/Transformation. Final Adopted Specification (2005)
21. Giorgini, P., Rizzi, S., Garzetti, M.: GRAnD: A goal-oriented approach to requirement analysis in data warehouses. DSS 45(1), 4–21 (2008)
22. Paige, R., Olsen, G., Kolovos, D., Zschaler, S., Power, C.: Building model-driven engineering traceability classifications. In: ECMDA-TW, pp. 49–58 (2008)
23. Ramesh, B., Jarke, M.: Toward reference models for requirements traceability. IEEE Transactions on Software Engineering 27(1), 58–93 (2001)
24. Ramesh, B., Stubbs, C., Powers, T., Edwards, M.: Requirements traceability: Theory and practice. Annals of Software Engineering 3(1), 397–415 (1997)
25. Samia Kaabi, R., Souveyet, C., Rolland, C.: Eliciting service composition in a goal driven manner. In: ICSOC, pp. 308–315 (2004)
26. Spanoudakis, G., Zisman, A.: Software traceability: a roadmap. Handbook of Software Engineering and Knowledge Engineering (2005)
27. Vassiliadis, P.: Data Warehouse Modeling and Quality Issues. Ph.D. thesis, Athens (2000)
28. Walderhaug, S., Stav, E., Johansen, U., Olsen, G.K.: Traceability in Model-Driven Software Development. Designing Software-Intensive Systems: Methods and Principle, 133–159 (2008)
29. Winkler, S., von Pilgrim, J.: A survey of traceability in requirements engineering and model-driven development. Software and Systems Modeling 9, 529–565 (2010)
30. Yu, E.S.K.: Modelling strategic relationships for process reengineering. Ph.D. thesis, Toronto, Ont., Canada (1995)
31. Yu, Y., Jurjens, J., Mylopoulos, J.: Traceability for the maintenance of secure software. In: ICSM 2008, pp. 297–306. IEEE, Los Alamitos (2008)
32. Yu, Y., Niu, N., Gonzalez-Baixauli, B., Candillon, W., Mylopoulos, J., Easterbrook, S., do Leite, J., Vanwormhoudt, G.: Tracing and validating goal aspects. In: RE 2007, pp. 53–56. IEEE, Los Alamitos (2007)

Ontological Foundations for Conceptual Part-Whole Relations: The Case of Collectives and Their Parts

Giancarlo Guizzardi

Ontology and Conceptual Modeling Research Group (NEMO),
Federal University of Espírito Santo (UFES), Vitória (ES), Brazil
gguizzardi@inf.ufes.br

Abstract. In a series of publications, we have employed ontological theories and principles to evaluate and improve the quality of conceptual modeling grammars and models. In this article, we advance this research program by conducting an ontological analysis to investigate the proper representation of types whose instances are *collectives*, as well as the representation of part-whole relations involving them. As a result, we provide an ontological interpretation for these notions, as well as modeling guidelines for their sound representation in conceptual modeling. Moreover, we present a precise qualification for the parthood relations of *member-collective* and *subcollective-collective* in terms of formal mereological theories of parthood, as well as in terms of the modal meta-properties of *essential* and *inseparable* parts.

Keywords: ontological foundations for conceptual modeling, part-whole relations, representation of collectives.

1 Introduction

In recent years, there has been a growing interest in the application of Foundational Ontologies, i.e., formal ontological theories in the philosophical sense, for providing real-world semantics for conceptual modeling languages, and theoretically sound foundations and methodological guidelines for evaluating and improving the individual models produced using these languages.

In a series of publications, we have successfully employed ontological theories and principles to analyze a number of fundamental conceptual modeling constructs such as Types, Roles and Taxonomic Structures, Relations, Attributes, among others (e.g., [1,2]). In this article we continue this work by investigating a specific aspect of the representation of part-whole relations. In particular, we focus on the ontological analysis of *collectives* and the part-whole relations involving them. The focus on collectives is timely given the increasing recognition of the importance of finding well-founded manners to represent collectives in domains such as bioinformatics in which collectives and their parts abound [3,4].

Parthood is a relation of fundamental importance in conceptual modeling, being present as a modeling primitive in practically all major conceptual modeling languages. Motivated by this, a number of attempts have been made to employ theories of different sorts to provide a foundation for part-whole relations. These initiatives

H. Mouratidis and C. Rolland (Eds.): CAiSE 2011, LNCS 6741, pp. 138–153, 2011.
© Springer-Verlag Berlin Heidelberg 2011

fall roughly in three different classes: (i) proposals that employ classical ontological theories of parthood (*Mereologies*). In this class, there is a number of works in the literature that employ the ontological theory put forth by the philosopher Mario Bunge [5] typically accessed through its most popular adaptation termed the BWW ontology [6,7]; (ii) proposals that are based on research from linguistics and cognitive science in which different sorts of *Meronymic* relations are elaborated. Most contributions in this class are based on the theory developed by Winston , Chaffin and Herrmann (henceforth WCH) [8]. An example of a pioneering article in this class is [9]; (iii) proposals which define a number of so-called *secondary properties* which have been used to further qualify parthood relations [6]. These include distinctions which reflect different *modal aspects* of parthood reflecting different relations of dependence (e.g., generic versus existential dependence). An example is the distinction between *essential* and *inseparable* parthood in [2].

Despite their important contributions, there are significant shortcomings in the current scenario considering the aforementioned approaches. On the one hand, accounts of parthood solely based on WCH suffer from many difficulties inherited from the original theory. As discussed in [10,11], WCH's original taxonomy turned out to be overly linguistically motivated, focusing on the linguistic term *part-of* (and its cognates). In fact, as demonstrated by these authors, the six linguistically-motivated types of part-whole relation originally proposed in WCH give rise to only four distinct ontological types, namely: (a) *subquantity-quantity*: modeling parts of an amount of matter (e.g., alcohol-wine, gin-Martini, Chocolate-Toddy); (b) *member-collective*: modeling a collective entity in which all parts play an equal role w.r.t. the whole (e.g., tree–forest, card–deck, lion-pack); (c) *subcollective-collective*: modeling a relation between a collective and the subcollectives that provide further structure to the former (e.g., the north part of the black forest-black forest, the underage children of John-the children on John); (d) *component – functional complex*: modeling an entity in which all parts play a different role w.r.t. the whole, thus, contributing to the functionality of the latter (e.g., heart-circulatory system, engine – car).

On the other hand, conceptual modeling accounts of parthood based on BWW inherit the limitations of Bunge's original treatment of parthood in its most basic core. Mereology is a mature discipline with well-defined and formally characterized theories. These in fact form a lattice of theories such that there is not one single formal meaning of part in mereology but several alternative axiomatizations of parthood that extend each other. Mapping modeling primitives representing part-whole relations to these theories can indeed provide an important contribution to conceptual modeling. Firstly, in a direct manner because these theories can provide sound and fully characterized formal semantics for these relations. But, also because nowadays several authors have proposed codifications of different mereological theories by mapping them to different Description Logics, hence, providing a mechanism for automated reasoning with *partonomies* in conceptual models [12,13]. The negative point here is that Bunge's theory of parthood corresponds to the weakest theory in mereology. In fact a theory which is even considered to be too weak to count as a characterization for a true part-whole relation [14].

Finally, most current approaches limit themselves to analyze the relation between the whole and its parts. However, as discussed in [15], a conceptual theory of parthood should also countenance a *theory of wholes*, in which the relations that tie the parts of a whole together are also considered. To put it simply, the composite objects in which we are interested in conceptual modeling are not mere aggregations of arbitrary entities but complex entities suitably unified by proper binding relations.

This paper should then be seen as a companion to the publications in [2,16] and [17]. In this research program, we have managed to show that the three classification schemes aforementioned, namely, the linguistic-cognitive meronymic distinctions, the mereological theories of parthood, and the so-called secondary properties are not orthogonal. In fact, each particular meronymic distinction in the first scheme commits to basic mereological properties, secondary properties, and even requires binding relations of specific kinds to take place between their parts. In [17], we have managed to show the interconnection between these classification schemes for the case of the *subquantity-quantity* relation. In a complementary form, we did the same in [2,16] for the case of the *component – functional complex* relation. The objective of this paper is to follow the same program for the case of part-whole relations involving collectives, namely, the *member-collective* and the *subcollective-collective* relations. This paper is, thus, a substantial extension to the preliminary work reported in [18] in which only the *member-collective* relation is analyzed and in some of its aspects.

The remainder of this article is organized as follows. Section 2 reviews the theories put forth by classical mereology as well as its connection with modal secondary properties of parts and wholes. The section also discusses how these mereological theories can be supplemented by a *theory of (integral) wholes*. In section 3, we discuss collectives as integral wholes and present some modeling consequences of the view defended there. Moreover, we elaborate on some ontological properties of collectives that differentiate them not only from their sibling categories (quantities and functional complexes), but also from sets (in a set-theoretical sense). The latter aspect is of relevance since collectives as well as the member-collective and subcollective-collective relations are frequently taken to be identical to sets, set membership and the subset relation, respectively. In section 4, we promote an ontological analysis of two part-whole relations involving collectives, clarifying on how these relations stand w.r.t. to basic mereological properties (e.g., transitivity, weak supplementation, extensionality) as well as regarding the modal secondary properties of essential and inseparable parthood. As an additional result connected to this analysis, we outline a number of metamodeling constraints that have been used to implement a UML modeling profile for representing collectives and their subparts in conceptual modeling. Section 5 presents final considerations of this paper.

2   A Review of Formal Part-Whole Theories

In practically all philosophical theories of parts, the relation of (proper) parthood (symbolized as <) stands for a strict partial ordering, i.e., an asymmetric (2) and transitive relation (3), from which irreflexivity follows (1):

$$\forall x \ \neg(x < x) \tag{1}$$

$$\forall x,y \ ((x < y) \rightarrow \neg(y < x)) \tag{2}$$

$$\forall x,y,z \ ((x < y) \land (y < z) \rightarrow (x < z)) \tag{3}$$

These axioms amount to what is referred in the literature by the name of *Ground Mereology (M)*, which is the core of any theory of parts, i.e., the axioms (1-3) define the minimal (partial ordering) constraints that every relation must fulfill to be considered a parthood relation. As previously mentioned, Mario Bunge's mereological theory (*Assembly Theory*) corresponds to the axiomatization of Ground Mereology (with the only difference of assuming the existence of a null individual which is supposed to be part of everything else) [5]. Although necessary, these constraints are not sufficient, i.e., it is not the case that any partial ordering relation qualifies as a parthood relation. Most authors require an extra axiom termed the *weak supplementation principle (WSP)* (4) as constitutive of the meaning of part and, hence, consider (1-3) plus (4) (the so-called *Minimal Mereology* (MM)) as the minimal constraints that a mereological theory should incorporate [14,19]:

$$\forall x,y \ ((y < x) \rightarrow \exists z \ (z < x) \land \neg\mathrm{overlap}(z,y)) \tag{4}$$

Figure 1.a below illustrates this notion of weak supplementation. It shows that if y is a part of x then there must exist another part of x which is disjoint from y (the "missing" part of x). Without this "missing" part, what differentiates y and x? Notice that x and y are supposed to be different given that parthood is irreflexive (1). From a practical point of view, without WSP, models such as the one in figure 1.b cannot be deemed incorrect. Now, following that model, suppose an event E which is composed of one single subevent. Isn't this alleged part identical to the event E itself? In a sound model, events are either atomic or are composed of at least two disjoint subevents.

Fig. 1. Invalid situation (a) and invalid conceptual model (b) according to Minimum Mereology

There is an extension to MM that has then been created by strengthening the supplementation principle represented by (4). In this system, (4) is thus replaced by something termed the *stronger supplementation principle (SSP)*. The resulting theory is named *Extensional Mereology (EM)*. A known consequence of the introduction of SSP is that in EM we have that two objects are identical iff they have the same parts, i.e., SSP entails a mereological counterpart of the *extensionality principle* (of identity) in set theory. As a consequence, if an entity is identical to the (mereological) sum of its parts, thus, changing any of its parts changes the identity of that entity. Ergo, an entity cannot exist without each of its parts, which is the same as saying that all its parts are *essential parts*.

Essential parthood can be defined as a case of *existential dependence* between individuals, i.e., x is an *essential part* of y iff y cannot possibly exist without having that specific individual x as part [2]. This specific mode of existential dependence can also be defined from the part x to the whole y. We say that x is an *inseparable part* of y iff x cannot possibly exist without being a part of that specific individual y [2]. A stereotypical example of an essential part of a car is its chassis, since that specific car cannot exist without that specific chassis (changing the chassis legally changes the identity of the car); A stereotypical example of an inseparable part of a living cell is its membrane, since the membrane cannot exist without being part of that particular cell. As discussed in depth in [2], essential and inseparable parthood play a fundamental role in conceptual modeling. However, it is not the case for all types of entities that all their parts are essential. In other words, although EM describes the basic meaning of parthood for some types of entities (e.g., quantities [17] and events [14]), this is not the case for entities of all ontological categories. In particular, as we have shown in [16], for functional complexes while some of their parts are essential (inseparable), not all of them are essential (inseparable). As discussed in section 4, EM is too strong a theory in this sense also for the case of the member-collective and the subcollective-collective relations.

Classical mereological theories focus solely on the relation from the parts to the wholes. Thus, just like in set theories we can create sets by enumerating any number of arbitrary entities, in classical mereologies one can create a new object by summing up individuals that can even belong to different ontological categories. For example, in these systems, the individual created by the aggregation (termed *mereological sum*) of Noam Chomsky's left foot, the first act of Puccini's Turandot and the number 3, is an entity considered as legitimate as any other. However, as argued by [10], humans only accept the aggregation of entities if the resulting mereological sum plays some role in their conceptual schemes. To use an example: the sum of a frame, a piece of electrical equipment and a bulb constitutes a whole that is considered meaningful to our conceptual classification system. For this reason, this sum deserves a specific concept in cognition and a name in human language. The same does not hold for the sum of bulb and the lamp's base.

According to Simons [14], the difference between purely formal mereological sums and, what he terms, *integral wholes* is an ontological one, which can be understood by comparing their existence conditions. For sums, these conditions are minimal: the sum exists just when the constituent parts exist. By contrast, for an integral whole (composed of the same parts of the corresponding sum) to exist, a further *unifying condition* among the constituent parts must be fulfilled. A unifying condition or relation can be used to define a closure system in the following manner. A set B is a closure system under the relation R, or simply, R-closure system iff

$$\text{cs} \langle R \rangle \, B =_{\text{def}} (\text{cl} \langle R \rangle \, B) \wedge (\text{con} \langle R \rangle \, B) \tag{5}$$

where **(cl $\langle R \rangle$ B)** means that the set B is closed under R (R-Closed) and **(con $\langle R \rangle$ B)** means that the set B is connected under R (R-Connected). R-Closed and R-Connected are then defined as:

$$\text{cl} \langle R \rangle \, B =_{\text{def}} \forall x \, ((x \in B) \rightarrow ((\forall y \, R(x,y) \vee R(y,x) \rightarrow (y \in B))) \tag{6}$$

$$\text{con} \langle R \rangle \, B =_{\text{def}} \forall x \, ((x \in B) \rightarrow (\forall y \, (y \in B) \rightarrow (R(x,y) \vee R(y,x)))) \tag{7}$$

An integral whole is then defined as an object whose parts form a closure system induced by what Simons terms a *unifying* (or *characterizing*) *relation* R.

3 What are Collectives?

According to WCH, the main distinction between collectives and quantities is that the latter but not the former are said to be *homeomeros* wholes [8]. In simple terms, homeorosity means that the entity at hand is composed solely of parts of the same type (*homo*=same, *mereos* = part). The fact that quantities are homeomeros (e.g., all subportions of wine are still wine) causes a problem for their representation (and the representation of relationships involving them) in conceptual modeling. In order to illustrate this idea, we use the example depicted in figure 2.a below. In this model, the idea is to represent that a certain portion of wine is composed of all subportions of wine belonging to a certain vintage, and that a wine tank can store several portions of wine (perhaps an *assemblage* of different vintages). However, since Wine is homeomeros and infinitely divisable in subportions of the same type, if we have that a Wine portion x has as part a subportion y then it also has as part all the subparts of y [17]. Likewise, a wine tank storing two different "portions of wine" actually stores all the subparts of these two portions, i.e., it actually stores infinite portions of wine. In other words, maximum cardinality relations involving quantities cannot be specified in a finite manner. As discussed, for instance in [20], *finite satisfiability* is a fundamental requirement for conceptual models which are intended to be used in information systems. This feature of quantities, thus, requires a special treatment so that they can be property modeled in structural conceptual models. A treatment that does not take quantities to be mere aggregations (*mereological sums*) of subportions of the same kind but integral wholes unified by a characterizing relation of *topological maximal self-connectedness* [17].

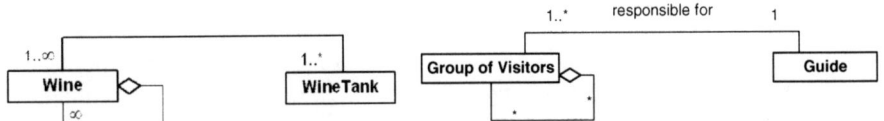

Fig. 2. Representations of a Quantity (a-left) and a Collective (b-right) with their respective parts in UML conceptual Models

As correctly defined by WCH, collectives are not homeomeros. They are composed of subparts parts that are not of the same kind (e.g., a tree is not forest). Moreover, they are also not infinitely divisible. As a consequence, a representation of a collection as a simple aggregation of entities (analogous to an enumerated set of entities) does not lead to the same complications as for the case of quantities. Take, for instance, the example depicted in figure 2.b, which represents a situation analogous to the one of figure 2.a. Different from the former case, there is no longer the danger of an infinite regress or the impossibility of specifying finite cardinality constraints. In figure 2.b, the usual maximum cardinality of "many" can be used to express that a group of visitors has as parts possibly many other groups of visitors and that a guide is responsible for possibly many groups of visitors.

Nonetheless, in many examples (such as this one), this model of figure 2.b implies a somewhat counterintuitive reading. In general, the intended idea is to express that, for instance, John as a guide, is responsible for the group formed by {Paul, Marc, Lisa} and for the other group formed by {Richard, Tom}. The intention is not to express that John is responsible for the groups {Paul, Marc, Lisa}, {Paul, Marc}, {Marc, Lisa}, {Paul, Lisa}, and {Richard, Tom}, i.e., that being responsible for the group {Paul, Marc, Lisa}, John should be responsible for all its subgroups. A simple solution to this problem is to consider groups of visitors as maximal sums, i.e., groups that are not parts of any other groups. In this case, depicted in figure 3, the cardinality constraints acquire a different meaning and it is no longer possible to say that a *group of visitors* is composed of other *groups of visitors* in this technical sense.

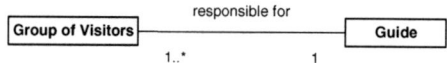

Fig. 3. Representation of Collections as Maximal Sums

The solution above is similar to taking the meaning of a quantity K to be that of a maximally-self-connected-portion of K [17]. However, in the case of collections, topological connection cannot be used as a *unifying or characterizing relation* to form an integral whole, since collections can easily be spatially scattered. Nonetheless, another type of connection (e.g., social) should always be found. A question begging issue at this point is: why does it seem to be conceptually relevant to find *unifying relations* leading to (maximal) collections? As discussed in the previous section, collections taken as arbitrary sums of entities make little cognitive sense: we are not interested in the sum of a light bulb, the North Sea, the number 3 and Aida's second act. Instead, we are interested in aggregations of individuals that have a purpose for some cognitive task. So, we require all collectives in our system to form closure systems unified under a proper characterizing relation. For example, a group of visitors of interest can be composed by all those people that are attending a certain museum exhibition at a certain time. Now, by definition, a closure system is maximal (see formula (5)), thus, there can be no group of visitors in this same sense that is part of another group of visitors (i.e., another integral whole unified by the same relation).

Nonetheless, it can be the case that, among the parts of a group of visitors, further structure is obtained by the presence of other collections unified by different relations. For example, it can be the case that among the parts of a group of visitors A, there are collections B and C composed of the English and Dutch speaking people in that group, respectively. Now, neither the English speaking segment nor the Dutch speaking segment are *groups of visitors* in the technical sense just defined, since the latter has properties lacking in both of them (e.g., the property of having both English and Dutch segments). Moreover, the unifying relations of B and C are both specializations of A's unifying relation. For example, A is the collection of all parties attending an exhibition and the B is the collection of all English speakers among the parties attending that same exhibition. We return to this point in section 4.2.

By not being homeomeros and infinitely divisible, collectives actually bear a stronger similarity to functional complexes than to quantities in the classifications of [10,11]. In [11], for instance, the authors propose that the difference between a collective and a functional complex is that whilst the former has a *uniform* structure, the latter has a *heterogeneous* and *complex* one. We propose to rephrase this statement in other terms. In a collective, *all member parts play the same role type w.r.t. the whole*. For example, all trees in a forest can be said to play the role of a forest member. In complexes, conversely, a variety of roles can be played by different components. For example, if all ships of a fleet are conceptualized as playing solely the role of "member of a fleet" then it can be said to be a collection of ships. Contrariwise, if this role is further specialized in "leading ship", "defense ship", "storage ship" and so forth, the fleet must be conceived as a functional complex.

Finally, we would like to call attention to the fact that collectives are not sets and, thus, the *member-collective* and the *subcollective-collective* relations are not the same as the *membership* ($\in$) and *subset* ($\subset$) relations, respectively. Firstly, collectives and sets belong to different ontological categories: the former are concrete entities that have spatiotemporal features; the latter, in contrast, are abstract entities that are outside space and time and that bear no causal relation to concrete entities [1]. Secondly, unlike sets, collectives do not necessarily obey an extensional principle of identity, i.e., it is not the case that a collective is always completely defined by the sum of its members. We take that some collectives can be considered extensional by certain conceptualizations; however, we also acknowledge the existence of *intentional collectives* obeying non-extensional principles of identity [21]. Thirdly, collectives are integral wholes unified by proper characterizing relations; sets can be simply postulated by enumerating their members. This feature of the latter is named *ontological extravagance* and it is a feature to be ruled out from any ontological system [19]. Finally, contrary to sets, we do not admit the existence of empty or unitary collectives. As a consequence, we eliminate a feature of set theory named *ontological exuberance* [19]. Ontological exuberance refers to the feature of some formal systems that allows for the creation of a multitude of entities without differentiation in content. For instance, in set theory, the elements a, {a}, {{a}}, {{{a}}}, {...{{{a}}}...} are all considered to be distinct entities. We shall return to some of these points in the next section.

4 Parthood Relations Involving Collectives

4.1 The Member-Collection Relation

According to [22], classical semantic analysis of plurals and groups distinguish between *atomic entities*, which can be *singular* or *collectives*, and *plural entities*. From a linguistic point of view, the *member-collection* relation is considered to be one that holds between an *atomic entity* (e.g., John, the deck of cards) and either a *plural* (e.g., {John, Marcus}) *or a collective term* (e.g., the children of Joseph, the collection of ancient decks).

Before we can continue, a formal qualification of this notion of atomicity is needed. Suppose an integral whole W unified under a relation R. By using this unifying (or characterizing) relation R, we can then define a composition relation $<_R$ such that $(x <_R W)$ iff: (i) there is a set B such that $\mathbf{cs} \langle \mathbf{R} \rangle \mathbf{B}$; (iii) $(x < W)$ and $(x \in B)$. Intuitively, this relation captures the idea that there is indeed a genuine connection between a part x and the whole W (as opposed to a merely formal one). Now, one important thing to highlight is that if $(x <_R W)$ then there is no y such $(y <_R x)$. In other words, the closure set defined by relation R are the R-atoms of W. This is because, the whole W unified under R is maximal under this relation (by the definition of an R-closure system). The fact that no R-part of W can be unified under the same relation R of course does not imply that these R-parts need to be atomic in an absolute sense. In fact, given an element x such that $(x <_R W)$, x itself can be an integral wholes unified by a different relation R'. However, it should be clear by now that the sets of R'-atoms of x and the set of R-atoms of W (of which x is a member) are disjoint.

An example of a relation that takes place between an atom under relation R and an integral whole unified under that relation is the *member-collective* relation (symbolized as *M(part,whole)*). Following the above discussion, we have that these relations are never transitive, i.e., they are intransitive. Thus, if M(x,W) then x is atomic for W, and if we have M(y,x), we also have necessarily that ¬M(y,W). In other words, for the case of the member-collective relation, to say that a member must be a *singular* entity coincides with this entity being an *atom* in the sense just discussed, i.e., an atom w.r.t. to a characterizing relation unifying that specific whole.

The following example illustrates the intransitivity of the member-collection relation: *"I am member of a club C (collective) and my club is a member of an International Association of clubs C' (collective). However, it does not follow that I am a member of this International Association of Clubs C' since this only has clubs as members, not individuals"*. However, an even more general statement about the intransitivity of this relation can be made. Since members of a collective are considered to be atomic w.r.t. the context in which the collective is defined, if an individual x is a part of (member of) a collection y, then for every z which is part of (member of, functional part of, sub-collection of) x, then z is not a part of (member of) y. In other words, the member-collective relation causes the part to necessarily be seen as atomic in the context of the whole, hence, "blocking" a possible transitive chain of part-whole relations. Thus, for instance, although an individual John can be part of (member of) a Club, none of John's parts (e.g., his heart) is part of that Club.

Regarding the weak supplementation axiom, some authors claim that this axiom is too hard a constraint to be imposed to the member-collective relation [4]. From a formal point of view, this view implies that we accept *reflexive characterizing relations* for collectives as integral wholes. Such an approach seems at first to be somehow afforded by common sense. For instance, we can conceive a book of poems composed of a single poem, a CD composed of a single track, a purchase order composed of single order item, or a journal issue composed by a single article. Now, are there disadvantages to such an approach? We can foresee two of them.

Firstly, abandoning weak supplementation would set this relation apart from all the other parthood relations that we have considered, since this axiom (considered to be constitutive of the very meaning of part) is assumed by the relations of *component-functional complex* [16], *subquantity-quantity* [17], and *subcollective-collective* (section 4.2). Secondly, this choice opens the possibility for the creation of collectives with one single member. But what then would be the difference between John, {John}, {{John}}, {...{{John}}...}, etc? If entities such as these are generally adopted, then our system can face the objection of *ontological extravagance*, and we should be reminded that avoiding this feature was one of the motivations of mereology in the first place [19]. Given these two reasons, we adopt in this paper the view that weak supplementation should be part of the axiomatization of the member-collective relation. This obviously does not imply that we cannot have single-track CD's or single-article journal issues. Following [1], in these cases, we consider the relation between, for instance, the tracks and the CDs to be a relation of *constitution* as opposed to one of parthood. Relations of constitution abound in ontology. An example is the relation between a marble statue and the (single portion of marble) that constitutes it [1]. In fact, a more detailed analysis of WCH initial proposal showed that some of their original meronymic relations are in fact cases of constitution [1,11].

Finally, let us take the case of the secondary (modal) property of essentiality. As we have previously discussed, unlike sets and mereological sums, collectives do not necessarily have an extensional criterion of identity. That is, whereas for some collectives the addition or subtraction of a member renders a different individual, it is not the case that this holds for all of them. However, when this is the case, all member-collective relations that the extensional collective participates as a whole are relations of essential parthood. This is because, since a collective (by definition) has a uniform structure, then all members of a collective must be indistinguishable w.r.t. the whole. As a consequence, it cannot be the case that some members of a collection are essential while others are not. In summary, member-collective relations are only relations of essential parthood if the collective in the association end connected to the whole is an extensional individual. In the converse reading, if a collective is extensional then all its parts (members) are essential.

4.2 The Subcollective-Collective Relation

In contrast with the member-collective relation, from a linguistic point of view, the *subcollective-collective* is a relation that holds between two plural entities, or collectives constituted by such plural entities, such that all atoms of the first are also atoms of the second [22].

Let us start with an example. Figure 4 depicts an integral whole termed the DSRG (Distributed Systems Research Group) unified by the relation of *carrying out research in the area of distributed systems at University X (UtX)*. The R-atoms of DSRG are them {John, Mary, Peter, Mark}. The fact that no R-part of W can be unified under the same relation R also does not imply that these R-parts cannot be further structured to form new wholes. In other words, for example, we can take two different relations R'and R'' which are specializations of R, such that they can be used to form new closure systems among the R-parts of DSRG. Let R' be the relation of *carrying out research in the same sub-area of modeling of distributed systems at UtX*, and R' be the relation of *carrying out research in the same sub-area of performance of distributed systems at UtX*. This situation is depicted in figure 5.

Indeed, the fact that R' is a specialization of the condition R implies that the possible relata of R' are the R-atoms of R, i.e., R'⊆ R. When this is the case, we name the integral whole W' unified under R' a *subcollective* of the whole W unified under R. Let us name the relation between W' and W the relation of *subcollective-collective*, symbolized as C(W',W). We then have that C(W',W) iff: (i) all formal parts of W' are formal parts of W; (ii) the characterizing relation R' of W' is a specialization of the characterizing relation R of W. Now, suppose an integral whole W'' unified by relation R'' and C(W'',W'). By the above definition of the C-parthood relation, we

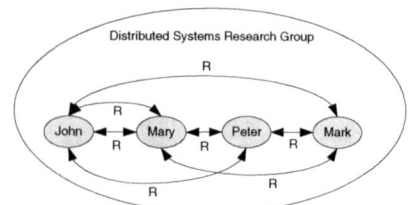

Fig. 4. Examples of an integral whole (collective) and its members

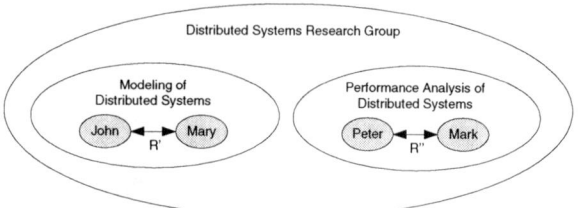

Fig. 5. Examples of a collective and its subcollectives

have that $R'' \subseteq R'$, and that all parts of W'' are also parts of W'. Due to the transitivity of both formal parthood (<) and the subset relation ($\subseteq$), we have that $R'' \subseteq R$, and also that all formal parts of W'' are also formal parts of W. Again, by definition, we conclude that $C(W'',W)$ holds. In other words, the *subcollective-collective* relation is always transitive.

A second property we would like to demonstrate is the following. Suppose we have that $M(y,x)$ and $C(x,W)$, and that R' and R are the characterizing relations of x and W, respectively. Since $M(y,x)$, we have both that $(y < x)$ and that y is an R'-atom of x. From $C(x,W)$, we have that all formal parts of x are formal parts of W, but also that $R' \subseteq R$. Again, due to the transitivity of both < and $\subseteq$, we have both that $(y < W)$ and that y is a R-atom of W. From this, we conclude that $M(y,W)$. In other words, transitivity always holds across a member-collective relation combined with a subcollective-collective relation.

Now, how does the subcollective-collective relation stand w.r.t. weak supplementation? Suppose that we have two collectives x and y such that $C(y,x)$. As we have previously discussed, x is closure system unified by relation R, and y must be a closure system unified by a specialization of this condition R'. Since the subcollective-collective relation is irreflexive, we have that R' is necessarily a proper subset of R, i.e., there are R-atoms of x which are not R'-atoms of y. This, at first, seems to imply that we can always have an integral whole z which is unified by another specialization R'' of R (the complement of R w.r.t. R'). However, the fact that there are members of x which are not part of y does not mean that these members can define a genuine integral whole. In other words, it can be the case that the only relation in common between these entities is that they obey the condition implied by R and the negation the condition implied by R'. As discussed in [1,5], characterizing entities based on negative properties is a poor ontological choice. For this reason, if we have that $C(y,x)$ we do not require that there is a z different from y such that $C(z,x)$, but we do require that there is a z such that $\neg overlap(z,y)$ and $M(z,x)$.

In the previous sections, we have discussed that collectives are not necessarily extensional and that, if a member of a collective is essential to the collective, then the collective is an extensional entity, i.e., all its members are essential. How do subcollectives stand w.r.t. to this secondary property? Suppose that we have a collective W composed of the subcollective W' and W'' such that the former is an essential part of W but not the latter. Now, as we have discussed, the structure of collectives is defined via specialization of the collective's unifying relation, i.e., the members of W' and W'' are also members of W. This implies that all subcollectives of W are *inseparable parts* of it, i.e., W' and W'' come to existence by refining the structure of W and by grouping the specific members of W. As a consequence, they cannot exist without that whole. A second observation we can make is that if there is an x which is a member of W', and x is an essential member of W then x must also be an essential member of W'. The argument can be made as follows. If x is an essential part of W then W cannot exist without x; If W' is an inseparable part of W then W' cannot exist without W; due to the transitivity of existential dependence [14], we have that W' cannot exist without x, ergo, x is an essential part of W'. Finally, since we are admitting that collectives are not necessarily extensional entities, it is conceivable that a whole W has an essential part W' composed of members which are not essential for either W or W'. For instance, suppose that by law, all juries must have at least two members which are older than sixty years old. Although this subcollective would be essential to the whole, it is conceivable that its individual members are exchangeable. By the same reasoning, one could admit a particular subcollective to be essential to a whole, without requiring the other collectives of that whole to be likewise essential.

4.3 Towards a UML Profile for Modeling Collectives and Their Parts

We summarize the results of these sections in a proposal that has been incorporated in a UML profile for representing the member-collective and the subcollective-collective relations (table 1). Since a profile is constituted by syntactical constraints and, since UML conceptual models are always defined at the type level, the meta-properties of irreflexivity, anti-symmetry and transitivity (at instance level) cannot be captured by profile constraints. We have included a constraint to guarantee weak supplementation for these relations taking into consideration the type-level nature of a UML class diagram, i.e., taking into consideration the minimum cardinality constraints of all parthood relations connected to the same type representing a whole.

Metaclass	Description
«collective» **A**	A «collective» represents a type whose instances are collectives, i.e., they are collections of entities that have a uniform structure. Examples include a deck of cards, a forest, a group of visitors, a pile of bricks.
subcollective-collective	This parthood relation holds between two collectives. Examples include: (a) the north part of the Black Forest is part of the Black Forest; (b) The collection of Jokers in a deck of cards is part of that deck; (c) the collection of forks in a cutlery set is part of that cutlery set. We use the symbols ◇——— and ◆——— to represent the shareable and non-shareable (see http://www.uml.org/) versions of this relation, respectively.

| member-collective | This is a parthood relation between a functional complex or a collective (as a part) and a collective (as a whole). Examples include: (a) a tree is part of forest; (b) a card is part of a deck of cards; (c) a club member is part of a club. We use the symbols ◆━━━━ and ◆━━━━ to represent the shareable and non-shareable versions of this relation, respectively. |

General Constraints

1. <u>Weak Supplementation:</u> Let T be a type whose instances are wholes and let $\{T_1...T_2\}$ be a set of types related to T via the subcollective-collective or member-collective relations. Let $lower_{Ci}$ be the value of the minimum cardinality constraint of the association end connected to C_i in the aggregation relation. Then, we have that

$$(\sum_{i=1}^{n} lower_{Ci}) \geq 2;$$

Constraints applied to the subcollective-collective relation

1. This relation only holds between collectives, i.e., they must be either stereotyped as «collective» or be a subtype of a type stereotyped as «collective»;

2. Collectives are maximal entities. For this reason, it is not the case that a collective can have as a part another collective of the same type (i.e., unified by the same relation). As a consequence, these relations are irreflexive at the type level. In UML terms, the two association ends of this relation must be connected to classes of different types (albeit both stereotyped as «collective»);

3. Also because collectives are maximal entities, a collective can have at maximum one subcollective of a given type. For this reason, the maximum cardinality constraint in the association end connected to the part in this relation must be one (in UML terms, self.target.upper = 1);

4. All subcollective-collective relations are relations of *inseparable parthood*. These relations are marked with a tagged value {insperable} and the association end connected to the whole must be immutable (in UML terms, self.source.readOnly = true);

5. This relation conforms to the axiomatization of Minimum Mereology (MM), i.e., it is an Irreflexive, Asymmetric and Transitivity relation which obeys the Weak Supplementation axiom. Moreover, if a collective W has one single direct subcollective W', then it must have members which are disjoint from W';

Constraints applied to the member-collective relation

1. This relation can only represent *essential parthood* if the object representing the whole on this relation is an extensional individual. In this case, all parthood relations in which this individual participates as a whole are essential parthood relations. These relations are marked with tagged value {essential} and the association end connected to the part must be immutable (in UML terms, self.target.readOnly = true);

2. The class connected to association end relative to the whole individual must be a type whose instances are collectives, i.e., they must be either stereotyped as «collective» or be a subtype of a type stereotyped as «collective»;

3. This is an Irreflexive and Asymmetric relation which obeys the Weak Supplementation axiom. However, it is also an Intransitive relation. Although transitivity does not hold across two member-collective relations, a member-collective relation followed by

> subcollective-collective relation is transitive. That is, for all a,b,c, if M(a,b) and M(b,c) then ¬M(a,c), but if M(a,b) and C(b,c) then M(a,c).
>
> 4. Asides from being intransitive, a member x of a collective W is atomic w.r.t. the collective. This means that for if an entity y is part of x then y is not a member of W.

For the sake of illustration, we revisit in figures 6.a and 6.b two of the examples discussed in this paper, explicitly representing them with the modeling primitives proposed in table 1. As one can observe, we decorate the standard UML symbol for aggregation with a C and an M to represent a subcollective-collective and member-collective relations, respectively.

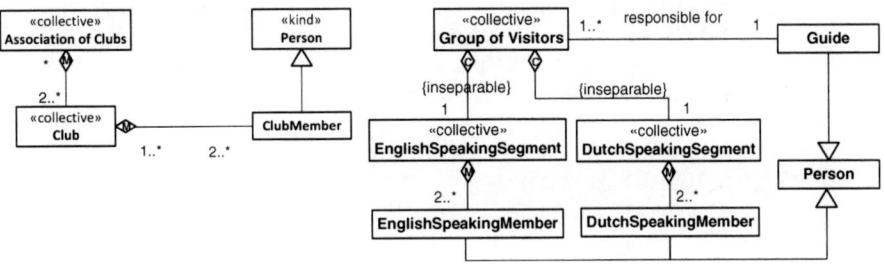

Fig. 6. Examples of subcollective-collective and member-collective part-whole relations

5 Final Considerations

The development of suitable foundational theories is an important step towards the definition of precise real-world semantics and sound methodological principles for conceptual modeling languages. This article concludes a sequence of papers that aim at addressing the three fundamental types of wholes prescribed by theories in linguistics and cognitive sciences, namely, functional complexes, quantities, and collectives. The first of these roughly correspond to our common sense notion of object and, hence, the standard interpretation of objects (or entities) in the conceptual modeling literature is one of functional complexes. The latter two categories, in contrast, have traditionally been neglected both in conceptual modeling as well as in the ontological analyzes of conceptual modeling grammars.

In this paper, we conduct one such ontological analysis to investigate the proper representation of types whose instances are collectives, as well as the representation of parthood relations involving them. As result, we were able to provide a sound ontological interpretation for these notions, as well as modeling guidelines for their proper representation in conceptual modeling. In addition, we have managed to provide a precise qualification for the relations of *member-collective* and *subcollective-collective* w.r.t. to both classical mereological properties (e.g., transitivity, weak supplementation, extensionality) as well as modal secondary properties that differentiate essential and inseparable parts. Finally, the results advanced here contribute to the definition of concrete engineering tools for the practice of conceptual modeling. In particular, the metamodel extensions and associated constraints outlined here have been implemented in a Model-Driven Editor using available UML metamodeling tools [23].

Acknowledgements. This is research is funded by FAPES (grant # 45444080/09) and CNPq (grant # 481906/2009-6) as well as a CNPq research productivity grant.

References

[1] Guizzardi, G.: Ontological Foundations for Structural Conceptual Models, Telematica Institute Fundamental Research Series, The Netherlands (2005) ISBN 90-75176-81-3

[2] Guizzardi, G.: Modal aspects of object types and part-whole relations and the shape de re/de dicto distinction. In: Krogstie, J., Opdahl, A.L., Sindre, G. (eds.) CAiSE 2007 and WES 2007. LNCS, vol. 4495, pp. 5–20. Springer, Heidelberg (2007)

[3] Rector, A., Rogers, J., Bittner, T.: Granularity, scale and collectivity: When size does and does not matter. Journal of Biomedical Informatics 39(3), 333–349 (2006)

[4] Bittner, R., Donelly, M., Smith, B.: Individuals, Universals, Collections: On the Foundational Relations of Ontology. In: 3rd Intl. Conf. on Formal Ontology in Inf. Systems (FOIS 2004), Torino, Italy (2004)

[5] Bunge, M., Ontology I.: The Furniture of the World. Springer, Heidelberg (1977)

[6] Opdahl, A., Henderson-Sellers, B., Barbier, F.: Ontological Analysis of whole-part relationships in OO-models. Information and Software Technology 43, 387–399 (2001)

[7] Wand, Y., Storey, V.C., Weber, R.: An ontological analysis of the relationship construct in conceptual modeling. ACM Transactions on Database Systems 24(4), 494–528 (1999)

[8] Winston, M.E., Chaffin, R., Herrman, D.: A Taxonomy of Part-Whole relations. Cognitive Science (1987)

[9] Odell, J.J.: Six Different Kinds of Composition. Journal of Object-Oriented Programming 5/8 (1994)

[10] Pribbenow, S.: Meronymic Relationships: From Classical Mereology to Complex Part-Whole Relations, The Semantics of Relationships. Kluwer Academic Publishers, Dordrecht (2002)

[11] Gerstl, P., Pribbenow, S.: Midwinters, End Games, and Bodyparts. A Classification of Part-Whole Relations. Intl. Journal of Human-Computer Studies 43, 865–889 (1995)

[12] Bittner, T., Donnelly, M.: Logical properties of foundational relations in bio-ontologies. Artificial Intelligence in Medicine 39, 197–216 (2007)

[13] Keet, C.M., Artale, A.: Representing and Reasoning over a Taxonomy of Part-Whole Relations. Applied Ontology 3(1-2), 91–110 (2008)

[14] Simons, P.M.: Parts. An Essay in Ontology. Clarendon Press, Oxford (1987)

[15] Gangemi, A., Guarino, N., Masolo, C., Oltramari, A.: Understanding top-level ontological distinction. In: Proceedings of IJCAI 2001, Workshop on Ontologies and Information Sharing (2001)

[16] Guizzardi, G.: The problem of transitivity of part-whole relations in conceptual modeling revisited. In: van Eck, P., Gordijn, J., Wieringa, R. (eds.) CAiSE 2009. LNCS, vol. 5565, pp. 94–109. Springer, Heidelberg (2009)

[17] Guizzardi, G.: On the Representation of Quantities and their Parts in Conceptual Modeling. In: 6th International Conf. on Formal Ontology and Information Systems (FOIS 2010), Toronto, Canada (2010)

[18] Guizzardi, G.: Representation of Collectives and their Members in UML Conceptual Models: An Ontological Analysis. In: Proc. of the 6th FP-UML International Workshop, Vancouver, Canada (2010)

[19] Varzi, A.C.: Parts, wholes, and part-whole relations: The prospects of mereotopology. Journal of Data and Knowledge Engineering 20, 259–286 (1996)

[20] Cadoli, M., Calvanese, D., De Giacomo, G., Mancini, T.: Finite satisfiability of UML class diagrams by constraint programming. In: Workshop on CSP Techniques with Immediate Application (2004)

[21] Botazzi, E., Catenacci, C., Gangemi, A., Lehmann, J.: From Collective Intentionality to Intentional Collectives: An Ontological Perspective, Cognitive Systems Research, Special Issue on Cognition and Collective Intentionality (2006)

[22] Vieu, L., Aurnague, M.: Part-of Relations, Functionality and Dependence, Categorization of Spatial Entities in Language and Cognition. John Benjamins, Amsterdam (2007)

[23] Benevides, A.B., Guizzardi, G.: A model-based tool for conceptual modeling and domain ontology engineering in ontoUML. In: Filipe, J., Cordeiro, J. (eds.) Enterprise Information Systems. Lecture Notes in Business Information Processing, vol. 24, pp. 528–538. Springer, Heidelberg (2009)

Product-Based Workflow Design for Monitoring of Collaborative Business Processes

Marco Comuzzi and Irene T.P. Vanderfeesten

School of Industrial Engineering, Eindhoven University of Technology
P.O. Box 513, 5600MB Eindhoven, The Netherlands
{m.comuzzi,i.t.p.vanderfeesten}@tue.nl

Abstract. Monitoring of cross-organizational processes requires the definition and implementation of monitoring processes that can deliver the right information to the right party in the collaboration. Monitoring processes should account for the temporal and aggregation dependencies among the monitoring information made available by the set of collaborating parties. We solve the problem of designing monitoring processes in collaborative settings using Product-Based Workflow Design (PBWD). We first discuss a methodology to apply PBWD in this context and then propose an architecture to implement the methodology using a service-oriented approach.

Keywords: Monitoring, business process design, cross-organizational processes.

1 Introduction

Continuous monitoring of a business process can be defined as the set of methodology and tools to collect and disseminate relevant information about the process execution to interested stakeholders simultaneously with, or within a reasonably short period after, the occurrence of relevant events in the process [6]. Continuous monitoring has straightforward benefits, such as the opportunity for process providers to detect anomalies in (almost) real time and apply control actions on-the-fly.

Research on (continuous) monitoring in cross-organizational processes has usually taken an *information-centric* perspective, focusing on the definition of monitoring requirements for the collaborating parties and their evolution [6,11], the design of architectures and tools to capture monitoring information [17], the detection of contract violations, given the available monitoring information [3], or the verification of the compliance of execution logs to a process specification [18].

Although the information-centric view can suffice for intra-organizational process monitoring, where all monitoring information is produced in a given business domain, in cross-organizational settings researchers stress the importance of process- and communication-oriented mechanisms to transmit relevant information to interested parties across the collaborative network [7,11]. In other words, once the monitoring information is captured and made available by the

H. Mouratidis and C. Rolland (Eds.): CAiSE 2011, LNCS 6741, pp. 154–168, 2011.

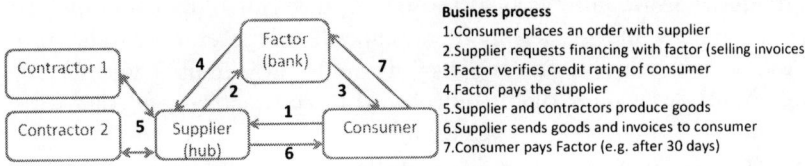

Fig. 1. Factoring example

collaborating parties, a process must be built to allow a specific party to retrieve (or be delivered) the monitoring information in the right way. Such monitoring process should account for the temporal and aggregation dependencies among monitoring information.

Let us consider the running example of a business network for factoring in the manufacturing industry (see Fig. 1), constituted by a supplier, a set of contractors (two in our case), a consumer, and a factor. The supplier, in particular, acts as the coordinating hub of the set of contractors, which execute the process required to deliver the product ordered by the consumer. Factoring is a financial transaction whereby a business (supplier) sells its account receivables (invoices) to a third party financial institution (factor) in exchange for immediate payment. Factoring allows consumers to obtain financing at an interest rate, i.e. the one provided by the factor, lower than the one they could obtain directly from the supplier [10]. In this context, continuous monitoring may help reducing the risks associated to the collaboration, e.g. the risk that the supplier will not deliver the goods as promised or the risk that the consumer will not be able to pay. Therefore, continuous monitoring may help to further increase the benefits that the involved parties achieve through the collaboration.

In this scenario, the supplier wants to monitor when the consumer has made the payment to the factor and when the factor has received such payment. If the consumer does not pay, in fact, the supplier should not agree to further transactions in the future involving the same consumer. Note that (i) payment confirmations are not conveyed to the supplier in the process depicted in Fig. 1 and (ii) a temporal dependency exists between monitoring information, i.e. if the supplier checks the factor acknowledgment of the payment without knowing if the consumer has actually sent out the payment, he or she may have an inconsistent view on the process (and may take wrong corrective actions accordingly).

The factor, similarly, wants to monitor the progress and quality of the process on the supplier side to reduce its own risk. The consumer, in fact, may not be satisfied with the goods, e.g. because of late delivery or poor quality, and, as a consequence, may not be willing to pay the factor according to established terms. This information may be delivered either by the supplier, or being reconstructed through more detailed progress information made available by the contractors and aggregated correctly. Again, note that progress information is not conveyed to the factor in the business collaboration depicted in Fig. 1.

The latter example also shows that there could be different alternatives for a party to obtain the required monitoring information and each alternative may be characterized in non-functional terms, e.g. in terms of cost and quality. For instance, progress information made available by contractors may be of higher quality and cost, whereas the supplier may only have limited visibility on the progress of an order once this is outsourced to contractors, and therefore may provide such information at a cheaper price.

In this paper, we propose a methodology to design monitoring processes in collaborative business settings. The methodology considers as input the monitoring information made available by the collaborating parties and builds monitoring processes embedding temporal and aggregation dependencies among monitoring information. Moreover, monitoring information in our methodology can be described also in non-functional terms, e.g. by cost, quality, and availability. Among the set of possible alternatives, the proposed methodology allows the selection of the monitoring process satisfying also the party non-functional requirements, e.g. the minimum cost monitoring process or the highest quality process, given a budget constraint.

The design of the monitoring process for cross-organizational business processes is framed as a PBWD (Product-Based Workflow Design, [16,20,22]) problem. PBWD is an analytical method for automatically deriving business process specifications from the set of information products involved in the process and their dependencies. In the monitoring of collaborative processes, information products are represented by the monitoring information made available by the actors involved in the collaboration.

The paper is organised as follows. Related work is discussed in Section 2, while Section 3 introduces the background on PBWD and explains its novel application to the problem of cross-organizational process monitoring. The architecture to integrate the PBWD design of monitoring processes in a service-oriented environment is presented in Section 4, while conclusions are eventually drawn in Section 5.

2   Related Work

Monitoring of cross-organizational business processes has been investigated, from a requirements engineering perspective, in [7] and [11]. In order to achieve a successful collaboration, both papers stress the importance of process- and communication-oriented mechanisms to transmit relevant information to interested parties across the network. Similarly, [5] considers the need to define external information requirements, i.e. information required by a consumer from its providers to correctly monitor and enforce a multiparty contract.

From the design and implementation perspectives, the CrossFlow [8] and CrossWork [9] projects consider architectural support for cross-organizational business processes. Both projects investigate issues such as the design of flexible architecture to support monitoring [9] and the definition of specific monitoring points in electronic contracts [8]. Monitoring, however, is still considered information-centric, i.e. the dependency among monitoring information products produced by various sources or the aggregation of monitoring information from several parties are not considered.

Research on Web service-based business processes has also extensively investigated the issue of monitoring. Web service-based processes are intrinsically cross-organizational, since each orchestrated Web service can in principle be exposed by a different organization. In this context, we can distinguish between *intrusive* and *non-intrusive* monitoring [2,13]. The former involves the interleaving of service and monitoring activities at runtime, whereas the latter separates the business from the monitoring logic, since information relevant for monitoring can be captured non-intrusively while a process is executing, e.g. intercepting service operation calls and responses or from the log of the process engine [13]. Cross-organizational settings require non-intrusive monitoring, since it would not be feasible to implement a different instrumentation to satisfy the monitoring requirements of each different party with which an organization has to interact. Furthermore, while the aforementioned approaches only consider the monitoring of performance variables, such as response time and availability, or simple conditions describing the behavior of a service, in this paper we take a business perspective, since PBWD considers business-related information that is deemed relevant by the user of a process.

The innovation of the approach presented in this paper concerns also the aggregation, according to user-specific dependencies, of monitoring information to design a customized monitoring process. Web service-based process monitoring considers also aggregation of monitoring information from multiple sources [2,13]. However, monitoring information, such as the timestamps of service calls or process variables, are meaningful only at a technical level, and they require further translation before becoming meaningful to and, therefore, relevant for a process user [12]. PBWD considers only informational products meaningful at a business level and, therefore, enables us to design monitoring processes using informational products that do not need translation.

3   Using PBWD for Collaborative Process Monitoring

This section introduces some background on the PBWD approach and shows how PBWD can be used to generate monitoring processes. PBWD is a scientifically grounded method for business process (re)design. The focus of this method is on the design of processes that deliver informational products, the so-called *workflow processes*. The PBWD methodology takes the structure of the *informational product*, which is described in a Product Data Model (PDM), as a starting point to derive a process model. Informational products are, for instance, a decision on an insurance claim, the allocation of a subsidy, or the approval of a loan. Based on the input data provided by the client or retrieved from other systems, the end (informational) product is constructed step-by-step. In each step new information is produced based on the specific data available for the case.

Over recent years, PBWD has shown to be a successful business process (re)design method [15]. For instance, the annual reporting process for mutual funds at a large Dutch bank was successfully redesigned using PBWD. The insights in the informational product, achieved by PBWD, led to a 50% decrease in throughput time [20].

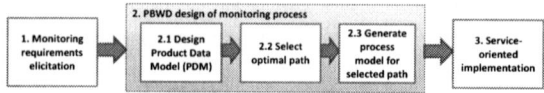

Fig. 2. Methodology for PBWD of monitoring processes

PBWD is particularly well-suited for achieving a process-centric view on monitoring in cross-organizational processes because of two reasons:

– *Clean-sheet approach to process design*: PBWD builds process models directly from the specification of the informational products involved in the process and their dependencies. This approach is a perfect fit for monitoring cross-organizational business processes, since the monitoring process must be built from the monitoring information made available by the collaborating parties. Note that, in highly dynamic collaborations, partner selection is *late-bound*, and collaborating parties are selected dynamically as they become available [9];
– *Cost- and quality-aware process design*: PDMs, i.e. available informational products and their dependencies, can be enriched with information about the cost of producing an information product or its quality for the interested stakeholders. PBWD can then derive various process specifications for the same PDM that differ for their overall costs and quality. The possibility to tune the costs and the quality of the process is an essential feature to derive the most suitable monitoring process for a given stakeholder. When a process is not mission critical, for instance, the monitoring process can be designed by maximizing its quality given a budget constraint, whereas in contexts characterized by severe quality requirements, for instance in highly regulated industries, such as healthcare, monitoring processes can be designed by minimizing the monitoring costs while guaranteeing a given required level of quality.

Fig. 2 shows the steps of the methodology for creating monitoring processes using PBWD for a specific stakeholder in the collaboration. After having analyzed the monitoring requirements of the stakeholder, the application of PBWD to monitoring processes design involves three steps.

First, the Product Data Model (PDM) of the stakeholder is designed. The PDM, which is the starting point for the PBWD method, is similar to the concept of a Bill-of-Materials (BoM) [14] used in manufacturing environments to manage and control production processes. Since (digital) information is more flexible than physical products, however, the PDM contains more complex structures than a BoM, such as re-use of information or alternative paths to produce an information element.

Second, among the set of all paths in the PDM that can lead to the correct production of monitoring information, the path which satisfies the non-functional requirements of the stakeholder is selected. As discussed later, we consider the cost, (data) quality, and availability dimensions to specify non-functional requirements.

In the third step, a process model is generated for the chosen path.

Eventually, the last step in the methodology concerns the implementation of the monitoring process. In this regard, we discuss an architecture for the implementation of the methodology in a service-oriented environment. This discussion is made in Section 4.

From now on, the paper will focus on the application of PBWD to design monitoring processes to satisfy the monitoring requirements of the consumer (stakeholder) in the running example depicted in Fig. 1. In order to decrease its own risk, the consumer is interested in monitoring (i) the status of the payment and (ii) the progress of requests on the supplier side. Monitoring information on the payment can be reconstructed as a combination of information on when the factor sent out the payment and information on when the supplier received the payment. If for instance, the factor has sent out the payment, but this has not been acknowledged by the supplier in a reasonably short period, then the payment may not have been successful.

Monitoring information on the progress of an order can be provided either by the supplier or directly by the contractors. The supplier has only a low-quality view on the order progress, e.g. the supplier may only report that the order has been sent to contractor 1, but he cannot access the details of the internal enactment of contractor 1's process. More detailed monitoring information can be provided directly by contractors.

3.1   The Product Data Model

Fig. 3 summarizes in a PDM the information products available to satisfy the monitoring information requirements of the consumer. A PDM is constituted by information products and operations. Information products in the PDM are depicted by circles, while the operations performed on the input elements to produce the output are represented by hyperarcs. Each operation has zero or more input elements and has exactly one output element, i.e. the information product obtained through the combination of its input elements. A PDM may contain alternative paths to produce a certain information element. Hence, we define a *path* as any sub-graph of the PDM. A *complete* path is a sub-graph of the PDM containing the root element.

In our example, the correct monitoring information for the consumer *(MON)* is obtained combining information on the delivery of the payment *(PAY)* and information on the progress of the request *(PRO)*. The monitoring information *PAY* can be obtained either as information on when the factor has sent out the payment *(PF)*, information on when the supplier has received the payment *(PS)*, *or* a combination thereof. The progress report *(PRO)* can be obtained either from the supplier *(SPx)* or from the contractors *(Eyx)*. Note that x represents the quality of the provided monitoring information, i.e. High or Low $(x \in \{H, L\})$, whereas y identifies the contractor $(y \in \{1, 2\})$. The supplier and contractors 1 and 2 can provide two different types of monitoring information, that is, more or less accurate (see *SPH* and *SPL* or *E1H* and *E1L*).

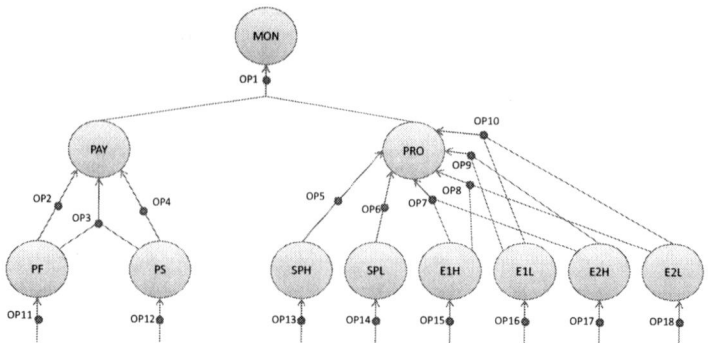

Fig. 3. Factoring monitoring product data model

Table 1. Constraints on example monitoring product data model

Operation	C	Q	A	Operation	C	Q	A
Op01	0	1	1.0	Op10	0.5	0.5	1.0
Op02	0.6	0.3	1.0	Op11	0.2	1	0.9
Op03	0.8	0.8	1.0	Op12	0.7	1	0.7
Op04	0.1	0.7	1.0	Op13	1	1	0.7
Op05	0.1	0.4	1.0	Op14	0.8	1	0.9
Op06	0.1	0.2	1.0	Op15	0.5	1	0.63
Op07	0.8	1	1.0	Op16	0.3	1	0.7
Op08	0.6	0.7	1.0	Op17	0.3	1	0.7
Op09	0.6	0.7	1.0	Op18	0.1	1	0.8

Operations in the PDM signify the production of an information product. In particular, *leaf* operations (*Op11* to *Op18*) signify the production of monitoring information made by the supplier, contractors, and the factor, while the other operations signify the combination of data elements to produce the parent. In this respect, for instance, *Op07* signifies the combination, made by the consumer, of high quality monitoring information from contractors (*E1H* and *E2H*) to produce the *PRO* data element, whereas *Op03* signifies the combination of payment information from the factor (*PF*) and the supplier (*PS*) to produce the payment monitoring information *PAY*.

We express a path by listing its set of operations. For instance, the path {*Op11, Op02, Op13, Op05, Op01*} is complete, since it leads to the production of the root element *MON*.

Apart from the functional structure, a PDM also contains additional information concerning the non-functional aspects of operations. Three dimensions characterize an operation from the non-functional point of view:

- **Cost (C).** Represents the cost of executing the operation. In other words, for leaf operations, it is the cost sustained by an actor to provide the correspondent monitoring information product, whereas for other operations it is the cost for the monitoring stakeholder (consumer in our case) to combine the leaf information products to obtain the parent product. Costs are summative, i.e. the cost of an information product is the sum of the costs of operations executed to obtain such an information product [1];

- **Quality (Q).** Represents the quality of the information product perceived by the *stakeholder* of the monitoring product model. Quality of monitoring information should be intended as *fit for use*, i.e. the ability of a piece information to satisfy the monitoring information requirements of its user [23]. Fit for use quality of an information product is aggregated using the minimum value of data quality over all operations required to create the information product [1];

- **Availability (A).** Represents the probability that an operation produces its output element(i.e., the output element becomes available after the execution of the operation). Availability is aggregated multiplying the availabilities of all considered operations [1].

Table 1 shows the values assigned to the non-functional dimensions for the PDM of Fig. 3. Note, for instance, that the high quality progress status information, i.e. obtained through operation *Op13* or *Op15*, is more costly than the corresponding low quality information, i.e. the one obtained through *Op14* and *Op16*, respectively. This because a supplier needs to capture more information from the infrastructure executing the process to provide high quality monitoring information. Note also that the cost for the consumer of combining monitoring information produced by other parties is very low. We assume, in fact, that monitoring information is made available digitally and, therefore, its aggregation is almost costless. Concerning availability, note that aggregation operations executed by the consumer, e.g. *Op07* or *Op03*, have always highest availability, whereas the availability of the leaf operations may not be optimal, since it depends on the availability of the infrastructure in which monitoring information is captured by the supplier(s) or the factor.

Apart from the cost, quality, and availability, other properties can be also considered. Ardagna and Pernici [1], for instance, consider execution time and reputation in addition, while Vanderfeesten et al. [21] also consider duration of execution.

3.2   Optimal Path in the PDM

After the design of the PDM, the next step in our methodology (see Fig. 2) is the selection of the optimal path. As explained in the previous section, the PDM may accommodate several alternative paths to produce the end product. The objective of this step is therefore to select a *complete* path that satisfies the requirements of the considered stakeholder, in terms of cost, quality, or availability of the monitoring information, or a combination thereof.

In order to select the optimal path, different constraints may be considered. In this paper we use the following scenarios to illustrate our approach: (i) optimal path considering the availability dimension only (A-path), (ii) optimal path minimizing costs given a minimum quality level (C_q-path), and (iii) optimal

Table 2. Values for the different dimensions (Cost, data quality, availability) for each path in the example product data model

Path	Operations	Total Cost	Total Quality	Total Avail- ability
1	Op01 Op02 Op11 Op06 Op14	1,7	0,2	0,81
2	Op01 Op02 Op11 Op05 Op13	1,9	0,3	0,72
3	Op01 Op02 Op11 Op08 Op15 Op18	2,0	0,3	0,43
4	Op01 Op02 Op11 Op10 Op16 Op18	1,7	0,3	0,50
5	Op01 Op02 Op11 Op07 Op15 Op17	2,4	0,3	0,38
6	Op01 Op02 Op11 Op09 Op16 Op17	2,0	0,3	0,44
7	Op01 Op03 Op11 Op12 Op06 Op14	2,6	0,2	0,57
8	Op01 Op03 Op11 Op12 Op05 Op13	2,8	0,4	0,50
9	Op01 Op03 Op11 Op12 Op08 Op15 Op18	2,9	0,7	0,30
10	Op01 Op03 Op11 Op12 Op10 Op16 Op18	2,6	0,5	0,35
11	Op01 Op03 Op11 Op12 Op07 Op15 Op17	3,3	0,8	0,26
12	Op01 Op03 Op11 Op12 Op09 Op16 Op17	2,9	0,6	0,31
13	Op01 Op04 Op12 Op06 Op14	1,7	0,2	0,63
14	Op01 Op04 Op12 Op05 Op13	1,9	0,4	0,56
15	Op01 Op04 Op12 Op08 Op15 Op18	2,0	0,7	0,17
16	Op01 Op04 Op12 Op10 Op16 Op18	1,7	0,5	0,39
17	Op01 Op04 Op12 Op07 Op15 Op17	2,4	0,7	0,15
18	Op01 Op04 Op12 Op09 Op16 Op17	2,0	0,6	0,34

path maximizing quality given maximum costs and minimum availability level ($Q_{c,a}$-path). We chose these type of constraints to exemplify our methodology. In the general case, however, the selection of the optimal path should be seen as the optimization of a utility function. Stakeholders can define the utility function, for instance, as the weighted sum of partial utility functions on individual dimensions [1].

From the PDM of Fig. 3 and the operation properties of Table 1, we derive 18 *complete* paths for the monitoring process that may all serve the need for monitoring the consumer order fulfillment process. These paths are reported in Table 2. Among the available complete paths, we now discuss the above mentioned optimal ones:

A-path. The path with the highest availability, i.e. the highest probability of delivering the required end product in the form of monitoring information (MON), is path 1. It produces the end product (MON) by executing operations $Op01$, $Op02$, $Op11$, $Op06$, $Op14$. The total availability of this plan is 0.81.

C_q-path. A cost optimal path given a minimum quality level is the monitoring path with the lowest cost that satisfies a quality requirement set by the consumer. Suppose the consumer sets the threshold for the quality level to 0.5. Then, paths 9, 10, 11, 12, 15, 16, 17, and 18 are to be considered. The path with the lowest cost is selected from this subset. The cost optimal path given a minimum data quality level therefore is path 16, with quality 0.5 and cost 1.7.

$Q_{c,a}$-path. The third scenario concerns the determination of the quality optimal path given maximum costs and a minimum level of availability. If the consumer has a budget constraint of at most 2.0 and wants to be for at least 50% sure that the monitoring information is delivered, then the highest quality possible

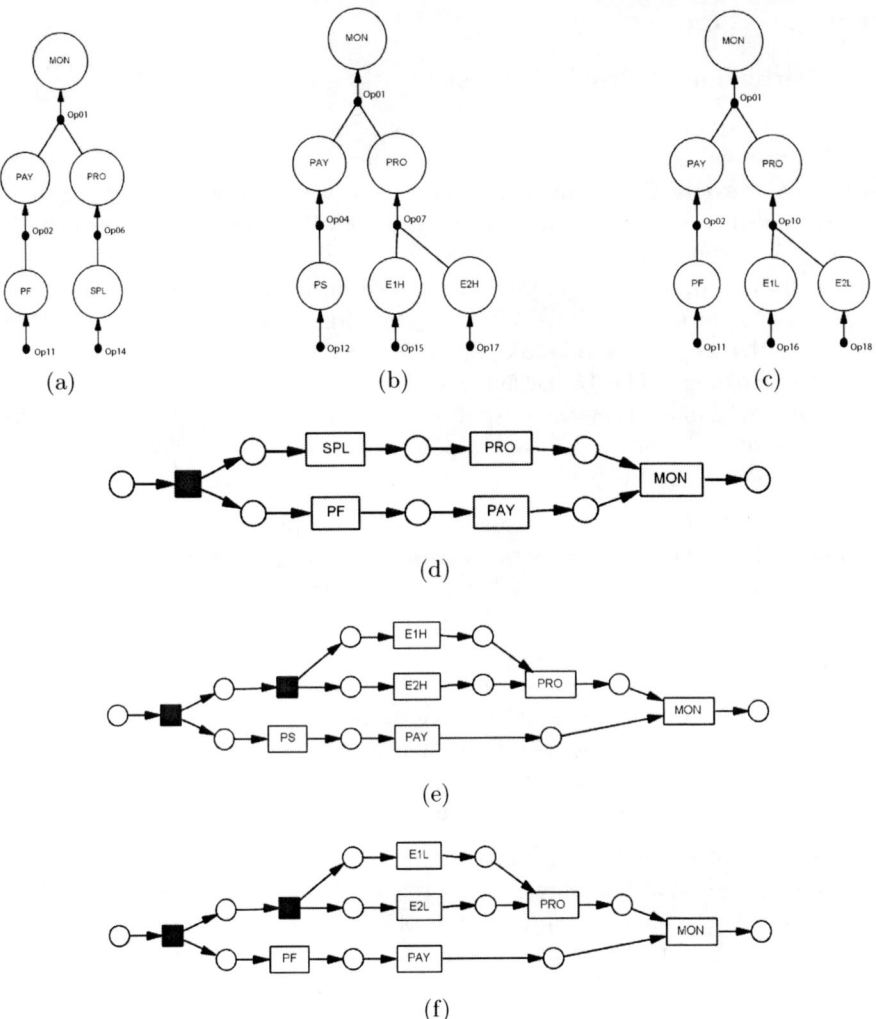

Fig. 4. (a) PDM of the A-path (path 1); (b) PDM of the C_q-path (path 16); (c) PDM of the selected $Q_{c,a}$-path (path 4); (d) Process model for path 1; (e) Process model for path 16; (f) Process model for path 4

is achieved by paths 2 and 4. Since there are two optimal paths and we want to have just one prescriptive process model, it has to be decided which one of the paths is best. In general, if more than one path satisfy the given constraints, then several approaches can be chosen to select an optimal path. These include: (i) random selection of one path among the identified optimal ones, (ii) selection (by the monitoring stakeholder) of one of the optimal paths by ranking the criteria, or (iii) selection of the path involving the lowest number of operations. As an illustration, we choose the second case and we define the cost as being the second most important criterion after the quality. From paths 2 and 4 we now select the one with the lowest cost (path 4).

3.3 Derivation of Process Models

After the determination of the optimal path in the PDM with respect to a certain criterion, a process model can be generated for the monitoring process (see Fig. 4). Several algorithms have been proposed to transform the PDM into a process model executable, for instance, by a workflow management system [22]. We use the algorithm described in [19] to automatically generate a process model for the optimal path of the PDM. This algorithm is implemented in the ProM framework for process analysis [20]. The resulting process models of our three optimal paths are discussed below. These process models are represented in the Petri Net language. The transitions (squares) represent the operations in the PDM and are named after the output element of the operation, e.g. transition MON indicates the operation that produces data element MON based on the input elements PAY and PRO.

A-**path.** Fig. 4(d) shows the process model for the optimal path with respect to the availability. There are two parallel branches in the process model that can be executed concurrently: (i) a branch in which first the element PF is determined followed by the element PAY, and (ii) a branch in which SPH is determined followed by PRO. Once both elements PAY and PRO are determined, the final element MON can be produced. Note that the black activity at the left hand side of the process model is a 'silent' activity, i.e. it is added only for routing purposes but does not process information. Using this process, the consumer retrieves monitoring information on the progress of its request only from the supplier and information on the payment only from the factor.

C_q-**path.** Fig. 4(e) depicts the process model for the optimal path C_q. Again there are two parallel branches in the process model, both providing input to produce MON. On the one hand, PAY is obtained using PS first. On the other hand, PRO is determined by $E1H$ and $E2H$, which can be determined in parallel as well.

$Q_{c,a}$-**path.** Fig. 4(f) shows the process model for the optimal path $Q_{c,a}$. It is similar in structure to the process model of Fig. 4(e), but it uses elements PF, $E1L$, and $E2L$ in place of PS, $E1H$, and $E2H$, respectively.

4   Implementing the Methodology

In this section we discuss the last step of the methodology proposed in this paper (see Fig. 2), i.e. its implementation in a service-oriented environment.

We take a service-oriented approach to the definition of the product data and the implementation of monitoring processes (see the architecture in Fig. 5(a)). The combination of PBWD and service orientation for design and execution of monitoring processes, respectively, can address the need for structural and operational dynamicity in business networks [9]. Concerning network structure, the

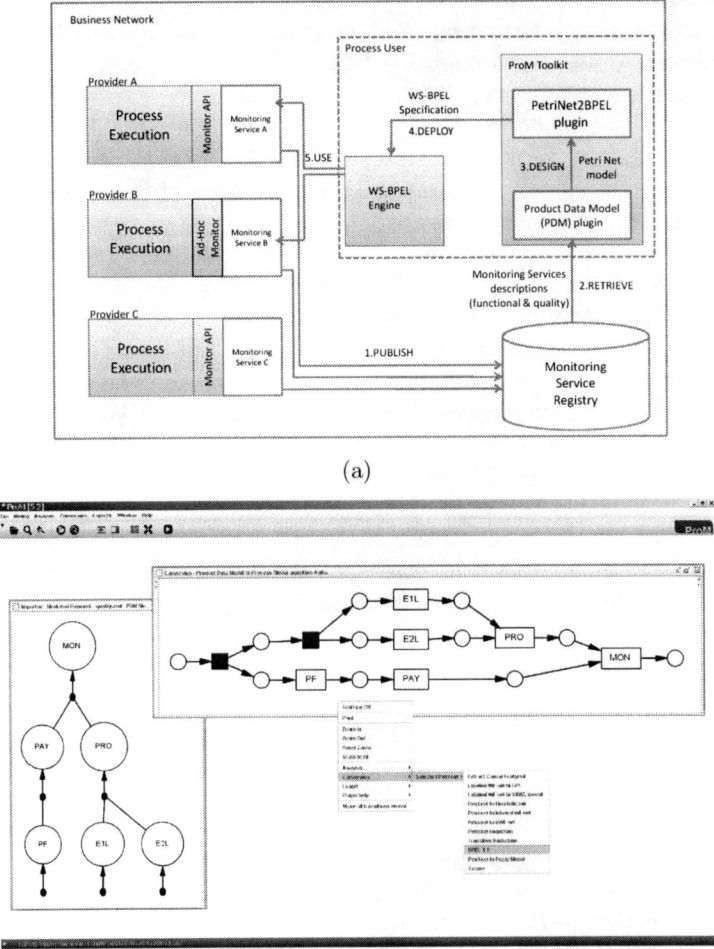

(a)

(b)

Fig. 5. (a) PBWD-based monitoring process creation architecture; (b) The ProM toolkit, showing a PDM, the relative process model, and the WS-BPEL export functionality

actors in the network can be dynamically replaced. Suppliers of spare parts in an automotive industrial district, for instance, can be dynamically substituted by the car manufacturer as new suppliers providing more convenient or higher quality options become available. Concerning network operations, the network business processes can be reconfigured on-the-fly as new business opportunities arise. Insurance companies, for instance, may decide to outsource part of their claim management process only on a temporary basis, e.g. to manage an abnormal amount of claims resulting from a possible fraud. Both scenarios require the dynamic set-up or update of monitoring processes as the network dynamically evolves. This is addressed by our methodology at design time, by adopting

PBWD, and at runtime, by using a service oriented approach, where services providing monitoring information can be dynamically orchestrated in the monitoring processes obtained through PBWD.

The elementary monitoring information products are the monitoring information that can be made available by actors in the business network to other actors for monitoring purposes. e.g. the progress information made available by the supplier or the factor in our running example. In the information system or process engine executing a process to be monitored, monitoring information can be captured from various sources and through different mechanisms, such as (i) native APIs of the actor's ERP system, e.g. SAP monitoring architecture, workflow or BPEL engine and (ii) ad-hoc instrumentation, e.g. through the development of event captors or other online process inspection techniques [2].

Irrespectively of how monitoring information is captured, the process provider can make such information available to users by exposing a Web service [6] implementing, for instance, a different operation for each leaf element in the monitoring product data model. The cost, quality, and availability of the monitoring information (see Table 1) can be then specified in a policy document [4], that can be attached to such service before its publication.

In a service-oriented architecture, process providers publish their monitoring services in a service registry (STEP 1 in Fig. 5(a)) and process users browse the registry to get service descriptions and build their monitoring PDM (STEP 2). Note that, in Fig. 5(a), process *provider* and *user* should be intended as roles, since an actor in the business network can act at the same time as a provider of processes and a user of processes contributed by other actors.

The architecture depicted in Fig. 5(a) supports also the creation of an executable monitoring process. Specifically, a process user in the business network interested in building a monitoring process retrieves the required monitoring service description from the registry. Service descriptions are then used to build a monitoring product data model using the ProM toolkit (STEP 3). Currently, as discussed in the previous sections, the algorithms for obtaining monitoring process models (STEP 3), expressed as Petri nets and satisfying given quality constraints, have been implemented as plugins of the ProM toolkit. ProM also provides a plugin for the translation of models from Petri nets to (abstract) WS-BPEL specifications (see Fig. 5(b)). The abstract WS-BPEL specification is bound by the monitoring stakeholder to the required monitoring services in the registry, in order to make it executable, and deployed in a process engine (STEP 4). When in execution, a monitoring process will use the monitoring services originally published by actors in the business network, according to the aggregation and dependency constraints specified in the monitoring PDM and the derived monitoring process model (STEP 5).

In respect of the scenario depicted in Fig. 5(a), most of the steps of our methodology, such as the definition and retrieval of monitoring services from the registry, the binding of the WS-BPEL specification obtained through ProM to actual services in the registry, and the deployment of the WS-BPEL specification in the process engine, are still executed manually. Future work will concern the implementation of an integrated approach in which the aforementioned activities could be fully automated.

5   Conclusions

This paper presents an innovative application of PBWD, that is, the product-based design of monitoring processes in collaborative business networks. The innovation brought about by this paper is twofold. On the one hand, given a collaborative scenario and available monitoring information, we propose a methodology to design from scratch a monitoring process that matches the process user's requirements, in terms of cost, (data) quality, and availability of monitoring information. On the other hand, we discussed an architecture for implementing the proposed methodology.

A first area of improvement for this work concerns the implementation of the architecture reported in Fig. 5(a) and, specifically, the connections between the implemented ProM's plugins and the service-oriented process execution environment. Future work will also concern the extension and refinement of the product-based generation of monitoring processes. In particular, we plan to consider additional non-functional dimensions, such as reputation, and more complex utility functions for capturing the monitoring stakeholder requirements. Constraints describing monitoring products can also become dependent on the type of monitoring information and the type of stakeholder requiring it. Hence, we want to investigate the issue of provider and user profiling for automatically designing more customized monitoring processes. Also, while this paper considers the monitoring requirements for a stakeholder at the process level, we are planning to consider also instance-level monitoring requirements, i.e., monitoring requirements that can change with every different instance involving a given stakeholder. Finally, from the modeling perspective, we want to investigate the opportunity of specifying monitoring processes as choreographies, e.g. in BPMN 2.0, for capturing more complex dependencies among monitoring information.

References

1. Ardagna, D., Pernici, B.: Adaptive Service Compostion in Flexible Processes. IEEE Transactions on Software Engineering 33(6), 369–384 (2007)
2. Baresi, L., Guinea, S., Nano, O., Spanoudakis, G.: Comprehensive monitoring of BPEL processes. IEEE Internet Computing 14(3), 50–57 (2010)
3. Baresi, L., Guinea, S., Pistore, M., Trainotti, M.: Dynamo + Astro: An integrated approach for BPEL monitoring. In: Proc. ICWS (2009)
4. Cappiello, C., Comuzzi, M., Plebani, P.: On automated generation of web service level agreements. In: Krogstie, J., Opdahl, A.L., Sindre, G. (eds.) CAiSE 2007 and WES 2007. CAiSE, vol. 4495, pp. 264–278. Springer, Heidelberg (2007)
5. Chiu, D.K.W., Karlapalem, K., Li, Q., Kafeza, E.: Workflow view based E-contracts in a cross-organizational E-services environment. Distrib. Parallel. Dat. 12, 193–216 (2002)

6. Comuzzi, M., Vonk, J., Grefen, P.: Continuous monitoring in evolving business networks. In: Proc. CoopIS, pp. 168–185 (2010)
7. Daneva, M., Wieringa, R.: A requirements engineering framework for cross-organizational erp systems. Requirements Engineering 11, 194–204 (2006)
8. Grefen, P., Aberer, K., Hoffner, Y., Ludwig, H.: CrossFlow: cross-organizational workflow management in dynamic virtual enterprises. Comput. Syst. Sci. & Eng. 5, 277–290 (2000)
9. Grefen, P., Eshuis, R., Mehandijev, N., Kouvas, G., Weichart, G.: Internet-based support for process-oriented instant virtual enterpises. IEEE Internet Comput., 30–38 (2009)
10. Kappler, L.: The role of factoring for financing small and medium enterprises. Journal of Banking and Finance 30, 3111–3130 (2006)
11. Kartseva, V., Hulstijn, J., Gordijn, J., Tan, Y.-H.: Control patterns in a health-care network. European Journal of Information Systems 19, 320–343 (2010)
12. Kotsokalis, C., Winkler, U.: Translation of service level agreements: A generic problem definition. In: Dan, A., Gittler, F., Toumani, F. (eds.) ICSOC/ServiceWave 2009. LNCS, vol. 6275, pp. 248–257. Springer, Heidelberg (2010)
13. Moser, O., Rosenberg, F., Dustdar, S.: Non-intrusive monitoring and service adaptation for ws-bpel. In: Proc. WWW (2008)
14. Orlicky, J.: Structuring the Bill of Materials for MRP. In: Production and Inventory Management, pp. 19–42 (December 1972)
15. Reijers, H.A. (ed.): Design and Control of Workflow Processes. LNCS, vol. 2617. Springer, Heidelberg (2003)
16. Reijers, H., Limam, S., van der Aalst, W.: Product-Based Workflow Design. Jorunal of Management Information Systems 20(1), 229–262 (2003)
17. Robinson, W.: A requirements monitoring framework for enterprise systems. Requirements Engineering 11, 17–41 (2006)
18. Rozinat, A., van der Aalst, W.: Conformance checking of processes based on monitoring real behavior. Information Systems 33(1), 64–95 (2008)
19. van der Aalst, W.: On the Automatic Generation of Workflow Processes based on Product Structures. Computers in Industry 39, 97–111 (1999)
20. Vanderfeesten, I.: Product-Based Design and Support of Workflow Processes. PhD thesis, Eindhoven University of Technology, Eindhoven, the Netherlands (2009)
21. Vanderfeesten, I., Reijers, H., van der Aalst, W.: Product-Based Workflow Support. Information Systems (2011) (to appear)
22. Vanderfeesten, I., Reijers, H., van der Aalst, W., Vogelaar, J.: Automatic Support for Product Based Workflow Design: Generation of Process Models from a Product Data Model. In: OTM 2010 Workshops, pp. 665–674 (2010)
23. Wang, R.: A product perspective on total data quality management. Communications of the ACM 41(2), 58–65 (1998)

Modeling Design Patterns with Description Logics: A Case Study

Yudistira Asnar, Elda Paja, and John Mylopoulos

Department of Information Engineering and Computer Science
University of Trento, Italy
{yudis.asnar,paja,jm}@disi.unitn.it

Abstract. Design Patterns constitute an effective way to model design knowledge for future reuse. There has been much research on topics such as object-oriented patterns, architectural styles, requirements patterns, security patterns, and more. Typically, such patterns are specified informally in natural language, and it is up to designers to determine if a pattern is applicable to a problem-at-hand, and what solution that pattern offers. Of course, this activity does not scale well, either with respect to a growing pattern library or a growing problem. In this work, we propose to formalize such patterns in a formal modeling language, thereby automating pattern matching for a given problem. The patterns and the problem are formalized in a description logic. Our proposed framework is evaluated with a case study involving Security & Dependability patterns specified in Tropos SI*. The paper presents the formalization of all concepts in SI* and the modeling of problems using OWL-$\mathcal{DL}$ and SWRL. We then encode patterns as SPARQL and SQWRL queries. To evaluate the scalability of our approach, we present experimental results using models inspired by an industrial case study.

Keywords: design patterns, description logics, pattern matching.

1 Introduction

Design patterns represent recurring design problems and how to solve them. Design patterns gained prominence initially in Architecture [1], and within Computer Science with the widely-known and used design patterns for object-oriented design [2]. Today, there are dozens of proposals for design patterns covering a range of design domains, such as: requirements, software architectures, business processes, workflows etc.

Generally speaking, a design pattern consists of a triple *(context, problem-to-solve, solution-pattern)* [1]. To use a pattern one needs to first match the design problem-at-hand to the context, (if successful), match the problem-at-hand to the pattern context (thereby creating mapping for pattern variables), and revise the problem at-hand by using the solution-pattern. However, design patterns often are represented and documented informally in natural language (for an example [2]). This means that it is up to users of a pattern library to determine which patterns are relevant, and also exactly how they apply to the design problem-at-hand. Unfortunately, this approach does not scale with respect to the size of the pattern library, the problem-at-hand, or the expertise of the designer. Indeed, there is plenty of evidence that pattern libraries have low acceptability rates, especially so among non-expert users by now [3]. Some of the prominent

H. Mouratidis and C. Rolland (Eds.): CAiSE 2011, LNCS 6741, pp. 169–183, 2011.

reasons are (i) finding patterns relevant to the problem-at-hand is hard [3], often because pattern libraries are unstructured; (ii) understanding patterns requires some expertise [4] that users often do not have.

The main objective of this paper is to address this situation by formalizing design patterns in a language that has a built-in matching operation, thereby offering automated support for the identification of applicable patterns for a given design problem, also for formulating a design solution. In this work, we formalize the pattern library as well as the design problem as a knowledge base using description logics [5], and represent the context of the patterns as a query to the knowledge base (SPARQL and SQWRL). To evaluate the proposed framework, we use a case study coming from SERENITY Security and Dependability (S&D) patterns [6]. Our evaluation is conducted along two fronts. Firstly we want to see to what extent a description logic can accommodate patterns expressed in an ontologically rich pattern language with built-in concepts such as goal, agent, strategic dependencies among agents, and more. After all, description logics are ontologically simple in the sense that they usually come only with two primitives: concepts and roles (binary relationships). Secondly, we want to know if our proposed solution scales as the size of the design problem to be matched against patterns in the pattern library grow.

The rest of the paper is structured as follows. We outline the research baseline (description logic and OWL-$\mathcal{DL}$) in Section 2. Section 3 presents the case study, while Section 4 details the formalization of patterns. In Section 5, we explain the experimental setup we used and its results. We then offer a discussion of related work in Section 6 and concluding remarks in Section 7.

2 Baseline

Description Logics (DL).   [5] are used to formalize design patterns and also the problem-at-hand in terms of concepts, roles, individuals and their relations. In DL there is no explicit use of variables. Instead, operators are exploited to define complex concepts and roles starting from atomic ones. The set of operators is restricted to ensure tractability for reasoning, such as deciding if a concept is a generalization (subsumes) another concept. A DL knowledge base is composed of two components: 1) a terminological box (TBox) consisting of definitions for concepts and roles, and 2) an assertional box (ABox) consisting of individuals and true facts about them. For instance, if we consider a knowledge base about persons, the TBox would contain concepts such as Person, Male, Female, Parent, Child etc., and roles such as hasChild, and axioms of the form:

$$Person \sqsubseteq Male \sqcup Female$$
$$Man \equiv Person \sqcap Male$$
$$Father \sqsubseteq Man$$
$$Father \equiv Man \sqcap \exists hasChild.Person$$

The ABox, on the other hand, contains assertions regarding individuals, such as axioms of the form: $Father(Tom)$.

OWL-$\mathcal{DL}$ Among DLs, we chose OWL-$\mathcal{DL}$ because it is state-of-the-art as far as DL go, it is part of W3C standard, and is readily available with several possible implementations to choose from. Our use of description logics is as follows: we formalize the modeling language concepts using OWL-$\mathcal{DL}$(creating the TBox), and we use this as a basis for representing formally the context of the patterns. We perform matching at the instance level (ABox), that is why we represent patterns as queries. However, the expressiveness of OWL-$\mathcal{DL}$ alone is weak to represent some constraints, such as the ones related to individuals. We solved the problem by adopting the SWRL rule language [7]. This allows us to enrich the formal pattern description with inferred knowledge, thereby ensuring better pattern matching for the problem-at-hand.

3   Case Study

Several works have been proposed in the literature on S&D patterns (e.g., fault-tolerant patterns [8], security patterns [9], SERENITY patterns [6]). In this work, we use SERENITY patterns, developed within the EU SERENITY project, it is state-of-the-art for its intended application domain and we had expertise on both the patterns and their uses.

SERENITY Patterns.   [6] are represented using Alexander's pattern language as triples: ⟨Context, S&D Requirements, S&D Solution⟩ The *Context* defines the state-of-affairs the problem/situation where the pattern could be applied, which is depicted in terms of the minimum set of actors and relationships, where the S&D Requirements are not fulfilled. *S&D Requirements* specify the required S&D Properties that must be satisfied in the model (representing the problem). *S&D Solution* describes the modifications that need to be performed to the context in order to meet S&D Requirements. The description of SERENITY patterns is enriched with additional description about when, how, and for what the patterns are intended for.

In [6], patterns are identified in scenarios extracted from business cases (e.g., Air Traffic Management, e-Business, Online-Tax, Smart-home) and then described in natural language. Patterns, then, are represented formally; *Context* and *S&D Solution* are represented in terms of SI* models, whereas the *S&D Requirements* in ASP (answer set programming - an extension of DATALOG). The pattern library is composed of 29 SERENITY patterns (4 legal, 3 privacy, 11 security and 11 dependability patterns) [6]. Table 1 presents some of the SERENITY patterns described in natural language. For an illustrative example, we use pattern DP2.1 on *Collaboration in Small Groups for Risky Activities*. In addition, the SERENITY pattern library was used for evaluating the performance of our implementation (Section 5).

Tropos SI*.   [10,11] is a modeling language for security requirements. The language offers primitive concepts such as actor, goal, task, as well as various kinds of relationships among actors. This modeling language is used for representing *Context* and *S&D Solution* of SERENITY patterns.

Fig. 1 depicts one of the SERENITY scenarios (e.g., Air Traffic Management - ATM). SI* considers *intentional Actor* as basic concept (e.g., Executive Controller,

Table 1. SERENITY S&D Patterns in Natural Language

Pattern Name	Natural Language Description
SP1. Proof of Fulfillment for Ensuring Non-Repudiation	To prevent repudiation, the executor needs to provide evidence of performing the action to the benefitor, in addition to performing the action.
SP4. Artefact Generation as an Audit Trail	To prevent repudiation of some actions upon a shared resource, a group of agents needs to keep a common audit trail.
DP2.1. Collaboration in Small Groups for Risky Activities	To cope with an activity where a tight coordination among agents is crucial, a failure on a risky sub-activity may compromise the team goal. However, one team member might have a capability to mitigate the risk of the risky sub-activity therefore that team member must mitigate the risk for the team success.
DP6. Reinforcing Overlapping Responsibilities for Robustness	A critical task must be completed successfully most of the time. Therefore, several team members are responsible to perform the task.
PP1. Sign an Agreement to Address Lack of Trust on the Use of Private Data	Sometime a customer does not trust an organization accessing its data. Therefore a representative agent of the organization needs to ask for customer consent before accessing customer's data.

Supervisor, Alice) that wants to achieve goals (ensure traffic safety in its sector, form team sectors). Actors are equipped with certain abilities (e.g., resolve traffic conflict), have beliefs, etc. They are further specialized into *roles* (e.g., Supervisor, Executive Controller, and Team sector) as abstract actors in an organization that are played by *agents* (e.g., Bob, Alice, Dan), which are concrete actors.

Actors (e.g., executive controllers) intend to achieve/satisfy their business goals (manage traffic in the sector) by relying on their capabilities and those of other actors (e.g., resolve traffic conflict). The term Business Object refers to a *goal*, a *task*, or a *resource*. *Goal* represents a state-of-affair that an actor intends to achieve (manage traffic in sector). *Task* is a course of actions performed by an actor to achieve a desired goal (give airway commands). *Resource* refers to physical or informational entities required to achieve goals (flight progress strips) or to perform tasks (air situation display). However, the fulfillment of these business objects is affected by uncertain *events*. Events that can cause a goal failure are risks (overload traffic), while events that can help in the fulfillment of a goal are treated as opportunities (deployment new system for air conflict prediction).

In addition to capturing the strategic rationale of an actor, SI* captures strategic inter-dependencies among actors in an organization. Inter-dependencies can be either *delegation* and *trust* relationships among actors. Delegations also come in two flavours: 1) execution of business objects and 2) permission/entitlement on business objects. *Delegation* refers to the transfer of responsibilities (*Delegation on Execution*) or rights (*Delegation on Permission*). In Fig. 1, team sector delegates the execution of manage inbound traffic to another actor - planning controller. **Trust** refers to the belief and expectation of an actor that another actor (trustee) will fulfill its commitments (will execute all assigned business objects) and will respect its permissions. For an example, Alice trusts Bob to fulfill managing traffic in Sector SU1.

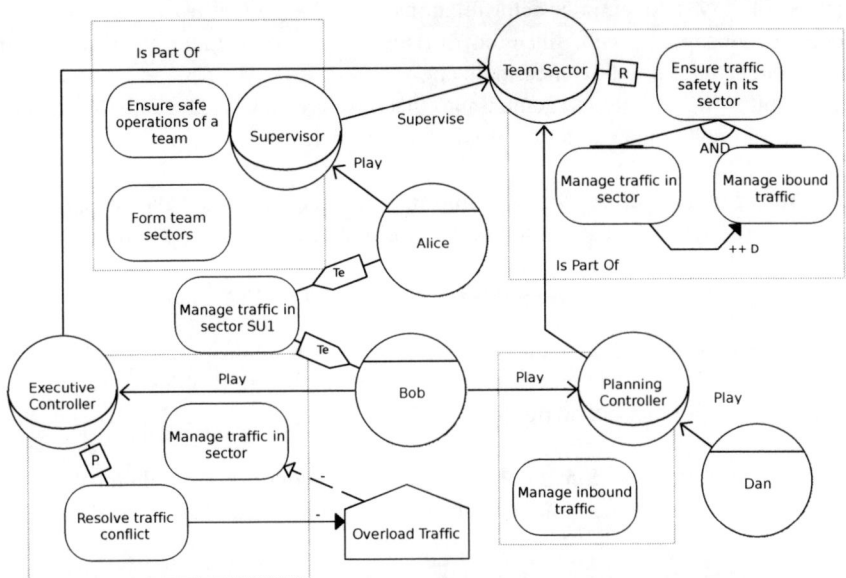

Fig. 1. A SI* Diagram from a fragment of the Air Traffic Management scenario

A SI* model captures relationships between concepts using several basic relations: 1) *AND/OR-decomposition* to refine a goal, 2) *contribution* to capture the effects of a goal to another, 3) *impact* to model the impact of an uncertain event to a business object. Fig. 1 depicts the goal ensure traffic safety in its sector is AND-decomposed into manage traffic in sector and manage inbound traffic, where the achievement of both subgoals are necessary to achieve the up-level goal. Moreover, the achievement of the latter goal contributes positively to the success of the former one. In ATM, we consider the effect of overload traffic event to the goal manage traffic in sector, and it can be mitigated with the capability of an actor in resolve traffic conflict.

4   Formalizing Patterns

Our formalization process includes four steps: (i) Formalize the SI* language by defining non-overlapping OWL-$\mathcal{DL}$ concepts for all SI* primitives and one or more roles for every primitive SI* relationship; (ii) Formalize the context of each pattern using the concepts and roles introduced in step (i); (iii) Enrich the formal pattern descriptions with implicit knowledge [1]; (iv) Represent the problem-at-hand in the ABox by instantiating the concepts and roles of step (i).

4.1   Formalizing SI* Primitives

In general, we represent nodes (e.g., goal, task, resource, event) in a SI* model as concepts and binary relations (e.g., actor's associations, contributions, decompositions, im-

---

[1] Availability of a domain expert is essential here because this implicit knowledge (constraints, alternatives, and more) is often missing from the informal pattern description.

pacts) as roles. Moreover, inter-actor relations (e.g., delegation, trust, monitoring) are encoded as concepts as well, since they are ternary relations. Later, we discuss some considerations that underlie our formalization.

In the following, we discuss some issues that have arisen while formalizing concepts/relations of the language in terms of DL concepts/roles.

Role versus Subsumption. The relationship between a goal and its subgoals could have been represented as a subsumption relationship or a role, say $hasSubgoal$:

$$Subgoal \sqsubseteq Goal \quad vs. \quad hasSubgoal.Goal$$

But instances of a subgoal do not need to also be instances of the parent goal (for example, consider goal "schedule meeting" and subgoal "collect timetables"). Accordingly, we chose to go with the second option.

Ternary/n-ary relations. Since OWL-$\mathcal{DL}$ does not support N-ary relations, $N \geq 3$, we decided to represent such relationships in terms of a concept and several roles. For example,

$$DelegationOnExecution \equiv DelegationOnExecution \sqcap (hasDelegator = 1) \sqcap$$
$$(hasDelegatum = 1) \sqcap (hasDelegatee = 1)$$

Consider Fig. 1, where Team Sector delegates execution of manage traffic in the sector to Executive Controller. In the ABox, a delegation on execution in such a setting is represented as follows:

$$DelegationOnExecution(\text{Del-exec}1)$$
$$hasDelegator(\text{Del-exec}1, \text{Team Sector})$$
$$hasDelegater(\text{Del-exec}1, \text{Executive Controller})$$
$$hasDelegatum(\text{Del-exec}1, \text{Manage traffic in sector})$$

4.2   Understanding and Formalizing a Pattern as a Query

Once designers specify the pattern language in the DL TBox, the next step is understanding the essence of the pattern description and formalizes the context in terms of OWL queries (i.e., using terms specified in the DL TBox). Designers need to be aware that some patterns are very generic and sometimes vague, (e.g., patterns described in natural language in [8]), and others are rather restrictive because of the limitation of the pattern language (e.g., patterns described using a modeling language, such as [6]). In this phase, we leave it in the hands of the designers to decide how much detail they want to put into the patterns or how generic the pattern should be.

In [6], the context of the patterns is modeled in terms of an "abstract" SI* model that includes variables (denoted by identifiers with capitalized letters). For instance in

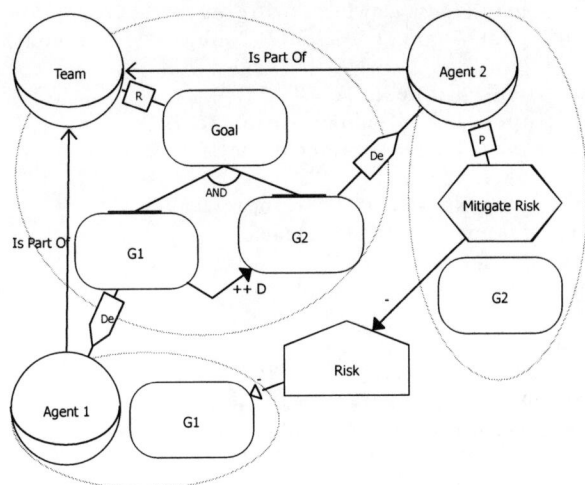

Fig. 2. The "Collaboration in Small Groups for Risky Activities" pattern

Fig. 2, the pattern indicates **TEAM delegates execution of G1 to AGENT1** will match elements that involve two roles, where the first role delegates execution of a goal to the second role. Moreover, in some patterns (e.g., GoF [2]) pattern contexts are left implicit and designers need to fill the details. A pattern is applicable to the problem-at-hand if all constructs in the pattern match corresponding elements of the problem. When such a match is found, the reasoner returns not only true, but also mappings for pattern variables. Patterns are formalized in two parts: (i) the *mandatory part* must be matched in the problem-at-hand for the pattern match to succeed; (ii) the *optional part* can bring about useful mappings for variables, but do not affect the outcome of a pattern match. In our approach, matching is performed at the individual level (ABox) of the knowledge base. Therefore, the reasoner checks for individuals present in the problem and the relationships among them, and lists all individuals that match the pattern context. Note that reasoners can return more than one resultset, since it is likely there are several parts of the model that match a given pattern context.

In the context of security and dependability, the *optional part* is only used to define a new actor (including its capabilities & responsibilities) that is not necessarily present in the problem. For instance, A client needs to buy a house from Company A, but he does not trust Company A. Based on [6], to ensure security, the designers can "patch" the trust issue by having a contract arranged by a lawyer. However, in most cases the lawyer does not exist in the "current" statement of the problem, hence the need for optional elements during a pattern match. Application of the pattern basically introduces, a new role - lawyer (i.e., a trusted $3^{rd} party$).

OWL-$\mathcal{DL}$ models can be queried using two languages: SPARQL [12], and SQWRL [13]. The SPARQL is a W3C Recommendation [12] for querying RDF. RDF essentially offers directed labeled graph data format, built out of triples. Thus, SPARQL queries are expressed in terms of triple patterns, consisting of a subject, predicate, and object. The

```
prefix  tropos:  <http://www.owl-ontologies.com/Tropos_Ontology.owl#>
SELECT    ?team ?goal ?agent1 ?agent2 ?subgoal1 ?subgoal2
WHERE  {
          ?team  tropos:request  ?goal.
          ?goal  tropos:isAndDecompositionOf  ?g1.
          ?goal  tropos:isAndDecompositionOf  ?g2.
     FILTER  (?subgoal1 != ?subgoal2).
          ?agent1  tropos:isPartOf  ?team.
          ?agent2  tropos:isPartOf  ?team.
     FILTER  (?agent1 != ?agent2).
          ?team  tropos:hasDelegation  ?d1.
          ?d1  tropos:hasDelegatee  ?agent1.
          ?d1  tropos:hasDelegatum  ?g1.
          ?team  tropos:hasDelegation  ?d2.
          ?d2  tropos:hasDelegatee  ?agent2.
          ?d2  tropos:hasDelegatum  ?g2.
          ?agent2  tropos:provides  ?mitigateRisk.
          ?g1  tropos:hasNegDContribution  ?g2.
          ?task  tropos:hasNegContribution  ?risk.
          ?risk  tropos:hasNegImpact  ?g1.
     OPTIONAL{?agent2  tropos:requests  ?mitigateRisk.}
}
```

Fig. 3. SPARQL representation of DP 2.1

Turtle data format [2] is used to represent triple patterns. The query attempts to match the triples on the graph pattern against the model [14]. SPARQL just queries the model and does not support inference [12], nor does it modify the RDF dataset. However, some frameworks (e.g., JENA) and rule engines [15], have the capacity to perform inference and update the dataset by performing OWL reasoning.

Alternatively, a more expressive query language that is founded on DL semantics and supports comprehensive querying of OWL is SQWRL [13]. SQWRL is a SWRL-based query language [7]. SQWRL provides SQL-like operations to retrieve knowledge from an OWL ontology. Similarly to SPARQL, in SQWRL we try to capture all concepts and relationships present in a pattern. Since SQWRL understands the semantics of OWL and SWRL rules, it understands not only the explicit, but also the inferred knowledge. For example, the DP2.1 of SERENITY pattern (Fig. 2 described in Table 1), can be translated into a SPARQL Query (Fig. 3). Each node of the pattern context is a variable in the query and each edge is an RDF triplet. For a SQWRL Query, the DP2.1 translational is shown in Fig. 4.

4.3   Enriching DL T-Box with Implicit Knowledge

Often details of the patterns are described in natural language, due to the expressivity limitation of the pattern language. This was certainly the case with our case study.

Back to our example in DP2.1, in SI* the notion of "request" means that an actor intends to achieve a particular goal. However, based on DP2.1's description the intent

---

[2] Turtle: http://www.w3.org/TeamSubmission/turtle/

1 : requests(?*team*, ?*goal*) ∧ isPartOf(?*agent1*, ?*team*) ∧ isPartOf(?*agent2*, ?*team*) ∧
2 : hasSubgoal(?*goal*, ?*g1*) ∧ hasSubgoal(?*goal*, ?*g2*) ∧
3 : hasDelegation(?*team*, ?*d1*) ∧ hasDelegatee(?*d1*, ?*agent1*) ∧ hasDelegatum(?*d1*, ?*g1*) ∧
4 : hasDelegation(?*team*, ?*d2*) ∧ hasDelegatee(?*d2*, ?*agent2*) ∧ hasDelegatum(?*d2*, ?*g2*) ∧
5 : hasNegDContribution(?*goal1*, ?*goal2*) ∧ provides(?*agent2*, ?*mitigateRisk*) ∧
6 : hasNegImpact(?*risk*, ?*g1*) ∧ hasNegContribution(?*mitigateRisk*, ?*risk*) →
7 : *sqwrl* : select(?*team*, ?*goal*, ?*agent1*, ?*g1*, ?*agent2*, ?*g2*, ?*risk*, ?*mitigateRisk*) ∧
8 : *sqwrl* : *columnNames*("*team*", "*goal*", "*agent1*", "*g1*", "*agent2*", "*g2*", "*risk*", "*mitigateRisk*")

Fig. 4. SQWRL representation of DP 2.1

aim(?*a*, ?*goal*)   ← requests(?*a*, ?*goal*)
aim(?*a2*, ?*goal*) ← requests(?*a1*, ?*goal*) ∧ hasDelegation(?*a1*, ?*d*)∧
$\qquad$ hasDelegatee(?*d*, ?*a2*) ∧ hasDelegatum(?*d*, ?*goal*)
aim(?*a2*, ?*goal*) ← aim(?*a1*, ?*goal*) ∧ hasDelegation(?*a1*, ?*d*) ∧ hasDelegatee(?*d*, ?*a2*)∧
$\qquad$ hasDelegatum(?*d*, ?*goal*)

Fig. 5. Relaxing "request" on SI* in SQWRL

of "request" is more relaxed – direct request (i.e., the actor "requests" fulfillment of a goal) or indirect request (i.e., another actor delegates the execution of a goal to him/her, to the actor). Accordingly, we decided to extent the DL TBox and revise the pattern formalization using those new concepts/roles. However, this extension can only be done when we use SQWRL and **not** SPARQL. In Fig. 5, we illustrate an example of the extension of the "request" relation in SI*, namely "aim". To be closer with the DP2.1's description one needs to replace line 3-4 of Fig. 4 with the following:

$$\text{aim}(?agent1, ?g1) \wedge \text{aim}(?agent2, ?g2)$$

4.4   Representing the Problem in the ABox

Finally, designers need to represent the problem-at-hand in terms of instances of concepts and roles in the ABox.

Concept versus Individual. Individuals have a unique identity, and their description can be modified by adding more assertions in the ABox. Conversely, the definition of concepts cannot be modified [16].

The first alternative (i.e., subclasses of the DL TBox) allows us to reason whether a pattern appears anywhere in the problem, but it cannot provide the mapping between construct in the pattern's solution and the problem-at-hand. The second alternative (i.e., individuals in the DL ABox) on the other hand can provide such mappings, but it will not allow us to reason in a situation where the problem contains both abstract and concrete entities in the real world, because both entities will be encoded as individuals and the reasoner will treat them equally. Since we deal with the problem at the design level where mostly models capture the class level instead of the object one, we have chosen the second alternative (i.e., as series of individuals) as the most suitable to our needs.

Moreover, providing mapping between the pattern and the problem is a critical feature to support designers in applying the patterns to resolve their problem.

Here is a fragment of the representation of the problem-at-hand in Fig. 1 in terms of concept and role instances in the ABox:

- Role(Team Sector)
- Role(Executive Controller)
- Role(Planning Controller)
- Agent(Bob)
- Goal(Ensure traffic safety in its sector)
- Goal(Manage traffic in sector)
- Goal(Manage inbound traffic)
- Goal(Resolve traffic conflict)
- play(Bob, Executive Controller)
- play(Bob, Planning Controller)
- isPartOf(Executive Controller, Team Sector)
- isAndDecompositionOf(Ensure traffic safety in its sector, Manage traffic in sector)
- isAndDecompositionOf(Ensure traffic safety in its sector, Manage inbound traffic)
- hasPosContribution(Manage inbound traffic, Manage traffic in sector)
- provide(Executive Controller, Resolve traffic conflict)
- DelegationOnExecution(Del-exec1)
- hasDelegator(Del-exec1,Team Sector)
- hasDelegater(Del-exec1,Executive Controller)
- hasDelegatum(Del-exec1,Manage traffic in sector)

4.5   System Architecture

Fig. 6 depicts the architecture of our implemented system. Though this work supports two types of queries (SPARQL and SQWRL), most system components and artifacts are common for both inputs (normal line). The ones with thick lines refer to parts for SPARQL, while dashed lines to SQWRL. In both cases, the implemented system requires the same input SI* model representing the problem-at-hand and a set of SI* models representing patterns.

Since we need some inference capabilities to deduce implicit facts, we use a rule engine (i.e., JESS) that is integrated in Protégé. Essentially, a rule engine takes an input (rules and facts) and produces a model. In SQWRL setting, the input consists of the TBox and ABox defined so far, patterns in SQWRL, and SWRL rules [3]. In this system, the model (produced by the rule engine) contains the resultset of the matching. In SPARQL setting, the input to the rule engine only contains TBox, ABox, and SWRL rules. The rule engine produces a model containing inferred knowledge from available facts and rules. Using the Model-to-OWL library in Protégé, inferred knowledge is added to the knowledge base (TBox and ABox). By means of the OWL-$\mathcal{DL}$ reasoner (e.g., Pellet, JENA), we can query the revised knowledge base to find a match to a pattern. In both settings, if the length of resultset is zero, then there is no match found in

---

[3] available at:http://www.w3.org/Submission/SWRL/swrl.owl

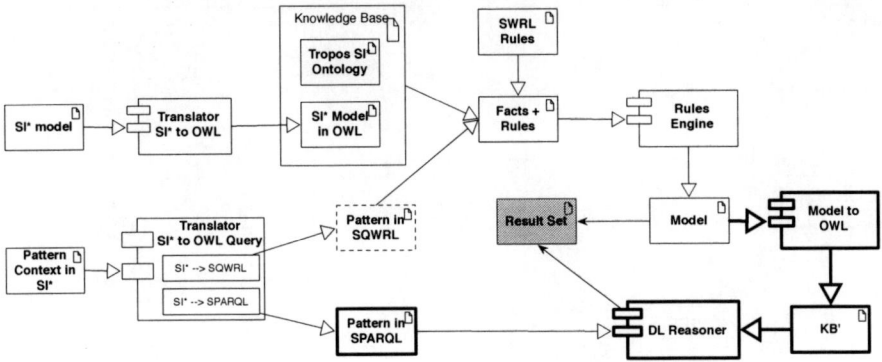

Fig. 6. System Architecture for Pattern Matching

the problem. The resultset will contain several sets when a pattern matches to several parts of the problem. Moreover, each set will provide a mapping from a pattern's constructs to the problem. We have implemented this approach using Java Platform v1.6 along some features from Protégé libraries.

5   Experimental Results

Design of experiments. To evaluate this approach and its implementation, we have conducted experiments using a laptop Intel Core2 Duo T7300 2.0GHz, 2Gb DDR2 667. Through these experiments we intend to assess the performance of our implementation, and investigate how performance (i.e., execution time) is affected by an increase in problem size. To make the experiment realistic, we consider the ATM Scenario [17] as the problem-at-hand. First, the SI* model of the problem is translated to a corresponding model in OWL-$\mathcal{DL}$. Similarly, we translate SERENITY S&D patterns (21 patterns), defined in [6], into OWL-$\mathcal{DL}$ queries in SPARQL and SQWRL. The model is then queried using SPARQL and SQWRL queries to find matches to those patterns. Bigger models are obtained by cloning the OWL-$\mathcal{DL}$ model of ATM scenario facilitated by the "deep copy" feature of Protégé. Originally, the model of ATM scenario has the size of 472 elements composed of 83 nodes and 389 relations [4]. Cloning was performed on the ATM model by cloning nodes and their respective relations. The cloning process was not linear; as we could not control the number of relations a node participates in. Seven models were obtained through this process, starting from a model size of 832 (136 nodes) up to the biggest model with size 6203 (941 nodes).

In the experiment, each pattern is matched against 8 different models. To ensure stability of "execution time", we perform 20 executions for each pair (pattern, model) and used the average of each execution time. Moreover, a manual verification has been performed to validate the correctness of each pattern match.

---

[4] All datasets can be found at http://disi.unitn.it/~yudis/lib/exe/fetch.php?media=files:dataset.rar

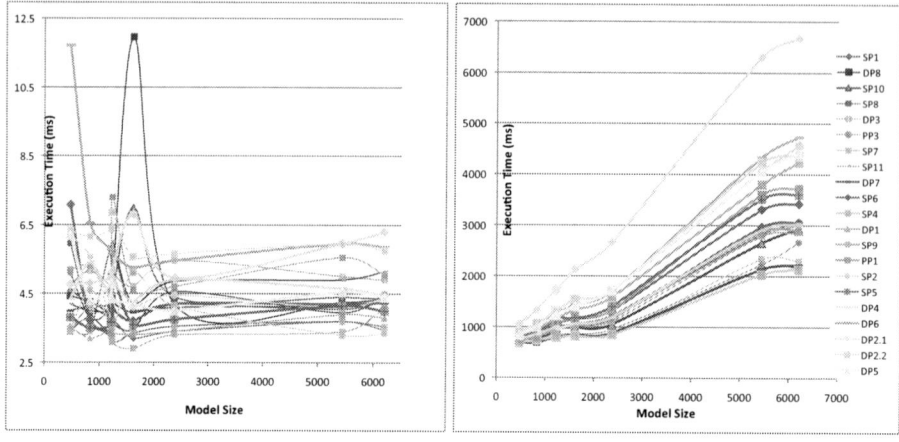

(a) Execution time in SPARQL   (b) Execution time in SQWRL

Fig. 7. Performance of Pattern Matching

Results. After running the queries (patterns) over the ATM model we found that there are four applicable patterns (e.g., SP1, SP8, DP2.1, DP6). In Fig. 7, we present the performance in milliseconds, of our implementation for both: SPARQL and SQWRL.

In general, there are significant differences between the two query representations. In particular, the worst performance in SPARQL (11.7ms) is much faster then the best performance in SQWRL (713ms). The main reason is that in the case of SPARQL queries, the inferred model is computed only once before the matching starts and used throughout all the queries. Thus, the inference time is not taken into account in the SPARQL execution time. In the SQWRL case, execution time is highly affected by the inference time.

Considering Fig. 7(b), it is an almost linear correlation between the size of the problem model and execution time. However, Fig. 7(a) indicates that the SPARQL performance is constant after a certain model size (model size $\geq$ 2378) [5]. Even though the SPARQL case outperforms the SQWRL case, designers need to be aware on the fact that SPARQL engine does not exploit the semantics of OWL-$\mathcal{DL}$. Moreover, SPARQL is meant to be used for querying RDF and OWL-$\mathcal{DL}$ needs to be serialized before it can be queried. This serialization of OWL-$\mathcal{DL}$ to RDF is vendor specific therefore it could be the case that the same OWL-$\mathcal{DL}$ has several representations in RDF and consequently different SPARQL queries. However, this dilemma does not hold for SQWRL.

6   Related Work

The growing size of pattern libraries has spawned the following challenges: 1) finding a relevant pattern in the pattern library, 2) selecting a pattern that is suited with the problem-at-hand, and 3) applying a pattern. For the first challenge, though there is no central index as mentioned in [3] several initiatives are trying to collect software design

---

[5] We acknowledge some irregularities on the execution time for the 3 smallest models.

patterns (e.g., Pattern Forge [6], Net Objectives [7], Portland Pattern Repository [8]). In comparison to our approach, such initiatives receive a pattern contribution in natural language without formalizing it. In addition to disadvantages presented in Section 4.2, users might have difficulties in finding relevant patterns in the library because textual matching does poorly without domain knowledge.

To improve the finding and the selection phase, several works use (semi-)formal languages for representing patterns and selection mechanisms of such representation. In [18], Mens et al. use DL to detect inconsistencies between UML models in evolving systems. In that work, the authors take advantage of the underlying DL representation and reason about UML models exploiting the DL reasoning engine (e.g., Racer, Loom). In a nutshell, this work takes a similar approach to ours where the TBox formalizes the UML meta-model and the ABox represents instances in the designers' model. In our work, the query represents the context of a pattern, while in this work the query represents the rules characterizing model constraints. In [19], the authors describe how to use a meta-model to obtain a representation of a pattern at the code level. The meta-model consists of a set of entities and interaction rules between them, and defines pattern semantics. The meta-model is further specialized by adding structural and behavioral constituents, thereby obtaining an abstract representation of patterns. These are gathered in a repository and used to generate code automatically.

Some works facilitate the selection phase by structuring the pattern library in a certain manner. In [20,21], the authors proposed a structure to organize patterns. Moreover, other authors [22,23] provide systematic and automated reasoning to select a pattern. In these works, the authors do not formalize the pattern itself, but rather formalizing the structure and relationships among patterns. Conversely, our approach formalizes a pattern and does not prescribe a particular structure on the pattern library. In our approach, we aim to find an applicable pattern and provide a mapping, while these works intends to limit the solution space so that the pattern users need only evaluate a small number of patterns. In other words, these approaches require less efforts in contributing a new pattern because they only require where a pattern should be categorized and its relationships with other patterns, while in our approach "the formalization" of a pattern defines the performance, in term of correctness, of the system. Note that these approaches does not guarantee the resulted pattern will be applicable to the problem-at-hand, while ours

Some works concentrate on how to apply the patterns in the problem-at-hand. For instance, Eden et al. [24] represent patterns as meta-programs that modify other code (i.e., the problem-at-hand). The authors have implemented a prototype that supports design pattern specification and realization in a given program and this approach allows programmers to edit the source code at any time in the process. In comparison to our approach, this work aims at modifying the problem before implementing a chosen pattern, whereas ours aims to find the pattern(s) that are applicable to solve a given problem.

In the area of Model-Driven Engineering, several frameworks have been proposed to support model transformations [25]. In comparison to ours, their approaches are more expressive in describing how a pattern is to be applied. However, these frameworks

---

[6] http://www.patternforge.net/
[7] http://www.netobjectives.com/PatternRepository
[8] http://c2.com/ppr/

have some limitations in finding a match because matching is based on graph similarity techniques only, rather than inference in a DL. More generally, our framework can leverage reasoning provided by DL to support pattern matching and pattern application.

7 Final Remarks

We have presented an approach to formalize problems and patterns using Description Logics, so that, given a problem, we can find applicable patterns from a pattern library. Moreover, when a pattern match succeeds, it provides mapping between elements of the problem and variables in the pattern. These mapping are useful in determining how to apply the pattern to a given problem. Our proposal has been evaluated in terms of a case study using the SERENITY pattern library. Our experiences suggest that description logics do constitute a viable solution to formalize patterns, and the problem represented by a rich modeling language such as SI* can be accommodated in a description logic using its concept definition facilities. A corollary of our case study is that there is an important trade-off in formalizing patterns between making them too generic or too specific. Generic patterns match in many contexts but offer vanilla solutions. Conversely, specific ones match few concepts but offer insightful solutions. Pattern designers need to tread carefully as they navigate between these alternatives.

Our future work includes applying our framework to other pattern libraries. In addition, we propose to conduct a controlled experiment to empirically evaluate our approach with pattern designers and pattern users.

Acknowledgments

The research leading to these results has received funding from the EU FP7 under grants no. 216917 MASTER, no. 256980 NESSoS, and no. 257930 ANIKETOS.

References

1. Alexander, C., Ishikawa, S., Silverstein, M.: A pattern language. Oxford Press (1977)
2. Gamma, E., Helm, R., Johnson, R., Vlissides, J.: Design Patterns: Elements of Reusable Object-Oriented Software. Addison-Wesley Longman Publishing Co., Inc., Reading (1995)
3. Manolescu, D., Kozaczynski, W., Miller, A., Hogg, J.: The growing divide in the patterns world. IEEE Software 24(4), 61–67 (2007)
4. Sommerville, I.: Software Engineering, 7th edn. Addison Wesley, Reading (May 2004)
5. Baader, F., Calvanese, D., McGuinness, D.L., Nardi, D., Patel-Schneider, P.F. (eds.): The description logic handbook: theory, implementation, and applications. Cambridge University Press, New York (2003)
6. Asnar, Y., Bryl, V., Dalpiaz, F., El-Khoury, P., Felici, M., Halas, H., Krausová, A., Li, K., Riccucci, C., Saidane, A., Séguran, M., Yautsiukhin, A.: Final set of S&D Patterns at Organizational Level. Project Deliverable A1.D3.3, SERENITY Consortium (January 2009)
7. Horrocks, I., Patel-Schneider, P.F., Boley, H., Tabet, S., Grosof, B., Dean, M.: SWRL: A Semantic Web Rule Language Combining OWL and RuleML (May 2004), http://www.w3.org/Submission/2004/SUBM-SWRL-20040521/

8. Hanmer, R.: Patterns for Fault Tolerant Software. Wiley, Chichester (2007)
9. Schumacher, M., Fernandez-Buglioni, E., Hybertson, D., Buschmann, F., Sommerlad, P.: Security Patterns: Integrating Security and Systems Engineering, 1st edn. Wiley, Chichester (2006)
10. Giorgini, P., Massacci, F., Mylopoulos, J., Zannone, N.: Requirements Engineering for Trust Management: Model, Methodology, and Reasoning. IJIS 5(4), 257–274 (2006)
11. Asnar, Y., Giorgini, P., Mylopoulos, J.: Goal-driven risk assessment in requirements engineering. REJ, 1–16 (2010)
12. Prud'hommeaux, E., Seaborne, A.: SPARQL Query Language for RDF (February 2004), http://www.w3.org/TR/rdf-sparql-query/ (lastchecked: March 7, 2010)
13. O'Connor, M., Das, A.: SQWRL: a query language for OWL. In: Proc. of OWLED 2009 (2009)
14. McCarthy, P.: Search RDF data with SPARQL: SPARQL and the Jena Toolkit open up the semantic Web (May 2005), http://www.ibm.com/developerworks/library/j-sparql/ (lastchecked: March 1, 2010)
15. Friedman-Hill, E.: Jess Rule Engine, http://www.jessrules.com/ (lastchecked: December 2009)
16. Görz, G.: Description Logics, Knowledge Bases, Formal Ontologies and Data Bases: Content. Lecture Notes (2008)
17. Asnar, Y., Giorgini, P., Massacci, F., Saidane, A., Bonato, R., Meduri, V., Riccucci, C.: Secure and Dependable Patterns in Organizations: An Empirical Approach. In: Proc. of RE 2007 (2007)
18. Mens, T., Van Der Straeten, R., Simmonds, J.: Maintaining consistency between UML models with description logic tools. In: Proc. of UML 2003, Workshop on Consistency Problems in UML-based Software Development II (2003)
19. Albin-amiot, H., gaël Guéhéneuc, Y., Kastler, R.A.: Meta-modeling design patterns: Application to pattern detection and code synthesis. In: Proc. of ECOOP 2001 Workshop Automating Object-Oriented Software Development Methods, pp. 1–35 (2001)
20. Manolescu, D., Kozaczynski, W., Miller, A., Hogg, J.: The growing divide in the patterns world. IEEE Software 24(4), 61–67 (2007)
21. Zdun, U.: Systematic pattern selection using pattern language grammars and design space analysis. Software: Practice and Experience 37(9), 983–1016 (2007)
22. Gross, D., Yu, E.: From Non-Functional requirements to design through patterns. Requirements Engineering 6(1), 18–36 (2001)
23. Weiss, M., Mouratidis, H.: Selecting security patterns that fulfill security requirements. In: 16th IEEE International Requirements Engineering, RE 2008, pp. 169–172 (2008)
24. Eden, A.H., Yehudai, A., Gil, J.: Precise specification and automatic application of design patterns. In: Proc. of ASE 1997, pp. 143–152 (1997)
25. Czarnecki, K., Helsen, S.: Classification of model transformation approaches. In: Proc. of OOPSLA 2003 Workshop on Generative Techniques in the Context of the MDA (2003)

Interactively Eliciting Database Constraints and Dependencies

Ravi Ramdoyal and Jean-Luc Hainaut

Laboratory of Database Application Engineering - PReCISE Research Centre
Faculty of Computer Science, University of Namur
Rue Grandgagnage 21 - B-5000 Namur, Belgium
{rra,jlh}@info.fundp.ac.be
http://www.fundp.ac.be/precise

Abstract. When designing the conceptual schema of a future information system, it is crucial to define a set of constraints that will guarantee
the consistency of the subsequent database once it is implemented and
operational. Eliciting and expressing such constraints and dependencies
is far from trivial, especially when end-users are involved and when there
is no directly usable data to play with. In this paper, we present an interactive process aimed to elicit hidden constraints such as value domains,
functional dependencies, attribute and role optionality and existence constraints. Inspired by the principles of Armstrong relations, it attempts to
acquire minimal data samples in order to validate declared constraints,
to elicit hidden constraints and to reject irrelevant constraints in conceptual schemas. This process is part of the RAINBOW approach, destined
to develop the data model of an information system based, among others,
on the reverse engineering of user-drawn form-based interfaces.

Keywords: Information Systems Engineering, Requirements Engineering, Database Engineering, Electronic Forms Reverse Engineering,
Constraint Discovery.

1   Introduction

In the realm of Requirements engineering, Database engineering focuses on data
modelling, where the static data requirements are typically expressed by means
of a conceptual schema, which is an abstract view of the static objects of the
application domain. There are numerous types of constraints and dependencies
that can be established for such a schema. They can concern individual elements,
their components, or even how (the components of) an element can affect (the
components of) other elements. Traditional database elicitation techniques, such
as the analysis of corporate documents and interviews of stakeholders, usually
yield many relevant constraints during the design of the conceptual schema,
however some constraints may be forgotten, typically because the domain experts
were not aware of them, or (more probably) because they are part of some tacit
knowledge.

H. Mouratidis and C. Rolland (Eds.): CAiSE 2011, LNCS 6741, pp. 184–198, 2011.

Though the necessity to associate end-users of the future system with its specification and development steps has long been advocated [1], several approaches rather propose to deal with the discovery of such constraints by analysing the content of a related database (or at least, a set of relevant data samples). However, they rely on the preexistence of large sets of data, which can obviously be problematic in the process of designing a new information system: there might be no usable legacy database, or gathering and reencoding a significant amount of data would be unrealistic.

In this context, the RAINBOW approach [2] provides an alternative and interactive process based on the analysis of a limited set of user-provided data samples in order to elicit and suggest database constraints and dependencies for a given schema. RAINBOW is a collaborative and interactive user-oriented approach to develop the static data model of an information system based on the reverse engineering of user-drawn form-based interfaces. It relies on the adaptation and integration of principles and techniques coming from various fields of study, ranging from Database Forward and Reverse Engineering to Prototyping and Participatory Design.

In this paper, we present how to use the RAINBOW approach to discover constraints and dependencies. In particular, we focus on the elicitation of functional dependencies, which are a fundamental and critical aspect of conceptual modelling that has proved difficult to apprehend. The remainder of the paper is structured as follows. Section 2 delineates the research context, while Section 3 describes the related works. The main principles of the proposal are detailed in Section 4. Section 5 discusses the evaluation of this process, while Section 6 discusses the proposal and concludes this paper.

2   Research Context

2.1   The RAINBOW Approach

The RAINBOW approach is a collaborative and interactive user-oriented approach to design database conceptual schemas in the context of Information System engineering [2]. It exploits the expressiveness of user-drawn form-based interfaces and prototypes, and specialises and integrates standard techniques to help acquire and validate data specifications from existing artefacts in order to use such interfaces as a two-way channel to communicate static data requirements between end-users and analysts. The approach is formalised by a semi-automatic seven-step process dealing with the progressive modelling of the application domain:

1. *Represent*: the end-users are invited to draw and specify a set of form-based interfaces to perform usual tasks of their application domain;
2. *Adapt*: the forms are "translated" into data models, which basically consists in extracting a data model from each interface using mapping rules;
3. *Investigate*: the data models are cross-analysed to highlight and arbitrate semantic and structural similarities and produce a pre-integrated schema;

4. *Nurture*: using the interfaces that they drew, the end-users are invited to provide data examples that are analysed to infer and arbitrate possible constraints and dependencies;
5. *Bind*: the pre-integrated schema is completed and refined into a non redundant integrated conceptual schema;
6. *Objectify*: from the integrated conceptual schema, the artefacts of a prototypical data manager application are generated;
7. *Wander*: finally, the end-users are invited to play with the prototype in order to refine and ultimately validate the integrated conceptual schema.

In order to position end-users as major stakeholders throughout the data requirements process, the approach uses form-based interfaces as a controlled basis for joint development, analysis and discussion. In particular, in order to make the development of the interfaces more accessible and focus the drawing on the substance rather than (ironically) the form, the available graphical elements are restricted to the most commonly used ones (`forms`, `fieldsets` and `tables`, `inputs`, `selections` and `buttons`) and limited the layout of forms as a vertical sequence of elements, which also simplifies the transition from the form model and the ER model. This drawing phase is supported by the RAINBOW Toolkit, which is the dedicated and integrated tool support intended to assist end-users and analysts during the different RAINBOW processes.

The interfaces being drawn by non experts and possibly multiple end-users increases the possible inconsistencies among the individual labels and the structures used in the forms [2]. The semantic and structural similarities are therefore analysed to manage commonality and standardise the form constructs and their underlying data counterpart. Semantic similarities arise due to the richness of written natural language, which can lead to spelling and meaning ambiguities. Structural similarities which occurs when two entity types share a pattern, which is a bijection between two sets of attributes belonging to different entity types. RAINBOW deals with the elicitation and subsequent unification of semantic similarities using String Metrics, Ontologies and dictionaries. Though the forms and their underlying data models have a tree-like arborescence, their structure is simple and does not necessitate using complex techniques such as tree mining approaches to discover structural similarities. The shared patterns are instead elicited by comparing the different entity types, and the structures are subsequently unified according to the meaning of the pattern (equality, union, comprehension, complementarity, composition or difference).

The RAINBOW approach also deals with the integration of each schema corresponding to a form into a single normalised schema representing the domain of application, before generating and testing the resulting associated applicative components. However, before leading these processes, an important step consists in eliciting the relevant constraints and dependencies of the domain of application.

2.2   Constraints and Dependencies in Conceptual Schemas

When designing a conceptual schema, it is indeed important to define a set of constraints that will guarantee that once the subsequent database is implemented and operational, any change made to its content by authorised users will maintain its consistency. Typically, inserting, modifying or deleting values from the database should not result into data anomalies or unnecessary redundancies. For this purpose, let us introduce the *relational model* of a database according to the *First normal form* (1NF), which is a database model based on first-order predicate logic, first formulated and proposed by Codd [3]. In the relational model, all the data is represented through *relations* (also known as *tables*). A relation is composed of *attributes* (a.k.a. as *columns*), each of which is defined on a *domain*, which is a given set of values. A *tuple* (a.k.a. *row*) contains all the data of a single instance, that is, a value for each attribute of the relation. Relations and attributes of the relational model will be used to model entity types and attributes in the GER model[1], as illustrated in Figure 1.

The transition between conceptual and relational models is facilitated in the RAINBOW approach since the complex conceptual schemas obtained from the form-based interfaces are recursively transformed until they reduce to simple schemas including flat entity types (with elementary attributes) and binary relationship types. Coping with these relationship types in the 1NF relational modeling requires a little trick; the roles are represented through *role attributes*, so that they appear as attributes for which the value belongs to the possible tuples of the entity type associated through the binary relationship type. In addition, when no value is provided in a form for a given attribute, a default "empty" value (noted ∅) is encoded, so that null values are avoided.

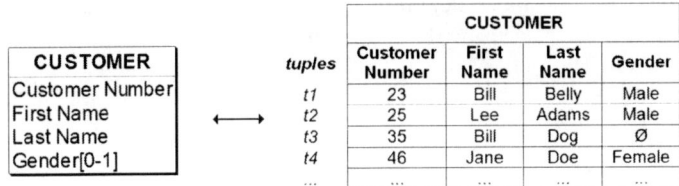

Fig. 1. The representation of a Customer using the GER and relational model

Among the many *constraints* usually found in database schema, we have selected the following ones. *Domains of values* restrict the possible values of given attributes, for instance using domain types, sets or ranges of (un)authorised values, rule-based formulas for the values, and so on. For instance, a `Customer Number` may be restricted to integer values, and the `Gender` limited to the set of values {*Female, Male*}. *Cardinality constraints* define the minimal (typically

---

[1] The Generic Entity-Relationship (GER) model is a wide-spectrum model used to describe database schemas in various abstraction levels and paradigms [4].

zero or one) and maximal numbers (typically one or infinite) of occurrence(s) of given attributes and roles. *Existence constraints* define how optional components (attributes and roles) may influence each other. For two components A and B, these constraints can be:

- *coexistence*, which implies that A and B must always be not null simultaneously;
- *exclusion*, which implies that A and B cannot be not null simultaneously;
- *at-least-one*, which implies that A and B cannot be null simultaneously;
- *exactly-one*, which implies that if A is not null, then B should be null, and vice-versa;
- *implication*, where A implies B means that A can be not null only if B is not null itself;

For a relation, an *identifier* (a.k.a. *candidate key*) is a set of attributes so that, when considering all the possible tuples of the relation, there cannot be more than one tuple having the same combination of domain values for these attributes. For instance, from the tuples visible in Figure 1, we could assume that `Customer Number` forms an identifier for relation `Customer`, since there are no two tuples with the same value of this attribute.

A similar notion is the concept of *functional dependencies*, which are materialised by the explicit or implicit constraints between two sets of attributes in a relation. Given relation R, a set of attributes $X \subseteq R$ is said to functionally determine another set of attributes $Y \subseteq R$, if and only if all the tuples with the same combined values of X also have the same combined values of Y. This functional dependency is written $R : X \rightarrow Y$, with X and Y respectively called the *left-hand side* (a.k.a. *determinant*) and *right-hand side* (a.k.a. *dependent*) of the functional dependency f. For instance, from the tuples visible in Figure 1, it seems that the functional dependency `Customer:First Name` $\rightarrow$ `Last Name` does not hold, since there are two persons named "Bill" but with a different family name. On the other hand, the functional dependency `Customer:First Name, Last Name` $\rightarrow$ `Gender` could hold, but would need to be validated.

3   State of the Art in Constraints and Functional Dependencies Mining

Analysing the content of a database or a subset of data samples can intuitively lead to make plausible assumptions on, e.g., the domains of values, the cardinalities of the attributes, their existence constraints and possibly their identifiers. Let $t[\mathcal{C}]$ be the restriction of a tuple t to the set of components $\mathcal{C}$ (called *projection* of t onto $\mathcal{C}$), and $t[C]$ be the restriction of t onto a given component C. Now consider for instance an optional textual attribute A: if for any tuple t_i, we observe that $t_i[A]$ is never null and always composed of a number, we could easily wonder if A is not actually a mandatory numeric attribute. Moreover, if all the t_i have different values A, this could suggest that A is in fact an identifier. The same kind of *induction* can be applied on optional attributes to assess their possible existence constraints. However, functional dependencies mining is far less trivial.

Back in 1995, Ram presented four types of approaches to derive functional dependencies from an existing conceptual ER schema [5]. The first category consists in using keyword analysis to identify intra-entity functional dependencies: typically, attributes bearing a suffix or prefix such as "id" or "number" should be considered potential determinants, while attributes bearing a suffix or prefix such as "maximum", "minimum", "average" or "total" should be considered potential dependent attributes. The second category consists in analysing the cardinalities of binary relationships to identity inter-entity functional dependencies, typically between their identifiers. The third category is similar, but concerns N-ary relationships. And finally, the fourth category consists in analysing sample data to elicit undiscovered functional dependencies. These principles were supported by the FDExpert tool.

The first three categories rely on the analysis of the schema itself, while the latter category, known as the *dependency discovery problem*, focuses on the content of the database. The latter category is a standard issue, especially in data mining, database archiving, data warehouses and Online Analytical Processing (OLAP). The most prominent existing algorithms dealing with this issue can be classified in three categories, that are difficult to compare qualitatively [6,7].

The first two categories basically try to explore the search space (i.e. the possible combinations of the attributes for a given relation) in the most efficient way possible, in order to test the associated functional dependencies using a stripped partition database computed from that relation. The *candidate generate-and-test* approach progressively explores and prunes the search space in a levelwise manner, while partitioning the database using attribute-based equivalence classes, as in Huhtala et al.'s TANE [8], Novelli and Cicchetti's FUN [9], or Yao and Hamilton's FD_Mine [7]. The *minimal cover* approach structures the search space using hypergraphs that are explored to discover the minimal cover of the set of FDs for a given database, i.e. the minimal set of FDs from which the entire set of FDs can be generated using the Armstrong axioms, as in DepMiner, proposed by Lopes et al. [10] and FastFDs, proposed by Wyss et al. [11].

Finally, *Formal concept analysis* (FCA) has also been used recently to find and represent logical implications in datasets [12], mainly through a closure operator from which concepts (closed sets) can be derived. For instance, Baixeries uses Galois connections and concept lattices as a framework to find functional dependencies [13], while Rancz et al. optimise an existing method introduced by [14], which provides a direct translation from relational databases into the language of power context families, in order to build inverted index files to optimise the elicitation the functional dependencies in a relational table through the construction of their formal context [15]. The latter authors also developed the subsequent FCAFuncDepMine software to detect functional dependencies in relational database tables [16]. Similar principles were also used in Flory's method, which was based on the definition and analysis of a *matrix* and its associated *graph* of functional dependencies [17].

4 An Interactive Process to Elicit Constraints and Functional Dependencies

4.1 Overview

As we have seen, traditional elicitation techniques may neglect constraints, while dependency discovery approaches rely on massive pre-existing data sets, which is problematic when there is no data samples available, or when their re-encoding would be too expensive. To tackle this problem, the RAINBOW approach proposes to use form-based interfaces that were previously drawn by the end-users themselves in order to let them provide a limited set of data samples from which constraints and dependencies could be inferred and suggested. Though such constraints can be provided directly, it appears that the interactive acquisition and processing of data samples is useful and more natural in this process, as it also helps to visualise the implications of existing constraints.

In this section, we present an interactive process inspired by the principles of *Armstrong relations*, which are relations that satisfy each functional dependency implied by a given set of functional dependencies, but no functional dependency that is not implied by that set [18]. This twofold process focuses on eliciting the constraints and dependencies mentioned in Section 2.2, i.e. domains of value, cardinalities, existence constraints, identifiers and functional dependencies. On the one hand, data samples are acquired to restrict the potentially "hidden" constraints, and on the other hand, potential constraints are arbitrated to conversely restrict the tuples that can be encoded. Some of these properties can be trivial and may be expressed directly, or have been expressed during the preliminary drawing of the form-based interfaces. For instance, in the form of Fig. 2, the Last Name of a Person is mandatory, while its Title appears to be optional. Likewise, the Birth date has been encoded as a date value, while the Zip code of the Contact may have been encoded as a textual value. However, the specified properties may need to be refined (for instance, the Zip code may prove to be numeric), and there may be some unsuspected constraints and dependencies among the elements of the schema.

We therefore propose to start by envisaging initial potential constraints and dependencies. Then, using user input, we progressively *validate* or *discard* them, and generate alternatives until they are all arbitrated. This process hence relies on several sub processes:

- the initialisation of all the currently *declared* (explicitly expressed during the drawing step) and *potential* (implicitly verified by the present set of tuples) constraints and dependencies;
- the acquisition and analysis of new valid data samples in order to automatically discard the invalid potential constraints and dependencies, and possibly generate acceptable potential alternatives;
- the arbitration of potential constraints and dependencies through user *validation* or *discardure*, and the subsequent generation of new potential alternatives;

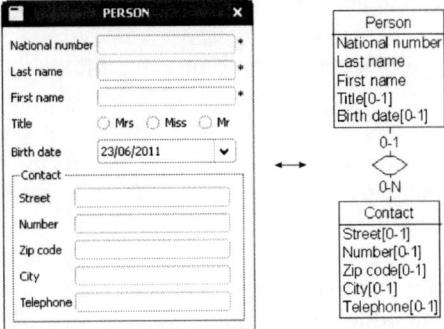

Fig. 2. A form-based interface describing a "Person" and its associated conceptual schema, using the GER representation[2]

- the processing of the validated constraints and dependencies, once there are no other potential constraints or dependencies left.

The acquisition of data samples progressively restricts the set of potential constraints, while conversely, validating constraints also restricts the future data samples that will be encodable.

4.2 Initialisation

Before beginning the interaction with the end-users, we start by initialising an empty set of tuples and defining the initial sets of validated, potential and discarded constraints and dependencies for each entity type associated with a given form. The *validated* constraints are initially the same than the previously *declared* constraints. From these initial validated constraints, the *potential* domains of value, cardinalities and existence constraints are initialised using *induction*. Typically, if a given component is considered optional so far, it could actually be mandatory if there is no tuple with this component empty (whereas the opposite is not possible). Similarly, an attribute declared as textual could be of any other type, while an attribute declared as real could only be restricted to the type integer. In the same way, any subset of optional components for which no existence constraint has been declared should be submitted to existence constraint elicitation. Consider for instance that the attribute `Zip code` has been declared optional and textual in Fig. 2. Further examination should therefore check whether this attribute is not mandatory and whether its value domain is not restricted to integers, reals, dates, ...

Regarding functional dependencies, the ideal process should lead us to build a set of data samples and dependencies so that each entity type of the underlying conceptual schema becomes an *Armstrong relation*. Reaching such a state is

---

[2] Note that the GER notation uses the *participation* interpretation rather than the *look-across* interpretation (as in UML). In the given example, the notation therefore indicates that same contact details may apply to more than one person.

obviously not trivial *per se*, and these principles are here inapplicable due to the requirement of user involvement. However, we can try to near it by progressively narrowing the functional dependencies. Since the number of possible functional dependencies for each entity types can be very high, we start from the set of *strongest* dependencies, through which each component of a given entity type determines the other components. For instance, the form of Fig. 2 induces the initial functional dependencies F_1, F_2, F_3, F_4, F_5 and F_6 of Fig. 3. From these dependencies, we will be able to recursively generate *weaker* functional dependencies to cover all the existing ones, by progressively reducing the right-hand sides and enlarging the left-hand sides. The objective is to favour functional dependencies with minimal left-hand sides and maximal right-hand sides.

F1: National number → Last name, First name, Title, Birth date, Contact
F2: Last name → National number, First name, Title, Birth date, Contact
F3: First name → National number, Last name, Title, Birth date, Contact
F4: Title → National number, Last name, First name , Birth date, Contact
F5: Birth date → National number, Last name, First name, Title, Contact
F6: Contact → National number, Last name, First name, Title, Birth date

~~F2: Last name → National number, First name, Title, Birth date, Contact~~
F21: Last name, National number → First name, Title, Birth date, Contact
F22: Last name , First name → National number, Title, Birth date, Contact
F23: Last name , Title → National number, First name, Birth date, Contact
F24: Last name, Birth date → National number, First name, Title, Contact
F25: Last name, Contact → National number, First name, Title, Birth date

Fig. 3. The initial functional dependencies F_1, F_2, F_3, F_4, F_5 and F_6 for form of Fig. 2, as well as alternatives for F_2

Finally, potential *unique constraints* are induced from validated and potential functional dependencies, using the fact that the left-hand side of a given functional dependency $f : X \rightarrow Y$ is a potential identifier for an entity type having the set of components $\mathcal{C} = X \cup Y$.

4.3   Analysing New Data Samples to Suggest Constraints and Dependencies

Once the sets of constraints and dependencies have been initialised, we take advantage of user input to acquire data samples that will progressively reduce the set of potential constraints and dependencies. To be consistent with the previously validated constraints and dependencies, any new tuple must respect the latter to be accepted. Once a new tuple is acceptable, we proceed with its analysis to determine which potential constraints and dependencies do not hold any more. The invalidated constraints are discarded, while the invalidated functional dependencies are replaced by alternative dependencies. Let us explicit this process for each type of constraint and dependency when adding a new valid tuple. Fig. 4 illustrates three data samples that could be encoded for the form of Fig. 2, and Fig. 5 illustrated the underlying relation **Person** after the acquisition of these data samples. Despite the apparent structure of attribute **Contact**, this relation is in 1NF. Indeed, the compound value must be considered as a whole whose unique goal is to reference a target tuple in the **CONTACT** table.

First of all, discarding the potential *domains of value* and *cardinalities* that do not hold any more is relatively straightforward, since it consists in removing

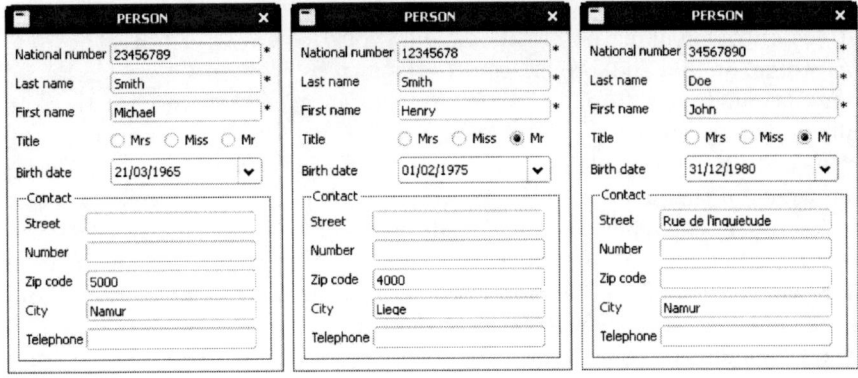

Fig. 4. Three data samples for the form of Fig. 2

	PERSON					
tuples	**National Number**	**Last Name**	**First Name**	**Title**	**Birth date**	**Contact**
t1	23456789	Smith	Michael	Ø	21/03/1965	< Ø, Ø, 5000, Namur, Ø >
t2	12345678	Smith	Henry	Mr	01/02/1975	< Ø, Ø, 4000, Liege, Ø >
t3	34567890	Doe	John	Mr	31/12/1980	< Rue de l'inquiétude, Ø, Ø, Namur, Ø >

Fig. 5. The relation `Person` after the acquisition of the data samples of Fig. 4. The compound value of attribute `Contact` must be considered as a whole whose unique goal is to reference a target tuple in the corresponding `CONTACT` table.

the constraints with which the tuple does not agree. Regarding the cardinalities, we remove the possible mandatory constraints for components that are empty, and we remove the value type constraints that are not compatible with the value provided for each attribute and replace the value size if the provided value is longer. For instance, adding the first data sample confirms that the `Title` is definitely optional, while the `Birth date` potentially remains a mandatory attribute. The `Zip Code` of a `Contact` can now only be validated as integer, real or textual. The second data sample still supports `Birth date` being a possibly mandatory attribute.

Secondly, discarding the potential *existence constraints* that do not stand any more also consists in removing the constraints with which the tuple does not agree. Consequently, coexistence constraints are removed if their set of components is different from the set of non empty optional components of the tuple. Exactly-one, exclusion and at-least-one constraints are respectively removed if there is not one and only one, more than one or less than one of their components that is not null among the set of non empty optional components of the tuple. Finally, we remove all the implication constraints for optional components if the suggested prerequisite components are not part of the non empty components of the tuple. The first data sample for instance suggests that there could be at-least one, at-most one or exactly one value of `Title` or `Birth date`, or that the former could require the latter (implication). The second data sample implies that the only remaining potential configurations for `Title` or a `Birth date` is at-least-one, or that the former requires the latter.

We also analyse each potential *functional dependency* to check if there is a conflictual tuple among the previously provided tuples, i.e. if an existing tuple has the same left-hand side but a different right-hand side when considering the components of the functional dependency. If such a conflictual tuple exists, the functional dependency is discarded and alternatives are recursively generated. First of all, this implies that the right-hand side may be too large with respect to left-hand side, and we therefore consider smaller right-hand sides by removing a component. The removed component may be purely dismissed, or added to the left-hand side to consequently generate two alternatives per component. For instance, the first data sample doesn't jeopardize the potential functional dependencies, but the second data sample discards the FD F_2 and generates the alternatives F_{21}, F_{22}, F_{23}, F_{24} and F_{25} of Fig. 3. The second data sample then discards the FD F_4 and generates its subsequent alternatives.

Understanding the implications of a functional dependency is not always trivial and easy to grasp. Presenting the end-users with automatically generated data samples that would contradict the validity of existing functional dependencies therefore helps them to visualise the relevance of these dependencies, while reducing the number of tuples that they would need to provide by themselves. As we can observe, a tuple t is actually problematic for the functional dependency $f : X \rightarrow Y$ and the existing set of tuples $\mathcal{T}$ if $\exists\, t' \in \mathcal{T} : t'[X] = t[X] \wedge t'[Y] \neq t[Y]$. If we already have several tuples in the tuples set of a given entity type, we generate problematic data samples for a given dependency by putting together previously provided data samples. Accepting such a generated data sample would imply discarding the associated functional dependency and generate alternatives. For instance, considering the functional dependency F_{23}, we generate the problematic tuple illustrated in Fig. 6 from the composition of the second and third data samples of Fig. 4.

Finally, potential *unique constraints* are again induced from validated and potential functional dependencies, using the same principle than during the initialisation of the process, i.e. the left-hand side of a functional dependency involving all the components of a given entity type is a potential identifier for that entity type.

4.4   Acquiring Constraints and Dependencies

Another way to take advantage of user input is to directly acquire validated or discarded constraints and dependencies, whenever they are trivial and easy to express for the participants, and to invite them to arbitrate the potential constraints and dependencies that could be suggested after the acquisition of multiple data samples. The end-users should indeed be able to directly specify validated or discarded constraints and dependencies, even without looking at possible suggestions. To be accepted as validated, a given constraint or

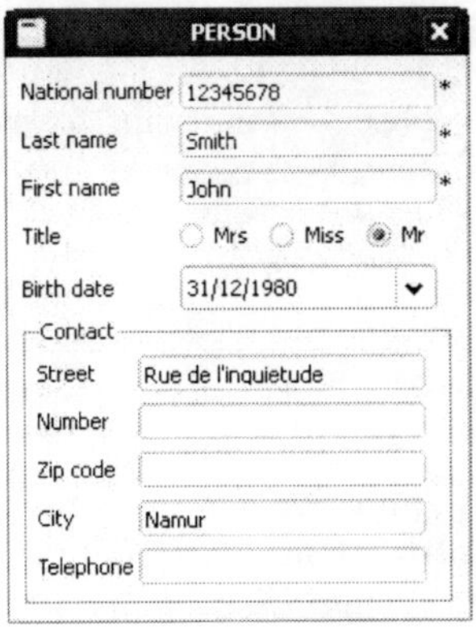

Fig. 6. A problematic data sample for the FD `Title` → `National number`, `Last name`, `First name`, `Birth date`, `Contact`, given the valid data samples of Fig. 4

dependency must be satisfied by the existing set of tuples associated with the considered entity type.

Alternatively, the participants can also take advantage of the potential constraints and dependencies to arbitrate them, i.e. to validate or discard them. The advantages of this approach are that the participants do not have to imagine all the possible constraints and dependencies for each entity type, and that we directly know that each candidate constraint or dependency is currently potential for the given entity type. One can suspect that validating or discarding a constraint or a dependency may impact on the constraints or dependencies of other types. Such a correlation actually exists between functional dependencies and unique constraints. Indeed, discarding a potential functional dependency may change the potential unique constraints, whereas validating a unique constraint automatically validates its underlying functional dependency and discards others. When these cases occur, the relevant sets must therefore be updated.

Besides, it obviously appears that the number of suggested constraints and dependencies can eventually become very high. It is therefore crucial to organise these suggestions in an accessible fashion, so that the end-users do not feel overwhelmed. Besides, this underlines the importance of the analyst to guide the end-users through this collaborative process, by assessing the relevance of these suggestions. This observation is especially true regarding the elicitation of the functional dependencies, since the number of suggestions can increase dramatically.

We therefore propose to filter the potential functional dependencies in order to limit the number of relevant suggestions, while privileging the "stronger" functional dependencies (i.e. the dependencies with smaller left-hand side and larger right-hand side, as previously explained). For this purpose, we therefore propose to "hide" dependencies that can be obtained from other potential dependencies using *Armstrong's axioms*, which are a set of inference rules used to infer all the functional dependencies on a relation [19]. Hiding these functional dependencies does not mean discarding them. Indeed, they are still potential, and may eventually become visible again with the progressive arbitration of the other dependencies. Still, we observed that this filtering helps keeping the focus of the end-users during this elicitation process.

5   Evaluation

To experiment and evaluate the RAINBOW approach, a validation protocol was defined based on the *Participant-Observer* principles to monitor the use of the RAINBOW approach, and the *Brainstorming/Focus group* principles to analyse the resulting conceptual schemas, as defined in [20]. This protocol was used for a first series of experiments where pairs of end-users and analysts were asked to jointly define the conceptual schema of a future information system, including constraints and dependencies, using the RAINBOW approach and its tool support.

For each project, the first task consisted in preparing the experimentation by defining the subject based on real-life concerns of the end-users, then training the participants to understand the method and to use the tools. Secondly, the end-users and analysts were asked to apply the approach on their project and focus on the five first phases, while observers took notes. In particular, for the *Nurture* phase which dealt with the elicitation of constraints and dependencies, the participants were asked to progressively provide data samples and constraints, while arbitrating the candidate constraints suggestions. Thirdly, the observations on the efficiency of the approach were analysed, and finally, the quality of the produced schemas was debated, taking in account schemas that were designed by the analysts independently of the approach.

The analysis of these experiments notably highlighted that the RAINBOW approach and tool support did help end-users and analysts to communicate static data requirements to each other, inclusive of constraints and dependencies. Though all the requirements could not be expressed through the toolkit, the latter did serve as a basis for discussion and modifications. Since the validation aspect of the proposed approach cannot be addressed more extensively in this paper, the interested reader may refer to [21] for further details on the validation process and methodology.

6   Conclusion

In this paper, we extend the user-oriented RAINBOW approach presented in [2], and describe how it can be used to interactively elicit database constraints and dependencies, and more specifically domains of values, optionality, existence constraints, identifiers and functional dependencies. The process, inspired by the principles of Armstrong relations, uses form-based interfaces that were previously drawn by the end-users themselves in order to let them provide a limited set of data samples that will restrict the potential implicit constraints of the underlying conceptual schema. Conversely, end-users are invited to arbitrate potential constraints that will in turn restrict the tuples that can be encoded. Such a process prevents the development of unsatisfiable systems of constraints and dependencies, since such set of constraints will never accept the introduction of new tuples. Whereas usual dependency discovery approaches rely on extensive data sets, this specific *modus operandi* is particularly adapted when engineering information systems with no legacy data samples available, or when their re-encoding would be too expensive. The application of this approach to different case studies have proved that such intensive end-user involvement with inter-active support is particularly fruitful and sustainable, while merely providing, without support, significant amount of data samples is a particularly tedious and time-consuming process, and in most situations, unrealistic. Besides, manipulat-ing form-based interfaces to encode data samples leads to directly expressing trivial constraints, while inducing further discussion and reflection on their un-derlying conceptual schema. Though this approach relies on a set of pre-existing form-based interfaces, its principles are easily generalisable to any given concep-tual schema. Indeed, the constructs of the schema can first be transformed to comply with the structures used in the RAINBOW approach [4]. Subsequently, a set of form-based interfaces can then be generated from this transformed schema [22], hence enabling the encoding of data samples and the application of the pro-posed approach.

References

1. Rosson, M.B., Carroll, J.M.: Usability Engineering: Scenario-Based Development of Human-Computer Interaction (Interactive Technologies). Morgan Kaufmann, San Diego (2001)
2. Ramdoyal, R., Cleve, A., Hainaut, J.-L.: Reverse engineering user interfaces for interactive database conceptual analysis. In: Pernici, B. (ed.) CAiSE 2010. LNCS, vol. 6051, pp. 332–347. Springer, Heidelberg (2010)
3. Codd, E.F.: A relational model of data for large shared data banks. Communica-tions of the ACM 13(6), 377–387 (1970)
4. Hainaut, J.-L.: The transformational approach to database engineering. In: Läm-mel, R., Saraiva, J., Visser, J. (eds.) GTTSE 2005. LNCS, vol. 4143, pp. 95–143. Springer, Heidelberg (2006)
5. Ram, S.: Deriving functional dependencies from the entity-relationship model. Communications of the ACM 38(9), 95–107 (1995)
6. Lopes, S., Petit, J.-M., Lakhal, L.: Functional and approximate dependency min-ing: database and FCA points of view. Journal of Experimental and Theoretical Artificial Intelligence (JETAI) 14(2-3), 93–114 (2002)
7. Yao, H., Hamilton, H.J.: Mining functional dependencies from data. Data Mining and Knowledge Discovery 16(2), 197–219 (2008)
8. Huhtala, Y., Kärkkäinen, J., Porkka, P., Toivonen, H.: TANE: An efficient algo-rithm for discovering functional and approximate dependencies. Computer Jour-nal 42(2), 100–111 (1999)
9. Novelli, N., Cicchetti, R.: FUN: An efficient algorithm for mining functional and embedded dependencies. In: Van den Bussche, J., Vianu, V. (eds.) ICDT 2001. LNCS, vol. 1973, pp. 189–203. Springer, Heidelberg (2000)

10. Lopes, S., Petit, J.-M., Lakhal, L.: Efficient discovery of functional dependencies and armstrong relations. In: Zaniolo, C., Grust, T., Scholl, M.H., Lockemann, P.C. (eds.) EDBT 2000. LNCS, vol. 1777, pp. 350–364. Springer, Heidelberg (2000)

11. Wyss, C.M., Giannella, C., Robertson, E.L.: Fastfds: A heuristic-driven, depth-first algorithm for mining functional dependencies from relation instances. In: Kambayashi, Y., Winiwarter, W., Arikawa, M. (eds.) DaWaK 2001. LNCS, vol. 2114, pp. 101–110. Springer, Heidelberg (2001)

12. Priss, U.: Establishing connections between formal concept analysis and relational databases. In: Common Semantics for Sharing Knowledge: Contributions to ICCS 2005, pp. 132–145 (2005)

13. Baixeries, J.: A formal concept analysis framework to mine functional dependencies. In: Proceeding of Mathematical Methods for Learning 2004: Advances in Data Mining and Knowledge Discovery (2004)

14. Correia, J.H.: Relational scaling and databases. In: Proceedings of the 10th International Conference on Conceptual Structures (ICCS 2002), Borovets, Bulgaria, July 15-19, pp. 62–76 (2002)

15. Rancz, K.T.J., Varga, V.: A method for mining functional dependencies in relational database design using FCA. Studia Universitatis Babes-Bolyai Cluj-Napoca, Informatica LIII(1), 17–28 (2008)

16. Rancz, K.T.J., Varga, V., Puskas, J.: A software tool for data analysis based on formal concept analysis. Studia Universitatis Babes-Bolyai Cluj-Napoca, Informatica LIII(2), 67–78 (2008)

17. Flory, A.: Bases de données: conception et réalisation. In: ECONOMICA, Paris (1982)

18. Lopes, S., Petit, J.-M., Lakhal, L.: Efficient discovery of functional dependencies and armstrong relations. In: Zaniolo, C., Grust, T., Scholl, M.H., Lockemann, P.C. (eds.) EDBT 2000. LNCS, vol. 1777, pp. 350–364. Springer, Heidelberg (2000)

19. Armstrong, W.W.: Dependency structures of data base relationships. In: IFIP Congress, pp. 580–583 (1974)

20. Singer, J., Sim, S.E., Lethbridge, T.C.: Software engineering data collection for field studies. In: Shull, F., Singer, J., Sjøberg, D.I. (eds.) Guide to Advanced Empirical Software Engineering, pp. 9–34. Springer, Heidelberg (2008)

21. Ramdoyal, R.: Reverse Engineering User-Drawn Form-Based Interfaces for Interactive Database Conceptual Analysis. PhD thesis, University of Namur, Namur, Belgium, (December 2010) Electronic version available from http://www.info.fundp.ac.be/libd/rainbow

22. Pizano, A., Shirota, Y., Iizawa, A.: Automatic generation of graphical user interfaces for interactive database applications. In: CIKM 1993: Proceedings of the Second International Conference on Information and Knowledge Management, pp. 344–355. ACM, New York (1993)

A Conceptual Model for Integrated Governance, Risk and Compliance

Pedro Vicente and Miguel Mira da Silva

Instituto Superior Técnico, Universidade Técnica de Lisboa,
Avenida Rovisco Pais, 1, 1049-001 Lisboa, Portugal
{pedro.vicente,mms}@ist.utl.pt

Abstract. As integrated Governance, Risk and Compliance (GRC) becomes one of the most important business requirements in organizations,
the market is incongruously struggling to satisfy organizations' needs.
The absence of scientific references regarding GRC is leading to a dis-
persion of concepts involving this topic. Without boundaries and correct
domain definition, poor implementation of GRC solutions can lead to low
performances and high vulnerabilities for organizations. This paper proposes a set of high level concepts covering the GRC domain. Through
literature review and framework research we propose key functions of
governance, risk and compliance and their associations, resulting in a
reference conceptual model for integrated GRC. The model was evaluated by comparing the GRC capability model from OCEG with a quality
model evaluation framework. We concluded that the proposed model is
valid and complete.

Keywords: governance, risk, compliance, conceptual model, integrated.

1 Introduction

Some research is starting to finally arise in the study of governance, risk and compliance as an integrated concept. Since PricewaterhouseCoopers introduced the
term GRC in 2004 [1], a bewildering amount of definitions have been presented,
distinguishing in terms of scope and levels of integration.

The first scientific definition for integrated Governance, Risk and Compliance
(GRC) was proposed by Racz et al. [2] and states that: *"GRC is an integrated,
holistic approach to organization-wide governance, risk and compliance ensuring
that an organization acts ethically correct and in accordance with its risk appetite,
internal policies and external regulations, through the alignment of strategy, processes, technology and people, thereby improving efficiency and effectiveness."*

However, if you ask 10 organizations to describe governance, risk and compliance, probably you will get at least 20 definitions [3]. Therefore, there is
not a common understanding of what GRC is. Instead, there are very different
perspectives [4].

Just like Enterprise Resource Planning (ERP), GRC is becoming one of the
most important business requirements of an organization [5], mainly due to the

H. Mouratidis and C. Rolland (Eds.): CAiSE 2011, LNCS 6741, pp. 199–213, 2011.

rapid globalization, increasing regulations like BASEL II, the Sarbanes-Oxley Act (SOX), Anti-Money Laundering (AML), etc., and growing demands of transparency for companies [5].

Traditionally, governance, risk and compliance activities were scattered in silos all over the organization, which has a negative impact on transparency and decision making. GRC activities are important in organizations, not only to boost their performance, but above all, to protect organizations from the inside and the outside. To accomplish this objective, organizations need to shift these activities from niche groups to business units [5] in order to improve these same activities.

Although many organizations agree on the benefits that arise from integrating GRC processes, there is no congruence between software vendors, organizations and market research [4].

In this paper we use conceptual modelling to define the domain of integrated GRC. It is widely accepted that conceptual models are a prerequisite for successfully planning and designing complex systems, particularly information systems [6,7,8,9]. Over the last decades, conceptual modelling has been employed to facilitate, systematize, and aid the process of information system engineering [8].

Based on the four design artefacts produced by design science research in information systems - *constructs*, *models*, *methods* and *instantiations* - we will focus on constructs and models. Constructs are necessary to describe certain aspects of a problem domain and allow the development of the research project's terminology [10]. In other words, they provide the language in which problems and solutions are defined and communicated [11]. Models use constructs to represent a real world situation, the design problem and the solution space [12].

A conceptual reference model, a specific type of conceptual models, is a "claim that the model comprises knowledge that is useful in the design of specific solutions for a particular domain" [10]. A conceptual model is a typically graphical representation, hence can provide limited vocabulary [10], constructed by IS professionals of someone's or some group's perception of a real-world domain [13].

Conceptual modelling may be used to ease the implementation of an information system or to provide a common understating between the organization's needs and an enterprise application [13]. It is also suitable to systematize knowledge, provide guiding research and map a portion of reality [14].

In this paper, we use conceptual modelling to supply a reference model to the scientific community that can lead to a common understanding of what constitutes the universe of integrated GRC. Currently, the most complete and recognized framework for integrated GRC was developed by the "Open Compliance & Ethics Group"(OCEG). OCEG is a non-profit organization that uniquely helps other organizations to enhance corporate culture and integrate governance, risk management, and compliance processes. The GRC Capability Model [15] is the central piece of the OCEG framework and describes practices to implement and manage GRC activities.

Our approach is to design a conceptual model that contains domain level concepts, representing a high level of integration between the following sub-domains:

governance, risk management and compliance. The higher the semantic content of those concepts, the better the integration [7]. Although it may seem impossible to find general and meaningful concepts for the entire domain of integrated GRC, it is better to adopt the so-called "constructive" research strategy [7].

2   Methodology

The methodology applied is divided according to the two processes of design science research in information system, *build* and *evaluate* [16]. The build process is composed by two stages whereas and the evaluation process is composed by only one stage (Fig. 1).

Build		Evaluate
Construct Definition	Conceptual Model Construction	Evaluation
- Conceptual definition - Domain definition - Categorization of concepts	- Analysis of relations between concepts - Integration of the three domains	- OCEG Capability Model - Quality Assessment

Fig. 1. Research Methodology

The first stage, construct definition, has two main milestones: conceptual domain establishment and conceptual definition within the set up boundaries established. In this stage we have proceeded with literature study and benchmarking of integrated GRC solutions in the market. Throughout it, we have come to support the observations made by Racz et al. [2]: "there is basically no scientific research on GRC as an integrated concept", "software vendors, analysts and consultancies are the main GRC publishers" and "software technology is the prevailing primary topic". Hence, gathering solid information was a hard task due to the lack of scientific research. Also, at this stage, we began to categorize the concepts that we will present in Sect. 3.

According to Hevner et al. [17], the results from this stage can be called constructs. "Constructs provide the vocabulary and symbols used to define problems and solutions" within an outlined domain. To favour the boundary definition of the domain, we used the design science research pattern proposed by Vaishnavi and Kuechler [18], *building blocks*, which consists in dividing "the given complex research problem into smaller problems that can form the building blocks for solving the original problem". Especially in this case, we divided the domain in G, R and C areas so as to simplify it and the concepts involved.

In the second stage the concepts were separated according to their most evident domain. For example, risks are more likely to belong to the risk domain (R in GRC). However, this does not imply that they could not be represented in governance and compliance domains for they might maintain relations with other concepts. One of the goals of this phase was to identify the concepts duplicated among domains. This way we could determine the integration points between the three areas. Also, by having concepts divided into smaller domains, it became simpler to define the relations between them.

Still at this stage, three conceptual models were built, one for each area, G, R and C (Sects. 3.1, 3.2 and 3.3). In Sect. 3.4 we present the domain of integrated GRC with concepts and relations adjusted to the integrated context.

Even though little is known about how to validate conceptual models effectively and efficiently [13], in the final stage, we proceeded with the evaluation of the final conceptual model, by mapping the relations between concepts with the eight components of the GRC Capability Model presented by OCEG [15]. We used this mapping to evaluate the quality of the conceptual model according to its syntactic and semantic quality, using the Conceptual Model Quality Framework proposed by Moody et al. [19].

3   Conceptual Model

Information integration is one of the core problems in cooperative information systems [20]. Also, GRC functionalities have shown to overlap themselves [15,21] making integration difficult. Governance, risk and compliance as separate concepts are nothing new [1] and many researchers have addressed each area. The proposed model describes GRC functionalities and information that are considered to be within the scope of each of the three areas (G, R and C).

The components of the model. Before we begin describing each of the three scopes, a proper explanation concerning the model is required. The model has three types of concepts, represented by different colours and different shapes. The rectangular concepts, coloured orange, stand for what we propose to be the GRC main functionalities:

1. Audit Management
2. Policy Management
3. Issues Management
4. Risk Management

We have chosen the four functionalities for three reasons. First, a study performed by Racz et al. [4] concluded that Risk Management, Policy Management and Audit Management were mentioned seven times by GRC vendors as GRC functionalities. Issues Management was mentioned six times. Second, we decided to propose these four core functionalities to maintain the conceptual model simple without withdrawing GRC capabilities. Finally, although there are diverse opinions, the benchmarking performed supports these functionalities. The importance and role of each one will be described in the next sections.

Additionally, rectangular concepts, coloured grey (Reporting, Dashboards and Monitoring), also represent imperative functionalities to access and deliver important information in real-time through an automated manner. It is arguable that the four main functionalities presented implicitly cover reporting, dashboards and monitoring but we opted to include them since they represent essential functions for GRC to perform in an adequate, efficient and effective basis [22].

For this reason, they are explicitly represented. We have distinguished these four from the key functions, because they represent horizontal functionalities available through the three areas.

The concepts, in a blue round shape, represent information that is managed by these functionalities or are presented as a responsibility of the G, R or C areas. As stated before, G, R and C areas overlap [15,21], and some information is managed by different areas simultaneously. One way to observe the points of integration of GRC is through the information that is used collaboratively between governance, risk management and compliance.

Next, we address governance, risk and compliance separately and in more detail.

3.1 Governance

OCEG states that "governance is the culture, values, mission, structure, layers of policies, processes and measures by which organizations are directed and controlled" [15]. According to this definition, one of the most important responsibilities of governance is to determine guidelines, which are translated into policies composed by culture, values, mission, objectives and supported by procedures (see Fig. 2).

Policy Management, a key functionality, can be said to be an important activity with direct governance responsibility. Policy management must "develop, record, organize, modify, maintain, communicate, and administer organizational policies and procedures in response to new or changing requirements or principles, and correlate them to one another" [23].

Policies play an essential role at GRC, because they represent the board and top management's point of view on how the organization should be driven. It can be said that governance defines an interface, and the rest of the organization implements it to operate according with what is established. Once agreed upon, policies have to be transmitted across the organization. It is also important that they be reviewed and preserved. It is all part of the policy life cycle that must be set up (Fig. 2).

Since governance defines how the organization should perform, describing through policies what is acceptable and unacceptable, compliance is the area responsible for inspecting and proving that they are: adequate, being implement and followed. In Sect. 3.3 we will address the influence of compliance in policy management in more detail.

Governance is also responsible for risk and compliance oversight, as well as evaluating performance against enterprise objectives [21]. "The board acts as an active monitor for shareholders' and stakeholders' benefit, with the goal of Board oversight to make management accountable, and thus more effective" [15]. Accordingly, governance should be able to understand and foresee the organization's vulnerabilities and, hence make decisions to reduce them.

Also, governance should distribute power to provide insight and intelligence, at the right time, so that the right people in the management can make risk-aware decisions in accordance with key business objectives. Risk-awareness is possible

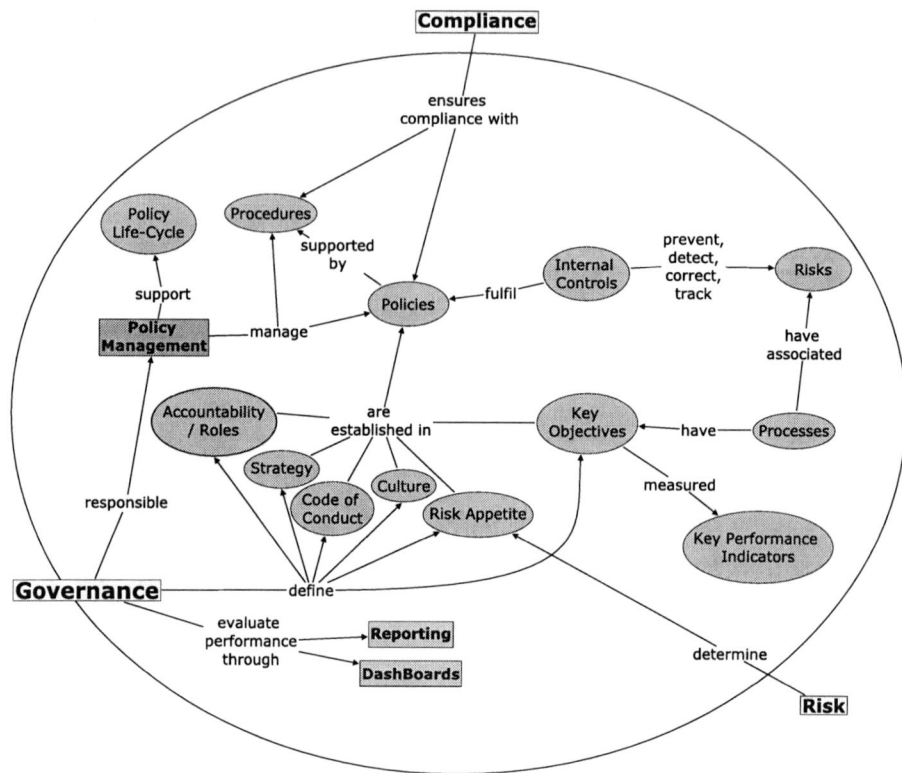

Fig. 2. Conceptual Model for Governance

through the close proximity that governance should have with risk management, which may provide very useful information in strategy setting and decision making. We will address the relation with risk management in Sect. 3.2.

Controlling the organization over intelligent, reliable and real-time information that is available through dashboards, appropriate reporting and monitoring mechanisms, provides C-level executives a paramount tool for an effective and efficient supervision of the performance of all GRC activities.

3.2   Risk Management

Risk management is more than to just identify and respond to risks. Risk management enables us to predict and avoid risk taking consequently decreasing the possibility of unexpected events to occur. A well-structured risk management must be aligned and linked with both governance and compliance information in order to attain advantages (Fig. 3).

According to OCEG [15], risk management is "the systematic application of processes and structure that enable an organization to identify, evaluate, analyse, optimize, monitor, improve, or transfer risk while communicating risk and risk

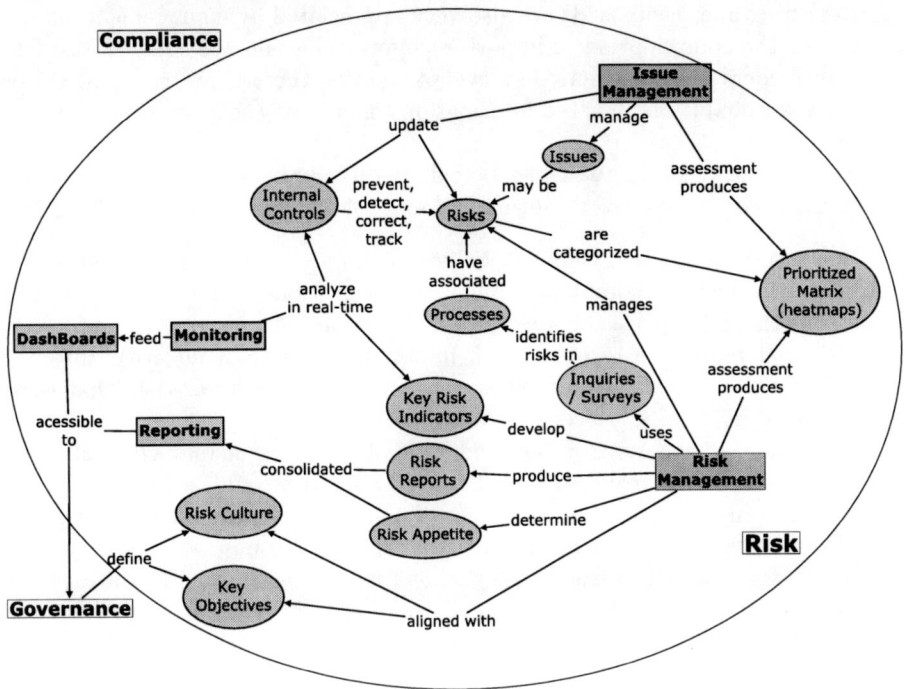

Fig. 3. Conceptual Model for Risk Management

decisions to stakeholders". A strong risk management structure can provide for a better decision making and strategy setting.

Nowadays, risk management itself cannot take full advantage of its features. It needs structured governance and compliance management in order to better align business aims with risks and assist audit management in improving controls which in turn will help detect and prevent risks. This way the organization as a whole can benefit from all risk management capabilities.

So, in order to make risk management more effective in detecting and mitigating risks that can compromise the achievement of business goals, risk identification should be based on a holistic top-down approach by aligning risk management with key corporate objectives defined by governance (see Fig. 3). This approach enables risk management to be infused into the corporate culture, quickly identifying gaps, while maintaining a proactive approach [24]. Accordingly, risk appetite must be seen as a component of both the culture and strategy of organizations.

By identifying information that is mutual or has influence between governance and risk management, we can identify several specific points of integration:

1. The defined corporate objectives should be taken into consideration in the identification of risks, adopting a top-down approach while avoiding an expensive and ineffective bottom-up approach;

2. Reporting and dashboards are also very appreciated by management, allowing for the consolidation of important information, in real-time. It also lets stakeholders reach an increased level of trust on the organization since they possess valuable and trusted information concerning the level of exposure to risks;
3. The level of risk appetite must be collaboratively defined in order to make governance and business performance more risk-aware in decision making [15].

Another important aspect that can be very helpful in risk identification is the information concerning complaints, incidents, suggestions, etc., that are reported when something happens. This we present as issues. An issue is a nonroutine stimulus that requires a response [25]. It may be positive or negative, internal or external to the organization. Issues can be risks that occur or risks that were not identified in the first place.

As risk management acts on the prediction of events, issue management identifies threats that occurred and need to be categorized and addressed. Additionally, it is in the organization's interest not only to correct what is wrong, but also to have a mechanism in place that could help improve the organization itself, for example, through suggestions from clients. By integrating this functionality in the GRC system, the information from issues management can be helpful in identifying new sources of risk and improve the activities of the organization.

Monitoring plays a crucial role on the efficiency of risk management, since it provides the capability to effectively and efficiently identify potential risks and issues. Therefore, it gives the organization the key to identify opportunities and mitigate "risks in the context of corporate strategy and performance" [24]. Internal Controls can be seen as a monitoring tool, since their role in risk management is to help prevent, detect, correct and also track risks.

Monitoring, reporting and dashboards are essential in risk and issue management because they allow organizations to answer very important questions: What are our top 10 risks? What is the percentage of issues that were identified as risks? What are the impacts of those risks and what is their status? Which risks can our organization endure? What objectives are compromised?

3.3   Compliance

Compliance must assure that the organization is following all its obligations, and thus is operating within the defined boundaries. According to OCEG, "compliance is the act of adhering to, and the ability to demonstrate adherence to, mandated requirements defined by laws and regulations, as well as voluntary requirements resulting from contractual obligations and internal policies" [15]. Through this definition, the relation between governance and compliance becomes clearer.

Compliant organizations need an effective approach to verify that they are in conformity with external (standards, regulations) and internal (internal policies) rules. This approach is assisted by risk management, which must identify and

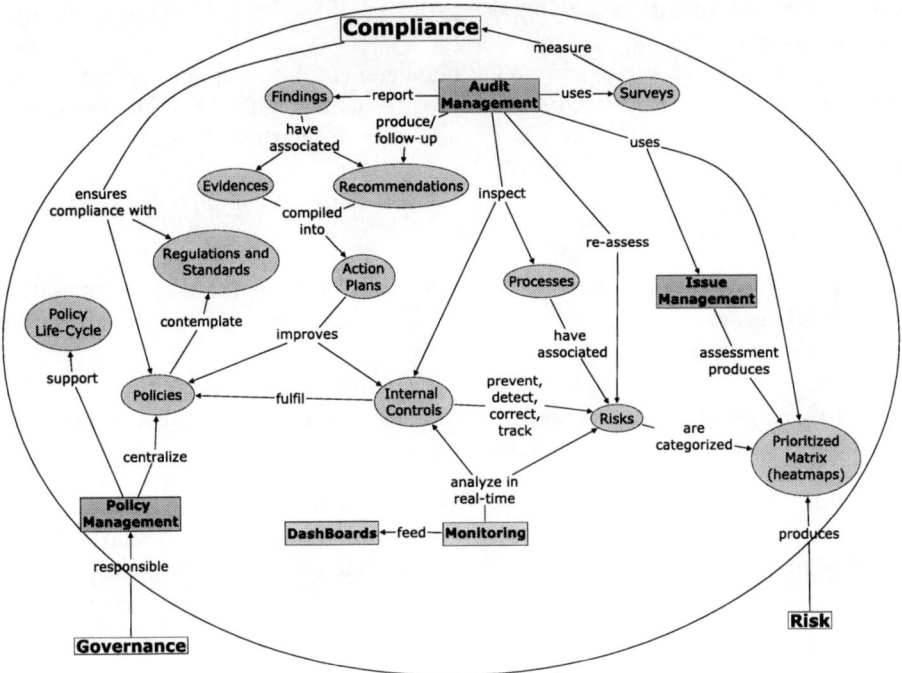

Fig. 4. Conceptual Model for Compliance

prioritize risks that are already aligned with corporate objectives defined by governance (Fig. 4).

This way, audit management, one of the key components of GRC, is responsible for auditing the processes or departments of the organization in which risks that menaced and compromised the achievement of goals were identified. By having risks aligned with objectives, audit teams can address the most important threats that place organizations' compliance under risk. Audit management is responsible for internal controls testing and policies review [22] in order to report findings and produce recommendations that will subsequently improve controls and policies (Fig. 4). Findings and issues are very similar. Organizations, therefore, need to pay close attention to them to know what needs to be fixed, who is responsible and what is the progress in accomplishing it [22].

Although audit management is very important and a crucial piece of the puzzle, it must be presented as an independent and neutral component [21], so as to preserve reliable conclusions and results that can be translated into important improvements. Consequently, compliance is responsible for defining the tactical approach that the organization should follow in order to be compliant with standards and regulations and translate it to policies and procedures. By tactical approach, we mean implementing communications so that everyone knows about the compliance problems [21], through training, surveys and self-assessments.

This is very much related to policy management, as compliance must determine if the organization is conforming to its defined policies. If it is not, the organization must take the necessary measures to upgrade the current policies and, thus influence the policy life-cycle.

Summarizing, we can identify more relations between compliance, governance and risk areas:

1. Risk categorization is used to schedule and prioritize audits. Consequently, investigations and recommendations have an impact on risks due to the improvement of controls;
2. Policies are reviewed and improved by compliance, mirroring the impact of external regulations, standards and audits, and thus has an influence on policy management and the inherent life-cycle of policies.

Real-time monitoring also provides the opportunity to eliminate or greatly reduce sample-based audits [26]. This way, through continuous monitoring, auditors can rely in the existence of automated controls as evidence of compliance [26].

3.4   Integrated GRC Conceptual Model

In this section we present an integrated view of the three scopes presented(Fig. 5). The points of integration that we specified in each section are now combined in an integrated model. We opted not to include monitoring, dashboards and reporting to remove further complexity from the model.

As previously stated, internal controls are paramount in this model since they are crucial for governance, risk and compliance activities [15]. Controls are clearly a common thread among the GRC components (Fig. 5). An organization should, then, develop and implement adequate controls that mirror policies and procedures' objectives.

According to the Committee of Sponsoring Organizations of the Treadway Commission (COSO), controls are also indispensable to achieve key business objectives through the mitigation of risks that menace the same objectives, and thus have a tremendous impact on effective risk management. Compliance manages controls through audit management, which is responsible for testing and improving controls based on findings and respective recommendations, a travail of auditors' work. By having adequate, effective and efficient controls, organizations are not only better prepared and safeguarded from external audits, but also guarantee organizations' health.

Risks and processes are also presented with a central role in integrated GRC, because they are linked to everything. In all activities, there are processes and subsequently, risks. In order to successfully and proficiently manage all GRC activities, processes must be associated with risks, and risks have to be linked with controls. This way, all information is organized, making it highly manageable and traceable.

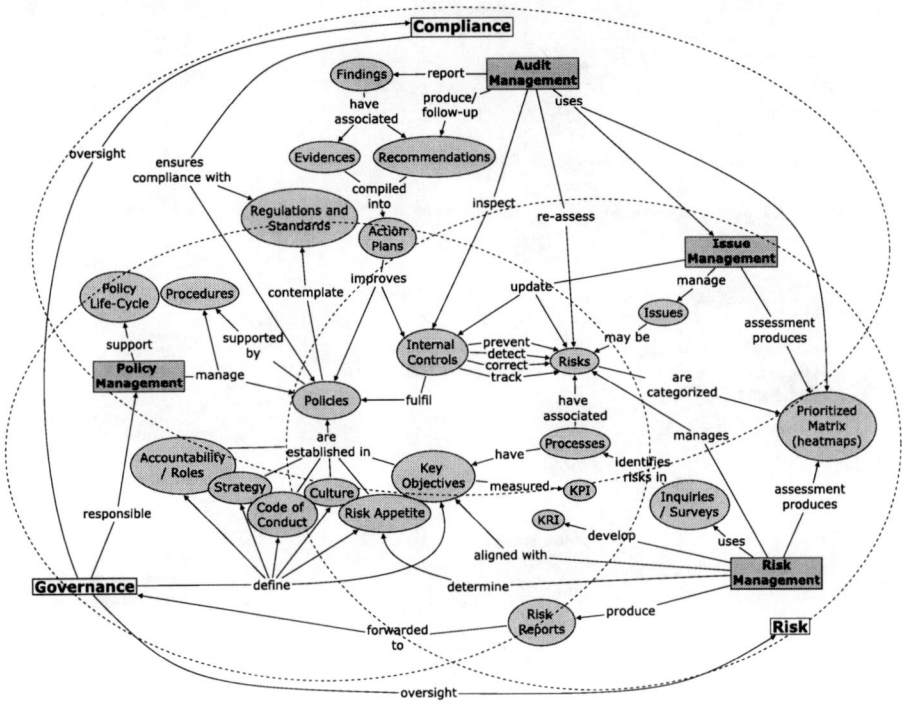

Fig. 5. Integrated GRC Conceptual Model

Finally, we opted to include policies into this crucial group that represents the integration of the three areas. On the one hand, because they are linked to controls that help ensure the fulfilment of policies, and on the other hand, because policies articulate culture and accountability at the level of governance, risk and compliance, consequently having an impact across the entire organization.

The integrated conceptual model in Fig. 5 shows the information with central roles in integrated GRC, thus it should be centralized and properly associated.

4   Evaluation

4.1   OCEG Capability Model

We opted to map the relations between the concepts of the model with OCEG Capability Model components (Fig. 6), a recognized framework that provides eight components that gather detailed practices (Fig. 7).

The components contain 32 associated elements with 132 practices. The relations that cover elements and practices of the component have been coloured with the according shade attributed to the component(Fig. 7).

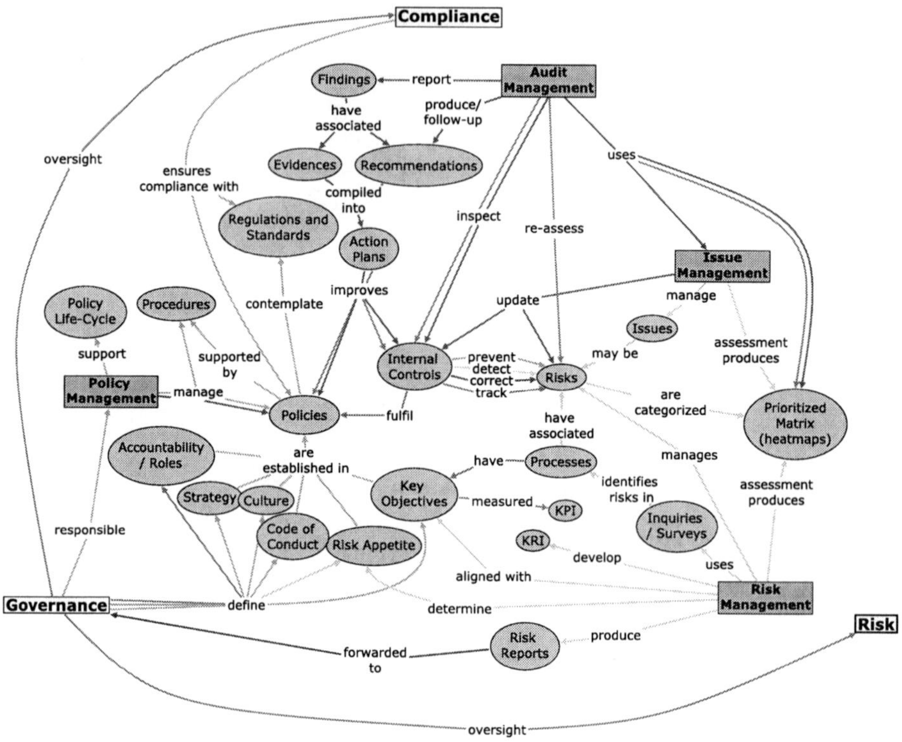

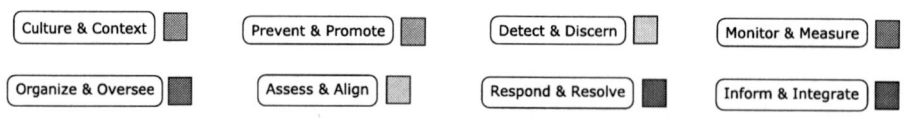

Fig. 6. Mapping between the Reference Model and the OCEG Capability Model

Culture & Context

Prevent & Promote

Detect & Discern

Monitor & Measure

Organize & Oversee

Assess & Align

Respond & Resolve

Inform & Integrate

Fig. 7. GRC Capability Model Components

4.2   Conceptual Model Quality

The quality framework used to assess the conceptual model (Fig. 8) presents four components (Interpretation, Domain, Language and Model) and three quality categories (Syntactic, Semantic and Pragmatic quality) [19].

A model has syntactic correctness if there are no statements included in the model that are not a part of the language [19]. Syntactic quality is the relationship between the model and the language while semantic quality is the relationship between the model and the domain, and it is divided into two goals: Validity and Completeness. A model is valid if there are no statements in the model that are not correct and relevant about the domain [19]. A model is complete if there are no statements that are correct and relevant about the domain, but are not included in the model [19].

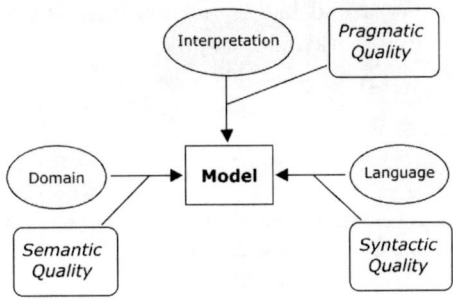

Fig. 8. Conceptual Model Quality Framework - adapted from [19]

The model presented in Fig. 6, shows that every relation is signalled with a colour, proving the validity of the model. Concerning the model's completeness, this attribute is not entirely fulfilled, because some elements of the components were not shown in the conceptual model. Since the language used to create the model was ad-hoc, we will not consider syntactic quality.

The completeness of the model can be measured by calculating the relation between the number of elements and practices covered by the conceptual model and the total number of elements and practices of the OCEG Capability Model. After an analysis of the elements presented in the capability model, we have identified 100 practices and the corresponding 24 elements that our model fulfils, with a result of approximately 76% of coverage (75,75%).

Pragmatic quality is the relationship between the model and the audience's interpretation and has not been accomplished in this research.

5   Conclusion

In this paper, we developed and evaluated a high-level conceptual model for integrated GRC and thus providing new research concerning the topic. The conceptual model was built from the integration of the three domains - governance, risk Management and compliance - but always maintaining an integrated context.

Through the identification of the concepts of each domain, the conceptual models were merged through common concepts and relations between G, R and C, resulting in a conceptual model for integrated GRC. The evaluation was performed by combining two frameworks: the OCEG capability model [15] and a conceptual model quality framework [19].

However, the evaluation is not yet complete. The pragmatic quality of the conceptual model needs to be assessed. As a future research, we will conduct surveys to obtain critical enhancements from GRC professionals in order to improve the model, and thus feed the build and evaluate loop of design science research.

Acknowledgments. We would like to acknowledge the support provided by Methodus to our research work in the scope of an innovation project partly financed by QREN.

References

1. PricewaterhouseCoopers: 8th annual global CEO survey (2004), http://www.grc-resource.com/resources/pwc_integritydrivenperformance.pdf
2. Racz, N., Weippl, E., Seufert, A.: A Frame of Reference for Research of Integrated Governance, Risk and Compliance (GRC). In: De Decker, B., Schaumüller-Bichl, I. (eds.) CMS 2010. LNCS, vol. 6109, pp. 106–117. Springer, Heidelberg (2010)
3. Hagerty, J., Kraus, B.: GRC in 2010: $29.8B in Spending Sparked by Risk, Visibility, and Efficiency (2009)
4. Racz, N., Weippl, E., Seufert, A.: Governance, Risk & Compliance (GRC) Software An Exploratory Study of Software Vendor and Market Research Perspectives. In: Proceedings of the 44th Hawaii International Conference on System Sciences (2011)
5. Gill, S., Purushottam, U.: Integrated GRC - Is your Organization Ready to Move? In: Governance, Risk and Compliance. SETLabs Briefings, PP. 37–46 (2008)
6. Moody, D.L., Shanks, G.G.: Improving the Quality of Data Models: Empirical Validation of a Quality Management Framework. Inf. Syst. 28, 619–650 (2003)
7. Frank, U.: Conceptual Modelling as the Core of the Information Systems Discipline: Perspectives and Epistemological Challenges. In: Proceedings of the Fifth America's Conference on Information Systems (AMCIS 1999), Milwaukee, Association for Information Systems, pp. 695–698 (1999)
8. Recker, J.C.: Conceptual Model Evaluation. Towards more Paradigmatic Rigor. In: Halpin, T., Siau, K., Krogstie, J. (eds.) Proceedings of the Workshop on Evaluating Modeling Methods for Systems Analysis and Design (EMMSAD 2005), Held in Conjunctiun with the 17th Conference on Advanced Information Systems (CAiSE 2005), Porto, Portugal, EU, FEUP (2005)
9. Jeusfeld, M.A., Jarke, M., Nissen, H.W., Staudt, M.: ConceptBase: Managing Conceptual Models about Information Systems. In: Bernus, P., Mertins, K., Schmidt, G. (eds.) Handbook on Architectures of Information Systems. International Handbooks Information System, pp. 273–294. Springer, Heidelberg (2006)
10. Schermann, M., Böhmann, T., Krcmar, H.: Explicating Design Theories with Conceptual Models: Towards a Theoretical Role of Reference Models. In: Becker, J., Krcmar, H., Niehaves, B. (eds.) Wissenschaftstheorie und Gestaltungsorientierte Wirtschaftsinformatik, pp. 175–194. Physica-Verlag, HD (2009)
11. Schon, D.A.: The reflective practitioner: how professionals think in action. Basic Books, New York (1983)
12. Simon, H.A.: The Sciences of the Artificial - 3rd Edition, 3rd edn. The MIT Press, Cambridge (1996)
13. Shanks, G., Tansley, E., Weber, R.: Using Ontology to Validate Conceptual Models. Commun. ACM 46, 85–89 (2003)
14. Järvelin, K., Wilson, T.D.: On Conceptual Models for Information Seeking and Retrieval Research. Information Research 9 (2003)
15. OCEG: GRC Capability Model (2009), http://www.oceg.com
16. March, S.T., Smith, G.F.: Design and natural science research on information technology. Decis. Support Syst. 15, 251–266 (1995)

17. Hevner, A.R., March, S.T., Park, J., Ram, S.: Design Science in Information Systems Research. MIS Quarterly 28, 75–106 (2004)
18. Vaishnavi, V.K., Kuechler, W.: Design Science Research Methods and Patterns: Innovating Information and Communication Technology, 1st edn. Auerbach Publications, Boca Raton (2008)
19. Moody, D.L., Sindre, G., Brasethvik, T., Sølvberg, A.: Evaluating the Quality of Information Models: Empirical Testing of a Conceptual Model Quality Framework. In: Proceedings of the 25th International Conference on Software Engineering. ICSE 2003, pp. 295–305. IEEE Computer Society, Los Alamitos (2003)
20. Calvanese, D., de Giacomo, G., Lenzerini, M., Nardi, D., Rosati, R.: Information Integration: Conceptual Modeling and Reasoning Support. In: IFCIS International Conference on Cooperative Information Systems, P. 280 (1998)
21. Mitchell, S.L.: GRC360: A Framework to help Organisations drive Principled Performance. International Journal of Disclosure and Governance 4, 279–296 (2007)
22. Tarantino, A.: Governance, Risk and Compliance Handbook: Technology, Finance, Environmental and International Guidance and Best Practices. John Wiley & Sons, Hoboken (2008)
23. Rasmussen, M.: Defining a Policy Management Lifecycle. (2010),
    http://www.corp-integrity.com/compliance-management/
    defining-a-policy-management-lifecycle
24. Chatterjee, A., Milam, D.: Gaining Competitive Advantage from Compliance and Risk Management. In: Pantaleo, D., Pal, N. (eds.) From Strategy to Execution, pp. 167–183. Springer, Heidelberg (2008)
25. Brache, A.P.: How Organizations Work: Taking a Holistic Approach to Enterprise Health. Wiley, Chichester (2001)
26. Rasmussen, M.: Achieve GRC Value: Efficient Business Process and Application Monitoring (2010),
    http://www.corp-integrity.com/wp-content/uploads/2010/12/Achieve-GRC-
    Value-Efficient-Business-Process-and-Application-Monitoring.pdf

Using Synchronised Tag Clouds for Browsing Data Collections

Alexandre de Spindler, Stefania Leone, Michael Nebeling,
Matthias Geel, and Moira C. Norrie

Institute for Information Systems, ETH Zurich
CH-8092 Zurich, Switzerland
{despindler,leone,nebeling,geel,norrie}@inf.ethz.ch

Abstract. Tag clouds have become a popular means of visualising and browsing data, especially in Web 2.0 applications. We show how they can be used to provide flexible and intuitive interfaces to web search services over data collections by using multiple synchronised tag clouds to browse that data. A data collection can have alternative tag clouds and a tag cloud alternative visualisations, with the choice of tag cloud and visualisation at any time controlled by a combination of user selection, developer specification and default system behaviour. A search interface is defined by an augmented data model that specifies the viewer classes, their associated tag clouds and the visualisations of these tag clouds. We demonstrate the approach by describing how we implemented a web application to browse data related to researchers and their publications.

Keywords: search service, tag clouds, data browsing, data visualisation.

1 Introduction

Tag clouds and faceted browsing have been used to address the challenge of providing users with intuitive interfaces to web search services. They offer visualisations of data collections that allow users to construct search queries through simple data selection. While faceted browsing allows complex search queries over a data collection to be constructed in a multi-step refinement process, tag clouds typically support only simple selections. However, the advantage of tag clouds is their capability to represent multiple features of a data collection within a single visualisation.

In this paper, we show how we have extended the use of tag clouds to allow the formulation of complex search queries by developing a browser that can offer multiple synchronised tag clouds to visualise the data stored in one or more data collections. By supporting alternative tag cloud representations for selected data collections within a database as well as alternative visualisations, we are able to combine features of tag clouds and faceted browsing.

The application developer can configure the browser through an extension of the data modelling language that is used to specify the view model of the database. We present an extension of SQL used to define the view model and the

H. Mouratidis and C. Rolland (Eds.): CAiSE 2011, LNCS 6741, pp. 214–228, 2011.

process of generating a browser from this model. To demonstrate the approach, we describe how a web application to browse data about researchers and their publications has been implemented.

We begin in Sect. 2 with a more detailed discussion of the background to this work and related research before going on to describe our approach in Sect. 3. Sect. 4 then introduces the SQL extension used to define the view model and shows how it can be used to specify a browser for a particular web search service. Sect. 5 provides details of the system architecture and the process of generating a browser from the view model definition. Implementation details are then given in Sect. 6. We discuss the contributions of our work in Sect. 7, and concluding remarks are given in Sect. 8.

2   Background

Tag clouds have become an extremely popular way of providing visual summaries of data collections and are nowadays used in many Web 2.0 sites to provide a basic search service based on user-generated tags. For example, both Flickr[1] and Del.icio.us[2] provide search services based on collaborative tagging. Although very simple, tag clouds can be used to support search, browsing and recognition as well as forming and presenting impressions [1,9]. In previous work, we have shown that tag clouds can also be used as the basis for a generic database browser [6].

The presentation and layout of tags can be controlled so that features such as the font size, type and colour can be used to give some measure of the importance of a given tag, while the positioning of tags may be based on pure aesthetics, alphabetical sorting or some form of relationship between tags. Studies have experimented with such features and their impact on users, concluding that font size, font weight and intensity are the most important features [7,1]. A study on search performance [8] found that topic-based layouts produced better results than random arrangements, but alphabetic layouts were best.

More recently tag clouds have been proposed as a means of summarising and refining the results of keyword searches over structured as well as unstructured data [5,3,4]. In [5], tag clouds are used to summarise query results of the PubMed biomedical literature database based on words extracted from the abstracts returned by a query. The term *data cloud* is used in [3,4] to refer to their particular adaptation of tag clouds for summarising keyword search results. Data clouds were implemented as part of CourseRank, an application that enables students to search for classes, give comments and ratings, and also organise their classes into a personalised schedule. The developer of a data cloud application specifies how application entities can be composed from the relations in the database in order that keyword search can be applied to entities rather than simple attributes or tuples. The keyword search is based on a traditional information retrieval

---

[1] http://www.flickr.com
[2] http://www.delicious.com

approach where entities are considered as documents and attribute values as weighted terms.

At the same time, the use of faceted browsers for web search services has become widespread. Faceted browsing allows items in a data collection to be filtered based on the selection of values of one or more properties of these items. For example, the Flamenco image browser [12] provides facets such as shape, colour, location and date to search and browse image collections. Unlike a simple hierarchical scheme, faceted browsing gives users the ability to find items based on more than one dimension. Another example is yelp.com, a local search and reviewing platform, where users can browse for information of interest using a mix of keyword search and filtering. A user might initially search on keyword and then refine the result collection based on multiple facets which are then refined with every filter selection.

Faceted browsers come in various flavours and have been extended with various features. Facets are usually visualised as ordered lists of possible values, where each value is followed by the number of items associated with that value. Non-directional browsers such as Flamenco [12] and the faceted browser for DBLP[3] offer multiple such lists, from which users can select values as part of the filtering process. In the case of Flamenco, the selection of a value in one facet filters the values of all other facets, but DBLP does not offer such synchronisation of facets. Directional browsers, such as Apple's iTunes, have a specific order of facets, most often represented as columns, where the browsing process goes from left to right. The selection of a facet value in one column, triggers a filtering action on facet values of all subsequent columns. In [11], they extend the column representation of facets with the concept of backward highlighting, where the selection of an item or facet value highlights all possible facet values of precedent columns associated with the current selection and that could have led to that selection. In [10], the information presentation is extended with so called *elastic lists*[4] that visualise the weight cardinality of the facet values. Facets are presented in the form of an ordered list where the size of a facet value indicates the cardinality of information items associated with that value.

Both faceted browsing and tag clouds simplify search processes for users, but have limitations in terms of how they are usually used. Often they only support searches over one particular data collection such as products in the case of online stores or publications in the case of DBLP. While tag clouds offer richer visualisation in terms of being able to encode different properties of a data collection in a single visualisation, clearly only limited information can be visualised at one time and usually they support only very simple selection processes. We propose an approach that combines the features of facets and tag clouds, and extends their use to support more general search services over multiple data collections. Our approach has been inspired by [2], where they provide a search tool to summarise, browse and compare search results over clinical trial data that combines faceted browsing with tag cloud visualisation.

---

[3] http://dblp.l3s.de

[4] http://well-formed-data.net/experiments/elastic_lists

3   Approach

Before introducing our view model that developers can use to specify a browser interface for a particular application, we will present an overview of the publications application that we developed in order to explain the main ideas behind the approach. In contrast to our previous work where we aimed to realise a generic database browser based on tag clouds [6], we now focus on providing developers with a framework that enables them to provide users with simple data browsers tailored to a particular application domain.

Fig. 1 shows the initial screenshot of the application when started. The interface comprises three tag clouds—one showing ranges of author names, one showing keywords associated with publications and one showing the names of conferences where papers have been published. In all three cases, the size of the tag is relative to the number of associated publications. Each tag cloud is labelled with the name of the data collection that it visualises, followed by the properties used for the tag cloud visualisation.

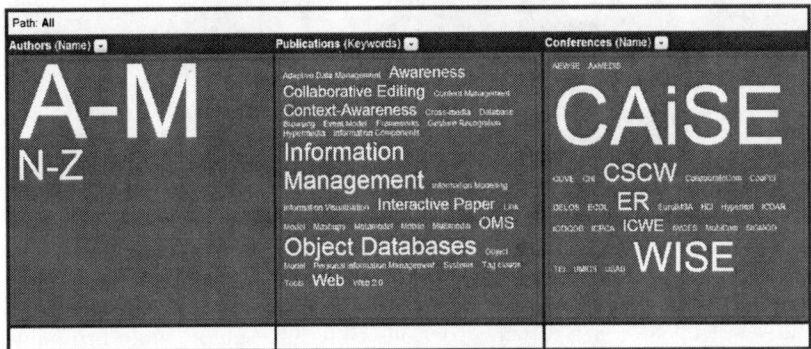

Fig. 1. Initial browser view

In contrast to the conference names and publication keywords, the authors are not represented by their names but by ranges of names instead. This happens if the number of tags shown in a tag cloud exceeds what can be displayed in a browser. In the example, *Authors* initially contains tags for alphabetical ranges "A..M" and "N..Z", where the size is relative to the number of publications aggregated for the respective range of authors. When a user selects one of the ranges, the author names contained in the range are displayed. The intervals used for the ranges can be controlled by the application developer who can specify a threshold limiting the number of tags shown in a tag cloud.

The three tag clouds are synchronised and users can filter data by clicking on a tag in any of the three clouds and the effect will be that values are filtered accordingly in all three clouds. So a user could, for example, click on the tag *CAiSE* in *Conferences* and the *Authors* cloud would then show only the names of people who have authored at least one CAiSE paper, the *Publications* cloud would show only the keywords associated with CAiSE papers and the *Conferences* cloud would show only *CAiSE*.

Presenting single items in one of the tag clouds is useful in terms of showing the context in which the other tag clouds should be interpreted. However, once a tag has been selected there are no further selections possible in that cloud and this is one of the limitations of using tag clouds that we alluded to before when we stated that they can only support simple, single step searches. To support further filtering of *Conferences* based on other properties, an application developer could specify alternative tag clouds for that data collection and these would be available through a dropdown menu. As illustrated in Fig. 2, we also offer a tag cloud for *Conferences* showing the year of conference as well as one showing the title instead of keywords for *Publications*.

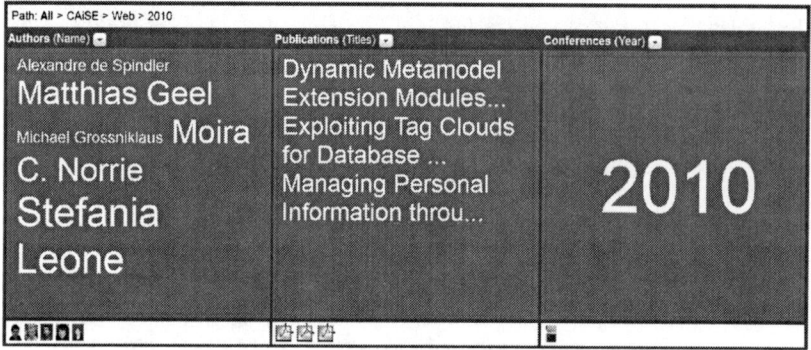

Fig. 2. Browsing example

For users to keep track of the selections they have made while browsing the database, we show the navigation path in terms of the tags that have been selected from the tag clouds at the top of the browser. In the example of Fig. 2, the user has first selected the *CAiSE* conference and the keyword *Web* before switching to an alternative view for *Conferences* to select the year *2010*. The user has then also switched the *Publications* view to show the titles instead of the keywords of the resulting three publications. Users can use this kind of breadcrumb navigation to go one or more steps back in the sequence of tag selections or click on *All* to return to the initial browser view shown in Fig. 1.

For each tag cloud, the items resulting from the filtering process are visualised in the designated areas beneath the respective tag cloud if the number of results does not exceed a certain threshold also defined by the application developer. A user can view these items by simply clicking on them. In the case of the *Authors* tag cloud, the author information of a specific author can be accessed. For the *Publications* cloud, publications can be accessed as PDF files and, in the case of the *Conferences* cloud, the conference proceedings can be viewed.

We offer different visualisations for tag clouds so that the developer may choose one appropriate to the information to be displayed and even the task at hand. For example, if the titles of a collection of publications are to be displayed, then it is more appropriate to display these as an ordered list rather than as the sorts of tag clouds one typically sees where several tags are displayed in a single line. We therefore provide a set of basic visualisation types, illustrated in Figure 3, from which a developer may choose. We will describe each of these working from left to right.

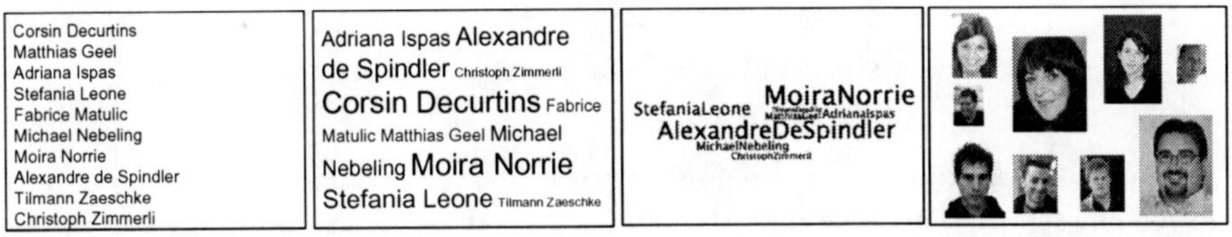

Fig. 3. Visualisation modes for authors

The first visualisation is a simple list of author names, sorted alphabetically by surname and with no variable visual features. The second visualisation is a tag cloud where the tags are aligned horizontally, sorted alphabetically by forename, and their size indicates the number of publications of the author. We refer to this as a line-based visualisation. The third visualisation is similar to ones produced by tools such as Wordle[5] where an advanced algorithm is used to align tags with aesthetics in mind and it can also be used to visualise various forms of relationships between tags. We refer to this as a spiral visualisation since the tag cloud is formed working from a central point and then positioning tags around that point while moving outwards. The fourth visualisation shows that we also support non-textual tags such as images. Any of the three basic visualisation types—list, line-based and spiral—can be used with both textual and non-textual tags.

4   Model and Specification

Our approach builds on the model of a browser that can define multiple synchronised tag clouds to visualise the data stored in one or more data collections. The application developer can configure such a *browser* as a search interface through an augmented data model that specifies the *viewer* classes, their associated *views* in the form of tag clouds and the *visualisations* of these tag clouds. Figure 4 illustrates this concept and how it extends the data model stored in a relational database with a view model based on the shared concept of views. Note that we use the relational model and later SQL, since the majority of web sites build on relational database systems, such as MySQL, for the storage and retrieval of data. However, we note critically that our approach is based on general database principles, such as data tuples and views, and is therefore not tied to a particular database system or modelling language.

---

[5] http://www.wordle.net

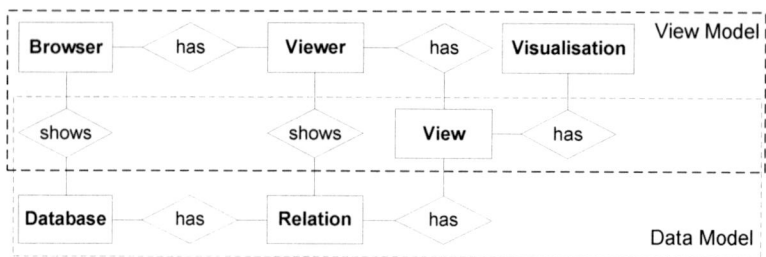

Fig. 4. Extension of the data model with a view model

iN a relational database, the concept of a browser translates to one or more viewer classes used to visualise the tuples stored in a relation. The concept of views is shared by both the data and view model so that viewer classes can associate them with different visualisations, such as simple vertical and horizontal line-based as well as spiral tag clouds. By building directly on the database to specify the view model, many aspects of the visualisations presented in the previous section can be derived directly from the data. For instance, the attribute types of data tuples decide how tags are formatted and displayed, e.g. as text or images, and the number of occurrences of a tag within a data collection determines its size in the visualisation.

To give a concrete example of how a given data model can be augmented to define a search interface using a combination of faceted browsing and tag clouds, Fig. 5 shows a simple domain model for the management of conferences, publications and authors. We will show how the browser and viewer classes corresponding to the interface illustrated in Figures 1 and 2 were defined. The domain model translates to the following relational schema.

Fig. 5. Example of a simple domain model

Authors (<u>id</u>, first_name, last_name, image)
Publications (<u>id</u>, title, keywords, abstract, image)
Conferences (<u>id</u>, name, year, image)
Authored (<u>author_id, publication_id</u>)
Published (<u>publication_id, conference_id</u>)

The relational schema uses a separate relation to store the tuples, not only for each entity defined in the model above, but also for the foreign key relationships, i.e. Authored and Published, between authors and publications or publications and conferences, respectively. It also covers attributes, such as image, which are primarily used for visualisation when showing the results. In our example from the previous section, the image attribute for authors is used to show a photo, while publications and conference proceedings are represented by a general PDF icon or thumbnail of the front page. Based on this relational schema, Listing 1 now defines the necessary browser and viewer classes using an enhanced version of SQL, which we will describe in more detail later.

```
CREATE VIEWER VR_AUTH (
    "Name"
        (SELECT CONCAT(a.first_name, " ", a.last_name) AS tag, COUNT(ap.
            publication_id) AS count, a.id, ap.author_id FROM Authors AS a,
        Authored AS ap WHERE a.id = ap.author_id GROUP BY ap.publication_id
            RANGE 15)
        LINE MULTISELECT,
);

CREATE VIEWER VR_PUB (
    "Titles"
        (SELECT title AS tag FROM Publications LIMIT 30)
        LIST,
    "Keywords"
        (SELECT SPLIT(keywords) AS tag FROM Publications LIMIT 100 ORDER BY tag
            ASC)
        LINE,
);

CREATE VIEWER VR_CONF (
    "Name"
        (SELECT c.name AS tag, COUNT(p.id) AS count, c.id, p.conference_id FROM
            Conferences AS c, Publications AS p WHERE c.id = p.conference_id
            GROUP BY p.id ORDER BY c.name ASC)
        LINE,
    "Year"
        (SELECT c.year AS tag, COUNT(c.year) AS count, c.id, p.conference_id FROM
            Conferences AS c, Publications AS p WHERE c.id = p.conference_id
            GROUP BY c.year ORDER BY c.year DESC)
        SPIRAL,
);

CREATE BROWSER B_PUBLICATIONS (
    "Publications" VR_PUB,
    "Authors" VR_AUTH,
    "Conferences" VR_CONF,
);
```

Listing 1. Example browser in SQL

The first viewer class, VR_AUTH, defines a single view over all authors with each tag built using SQL's standard CONCAT function to combine the first and last names. The tag size is calculated using an SQL count over the authored publications. This view is then associated with a horizontal, line-based tag cloud visualisation that will use count to make the size of an author's name dependent on the number of publications that they have. The viewer also allows for multiple selections of authors so that publications authored or co-authored by selected authors will be shown. Additionally, we use ranges in the case that more than 15 authors are displayed in the cloud. The next viewer, VR_PUB, associates a view over all titles of the publications with a vertical list visualisation. An alternative view of the publications is defined as the top 100 keywords in a line-based tag cloud. Here we use a non-standard SQL function SPLIT that we have defined to parse a comma-separated VARCHAR value and return the set of tokens. VR_CONF defines a primary view for conferences by name, where the size of the tag is relative to the number of publications in that conference. A variation here is to use an advanced, spiral tag cloud visualisation to show the publications by year starting from the latest conference, where the size of the tag is relative to the number of publications in the year. Finally, the browser B_PUBLICATIONS defines the search interface with the three viewer classes.

```
CREATE VIEW <view_name> (tag, [count , <other_columns>]) AS (SELECT <column>
    AS tag[, COUNT(<column>) AS count , <other_columns>] FROM <table_1>[, <
    other_tables>] [GROUP BY <column>] [ORDER BY <column> ASC|DESC] [LIMIT <
    number>] [RANGE <number>]);

CREATE VIEWER <viewer_name> (
    "View Name" <view_name>|<inner_view_definition> LIST|LINE|SPIRAL|<
        other_visualisations> [MULTISELECT],
    [<other_views>]
);

CREATE BROWSER <browser_name> (
    "Viewer Name" <viewer_name>,
    [<other_viewers>]
);
```

Listing 2. Extended SQL to specify browsers and viewer classes based on views

Listing 2 gives more details of the extended SQL syntax used above to define the view model. In SQL, views are essentially named SELECT statements that represent stored queries in the form of a virtual table composed of the respective result sets. We build on this concept of views to enable different visualisations of the data. Note that views can either be defined as an inner view as part of the viewer class or referenced by name, which enables re-use and combinations of views. For the proposed visualisations, the developer is required to specify a reserved column tag that will represent the tuples used to display the tags in the tag cloud visualisation. If the tags are to be displayed in different sizes according to certain criteria, then a second reserved column count is required that can build on SQL's COUNT aggregator function to count the occurrences of different values for a given column. In that case, also the GROUP BY statement is required to group the result set by the aggregated values. Note that other SQL statements such as ORDER BY and LIMIT can be used to sort tags in ascending or descending order as well as to limit the amount of tags displayed. Additionally, we define the RANGE statement to display the range of values rather than all retrieved values if the number of tags returned for the query exceeds the specified number. This can be helpful to navigate through large amounts of tags within a single view, e.g. by first showing ranges A..E, F..J and so forth, and, upon selection, showing the names of the respective subset of authors. Such ranges can be built by a custom SQL function that we defined to first sort all retrieved tags and then divide them into categories. For example, in the case of type VARCHAR, ranges could be built from only the first letters of all tags. Finally, the set of values used to display and size the tags typically comes from different columns and not necessarily from the same entity, e.g. to use a larger font for authors the more publications they have. While other_columns will then be required for joining associated relations, they will be ignored by the default tag cloud visualisations. On the other hand, additional columns offer a simple way of allowing for extensions and refinements. For example, new reserved columns, such as color, could be introduced to extend the proposed visualisations and visually group the result set by a specified range of colours.

In addition to this augmentation of view definitions, we further extend SQL with VIEWER and BROWSER definitions, respectively. A viewer class defines a set of alternative views, each of which is associated with a name displayed for the user to switch between visualisations, and a combination of parameters LIST, LINE or SPIRAL and MULTISELECT. The first three determine which of the visualisations shown in Fig. 3 will be used, where LIST represents the vertical alignment of tags, LINE a horizontal, line-based visualisation and SPIRAL the advanced tag cloud visualisation. Again, other visualisations could be supported by introducing new parameters that represent the respective visualisations. If the optional parameter MULTISELECT is provided, then the associated visualisation must allow for multiple selection of tags. With multiple selection, a combination of conjunctive queries between and disjunctive queries within views can be supported. Finally, a BROWSER defines a set of viewer classes and also provides a display name for each of them.

By using these augmentations of SQL, we can build on established database concepts and directly benefit from the rich support for SQL expressions and functions such as COUNT. Moreover, the caching strategies and high performance of query execution in many database management systems, such as MySQL, makes it optimal for web search interfaces. The way in which the final presentation of the tag clouds is generated as well as how the synchronisation between views on selection of a particular tag works are discussed in the next section.

5   Framework

Having described our language extension, we now present a framework that can process such browser specifications and generate a browser interface to search and browse specific data collections. The framework is shown in Fig. 6 in terms of its main components and their interactions. The browser defined in terms of an extended SQL specification is provided as input to the framework ①. The framework processes this specification as follows. First, a document template representing the browser's web interface is generated ②. This document contains one designated placeholder for each viewer, in which the tag clouds will be inserted at application runtime. Second, a browser-specific SQL view manager is created ③ based on the viewer specifications that will create the SQL views specified for each viewer in the database. At run-time, the SQL view manager queries these views ④ to retrieve the tags and their sizes which then provide the necessary input for the tag cloud generator that is called to create the tag clouds and the associated visualisations ⑤. In a final step, these generated tag clouds are inserted into the placeholders of the document template ⑥, which yields the final browser interface presented to the user.

When a user selects a tag in one of the tag clouds, the current view associated with that tag cloud as well as the associated views of all other tag clouds are

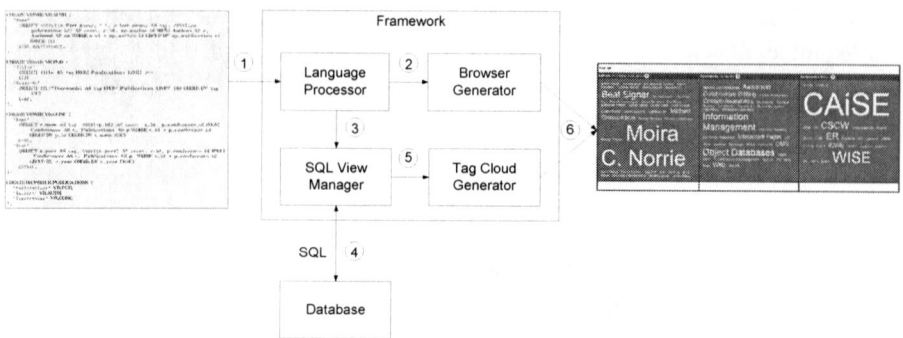

Fig. 6. Framework architecture and workflow

synchronised according to the selection. For this purpose, the view manager first restricts the current view by temporarily extending the **WHERE** clause in order to reflect the user selection. This updated view is then used as the starting point for the **PROPAGATE-UPDATE** function shown in Fig. 7, which implements an algorithm propagating the tag selection to associated views in order to keep them synchronised.

PROPAGATE-UPDATE(*View*)
1   $N \leftarrow \emptyset$
2   $N \leftarrow$ GET-ASSOCIATED-VIEWS(*View*)
3   **for** $\forall\, n \in N$
4   **do** ALTER-VIEW(*View*,n)
5   PROPAGATE-UPDATE(n)

Fig. 7. Update Propagation Algorithm

For the view passed as the argument, the set N of all associated views are retrieved using the **GET-ASSOCIATED-VIEWS** function. For every view $n \in N$, the view creation statement is extended by a join operation with respect to the view argument. Such extensions are carried out by the **ALTER-VIEW** function. Then, the **PROPAGATE-UPDATE** function is invoked recursively, in order to propagate the selection to all views related to the one currently processed. Note that, if multiple tags are selected subsequently, the algorithm is executed for each affected entity.

As a result, the selection of a tag is propagated along the relationships among database relations and, therefore, all related tag clouds are synchronised. For example, if a conference tag is selected, the view associated with the tag cloud is extended in order to filter the selected conference. Next, all associated views are determined, which, in our publication browser example, would be the publications and authors views. Then, the publications view would be filtered for those publications that were published in the selected conference. Finally, the authors view is extended in order to retrieve only those authors who have a publication at the selected conference.

In general, the sequence of views to be extended is determined by starting with the view in which the tag selection occurred and then following the relationships in a breadth-first manner. Note that this propagation algorithm was designed to work with data models that can be represented as connected and acyclic graphs. However, if there were cycles, endless loops are avoided because the framework keeps track of the views already extended. If there is a viewer showing a database relation not connected to any other, this viewer is independent and therefore cannot be synchronised.

As users continue selecting tags, the cumulated selections are propagated individually and in the same order as they were made by the users. Finally, if users make multiple disjunctive selections at once, the WHERE clause of the respective view is extended with all selection criteria combined in a disjunctive manner. Similarly, if a tag representing a range is selected, all values contained in this range are taken as disjunctive selection criteria.

6   Implementation

We now present how the framework was implemented in the form of a web application. We used a standard Client/Server setup consisting of HTML, CSS and JavaScript on the client side and PHP and MySQL on the server side. The web application provides an administration page where developers can input and execute a browser specification in extended SQL to generate a new browser that is then available from a new URL. Users may then interact with the browser as described in the previous sections.

Figure 8 shows the PHP classes which are involved in the creation of a browser as well as in processing user interactions at runtime. The specification of a browser in extended SQL is handled by the method **generateBrowser** declared in the **LanguageProcessor** class as follows. First, the information required to generate the HTML document template is extracted, which includes the number of viewers, their names and contained views. This information is passed on to the **generate** method defined in the **BrowserGenerator** class in order to create the client-side browser interface. It consists of the top bar for the breadcrumb navigation, the viewers with their names, the views and the dropdown menus for the selection of alternative views, the placeholders for the tag clouds and the bottom bar for the result sets.

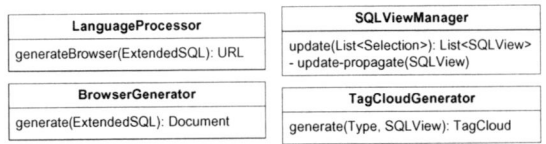

Fig. 8. Classes implementing the framework

Second, the SQL view definitions are extracted from the extended SQL. The respective views are created in the database and their names are stored in a separate database relation from where they can be accessed at application runtime. Finally, a new folder is created on the server, containing the generated browser interface and a PHP script index.php responsible for processing user tag selections and returning viewer contents where tag clouds reflecting user selections are dynamically updated at runtime. The URL returned by the `generateBrowser` method in the `LanguageProcessor` class points to this folder. The generated browser interface can then be tailored and styled according to specific application requirements.

The initial tag clouds presented to the users consist of tags that are HTML links. These links point to the index.php file created for the current browser, and the selection to be carried out when a particular link is chosen by the user is appended as a query string. The following example URL is the target of a link associated to the *CAiSE* tag in the *Conferences* viewer.

```
index.php?Conferences=CAiSE
```

Such a request is processed on the server side by the `SQLViewManager` class. Its method `update` takes the selection contained in the query string as a parameter and performs the update propagation algorithm described in the previous section. As a result of this update propagation, all extended SQL views reflecting the user selection are created. For each of these extended SQL views, the `generate` method in the `TagCloudGenerator` class creates an updated tag cloud which is merged with the document template and returned to the client. The URLs in the links of these updated tag clouds contain the previous selection in the query string as well as the subsequent selection they represent. For example, the URL of a link associated to the author Matthias Geel would be written as follows.

```
index.php?Conferences=CAiSE&Authors=Matthias%20Geel
```

For each tag selection specified by a user, the URLs of the links in the updated tag clouds are extended in order to contain all previous selections as well as the one to be carried out if the link was followed.

Similarly, the breadcrumb navigation consists of links pointing to the URLs previously requested. Due to the fact that our implementation follows a stateless approach, the implementation of the breadcrumb navigation is a simple manner of creating URLs including the respective query strings.

In order to support multiple disjunctive tag selections at once, the user can switch to a *multi select* mode. In this mode, the selection of a tag does not immediately initiate a request to the server. Instead, a search button is added to the browser interface which triggers the request to the server when the user is finished selecting tags.

7   Discussion

We have presented a general framework that supports the configuration of search interfaces for browsing and querying data collections using multiple synchronised tag clouds. We have illustrated its use based on the example of browsing a publication collection. Such interfaces could support a web search service either of a single research group's publications or over an entire digital library—simply by adapting the specification. While there are faceted search interfaces to publication collections, such as DBLP, our approach is much more flexible, since it not only supports searching for publications, but users can also shift their search focus to other entities of interest, such as authors or conferences. Furthermore, the selection of the visualised entities, their relationship and alternative tag cloud visualisations are configurable based on a combination of user selection, developer specification and default system behaviour.

Our approach is not dissimilar to the one taken by [2], where they provide a domain-specific tool for searching semi-structured clinical trial data where a set of predefined categories are represented using tag clouds. As with standard faceted browsing, users can start with a keyword search and the number of relevant documents are returned as a list, which can be further refined using the tag clouds. The selection of a tag in one dimension triggers the synchronisation of the tag clouds representing all other dimensions, as well as the filtering of the search result. While our approach could be seen as a generalisation of their work as we propose an augmented data model and a framework that supports the configuration of search interfaces for a domain of choice, it is also important to highlight the differences. Their interface consists of a set of predefined facets represented as a tag cloud, while we offer configurability at the interface level through dropdown menus that allow the selection of other tag cloud representations of the same entity. Furthermore, their data model corresponds to a typical data model underlying faceted browsing that is often based on star or multi-dimensional schemas, while our synchronised tag clouds do not evolve around a particular pivot entity. This means that there is no central entity that all other dimensions depend upon. In addition, with our approach, the tag size can be configured to represent dependencies to other entities of interest or simply the occurrence of a specific term, while with their approach the tag size always refers to the number of occurrences of a term in relation to the entity of interest, which in their case is clinical trial data. However, there are also some restrictions to the database schemas we support. The schema has to be a connected acyclic graph in order for our propagation algorithm to calculate the tags for each viewer correctly. While with cyclic structures, the propagation algorithm simply uses a shortest path approach, we could extend our framework so that a developer could configure the algorithm to achieve a different behaviour, if desired.

We note that our current implementation follows a stateless approach. This has some implications on system performance. Users can always choose to navigate to a breadcrumb, which is a bookmark to an individual search and allows a user to continue from there. With our current approach, these queries are executed again, invoking the propagation algorithm to adapt all adjacent tag clouds, while with a stateful approach these views could simply be cached. However, such an approach would be memory-intensive since it requires these views to be materialised.

8 Conclusions

We have presented an approach for browsing and searching data collections based on an extended data model that supports the configuration of a synchronised tag cloud browser for a domain of choice and we have illustrated its use through a publication browser. The generation of the browser is automated and its configuration is a mix of developer configuration using the extended SQL syntax, system default behaviour and user selection. We are also planning a user study to compare our approach to regular web search interfaces as well as faceted browsers.

References

1. Bateman, S., Gutwin, C., Nacenta, M.: Seeing Things in the Clouds: The Effect of Visual Features on Tag Cloud Selections. In: Proc. ACM Conf. on Hypertext and Hypermedia, HT 2008, pp. 193–202 (2008)
2. Hernandez, M.E., Falconer, S.M., Storey, M.A., Carini, S., Sim, I.: Synchronized Tag Clouds for Exploring Semi-Structured Clinical Trial Data. In: Proc. Conf. of the Center for Advanced Studies on Collaborative Research (CASCON 2008), pp. 42–56 (2008)
3. Koutrika, G., Zadeh, Z.M., Garcia-Molina, H.: Data Clouds: Summarizing Keyword Search Results over Structured Data. In: Proc. Intl. Conf. on Extending Database Technology (EDBT 2009), pp. 391–402 (2009)
4. Koutrika, G., Zadeh, Z.M., Garcia-Molina, H.: CourseCloud: Summarizing and Refining Keyword Searches over Structured Data. In: Proc. Intl. Conf. on Extending Database Technology (EDBT 2009), pp. 1132–1135 (2009)
5. Kuo, B.Y.L., Hentrich, T., Good, B.M., Wilkinson, M.D.: Tag Clouds for Summarizing Web Search Results. In: Proc. Intl. Conf. on World Wide Web (WWW 2007), pp. 1203–1204 (2007)
6. Leone, S., Geel, M., Müller, C., Norrie, M.C.: Exploiting tag clouds for database browsing and querying. In: Information Systems Evolution. LNBIP, vol. 72, pp. 15–28 (2011)
7. Rivadeneira, A.W., Gruen, D.M., Muller, M.J., Millen, D.R.: Getting our Head in the Clouds: Toward Evaluation Studies of Tag Clouds. In: Proc. Intl. Conf. on Human Factors in Computing Systems (CHI 2007), pp. 995–998 (2007)
8. Schrammel, J., Leitner, M., Tscheligi, M.: Semantically Structured Tag Clouds: An Empirical Evaluation of Clustered Presentation Approaches. In: Proc. Intl. Conf. on Human Factors in Computing Systems (CHI 2009), pp. 2037–2040 (2009)
9. de Spindler, A., Leone, S., Geel, M., Norrie, M.C.: Using Tag Clouds to Promote Community Awareness in Research Environments. In: Luo, Y. (ed.) CDVE 2010. LNCS, vol. 6240, pp. 3–10. Springer, Heidelberg (2010)
10. Stefaner, M., Muller, B.: Elastic Lists for Facet Browsers. In: Wagner, R., Revell, N., Pernul, G. (eds.) DEXA 2007. LNCS, vol. 4653, pp. 217–221. Springer, Heidelberg (2007)
11. Wilson, M.L., André, P., Schraefel, m.c.: Backward Highlighting: Enhancing Faceted Search. In: Proc. ACM Symposium on User Interface Software and Technology (UIST 2008), pp. 235–238 (2008)
12. Yee, K.P., Swearingen, K., Li, K., Hearst, M.: Faceted Metadata for Image Search and Browsing. In: Proc. ACM Intl. Conf. on Human-Computer Interaction (CHI 2003), pp. 401–408 (2003)

Revisiting Naur's Programming as Theory Building for Enterprise Architecture Modelling

Balbir S. Barn and Tony Clark

Middlesex University, Hendon, London, UK, NW4 4BT
b.barn@mdx.ac.uk, t.n.clark@mdx.ac.uk

Abstract. The recent burgeoning interest in Enterprise Architecture and its focus on artifact driven methods is taken as a motivation for the re-appraisal of Peter Naur's notion of "programming as theory building". Naur strongly disputes the value of the role and orientation of the IS discipline around artifacts and argues that algorithmic methods do not lead to a theory of a domain. Such a viewpoint provides an alternative lens with which to view current developments and may lead to additional insights by providing the reader with a source for questioning and reflecting critically on the re-focusing of method design on conversation rather than artifact production . It is suggested that such a conversational framework based on Toulmin and Pask may provide a means to establish and test theory building views of enterprise architecture.

1 Introduction

This account sets out to re-appraise Peter Naur's influential paper on Programming as Theory Building [11] in the context of model building and the recent focus on Enterprise Architecture. It is the intention of this paper to evaluate how theory building can play an important role in helping organizations make more use of their enterprise architecture activity and in particular how theory building may influence methods, techniques and tools to support enterprise architecture by focusing on conceptual modelling as a conversation process.

The starting point for this work has been triggered by the extent of activity that is currently surrounding Enterprise Architecture. As systems supporting business become increasingly more significant and complex an important approach to management and planning of systems that has gained prominence is model-based Enterprise Architecture (EA). EA has its origins in Zachman's original EA framework [21] while other leading examples include the Open Group Architecture Framework (TOGAF) [17] and the framework promulgated by the Department of Defense (DoDAF) [19]. In addition to frameworks that describe the nature of models required for EA, modeling languages specifically designed for EA have also emerged. One leading architecture modelling language is Archimate [7].

Central to enterprise architecture is the notion of development and presentation of models. Given the plethora of models available two concerns of note arise: Firstly, given the range of models available, it is difficult to ascertain why a particular model is relevant and preferable over others. This arises from a

H. Mouratidis and C. Rolland (Eds.): CAiSE 2011, LNCS 6741, pp. 229–236, 2011.

lack of clarity of the link between the contents and structure of a model on one hand and its purpose on the other [4]. Secondly, evaluation of quality of models in general, and therefore EA models, is relatively under researched. While there are international standards for software systems there is "little agreement among experts as to what makes a "good" model" [8]. Empirical measurements of the goodness of an EA model are generally lacking in the literature. Is that because we want to evaluate the final outputs of the modelling process rather than the success of the modelling process? Would a re-appraisal of Naur's ideas will provide a new insight and approach to questioning the "why" of a model? Similarly, would a re-appraisal provide insight to the "goodness" of a model for assessing the efficacy of a model in representing knowledge of a domain? These questions are the subject of this paper.

The remainder of the paper is structured as follows: Section 2 outlines the main hypotheses posed by Naur. Section 3 and 4 presents a more detailed analysis of aspects of the hypotheses (programs as models and methods). Section 5 provides an alternative view of how methods for EA should structured and presents two underlying philosophical and psychological theories and their conceptual integration as the basis for a conversation framework approach to EA method design rather than the algorithmic artifact oriented views that are currently prevalent. Section 6 concludes with an overview of the implications arising from this re-appraisal of Naur's seminal paper in the context of conversational processes for EA modelling.

2   Programming as Theory Building

Peter Naur wrote "Programming as Theory Building" in 1985, it was reprinted later in his collection of works, Computing: A Human Activity in 1992 [10]. The paper presents a discussion that contributes to what programming is. While it is tempting to assume from the title of the paper that Naur is focused on the minutiae of programming, he is specific in that programming denotes the "whole activity of design and implementation" and thus his theory applies to the field of software engineering. The fundamental premise asserts that programming should be regarded as an activity by which programmers achieve a certain insight or theory of some aspect of the domain that they are addressing. The knowledge, insight or theory that the programmer has come into possession of is a theory in the sense of Ryle [15]. That is, a person who has a theory knows how to do certain things and can support the actual doing with explanations, justifications and responses to queries. That insight or theory is primarily one of building up a certain kind of knowledge that is intrinsic to the programmer whilst any auxillary documentation remains a secondary product. Of particular interest, is how Naur explains the life-cycle of a program. Programs are created by the establishment of a theory, the maintenance of a program is dependent on the theory being transferred between programmers; and the program dies when the theory has decayed. Program revival is described as re-establishing the theory behind the program which cannot be done merely from documentation and should only be considered in exceptional circumstances as the cost of theory revival is prohibitive and the resulting theory may be different from that originally conceived.

In addition to the theory view of a program life cycle, he also directed criticism at the then significant emphasis of methods for program development. He claimed that the methods based on sequences of actions of certain kinds cannot lead to the development of a theory of the program because the intrinsic knowledge held by a human has no inherent division into parts nor an inherent ordering. Instead the person possessing the knowledge is able to present multiple viewpoints as responses to requests. Where methods were supplemented with notations or formalizations then these were treated as secondary items as the theory of a program is intrinsic and cannot be expressed. Thus: "...there can be no right method".

Having outlined the basic hypotheses of Naur's paper, the remainder of this account continues the critique of Naur's ideas and applies them to modeling and Enterprise Architecture.

3   Programs as Models

Naur was concerned with programs, but Enterprise Architecture is concerned with the production of models of interconnected systems or components. Thus we need to explore the relationship between programs and models and use that as a basis for analysing the applicability of Naur's hypothesis in the context of Enterprise Architecture.

A major activity in software engineering and computer science in general is modelling and as Fetzer [1] has noted "the role of models in computer science appears to be even more pervasive than has been generally acknowledged..". A key feature of modelling is the existence of an isomorphic relationship between the parts of the model and the parts of the thing modelled at some level of abstraction. Smith [16] whilst noting these different types of models emphasizes the nature and importance of "representation":

> "To build a model is to conceive of the world in a certain delimited way... Computers have a special dependence on these models: you write an explicit description of the model down inside the computer...".

Smith suggests this feature distinguishes computers from other machines because they run by manipulating representations. "Thus there is no computation without representation" [16, p, 360] If we pursue this analysis further: From Naur we can state that the program is a theory; from general computation principles, we can state: the program is a model. This leads to the notion that there is an equivalence between program = theory = model. We might moderate this further by noting that a program is a representation of a slice of "the" theory. In general though, this blurring between programs, theories and models is confusing and inaccurate. While models may exhibit an isomorphic relationship with their subject matter, this relationship may not reveal the theoretical connections that allow the theory to be defended in the form of Ryle's definition of a theory. Ideally, then, the theory must be statable independent of the computer model. In an essay that predates Naur's paper but still based on a prevailing view of the time that programs are theories, James Moor notes:

> "My claim is that this is rarely, if ever, the correct response. Even if there is some theory behind a model, it cannot be obtained by simply examining the computer program. The program will be a collection of instructions which are not true or false, but the theory will be a collection of statements which are true or false. Thus, the program must be interpreted in order to generate a theory. Abstracting a theory from the program is not a simple matter for different groupings of the program can generate different theories. Therefore, to the extent that a program, understood as a model, embodies one theory it may well embody many theories."[9, p.221]

From this analysis arises two key concerns. Firstly, programs and models may have multiple theories and a program or model may not refer to the same theory. Secondly these theories must be state-able independent of the program or the model. Then, there is an additional dichotomy: Is a program a representation of one view of an aggregate theory or is the program a representation of a component theory of the aggregate theory? These complexities, in the case of Naur's Programming as theory perspective have implications, because if the program is the only vehicle through which a theory can demonstrate that requirements of the intended system have been met, then, that theory testing process comes too late in the system life cycle.

4   On Methods and Theory Building

Earlier we noted that Naur had reserved considerable criticism for methods. We develop this discussion further in this section. The tendency of methods research in the IS discipline is to propose algorithmic steps to analysing and designing solutions to problems. As Naur notes: "A method implies a claim that program development can and should proceed as a sequence of actions leading to a particular kind of documented result". In contrast, a theory building view holds that a theory "held by a person has no inherent division into parts and no inherent ordering". At large, IS/SE research is embarked on a journey based on epistemological foundations and as a consequence has mostly neglected *techne* (the technical know how of getting things done) and *phronosis* (wisdom derived from socialised practices) [20]. In a more generalised form, this has correspondence to the distinctions between explicit and tacit knowledge [12] and Naur would seem to be arguing the case for methods research that suggests more attunement with the effects that methods may have in the education of programmers. That is, the creation and embedding of tacit knowledge rather than the production of artifacts representing explicit knowledge through an algorithmic process.

Naur cites a study of five different methods by Floyd et al (cited here for completeness [2])where the key result that a system of rules will lead to good solutions is an illusion, what remains is *the effect of methods on the education of programmers*. Thus the use of methods may themselves not lead to a good design but the practice of the method may lead to a better innate ability for theory building.

5   Theory Building and Testing as a Conversation Process

The act of constructing a conceptual model that describes an enterprise architecture is essentially all about communication. For example, when we engage in a discussion of budgetary requirements, we are requiring the architecture description to provide us with a theory of budgetary models. A description of the communication of how that theory is explicated is at the heart of that architecture description. In a modelling process, participants such as the domain expert and the systems analyst (who may have no knowledge of the domain) engage in a conversational process through which concepts understood by the domain expert are formalised by the systems analyst through some dialogue document in a controlled language[3]. The goal of the modelling process is to reach a state where all participants agree that they have some degree of common understanding[14].

When Naur describes theory building amongst teams of programmers who share the same theory he would appear to be alluding to a similar socialisation process. More recently and in line with what we propose in this section, Kruchten [5] provides a critique of software architecture from a knowledge management perspective where architectural knowledge is a composite of the architecture (design) and a rationale for design decisions. The support for the rationale comes through a socialisation process framed by the SECI (socialisation, externalisation, combination, internalisation), model [12]. Significantly, though, the artifact remains central albeit augmented by more human centred activity.

In the development of a theory, Naur also suggests three tests to check if the programmers knowledge transcends the written documentation consistent with Ryle's notions that a theory should be defensible and justifiable by the presentation of evidence. These are: the programmer can explain how the solution relates to the affairs of the worlds that it helps to handle; the programmer can explain why each part of the program is what it is, in other words, is able to support the actual program text with a justification of some sort; and the programmer is able to respond constructively to any demand for a modification.

Here we propose that the first test can be addressed by consideration of an integration of two other philosophical theories in this field: Toulmin's informal argumentation model [18] and Pask's conversation theory [13] and to suggest that theory testing can be achieved by constrained conversations using models as the subject. More pertinently, it may help us to address the "why" of an enterprise architecture.

5.1   Conversation Theory

A conceptual model represents the arrival of a shared understanding of a subject area between two different actors – the domain expert and the systems designer. One way of viewing the process of understanding is through the lens of Conversation Theory (CT) [13]. As theory of exposition and defence, CT can be summarised as follows: one participant (say, the domain expert) describes a body of knowledge to a second participant (the Systems Analyst). Both these participants are a type of organization – the psychological (p-) individual. A p-individual is a stable closed system comprising memory (facts), rules for interpreting the memory (concepts), rules for structuring the derivation of concepts – "how to" understand concepts and rules for understanding how topics in the memory relate to each other. In a basic conversation ("skeleton of a conversation"), there are two levels – the "how" and "why". The "how" level describes how to do a topic for example, recognizing, constructing and maintaining a topic, while the "why" level is focused on explaining or justifying the topic perhaps in terms of other topics. The basic conversation is provocative, that is participants are provoked into constructing understandings of each others' beliefs. A "modeling facility" provides the medium in which concepts are understood between individuals.

A key aspect of CT is the embodiment of knowledg (e.g. the workings of the combustion engine, finite state machines or any other coherent whole) which is viewed as a set of topics or facts that are related to each other. Relations between topics are either decompositional (hierarchical) or analogous (heterarchical), when such relationships and topics are static then that static representation is called an entailment structure. When a topic is understood by a learner (via a reproducible procedure) then the topic also exists as a concept for potential sharing with another p-individual.

5.2   Argumentation Theory

A person who has or possesses a theory knows how to do certain things and can support those actions with explanations, justifications and answers to queries. This is similar to Toulmin's argumentation model [18] - a logical structure for reasoning about the validity of arguments, the structure of which are described below:

Claim. A proposition representing a claim being made in an argument;
Grounds. One or more propositions acting as evidence justifying the Claim;
Warrant. One or more rules of inference describing how the Grounds contribute to the Claim;
Backing. The knowledge establishing the Grounds for believing the Warrant;
Qualifier. A phrase qualifying the degree of certainty in the argument for the Claim;
Rebuttal. One or more propositions challenging the validity of the Claim.

An example of a Toulmin argumentation model might be as follows: Object oriented modelling is a more natural way for most business analysts to capture requirements. Such a statement is a claim that includes a qualifier - most. The grounds for this statement might refer to hard facts or evidence that supports this claim. The warrant might indicate how object concepts provide a closer correspondence to objects in the real world. The backing for the claim might be: because object modeling is derived from entity modeling and entity relationship modeling has considerable history of efficacy in requirements capture. A rebuttal is a counter claim and has its own argumentation model.

Taken together, the two theories present a potential opportunity to review how we design methods and their supporting tools. The argumentation model presents a conversational framework which allows the theory builder to create an orderly and intelligible conversation - a discussion of the theory. But because such discourse analysis has the potential to generate large amounts of data by utilizing a limited set of concepts derived from the domain (the topics in the entailment mesh from conversation theory) it is possible to make the resulting analysis more amenable.

6   Implications for Enterprise Architecture

Enterprise Architecture (unlike programming) has no target theory. The execution of a program can be used to validate the quality of the theory that a programmer constructed but mechanisms for executing enterprise architectures are still largely an area of research focus. Prevailing methods and languages for EA (and using ArchiMate as a canonical example) have focused on developing artifacts and models for explicit knowledge [6, p.75] and so are subject to Naur's criticisms. EA frameworks such as TOGAF provides an exhaustive set of activities, phases of activities, ordering of activities and artifacts to be produced by activities with the intent of capturing in its entirety a theory of the EA. Accepting the theory building view forces us to reject firstly that such an exhaustive methodological approach can lead us to a universal theory of Enterprise Architecture for a domain. Secondly the focus on explicit knowledge does not allow us to extract from the plethora of method the essence of "why". Instead, a model of incremental, modular theory building which involves the real world thorugh a conversational process as a source of knowledge and validation may unlock the real value of an enterprise architecture.

This takes us then to a more fundamental re-thinking of method development. A method for EA should not (algorithmically) take us to a model of EA (because no one model exists), instead a method should instill in the practitioner, the cognitive processes for constructing theories about the enterprise architecture. The conversational approach outlined earlier is one such candidate basis for such cognitive processes as it enables both the testing of a theory and the collaborative development of a theory. Indeed it might allow us to measure the efficacy of a method not by how a solution is designed or quality of solution but by how the engineer has modified their psychological processes for theory building and so the corresponding implications for software engineering education.

References

1. Fetzer, J.H.: The role of models in computer science. The Monist 82(1), 20–36 (1999)
2. Floyd, C.: Eine untersuchung von software-entwicklings-methoden. In: Morgenbrod, H., Sammer, W., Tagung, I. (eds.) Programmierumgebugnen und Compiler, Tuebner Verlag (1984)
3. Hoppenbrouwers, S., Proper, H.A., der Weide, T.P.: A fundamental view on the process of conceptual modeling. In: Delcambre, L.M.L., Kop, C., Mayr, H.C., Mylopoulos, J., Pastor, Ó. (eds.) ER 2005. LNCS, vol. 3716, pp. 128–143. Springer, Heidelberg (2005)
4. Johnson, P., Ekstedt, M., Silva, E., Plazaola, L.: Using enterprise architecture for cio decision-making: On the importance of theory. In: The Proceedings of the 2nd Annual Conference on Systems Engineering Research, CSER (2004)
5. Kruchten, P.: Documentation of Software Architecture from a Knowledge Management Perspective–Design Representation. Software Architecture Knowledge Management, 39–57
6. Lankhorst, M.: Enterprise architecture at work: Modelling, communication and analysis. Springer-Verlag New York Inc., Heidelberg (2009)
7. Lankhorst, M.M., Proper, H.A., Jonkers, J.: The Anatomy of the ArchiMate Language. International Journal of Information System Modeling and Design 1(1)
8. Moody, D.L.: Theoretical and practical issues in evaluating the quality of conceptual models: current state and future directions. Data & Knowledge Engineering 55(3), 243–276 (2005)
9. Moor, J.H.: Three myths of computer science. British Journal for the Philosophy of Science 29(3), 213–222 (1978)
10. Naur, P.: Computing: a human activity. ACM, New York (1992)
11. Naur, P.: Programming as theory building. Microprocessing and Microprogramming 15(5), 253–261 (1985)
12. Nonaka, I., Takeuchi, H.: The knowledge-creating company, New York, vol. 1 (1995)
13. Pask, G.: Conversation, cognition and learning. Elsevier, Amsterdam (1975)
14. Pohl, K.: The three dimensions of requirements engineering: A framework and its applications* 1. Information Systems 19(3), 243–258 (1994)
15. Ryle, G.: The concept ofmind, London, Hutchinson (1949)
16. Smith, B.C.: Limits of correctness in computers. Academic Press Professional, Inc., London (1991)
17. Spencer, J., et al.: TOGAF Enterprise Edition Version 8.1 (2004)
18. Toulmin, S.E.: The uses of argument. Cambridge Univ Pr., Cambridge (2003)
19. Wisnosky, D.E., Vogel, J.: DoDAF Wizdom: A Practical Guide to Planning, Managing and Executing Projects to Build Enterprise Architectures Using the Department of Defense Architecture Framework, DoDAF (2004)
20. Wyssusek, B.: A philosophical re-appraisal of peter naur's notion of programming as theory building. In: European Conference on Information Systems, ECIS (2007)
21. Zachman, J.A.: A framework for information systems architecture. IBM Systems Journal 38(2/3), 454–470 (1999)

A DSL for Corporate Wiki Initialization

Oscar Díaz and Gorka Puente

Onekin Research Group, University of the Basque Country, Spain
{oscar.diaz,gorka.puente}@ehu.es

Abstract. Some wikis support virtual communities that are built around the wiki itself (e.g., *Wikipedia*). By contrast, corporate wikis are not created in a vacuum since the community already exists. Documentation, organigrams, etc are all there by the time the wiki is created. The wiki should then be tuned to the existing information ecosystem. That is, wiki concerns (e.g., categories, permissions) are to be influenced by the corporate settings. So far, "all wikis are created equal": empty. This paper advocates for corporate wikis to be initialized with a "wiki scaffolding": a wiki installation where some categories, permissions, etc, are initialized to mimic the corporate settings. Such scaffolding is specified in terms of a Domain Specific Language (*DSL*). The DSL engine is then able to turn the DSL expression into a *MediaWiki* installation which is ready to be populated but now, along the company settings. The DSL is provided as a *FreeMind* plugin, and DSL expressions are denoted as mindmaps.

Keywords: wiki, dsl, MDE, information system.

1 Introduction

Wiki's pioneer, Ward Cunningham, defines wikis as "the simplest online database that could possibly work"[4]. Nowadays, wikis are becoming a favourite approach for collaborative knowledge formation and knowledge sharing [12]. So far, most studies are conducted for public-access wikis or wikis for supporting learning activities [13]. However, companies are increasingly realizing the benefits of wikis [3]. Indeed, the Intranet 2.0 Global Survey reports that around 47% of the respondent companies were somehow using wikis [10]. Based on these figures, we can expect an increasing adoption of wikis among companies.

As any other Information System, the interplay of technology, work practice, and organization is paramount to achieve successful wiki deployments. Therefore, we can expect differences when wikis are deployed to sustain open communities (e.g., *Wikipedia*), offered within a learning organization [13] or are deployed at a company [8]. The peculiarities of each organization will certainly percolate the wiki itself. Indeed, unlike other settings, companies provide an existing infrastructure that frames the wiki. Users, roles, permissions, terminology, documents, templates or project milestones are already there before the wiki is created. This is not the case (or at least not to the same extent) in open-access wikis (e.g., *Wikipedia*) where the community originates around the wiki itself.

H. Mouratidis and C. Rolland (Eds.): CAiSE 2011, LNCS 6741, pp. 237–251, 2011.

Educational settings sit in-between since they offer some pre-existing context but with less demanding constraints than companies.

Consequently, our premise is that, unlike other environments, corporations have all, personal organigrams, documentation practices and task schedules that frame both wiki users and wiki editing. The term **"wiki scaffolding"** is introduced to denote a wiki installation where some categories, templates, permissions, etc are initialized at the outset to mimic the corporate background. This includes structural concerns (e.g., how are wiki pages arranged along which categories), communication means (who is going to be notified of what), permission needs (e.g., who is allowed to do what), etc. So far, this background is patiently replicated by wiki users that, in some cases, are forced to go down to code.

Wiki scaffolding implies not only being knowledgeable about the wiki engine (e.g., *MediaWiki*) but also installing third-party extensions. This can certainly discourage users. Lawyer, architects, medical doctors are all profiting from wikis. Hence, *our aim is for wiki scaffolding to be made accessible to non technical people (**who**) that collaboratively agree (**how**) on a blueprint for the wiki (**what**)*. To this end, we propose the use of a *Domain Specific Language* (DSL). DSLs are reckoned to enhance the quality, productivity, maintainability and portability while permitting domain experts understand, validate and develop the DSL programs themselves [7]. Additionally, collaboration and easy sharing can be promoted by using a *graphical DSL* (as opposed to a *textual DSL*). Specifically, the collaborative mandate suggests capitalizing on existing tools for supporting brainstorming. A common way of recording and expressing brainstorming sessions are *mind maps*. A mindmap is a diagram to express ideas around a central topic. Now, this central topic is "wiki scaffolding", and mindmaps constructs are reinterpreted to denote scaffolding concerns.

This paper presents the *Wiki Scaffolding Language (WSL)* (pronounced "whistle"). *WSL* is built on top of *FreeMind* [1], a popular, open source tool to create mindmaps. Hence, *WSL* expressions are mindmaps. Our bet is that users might already been exposed to mindmaps and even to *FreeMind*, hence reducing the learning curve for *WSL*. These maps (i.e., *WSL* expressions) are then compiled into a set of *MediaWiki* directives whose execution generates the wiki scaffold. *MediaWiki* is one of the most popular wiki engines [2].

This paper is organized along the design and use of *WSL*: *WSL* analysis (Section 2), *WSL* design (Section 3), *WSL* realization (Section 4), *WSL* verification (Section 5) and *WSL* enactment (Section 6). Discussion through related work is presented in Section 7. Some conclusions 8 end the paper.

2 WSL Analysis

This section identifies the scope and main abstractions behind *Wiki Scaffolding* (WS). The aim is to capture the company's work practice and settings as long as their impact on wiki operation. A main outcome of this analysis is a *feature diagram* that describes the commonalities and variabilities of domain concepts

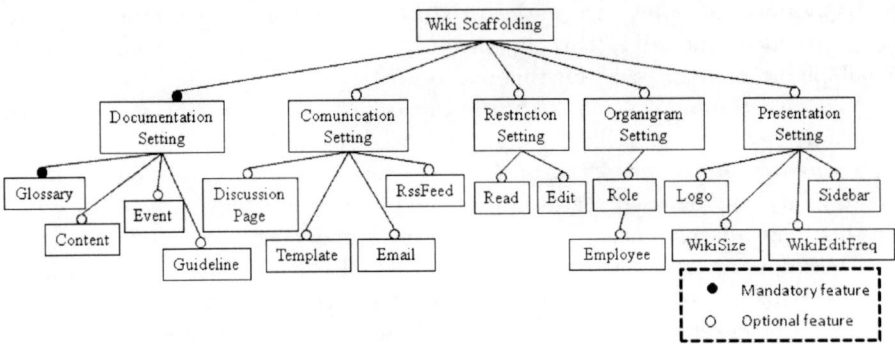

Fig. 1. Feature diagram

and their interdependencies [6]. Fig. 1 depicts the feature diagram for WS. The diagram states that a WS expression captures the company settings in terms of existing documentation practices, communication means, restrictions, the existing organigram and finally, presentation concerns. Next paragraphs delve into these notions.

Documentation Setting. A common problem for open communities is that of fixing a common terminology and understanding. This is easier in the case of corporate wikis where glossaries, documentation guidelines or even, some content might already exists. This setting needs to be captured in wiki terms. A basic classification of wiki pages is that of *"articles", "categories"* and *"templates"*. Articles stand for the *content* that is progressively and socially edited. Next, categories are commonly used as tags to easily locate, organize and navigate among articles. *Glossaries* can help to identify initial wiki categories. Finally, templates provide content to be embedded in other pages. Through parameterization, they permit to reuse and ensure a formatted content along distinct pages. Corporate *guidelines* can then be re-interpreted as wiki templates that guide article editing.

Fig. 1 depicts *"glossary", "content"* and *"guideline"* as three features of the company's documentation setting that can impact the wiki Moreover, wikis frequently support living projects where project milestones might need to be accounted for by the wiki. This does not apply to other settings where content is the result of free-willing participation and hence, contribution is not tight to pressing schedules. Wiki wise, this implies that *"event"* is a semantically meaningful piece of data, and so should it be markuped and rendered (e.g., through a calendar).

Communication Setting. Wikis are an effective mechanism to support knowledge building through collaboration. This implies the existence of coordination and conflict resolution strategies. When wikis are deployed in an existing organization, wikis become an additional means that should be integrated with existing communication channels. This poses a range of questions: who is going to be notified of what? Does the existing organizational structure need to be mirrored in the wiki? How is currently achieved such communication? Is email/phone/chatting used for this purpose?

Wiki wise, communication can be internal or external. Internal communication is achieved within the wiki. At this respect, two mechanisms are considered: *"discussion pages"* and *"templates"*. Discussion pages (a.k.a. *"talk"* pages in *MediaWiki*) can be used for discussion and communicating with other users. In this way, discussions are kept aside from the content of the associated page. Templates have also been identified as effective means to deliver fixed messages (e.g., warnings, to-do reminders, etc). On the other hand, external communication refers to the ability to notify wiki changes outside the wiki itself (e.g., through **"RSS feeds"** any *rss* client can be used).

Organigram Setting. Companies tend to be organized somehow, what is indicated with the distinct *"roles"* that the *"employees"* adopt in projects.

Restriction Setting. Unlike public-access wikis, corporate wikis normally limit access to employees. Permissions are counterintuitive in a wiki setting where openness is a hallmark. Indeed, *MediaWiki* natively supports a basic mechanism where the scope of permissions is the whole wiki: you can either edit the whole set of wiki pages or not (e.g., anonymous users cannot read pages). By default, wiki pages can be freely operated. However, permission demands are more stringent in a company setting. Indeed, a study on the use of wikis in the enterprise reports that "power relationships and competition between stakeholders created a need to read access in the ResearchWiki" [5]. For the time being, two permissions are considered: *"read"* and *"edit"*. Additional permissions could be added in future releases if feedback so advises[1].

Presentation Setting. Most companies project a unified image in terms of rendering and presentation. Wikis resort to "skins"[2] for rendering. These skins are engine specific. However, we do not expect our target audience to know about skins. We should strive to capture presentation concerns in abstract terms, better said, through domain criteria that could later be used by the *DSL* engine to determine the most appropriate skin. Specifically, we consider *"wikiSize"* and *"wikiEditFreq"*. Based on the expected size and edit frequency of the wiki, heuristics can make an educated guess about the wiki skin. In this way, the *DSL* engine frees stakeholders from being knowledgeable about presentation issues, offering good-enough outputs. Notice that the wiki administrator can latter change this automatically-selected skin. Additionally, the *"logo"* and *"sidebar"* features are introduced for customizing both headers and index panes which are available for speeding up wiki access.

---

[1] *MediaWiki* permissions include *"read"*, *"edit"*, *"createpage"*, *"createtalk"*, *"upload"*, *"delete"*, *"protect"* (i.e., allows locking a page to prevent edits and moves), etc.

[2] A skin is *"a preset package containing graphical appearance details"*, used to customise the look and feel of wiki pages.

3   WSL Design

The aforementioned concerns are now captured through a *DSL*. *DSL* design implies first to set the abstract syntax, and next, select one of the possible concrete syntaxes [11]. Based on the feature diagram, the **abstract syntax** describes the concepts of the language, the relationships among them, and the structuring rules that constrain the model elements and their combinations in order to respect the domain rules. This is expressed as the *DSL* metamodel. Fig. 2 depicts the abstract syntax for *WSL*. A *scaffolding* model includes four main model classes, namely:

(1) The **Content** class, which is a graph described along **Items** and **Links**. *Items* capture the different kinds of data existing in the company that need to be also available at wiki inception. As identified in section 2, this content includes glossary terms (*"category" itemType*), content ready to be available as a wiki article (*"article" itemType*), guides for content structure (*"template" itemType*) or events to capture scheduling milestones (*"event" itemType*). Next, **Links** relate these *Items* together. *Links* are also typed based on the type of the related *items*: *"relatedWith"* link (a general *item-to-item association*); *"belongsTo"* link (to associate a *category* to an *item*); *"templatedBy"* link (to associate a *template* to an *item*); *"scheduledFor"* link (to associate an *event* to an *item*). Items also hold three boolean properties: **discussion** (to indicate whether this *item* is subject to discussion), **rssFeed** (to specify the availability of a feed subscription for this *item*) and **indexPaneEntry** (to capture that the *item* is to be indexed in the sidebar).

(2) The **Organigram** class, which captures a basic arrangement of **Employees** in terms of **Roles**.

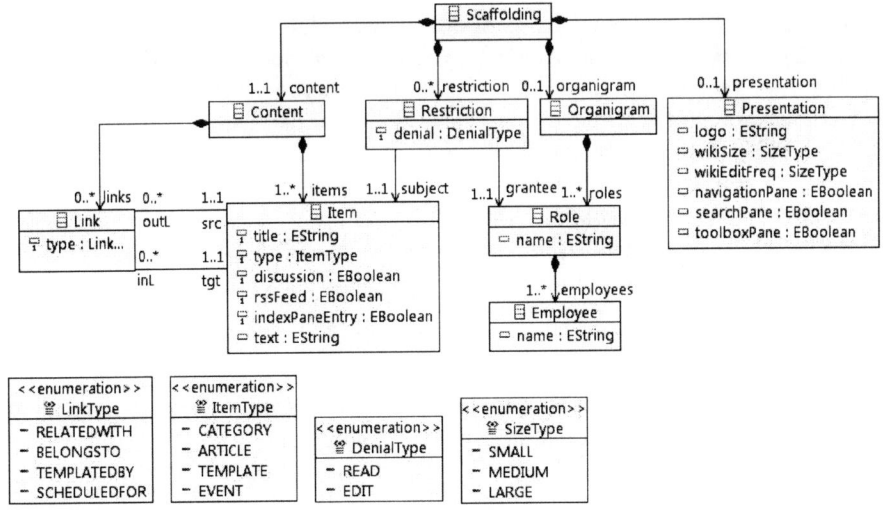

Fig. 2. *WSL* metamodel (abstract syntax)

(3) The **Restriction** class, which binds together three elements: a permission subject (i.e., an *Item*), a permission grantee (i.e., a *Role*) and a **denial** (i.e., *"read"* and *"edit"*).

(4) The **Presentation** class, which holds properties to guide the rendering of the wiki. Specifically, index requirements are captured through four common indexing schemas: *toolboxPane* (entries include *"what links here"*, *"Upload file"*, *"printable version"*, etc), *navigationPane* (entries include *"recent changes"*, *"help"* *"main page"*, etc), *indexPane* (where entries are set by the designer through the *Item*'s *indexPaneEntry* attribute), and *searchPane* (as a search facility to locate articles based on content). The logo is also captured here.

Next, this abstract syntax is realized through a concrete syntax. This implies a mapping between the metamodel concepts and their textual or visual representation. Preliminary feedback indicates that a visual syntax would be more suitable. The user profile (i.e., domain experts) as well as the collaboratively way of obtaining the wiki blueprint, advise to go for a visual DSL. Rather than developing our own visual language, we decide to capitalize on an existing one: *FreeMind*. With over 6,000 daily downloads, *FreeMind* is one of the most popular tools for mindmap drawing. Fig. 3 shows a snapshot of a *FreeMind* map. The main advantage of this tool is the easiness to play around to capture your mental model (e.g., nodes, and their descendants, can be easily moved around; branches can be collapsed, etc). This decision not only speeds up development but, more importantly, it will hopefully facilitate *WSL* adoption among end users. Next section introduces *WSL* as a visual language on top of *FreeMind*.

4   WSL Realization

WSL is a visual language on top of *FreeMind*. That is, a *WSL* expression is a compliant *FreeMind* map. However, the opposite does not hold. Some maps might not deliver a compliant wiki scaffolding, where compliance is determined by the abstract syntax in fig. 2. Therefore, *WSL* maps are a subset of *FreeMind* maps. *FreeMind* maps are internally represented as *XML* files along an *XML* schema. On top of it, *WSL* imposes an additional set of constraints that ensures that maps account for compliant scaffoldings (i.e., conform to the *WSL* abstract syntax). Before delving into how *WSL* constructs are mapped into *FreeMind* elements, next subsection introduces an example.

4.1   WSL Example

Consider the use of wikis to support software projects. The scattering of stakeholders, the need for collaboration and tracking, and the iterative manners that characterize software projects make wikis an attractive platform [9]. Fig. 3 provides an example for the *Purchase Project* as a *WSL* mindmap.

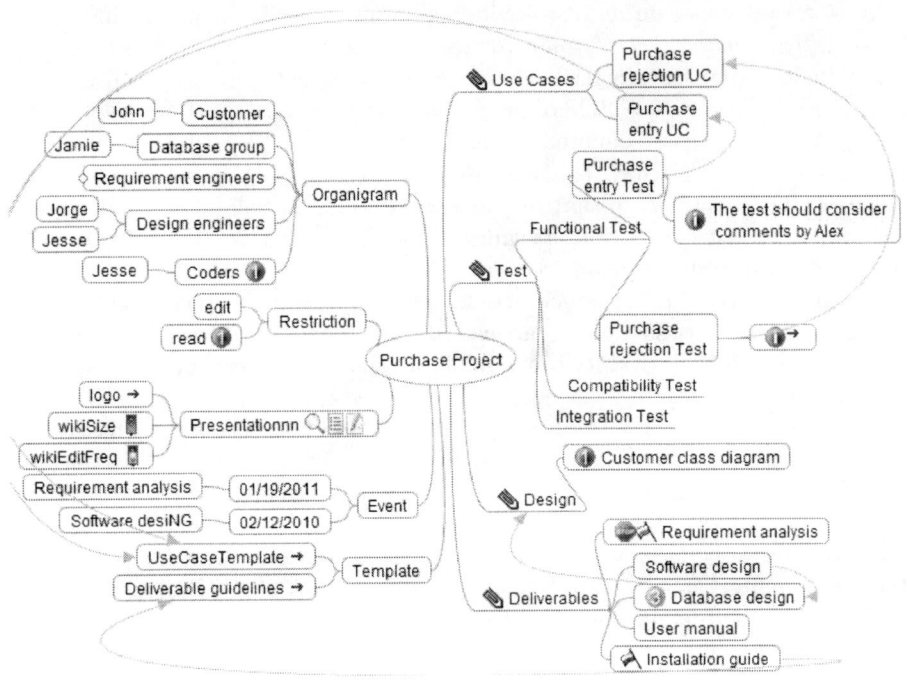

Fig. 3. *Purchase Project* Scaffolding

FreeMind depicts ideas and their relationships as nodes and edges that follow a radial distribution. In our example, the *Organigram* branch captures the existing roles as well as the employees assigned to these roles. The *Restriction* branch lists limitations in terms of wiki operations. The *Event* branch captures two milestones attached to pages *"Requirement analysis"* and *"Software desiNG"* at the onset. Next, the company already has some guidelines to capture use cases and document deliverables. Such practices should also be adhered to when in the wiki. The *Template* branch refers to two such guidelines through the *"UseCaseTemplate"* node and the *"DeliverableGuidelines"* node. The *Presentation* branch will impact on the rendering of the wiki based on the expected *"wikiSize"* and *"wikiEditingFreq"*. A *"traffic light"* icon 🔋 is used to indicate the three possible values of these properties: large (red light), medium (yellow light) and small (green light). As for the sidebar, this node includes a navigation pane (denoted by the *"list"* icon 📋) and a search pane (denote by the *"magnifier"* icon 🔍). The sidebar is finally completed with an index pane (denoted by the *"look here"* icon 🔖 on categories *"Use Cases"*, *"Test"*, etc). Regarding to restrictions, *"priority"* ⬤ icon sets a restriction whereby *"Coders"* (i.e., the role) are restricted from *reading* (i.e., the denial) the article *"Customer class diagram"* (i.e., the item).

As for the corporate glossary, common terms already in use include *"Use Cases"*, *"Functional Test"*, *"Compatibility Test"* etc. These terms find their way as wiki categories. Hierarchical relationships among categories are captured by describing a category as a child of the parent category (e.g., *"Test"* ← *"Functional Test"*). Wiki articles are denoted as *bubbled* nodes (e.g., *"Requirements analysis"* stands for an article which is categorised as *"Deliverable"*). The title of a node behaves as an identifier, so that two *FreeMind* nodes placed differently but with the very same title, stand for the same notion. This permits the *Content* graph to be flattened as a *FreeMind* tree.

It can look odd to introduce articles at wiki inception since wiki's *raison d'etre* is precisely collaborative article editing. Indeed, we do not expect too many articles to be introduced at scaffolding time. However, the need to come up with some articles might be known from the very beginning. The scaffolding permits so by introducing a node whose title becomes the title of the wiki article. For instance, the node *"Software design"* yields a wiki article with the namesake title. Even more, some relationships might be known at the outset. For instance, trace requirements made advisable to keep a hyperlink between the *"Purchase entry test"* and the *"Purchase entry UC"*. This is depicted as an arrow between the node counterparts.

Based on preliminary user feedback, we also consider article content to be known at scaffolding time. This is realized as a child of the given article (together with the *"info"* icon ⬤). Fig. 3 illustrates the two options. The content of *"Purchase entry test"* is explicitly provided as the text of its child node. By contrast, the content of *"Purchase rejection test"* is already available at the company as a *Word* document. *FreeMind* permits to introduce hyperlinks as node content (denoted through a small red arrow). This facility is used to our advantage to link *"Purchase rejection test"* to the external document holding its content. Likewise, corporate guidelines can find their way as wiki templates. So far, *WSL* only supports *Word* documents (exported as *XML*). At deployment time (i.e., when the *WSL* map is enacted), these external documents are turned into either, article content or wiki templates. Fig. 6 provides a screenshot of the main page as generated by the WSL engine. The rest of this section provides a detail account of *WSL* expressivity.

4.2   WSL Concrete Syntax

WSL abstract syntax is realized as a *graphical* concrete syntax. A mapping is then set between elements of the abstract syntax and their visual counterparts in *FreeMind*. These "visual counterparts" are set by the *FreeMind* metamodel. Therefore, a set of mappings between elements of the *WSL* metamodel and elements of the *FreeMind* metamodel is realized. Additionally, some constraints need to restrict the expressiveness of *FreeMind* to result in valid "scaffolding maps" (i.e., compliant with the *WSL* metamodel).

FreeMind metamodel (see fig. 4). A **Map** is a compound of **Nodes**. Nodes have a **title** and might hold a *link* to an external document (local or remote) as well as a set of properties mainly referring to rendering concerns. For instance,

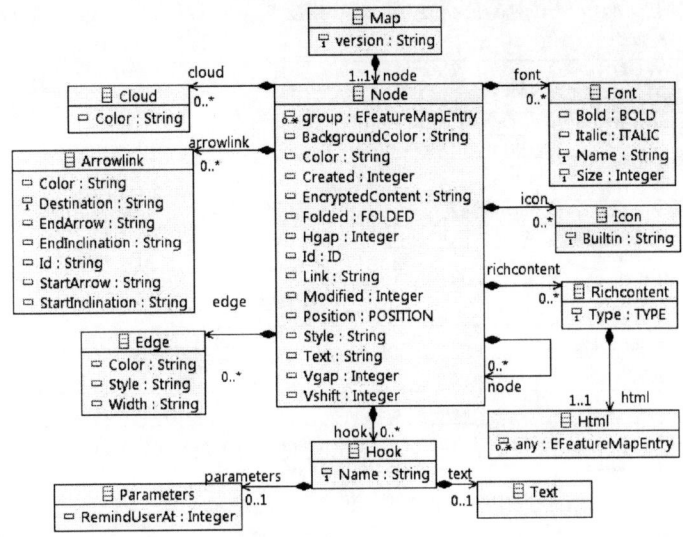

Fig. 4. *FreeMind* metamodel: primitives for mindmap drawing

the **Style** property can be **fork** and **bubble** and determines the look of the node as a tagged line tag or a bubble, respectively. Next, nodes are basically arranged in a tree-like way. A central node serves as the common root. Tree structures are constructed using **Edges**. An Edge is a connector that relates a node with its parent node. Additionally, **Arrowlinks** are also connectors but in this case, the connection is between two arbitrary nodes. Finally, **Icons**[3] and **Fonts** can be associated with nodes in an attempt to reflect the underlying semantics of the node. Of course, this semantics resides in the users' head.

WSL-to-FreeMind mapping (see Table 1, first two columns). Once *FreeMind* visual symbols are introduced, the next step is to indicate a mapping between the *WSL* abstract syntax and these symbols:

– *Scaffolding* class. The *root node* is the *FreeMind* counterpart of this class.
– *Organigram* class. A bubble node with title *"Organigram"* denotes the origin of the organigram hierarchy. Nodes having *"Organigram"* as parent denote roles. Likewise, nodes having *"Organigram"* as grandparent are interpreted as employees.
– *Presentation* class. A bubble node with title *"Presentation"* denotes this class. Boolean properties are captured as icons on *"Presentation"*. Value-based properties are represented as children nodes: *logo* (captured as a link to an image file), *wikiSize* and *wikiEditFreq*. The latter are decorated with traffic-light icons to account for their values.

---

[3] *Freemind* provides a fixed set of icons. In the last version, users can introduce their own icons, though it is not recommended for interoperability reasons.

Table 1. *WikiScaffolding-to-FreeMind* mapping & *FreeMind-to-MediaWiki* mapping

WSL	FREEMIND	MEDIAWIKI
Scaffolding	root node	main page[4]
Organigram	*"Organigram"* bubble node	n.a.
Role	child of *"Organigram"* node	wiki group
Employee	grandson of *"Organigram"* node	wiki user & user page
Presentation	*"Presentation"* bubble node	wiki skin[5]
logo	logo node	wiki logo
wikiSize	*wikiSize* node	wiki skin
wikiEditFreq	*wikiEditFreq* node with *"traffic light"* icons	wiki skin
navigationPane	*"list"* icon	navigation in sidebar
searchPane	*"magnifier"* icon	search in sidebar
toolboxPane	*"refine"* icon	toolbox in sidebar
indexPane entry	*"look here"* icon at *Item*	element in the navigation bar
Restriction	*"Restriction"* bubble node and *"priority"* icons	blacklisted pages for groups[6]
denial	child of *"Restriction"* node	wiki permission.
Item		
title	node text	page title
category Item	fork node	category page
article Item	bubble node	article page
template Item	child of *"Template"* node	template page
event Item	child of *"Event"* node	calendar extension[7]
discussion	*"stop-sign"* icon	talk page for that page
RSSfeed	*"flag"* icons	RSS generator for that page[8]
text	child with *"info"* icon or linked files	page content
Link		
relatedWith Link	arrowLink connector	inter-page hyperlink *[[page]]* *[[:Category:parentCat]]*
belongsTo Link	edge connector	page-category hyperlink *[[Category:parentCat]]*
templatedBy Link	arrowLink connector	template-page hyperlink *{{template}}*
scheduledFor Link	edge connector	event-to-page link in the calendar widget

- *Restriction* class. A bubble node with title *"Restriction"* denotes this class. A restriction is a triplet: subject (i.e., an *Item* node), grantee (a *Role* node), and the denial type (i.e., *read* or *edit*). We resort to *priority* icons to denote those elements that conform to a restriction unit. That is, map nodes decorated with the same *priority* icon belong to the same *restriction*. Due to icon availability, *permissions* are limited to ten (*"priority"* icon ..).

---

[4] CategoryTree extension: www.mediawiki.org/wiki/Extension:CategoryTree
[5] MediaWiki skins include monobook (default), vector (e.g., used by Wikipedia), etc. WSL completes the offer with cavendish, rilpoint, guMax, guMaxDD and guMaxv.
[6] Blacklist extension at www.mediawiki.org/wiki/Extension:Blacklist
[7] Barrylb extension at www.mediawiki.org/wiki/Extension:Calendar_(Barrylb)
[8] WikiArticleFeeds extension at www.mediawiki.org/wiki/Extension:WikiArticleFeeds

- *Content* class. There is not a *FreeMind* counterpart for the *Content* class as such. Rather all nodes in the map except for *"Organigram"*, *"Presentation"*, *"Restriction"*, *"Event"* and *"Template"* nodes (and descendants) stand for *Content Items*. The node title behaves as an identifier, so that two *FreeMind* nodes placed differently but with the same title, stand for the same *Item*. This allows the *Content* graph to be flattened as a *FreeMind* tree.
- *Item* class. *Items* are typed as *"category"*, *"article"*, *"template"* and *"event"*. *Category Items* are denoted as fork nodes (i.e., nodes with the "fork" style). *Article Items* are captured as bubble nodes. Next, *Template Items* are children of the *"Template"* node. These nodes can either hold the page text content (i.e., text attribute) themselves as a child with the *"info"* icon ⬤ or point to external documents from where the content is obtained at compile time (only txt and word as xml exported files in the current version) Finally, *Event Items* are children of the *"Event"* node. As for the boolean properties, *discussion, rssFeed* and *indexPaneEntry*, the affected *Items* (regardless of their type) are decorated with the *"stop sign"* icon ⬤, a *"flag"* icon ⚑ and *"look here"* icon ✎, respectively.
- *Link* class. *Links* are classified as *relatedWith, belongsTo, templatedBy* and *scheduledFor*. *FreeMind* offers two kinds of connectors: *Edges,* which are the default arcs connecting a node with its child, and *ArrowLinks*, which are arcs connecting two nodes anywhere in the map. *Edges* are interpreted as *belongsTo* links when they connect an *Item* to a *category Item* (e.g., fig. 3, arc from *"Database design"* to *"Deliverables"*) and as *scheduledFor* when they connect an *Item* to an *event Item* (e.g., fig. 3, edge from *"Requirement analysis"* to *"01/19/2011"*). As for *ArrowLinks*, they sustain (1) *RelatedWith* links when they relate an *Item* to another *Item* (e.g., fig. 3, arc from *"Software design"* to *"Database design"*) and *(3) TemplatedBy* links when the ingoing node stands for a *template Item* (e.g., fig. 3, arc from *"Purchase entry UC"* to *"UseCaseTemplate"*).

5   Verification of WSL Maps

WSL maps are a subset of *FreeMind* maps, i.e., *WSL* metamodel imposes additional constraints on top of the *FreeMind* metamodel. Such constraints can be verified on user request or at deployment time. Fig. 5 provides a snapshot of the *"Tools"* menu now extended to address *WSL* maps: *"WSL configuration"* permits to configure parameters for the *MediaWiki* installation; *"WSL deployment"* causes the generation of the wiki instance from the *WSL* specification; *"WSL Skeleton"* provides a *FreeMind* map with the basic *WSL* nodes (e.g., *Organigram, Restriction,* etc) so that misspells are prevented; and finally, *"WSL Map Checking"* triggers *WSL* map verification.

Fig. 5 depicts the verification outcome for our sample problem (see fig. 3). Messages can be warnings and errors. For our sample, two warnings are noted.

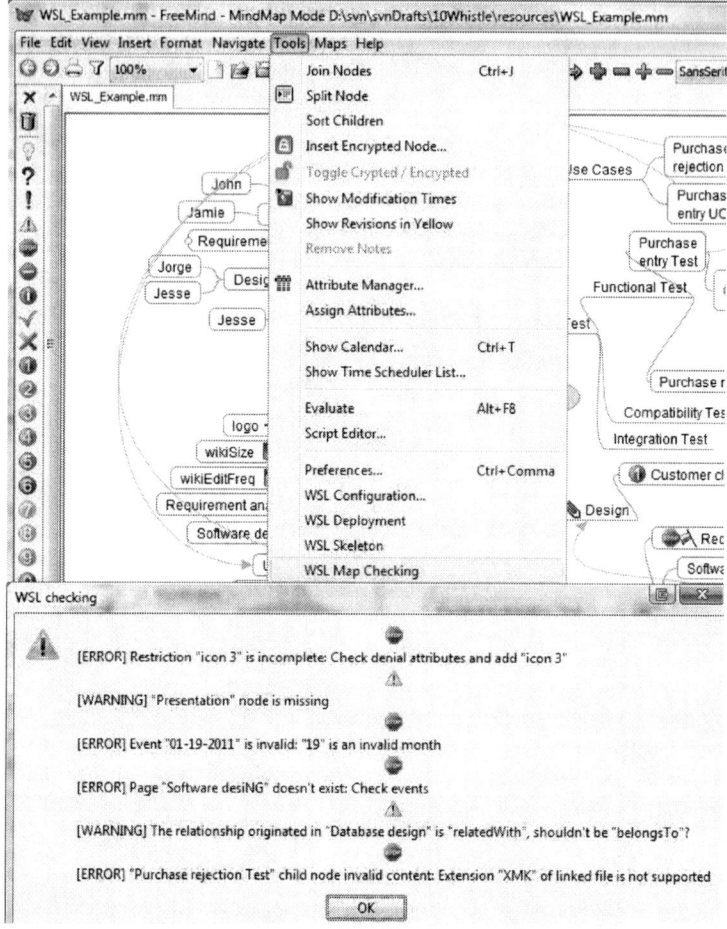

Fig. 5. Verifying the *WSL* map at fig. 3

One informs about the lack of the optional *Presentation* node which, in this example, is due to a misspelling (*"Presentationnn"*). The other warning notifies about a common mistake in wiki construction: setting a *relatedWith* relationship between an article and a category. This is an odd situation that could be mistaken with the *belongsTo* relationship, and so is it indicated. As for errors, they prevent the wiki from being generated. For our sample case, these errors include: a misspelling of an event date (e.g., *"01/19/2011"*); referring to a non-existent node (e.g., *"Software desiNG"*); partial definition of a restriction where either the denial, the employee or the article is missing (e.g., restriction ③); unsupported document extension (e.g., extension *"XMK"* is not supported; so far, only XML and TXT files can become page content).

6   Enactment of WSL Maps

By selecting the *"WSL deployment"* option of the *Tool* menu (see fig. 5), the current map is turned into a wiki installation in *MediaWiki*. This means that around 400 LOC (mainly *SQL* statements) are automatically generated for the current example. Figs. 6 and 7 provides three screenshots of the generated pages: the main page (illustrating the use of the *CategoryTree* and *Calendar* extensions), the *"Purchase rejection Test"* article page (which is obtained from a *Word XML* document) and the *"Purchase Rejection UC"* (which follows the *"UseCaseTemplate"* also externally obtained). Space limitations prevent us from giving a detail account of this generation process. For the purpose of this paper, it is enough to show the mapping between *FreeMind* constructs and *MediaWiki* primitives. The last two columns in Table 1 indicates such mappings.

It is important to notice that some scaffolding features require additional *MediaWiki* extensions (e.g., *CategoryTree*). The *WSL* engine builds upon *MediaWiki* version 1.16 and the extensions have been tested against it. Such

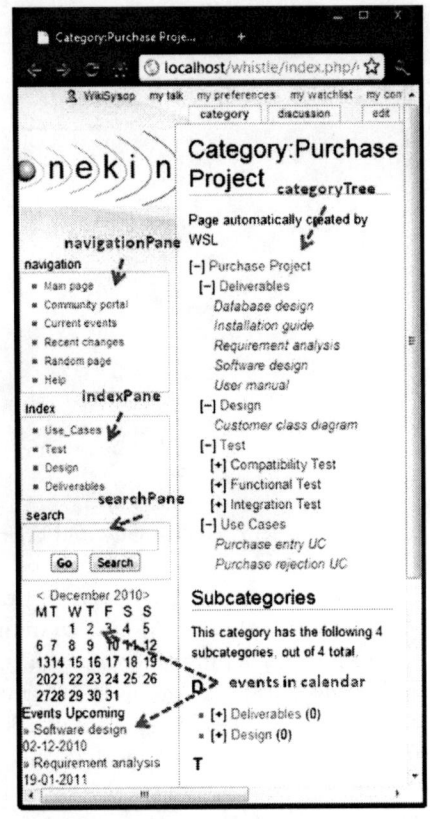

Fig. 6. *"Purchase Project"* wiki main page as generated by *WSL*

composition is provided as a unit by *WSL*. This raises the issue of platform evolution, i.e., new versions of *MediaWiki* (or its extensions) might impact the *WSL* engine. This is certainly true. But, how real is this threat? First, *MediaWiki* is a stable platform backed by thousands of installations. And second, wikis can be upgraded once deployed. That is, *WSL* can be used to generate the wiki scaffold, and next, the user can upgrade to the newest version (just two clicks away). This makes us confident about the lifespan of *WSL*.

7   Discussion through Related Work

Mindmaps have long been recognized as a useful technique for brainstorming. Recently, enhancements have been proposed to improve the efficacy of mind maps (e.g., use of pictorial stimuli [14]). Although benefits are reported, these extensions decrease simplicity, and jeopardize interoperability. There certainly

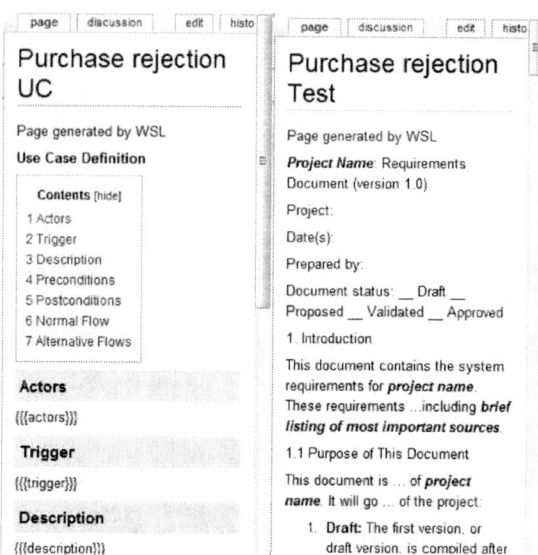

Fig. 7. Template and article pages as generated by *WSL*

exists more sophisticated tools for brainstorming than *FreeMind,* but we value popularity, simplicity and cost as main selection criteria.

Another important remark is that of scalability. Although it is not the aim of scaffolding to offer a complete wiki map but just a blueprint, large projects can require large scaffoldings. This can lead to cluttered *WSL* maps. Fortunately, *FreeMind* offer view-like mechanisms that permit to filter map nodes based on content and relationships. Testing stakeholders can filter those nodes based on containing the string *"test"*, whereas template-minded stakeholders can restrict the view to those nodes related with a template.

As for visual DSLs, they are still scarce compare with textual DSL ([11] for an overview). Our insight here is that the context where the DSL is to be deployed is generally overlooked in DSL publications. Our experience is that DSL success not only depends on finding the right abstractions but also on producing minimum disturbance to existing practices. *FreeMind* was chosen on these grounds.

8   Conclusions

We introduced the notion of "wiki scaffolding" as a way to capture the contextual setting for wikis deployed in an existing organization. While wikis for virtual communities create such setting as they go along, corporate wikis know this context at the onset. We introduced a *DSL* for wiki scaffolding that abstracts from the technicalities that go in setting those parameters down to wiki code. By capitalizing on *FreeMind* as the conduit for *WSL* concrete syntax, we expect non-technical communities to benefit from the scaffolding. Anecdotical evidences suggest that the benefits of the *DSL* go beyond speeding up wiki deployment or promoting user participation. Knowledge retention is achieved by the *DSL* engine embedding good practices about both presentation and structure. This helps introducing wikis in organizations without wiki experience. As for the expressiveness of *WSL*, current constructs are based on a literature survey about the use of wikis in companies. Social conventions and incentives will emerge and evolve to guide contributors, resolve disputes and help manage wiki deployments in organizations. As these issues find support in wiki engines, *WSL* constructs will need to be extended.

Acknowledgments. This work is co-supported by the Spanish Ministry of Education, and the European Social Fund under contract TIN2008-06507-C02-01/TIN (MODELINE), and Consejería de Educación y Ciencia of Castilla-La Mancha under contract PAC08-0160-6141 (IDONEO). Puente has a doctoral grant from the Spanish Ministry of Science & Education.

References

1. Freemind. Online, `http://freemind.sourceforge.net` (accessed November 25, 2010)
2. Mediawiki. Online, `http://www.mediawiki.org` (accessed November 25, 2010)
3. Carlin, D.: Corporate Wikis Go Viral. Online, `http://www.businessweek.com/technology/content/mar2007/tc20070312_476504.htm` (accessed November 25, 2010)
4. Cunningham, W.: What is a Wiki. Online, `http://www.wiki.org/wiki.cgi?WhatIsWiki` (accessed November 25, 2010)
5. Danis, C., Singer, D.: A Wiki Instance in the Enterprise: Opportunities, Concerns and Reality. In: Computer Supported Cooperative Work, CSCW (2008)
6. Kang, K.C., Cohen, S.G., Hess, J.A., Novak, W.E., Peterson, A.S.: Feature-oriented domain analysis (foda) feasibility study. Technical report, Carnegie-Mellon University Software Engineering Institute (1990)
7. Kelly, S., Tolvanen, J.-P.: Domain-Specific Modeling: Enabling Full Code Generation. Wiley-IEEE Computer Society (2008)
8. Lee, H., Bonk, C.: The Use of Wikis for Collaboration in Corporations: Perceptions and Implications for Future Research. In: World Conference on E-Learning in Corporate, Government, Healthcare, and Higher Education (2010)
9. Louridas, P.: Using wikis in software development. IEEE Software 23(2), 88–91 (2006)
10. Prescient Digital Media. Intranet 2.0 Global Survey. Online, `http://intranetblog.blogware.com/blog/_archives/2009/5/15/4187339.html` (accessed November 25, 2010)
11. Mernik, M., Heering, J., Sloane, A.M.: When and How to Develop Domain-Specific Languages. ACM Computing Surveys 37(4), 316–344 (2005)
12. Raman, M.: Wiki Technology as A "Free" Collaborative Tool within an Organizational Setting. IS Management 23, 59–66 (2006)
13. Toker, S., Moseley, J.L., Chow, A.T.: There a Wiki in Your Future?: Applications for Education, Instructional Design, and General Use. Educational Technology Magazine, 6 (2008)
14. Wang, H.-C., Cosley, D., Fussell, S.R.: Idea Expander: Supporting Group Brainstorming with Conversationally Triggered Visual Thinking Stimuli. In: Computer Supported Cooperative Work (CSCW), pp. 103–106 (2010)

The REA-DSL: A Domain Specific Modeling Language for Business Models

Christian Sonnenberg[2], Christian Huemer[1], Birgit Hofreiter[2],
Dieter Mayrhofer[1], and Alessio Braccini[3]

[1] TU Vienna
last@big.tuwien.ac.at
[2] University of Liechtenstein
{first.last}@uni.li
[3] LUISS University
abraccini@luiss.it

Abstract. In the discipline of accounting, the resource-event-agent (REA) ontology is a well accepted conceptual accounting framework to analyze the economic phenomena within and across enterprises. Accordingly, it seems to be appropriate to use REA in the requirements elicitation to develop an information architecture of accounting and enterprise information systems. However, REA has received comparatively less attention in the field of business informatics and computer science. Some of the reasons may be that the REA ontology despite of its well grounded core concepts is (1) sometimes vague in the definition of the relationships between these core concepts, (2) misses a precise language to describe the models, and (3) does not come with an easy to understand graphical notation. Accordingly, we have started developing a domain specific modeling language specifically dedicated to REA models and corresponding tool support to overcome these limitations. In this paper we present our REA DSL which supports the basic set of REA concepts.

Keywords: Domain Specific Languages, Conceptual Modeling, Business Models, Accounting Information Systems.

1 Introduction

Analyzing the economic phenomena on which companies base their business may serve as a good starting point in the requirements elicitation phase when developing enterprise information systems. Business models specify - amongst other things - the main actors, their relationships and the values exchanged between them (cf. [1]).

We see three main ontologies to conceptualize business models: the Business Model Ontology (BMO) [2], the e3-value ontology [3], and the Resource-Event-Agent ontology (REA) [4]. BMO is easy to use by the domain expert because it focuses mainly on the categorization of aspects relevant for the delivery of products and services to fulfill customers' requests. It helps the domain expert to ask herself the right questions when developing a business model, but has

H. Mouratidis and C. Rolland (Eds.): CAiSE 2011, LNCS 6741, pp. 252–266, 2011.

a limited focus on conceptualizing the elements of the business model. In contrary, e3-value defines a conceptual model to describe the exchanges of value among actors in a network. e3-value comes with a graphical syntax that is easy understood by the domain expert. Furthermore, it allows the domain expert to perform financial assessment of the value exchanges.

The Resource-Event-Agent (REA) ontology has its roots in the accounting discipline and was originally developed as a reference framework to conceptualize economic phenomena in an enterprise. In its proposal in 1982, McCarthy already had the vision to facilitate the design of data structures of accounting information systems by means of REA [4]. Since this time the REA model has been further extended and evolved into a domain ontology [5]. All REA concepts are based on well established concepts of the literature in economic theory - which is certainly one of the strengths of REA. However, REA has no dedicated representation format and, consequently, no graphical syntax. Thus, users may struggle when describing the REA models leading to the impression that REA is a rather heavyweight approach. A dedicated graphical syntax - such as it exists for e3-value - may help in overcoming this problem and may lead to a much more significant adoption of REA. Accordingly, we have started the endeavor of developing a domain specific modeling language for REA.

Most domain-specific languages (DSL) are small textual and usually declarative languages. A DSL offers expressive power through appropriate notations and abstractions focused on – and usually restricted to – a particular problem domain [6]. Besides textual DSLs, we see an increasing interest in domain-specific modeling languages [7,8] based on dedicated meta-models and notations. van Deursenet et al. claim [6] the following benefits of a DSL approach: They allow solutions to be expressed at the level of abstraction of the problem domain. As a matter of fact, domain experts themselves can understand, validate and often modify DSL programs/models. The DSL programs/models are concise and self documenting to a large extent. They enhance productivity, reliability and maintainability. DSLs allow for validation and optimization at the domain level.

When developing our REA-DSL we followed methodological steps that have been suggested by Strembeck and Zdun for the systematic development of domain specific languages [9]. Amongst other variants, they describe the development process for extracting a DSL from an existing system, which is appropriate for our needs, because we extract the DSL from the existing REA ontology. Accordingly, we started with (1) the identification of elements in the REA ontology. Next, we underwent a number of revision cycles of (2) deriving the abstract syntax of the REA model including the core language model and the language model constraints and (3) defining the DSL behavior, i.e. determining how the language elements of the DSL interact to produce the intended behavior. Once we had reached a stable state, we defined the DSL concrete syntax (4). Finally, we implemented a modeling tool support for the DSL (5), but we skipped the last step described in [9], i.e. integrating the DSL into a software platform, since REA stays at the platform independent level.

The remainder of this paper is structured as follows: In Section 2 we give a basic introduction into the REA ontology including an illustrative example. The core of the paper in Section 3 presents our DSL for REA models. We elaborate on the meta model of our REA DSL for describing duality models and value chains. Furthermore, we use the same example as in Section 2 to illustrate our DSL. Section 4 reports on our evaluation by means of a REA DSL tool. A summary and remarks on future work conclude the paper in Section 5.

2   Resource - Event - Agent (REA)

2.1   The REA Ontology

The objective of the REA ontology is the conceptualization of common economic phenomena of a firm independent of application-specific demands. REA accounting information systems focus on economic exchanges as the central unit of analysis. Instead of representing these exchanges with double-entry bookkeeping artifacts (e.g. debits, credits, accounts), REA proposes concepts and patterns to derive semantic models of economic exchanges and transformations. The underlying assumption of REA is that all business enterprises operate in the same manner [10] according to an entrepreneurial script: acquiring financial resources, engaging in a chain of economic exchanges with other parties, each time giving up an economic resource in return for another resource of greater value [10]. After executing this script, the business generates a justifiable profit after having paid interests and creditors. This entrepreneurial script essentially discloses the entrepreneurial rationale of a business. Hence, the REA ontology is not only used to facilitate the design of accounting information systems but in particular for business modeling.

The basic REA ontology is a stereotypical representation of an economic exchange as a core economic phenomenon [5]. This exchange is executed between parties inside and outside of a firm's boundaries and follows a particular object pattern (cf. [5], see also Figure 1(a) and 1(b)). In order to conceptualize this pattern, the REA ontology suggests three concepts that constitute an exchange: *resource*, *event*, and *agent*. Resources are things being exchanged between participating agents. In an exchange, an agent (inside agent) usually gives up control of a resource to an outside agent in order to gain control over another resource. Events occur in the course of executing economic activities. In REA basically two types of events are distinguished: *increment* and *decrement* events. Extensions of the REA ontology [11] also distinguish between *transfer* (exchanges with external actors) and *transformation* (concerns value creation within the firm) events.

Furthermore, the following economic primitives (relationship types) are specified by the REA: *duality*, *stock-flow*, and *participation*. A duality relationship connects decrement events with corresponding increment events and thus provides the rationale of individual economic activities. Stock-flow relationships connect economic resources with economic events (decrement or increment events). Depending on the connected event type, the following stock-flow relationship

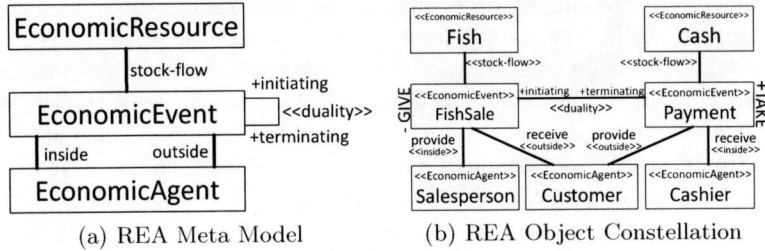

(a) REA Meta Model   (b) REA Object Constellation

Fig. 1. Basic REA Ontology

types are distinguished: *give* and *take* (transfer events), *use, consume*, and *produce* (transformation events). Participation relationships describe the involvement of an agent in an economic event. As such, the basic REA ontology not only conceptualizes economic exchanges but also relates to business process concepts. It captures *Who* is involved in an exchange (*economic agents*), *What* is being exchanged (*economic resources*), *When* (and under what conditions) do the components of an exchange occur (*economic events*), *Why* are the participants engaged in an exchange (*duality relationships, stock-flows*), and *How* do the exchanges materialize as economic activities or business processes (series of small events that move business process through to completion) (cf. [12]).

The basic REA ontology with its economic primitives is illustrated in Figure 1(a) by means of a UML class diagram. Figure 1(b) shows an instantiation of the REA ontology, a so called object constellation. Furthermore, the ontology defines three axioms that restrict the use of the concepts and primitives for conceptualizing economic exchanges (cf. [11]):

– **Axiom 1:** At least one take event and one give event exist for each economic resource (guarantees modeling of economic activities as a sequence of exchanges).
– **Axiom 2:** All events effecting a resource decrement must be eventually paired in duality relationships with events effecting an increment and vice versa (ensures correct enumerations of exchanges).
– **Axiom 3:** Each exchange needs an instance of both the "inside" and the "outside" agents(ensures presence of exchanges between parties with competing economic interests).

Axioms one to three apply for transfer events. Axiom two also holds for transformation events.

Since its proposal in 1982, some extensions to the basic REA ontology have been proposed (e.g. [5]). The extended REA ontology envisions a vertical and horizontal layering of economic exchanges. With regard to the vertical layering, a hierarchy consisting of three levels is proposed: (1) the *value chain* specification level, (2) the *duality* specification level, and (3) the *task or workflow* level ([5]). In this paper we focus on the upper two specification layers: the value chain and the duality . Thereby, we base our work on the diagram style used by Geerts and McCarty in [13], which has never been formalized.

The horizontal layering in REA enables the analysis of economic exchanges at different points in time on all three vertical layers described above. Therefore, the REA ontology considers an *accountability infrastructure* and a *policy infrastructure* [5]. The REA accountability infrastructure conceptualizes actual business events and captures "what has occurred" or "what is or has been". The policy infrastructure conceptualizes what "could be" or "should be" within the context of a defined portfolio of a firm's resources and capabilities [5]. In this paper we focus on the concepts associated with the REA accountability infrastructure. Concepts associated with the policy infrastructure like commitments, agreements, and type images are not covered here but are subject to future work.

In the following section the REA ontology is applied to a simplified example in order to demonstrate how the REA constructs can be used to specify an entrepreneurial script.

2.2 REA Ontology Example

The simplified example used to apply the REA ontology is taken from [13]. It is based on an actual company and is called *Sy's Fish* and is introduced below:

Sy's Fish is a distributor of seafood and provides his base of restaurant customers with over 50 types of fish which can be stored at all locations or stores. However, each store usually specializes in local favorites. Fish are purchased from local fishers, cleaned at the store, and then sold at restaurants to customers. Customers are allowed to buy on credit, and all pay on the last day of the month. Most employees are generalists and can perform many duties such as purchasing, cleaning, and delivering fish. They fill out time cards fortnightly upon which they may note the percentage of time devoted each day to buying, cleaning and selling fish. Non-generalist employees for the most part comprise of cashiers. Sy's also possesses a fleet of trucks, used to bring fish from the docks and to deliver fish to the restaurants. Both the truck and the employees involved in each purchase and sale of fish are noted. All trucks are leased on yearly contracts, and lease payments are made monthly. Cash receipts and disbursements are made to/from one of the multiple checking accounts of the firm.

The value chain of Sy's Fish (i.e. the entrepreneurial script) comprises the processes payroll, truck acquisition and store processing. The last process is

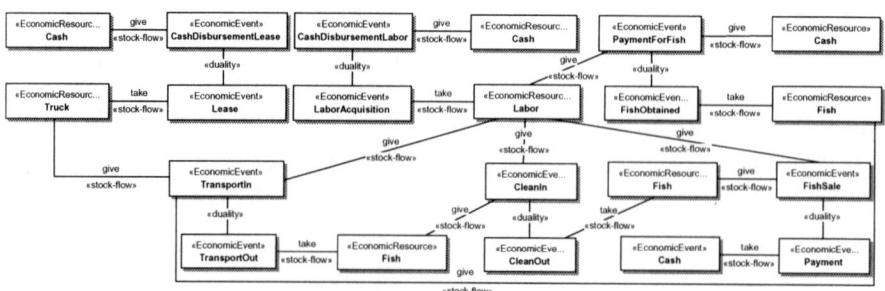

Fig. 2. REA Sy's Fish Example

further decomposed into the processes buying, cleaning, and selling. The process level specification is provided in 2 as a UML class diagram. It depicts the object constellations for the entire value chain. Due to space limitations, economic agents have been omitted from the process level specification. The process level specification with its sequence of duality relationships determines the rationale of how Sy's makes money. Initial outlays (cash disbursements) are followed by subsequent cash flows. These cash flows can be used to pay creditors and interests or to finance further economic activities.

2.3 Limitations of the REA Ontology

A first limitation of REA is given with regard to the clarity of ontological statements. Although the REA model has evolved into a domain ontology, REA leaves space for diverging interpretations of the relationships between core concepts. In particular, the multiplicities for individual relationships are not clearly defined. It remains unclear, whether a resource increment or decrement is caused by one event or by multiple events. Moreover, it is not clear if an event can be related to multiple stock-flows and agents. Furthermore, there is no axiom restricting event-resource (stock-flow) relationships. In its original specification, REA would allow to connect a single event with stock-flows to resources which are incremented and also to those which are decremented. However, this is not compliant with the duality concept which is established between an increment event(s) and decrement event(s) which are well distinguished from each other. Accordingly increment events are connected by stock-flow reletationships to resources which are incremented and decrement events to those which are decremented.

These ambiguities and potential semantic inconsistencies may be due to a lack of a dedicated specification language. In the current state no means have been proposed to specify precisely how to relate the REA concepts and how to enforce compliance of REA models with the axioms stated above. The compliance of REA models can only be checked manually. Thus, domain experts are responsible for including the REA pattern into particular enterprise information architectures. There is no conceptual facility that enables the development of interoperable REA models (a modeling language would help here).

A third limitation is the complexity of REA models which increases progressively even with very few processes to be modeled. Each process generally includes eight entities which have to be modeled (two events, two resources, two times an inside and an outside agent) [10]. A dedicated REA modeling notation would help to overcome the complexity. However, proposing a modeling notation necessitates resolving the inconsistencies and incompleteness of the ontological statements. This is subject to the following sections.

3   The REA Domain Specific Language

Given these limitations we started to develop a graphical domain specific language (DSL) for REA. We based the development of the REA DSL on The

Object Management Group's(OMG) metamodeling architecture called Meta-Object Facility (MOF) [14]. MOF comes with a meta-meta model (M3 layer) that allows us to define the structure, i.e. the abstract syntax, of the REA DSL as a meta-model (M2 layer). The resulting REA DSL meta-model comprises three interlinked views, of which we describe in detail the *duality* and the *value chain* perspective. Due to space limitations, we do not elaborate in detail on the third view on economic resources. However, it is important to note that economic resources - scarce objects having utility and under control of an enterprise [15] - may form a generalization hierarchy.

3.1   The Duality Model

The duality relationship is a core concept in REA. It links an increment in the resource set with a corresponding decrement; where increments and decrements must be members of two different event entity sets. In the REA ontology, the duality relationship is characterized by the unary relationship assigned to the concept of an economic event (see Figure 3(a)).

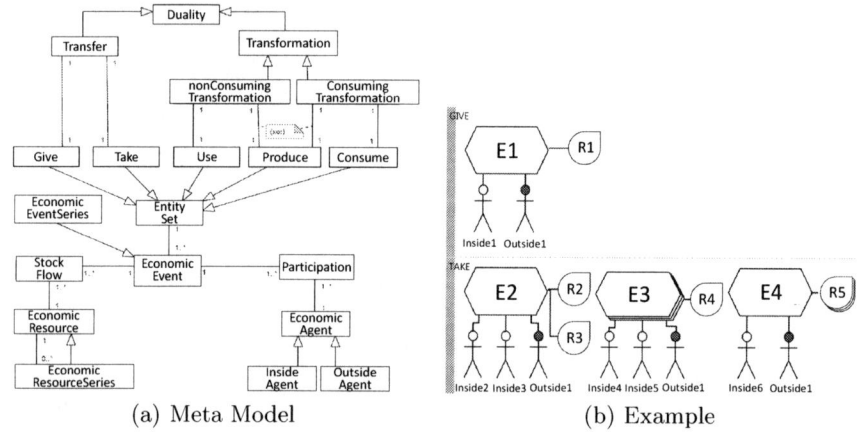

(a) Meta Model   (b) Example

Fig. 3. Duality

In the REA DSL, *duality* becomes a core model element that serves as a building block for further purposes. The duality meta model is depicted in Figure 3(a). Figure 3(b) presents a (rather abstract) example model - which should help understanding the meta model concepts and to introduce the concrete syntax and the corresponding stencils.

The *duality* concept applies to *transfers* (exchanges with external actors) and *transformations* (value creation inside the enterprise). In the case of *transformations* we distinguish between *resource-consuming* and *non-resource consuming transformation*. As a consequence, the meta model defines *consuming transformation* and *non-consuming transformation* as special kinds of *transformation* as well as *transfer* and *transformation* as specializations of *duality*.

No matter which kind of specialization it is, a *duality* always covers two distinct *entity sets*; one describing the increments in the resource sets and the other one the decrements in the resource sets. In the case of a *transfer* the decrement is called *give* and the increment is called *take*. The decrement of a *non-consuming transformation* is denoted as *use* and the one of a consuming transformation is denoted as *consume*. In both kinds of *transformation* the increment is called *produce*.

Figure 3(b) shows an abstract example of a *duality* model to illustrate the concepts and their stencils. An entity set is modeled as a partition. Accordingly, a duality model includes two partitions. The *duality* shown in the example is a *transfer*. It follows that one *entity set* is a *give* and the other one is a *take*. The kind of *entity set* is denoted in the upper left corner of the partition.

According to the meta model, an *entity set* covers at least one but up to multiple *economic events*. An *economic event* is considered as a class of phenomena reflecting changes in scare means [16]. An *economic event* is specific to the *entity set* it belongs to. Following the principles of *duality*, all *economic events* in the decrement *entity set* (*give, use, consume*) are counterbalanced by the *economic events* in the corresponding increment *entity set* (*take, produce*) of the same *duality* model.

An *economic event* is usually executed at a certain point in time. However, in certain cases the increment/decrement is realized by a series of economic events each of the same nature. For example, consider that the payment for goods is split into a number of partial payments. For this purpose we use the concept of an *economic event series*, which is defined as a specialization of the *economic event* in the meta model. Consequently, the concept of an *economic event series* may substitute an *economic event*. This means, instead of modeling each *economic event* of each partial payment, one may model a single *economic event series* of partial payments.

An *economic event* in a *give/use/consume entity set* decrements resource(s). Similarly, an *economic event* in a *take/produce entity set* increments resource(s). This relationship between an *economic event* and an *economic resource* is described by the concept of a *stock-flow*. A *stock-flow* models an association between exactly one *economic event* and exactly one *economic resource*. An *economic event* will affect most of the time one *economic resource* only, but it may affect multiple ones. Thus, an *economic event* may have one up to many *stock-flows* connected. An *economic resource* usually is affected by many different *economic events* (in different *entity sets* of different *duality* models). At a minimum an *economic resource* is affected by one *economic event* - otherwise it would not be worth considering the *economic resource* at all. Consequently, an *economic resource* is connected to one up to many *stock-flows*.

A similar concept to *economic event series* is defined for *economic resources*: the *economic resource series*. An *economic resource series* is used if an *economic event* affects a number of *economic resources* of the same kind. For example, on a high level of abstraction one may define raw material as an *economic resource* and an *economic event* may affect a number of *economic resources*. It is important to note that an economic resource series may substitute an economic resource in a stock-flow, but the economic resource series must not be used in economic resource generalization hierarchies. This is prohibited by a corresponding OCL constraint to the meta model. An *economic resource series* must always be based on exactly one existing *economic resource*. For an *economic resource* one may define zero to many *economic resource series* (used in different *stock-flows*).

An *economic event* involves *economic agents*. We distinguish between *outside agents*, i.e. trading partners outside the company, and *inside agents* who are accountable inside the company. The involvement of *economic agents* in *economic events* is denoted by the concept of *participation*. A *participation* is an association that connects exactly one *economic event* with one *economic agent*. An economic event is associated to at least one, but up to many economic agents. Hence, an economic event has one to many participation associations. An economic agent participates in at least one, but up to many economic events (in the same, but also in different entity sets of the same or different duality models). Thus, an economic agent has one to many participations connected.

In addition, there are further constraints assigned to the meta model to handle specifics of *transfers*. In case of a *transfer*, each *economic event* must be assigned to exactly one *outside agent* and, in addition, to at least one *inside agent*. All *economic events* of the same *transfer* (both in the *give* and the *take entity set*) must involve one and the same *outside agent*.

The relationships among *economic events*, *economic resources*, and *economic agents* within an *entity set* is also illustrated in the abstract example of Figure 3(b). *Economic events* are denoted by an hexagon. In the *give* partition, there is only one *economic event* E1. This *economic event* leads to a decrement of the only associated *economic resource* R1, notated by a drop. Since the *duality* model is a *transfer*, exactly one *outside agent* Outside 1 and an *inside agent* Inside 1 is involved. The symbol for *economic agents* is a stickman - *outside agents* have a black head, whereas *inside agents* have a white head.

The *take* partition includes three *economic events* E2, E3, and E4; one of which (E3) is an *economic event series*. The symbol of an *economic event series* is a pack of hexagons. All three *economic events* together compensate the single *economic event* E1 of the *give* partition. E2 is associated with two *economic resources* R2 and R3 that are incremented, whereas the series E3 increments only the *economic resource* R4. E4 leads to an increase of the *economic resource series* R5. An *economic resource series* is depicted by a pack of drops. Given the example of a *transfer*, all three events in the *take* partition must be associated with the same *outside agent* as the *economic event* of the *give* partition: Outside 1. In addition, E2 and E3 involve two *inside agents* and E4 only one.

3.2   The Value Chain

According to Geerts and McCarty [13], the duality relationships introduced in the previous chapter are the glue that binds a company's economic events together into rational economic processes, while "stock-flow" relationships weave these processes together into an enterprise value chain. In the latter case, they

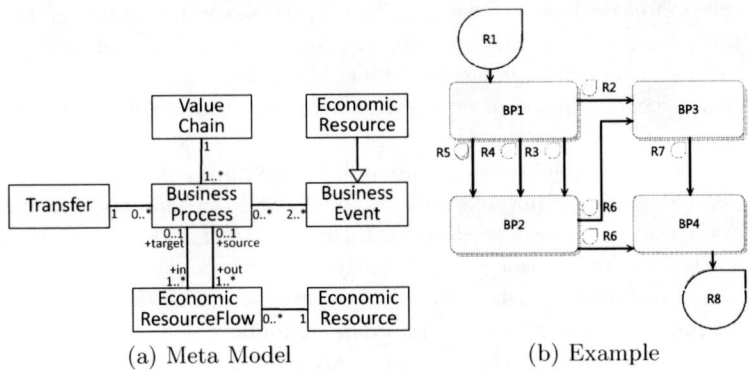

(a) Meta Model   (b) Example

Fig. 4. Value Chain

do not refer to the stock-flows within a single duality model. Rather, they mean that an increment in a resource as a result of one duality model will serve as a chance to decrement this resource in a subsequent duality model. In other words, a value chain is built by a well defined series of duality models. The transitions between the duality models reflect the resource dependencies between the duality models.

These general ideas are reflected in the value chain meta model of Figure 4(a). A *value chain* is built by a number of *business processes*. A *business process* is defined as a *transfer* or a *transformation*, and in addition the tasks needed to execute the *transfer/transformation*. It follows a *business process* points to an underlying *transfer/transformation* described by a *duality* model. Furthermore, a *business process* results in a number of *business events*, two of which are the *economic events* of the underlying *duality* model.

A *value chain* includes one to *many business processes*. A *business process* is used only once in one distinctive *value chain*. A *business process* points to exactly one *duality* model. A *duality* model is usually the basis of one business process, but may be referred to by multiple *business processes*. A *business process* involves at least two *business events*, these are the two *economic events* (which are considered as special kinds of *business events*). Usually, there will be more *business events* associated to a *business process* as a result of the tasks needed to execute the *transfer/transformation*.

Economic resources tie the *business processes* together. Thus, an *economic resource flow* - which points to exactly one *economic resource* - connects two *business processes*. An *economic resource flow* is a directed association that usually starts from a source *business process* and ends at a target *business process*. However, we also allow for *economic resource flows* that have either no source *business process* or no target *business process*. This allows for a partial analysis, when one considers a certain *economic resource* as given or when an *economic resource* is considered as final output of the *value chain*. Typically, cash is often assigned to such *economic resource flows*.

A *business process* has at least one, but up to many outgoing *economic resource flows*. Similarly, a *business process* has at least one, but up to many ingoing *economic resource flows*. In order to deliver a consistent *value chain* two important constraints on the business processes have to be considered: For each *economic resource* in the *give / use / consume entity set* of the underlying *duality* model, at least one corresponding incoming *economic resource flow* pointing to the same *economic resource* must exist. In analogy, for each *economic resource* in the *take / produce entity set* of the underlying *duality* model, at least one corresponding outgoing *economic resource flow* pointing to the same *economic resource* must exist. However, one may consider the substitutability concept of more general and more specific economic resources as defined in a resource generalization hierarchy. In other words, a duality model expecting an economic resource in *give / use / consume entity set* may also accept a more specific *economic resource* in the incoming *economic resource flow*.

In Figure 4(b) we depict an abstract example of a *value chain* to illustrate its concepts and their stencils. Our *value chain* example includes four *business processes* BP1, BP2, BP3, and BP4. The symbol for a *business process* is a rounded rectangle. *Economic resource flows* are depicted by an arrow with the *economic resource* assigned to this arrow. For example, the *economic resource* R2 flows from BP1 to BP3. However, for esthetic reasons we provide an alternative (but still similar) notation for *economic resource flows* that do not start from or do not end in a *business process*. In this case, the arrow may start from the *economic resource* or may lead to the *economic resource*, instead of assigning the *economic resource* to the arrow. Examples are the resource flows of R1 leading into BP1 and of R8 starting from BP4.

BP1 points to the *duality* model of Figure 3(b). In this *duality* model the *economic resource* R1 sits in the *give* partition; and the *economic resources* R2, R3, R4, and R5 - where the latter is a *economic resources series* - are included in the *take* partition. This is consistent with the *economic resource flows* to/from BP1 in the *value chain* model of Figure 4(b). BP1 receives R1 and delivers R2, R3, R4, and R5. It should be noted that the number of ingoing/outgoing transitions does not necessarily meet the number of *economic resources* in the *duality* model, since a *business process* may provide an *economic resource* or - more unlikely in practice - receive a resource from multiple business processes. For example, BP2 provides R6 to both BP3 and BP4.

Let us assume that the *duality* model on which BP2 is based requires the *economic resources* R3, R5, and R7. In this case one might think of an inconsistency since BP2 receives R3 and R5, but R4 instead of R7. However, if R4 is defined as a specialization of R7 in a resource generalization hierarchy, the model is consistent since in this case R4 may substitute R7 in the *duality* model of BP2.

3.3   REA DSL Example

Having introduced the meta models and rather abstract examples, we now demonstrate our REA DSL by means of a simple, but more realistic example.

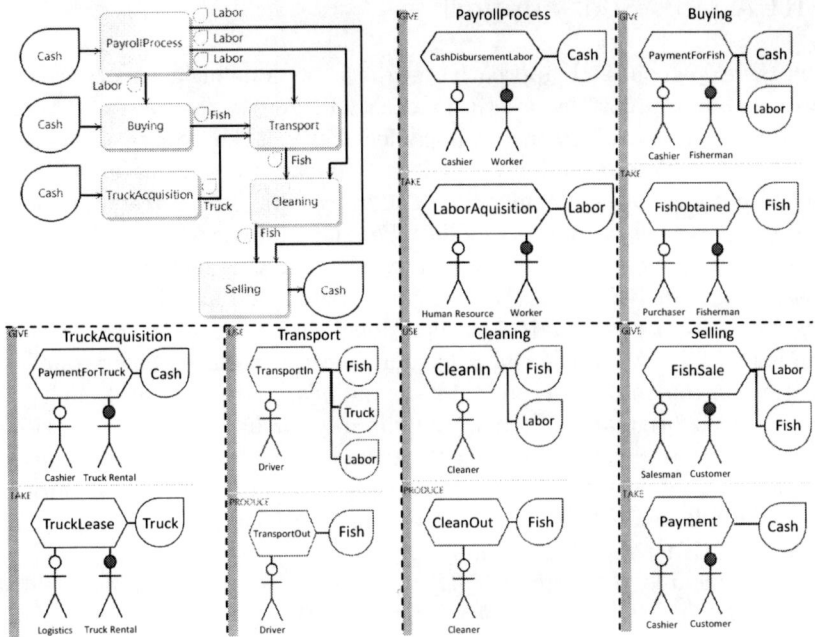

Fig. 5. Sy's Fish Value Chain and Dualities

For this purpose we use again the Sy's Fish example [13] of subsection 2.2. The resulting Sy's Fish value chain and duality models are depicted in Figure 5.

The value chain begins with the `payroll process` which receives some `cash` to realize a transfer. By the duality relationship of the economic events `cash disbursement` and `labor acquisition`, the economic resource `cash` is decremented and *labor* is incremented. The resulting labor is then input for the business processes `buying`, `transport`, `cleaning`, and `selling`.

`Buying` transfers `labor` and additionally `cash` to `fish`, by the dual economic events `payment for fish` and `fish obtained`. `Truck acquisition` is another transfer that turns `cash` into a `truck` that is leased by the duality of `payment for truck` and `truck lease`.

`Transport` uses all acquired resource - `labor`, `fish`, and `truck` - to deliver the fish to the company. The `transport` is done by the company itself, so it does not involve an outside actor. Accordingly, it is a transformation process that uses, but does not consume its economic resources in the economic event `transport in`. The dual economic event `transport out` results again in the `fish`, this time at the right place. `Cleaning` is another non-consuming transformation process that receives the `fish` and turns it into a cleaned `fish`.

The final transfer in the value chain is `selling`. `Labor` and `fish` are given in the `fish sale` in order to take `cash` as result of the `payment`.

4   REA DSL Tool Support

Having developed our DSL approach towards REA modeling in theory, we wanted
to evaluate our approach by practical means. Thereby, we focused on three major
steps. Firstly, we evaluated the technical feasibility of the DSL by an implementation based on Microsoft DSL tool kits. Thereby, we were able to eliminate
technical flaws in the meta model. Secondly, we wanted to test our tool if it
properly supports existing REA models. For this purpose we had 32 REA models available and successfully completed the reengineering task to represent these
models within our REA DSL. During our reengineering activities the strengths
of our rather strict meta model became evident - we were able to recognize flaws
in the existing REA models which have not been recognized before due to the
complexity in the ontology representation and missing tool support. Thirdly, we
approached the originator of REA - William McCarthy - to seek his advise and
to report inconsistencies in existing REA models. However, so far we have not
yet done any usability studies, nor have we used the tool set in real world case
studies. We will conduct these kinds of evaluations once we extend our approach
to cover the REA policy infrastructure.

We implemented the graphical REA-DSL tool based on *Microsoft's Visual
Studio 2010 Visualization & Modeling SDK (V&M SDK)*. Accordingly we used
the V&M SDK to create the meta models explained before in Section 3. Additional custom code enables to set further constraints, necessary for the validation
of the REA model. In a second step the designer - see Figure 6 - is created to
support the REA modeling.

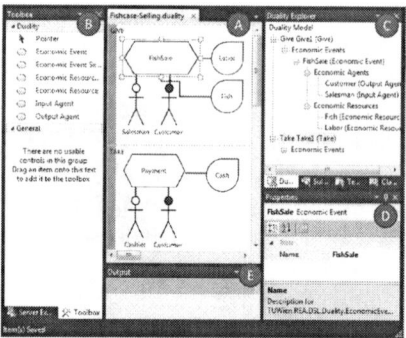

Fig. 6. REA-designer

The designer is separated into five major areas: The *modeling canvas* (A) the
toolbox (B), the *solution explorer* (C), the *property window* (D) and the *validation window* (E). By dragging the modeling elements from the *toolbox* on the
modeling canvas, a REA model can be assembled in a graphical representation.
The *solution explorer* provides a tree based overview of the elements of the currently displayed model as well as a file and directory structure to hold different

model instances. Properties of the selected model element can be changed in the *property window*. The *validation window* informs the user of any errors/warnings.

5   Summary and Future Work

In this paper we proposed a domain specific modeling language to support the conceptual modeling of economic events based on the REA ontology. Conceptual models based on our modeling language – the REA-DSL – aim at facilitating the requirements elicitation process during the design of accounting and enterprise information systems. The original REA ontology has a long history in accounting and is based on well grounded accounting concepts. However, REA leaves space for diverging interpretations of the relationships between core concepts. This has also been criticized by Gailly and Poels who have proposed a new conceptual representation of REA guided by proven ontology engineering principles [17]. They come up with a presentation based on UML class diagrams, which we feel results in a complex visual representation that is hardly understood by domain experts (cf. Figure 2). Similar to Gailly and Poels, we developed a representation format that does not leave space for divergent interpretations. In our case, the relationships between the core concepts are precisely defined by using OMG's Meta Object Facility (MOF) leading to a dedicated REA domain specific modeling language. The MOF-based approach enables the development of a graphical syntax that is dedicated to the needs of business modeling. Furthermore, our REA-DSL comes with a graphical syntax covering a set of stencils that facilitate the understanding of the domain expert. Beside proposing a meta model and a graphical language, we developed a REA-DSL tool as a proof of concept.

In this work, we concentrated on the basic REA principles [4] and the value chain perspective [13]. In future work, we will gradually extend the REA-DSL. Currently, we are working on an integration of the REA policy infrastructure [18] covering commitments, agreements and, furthermore, the typification of the operational concepts [5]. Next, we plan to extend the REA-DSL by concepts to derive a database design for enterprise information systems, which has been one of the original goals of REA. For this purpose, we have already introduced the concepts of economic event series and economic resource series in the current DSL, because they will affect the multiplicities in the database design. By including the policy infrastructure and the REA-driven database design, it will become more obvious that REA models are of structural nature rather and do not concentrate on the behavioral aspects of process models. Another intended REA-DSL extension addresses the perspective from which the REA models are described. Currently, we focus on the perspective of a single enterprise which exchanges value with other enterprises. In the Open-edi reference model (ISO/IEC 15944-4) REA concepts are used to describe the exchanges of value among enterprises from a neutral observer's point of view. We plan to integrate the observer's perspective into our REA-DSL and to support semi-automatic mappings between the perspectives.

References

1. Andersson, B., Bergholtz, M., Edirisuriya, A., Ilayperuma, T., Johannesson, P., Gordijn, J., Grégoire, B., Schmitt, M., Martinez, F.H., Abels, S., Hahn, A., Wangler, B., Weigand, H.: Towards a reference ontology for business models. In: Embley, D.W., Olivé, A., Ram, S. (eds.) ER 2006. LNCS, vol. 4215, pp. 482–496. Springer, Heidelberg (2006)
2. Osterwalder, A., Pigneur, Y., Tucci, C.L.: Clarifying Business Models: Origins, Present and Future of the Concept. Communications of the Association for Information Science (CAIS) 15, 751–775 (2005)
3. Gordijn, J., Akkermans, H.: E3-value: Designing and evaluating e-business models. IEEE Intelligent Systems 16(4), 11–17 (2001)
4. McCarthy, W.E.: The REA Accounting Model: A Generalized Framework for Accounting Systems in a Shared Data Environment. The Accounting Review 57(3) (1982)
5. Geerts, G.L., McCarthy, W.E.: An ontological analysis of the economic primitives of the extended-rea enterprise information architecture. International Journal of Accounting Information Systems 3(1), 1–16 (2002)
6. van Deursen, A., Klint, P., Visser, J.: Domain-specific languages: an annotated bibliography. SIGPLAN Not. 35(6), 26–36 (2000)
7. Kelly, S., Tolvanen, J.P.: Domain-Specific Modeling. Wiley-IEEE Computer Society Press (2008)
8. Greenfield, J., Short, K., Cook, S., Kent, S.: Software Factories. John Wiley, Chichester (2008)
9. Strembeck, M., Zdun, U.: An approach for the systematic development of domain-specific languages. Softw., Pract. Exper. 39(15), 1253–1292 (2009)
10. Geerts, G.L., McCarthy, W.E.: An Accounting Object Infrastructure for Knowledge-Based Enterprise Models. IEEE Intelligent Systems 14(4), 89–94 (1999)
11. Geerts, G.L., McCarthy, W.E.: The Ontological Foundations of REA Enterprise Systems (August 2000)
12. Motal, T., Schuster, R.: From e3-value to REA: Modeling multi-party eBusiness Collaborations. In: Proc. of the 11th IEEE Conference on Commerce and Enterprise Computing, pp. 202–208. IEEE CS, Los Alamitos (2009)
13. Geerts, G.L., McCarthy, W.E.: Modeling business enterprises as value-added process hierarchies with resource-event-agent object templates. In: Business Object Design and Implementation, pp. 94–113. Springer, Heidelberg (1997)
14. OMG: Meta Object Facility (MOF) Core Specification, Version 2.0 (January 2006)
15. Ijiri, Y.: Theory of accounting measurement. American Accounting Association, Sarasota (1975)
16. Yu, S.C.: The Structure of Accounting Theory. The University Presses of Florida (1976)
17. Gailly, F., Poels, G.: Ontology-driven business modelling: Improving the conceptual representation of the REA ontology. In: Parent, C., Schewe, K.-D., Storey, V.C., Thalheim, B. (eds.) ER 2007. LNCS, vol. 4801, pp. 407–422. Springer, Heidelberg (2007)
18. Geerts, G.L., McCarthy, W.E.: Policy-level specification in rea enterprise information systems. Journal of Information Systems 20(2), 37–63 (2006)

A Foundational Approach for Managing Process Variability

Matthias Weidlich[1], Jan Mendling[2], and Mathias Weske[1]

[1] Hasso Plattner Institute at the University of Potsdam, Germany
{Matthias.Weidlich,Mathias.Weske}@hpi.uni-potsdam.de
[2] Humboldt-Universität zu Berlin, Germany
Jan.Mendling@wiwi.hu-berlin.de

Abstract. A business process often shows different variations in a large organisation, due to different legal requirements in different countries, deviations in the IT infrastructure, or organisational differences. These variants are documented in separate independent process models. Management of these variants imposes various challenges. Invariant behaviour needs to be identified and redundancies among the variants have to be avoided. In this paper, we address these questions by defining a set-algebra for behavioural profiles. These profiles represent a behavioural abstraction of process models that can be computed efficiently. We trace back many questions of process variability management to set-theoretic operations and relations defined for behavioural profiles. As a validation, we apply our approach to an industry model collection.

1 Introduction

In large organisations, a business process often exists in many variations. Those stem from different legal requirements in different countries, deviations in the IT infrastructure, or organisational differences [1]. The existence of *process variants* is inevitable – a certain degree of variability is needed to meet the concerns of a specific organisational unit. Variants are often documented in independent process models. As the variability is not made explicit, synergies between variants are hard to explore and process harmonisation is impeded.

Recently, approaches to control the creation of process variants based on well-defined change patterns [2,3,4] or configurations [5,6] have been presented. However, enforcement of such a controlled variability of processes is hard to achieve once process variants are created independently in different organisational units. Therefore, we focus on the use case of managing decoupled process variants. To this end, identification of commonalities and differences between process variants is of central importance. Those have to be made explicit to allow for a comparison of the variants. This is a prerequisite for any process harmonisation effort that aims at reducing the amount of allowed process variability.

There are two fundamental challenges towards efficient management of process variants: the appropriate specification of formal operations to support reasoning with process variants, and the foundation of such operations upon an

H. Mouratidis and C. Rolland (Eds.): CAiSE 2011, LNCS 6741, pp. 267–282, 2011.

appropriate behavioural abstraction of processes. Throughout this paper, we will demonstrate that the essential variant management questions can be answered based on set algebraic operations of complementation, intersection, and union, and the relations of set equivalence and inclusion. Although there are several works on particular subsets of these aspects such as inclusion (or behaviour inheritance) [7,8,9] and union (or merging behaviour) [10,11,12,13], we currently miss an overarching set algebra which provides the means to efficiently calculate with behaviour.

There is a wide spectrum of behavioural abstraction available upon which algebraic operations could be defined. Existing work on behaviour inheritance [8,9,14] builds on an adapted notion of branching bisimilarity. That is, activities that are without counterpart can either be *blocked* or *hidden* when assessing behavioural equivalence. Such an approach has the drawback that the underlying notion of branching bisimilarity is computationally hard as it is based on state space analysis, cf., [15]. Therefore, we base our set algebra on behavioural profiles, a behavioural abstraction of process models that can be computed efficiently for a broad class of models. Behavioural profiles capture the essential behaviour of the set of traces of a process in terms of order constraints, and they are insensitive to skipping of an activity. Differences related to causal constraints are often observed among process variants, and behavioural profiles have been found to provide a suitable abstraction to reason about consistency of process variants [16].

Our contribution is a formal definition of a set algebra for calculating with behavioural profiles of process variants. Our approach is inspired by work on a set algebra for service interaction [17]. The concepts are applied to an industry model collection, for which we identify variant clusters and analyse behavioural commonalities. In this way, our work informs formal research on behavioural inheritance as much as engineering approaches towards merging of behaviour.

The remainder of this paper is structured as follows. The next section illustrate the use case in more detail and presents formal preliminaries. Section 3 defines a set algebra for behavioural profiles, which is applied in Section 4 to address various questions on managing process variability. We present findings from a case study with industry models in Section 5. Finally, Section 6 discusses our approach in the light of related work, before Section 7 concludes the paper.

2   Background

This section introduces the background of our work. First, Section 2.1 elaborates on the use case of managing decoupled process variants. Subsequently, Section 2.2 presents our formal model and Section 2.3 defines behavioural profiles.

2.1   Challenges for Managing Decoupled Process Variants

We illustrate our use case using the two process models depicted in Fig. 1. Both models capture the process of registering a newborn and are adapted versions of the models obtained in different Dutch municipalities, cf., [18]. The administrative processes of Dutch municipalities are rather similar due to legal regulations

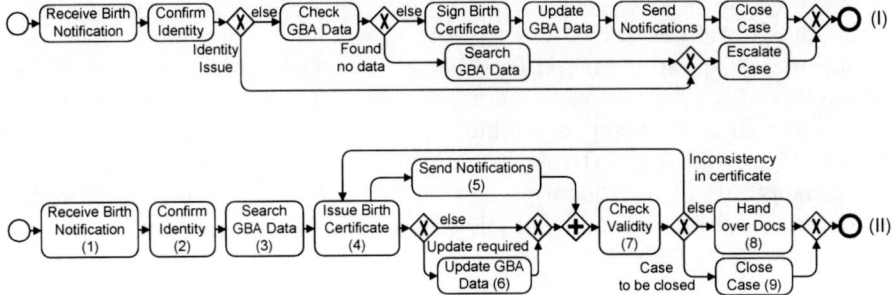

Fig. 1. Variants of the process of registering a newborn

and a non-normative reference model. Nevertheless, these processes show a certain degree of variation caused by, for instance, the size of the municipality or the used information systems, and evolve independently from each other.

Identification of best practices or reducing of the number of process variants among the municipalities can be seen as driven factors for process harmonisation. This requires the identification of semantically corresponding activities of different process models based on linguistic or structural analysis [19,20,21]. Those techniques may reveal that the activity *'Sign Birth Certificate'* in model (I) corresponds to the activity *'Issue Birth Certificate'* in model (II). For this paper, we assume correspondences between activities to be given.

Once we identify correspondences between activities of various pairs of models, we can define clusters of related models that capture variants of the same process. The following challenges must be addressed to support process harmonisation efforts for a particular cluster of process variants.

C1: Given a set of variants, what are their commonalities in terms of shared behaviour? This question relates to the challenge of identifying the behaviour that is invariably agreed on by all variants, i.e., the implemented invariant behaviour.

C2: Given a set of variants, what is the most general allowed behaviour? The challenge when answering this question is to integrate the behaviour defined by all process variants, such that the most general behaviour becomes visible. This is required to come to a notion of a configurable model known from [5,6], which subsumes the complete behaviour defined in single variants.

C3: Given a set of variants, what are their commonalities in terms of shared forbidden behaviour? Similar to the first question, also the shared forbidden behaviour is of interest for managing process variants.

C4: Given a set of variants, which variants are redundant in terms of the specified behaviour? The challenge behind this question is to identify variants for which the specified behaviour is completely subsumed by another process variant. In that case, the existence of the former variant may be challenged.

Although these questions may be approached on a structural level by investigating the set of shared activities, more detailed conclusions can only be drawn once the behavioural perspective is considered. For instance, the activities

'*Update GBA Data*' and '*Send Notifications*' show different behavioural constraints in the models in Fig. 1. To consider such differences, we approach the aforementioned questions based on a set algebra for behavioural profiles. Behavioural profiles are a behavioural abstraction that focusses on order constraints and is insensitive to causal constraints between activities, such as skipping an activity. Fig. 1 illustrates that such differences are often observed among process variants due to additional process entries and exits. Behavioural profiles have been found to provide a suitable abstraction to reason about consistency of process variants [16].

2.2   Formal Model

We use a notion of a process model that is based on a graph containing activity nodes and control nodes. It captures the commonalities of process description languages. For illustration purposes, we use a subset of BPMN and EPCs.

Definition 1. (Process Model)
A process model is a tuple $P = (A, s, e, C, F, T)$ where:
   o *A is a finite non-empty set of activity nodes,*
   o *C is a finite set of control nodes,*
   o *$N = A \cup C$ is a finite set of nodes with $A \cap C = \emptyset$,*
   o *$F \subseteq N \times N$ is the flow relation, such that (N, F) is a connected graph,*
   o *$\bullet n = \{n' \in N | (n', n) \in F\}$ and $n\bullet = \{n' \in N | (n, n') \in F\}$ denote direct predecessors and successors, we require $\forall a \in A : | \bullet a| \leq 1 \wedge |a \bullet | \leq 1$,*
   o *$s \in A$ is the only start node, such that $\bullet s = \emptyset$,*
   o *$e \in A$ is the only end node, such that $e\bullet = \emptyset$,*
   o *$T : C \rightarrow \{and, xor\}$ associates each control node with a type.*

The start and end activity nodes are assumed to carry no semantic meaning but indicate initialisation and termination of a process. Refactoring may be applied to arrive at these start and end activity nodes [22]. We assume trace semantics for process models. Execution semantics of a process model is defined by a translation into a Petri net following on common formalisations, cf., [23]. As our notion of a process model comprises a dedicated start and end activity nodes, the resulting Petri net is a workflow net (WF-net) [24]. All control nodes are of type *and* or *xor*, such that the WF-net is free-choice [24]. The translation into WF-nets defines the behaviour of a process model $P = (A, s, e, C, F, T)$, which is captured by a set of *traces* $\mathcal{T}_P$. It comprises a set of lists of the form $s \cdot A^*$, which represent the execution order of activities.

2.3   Behavioural Profiles

A behavioural profile captures behavioural characteristics of a process model by three relations between pairs of activity nodes. These relations are based on the notion of *weak order*. Two activities of a process model are in weak order, if there exists a trace in which one activity occurs after the other. Note that we require only the *existence* of such a trace.

Definition 2 (Weak Order Relation). *Let $P = (A, s, e, C, F, T)$ be a process model and $\mathcal{T}_P$ its set of traces. The* weak order relation *$\succ_P \subseteq A \times A$ contains all pairs (x, y), such that there is a trace $\sigma = n_1, \ldots, n_m$ in $\mathcal{T}_P$ with $j \in \{1, \ldots, m - 1\}$ and $j < k \leq m$ for which holds $n_j = x$ and $n_k = y$.*

Two activity nodes of a process model might be related by weak order in three different ways, which define the characteristic relations of the behavioural profile.

Definition 3 (Behavioural Profile). *Let $P = (A, s, e, C, F, T)$ be a process model. A pair $(x, y) \in A \times A$ is in one of the following relations:*
  o *The* strict order relation *$\rightsquigarrow_P$, if $x \succ_P y$ and $y \not\succ_P x$.*
  o *The* exclusiveness relation *$+_P$, if $x \not\succ_P y$ and $y \not\succ_P x$.*
  o *The* interleaving order relation *$\|_P$, if $x \succ_P y$ and $y \succ_P x$.*
The set $\mathcal{B}_P = \{\rightsquigarrow_P, +_P, \|_P\}$ of all three relations is the behavioural profile *of P.*

We illustrate the relations of the behavioural profile for the lower model in Fig. 1. For instance, it holds $(1) \rightsquigarrow (5)$ as both activities are ordered whenever they occur together in a trace. It holds $(8) + (9)$ as both activities will never occur in a single trace, and $(5)\|(6)$ due to the potential concurrent execution. The relations of the behavioural profile are mutually exclusive. Along with the reverse strict order $\rightsquigarrow^{-1} = \{(x, y) \in (A \times A) \mid (y, x) \in \rightsquigarrow\}$, the relations partition the Cartesian product of activities. An activity is either said to be *exclusive to itself* (e.g., $(1) + (1)$ in Fig. 1) or in *interleaving order to itself* (e.g., $(6)\|(6)$). The former holds, when an activity cannot be repeated, whereas the latter implies that the activity may be executed multiple times. The behavioural profile is a behavioural abstraction. Hence, there may be process models that show different trace semantics but have equal behavioural profiles. Moreover, behavioural profiles may be lifted from activities to labels of activities. Still, this may result in serious information loss as two occurrences of activities with equal labels would not be distinguished. The behavioural profile $\mathcal{B}_P$ of a process model $P = (A, s, e, C, F, T)$ may be restricted to a subset of activity nodes $A' \subset A$ by restricting the definition of the three relations $\rightsquigarrow_P, +_P$, and $\|_P$ to $A' \times A'$. We refer to this restricted profile as the behavioural profile over A'.

Computation of the behavioural profile of a process model is done efficiently under the assumption of soundness. Soundness is a correctness criteria that guarantees the absence of behavioural anomalies, such as deadlocks [25]. It has been defined for WF-nets, so that it can be directly applied to our notion of a process model. We are able to reuse techniques for the computation of behavioural profiles introduced for sound free-choice WF-nets [16]. Using these results, behavioural profiles are computed in $O(n^3)$ time with n as the number of nodes of the respective WF-net.

3   A Set Algebra for Behavioural Profiles

To introduce a set algebra for behavioural profiles, we first discuss a notion of strictness for profile relations in Section 3.1. Then, Section 3.2 introduces set-theoretic relations, while Section 3.3 turns the focus on set-theoretic operations.

3.1 Strictness of Behavioural Relations

The relations of the behavioural profile can be classified according to their *strictness*, as they allow different levels of freedom for the occurrences of activities in a trace. Interleaving order can be seen as the absence of any restriction on the order of po-

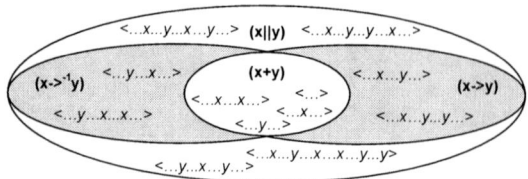

Fig. 2. Sets of traces induced by the relations of the behavioural profile for two activities x and y

tential occurrence – the activities are allowed to appear in an arbitrary order. In contrast, (reverse) strict order defines a particular order of execution for two activities, whereas exclusiveness completely prohibits the occurrence of two activities together in one trace. Therefore, we consider interleaving order to be the weakest relation, while exclusiveness is the strictest relation. For a dedicated pair of activities (x, y), this strictness is also reflected in the containment hierarchy of traces that show either none, one, or both activities, illustrated in Fig. 2. Here, all traces that conform to a specific behavioural relation are part of the encircled set of traces. A process model that defines interleaving order between two activities x and y allows for any trace, i.e., it may contain none, one, or both activities in any order. A model that imposes exclusiveness for both activities is most restrictive. It allows for traces that comprise none or only one of the activities. This set is a proper subset of the traces induced by interleaving order.

This notion of strictness is the foundation for the definition of a set algebra for behavioural profiles. Our operations and relations are *not* defined based on sets of traces, but on the relations of the behavioural profile. Still, the strictness of behavioural relations illustrated above with sets of traces is taken into account.

3.2 Set-Theoretic Relations

We start by introducing three set relations, i.e., equivalence, inclusion, and emptiness for behavioural profiles. Most use cases require the application of these concepts for behavioural profiles of different models for which a separate relation identifies corresponding pairs of activities. To keep the formalisation concise, we abstract from such correspondences and assume corresponding activities to be identical. As partially overlapping sets of activities do not impose serious challenges, we restrict the discussion to the behavioural aspects and assume identical sets of activities once more than one behavioural profile is considered.

Equivalence. Two behavioural profiles are equivalent, if they enforce equal behavioural constraints for the shared activities. Equivalence of behavioural profiles does not imply equal trace semantics for the shared activities, cf., [16].

Definition 4 (Equivalence). *Let* $\mathcal{B}_1 = \{\leadsto_1, +_1, \|_1\}$ *and* $\mathcal{B}_2 = \{\leadsto_2, +_2, \|_2\}$ *be two behavioural profiles over a set of activities* A. $\mathcal{B}_1$ *equals* $\mathcal{B}_2$, *denoted by* $\mathcal{B}_1 = \mathcal{B}_2$, *if and only if their relations are equal for all activities, i.e.,* $\leadsto_1 = \leadsto_2$, $+_1 = +_2$, *and* $\|_1 = \|_2$.

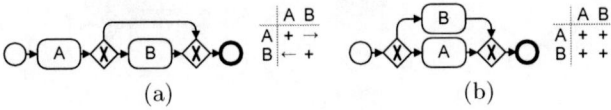

Fig. 3. The behavioural profile of (a) includes the one of (b)

Inclusion. An inclusion holds between two behavioural profiles, if one profile completely subsumes the behavioural constraints of another profile for shared activities according to the notion of strictness discussed in the previous section.

Definition 5 (Inclusion). *Let* $\mathcal{B}_1 = \{\leadsto_1, +_1, \|_1\}$ *and* $\mathcal{B}_2 = \{\leadsto_2, +_2, \|_2\}$ *be two behavioural profiles over a set of activities* A. $\mathcal{B}_1$ *includes* $\mathcal{B}_2$, *denoted by* $\mathcal{B}_1 \subseteq \mathcal{B}_2$, *if and only if for all pairs of activities* $(a_1, a_2) \in A \times A$ *it holds:*

- $a_1 +_1 a_2$ *implies* $a_1 +_2 a_2$,
- $a_1 \leadsto_1 a_2$ *implies* $a_1 +_2 a_2$ *or* $a_1 \leadsto_2 a_2$,
- $a_1 \leadsto_1^{-1} a_2$ *implies* $a_1 +_2 a_2$ *or* $a_1 \leadsto_2^{-1} a_2$.

If $\mathcal{B}_1 \subseteq \mathcal{B}_2$, *but not* $\mathcal{B}_1 = \mathcal{B}_2$, *we speak of proper inclusion, denoted by* $\mathcal{B}_1 \subset \mathcal{B}_2$.

Fig. 3 illustrates inclusion of behavioural profiles (relations for start and end nodes are omitted). The profile of model 3(a) includes the one of model 3(b) as the former is less restrictive. It allows for ordered execution of activities A and B, whereas model 3(b) forbids any occurrence of both activities in the same trace. Due to the assumed behavioural abstraction, inclusion of the behavioural profiles does not imply inclusion of the respective sets of traces. Still, the behavioural abstraction allows us to cope with variations as they are visible for activities *'Search GBA Data'* and *'Update GBA Data'* in our example in Fig. 1. Model (II) defines strict order for both activities, whereas model (I) completely disallows joint occurrence of both activities. Hence, the behavioural constraints imposed by model (I) are included in the constraints imposed by model (II). Any trace-based assessment will fail to address such cases.

Emptiness. A behavioural profile is empty, if it defines all pairs of activities, except for the start and end activity, to be exclusive. Such a profile forbids the execution of any activity other than the start and the end activity. As those activities are assumed to carry no semantic meaning but indicate initialisation and termination of the process, such a trace is considered to be empty.

Definition 6 (Emptiness). *Let* $\mathcal{B} = \{\leadsto, +, \|\}$ *be a behavioural profile over a set of activities* A *with* $s, e \in A$ *being start and end activities.* $\mathcal{B}$ *is empty, if and only if all activity pairs* $(a_1, a_2) \in (A \times A) \setminus \{(s, e), (e, s)\}$ *are exclusive,* $a_1 + a_2$.

3.3 Set-Theoretic Operations

We introduce three set operations for behavioural profiles, i.e., complementation, intersection, and union. Again, we abstract from a correspondence relation, assume that corresponding activities are identical, and focus on the constraints for shared activities once multiple behavioural profiles are considered.

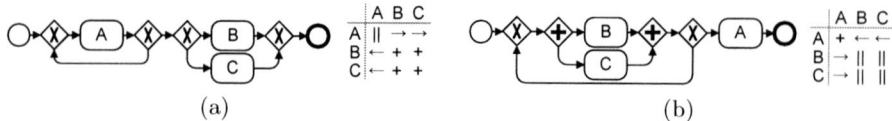

Fig. 4. Process models with complementary behavioural profiles

Complementation. The complement operation is defined for a single behavioural profile and returns a profile that specifies reverse relations for all pairs of activities of the original behavioural profile.

Definition 7 (Complement). *Let $\mathcal{B} = \{\leadsto, +, ||\}$ be a behavioural profile over a set of activities A. The complement $\overline{\mathcal{B}} = \{\leadsto_C, +_C, ||_C\}$ of $\mathcal{B}$ is a behavioural profile over A, such that for all pairs of activities $(a_1, a_2) \in A \times A$ it holds:*

- *$a_1 +_C a_2$ if and only if $a_1 \| a_2$,*
- *$a_1 \leadsto_C a_2$ if and only if $a_1 \leadsto^{-1} a_2$,*
- *$a_1 \|_C a_2$ if and only if $a_1 + a_2$.*

The complement operation is illustrated in Fig. 4. Model 4(a) and model 4(b) show complementary behavioural profiles (neglecting start and end nodes) – the profile of the left model is the complement of the profile of the right model, and vice versa. For instance, there is a strict order constraint between activities A and B in model 4(a), whereas both activities are in reverse strict order in model 4(b). Further, activity A may occur multiple times in model 4(a), whereas it is exclusive to itself in model 4(b).

Intersection. Given two behavioural profiles, the intersection operation yields a third behavioural profile that combines the strictest relations of the behavioural profile for all shared pairs of activities. Therefore, the intersection represents the behavioural constraints that are shared by both profiles.

Definition 8 (Intersection). *Let $\mathcal{B}_1 = \{\leadsto_1, +_1, ||_1\}$ and $\mathcal{B}_2 = \{\leadsto_2, +_2, ||_2\}$ be two behavioural profiles over a set of activities A. The intersection $\cap$ of these profiles is a behavioural profile $\mathcal{B}_3 = \{\leadsto_3, +_3, ||_3\}$ over A, denoted by $\mathcal{B}_1 \cap \mathcal{B}_2 = \mathcal{B}_3$, such that for all pairs of activities $(a_1, a_2) \in A \times A$ it holds:*

- *$a_1 +_3 a_2$ if and only if either $a_1 +_1 a_2$, $a_1 +_2 a_2$, $(a_1 \leadsto_1 a_2 \wedge a_1 \leadsto_2^{-1} a_2)$, or $(a_1 \leadsto_1^{-1} a_2 \wedge a_1 \leadsto_2 a_2)$,*
- *$a_1 \leadsto_3 a_2$ if and only if either $(a_1 \leadsto_1 a_2 \wedge (a_1 \leadsto_2 a_2 \vee a_1 \|_2 a_2))$ or $(a_1 \leadsto_2 a_2 \wedge (a_1 \leadsto_1 a_2 \vee a_1 \|_1 a_2))$,*
- *$a_1 \|_3 a_2$ if and only if $a_1 \|_1 a_2$ and $a_1 \|_2 a_2$.*

We illustrate the intersection of behavioural profiles with the models in Fig. 5. The lower model (c) shows a behavioural profile that corresponds to the intersection of the behavioural profiles of the upper two models (a) and (b). Consider, for instance, activities A and B. While model (a) allows for interleaving order between both activities, model (b) is more restrictive and enforces strict order. Hence, the intersection also defines strict order for both activities. Due to the assumed behavioural abstraction, again, model (c) does not represent the intersection of the

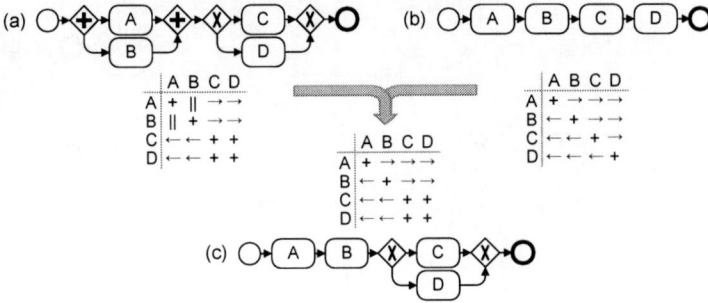

Fig. 5. The behavioural profile of model (c) corresponds to the intersection of the profiles of models (a) and (b)

sets of traces of models (a) and (b). Referring to our example in Fig. 1, the set of shared complete traces is empty for both models. The evident behavioural commonalities of both models are not revealed by a trace-based assessment. The intersection of behavioural profiles allows to address such scenarios.

Union. The union operation for two behavioural profiles yields a third behavioural profile that combines the weakest constraints of the two profiles given as input parameters for all pairs of activities. We trace back the definition of the union operation to the complement and intersection operations using De Morgan's rule. The union $\mathcal{B}_3$ of two behavioural profiles $\mathcal{B}_1$ and $\mathcal{B}_2$ over a set of activities A, denoted by $\mathcal{B}_3 = \mathcal{B}_1 \cup \mathcal{B}_2$ is defined as $\mathcal{B}_3 = \overline{\overline{\mathcal{B}_1} \cap \overline{\mathcal{B}_2}}$.

4   Managing Process Variability

In this section, we discuss how the set algebra for behavioural profiles is applied to address the questions raised in Section 2.1 regarding the management of decoupled process variants. Let $P_1, \ldots, P_n$ be a set of models that were found to capture different variants of a business process with A being the shared activities.

C1: Shared behaviour: Greatest Common Divisor. The shared behaviour may be referred to as the Greatest Common Divisor (GCD), cf., [14]. Given a set of process variants, the GCD is characterised by a behavioural profile $\mathcal{B}_{GCD}$ over A that is derived by computing the intersection of all profiles $\mathcal{B}_{P_1}, \ldots, \mathcal{B}_{P_n}$. Hence, the GCD integrates the constraints shared by all variants. The profile $\mathcal{B}_{GCD}$ may be checked for emptiness. If it is empty, all variants impose contradicting constraints for the shared activities. A model representing the GCD can be synthesized from the profile $\mathcal{B}_{GCD}$ following the approach introduced in [26]. Note that the synthesises imposes certain consistency requirements on the behavioural profile. Further, the behavioural commonality between a single variant P_i and the GCD is quantified as the relative share of activity pairs that have equal relations in $\mathcal{B}_{GCD}$ and $\mathcal{B}_{P_i}$. This measure quantifies how much additional behaviour the variant allows for, relative to the behaviour shared with all variants.

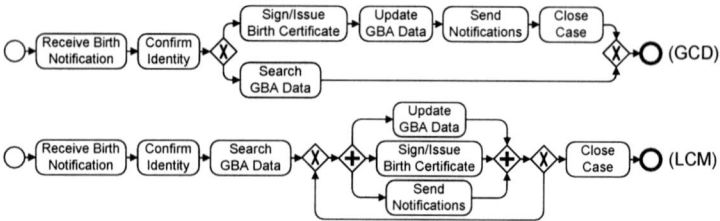

Fig. 6. Synthesised GCD and LCM for the scenario of Fig. 1

Fig. 6 depicts the GCD for the variants of our initial example in Fig. 1. The behavioural profile of the GCD comprises all constraints from model (I) as they are more restrictive than those imposed by model (II). The GCD visualises the basic ordering constraints of both variants. Still, it is not identical to model (I) as certain causalities are abstracted by the behavioural profile.

C2: Most general behaviour: Least Common Multiple. The most general behaviour is referred to as the Least Common Multiple (LCM) of a set of variants. It is characterised by a behavioural profile $\mathcal{B}_{LCM}$ over A that is derived by computing the union of all profiles $\mathcal{B}_{P_1}, \ldots, \mathcal{B}_{P_n}$. The LCM imposes solely the weakest constraints for a pair of activities in the set of variants. Again, a model is derived from $\mathcal{B}_{LCM}$ using the synthesis approach for behavioural profiles.

Fig. 6 depicts the LCM for the variants of our initial example. The parallel execution of three activities is caused by the interleaving order imposed by the profile of the LCM. As all of them may be executed multiple times (interleaving order as a self-relation), they are also part of a control flow cycle. Note that model synthesis for these activities includes various design decisions on how to represent interleaving order (by concurrency or by cyclic structures) [26]. Hence, the synthesis approach may be adapted so that an LCM with a different structure, but identical behavioural profile is created.

C3: Shared forbidden behaviour: Complementary LCM. In order to characterise the behaviour that is forbidden by all variants for shared activities, a behavioural profile $\mathcal{B}_{SFB}$ over A is created as the complement of the LCM, $\mathcal{B}_{SFB} = \overline{\mathcal{B}_{LCM}}$. However, this profile does not directly capture all constraints that are not implemented in any variant due to the strictness of behavioural relations, cf., Section 3.1. Conclusions are drawn solely from the (reverse) strict order and interleaving order relations of the profile $\mathcal{B}_{SFB}$. If $\mathcal{B}_{SFB}$ defines interleaving order between two activities, all variants show exclusiveness for these activities. Hence, the potentially arbitrary order implied by the interleaving order constraint is forbidden in all variants. Similar conclusions are drawn for (reverse) strict order constraints in $\mathcal{B}_{SFB}$.

We illustrate this concept with the activities *'Update GBA Data'* and *'Send Notifications'* of our initial example. For both activities, the LCM defines interleaving order, which yields exclusiveness in the complement. As exclusiveness is the strictest relation, we cannot draw any conclusions on shared forbidden behaviour. For the activities *'Confirm Identity'* and *'Update GBA Data'*, however, the LCM defines a strict order constraint from the former to the latter. There-fore, the reverse strict order in the complement is shared forbidden behaviour and prevents execution of activity *'Update GBA Data'* before *'Confirm Identity'*.

C4: Redundancy of variants: Inclusion of variants. Given the behavioural profiles of two process variants P_i and P_j, the question whether the behaviour of one variant is captured in the other variant for the shared activities is traced back to the inclusion of their behavioural profiles, i.e., $\mathcal{B}_{P_i} \subseteq \mathcal{B}_{P_j}$. In this case, all constraints imposed by P_i would be equal or less strict than the constraints imposed by P_j, so that the behaviour (according to the behavioural profile) of P_j is covered by P_i. That suggests integration of variant P_j into variant P_i.

Investigating the two models given in Fig. 1, the behavioural profile of model (II) includes the profile of model (I) for the shared activities. Hence, when aiming at reducing the number of variants, the existence of model (I) may be challenged.

5   Case Study

Our approach to managing process variability has been evaluated based on the SAP reference model [27]. This model collection describes the functionality of the SAP R/3 system and comprises 604 process diagrams, which are expanded to 737 models in EPC notation as some diagrams contain multiple disconnected EPCs. These EPC models capture different functional aspects of an enterprise, such as sales or accounting. However, the models are not fully orthogonal. Various models show an overlap, such that events and functions with identical labels occur in multiple models. These models, therefore, represent process variants.

For our analysis, we excluded all models that showed behavioural anomalies, such as deadlocks and livelocks [28], or ambiguous instantiation semantics [29]. We also normalised multiple start and end events, and replaced block-structured OR-split and OR-join connectors with AND connectors, which does not impact on the behavioural profile. These selections and transformations led to a set of 493 EPC models, that are grounded on our formal model introduced in Section 2.2. Hence, we were able to leverage the efficient techniques mentioned in Section 2.3, so that behavioural profiles were computed in milliseconds.

Table 1 gives on overview of the observed clusters of process variants in the SAP reference model. Given a threshold of shared nodes, we derived all process model clusters of maximal size for which the set of shared nodes was equal or

Table 1. Variant clusters in the SAP reference model

# Shared Nodes (Min)	4	6	8	10	12	14	16	18	20	22	24	26
# Variant Clusters	84	48	33	23	23	21	15	11	9	2	1	0
Avg Size of Clusters	3	3	3	3	3	2.6	2.4	2.3	2.3	2	2	0
Max Size of Clusters	10	9	8	8	7	6	4	4	4	2	2	0
# Models in Clusters	212	124	88	63	56	47	36	25	21	4	2	0
# Subsumed Models	127	73	55	41	35	28	20	14	12	2	1	0

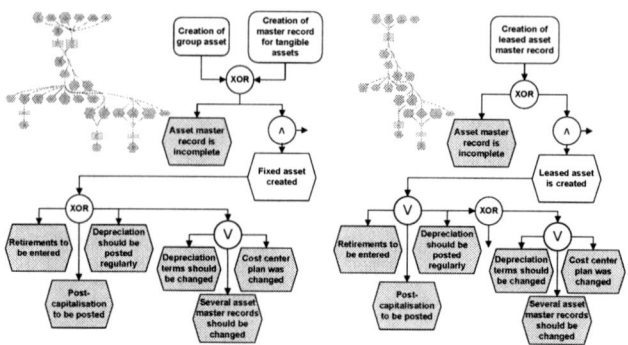

Fig. 7. Excerpts of two process variants in the SAP reference model

larger than the threshold. For these clusters, Table 1 depicts the number of clusters, their average and maximal size, and the number of considered models. For instance, requiring variants to share at least 12 nodes led to 23 clusters comprising 56 distinct models. The average size of the clusters is three models, the maximum size is seven models. The derived values witness the existence of a large number of process variants in the model collection.

In order to analyse the commonalities for all identified model clusters, we implemented the presented set algebra and derived the LCM. This analysis revealed that the behavioural profile of the LCM was often equal to the profile of at least one model in the cluster. Based on this observation, we checked for all clusters of variants whether the behavioural profile of one variant includes the profiles of all other variants for the shared nodes. In fact, this was the case for all but two clusters. The identified LCM models could be used as a starting point for harmonising the reference model by removing redundant variants and integrating their non-shared nodes into the former model. Consider, for instance, the excerpts of two encountered process variants in Fig. 7. Both models share 16 nodes, some of them are highlighted in the excerpt. For some shared nodes both models define different behavioural constraints. Still, the behaviour of the right model includes the behaviour of the left model for shared nodes, which suggests integration of the processing defined by the left model into the right one.

Using this approach, more than half of the identified variants could be removed from the collection as indicated by the last row in Table 1. Depending on the number of required shared nodes, the reference model could be reduced by up to 127 process models.

6   Related Work

Research related to our work is conducted mainly in three areas: comparison of behaviour, modelling of process variants, and behaviour synthesis.

The comparison of behaviour as defined by the operators in this paper is related to various notions of behavioural equivalence of the linear time – branching time spectrum [30]. Behavioural profiles provide an abstraction which approximates trace equivalence at the weaker end of this spectrum. Notions of inclusion are discussed in work on behavioural equivalence [7,8,9]. Recent works calculate a degree of behavioural similarity between process models based on linguistic, graph-matching, and state-based concepts, see [31] for an overview. Our work instead provides a precise definition of how to determine the behavioural intersection of two process models. There is also work that aims to determine the union of the behaviour captured by two process models [10,11,12,13]. None of these works has been integrated and extended towards the definition of an algebra.

The requirements for representing variants of a process have been addressed by different modelling approaches. In our work we build on the assumption that the union or intersection of two process models is again a process model. This approach is in line with work on the configuration of workflow models [6]. Other approaches use dedicated elements for capturing variation points on the level of the modelling language. Such languages include Configurable EPCs [5,32], aggregated EPCs [33], Provop [3], or variant rich process models [34], which pick up ideas and concepts from modelling of software product families [35] and feature diagrams [36]. These approaches typically assume that variation is identified and explicitly represented by a human modeller while our algebraic operations permit the calculation of intersection or union from two model variants.

Similar relations, but not exactly those of behavioural profiles, are used in a pre-processing step of approaches to synthesize a process model [37]. Please refer to [38] for a discussion of the conceptual differences between the relations used in [37] and those of the behavioural profile. Our general idea of defining an algebra for managing process variants is inspired by an algebra for operating guidelines, a formal concept for the synthesis of interaction partners for a process [17].

7   Conclusion

In this paper, we addressed two fundamental challenges of managing process variants, the specification of formal operations for reasoning with variants, and the foundation of such operations upon an appropriate behavioural abstraction. As a solution, we proposed a set algebra for behavioural profiles that enables conclusions on behavioural commonalities and differences using set-theoretic relations and operations. We showed how these concepts are used to address the problem of managing process variants. Our case study illustrated the potential of our approach for harmonisation of process model collections.

Having defined a complete algebra for the (abstracted) behaviour of processes allows for behavioural analysis way beyond similarity measurement. For instance, a process model may be used to encode forbidden behaviour. Using our algebraic operations, a model collection may then be analysed whether it allows for the forbidden behaviour. Our operations may also be used during the design of process models. As an example, subsumption of behavioural profiles may hint at the creation of model fragments that can be found in a model collection already.

We also have to reflect on some limitations of our approach. We focus on the control flow perspective and neglect data and resource assignments. We foresee that our set algebraic approach may be extended in these directions. As a starting point, the standard set theoretic operations may be applied to sets of data artefacts and resources. However, operations that consider the interplay between the perspectives (i.e., the intersection of data access constraints for a certain role) can be assumed to provide even more value. Another restriction is our assumption of elementary 1:1 correspondences between activities of two behavioural profiles. Differences in the abstraction level are likely to be observed in practice, in particular if models are created for different purposes (e.g., process analysis vs. process design). In future work, we aim at lifting our approach to the level of n:m correspondences between activities.

References

1. Wijnhoven, F., Spil, T., Stegwee, R., Fa, R.: Post-merger IT integration strategies: An IT alignment perspective. The Journal of Strategic Information Systems 15(1), 5–28 (2006)
2. Reichert, M., Rinderle, S., Kreher, U., Dadam, P.: Adaptive process management with ADEPT2. In: ICDE, pp. 1113–1114. IEEE Computer Society, Los Alamitos (2005)
3. Hallerbach, A., Bauer, T., Reichert, M.: Capturing variability in business process models: the provop approach. Journal of Software Maintenance 22(6-7), 519–546 (2010)
4. Weber, B., Reichert, M., Rinderle-Ma, S.: Change patterns and change support features - enhancing flexibility in process-aware information systems. Data Knowl. Eng. 66(3), 438–466 (2008)
5. Rosemann, M., van der Aalst, W.M.P.: A configurable reference modelling language. Inf. Syst. 32(1), 1–23 (2007)
6. Gottschalk, F., van der Aalst, W., Jansen-Vullers, M., Rosa, M.L.: Configurable workflow models. International Journal of Cooperative Information Systems (IJCIS) 17(2), 177–221 (2008)
7. Ebert, J., Engels, G.: Observable or Invocable Behaviour - You Have to Choose. Technical Report 94-38, Leiden University (December 1994)
8. Schrefl, M., Stumptner, M.: Behavior-consistent specialization of object life cycles. ACM Trans. Softw. Eng. Methodol. 11(1), 92–148 (2002)
9. Basten, T., van der Aalst, W.M.P.: Inheritance of Behavior. Journal of Logic and Algebraic Programming (JLAP) 47(2), 47–145 (2001)
10. Preuner, G., Conrad, S., Schrefl, M.: View integration of behavior in object-oriented databases. Data & Knowledge Engineering 36(2), 153–183 (2001)
11. Mendling, J., Simon, C.: Business Process Design by View Integration. In: Eder, J., Dustdar, S. (eds.) BPM Workshops 2006. LNCS, vol. 4103, pp. 55–64. Springer, Heidelberg (2006)
12. Gottschalk, F., van der Aalst, W.M.P., Jansen-Vullers, M.H.: Merging event-driven process chains. In: Chung, S. (ed.) OTM 2008, Part I. LNCS, vol. 5331, pp. 418–426. Springer, Heidelberg (2008)
13. Rosa, M.L., Dumas, M., Uba, R., Dijkman, R.M.: Merging business process models. In: Meersman, R., Dillon, T.S., Herrero, P. (eds.) OTM 2010. LNCS, vol. 6426, pp. 96–113. Springer, Heidelberg (2010)

14. van der Aalst, W.M.P.: Inheritance of business processes: A journey visiting four notorious problems. In: Ehrig, H., Reisig, W., Rozenberg, G., Weber, H. (eds.) Petri Net Technology for Communication-Based Systems. LNCS, vol. 2472, pp. 383–408. Springer, Heidelberg (2003)
15. van Glabbeek, R.J., Goltz, U.: Refinement of actions and equivalence notions for concurrent systems. Acta Inf. 37(4/5), 229–327 (2001)
16. Weidlich, M., Mendling, J., Weske, M.: Efficient consistency measurement based on behavioural profiles of process models. IEEE Transactions on Software Engineering (2010) (to appear)
17. Kaschner, K., Wolf, K.: Set algebra for service behavior: Applications and constructions. In: Dayal, U., Eder, J., Koehler, J., Reijers, H.A. (eds.) BPM 2009. LNCS, vol. 5701, pp. 193–210. Springer, Heidelberg (2009)
18. Gottschalk, F.: Configurable Process Models. PhD thesis, Eindhoven University of Technology, The Netherlands (December 2009)
19. Nejati, S., Sabetzadeh, M., Chechik, M., Easterbrook, S.M., Zave, P.: Matching and merging of statecharts specifications. In: ICSE, pp. 54–64. IEEE Computer Society, Los Alamitos (2007)
20. Dijkman, R.M., Dumas, M., García-Bañuelos, L., Käärik, R.: Aligning business process models. In: EDOC, pp. 45–53. IEEE Computer Society, Los Alamitos (2009)
21. Weidlich, M., Dijkman, R.M., Mendling, J.: The iCoP framework: Identification of correspondences between process models. In: Pernici, B. (ed.) CAiSE 2010. LNCS, vol. 6051, pp. 483–498. Springer, Heidelberg (2010)
22. Vanhatalo, J., Völzer, H., Leymann, F., Moser, S.: Automatic workflow graph refactoring and completion. In: Bouguettaya, A., Krueger, I., Margaria, T. (eds.) ICSOC 2008. LNCS, vol. 5364, pp. 100–115. Springer, Heidelberg (2008)
23. Lohmann, N., Verbeek, E., Dijkman, R.M.: Petri net transformations for business processes - a survey. TOPNOC 2, 46–63 (2009)
24. Aalst, W.: The application of Petri nets to workflow management. Journal of Circuits, Systems, and Computers 8(1), 21–66 (1998)
25. Aalst, W.: Workflow verification: Finding control-flow errors using petri-net-based techniques. In: van der Aalst, W.M.P., Desel, J., Oberweis, A. (eds.) BPM. LNCS, vol. 1806, pp. 161–183. Springer, Heidelberg (2000)
26. Smirnov, S., Weidlich, M., Mendling, J.: Business process model abstraction based on behavioral profiles. In: Maglio, P.P., Weske, M., Yang, J., Fantinato, M. (eds.) ICSOC 2010. LNCS, vol. 6470, pp. 1–16. Springer, Heidelberg (2010)
27. Curran, T.A., Keller, G., Ladd, A.: SAP R/3 Business Blueprint: Understanding the Business Process Reference Model. Prentice-Hall, Englewood Cliffs (1997)
28. Mendling, J., Verbeek, H.M.W., van Dongen, B.F., van der Aalst, W.M.P., Neumann, G.: Detection and prediction of errors in EPCs of the SAP reference model. Data Knowl. Eng. 64(1), 312–329 (2008)
29. Decker, G., Mendling, J.: Process instantiation. Data Knowl. Eng. 68, 777–792 (2009)
30. van Glabbeek, R.: The linear time - brancing time spectrum I. The semantics of concrete, sequential processes. In: Handbook of Process Algebra, pp. 3–99. Elsevier, Amsterdam (2001)
31. Dijkman, R., Dumas, M., van Dongen, B., Käärik, R., Mendling, J.: Similarity of business process models: Metrics and evaluation. Inf. Syst. 36(2), 498–516 (2011)
32. La Rosa, M., Dumas, M., ter Hofstede, A., Mendling, J.: Configurable multi-perspective business process models. Inf. Syst. 36(2), 313–340 (2011)
33. Reijers, H., Mans, R., van der Toorn, R.: Improved model management with aggregated business process models. Data Knowl. Eng. 68(2), 221–243 (2009)

34. Schnieders, A., Puhlmann, F.: Variability mechanisms in e-business process families. LNI, vol. 85, pp. 583–601. GI (2006)
35. Pohl, K., Böckle, G., Van Der Linden, F.: Software product line engineering: foundations, principles, and techniques. Springer New York Inc., Heidelberg (2005)
36. Schobbens, P.Y., Heymans, P., Trigaux, J.C., Bontemps, Y.: Generic semantics of feature diagrams. Computer Networks 51(2), 456–479 (2007)
37. van der Aalst, W.M.P., Weijters, T., Maruster, L.: Workflow mining: Discovering process models from event logs. IEEE Trans. Knowl. Data Eng. 16(9), 1128–1142 (2004)
38. Weidlich, M., Polyvyanyy, A., Mendling, J., Weske, M.: Efficient computation of causal behavioural profiles using structural decomposition. In: Lilius, J., Penczek, W. (eds.) PETRI NETS 2010. LNCS, vol. 6128, pp. 63–83. Springer, Heidelberg (2010)

Tangible Media in Process Modeling
– A Controlled Experiment

Alexander Luebbe and Mathias Weske

Hasso Plattner Institute, University of Potsdam, Germany
{alexander.luebbe,mathias.weske}@hpi.uni-potsdam.de
http://bpt.hpi.uni-potsdam.de

Abstract. In current practice, business processes modeling is done by trained method experts. Domain experts are interviewed to elicit their process information but typically not involved in actual modeling. We created a tangible toolkit for process modeling to be used with domain experts. We hypothesize that it results in more effective process elicitation.

This paper assesses nine aspects related to "effective elicitation" in a controlled experiment using questionnaires and video analysis. We compare our approach to structured interviews in a repeated measurement design. Subjects were 17 student clerks from a trade school.

We conclude that tangible modeling leads to more effective elicitation through activation of participants and validation of results. In particular, subjects take more time to think about their process and apply more corrections to it. They also report to get insights into process modeling.

Keywords: process elicitation, tangible media, controlled experiment.

1 Introduction

In business process management graphically depicted process models serve as communication vehicles about the working procedures of organizations. They are the basis for a shared understanding and process improvements. Moreover, process models are often used as requirements engineering artifacts for software implementation projects. Supporting processes with software offers great potential to save time, enhance reliability and deliver standardized output [9]. At the same time, misunderstandings in early stages lead to expensive change requests at later stages of the software project. Thus, the quality of communication between stakeholders is crucial to translate process requirements into software implementation.

In current practice, process models are created by trained method experts, typically external consultants. They gather the required information in interviews or workshops with the stakeholders of the process [1]. Afterwards, the method expert creates a business process model using notations such as EPC or BPMN. Creation of process models is done with dedicated software.

Domain experts provide information upfront but are typically passive while their knowledge is translated into a process model by the modeling expert. This translation step undertaken by the modeling expert de-couples the domain expert

H. Mouratidis and C. Rolland (Eds.): CAiSE 2011, LNCS 6741, pp. 283–298, 2011.

from the model. When asked for feedback, additional effort is needed to explain the models meaning and to resolve misunderstandings. This paper addresses this problem by introducing an approach to couple domain experts with process models using tangible objects.

We have developed the tangible business process modeling (t.BPM) toolkit. It is a transcribable set of plastic tiles that can be used to model processes on a table. It reflects the iconography of the Business Process Model and Notation (BPMN), see Fig. 1. It consists of shapes for activities, gateways, events and data objects. Control flow and roles are drawn on the table. In our opinion, it enables domain experts to actively shape their processes and allows the method expert to act as a facilitator rather than a translator. For the scope of our work, we consider domain experts to be the stakeholders of the project, i.e. clerks or managers. The method expert is either an external process consultant or an internal process expert who is trained in methods and notations to frame knowledge in process-oriented projects.

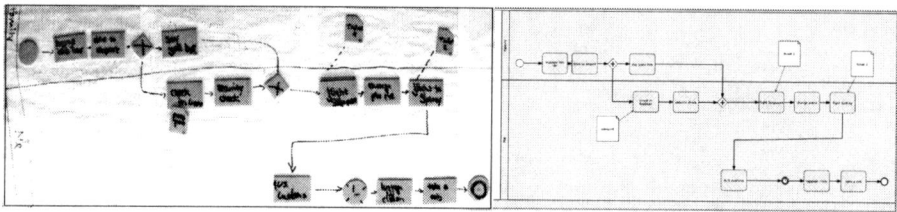

Fig. 1. Same process: modeled with t.BPM (left) and in a software modeling tool (right)

This paper reports on a controlled experiment in which we analyze one-to-one interview situations with respect of the effectiveness of process elicitation with or without t.BPM. It is a condensed version of a extensive technical report [15] on this experiment. Two hypotheses were cut out and discussed in detail in a separate publication [8]. Three more hypotheses were dropped for this paper as they did not hold and don't add value to the discussion here.

We review related research on process modeling in the next section. Afterwards, we explain the hypotheses, the experiment setup, the variables and the analysis procedures used in Section 3. The experiment execution is discussed in Section 4 and the data analysis is reported in Section 5. The results from the analysis are interpreted in Section 6 and the paper is concluded in Section 7.

2   Related Work

Empirical research on process modeling is typically focussed on the models that are produced with software tools and can be automatically analyzed, e.g. [5]. Only some research is turned towards the modelers in front of the screen and the process of model creation.

As examples, Sedera et al. [21] used case study research and survey methods to derive qualitatively a framework of factors that influence the success of process modeling efforts. Amongst others, they found tooling and participation to be key drivers. Participative model building was investigated by Persson [16]. She found that it leads to enhanced model quality, more stakeholder consensus and more commitment to results. The workshops are set up with a dedicated software tool operator to channel participant knowledge and create a common picture at the projector [23]. Rittgen developed software and guidance for modeling workshops in which the participants themselves use the software to create the model together [18]. Our approach uses intuitive tooling to remove software as a barrier for individuals to participate.

For individuals, Recker found that modeler performance is influenced by the complexity of the grammar [17], modeling experience and modeling background. Controlled experiments with individuals have been conducted e.g. by Weber et al. [24] to investigate the effect of events on process planing performance or by Holschke [10] to investigate the influence of model granularity on reusability of artifacts. To our best knowledge there is no controlled experiment that investigates the presence of an intuitive mapping tool for business process modeling.

The setup and execution of our controlled experiment was guided by Creswell [3] and Wohlin et al. [25]. We use literature from experimental software engineering [12] and statistics [6] to inform the structure of the paper and the level of reporting.

3   Experiment Planning

We outline all planning activities in this section. We start by deriving our hypotheses, talk about setup, the actual measurement of the hypotheses and the analysis procedures.

3.1   Goal and Hypotheses

The goal of this paper is to examine the effect of t.BPM on process elicitation with individuals. Therefore we compare t.BPM to structured interviews which are seen as the most effective requirements elicitation technique [4]. By 'effective' we mean that it produces a 'desired or intended result' [22]. In requirements engineering, more information is typically indicating more effective elicitation. But it was already shown that the presence of visual representations does not necessarily elicit more information [4]. We argue that effective process elicitation has more aspects such as user engagement and validated results. Fig. 2 visualizes how we refine our model towards hypotheses based on the following considerations:

User engagement is widely recognized as a key factor for success of IT projects [21]. Our approach uses tangible media which is seen as a key factor for task engagement, e.g. in HCI research [11]. In those cases, engagement is typically measured as the time spent on a problem, e.g. by Xie et al. [26]. Since tangible modeling consumes time to handle the tool itself (e.g, writing on tiles),

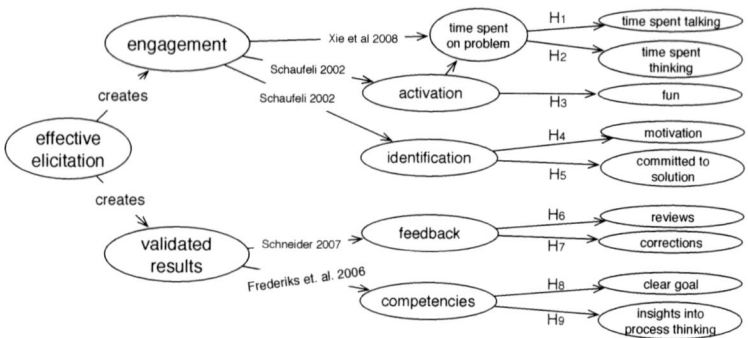

Fig. 2. Effective elicitation decomposed into nine hypotheses

we split up the time into more fine granular observations. We hypothesize that people will *spent more time talking (H₁)* about the process but also *spent more time to think (H₂)* about what they do.

Schaufeli segments engagement into the dimensions activation and identification [19]. While activation is already measured with the time spent on the task, we additionally hypothesize that people have *more fun (H₃)* as a further aspect of activation. The dimension of identification inspires us to hypothesize that people modeling with t.BPM will have *more motivation (H₄)* to accomplish the task and will be *more committed to the solution(H₅)* that they shaped.

The second key aspect that we see for effective elicitation is a validated result. Schneider [20] points out that validation cycles are a time consuming aspect of requirements elicitation projects. He proposes to create a model during the elicitation to trigger instant feedback and speedup validation. Validation cycles are characterized by reviews and adjustments to the model. We hypothesize that people will do *more reviews (H₆)* when using t.BPM and apply *more corrections (H₇)* to their process model.

Finally, validation in model building depends on the competencies of the participants. Frederiks [7] proposes that users validate models by deciding on the significance of information. We propose that this depends on a clear understanding of the modeling goal. We hypothesize that t.BPM provides a *clearer goal (H₈)* for the elicitation session. Frederiks also proposes that modeling experts guide the validation by grammatical analysis, in other words their modeling knowledge. We hypothesize that tangible modeling can create *insights into process thinking (H₉)* for the user and thereby support the validation process.

We note that the hypotheses are not a forced consequence of the identified aspects and their building involved interpretation. We come back to this decomposition when we assess the measurement validity in Section 5.3.

3.2   Experiment Setup and Sampling Strategy

We design the following experimental setup as illustrated in Fig. 3. Subjects get first conditioned to a certain level of BPM understanding. Afterwards they are

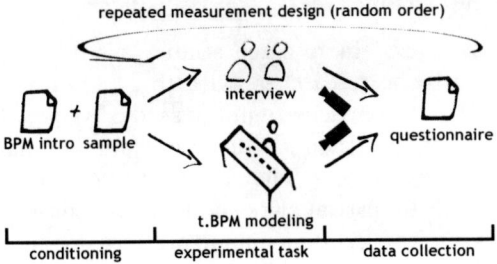

Fig. 3. Experiment Setup for this study

randomly assigned to do either interviews or model with t.BPM. The topic is randomly chosen between buying expensive equipment and running a call for tender. Two persons operate the experiment. One guides the subjects in the role of an interviewer, the other one observes the situation and ensures a stable treatment throughout the experiment. They randomly swap roles.

During the experimental task data is collected using video recording. Afterwards, a questionnaire is to be filled in by the subjects. In every step of the experiment, the time is tracked but time constrains are not imposed on subjects. After the first run, subjects rerun the experimental task using the other method and the other process to report on.

In other words, we use a randomized balanced single factor design with repeated measurements [25] also known as a within-subjects design. All subjects get both treatments assigned in different order. All subjects do interviews and process modeling. Finally, all subjects are rewarded for their participation with a chocolate bar and a cinema voucher.

3.3   Experimental Material

We briefly outline and explain the printed material used in this experiment. The original documents are appended to the technical report [15]. Like the experiment, the experimental material is in German.

- BPM introduction: Two pages explaining the terms Business Process Management, Business Process Modeling and process models.
- Sample model: One page depicting the process of "Making Pasta" and four pragmatical hints on process modeling such as verb-object style labeling.
- 2x task sheet: One paragraph explaining the process to report on. It explicitly sets the context, the start and the end-point of the process.
- Interview guide (for experimenter): Experimenters read out the same six questions in each experimental task. The sheet also contains standardized answers to potential questions from participants.
- Questionnaire: Items to be rated on a 5-point Likert scale. Details are explained in Section 3.5.

3.4   Participant Selection

The sample population used in research studies should be representatives of the population to which the researchers wish to generalize [2]. Thus, we want potential users of t.BPM to participate in the study. We got the opportunity to run an on-site experiment at a trade school in Potsdam (Germany) with graduate office and industrial clerk students. They all work in companies and study part time at the trade school. Industrial clerks do planing, execution and controlling of business activities. Office clerks do supporting activities in a department, e.g. as office managers. Both groups might be questioned in process-oriented projects by external consultants. Thus, they represent the target population that we like to address with t.BPM.

3.5   Operationalized Hypotheses

We operationalize the hypotheses presented in Section 3.1 by means of a questionnaire and video analysis. We define each hypothesis as H_x and its null hypothesis as H_{0x}.

Questionnaire-Based Hypotheses (H_3,H_4,H_5,H_8,H_9): Hypotheses which rely on perceived measures are tested using a questionnaire. On a five-point Likert scale, subjects rate their agreement to, in summary, fifteen statements. Three statements together represent one hypothesis. Two statements are formulated towards the hypotheses, one is negatively formulated. The level of agreement is mapped to the values [1..5] where 1 is no agreement and 5 is a strong agreement. The values are aggregated (negative statement is turned around by calculating $6 - value$) to retrieve the actual value to work with. The hypothesis holds if there is a significant difference according to the method immediately used before, t.BPM or interviews. We test the following hypotheses:

- H_3: Subjects report more fun in t.BPM sessions than in interviews.
  H_{03}: Subjects don't more fun in t.BPM sessions.
- H_4: Subjects report to be more motivated in t.BPM sessions than in interviews.
  H_{04}: Subjects don't report to be more motivated in t.BPM sessions.
- H_5: Subjects report to be more committed to the solution in t.BPM sessions than in interviews.
  H_{05}: Subjects report to be more committed to the solution in t.BPM sessions.
- H_8: Subjects report a clearer goal understanding in t.BPM sessions than in interviews.
  H_{08}: Subjects don't report a clearer goal understanding in t.BPM sessions.
- H_9: Subjects report to gain more new insights in process understanding from t.BPM sessions than from interviews.
  H_{09}: Subjects don't report to gain more new insights in process understanding from t.BPM sessions.

Video Hypotheses (H_1,H_2,H_6,H_7): We operationalize hypotheses related to time and actions taken during the experimental task using video coding analysis. We define the following coding schemes:

Time Slicing(H_1,H_2): The duration of the experimental task is sliced exclusively to belong to one of five categories. The use of t.BPM (Use_{tBPM}) such as labeling and positioning the shapes without talking, $Talk_{tBPM/int}$ is the time people talk about the process, $UseTalk_{tBPM}$ is talking and using t.BPM (to avoid overlap between Use_{tBPM} and $Talk_{tBPM}$). We define a code for the time spent silent ($Silence_{tBPM/int}$) when people do not talk and do not handle t.BPM. Finally, $Rest_{tBPM/int}$ captures remaining time such as interactions with the interviewer. The same coding scheme is used for both experimental tasks. However, Use and $UseTalk$ do not apply for interviews as there is no t.BPM to use.

Corrections and Reviews(H_6,H_7): Both are coded as distinct events. We code $Corrections_{tBPM/int}$ if the context of an already explained process part is explicitly changed. In t.BPM sessions this involves re-labeling or repositioning that impacts the process model meaning. In interviews explicit revisions of previously stated information is considered a correction. The $Reviews_{tBPM/int}$ are coded if subjects decide to recapitulate their process. This must involve talking about the process as we cannot account possibly silent reviews. This scheme is the same for both experimental tasks.

Using this coding scheme we operationalize the video hypotheses in the following way:

- H_1: Subjects talk more in t.BPM sessions than in interviews,
  i.e. $Talk_{tBPM} + UseTalk_{tBPM} > Talk_{int}$.
  H_{01}: Subjects don't talk more in t.BPM sessions,
  i.e. $Talk_{tBPM} + UseTalk_{tBPM} \not> Talk_{int}$.
- H_2: Subjects are more silent in t.BPM sessions than in interviews,
  i.e. $Silence_{t.BPM} > Silence_{int}$
  H_{02}: Subjects are not more silent in t.BPM sessions,
  i.e. $Silence_{t.BPM} \not> Silence_{int}$
- H_6: Subjects make more reviews in t.BPM sessions than in interviews,
  i.e. $Reviews_{t.BPM} > Reviews_{int}$
  H_{06}: Subjects don't make more reviews in t.BPM sessions,
  i.e. $Reviews_{t.BPM} \not> Reviews_{int}$
- H_7: Subjects make more corrections in t.BPM sessions than in interviews,
  i.e. $Corrections_{tBPM} > Corrections_{int}$.
  H_{07}: Subjects don't make more corrections in t.BPM sessions,
  i.e. $Corrections_{tBPM} \not> Corrections_{int}$.

3.6   Variables

The independent variable in this experiment setup is the method used for process elicitation. Subjects do either a structured interview or the same structured interview in the presence of t.BPM, the tangible modeling toolkit. The dependent variables are formed from the data collected during and immediately after a session. We use a notational convention for the data sets collected: $intention_{V/Qx}$. As an example, $talking_{V1}$ describes the set of talking times as measured with the video analysis for hypothesis 1. Likewise, fun_{Q3} is the set of all ratings collected with the fun related questionnaire items for hypothesis 3.

3.7   Analysis Procedures

For hypothesis testing, we use a one-way repeated-measures ANOVA (analysis of variances). It aims to determine the variation within subjects that is caused by the method. Additionally, we carry out a dependent t-test with acceptance level $p<.05$. It is used to get a different view on the data and to assess potentially confounding factors that might have influenced the performance of the subjects.

To assess reliability of the questionnaire, we use Cronbach's alpha. It determines the internal consistency of the three questionnaire items measuring one hypothesis. The video data is analyzed by two independent reviewers. They compare their results and (if needed) resolve conflicts by negotiation. Cohen's Kappa is used to determine the inter-rater agreement before negotiation to assess the quality of our coding guidelines.

4   Experiment Execution and Data Collection

The experiment design was executed in December 2009 at a trade school in Potsdam. Slots were offered to the students by short teasers given in the classes. All subjects were at the age of nineteen to twenty-one. Students could choose to swap one lecture unit for experiment participation (about 1h). We expected to test industrial clerks only, but only ten volunteered. Thus, we opened up the experiment to office clerks as well. We ended up testing 7 office clerks and 10 industrial clerks within the week.

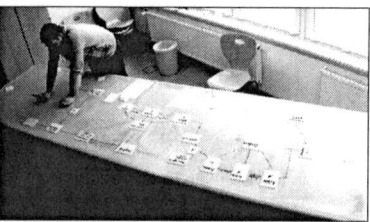

Fig. 4. Photos from the experiment execution. Subject giving interview (left) and modeling with t.BPM (right). Taped by the video cameras.

Each experiment run started with a short informal warm-up chat and afterwards followed the design as outlined in Section 3.2. One experimenter ran the experiment, the other one operated the cameras and observed the situation to ensure a stable treatment. Fig. 4 depicts the two experimental tasks as taped by the cameras. One video taping went wrong, leading to a sample size of sixteen for the video coding hypotheses.

5   Data Analysis

We explain the analysis techniques used and the results found in this section. We reason about the results in Section 6.

5.1   Descriptive Statistics

From seventeen students in two runs, we collected 34 questionnaires with 510 statements in total. The video analysis is based on 6,74 hours of video material. One t.BPM session taping went wrong. That results in N=16 for all hypotheses that rely on video analysis. Videos taken during t.BPM sessions took twenty minutes (19.52) on average ranging from ten (10.25) to almost forty minutes (38.98). On the other hand, interviews took about five minutes (5.42) on average ranging from three and a half (3.53) to ten minutes (9.68) at most.

5.2   Data Set Preparation

The data was tested with the Kolmogorov-Smirnov and Shapiro-Wilk test and is normally distributed. The original experiment evaluation involved two more video codings and three more questionnaire items. The related hypotheses did not hold and the data was therefore dropped for discussion in this paper due to the limited space. Apart from that, no collected data was excluded from the set.

5.3   Measurement Reliability and Validity

According to Kirk and Miller the reliability is the extent to which "a measurement procedure yields the same answer however and whenever carried out" ([13], p.19) while validity is the "extent to which it gives the correct answer".

We assess two aspects of measurement reliability. First we check the inter-rater agreement for the video codings using Cohen's kappa coeficient (κ). It compares both video codings before the negotiation process. The inter-rater agreement over all videos and all coding schemes is $\kappa = .463$ where $0.41 < \kappa < 0.60$ is a moderate agreement level [14]. Thus, we interpret our coding instructions as reasonably reliable and the results as moderately reproducible.

Furthermore, the reliability of the questionnaire is measured using Cronbach's alpha (α). It determines the degree to which the items related to one hypothesis coincide. In other words, whether they actually measure the same underlying concept, e.g. fun. In the literature [6] $\alpha > .8$ is suggested to be a good value for questionnaires, while $\alpha > 0.7$ is still acceptable. All our variables had $\alpha > .8$, except for $\alpha(motivation_{Q4}) = .702$ and $\alpha(clarity_{Q8}) = .687$. We keep those exceptions in mind but overall a high degree of reliability is indicated for the questionnaire.

Validating whether our variables correctly describe "effective elicitation" is not directly possible. We use effective elicitation as an umbrella term for the aspects of engagement and result validation. From there we derive variables to measure these aspects. In [15] we conducted a principal component analysis for validation. It is a technique to determine sets of strongly correlating variables which are approximated with one factor, the principal component [6]. Ideally, the variables would form two factors. Those that reflect the measures for engagement and those measuring result validation.

Using orthogonal (varimax) rotation, our nine dependent variables split up to three factors that do not match our hypothesis decomposition. Interestingly, all questionnaire based variables aggregate to one large principal component. These measures rely on self-perception of the subjects and therefore describe one side of the coin. Moreover, the time for $talking_{V1}$ and $silence_{V2}$ strongly correlate with the amount of $reviews_{V6}$ done. It indicates the degree to which people were involved with the task. Finally, $corrections_{V7}$ builds a single factor. Overall, the measurement validation calls for a more thought-out hypothesis decomposition and clever selection of measurement instruments.

5.4 Hypothesis Testing

We use the repeated-measures ANOVA to determine the effect of our independent variable (method) within each individual per dependent variable. In other words, to what extend did the method influence the performance of each individual? Fig. 5 illustrates how our data is partitioned. From the overall variability (SS_T), we identify the performance difference within participants (SS_W) and can further distinguish the variation caused by the treatment (SS_M) and the variation not explained by our treatment(SS_R).

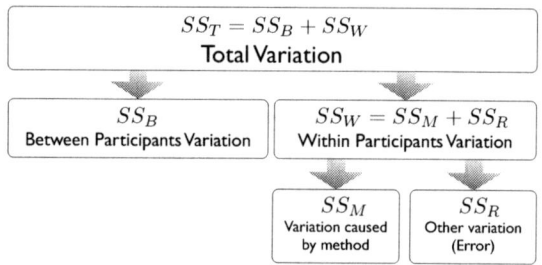

Fig. 5. Data partitioning for rep.-measures ANOVA. Drawing adopted from [6] p.463

The ratio of explained to unexplained variability in our dataset is described by $F = \frac{SS_M}{df_M} / \frac{SS_R}{df_R}$. Where df are the degrees of freedom calculated from the number of different methods (df_M=2-1=1) and the participant number (df_R=17-1=16). The critical ratio $F_{.05}(df_M, df_R)$ is the value to pass before the result is actually significant with an acceptance level of p<.05. For our variables collected in questionnaires $F_{.05}(1, 16) > 4.49$ is a significant result, for the video codings we only have N=16 thus $F_{.05}(1, 15) > 4.54$ is a significant ratio. In Table 1 we sorted the variables according to descending $F_{.05}$ ratios. We also report SS_B, SS_M, SS_R and η^2 (eta squared). The value of $\eta^2 = \frac{SS_M}{SS_W}$ describes the ratio of variation within the subjects that can be explained by the treatment method. It is an effect size measure.

Furthermore we conduct a dependent t-test to create a different view on the data, see Table 2. It compares the groups doing t.BPM and interviews by their mean scores (V=in minutes, Q=Likert scale [1..5]), the statistical significance of this difference (one-tailed with acceptance level p<.05) and the confidence interval. The upper and lower boundaries indicate that the real mean difference

Table 1. ANOVA result table based on $df_M=1$. Sorted by $F_{.05}$ ratios

dependend Variable	df_R	SS_T	SS_B	SS_M	SS_R	$F_{.05}$	η^2
$corrections_{V7}$	15	119.22	42.72	57.78	18.72	**46.30**	0.76
$silence_{V2}$	15	398.55	129.58	167.92	101.05	**24.93**	0.62
$insights_{Q9}$	16	18.24	14.9	0.84	2.50	**5.36**	0.25
$reviews_{V6}$	15	38.01	23.00	3.13	11.88	3.95	0.21
$talking_{V1}$	15	116.56	56.92	10.86	48.79	3.34	0.18
fun_{Q3}	16	18.31	15.03	0.55	2.73	3.24	0.17
$commitment_{Q5}$	16	24.68	20.90	0.33	3.45	1.52	0.09
$clarity_{Q8}$	16	32.78	25.78	0.12	6.88	0.27	0.02
$motivation_{Q4}$	16	10.90	9.46	0.05	1.39	0.23	0.04

Table 2. (one-tailed) t-test comparing groups by method. Ordered like Table 1

dependent variable	Effect Size		Significance	Confidence Intervals	
	t.BPM	interview		lower boundary	upper boundary
$corrections_{V7}$	3.00	0.31	**.000**	**1.85**	**3.53**
$silence_{V2}$	5.54	0.95	**.000**	**2.63**	**6.54**
$insights_{Q9}$	3.75	3.43	**.017**	**0.03**	**0.60**
$reviews_{V6}$	0.81	0.19	**.033**	-.046	1.30
$talking_{V1}$	4.65	3.49	**.044**	-0.19	2.52
fun_{Q3}	4.16	3.90	**.046**	-0.05	0.56
$commitment_{Q5}$	3.31	3.51	.118	-0.53	0.14
$clarity_{Q8}$	3.37	3.49	.304	-0.59	0.36
$motivation_{Q4}$	4.45	4.37	.225	-0.14	0.29

between the groups is in that range with 95 percent probability. It should not include zero to be sure about the effect between the groups.

From both tables we see, that all parameters for $corrections_{V7}$, $silence_{V2}$ and $insights_{Q9}$ meet scientific standards. For $reviews_{V6}$, $talking_{V1}$ and fun_{Q3} we see that they just missed acceptable standards in both tests. E.g. the difference between the groups is significant in Table 2 but the confidence intervals do not allow acceptance by rigor scientific standards. Finally, $commitment_{Q5}$, $clarity_{Q8}$ and $motivation_{Q4}$ did not hold.

5.5 Testing Potentially Influential Factors

We use a two-tailed dependent t-test to compare groups were two different influences were applied. For example, we had two processes to report on, two experimenters, and two different educational groups. Furthermore, each subject goes through the experimental task twice. Repetition effects might have influenced the performance of the subjects.

While the experimenters had no significant influence on the dependent variables, we found that the second experimental task led to significantly more $clarity_{Q8}$ about the goal (1st=3.1, 2nd=3.77, p=.001) and more $commitment_{Q5}$ to the solution (1st=3.2, 2nd=3.63, p=.004).

Participants' education had significant influence on $clarity_{Q8}$ and $insights_{Q9}$. In particular, office clerks reported to have a clearer goal understanding (o-clerks=3.98, i-clerks=3.05, p=.031) and get more new insights into process thinking (o-clerks=4.05, i-clerks=3.30, p=.022). In all cases, the confidence intervals left no doubt about the effect.

6   Interpretation of Results

We can identify three types of variables. Those that support their hypothesis, those that do not support their hypothesis, and those that just missed rigor scientific standards. We consider the latter ones as conditionally supportive and argue that a slightly larger sample set would have made the difference.

This claim is based on the t-test in Table 2. It indicates a significant difference for $talking_{V1}$, fun_{Q3}, and $reviews_{V6}$ due to method. The confidence intervals do not allow acceptance with scientific rigor. That means, we cannot rule out with 95 percent probability that the actual effect size is zero. For example, $talking_{V1}$ time is significantly higher (p=.044) in t.BPM sessions (t.BPM=4.65min, int=3.49min) but the confidence interval includes zero (lb=-0.19min, ub=2.52min). In this case we miss rigor acceptable levels by twelve seconds. The rest of the discussion is structured according to the hypothesis decomposition in Fig. 2.

The engagement variables to measure activation indicate a positive effect through method. Participants in t.BPM sessions did spend more time talking ($F_{0.5}(1, 15) = 3.34$, $\eta^2 = 0.18$) and significantly more time thinking ($F_{0.5}(1, 15) = 24.93$, $\eta^2 = 0.62$) about their process. They also report more fun ($F_{0.5}(1, 16) = 3.24$, $\eta^2 = 0.17$) in t.BPM sessions. We reject H_{02} and argue that H_{01} and H_{03} might be rejected with a bigger sample size.

The engagement variables measuring the dimension of identification did not hold. People did not report significantly more motivation or commitment to their solution due to the method used. We assume a ceiling effect for $motivation_{Q4}$. Participants got off from class, plus a chocolate bar and a cinema voucher for compensation. On a five point Likert scale we could not find a statistically relevant difference in $motivation_{Q4}$ due to the method applied (t.BPM=4.45, int=4.37). For $commitment_{Q5}$ we found in Section 5.5 that it significantly raises with repetition (p=.004). Thus, we assume that commitment (as operationalized by us) indicates self-confidence that raises with due to the learning effect. We do neither reject H_{04} nor H_{05}.

The variables that operationalize the aspect of validated results show a mixed picture. We note more reviews ($F_{0.5}(1, 15) = 3.95$, $\eta^2 = 0.21$) and significantly more corrections ($F_{0.5}(1, 15) = 46.3$, $\eta^2 = 0.76$) due to the method. We reject H_{07} and argue that H_{06} might be accepted with a sightly larger sample size. We conclude that t.BPM provokes more feedback in process elicitation sessions.

The competencies required for result validation rely on perceived measures. We see that people report significantly more insights into process thinking ($F_{0.5}(1, 15) = 5.36$, $\eta^2 = 0.25$) in t.BPM sessions but the goal clarity does not raise likewise ($F_{0.5}(1,15) = 0.27$, $\eta^2 = 0.02$). In Section 5.5 we found that goal clarity significantly raises with repetition (p=.001). We conclude that, similar to commitment, the goal clarity is determined by learning rather than the method. We reject H_{09} but not H_{08}.

In summary, we interpret the t.BPM method to be engaging through activation of subjects. We can not reason on the concept of identification which was determined by other effects in this experiment. The t.BPM method also leads to validated results through more feedback on the model. The competencies for result validation raise partially with the method and partially with learning through repetition.

6.1 Validity Threats

The **internal validity** was addressed by the experiment design. In particular, we use two processes and two experimenters assigned in random order. In Section 5.5 we assess potentially confounding variables for their influence. While experimenters and processes did not harm the results, we found learning effects due to the repeated measurements design on $clarity_{Q8}$ and $commitment_{Q5}$.

We found education to be influential on the reported $clarity_{Q8}$ and $insights_{Q9}$. In short, office clerks tend to report better scores while scoring worse in objective tasks [8]. While group heterogeneity is a threat to the internal validity, it also increases the **external validity** as both groups represent the population that we address with our tool. This is as important as choosing domain processes rather than artificial graphs to test with. We chose the domain processes in coordination with the school to ensure all students are equally familiar with them. However, we did not assess to which extend individuals are exposed to these processes in their companies. The **measurement instruments** were tested in one pre-study with ten computer science students. Small adjustments were made afterwards. To ensure quality standards for data analysis, we used two independent coders for the video analysis and we have split each questionnaire variable into three items, one poled negatively. Finally, we provide a longer version of this paper including more data and the experimental material in [15].

6.2 Generalizability of Findings

We think the findings about t.BPM can be generalized from the sample group to the general population. Besides their age (19-21years) the students represent exactly the group we address with the t.BPM tool.

We also think that the findings will hold for other tangible modeling approaches when compared to pure talking. Some aspects have also been reported for visual mappings of requirements such as instant feedback [20]. However, a different tests would be needed to determine exactly the aspects that lead to activation of participants.

6.3 Lessons Learned

If we had to start over again, we would put more effort into the reliability of our questionnaire items, in particular $clarity_{Q8}$. But we also learned that people may report a glorified self-image. Thus, we suggest to mix measurement instruments for each measured concept. In other words, complement perceived measures with external measures such as video codings. But we also had to learn that rigor video analysis is the most time-consuming evaluation task.

Besides all that, we think that the compact on-site experiment was a good idea. Instead of spreading it out over various weeks with changing conditions, we could collect the data in a compact week with a stable setup. Moreover, the two experimenters to review each others work did ensure a stable setup.

7   Conclusion

This paper reports on a controlled experiment which was conducted with 17 student clerks at a trade school. We investigate the process elicitation method as an independent variable. Subjects did structured interviews and t.BPM in a repeated measurement design. We claim that t.BPM enables more efficient process elicitation. We argue that efficient elicitation is not about the amount of information but about user engagement and validated results. We decompose these aspects into nine operationalized hypotheses. Three hypotheses did hold. Three more might hold with a larger sample set.

The results show strong support for user engagement through activation of participants and validated results through more feedback from participants. We think that these findings are reproducible with other tangible system modeling approaches when compared with interviews.

Our findings are limited by the measurement instruments and the small sample size (N=17). A future experiment with a larger group and better tested instruments might re-enforce our findings and also support H_1, H_3 and H_6. In other words, it would extend our rigor findings to more talking, more fun and more reviews with tangible media. For now we only showed significantly more thinking time (H_2), more corrections (H_7) and more insights into modeling (H_9) when using tangible media instead of interviews.

Acknowledgements

We are grateful to the students that helped setting up, running, and evaluating this experiment, namely Karin Telschow, Markus Güntert and Carlotta Mayolo. We'd also like to thank the reviewers for their valuable feedback. It led to a substantial revision of Sections 3.1 and 6.

References

1. Byrd, T., Cossick, K., Zmud, R.: A synthesis of research on requirements analysis and knowledge acquisition techniques. MIS Quarterly, 117–138 (1992)
2. Cooper, D., Schindler, P.: Business Research Methods, 10th edn. McGraw-Hill Higher Education, New York (2008)

3. Creswell, J.W.: Research design: Qualitative, quantitative, and mixed methods approaches. Sage Pubns, Thousand Oaks (2008)
4. Davis, A., Dieste, O., Hickey, A., Juristo, N., Moreno, A.: Effectiveness of requirements elicitation techniques: Empirical results derived from a systematic review. In: 14th IEEE International Conference Requirements Engineering, pp. 179–188 (2006)
5. van Dongen, B., van der Aalst, W., Verbeek, H.: Verification of ePCs: Using reduction rules and petri nets. In: Pastor, Ó., Falcão e Cunha, J. (eds.) CAiSE 2005. LNCS, vol. 3520, pp. 372–386. Springer, Heidelberg (2005)
6. Field, A.: Discovering statistics using SPSS. SAGE publications Ltd, Thousand Oaks (2009)
7. Frederiks, P.J.M., Van der Weide, T.P.: Information modeling: the process and the required competencies of its participants. Data & Knowledge Engineering 58(1), 4–20 (2006)
8. Grosskopf, A., Weske, M.: On business process model reviews. In: CAiSE 2010, pp. 31–42. Springer, Heidelberg (2010)
9. Hammer, M., Champy, J.: Reengineering the corporation: A manifesto for business revolution. Collins Business (2003)
10. Holschke, O., Rake, J., Levina, O.: Granularity as a cognitive factor in the effectiveness of business process model reuse. In: Dayal, U., Eder, J., Koehler, J., Reijers, H.A. (eds.) BPM 2009. LNCS, vol. 5701, pp. 245–260. Springer, Heidelberg (2009)
11. Ishii, H., Ullmer, B.: Tangible bits: towards seamless interfaces between people, bits and atoms. In: SIGCHI, pp. 234–241. ACM, New York (1997)
12. Jedlitschka, A., Ciolkowski, M., Pfahl, D.: Reporting experiments in software engineering. Guide to Advanced Empirical Software Engineering, 201–228 (2008)
13. Kirk, J., Miller, M.: Reliability and validity in qualitative research. Sage Publications, Inc., Newbury Park (1986)
14. Landis, J., Koch, G.: The measurement of observer agreement for categorical data. Biometrics 33(1), 159–174 (1977)
15. Luebbe, A., Weske, M.: The effect of tangible media on individuals in business process modeling - a controlled experiment. Tech. Rep. 41, Hasso-Plattner-Institute for IT Systems Engineering (2010),
http://bpt.hpi.uni-potsdam.de/Public/AlexanderGrosskopf
16. Persson, A.: Enterprise modelling in practice: situational factors and their influence on adopting a participative approach. Ph.D. thesis, Stockholm University (2001)
17. Recker, J., Rosemann, M.: The measurement of perceived ontological deficiencies of conceptual modeling grammars. Data & Knowledge Engineering (2010)
18. Rittgen, P.: Success factors of e-collaboration in business process modeling. In: Pernici, B. (ed.) CAiSE 2010. LNCS, vol. 6051, pp. 24–37. Springer, Heidelberg (2010)
19. Schaufeli, W., Salanova, M., González-Romá, V., Bakker, A.: The measurement of engagement and burnout: A two sample confirmatory factor analytic approach. Journal of Happiness Studies 3(1), 71–92 (2002)
20. Schneider, K.: Generating Fast Feedback in Requirements Elicitation. In: Sawyer, P., Heymans, P. (eds.) REFSQ 2007. LNCS, vol. 4542, pp. 160–174. Springer, Heidelberg (2007)
21. Sedera, W., Gable, G., Rosemann, M., Smyth, R.: A success model for business process modeling: findings from a multiple case study. In: PACIS, Shanghai (2004)
22. Stevenson, A.: Oxford Dictionary of English, vol. 24. Oxford University Press, Oxford (2010)

23. Stirna, J., Persson, A., Sandkuhl, K.: Participative Enterprise Modeling: Experiences and Recommendations. In: Krogstie, J., Opdahl, A.L., Sindre, G. (eds.) CAiSE 2007 and WES 2007. LNCS, vol. 4495, pp. 546–560. Springer, Heidelberg (2007)
24. Weber, B., Pinggera, J., Zugal, S., Wild, W.: Handling events during business process execution: An empirical test. In: ER-POIS at CAISE, pp. 19–30 (2010)
25. Wohlin, C., Runeson, P., Höst, M.: Experimentation in software engineering: an introduction. Springer, Netherlands (2000)
26. Xie, L., Antle, A.N., Motamedi, N.: Are tangibles more fun?: comparing children's enjoyment and engagement using physical, graphical and tangible user interfaces. In: Proceedings of TEI, pp. 191–198. ACM, New York (2008)

Experiences of Using Different Communication Styles in Business Process Support Systems with the Shared Spaces Architecture

Ilia Bider[1,2], Paul Johannesson[2], and Rainer Schmidt[3]

[1] IbisSoft ab/SU, Sweden
ilia@ibissoft.se
[2] SU, Sweden
pajo@dsv.su.se
[3] HTW-Aalen, Germany
Rainer.Schmidt@htw-aalen.de

Abstract. Though the concept of shared spaces had been known for quite a while, it did not become popular until the arrival of the internet and social software. Via this way, the concept has penetrated other IT-areas, including the area of Business Process Management, which brings about the needs of investigating the usage of shared spaces in connection to Business Process Support (BPS). The first part of this experience report describes the authors' experience of building, introducing in the operational practice, and using BPS systems based on the shared spaces architecture. It presents three examples of applications aimed at supporting collaboration/communication in the frame of business process instances. These systems use three different mechanisms for arranging communication/collaboration. The first system is based on collaborative planning; the second one is based on the specialized structure of the shared spaces, and the third one on changes in the status of the processes. The second part of the paper is devoted to the analysis of the examples from the first part in order to create a preliminary taxonomy of communication styles in systems with shared spaces architecture. For this end, the authors identified three binary parameters that characterize the way invitations to visit a shared space are issued. These parameters can be used for analyzing communication capabilities of BPS systems, as well as other types of computer systems, with the shared spaces architecture.

Keywords: business process, groupware, communication, shared space.

1 Introduction

The concept of shared spaces is well known [1][2] and has become widely used in the Internet era in connection with advances of social software. A blog, personal journal, and even a photo album are all examples of shared spaces, as they allow sharing information.

Now, shared spaces begin to penetrate the business world as a new generation of workers, brought up with computers, arrives to the labor market. Most of them are

H. Mouratidis and C. Rolland (Eds.): CAiSE 2011, LNCS 6741, pp. 299–313, 2011.
© Springer-Verlag Berlin Heidelberg 2011

digital natives [3] fully integrated into a multitude of social networks and continuously using shared spaces. The trend of using shared spaces for communication concerns the area of business processes in the highest degree [4].

We believe that proper employment of shared spaces in BPS systems requires good understanding of how such spaces enable communication and what kind of advantages and limitations they have. Therefore, we undertake an attempt to identify important styles of communication via shared spaces.

The goal of this paper is twofold. The first sub-goal is to present our experience of building, introducing in the operational practice, and using BPS systems based on the shared spaces architecture. The second sub-goal is to analyze this experience in order to create a preliminary taxonomy of various communication styles in systems with shared spaces architecture.

In order to create a proper foundation for the later considerations, we start with reviewing the role of shared spaces in BPS systems (Section 2). We also explain how a system that employs shared spaces differs from a traditional business process support system based on workflow.

The first sub-goal of the paper is achieved by discussing three examples of applications aimed at supporting collaboration/communication in the frame of business process instances (Sections 3-5). All three systems have been built based on the state-oriented view on business processes [5], and all of them use shared spaces to facilitate communication between process participants. The systems employ three different ways of using shared spaces for communication. The first system uses collaborative planning. In the second one, communication is based on the specialized structure of shared spaces. In the third one, the communication is based on changes in the status of processes.

The way of using shared spaces in each of the systems is not arbitrary but reflects the types of business processes each system supports. In the first case, the system supports loosely-structured processes that require much ad-hoc communication between people engaged in them. In the second case, the system supports relatively structured processes. In the third case, the system supports simple real-time processes with high requirements on the speed of communication.

The second sub-goal of the paper is achieved by analyzing the differences between the systems in respect of how the invitations to visit shared spaces are issued (Section 6). Based on this analysis, three binary parameters are introduced to differentiate communication styles in systems with shared spaces architecture.

In the last part of the paper, we give a short overview of related works (Section 7), and discuss our experience and draw plans for the future research based on it (Section 8). As the paper is an experience report, in the "Related works" section, we pay special consideration to list our own works that explain the theoretical background of our experience as well as give the possibility for the reader to learn about some parts of our experience in more details.

2   A Role of Shared Spaces in BPS Systems

There exist numerous definitions of what a business process is, each of them focusing on a particular property of business processes. For the sake of this paper, we take the following view on business processes:

"A business process is a way of combining efforts of several people for reaching a (well or not so well defined) goal".

Accepting this view, we consider only those business processes in which people play a role of the driving force behind the processes, leaving totally automated processes, and processes where people are used just for well-defined operations aside. Such processes require extensive communication between the participants of each process instance in order to reach the operational goal of the instance.

By a BPS system, we understand a system that helps process participants to run their process instances according to a process (type) definition. From the communication point of view, a BPS system with shared spaces differs from a pure workflow system by the kind of "information logistics" the system employs for providing process participants with instructions and information needed to complete their tasks in the frame of a process instance [6]. The workflow-type BPS systems use a so-called "conveyor belt" logistics [6] in which instructions, and information that is needed are sent to the "next in-line".

A BPS system with shared spaces employs a so-called "construction site" information logistics [6]. Such a system has no explicit data/information flow. A shared information space is created for each process instance to hold all information that is relevant to the process instance, e.g., documents received and sent, information on tasks planned and completed, reports on results achieved when completing these tasks, etc. All this information is easily available each time a process participant is invited to visit this space and complete some task related to it. Thus, a shared space is similar to a construction site where different kinds of workers are invited to complete their own tasks and leave the rest to the others.

The functioning of a BPS system based on shared spaces can be described in the following way:

– When a new process instance starts, a new shared space is created. It gets a unique name, an owner (responsible for the instance), and possibly, an instance team.
– When the process instance reaches its operational goal, the shared space is closed (sealed), but remains accessible for reading (an instance goes to the archive).
– A person who is assigned a task in the frame of the process instance "goes" to this instance's shared space to get information he/she needs for completing the task and reports the results achieved in the same space.

The "construction site" information logistics via shared spaces has certain advantages compared to the traditional communication schemes for those business processes in which instances can vary considerably from one to another. In such a situation, it is difficult to decide what and how much information needs to be sent to a person completing a certain task. When a person is invited to visit a certain part of a shared space, he/she oversees not only this local part, but also everything adjacent to it, and can use this additional information when completing his/her task without being explicitly told to do so. In other words, if using a traditional communication scheme you send one document to a person, and this is all he/she gets. If you send a person to work on a certain document placed in some corner of a shared space, he/she can access not only this document, but also other documents in this corner, or anywhere else in the whole shared space. More detailed justification of using shared spaces in BPS systems from the business point of view can be found in [7].

For the shared spaces technique to work efficiently in a BPS system, two conditions should be fulfilled:

- Shared spaces are properly structured. In a normal business environment, a person participates in many process instances, and, often, in parallel. For the shared space technique to work efficiently, he/she needs to understand the situation in a shared space he/she is visiting at a glance, and quickly find all information related to the task at hand.
- Invitations should give process participants a clear understanding on why they have been invited and what they are expected to do in each particular shared space.

Note that invitations to visit shared spaces in BPS systems have a different meaning from that in social software. In the latter, invitations are not binding; a person invited may not visit the shared space at all. In a BPS system, however, following an invitation is mandatory or at least strongly recommended; otherwise the whole communication /collaboration scheme will break down.

Naturally, getting an invitation does not constitute the only reason why a person would like to visit a shared space. He/she can do it in an arbitrary manner, or because some event happened in the frame of a process instance that is of value to be registered in its shared space.

In the next three sections, we will introduce three examples of BPS systems that use the principles outlined in this section.

3   A System with Collaborative Planning

3.1   Description

A system called ProBis was developed based on ideas from [5] for a Swedish interest organization (The Swedish Union of Tenants) in 2003-2006 as described in [8][9]. A shared space in ProBis is presented to the end-user as a window divided in several areas by using the tab dialogues technique, see Fig. 1.

Some areas of the window are standard, i.e. independent from the type of the business process; others are specific for each process type supported by the system. Standard areas comprise such attributes and links as:

- Name and informal description of a process instance
- Links to the owner, and, possibly, the process team
- Links to the relevant documents, created inside the organization, and received from the outside

The standard part of ProBis shared space includes also the task area (tab) that contains two lists, as in Fig. 1. The *to-do* list (to the left in Fig. 1) includes tasks planned for the given process instance; the *done* list (to the right in Fig.1) includes tasks completed in the frame of it. A planned task defines what and when something should be done in the frame of the process instance, as well as who should do it. All tasks planned for a given person from all process instances are shown in the end-user's personal calendar. From the calendar, the user can go to any shared space for which a task is assigned to him/her in order to inspect, change, or execute this task.

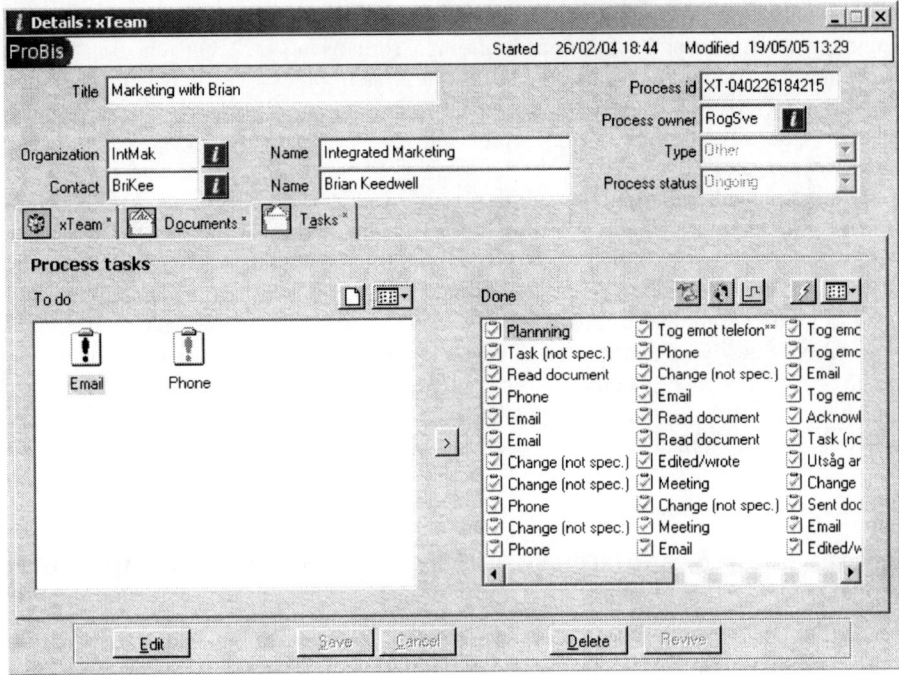

Fig. 1. View on the ProBis shared space

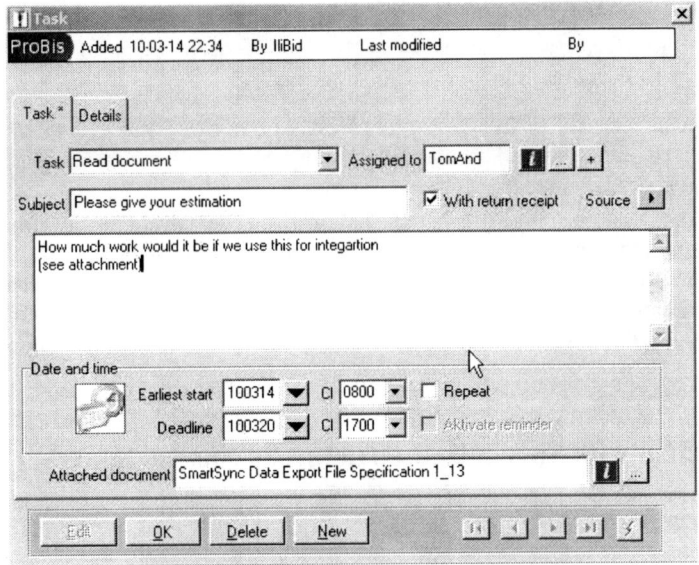

Fig. 2. Assigning a task to another user in ProBis

The only way of communicating via ProBis is by assigning a task to the communication partner. This is done by filling a form as in Fig. 2. One chooses the task from the list, assigns it to another user of the system, adds a textual description and some parameters, for example a document that is already registered in the process instance space. The task list is configurable and can be adjusted for each installation and process type.

To further facilitate communication, several more advanced features were added to ProBis. For example, there is a possibility to plan the same task to many users. Additional users can be added from the list with the "+" button (See Fig. 2), or can be fetched from a predefined group. Each user gets its own task in the calendar and will need to go and complete it independently of other users. Multi-user planning gives a possibility to easily raise attention of several people to some event that has happened in the process instance. Another advanced feature is the "Returned receipt" check-box which ensures that the planner gets a special "Attention" task planned as soon as the task he/she has assigned to somebody else has been completed.

3.2 Experience of Use

Based on our experience with ProBis, collaborative planning provides a very efficient way of communication/collaboration in the frame of business process instances. It is especially useful for:

- loosely structured processes, i.e. processes for which there are no predefined ways for handling each instance.
- processes driven by a professional team that knows how to use the system quite well.

There are, however, two drawbacks with the approach when using it for more structured processes that involve occasional users:

- The dynamic aspect of business processes is poorly visualized. One needs to go through the done-list and browse the history to get an understanding of how a given process instance is developing in time.
- Using the system puts some requirements on the user, as he/she needs to understand the general ideas built in the system and get some training. This means that the system is not very friendly for newcomers and casual users. Planning as a way of communication causes the major problem here, as it is considered to be counter-intuitive. Detailed planning is not as widespread in business life as one can imagine.

4   A System with Specialized Structure of Shared Spaces

4.1   Description

In a system with a specialized structure, shared spaces are structured according to the process map designed for a particular process type. In our case, such a map is designed with a tool called iPB [10]. Several systems have been built with the help of this tool. The biggest one is employed in the social office of one of the Swedish municipalities (municipality of Jönköping), where it helps to conduct investigations on suspected child abuse.

A process map in iPB is a drawing that consists of boxes placed in some order, see Fig. 3. Each box represents a step of the process, and the name of the step appears inside the box (no lines or connecters between the boxes). A textual description is attached to each step that explains the work to be done. Each process instance gets its own copy of the map that serves as a table of contents for its shared space, see Fig. 3. The map is used for multiple purposes: as an overview of the case, guidelines for handling the case, and a menu for navigating inside the shared space. The user navigates through the shared space by clicking on the boxes of the steps with which he/she wants to work. Not all boxes are clickable at the beginning; those that are grayed require that one or several previous steps are dealt with first, see Fig. 3. These constraints are defined with the help of so-called business rules.

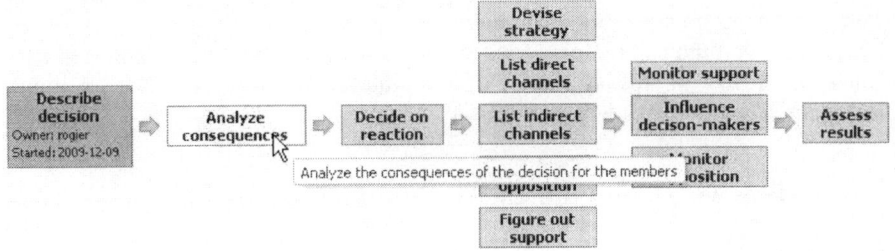

Fig. 3. A map used for structuring the shared space of a process instance

A click on a step box redirects the end-user to a web-form that assists him in completing the step. The form contains text fields, option menus and radio-buttons to make choices, checkboxes, as well as more complex fields. The form may also include "static" texts that explain what should be done before one can fill some fields.

From the shared spaces architecture point of view, the iPB solution can be interpreted as follows. The total process instance shared space is divided into a number of subspaces called process steps. The steps are graphically represented to the end-users as boxes. Subspaces may or may not intersect. The structure of a step subspace is represented to the end-users as a form to fill. Intersecting subspaces means that forms attached to different steps may contain the same field(s). Usually, in this case, the intersecting fields can be changed only in one form; they are made read-only in the second one.

The progress in filling the step forms is reflected in the map attached to the shared space via steps coloring. A gray box means that the step form has not been filled and cannot be filled for the moment. A white box means that the step form is empty but can be filled. A step with a half-filled form gets the green color, and additional information about when the work on it has been started, and who started it. A step with a fully filled form gets the blue color, and additional information about the finish date.

The primary way of "forcing" a person to visit a particular shared space in iPB is by assigning him/her to become an owner/co-owner of some step. Such an assignment results in an email message being delivered to this person, and the process to appear in his/her list of "My processes". When visiting a process shared space, a person can see directly on the map what step(s) are assigned to him, see a green box in Fig. 3.

4.2 Experience of Use

The communication possibilities in an iPB-based system may seem to be too limited. The only way of attracting a given person's attention to visit a particular shared space is by assigning him/her to be an owner/co-owner of some step in the given process instance. No clarification or explanation is given when making the assignment. This should be figured out by the person him/herself from the state of the process, i.e. from the partly filled step form.

However, in practice, this communication mechanism works quite well for relatively structured processes for which it is possible to identify steps. A system of this kind is quite easy to introduce in operational practice, which cannot be said about systems with a collaborative planning style. The communication works well even when participants do not know each other personally.

To extend communication possibilities, we added a rudimentary planning scheme similar to ProBis. In the systems currently introduced, however, this additional mechanism is not being widely used.

5 A System with Communication Based on Status Changes

5.1 Description

The system called eForm was developed for a large Swedish call center (Eniro 118118) to solve the daily staffing problems that can be defined as follows [11]. The scheduling software is run once per month. Staffing requirements may change from day to day (if not from hour to hour) due to changing volumes of inbound calls. In addition, unscheduled absences due to illness, traffic jams, snowfall, etc. make it impossible to totally rely on the pre-generated schedule even when the call volume follows the established pattern. Corrections in the schedule are constantly made to cope with fluctuations in volumes of calls and the number of agents not appearing for work. More agents need to be called in to deal with an increase in the volume of calls or increase in the number of absentees. Alternatively, fewer agents are required when the call volume decreases, or the sick rate due to a seasonable epidemics diminishes.

Effective dealing with fluctuations in the staffing level at a call center requires fast communication channels between agents and managers responsible for operative staffing of the call center. eForm is a web 2.0 system providing efficient channels of communication between three different categories of workers at a call center:

- Central staffing center, aka Control Tower, or just Tower
- Agents
- Coaches (managers)

The system works according to a simple scheme: a communication process starts when one of the participants fills an electronic form that serves as a shared space for this process. This form gets the status New and immediately appears in the list watched by another communication partner. The latter processes the form and sets its status to Finished or changes its status to something else that requires further processing, for example, Question. In the former case, the form disappears from the actual list; but it can still be found through search in the archive (if needed). In the second case, the form disappears from the actual list of this communication partner and appears in the list of some other communication partner, e.g. the one who originally filled the form.

Let us demonstrate how this scheme works on a particular example. Unscheduled absences are reported by agents to their coaches via direct phone calls. A coach communicates this information to the Tower via eForm with a couple of touches of the keyboard by selecting the agent's name, time, and reason for an unscheduled absence. While filling the form, the coach has access to the information about previous unscheduled absences of the same agent. Thus, he/she has a possibility to see a pattern of absences and discuss the matter with the agent. The coach has also a possibility to promptly fill an absence form that concerns more than one agent, for example, in case of an unscheduled training session.

As soon as the coach saves the form, it appears in the absentees list at the Tower. A tower worker makes corrections to the schedule accordingly, after which the form disappears from the actual list (but remains accessible via the archive search). In case of any uncertainty, the Tower quickly returns the form to the coach with a comment by changing its status to Questioned. The coach corrects the form, after which it again appears in the Tower's absentees list.

5.2 Experience of Use

In eForm, there are no explicit "calls" to visit a shared space. The form appears in the list of one of the participating partners dependent on the state of the process, more exactly on the value of one or more fields of the form. From our experience, this mechanism creates a very efficient communication channel, and the system is easy to learn, and introduce in operational practice. This is an important factor in the above business case, because the turnover of agents in a typical call center is, usually, quite high. The communication mechanism works very well for simple real-time processes with strong requirements on the speed of communication.

6   Identifying Communication Styles

Let us investigate differences between the three BPS systems types discussed in Sections 3-5. First of all, there is a difference in the structure of shared spaces. In ProBis – a shared space has a generalized, logical structure. Similar types of objects are gathered on the same tab. For example, there is a separate tab for documents, a separate tab for planned and completed tasks, etc. (see Fig. 1). This reflects the area of ProBis usage – loosely structured processes for which it is not possible to create a more exact structure. In iPB, a shared space is structured in steps according to the "dynamics" of a particular business process type (see Fig. 4). In eForm, shared spaces are quite simple and do not require complex structuring.

Secondly, these three systems implement, on the surface, completely different mechanisms of using shared spaces to facilitate communication/collaboration. ProBis uses collaborative planning. An iPB- based application uses assignment of owners/

co-owners to process steps. eForm uses a list management system to attract the attention of the end-users.

To compare the above mechanisms, we need to abstract from the technical details and consider each mechanism as a way of issuing invitations to visit a shared space. Then, the three mechanisms described in the previous sections can be interpreted as follows.

In ProBis, a task planned for a person represents a manually issued invitation for him/her to visit the process instance shared space with detailed instructions of what he/she is supposed to do there (whom to call, what document to read, etc.). In an iPB-based application, assignment to be owner/co-owner of a step represents a manually issued invitation for this person to visit a particular part of the process instance shared space (a form corresponding to the given step) without any instructions on what he/she is supposed to do there. In eForm, a form appearing in the given person's process list represents an automatically issued invitation to visit the process instance shared space without any instructions on what he/she is supposed to do there.

Generalizing the above, we suggest the following three parameters with binary values for identifying communication styles:

- Issuing technique (*Manual/Automatic*) – an invitation is issued manually by one of the process participants, or automatically by a system based on the state of the shared space.
- Invitation scope (*Global/Localized*) – a person is invited to visit the whole shared space or a particular part of it.
- Invitation instructiveness (*Non-instructive/Instructive*) – a person should him/herself figure out what to do in the shared space based on the state in which he/she finds it when he/she comes there, or an invitation may include instructions on what a person is supposed to do there.

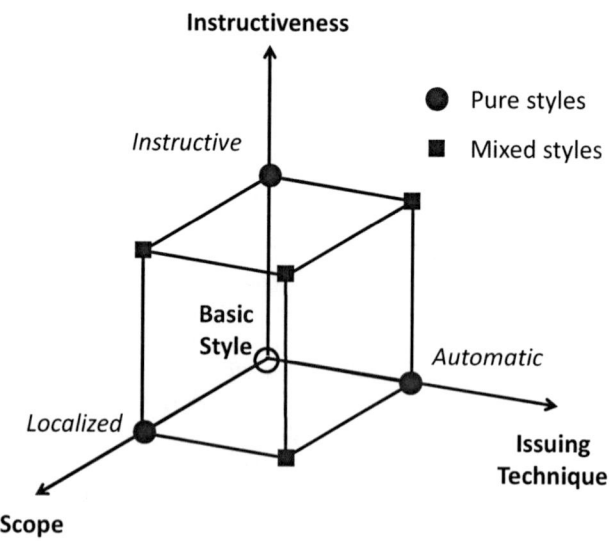

Fig. 4. Communication Styles

The above parameters formally give eight classes that we call communication styles. Let us consider *Automatic* as a more advanced feature than *Manual*, *Localized* as a more advanced feature than *Global*, and *Instructive* as a more advanced feature than *Non-instructive*. Then we can present the classification as a cube in a three-dimensional space where a "basic" style *Manual/Global/Non-instructive* constitutes a zero point, see Fig. 4.

Table 1 below presents styles employed in the systems discussed in this paper. As we can see, each system employs a style that includes only one advanced feature, which means that all styles employed in the systems lie on the axes of the cube in Fig. 4. We call such styles "pure" communication styles, as an opposite to other non-basic styles, which we call "mixed" styles.

Table 1. Communication styles employed in the systems discussed

	Issuing technique	Invitation scope	Invitation instructiveness
Collaborative planning (ProBis)	*Manual*	*Global*	***Instructive***
Specialized structure (iPB)	*Manual*	***Localized***	*Non-instructive*
Status change (eForm)	***Automatic***	*Global*	*Non-instructive*

In Section 2, we stressed that an invitation to visit a particular shared space should give the process participant a clear understanding on why he/she has been invited and what he/she is expected to do there. In the case of eForm, shared spaces are simply structured, thus neither localization nor instructiveness is needed for the invited person to understand what he/she is supposed to do.

Both ProBis and iPB allow quite complicated structures of shared spaces, thus some help is needed to find out what is required from the invited person. In ProBis, this help is provided by instructiveness, which compensates the absence of localization. In iPB, this help is provided by localization, which compensates the absence of instructiveness.

As we already mentioned, the systems discussed in this paper implement "pure" communication styles (only one advanced feature is present in each of them). It does not mean that a mixture is not possible or useful, but only that one advanced feature is sufficient for certain practical purposes as outlined in Sections 3.2, 4.2 and 5.2.

7 Related Works

Works on which the current paper is based are as follows. The earliest attempt to systematically build a BPS system based on the principles outlined in Section 2 (shared spaces along with collaborative and automated planning) was made in 1989-1990. The project, called "DealDriver", is described in [12], [13]. The ideas from the DealDriver project were first presented to the research audience in [14]. More detailed introduction into the state-oriented view on business processes that has been developed in connection to the DealDriver project can be found in [5]. For an overview of the practical experience in BPS systems development up to the time of building the first version of ProBis see [8]. The high-level theory underlying the state-oriented view on business processes is presented in [15][16]. Both DealDriver and ProBis belong to the category of case-handling systems later identified in [17].

Other research works that directly or indirectly belong to the topic of this paper are as follows. The concept of shared spaces is not a new one. It has been widely used in research literature for some time, see, for example, [1]. The simultaneous integration of face to face communication and the exchange of graphical information has been introduced by [2]. The integration of video and audio is proposed by [18]. Embodiments are used to enhance virtual shared spaces in [19]. In [20] media spaces are differentiated from collaborative virtual environments and spatial video conferences.

Using shared spaces for business collaboration was discussed in many research works. A task-oriented collaboration comparable to our collaborative planning can be found in [21]. The new media model introduced in [22] shows tight relationships to shared spaces. Media are spaces where agents can collect and represent information. Roles, describing rights and obligations determine the behavior of agents. In PRODNET [23], a federated database architecture has been used to support shared spaces. Gaia – a middleware infrastructure to enable spaces for cooperatively solving tasks - is presented in [24]. An introduction on how to manage multiple and collaborative tasks is give in [25]. An early system that incorporates some features of ProBis and iPB is OASIS [26]. It supports its users to cooperate for achieving a common goal.

8 Discussion and Future Plans

As was stated in the introduction, the goal of this paper was twofold. The first sub-goal was to present our experience in building domain-specific applications for supporting communication/collaboration in the frame of business process instances. This sub-goal was fulfilled by presenting examples of three systems. Each example contains the description of a system and the review of experience of its usage. In the latter, we outline types of business contexts for which each of the systems is best suited.

As any experience, ours is unique, though some features found in our systems can be found in others, as well. The main characteristic of our approach to building BPS system that differentiate us from others is that all our systems are systematically being built based on the principles of shared spaces architecture outlined in Section 2, and the state-oriented view on business processes from [7]. We are not aware of any other attempts of creating a line of on the surface dissimilar systems that implement these architectural and theoretical principles. In addition, our second example represents not a system, but a tool that can be (and already is) used by others for building BPS systems with the shared spaces architecture. The approach accepted for building BPS system implemented in iPB, as far as we know, is unique.

The second sub-goal was to introduce a taxonomy of communication styles for collaborative systems with shared spaces architecture. This was done based on the analysis of our experience and resulted in the introduction of three parameters with binary values: issuing technique, scope, and instructiveness.

The parameters proposed for the style identification can be applied for the analysis of publicly and commercially available systems that employ shared spaces for communication/collaboration. As an example, let us consider Google Sites [27], which is a shared spaces system created in the context of Google's E-Mail system Gmail [28]. A Google site is a tree-structured collection of pages that can be edited without an external tool using a browser-based interface. Other users may be invited to collaborate as a site owner, editor or reader. Furthermore, it is possible to share a site with everyone on the web. In the enterprise edition of Google Sites, it is possible to give read or write access to all users of the enterprise's domain. Google Calendar is used to delegate tasks. This can be done if the addressee of the delegation shares his/her calendar with the delegating person. The delegating person can create tasks in the calendar of the addressees and thus to delegate tasks to him/her.

According to the communication styles scheme suggested in section 6, Google Sites uses the manual issuing technique (through the calendars). An invitation always has a global scope, as it concerns the whole site. Invitations are instructive, as it is possible to specify what has to be done in the delegated task.

From the point of view of the parameters introduced, the three systems from our experience represent "pure" communication styles, which, of course, does not exclude creating a mixture of styles. In fact, we are in the process of adding the automatic issuing technique to both ProBis (based on [16]), and iPB.

The three parameters for style identification proposed in this paper give only basic characterization of communication capabilities of a system with the shared spaces architecture. More detailed classification is required for covering the nuances. For example, the automatic issuing technique can be divided into two subcategories: general rules, and instance rules. General rules ensure automatic issuing of invitation that covers all instances of the given process type. Instance rules mean a capability to ensure automatic invitations for a particular process instance/case. The latter are often expressed in the form of subscription to certain events in a shared space (see, for example, proposals for adding instance rules to ProBis in [16]).

As follows from the experience presented in the paper, different communication styles suit different kinds of business contexts. For example, high requirements on speed of communication, as in eForm, warrant automatic issuing technique. A complex structure of shared spaces, as in ProBis and iPB, requires either instructiveness or localization (or both).

To find a proper mixture of communication styles for a practical business case, the properties of each communication style should be understood, so that the styles are mixed based on the requirements of a particular business environment. Our current research in progress is devoted to this task. In this research, we analyze our experience from the point of view of business requirements that can be set on the communication/collaboration mechanisms. Here, we differentiate several groups of requirements: functional requirements (e.g., possibility of inviting an arbitrary person at any moment of time), security requirements (e.g., restricting a person's capability of viewing parts of the shared space), social requirements (e.g., support of week ties), and business process requirements (e.g., support for predefined tasks for the given process type).

Acknowledgments. This paper would have never been written without considerable efforts of the team of developers who have designed and implemented ProBis, iPB, and eForm. We are especially thankful to Tomas Andersson, Alexander Durnovo, Alexey Striy and Rogier Svensson.

References

[1] Takemura, H., Kishino, F.: Cooperative work environment using virtual workspace. In: Proceedings of the 1992 ACM Conference on Computer-Supported Cooperative Work, pp. 226–232 (1992)

[2] Ishii, H., Kobayashi, M., Grudin, J.: Integration of interpersonal space and shared workspace: ClearBoard design and experiments. ACM Transactions on Information Systems (TOIS) 11(4), 349–375 (1993)

[3] Prensky, M.: Digital natives, digital immigrants. On the Horizon 9(5), 1–6 (2001)

[4] Nurcan, S., Schmidt, R.: Introduction to the first international workshop on business process management and social software (BPMS2 2008). In: Ardagna, D., Mecella, M., Yang, J. (eds.) Business Process Management Workshops. Lecture Notes in Business Information Processing, vol. 17, pp. 647–648. Springer, Heidelberg (2009)

[5] Khomyakov, M., Bider, I.: Achieving Workflow Flexibility through Taming the Chaos. In: OOIS 2000-6th International Conference on Object Oriented Information Systems, pp. 85–92. Springer, Heidelberg (2000)

[6] Bider, I., Perjons, E., Johannesson, P.: In Search of the Holy Grail: Integrating social software with BPM. Experience Report. In: EB-PISM. LNBIP, vol. 50, pp. 1–13. Springer, Heidelberg (2010)

[7] Bider, I., Perjons, E., Johannesson, P.: A strategy for integrating social software with business process support. In: LNBIP, vol. 66, pp. 372–383. Springer, Heidelberg (2011)

[8] Andersson, T., Bider, I., Svensson, R.: Aligning people to business processes experience report. Software Process Improvement and Practice 10(4), 403–413 (2005)

[9] Bider, I., Striy, A.: Controlling business process instance flexibility via rules of planning. International Journal of Business Process Integration and Management 3(1), 15–25 (2008)

[10] iPB Reference Manual on-line documentation (Online), http://docs.ibissoft.se/node/3 (accessed June 20, 2010)

[11] eForm – a Staffing Communication System for running day-to-day operations at a call-center (Online), http://www.ibissoft.se/node/144 (accessed: June 20, 2010)

[12] Bider, I.: Developing tool support for process oriented management. In: Handbook of Systems Development, vol. 1999, pp. 205–222. CRC Press, Boca Raton (1998)

[13] Bider, I.: Object driver: a method for analysis, design, and implementation of interactive applications. In: Handbook of Systems Development, vol. 1999, pp. 81–96. CRC Press, Boca Raton (1998)

[14] Bider, I., Khomyakov, M.: Object-oriented model for representing software production processes. Object-Oriented Technologys, 319–322 (1998)

[15] Bider, I., Khomyakov, M., Pushchinsky, E.: Logic of change: Semantics of object systems with active relations. Automated Software Engineering 7(1), 9–37 (2000)

[16] Bider, I., Khomyakov: New technology - Great Opportunities. How to Exploit Them. In: Filipe, J. (ed.) Enterprise Information Systems, Kluver (2003)

[17] Van der Aalst, W.M.P., Weske, M., Grünbauer, D.: Case handling: a new paradigm for business process support. Data & Knowledge Engineering 53(2), 129–162 (2005)

[18] Bly, S.A., Harrison, S.R., Irwin, S.: Media spaces: bringing people together in a video, audio, and computing environment. Communications of the ACM 36(1), 46 (1993)

[19] Benford, S., Bowers, J., Fahlén, L.E., Greenhalgh, C., Snowdon, D.: User embodiment in collaborative virtual environments. In: Proceedings of the SIGCHI Conference on Human Factors in Computing Systems, pp. 242–249 (1995)

[20] Benford, S., Brown, C., Reynard, G., Greenhalgh, C.: Shared spaces: transportation, artificiality, and spatiality. In: Proceedings of the 1996 ACM Conference on Computer Supported Cooperative Work, pp. 77–86 (1996)

[21] Dillenbourg, P., Traum, D., Schneider, D.: Grounding in multi-modal task-oriented collaboration. In: Proceedings of the European Conference on AI in Education, pp. 401–407(1996)

[22] Schmid, B.F.: The concept of media. In: Workshop on Electronic Markets (1997)

[23] Garita, C., Afsarmanesh, H., Hertzberger, L.O.: The prodnet cooperative information management for industrial virtual enterprises. Journal of Intelligent Manufacturing 12(2), 151–170 (2001)

[24] Roman, M., Hess, C.K., Cerqueira, R., Ranganathan, A., Campbell, R.H., Nahrstedt, K.: Gaia: A middleware infrastructure to enable active spaces. IEEE Pervasive Computing 1(4), 74–83 (2002)

[25] Johnson, P., May, J., Johnson, H.: Introduction to multiple and collaborative tasks. ACM Transactions on Computer-Human Interaction (TOCHI) 10(4), 277–280 (2003)

[26] Martens, C., Woo, C.C.: OASIS: An Integrative Toolkit for Developing Autonomous Applications in Decentralized Environments. Journal of Organizational Computing and Electronic Commerce 7(2&3), 227–251 (1997)

[27] Google Sites, (Online), http://www.google.com/sites, http://sites.google.com/ (accessed June 18, 2010)

[28] Google Mail, (Online), http://www.google.com/mail (accessed June 18, 2010)

What Methodology Attributes Are Critical for Potential Users? Understanding the Effect of Human Needs

Kunal Mohan and Frederik Ahlemann

EBS Business School, Söhnleinstraße 8D, 65201 Wiesbaden
kunal.mohan@ebs.edu, frederik.ahlemann@ebs.edu

Abstract. Despite the overwhelming advantages of using IS development and management (ISDM) methodologies, organisations are rarely able to motivate their staff to use them. The resulting lack of methodology usage by individuals fails to deliver the expected advantages of better quality, control, less time, and less effort in IS development projects. We analyse the technical as well as non-technical aspects of an individual's use of ISDM methodologies, in order to enable organisations to engineer those that meet the needs of actual users and are actually used by them in a productive manner. We construct a conceptual model, based upon which, we posit that: technical methodology attributes such as relative advantage, complexity, compatibility, demonstrability, visibility, triability, and reinventability influence an individual's methodology usage behaviour. We also propose that the strengths of these relationships depend on non-technical, deeply rooted psychological needs of the people.

Keywords: Methodology acceptance, IS development, method engineering.

1 Introduction

In the search for ways to arrive at replicable, pragmatic, cost-effective, and timely solutions to real-world problems in systematic and predictable ways, organisations either adopt or customise and adaptively apply methodologies, which consist of tested bodies of methods, rules, and procedures. Despite the overwhelming advantages of using any methodology, only a handful of organisations are actually able to make their staff use such methodologies [18]. A software development project survey conducted by Russo et al. [41] shows that only 6% of organisations claim that their methodologies are always used as specified. Eva and Guilford [15] find that only 17% of respondents, in their survey of 152 organisations, claim to use a methodology in its entirety. Organisational theorists have long recognised that individual behavioural resistance against new methodology use is because they might not share the goals of the organisations in which they work and that exert pressure on them to use new methodologies [46]. The roots of this problem of methodology acceptance, which our research addresses, lies – among other factors – in the failure to understand individual attitudes towards a methodology's use, which ultimately leads to the development and implementation of a methodology that might be considered unsuitable and might be rejected by users [31].

H. Mouratidis and C. Rolland (Eds.): CAiSE 2011, LNCS 6741, pp. 314–328, 2011.
© Springer-Verlag Berlin Heidelberg 2011

Diffusions of innovation theory (DOI) has directed considerable attention towards understanding the diffusion of innovations [38]. Research in a vast array of academic disciplines such as anthropology, communication, geography, sociology, marketing, political science, public health, economics, social psychology, sociology, and political science has applied DOI to understand the process through which new ideas and technologies become accepted by people. Some of these studies have attempted to examine individual usage behaviour regarding IS methodologies from a technology adoption perspective. They view software development methodologies as technology innovations and make use of DOI and the technology acceptance model (TAM) (e.g., [22,37]). Others apply sociological models such as the theory of planned behaviour (TPB) and Triandis's theory of interpersonal behaviour to examine the development of individuals' intention to use methodologies (e.g., [21,26]).

Both approaches come to similar conclusions, and state that methodology characteristic *usefulness* is the single most important determinant of methodology acceptance and use by its actual users [22,37]. Subsequent research has therefore focused on this particular variable, but has neglected other potential crucial methodology attributes. Critics have also suggested that TAM and TPB are too parsimonious and need to be expanded by integrating variables specific to the innovation under investigation [49]. However, even when a handful of researchers seek to examine other methodology attributes, attributes are found to be either not significant, or of negligible effect (e.g., [22,37]), partly because these studies neglect to integrate other non-technical and non-economic variables from related theoretical perspectives [49], such as personality attributes like needs of individuals. As Warner [53] observes, the concept of adoption is a complex social phenomenon that involves both technical and non-technical factors, and sociologists would undoubtedly agree with this view. Unfortunately, the various disciplines, generally concentrating on their individual variables, have neglected to incorporate personality attributes in understanding the methodology acceptance problem. Little is known about the interactive effects of the attributes of methodologies and the non-technical personality characteristics, and it seems reasonable that variables from both sets are important in explaining the problem at hand [53].

Our study seeks to identify additional methodology attributes and to examine which of these attributes is more important for which type of person in which situations. Neglecting the impact of such complex relationships might lead to results that are not always valid [23]. Our study is a step toward filling the gap in the methodology development, adoption and implementation literature, which until now has not developed a theoretically and practically complete and relevant taxonomy of potential methodology characteristics and has also not studied the effect of personal traits such as needs on the effectiveness of the various methodology attributes. This leads us to fundamental questions regarding the impact of methodology attributes on an individual's usage behaviour: a) Which methodology attributes affect an individual's decision to use it? b) How do basic individual needs influence the predictive power of these different methodology attributes?

The remainder of the paper is organised as follows: Section 2 defines the research scope, and provides an overview of the conceptual foundations and the basic theories that provide the framework for our conceptual model. We also discuss prior research on the topic in order to clarify what has been done and what needs to be done. In Section 3, we present our research model and hypotheses, pointing out validated survey instruments that might be used to operationalize our constructs. In Section 4 we discuss the study's implications and contributions. Being research-in-progress, in section 5, we outline the next steps in our research to describe how we plan to execute the study's next phase.

2 Theoretical Foundations

We focus on examining the behaviour of *individual users* rather than an organisation as a whole because, although a particular methodology is developed and implemented by an organisation, the extent of its use is usually decided by the methodology's actual users [37,26]. We also focus only on the use of *methodologies* instead of methods/techniques (e.g., stakeholder analysis, earned value analysis, etc.) and tools (e.g., CASE tools, Word/Excel templates, project management software, etc.), because tools, techniques, and methods can be used in the absence of a formal methodology, and the use of a methodology represents a radical change [22]. Reasons why new methodology adoption and use might be so different from and so much more challenging than the adoption of specific methods and tools lies partly in the tacit organisational and individual problems caused by the new methodology introduction. For example, the stress associated with the learning of a new methodology, fear, the impact on self-esteem and identity associated with the organisational restructuring or re-engineering as well as the emotional costs of role conflict and ambiguity and/or workplace transformation might be serious inhibitors of methodology acceptance and usage [51].

DOI has been used over the past five decades to study how innovations diffuse and become adopted within wider social networks [38]. While early research using DOI concentrated on the diffusion and acceptance of products, the research community recently reached consensus on the fact that *ideas* and *practices* such as methodologies can also be regarded as innovations if they are perceived to be new by the potential adopter [38]. The foundations of DOI can be traced back to the three mainstream school of thoughts of Bass [5], Moore [30], and Rogers [38], with the Rogers gaining more attention and popularity. Bass [5] used mathematical methods to develop a model of innovations diffusions in 1969 in which he proposed five adopter categories depicting which type of person adopt innovations, and when: innovators, early adopters, the early majority, the late majority, and laggards. Moore [30] developed his own model of technology diffusion using the same adopter categories and the same terms as Bass [5].

The major difference between the two schools was that Moore's [30] work was based on the assumption of a discontinuous innovation process and focused only on organization, with a new technology adoption requirement. However, the best-known innovation diffusion theory was introduced by Rogers in 1962 in *Diffusion of Innovations*. Rogers classifies diffusion in his innovation adoption framework into five stages: innovators, early adopters, the early majority, the late majority, and laggards, with 2.5%, 13.5%, 34%, 34%, and 16% of the population respectively. According to Rogers, one of the most influential factors that determine an innovation's adoption rate is the innovation itself, i.e. its *characteristics*. Furthermore, differences between the adoption stages, i.e. how quickly an innovation gets adopted, depends on adopter characteristics such as socioeconomic status and *personality values* [53]. Another distinguishing feature of Rogers' theory, which makes it very attractive for our study, is that it can also be applied to individuals. As such, we find that Rogers' diffusion of innovations theory provides the most fertile theoretical foundation for our research.

A key aspect of diffusion theories relates to the perception of innovations by potential adopters. Based on DOI, a methodology's characteristics play a crucial role in how quickly it is accepted by potential users [9]. The more attractive the attributes of a methodology are perceived to be, the more swiftly it is accepted by potential users. Empirical studies related to Rogers' DOI theory have therefore focused on the identification and examination of innovations' characteristics. Rogers and Shoemaker's [40] comprehensive list includes relative advantage (e.g., profit, productivity, and the innovation's prestige-conferring qualities); compatibility with users skills and ways of working; complexity of use and understanding of the innovation; trialability (i.e. divisibility, as the ability to be tested by the potential adopter); and observability (the degree to which others see the results of use of an innovation; also called communicability) [53].

Extensive empirical research in the past has found that some of the elements are more important than others. After conducting a meta-analysis of 75 articles pertaining to innovation characteristics, Tornatzky and Klein [47] found that relative advantage, complexity, and compatibility were the only innovation characteristics consistently related to innovation adoption and implementation. Later, Moore and Benbasat [29] expanded Rogers and Shoemaker's [39] list to include image (enhancement of social image or status), result demonstrability (of tangible advantages), and voluntariness (free will to adopt). Although extensive empirical evidence in various fields suggests that these influences do hold [53] in the context of methodology adoption, except relative advantage, most of them have either been neglected or have been found to be insignificant. For example, in the study of Riemenschneider et al. [37], five theoretical models of individual intention to accept information technology tools were tested individually using least-square regression analysis, to understand why software developers accept or resist methodologies. They came to the following conclusions: *perceived usefulness* was the only significant variable across all five models (p < 0.001), *voluntariness* was found not significant (or was not included) in three models, *compatibility* was found not significant (or was not included) in four models, and result *demonstrability, complexity, observability,* and *image* were found to be not significant (or were not included) across all five models. In their study, Hardgrave et al. [23] also study software developers' intentions to use methodologies, and also find *usefulness* to significant (although comparatively weaker), *complexity* to not be significant, and *voluntariness* and *compatibility* to be significant but weak.

Seeing the large gap in the innovation attributes proposed by DOI and those studied in the context of methodology acceptance, we identify two areas in need of attention: a) examining which of the wide number of innovation characteristics apply to the methodology domain, and b) which of these different attributes are more important to what type of individuals. While, as mentioned earlier, DOI does provide a comprehensive list of attributes to examine the former issue (a), the latter problem (b) is virgin territory.

Recently, consumer research has acknowledged that personality-specific traits are of greater interest than demographic or psychographic influences, since they are "at the heart of consumer attitude formation and behavioural intentions" [11]. Over the past few decades, various categories of needs theories – e.g., Maslow's well-known hierarchy of needs [28] – have been developed so as to understand and predict human behaviour. They have become widely accepted in research studies, because they are considered to be the most enduring ways to understand an individual's motivation to act in a particular way [3]. According to the needs theories, an individual in an agent-target relationship is expected to be influenced by a certain user influence tactic (UIT), if this UIT corresponds to his or her desires and needs (e.g., the need for knowledge, affiliation with peers, etc.). Many definitions of basic needs have been proposed, of which Ryan and Deci's [42] is most consistent with the scope of this study. They indicate that "a basic need, whether it be a physiological need or a psychological need, is an energizing state that, if satisfied, conduces toward health and well-being but, if not satisfied, contributes to pathology and ill-being" [42]. This implies that, if an methodology attribute is not aligned with the potential adopters' basic needs, this might result in serious stress, anxiety, and depression, and that this discomfort might be visible in the weak effect of the particular attribute to motivate an individual to use the methodology. This view might help explain why previous methodology acceptance found inconsistent results (i.e. because they were void of personality factors such as needs of an individual).

Maslow's hierarchy of needs theory [27] is one of the most fundamental and influential needs theories. It suggests that there exists a hierarchy of needs and that certain lower needs must be satisfied in order for higher needs to be recognised as unfulfilled. However, critics of the theory state that: a) there is hardly any evidence of the existence of a definite hierarchy of needs or that fundamental human needs are non-hierarchical, and b) little evidence suggests that people satisfy only one motivating need at a time, except in situations where needs conflict (i.e. are mutually exclusive) [52]. Empirical research, finds that a) more than one need may motivate at any one time, and b) that different needs have different values for different people. For the purpose of our study, we therefore employ, Murray's theory of psychogenic needs [32], and Reiss' theory of 16 basic desires [36] as these are considered the most fundamental and comprehensive list of underlying psychological human needs and have been empirically tested in a number of studies.

3   Conceptual Model and Research Hypotheses

The decision as to whether or not to adopt a methodology often requires time, energy, and careful consideration by the potential adopter [38]. Based on the complementary use of DOI and needs theory in humanistic psychology, individuals are expected to use a methodology based on their perceptions that methodology attributes will enable them to fulfil their specific needs. Needs of an individual are therefore expected to play a *moderating* role (see Figure 1), and influence the explanatory power of the effect of different methodology attributes on an individual's use of a methodology. The reason why we focus on *perceptions* of methodology attributes, rather than *primary* attributes (intrinsic to a methodology, independent of the perceptions of potential adopters) is that individual behaviour is predicted by how one perceives these primary attributes [29]. Since different adopters might perceive primary attributes in different ways, their eventual behaviour might differ [29].

In our research, we specifically focus on moderating effects because, besides the examination of direct effects, scholars are increasingly seeking to understand complex relationships [23]. While the importance of taking moderation effects is emphasised repeatedly in the literature [10], its neglect has led to a lack of relevance, as "…relationships that hold true independently of context factors are often trivial" [23]. In the remainder of this section, we define each of the determinants, specify the role of key moderators, and provide theoretical justification for our hypotheses.

3.1 Attributes of a Methodology

Relative advantage (RA) is the degree to which a methodology is perceived as being superior to its precursor by potential adopters, which is either the previous way of doing things (if there is no current way), the current way of doing things, or doing nothing [9]. A methodology's superiority is not only measured in economic terms, but also in terms of reduced or increased status and other benefits (e.g., because of an increase in productivity and efficiency). The higher the relative advantage, the higher the rate of adoption, all other factors being equal. The expected favourable outcome or usefulness of a behaviour has emerged as a core construct in the field of MIS, driven largely by the use of the theory of planned behaviour (attitude) [1] and the technology acceptance model (TAM) (perceived usefulness) [12] in examining individual beliefs regarding performing a behaviour. A plethora of empirical research in various fields has confirmed that the favourable outcome or usefulness of a behaviour is the most important aspect in predicting it – e.g., [12,49,50]. Hardgrave et al. [22] state that "…usefulness generally has a beta (path coefficient) of around 0.60 in TAM studies". In the context of methodology adoption, Khalifa and Verner [26] find that better process and product quality have a substantial effect on a software developer's decision to use waterfall and prototyping methodologies. Riemenschneider et al. [37] apply five theoretical models and conclude that "…if a methodology is not regarded as useful by developers, its prospects of successful deployment may be seriously undermined". Consequently, we propose that relative advantage will have a positive effect of methodology use.

Complexity (CL) is the degree to which a methodology is perceived as difficult to understand and use. The more complex a methodology is perceived to be, the more resistance it is expected to generate. The *complexity* construct can be traced back to Bandura's [4] *self-efficacy* concept, which refers to the belief that one has the capability to perform certain actions in order to be able to use a methodology. Judgment of one's personal competence, reflected in one's self-efficacy, therefore not only determines *if* a person decides to use an methodology, but also how much *effort* he or she will expend to use it, how long he or she will *persevere* when confronting obstacles, and how *resilient* he or she will prove in the face of adverse situations [33]. The more complex a methodology is perceived to be, the more an individual doubts his or her own ability to be able to use a methodology properly. In technology adoption literature, *complexity* has been addressed through the *ease of use construct* (which is also based on the concept of self-efficacy), which refers to the degree to which a person believes that using a particular methodology would be a) free of physical and mental effort, and b) easy to learn [12]. Numerous empirical evaluations of *self-efficacy* and *ease of use* find them to be an important predictor of human behaviour (e.g., [12,48-50]) and therefore provide substantial justification for including the *complexity* construct in our model. As such, we propose that complexity will have a negative effect on methodology use.

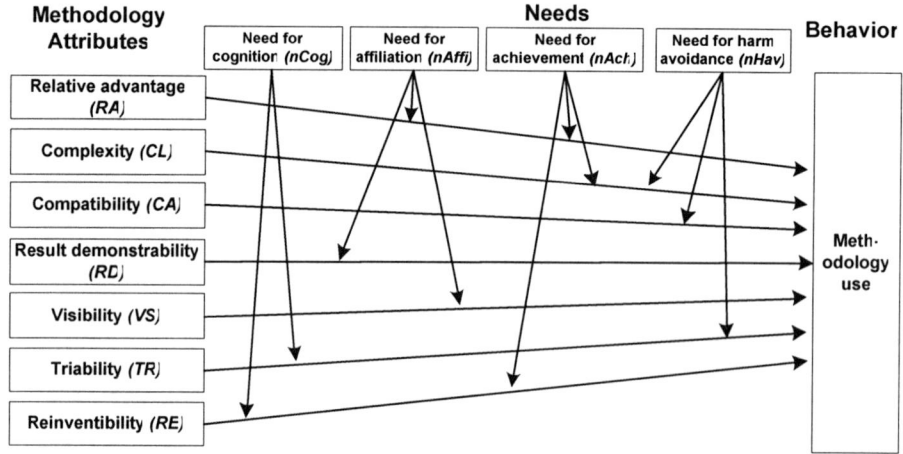

Fig. 1. Conceptual model

Compatibility (CA): is the degree to which a methodology is perceived to be consistent with existing social cultural values, and past experiences of potential adopters [9]. The higher the compatibility, the higher the desire to use the methodology. The roots of this lies in the understanding that individuals in organisations might be reluctant to change their habits, which they have learned unconsciously through past repetitions, and might therefore be unwilling to adopt new methodologies if they cause radical change. The more a methodology departs from the current work processes of an individual, the longer and harder he or she must strive to unlearn old routines and learn new ones [22]. In matters of radical change, as is the case with new methodology adoption (see Section 2), the methodology might not be compatible with the habits of potential users and would therefore activate negative feelings and emotions and, consequently, resistance. Past methodology acceptance research has found a significant positive but weak effect of compatibility on an individual's intention to use a methodology (e.g., [37]). However, research on this crucial construct is still relatively scarce, which calls for further attention. Based upon this discussion, we propose that compatibility will have a positive effect on methodology use.

Perceived observability is another attribute originally identified by Rogers [38]; it is the degree to which an individual believes that the results of using a methodology are visible [38]. However, this definition of observability indicates that the construct is complex and has been found to possess two unique dimensions [29]: a) the *demonstrability* of results to others, and b) the *visibility* of the innovation itself. While conducting a sorting exercise with a panel of judges, Moore and Benbasat [29] found that the items of the observability construct tapped into these two dimensions. As a result, they decided to split the construct into *visibility* and *demonstrability*. Similar to Moore and Benbasat [29], we hold that, in order to glean a deeper understanding of how observability influences methodology adoption, we need to study the effect of visibility and demonstrability individually. *Result demonstrability* (RD) is the degree to which the results of using a methodology are observable by others. While in some methodologies it is easy for others to see the results of a methodology's usage, this might not be the case with others (e.g., because of poor transparency or poor communication of outcomes achieved). Being able to show and communicate results achieved by using a methodology is important in acquiring tangible and intangible benefits such as praise, bonus, and promotion, since organisational incentive systems and management can only reward productive and efficient employees if they can observe these improvements. Furthermore, the easier it is for potential users to experience and see for themselves the positive effects of using a new methodology, the more confident they will be that they will also be able to realise the positive outcomes. In short, high observability acts as a motivator by reducing the risk in a potential adopter's mindset that using a methodology will be unfruitful (i.e. a high chance of the methodology not generating the promised positive results).

Result demonstrability might be especially critical to the methodology domain, because methodology outcome is characterised by the a) benefits not being realised swiftly [22], b) diffusion of potential benefits in a highly complex network of multiple actors. These characteristics make is hard to quantify and communicate total methodology usage benefits achieved to potential adopters (this is also a general problem in the IT field). *Visibility (VS)* is the degree to which a methodology is actually visible in the work environment. Visibility here implies that potential adopters can see their peers or seniors as they use methodologies, or know that they use them. Research has shown that mere exposure to objects (i.e. methodologies) is capable of rendering an individual's attitude towards these objects more positive. Extensive empirical research on human behaviour has shown that individuals would use a methodology because of their motivation to satisfy their notion of self-definition by doing what their peers (whom they want to be like) do [7]. Consequently, the more a potential adopter is able to observer his or her peers and seniors use a methodology, the more he or she will be inclined to use it. Consequently, we propose that result demonstrability and visibility will have a positive effect on methodology use.

Triability (TR) is the degree to which an individual believes he or she can experiment with a methodology on a limited basis prior to adoption. The chance to "test" a methodology prior to an individual making the final decision to adopt and use it helps to clear doubts relating to usefulness, complexity, and compatibility. High triability enables an individual to make a well-informed rational choice for himself, and is considered to be crucial for sceptical individuals who do not simply trust what they are told [16]. A number of studies also find strong evidence of the motivational effect of triability (e.g., [20,35]). Furthermore, prior testing of a methodology might also help a potential adopter to discover until then unknown or uncommunicated methodology benefits considered useful by the tester, since the evaluation of functionality is subjective and differs from person to person [16]. As such, we expect triability to have a positive effect on methodology use.

Reinventability (RE) is the degree to which a methodology is perceived to be modifiable by a potential user. If potential adopters can adapt, refine, or otherwise modify the methodology to suit their own needs and situation, it will be adopted more easily [19]. The concept of reinvention – the assumption that one size does *not* fit all – has been embraced by researchers, to the extent that a dedicated research stream, entitled *method engineering*, has developed (e.g., [40]). A number of empirical studies find strong empirical evidence of the motivational effect of reinventability on a person's decision to use an innovation (e.g., [38]). Although this construct has been widely neglected in methodology acceptance studies, based upon existing research findings, we consider it an important predictor in our model. We propose that reinventibility will have a positive effect on methodology use.

Although Moore and Benbasat [29] propose *voluntariness of use* and *image* as further attributes, we hold that they might not apply to our research, since enhancement of social image can be considered an aspect of usefulness, rather than a distinct methodology characteristic. Rogers [38] also includes the concept of *image* under *relative advantage*. Furthermore, in our research, we consider methodology use to be *voluntary*. Even though organisations can deploy obligatory methodologies, their actual use in a *productive manner* cannot be forced, and correct usage is ultimately a voluntary user act.

3.2 Personal Characteristics

Need for affiliation (nAffi) is the desire to achieve acceptance from one's social surroundings [32]. Individuals with a high need for affiliation tend to enjoy being with other people, making friends, and maintaining personal relationships. In a work environment, materialistic status symbols like promotion, higher salary, gifts, and praise from seniors have been found in a number of studies to be conveyors of, and an adequate substitute for, positive interpersonal relationships and feelings of acceptance [6]. Also, since individuals high in *nAffi* depend on approval from their work environment, it is critical to them that the results are visible to their peers and superiors. Based on this reasoning, *nAffi* is expected to have a moderating effect on the strength of the effect of relative advantage → methodology use, result demonstrability → methodology use, and visibility → methodology use.

Need for achievement (nAch) refers to an individual's desire to do things better, accomplish difficult tasks, overcome obstacles and become an expert, achieve high performance standards, or a need for significant task-related accomplishment [32]. Such individuals are focused on internal motivation and personal achievement, rather than on external rewards and recognition. They would be more inclined to use a methodology if they feel that it is useful and would help them to be more efficient and productive in their job. Furthermore, the more complex a methodology, the more gratification/satisfaction people high in *nAch* are expected to feel, since being successful at using methodologies which others fail to master symbolises and communicates personal competence. Individuals high in *nAch* are expected to expend *more effort* and *persevere longer* when confronted with obstacles, and show *resilience* in the face of adverse situations [36]. In order to achieve high performance and excel at using a methodology, individuals high in *nAch* will be interested in modifying and adapting the methodology to suit their own skills set and context of use. We therefore propose that *nAch* will have a moderating effect on the strength of the effect of relative advantage → methodology use, complexity → methodology use, and reinventability → methodology use.

Need for cognition (nCog) is the desire for knowledge, reasoning [32,36] as well as the need to explore and discover. Individuals high in *nCog* tend to naturally seek, acquire, think about, and reflect on information by experimenting and exploring, to make sense of a problem at hand [8]. Therefore, people high in *nCog* are more likely to want to try out methodologies, to better understand for themselves how proposed benefits are expected to be achieved, whether or not the promises are justified, and how their way of doing things changes. The desire to try out, extract, and process information *by oneself*, instead of simply "buying into" the anecdotes, demonstrations, reasoning tactics, and rational appeals of peers or experts is what characterises cognitive behaviour. Consequently, we expect *nCog* to have a moderating effect on the strength of the effect of triability → methodology use, and reinventability → methodology use.

Need for harm avoidance (nHav) is a personality trait characterised by excessive worrying, pessimism, fearfulness, and doubtfulness. Harm-avoidant individuals are biased in the direction of seeking to end behaviours that might involve worrying, fear of uncertainty, and increased risk of anxiety [32]. As such, individuals high in *nHav* attempt to pursue behaviour that helps them reduce any risk and uncertainty attached to new methodology use. Methodologies that are considered complex and hard to execute are avoided, since risk of failure increases with rising complexity. A methodology perceived to be compatible with old routines involves minimal change and is therefore considered safe, because it corresponds to previous experience. Similarly, individuals will seek to experiment and try out methodologies prior to making their final choice, in order to identify those that might be potentially risky and carry a high degree of outcome uncertainty. As a result, we propose that *nHav* will have a moderating effect on the strength of the effect of complexity → methodology use, compatibility → methodology use, and triability → methodology use.

Empirical research has shown that the above-mentioned needs are largely uncorrelated with one another [36,45]. Although the list of needs in the literature is extensive, we consider these four needs to be representative of the most fundamental high-level primary needs in the context of influence tactics, in the sense of being innate or "hardwired" [45]. Other secondary needs can be derived from these high-level primary needs. For example, Murray's *need for play, need for curiosity*, and *need for understanding* may be attributed to *nCog*, the *need for contrarience*, and the *need for acquisition* may be derived from *nAch*. The *need for family* – as proposed by Reiss [36] – and the *need for social recognition* may be attributed to *nAffi*, and the *need to compete or win* can also be derived from *nAch* [45]. Another reason to study fewer needs, rather than more, relates to the value of a parsimonious approach: as the list of needs increases, the utility of the approach diminishes. A long, unwieldy list of needs is precisely the reason why earlier needs-related theories fell out of favour [13].

The related research hypotheses are summarised in Table 1, which also provides an overview of some studies that have used validated instruments to operationalize the constructs of our conceptual model.

Table 1. Research hypotheses and prior operationalization of constructs

H1: RA[f] is positively associated with *methodology use*[a] (MU).
H2: CL[g] will be negatively associated with methodology use.
H3: CA[h] will be positively associated with methodology use.
H4: RD[i] will be positively associated with methodology use.
H5: VS[j] will be positively associated with methodology use.
H5: TR[k] will be positively associated with methodology use.
H6: RE[l] will be positively associated with methodology use.
H7: The influence of RA on methodology use will be moderated by *nAffi*[b] and *nAch*[e], such that the effect will be stronger for individuals with these specific needs.
H8: The influence of CL on methodology use will be moderated by *nAch* and *nHav*[c], such that the effect will be stronger for individuals with these specific needs.
H9: The influence of CA on methodology use will be moderated by *nHav*, such that the effect will be stronger for individuals with this specific need.
H10 The influence of RD on methodology use will be moderated by *nAffi*, such that the effect will be stronger for individuals with this specific need.
H11 The influence of VS on methodology use will be moderated by *nAffi*, such that the effect will be stronger for individuals with this specific need.
H12 The influence of TR on methodology use will be moderated by *nCog*[d] and *nHav*, such that the effect will be stronger for individuals with these specific needs.
H13 The influence of RE on methodology use will be moderated by *nCog* and *nAch*, such that the effect will be stronger for individuals with these specific needs.

[a] [49,50]; [b, e] [14,17]; [c] [34,54] ; [d] [8]; [f, g, h, i, j, k, l] [29].

4 Discussion and Implications

Our work seeks to further the research on acceptance and use of methodologies by individuals by unifying the theoretical perspectives on the *attributes of a methodology* and *needs of individuals* within a single model. Such a holistic approach for understanding why certain employees adopt a methodology while others reject it [25] is important, because people are not passive recipients of innovations. They actively seek new effective methodologies, "...experiment with them, evaluate them, find (or fail to find) meaning in them, develop feelings (positive or negative) about them, challenge them, worry about them, complain about them, 'work around' them, gain experience with them, modify them to fit particular tasks, and try to improve or redesign them—often through dialogue with other users" [19]. Only when we understand and acknowledge that such a diverse list of actions and feelings are typical of human behaviour, do we view the acceptance of new methodologies as a complex process and realise that research needs a holistic lens, integrating technical as well as non-technical factors.

Based on validated theories, we develop a conceptual model that holds that personal traits of individual – especially their needs – determine which technical methodology attributes has a larger effect on an individual's use of a methodology. The proposed multidimensionality of "what is a methodology" from a technical perspective represents a departure from traditional operationalization (which is devoid of human factors) and might reveal more complex and as yet unknown interaction effects on human decision-making, especially in regard to the use of new methodologies. Our findings might have significant implications not only for the MIS research community, but also for related fields as it might be able to explain how changes in needs change attitudes and preferences. Human needs have always played a key role in organisational development, and the proposed study is an attempt to "humanise" organisational methodologies [2], that is, to enable organisations to be more responsive to human concerns when developing and implementing new methodologies. Furthermore, by creating a theoretically and practically relevant and parsimonious taxonomy of attributes of methodologies, we present researchers as well as practitioners with a framework to help identify and understand the characteristics that their methodologies possess or should possess.

Our research also has significant implications for practitioners. Each of the proposed constructs reveals a different aspect of human behaviour and personality, and each can serve as a point of attack for organisations in their attempts to steer them in the desired direction [1] by means of tailor-made methodologies. Our findings could help organisations manage the selection, development, and implementation of new methodologies. We would like to propose that future research study the determinants of the constructs identified in this study as well as the interrelationships between them. For example, we still know very little about how an organisation perceives the needs of their employees, since misinterpretation might lead to misleading conclusions. Another very promising focus area is how culture influences the importance assigned by individuals to the specific attributes. Although the understanding of cultural influences is repeatedly emphasised by top journal editors (e.g., Straub [43]), this is seldom incorporated in research generally, because of the difficulty of data collection. If it is successful in collecting data that is sufficient for statistical analysis from a wide range of *different types* of cultures (categorised by Hofstede [24]), our study – as proposed – will further improve the generalisability of our findings as well as seek to reveal new avenues for future research. A better understanding of these determinants would enable us to design organisational interventions that would increase new methodology usage in order to improve productivity and quality as well as to reduce effort.

5 Future Research

The next steps in our research include developing a survey instrument to test the presented conceptual model. Regarding the operationalization of the proposed constructs, there might be a possibility that prior instruments may not be suitable to establish appropriate levels of construct validity in the context of our study; new scales might therefore need to be developed. In developing the initial set of items, we will follow the advice of Straub [44] and employ a rigorous step-by-step iterative process as well as utilise the existing literature (see Table 2 for an overview of the prior operationalisation of constructs). After obtaining the initial battery of items, two researchers will conduct expert interviews with six subject matter experts (three academics and three practitioners) to obtain specific information as to whether the initial items are comprehensible, valid, and complete [44]. To further improve content and construct validity, we will then conduct a Q-sorting and item ranking in two rounds. In the final step, we will subject the questionnaire to a pre-test based on a convenience sample with individuals representing the target population. The final survey instrument will be administered web-based to a diverse population of methodology users, to collect quantitative data needed for testing the model and hypotheses. To understand cultural influences, data will be collected from the USA, Germany, Austria, Switzerland, and India. We will seek to include more countries, especially developing and Asian nations such as Japan, China, and the African nations, as research based on Hofstede's cultural dimensions [24] shows that individuals from these nations, when compared to Western nations, are governed by different attitudes, preferences, and norms.

In conclusion, user acceptance of organisational methodologies remains a complex and elusive yet extremely important phenomenon. Past research has made some progress in unravelling some of its mysteries. The development and testing of our model seeks to advance theory and research on this fundamental matter.

References

1. Ajzen, I.: The theory of planned behavior. Organizational Behavior and Human Decision Processes 50(2), 179–211 (1991)
2. Alderfer, C.P.: Organizational Development. Annual Review of Psychology 28, 197 (1977)
3. Arnolds, C.A., et al.: Does higher remuneration equal higher job performance?: an empirical assessment of the need-progression proposition in selected need theories. South African Journal of Business Management 31(2), 53 (2000)
4. Bandura, A.: Social foundations of thought and action: a social cognitive theory. Social Foundations of Thought and Action: A Social Cognitive Theory (1986)
5. Bass, F.M.: A New Product Growth for Model Consumer Durables. Management Science 15(5), 215–227 (1969)
6. Belk, R.W.: Materialism: Trait Aspects of Living in the Material World. Journal of Consumer Research 12(3), 265–280 (1985)
7. Burnkrant, R.E., et al.: Informational and Normative Social Influence on Buyer Behavior. Journal of Consumer Research 2(3), 206–215 (1975)
8. Cacioppo, J.T., et al.: Dispositional differences in cognitive motivation: The life and times of individuals varying in need for cognition. Psychological Bulletin 119, 197–253 (1996)
9. Chigona, W., et al.: Using Diffusion of Innovations Framework to Explain Communal Computing Facilities Adoption Among the Urban Poor. Information Technologies & International Development 4(3), 57–73 (2008)
10. Chin, W.W., et al.: A Partial Least Squares Latent Variable Modeling Approach for Measuring Interaction Effects: Results from a Monte Carlo Simulation Study and an Electronic-Mail Emotion/Adoption Study. Information Systems Research 14(2), 189–217 (2003)
11. Dabholkar, P.A., et al.: An Attitudinal Model of Technology-Based Self-Service: Moderating Effects of Consumer Traits and Situational Factors. Journal of the Academy of Marketing Science 30(3), 184–201 (2002)
12. Davis, F.D.: Perceived Usefulness, Perceived Ease of Use, and User Acceptance of Information Technology. MIS Quarterly 13(3), 319–340 (1989)
13. Deci, E.L.: The Darker and Brighter Sides of Human Existence: Basic Psychological Needs as a Unifying Concept. Psychological Inquiry 11(4), 319–338 (2000)
14. Edwards, A.L.: Edwards personal preference schedule. Psychological Corporation, New York (1959)
15. Eva, M., et al.: Committed to a Radical approach? A survey of systems development methods in practice. In: Proceedings of the Fourth Conference of the British Computer Society Information Systems Methodologies Specialist Group, pp. 87–96 (1996)

16. Ford, G.T., et al.: Consumer Skepticism of Advertising Claims: Testing Hypotheses from Economics of Information. Journal of Consumer Research 16(4), 433–441 (1990)
17. Frs, R.H., et al.: A Validity Study of Scales to Measure Need Achievement, Need Affiliation, Impulsiveness, and Intellectuality. Educational and Psychological Measurement 32(1), 147–154 (1972)
18. Glass, R.L.: A Snapshot of Systems Development Practice. IEEE Softw. 16(3), 111–112 (1999)
19. Greenhalgh, T., et al.: Diffusion of Innovations in Service Organizations: Systematic Review and Recommendations. Milbank Quarterly 82(4), 581–629 (2004)
20. Grill, R., et al.: Evaluating the Message: The Relationship Between Compliance Rate and the Subject of a Practice Guideline (1992)
21. Hardgrave, B., et al.: Toward an information systems development acceptance model: the case of object-oriented systems development. IEEE Transactions on Engineering Management 50(3), 322–336 (2003)
22. Hardgrave, B.C., et al.: Investigating Determinants of Software Developers' Intentions to Follow Methodologies. Journal of Management Information Systems 20(1), 123–151 (2003)
23. Henseler, J., et al.: Testing Moderating Effects in PLS Path Models: An Illustration of Available Procedures. Handbook of Partial Least Squares, pp. 713-735 (2010)
24. Hofstede, D.G.: Culture's Consequences: Comparing Values, Behaviors, Institutions and Organizations Across Nations. Sage Publications, Inc., Newbury Park (2003)
25. Kanter, R.M.: Change Masters. Free Press, New York (1985)
26. Khalifa, M., et al.: Drivers for Software Development Method Usage. IEEE Transactions on Engineering Management 47(3), 360 (2000)
27. Maslow, A.H.: A Theory of Human Motivation. Psychological Review 50(4), 370–396 (1943)
28. Maslow, A.H.: Motivation and Personality. Harper & Brothers (1954)
29. Moore, G., et al.: Development of an Instrument to Measure the Perceptions of Adopting an Information Technology Innovation. Information Systems Research 2, 3, 222, 192 (1991)
30. Moore, G.A.: Crossing the Chasm: Marketing and Selling High-Tech Products to Mainstream Customers. HarperBusiness (1999)
31. Munns, A., et al.: The role of project management in achieving project success. International Journal of Project Management 14(2), 81–87 (1996)
32. Murray, H.A.: Explorations in Personality. John Wiley & Sons Inc., Chichester (1938)
33. Pajares, F.: Current directions in self-efficacy research. Advances in Motivation and Achievement 10, 1–49 (1997)
34. Pietrefesa, A.S., et al.: Moving Beyond an Exclusive Focus on Harm Avoidance in Obsessive Compulsive Disorder: Considering the Role of Incompleteness. Behavior Therapy 39(3), 224–231 (2008)
35. Plsek, P.E.: Complexity and the adoption of innovation in health care: For Accelerating Quality Improvement in Health Care Strategies to Speed the Diffusion of Evidence-Based. held in Washington, D.C. January 27-28, 2003. National Committee for Quality Health Care (2003)
36. Reiss, S.: Multifaceted Nature of Intrinsic Motivation: The Theory of 16 Basic Desires. Review of General Psychology 8(3), 179–193 (2004)
37. Riemenschneider, C.K., et al.: Explaining Software Developer Acceptance of Methodologies: A Comparison of Five Theoretical Models. IEEE Transactions on Software Engineering 28(12), 1135–1145 (2002)
38. Rogers, E.M.: Diffusion of Innovations, 5th edn. Free Press, New York (2003)
39. Rogers, E.M., et al.: Communication of Innovations: A Cross-Cultural Approach. Free Press, New York (1971)

40. Rossi, M., et al.: Managing Evolutionary Method Engineering by Method Rationale. Journal of the Association for Information Systems 5(9), 356–391 (2004)
41. Russo, N.L., et al.: The Failure of Methodologies to Meet the Needs of Current Development Environments. In: Proceedings of the British Computer Society's Annual Conference on Information System Methodologies, pp. 387–393 (1996)
42. Ryan, R.M., et al.: Self-Determination Theory and the Facilitation of Intrinsic Motivation, Social Development, and Well-Being. American Psychologist 55(1), 68 (2000)
43. Straub, D.W.: Creating Blue Oceans of Thought Via Highly Citable Articles. MIS Quarterly 33(4), iii-vii (2009)
44. Straub, D.W.: Validating Instruments in MIS Research. MIS Quarterly 13(2), 147–169 (1989)
45. Sun, R.: Motivational Representations within a Computational Cognitive Architecture. Cognitive Computation 1(1), 91–103 (2009)
46. Teodoro, M.P.: Bureaucratic Job Mobility and The Diffusion of Innovations. American Journal of Political Science 53(1), 175–189 (2009)
47. Tornatzky, L.G., et al.: Innovation Characteristics and Innovation Adoption-Implementation: A Meta-Analysis of Findings. IEEE Transactions on Engineering Management 29, 28–45 (1982)
48. Venkatesh, V.: Determinants of Perceived Ease of Use: Integrating Control, Intrinsic Motivation, and Emotion into the Technology Acceptance Model. Information Systems Research 11(4), 342 (2000)
49. Venkatesh, V., et al.: A Theoretical Extension of the Technology Acceptance Model: Four Longitudinal Field Studies. Management Science 46(2), 186 (2000)
50. Venkatesh, V., et al.: User Acceptance of Information Technology: Toward a Unified View. MIS Quarterly 27(3), 425–478 (2003)
51. Vickers, M.H.: Information technology development methodologies. Journal of Management Development 18(3), 255 (1999)
52. Wahba, M.A., et al.: Maslow Reconsidered: A Review of Research on the Need Hierarchy Theory. Organizational Behavior & Human Performance 15(2), 212–240 (1976)
53. Warner, K.E.: The Need for Some Innovative Concepts of Innovation: An Examination of Research on the Diffusion of Innovations. Policy Sciences 5(4), 433–451 (1974)
54. Wilson, R.S., et al.: Harm Avoidance and Disability in Old Age. Experimental Aging Research: An International Journal Devoted to the Scientific Study of the Aging Process 32(3), 243 (2006)

Exploratory Case Study Research on SOA Investment Decision Processes in Austria

Lukas Auer[1], Eugene Belov[2], Natalia Kryvinska[1], and Christine Strauss[1]

[1] Department of Business Administration, University of Vienna,
Bruenner Strasse 72, A-1210 Vienna, Austria
{lukas.auer,natalia.kryvinska,christine.strauss}@univie.ac.at
[2] FH Wien University of Applied Science of WKW,
Waehringer Guertel 97, A-1180 Vienna, Austria
eugene.belov@gmail.com

Abstract. Aligning information systems to financial key performance indicators and measure its returns has become one of the most important topics over the last years. However, despite the growing investments of many corporations into IT in general, an increasing number of questions and concerns have been arising to the effectiveness of these investments and their payoffs. Hence, an evaluation of an enterprise's IT architecture, as a part of IT investments, has been increasing its role throughout organizations worldwide. Numerous conceptual studies and tools to predict the business value of IT/SOA investment portfolios are being offered; nevertheless most of them substantially lack accuracy. For this purpose, we empirically investigate the application of SOA investment criteria in large Austrian corporations, which will be realized through a multiple case study collection reflecting current investment strategies and measurements.

Keywords: Business Value, IT, Empirical Study, Service-Oriented Architecture (SOA).

1 Introduction

The business value of information technologies in general and the adoption of different enterprise IT architecture paradigms in particular have been the cause of discussion for both academics and practitioners for ages [1, 2]. A particular challenge IT departments have been facing during the last several years is the question of how to measure the value of service-oriented architectures (SOA). The enterprise architectural model called SOA can be defined as a computing paradigm that utilizes services as the basic constructs to support the development of rapid, low-cost and easy composition of distributed applications even in heterogeneous environments [3]. However, IT infrastructure maintenance requires substantial financial resources; hence benefits should outweigh the costs and suffice expected returns. It is often argued that enterprises which made substantial investments in IT projects are often dissatisfied with their return on IT investments [4], as value assessments are frequently conducted inefficiently, using improper or not comprehensive methodologies, or are completely abandoned [5]. The challenge of whether or not to pursue SOA as a key element of a company's IT strategy was

H. Mouratidis and C. Rolland (Eds.): CAiSE 2011, LNCS 6741, pp. 329–336, 2011.
© Springer-Verlag Berlin Heidelberg 2011

primarily based upon the lack of clarity regarding potential benefits of this type of architecture. Various IT success models [6], SOA business value frameworks [2, 7] and SOA value drivers have been identified [8]. Yet a thorough literature review displayed that most empiric publications focus on either enhancing SOA and Web service concepts or exploring their adoption in practice [9]. It has been argued that an investment into SOA often cannot be properly evaluated because value potentials of this investment have not been made visible enough [10], in addition the consideration of risk is virtually absent in research on the returns on IT investment, even though the risks are widely recognized [11].

2 Research Design

This paper aims to empirically investigate SOA investment-decision making practices in large Austrian companies from a finance perspective, i.e. the underlying investment criteria rather than exploring organizational SOA effects [12], business process implications [13] or setting up a business case including detailed benefits and costs [14]. We also intend to examine the consideration of lag effects, i.e. the timing of IT investment decisions[1], which have been neglected in past literature reviews [15]. A total of six public and private companies (semi-structured interviews with 2 interviewees per case) have been analyzed and coded by 2 different interviewers from January to March 2010. After independent within-case and cross-case data analysis using an inductive approach [16], common patterns and experiences among the selected companies have been identified. Derived from the considerations explained so far, the research questions for this case study can be formulated as follows:

1. What is the strategy and expected business impact behind SOA initiatives?
2. What investment methodologies and investment criteria (qualitative and quantitative) are being applied by leading Austrian companies to evaluate initial and continuous SOA investments?

Based on a thorough literature review and the nature of the research questions, case study research for has been selected as the appropriate form of methodology [17], which needs to instantiate well-defined models before the data collection begins [18] and data collection methods used for this research had to follow a well-structured conceptual framework for this IT business value research [19], which is depicted in Figure 1. The categories covered include SOA strategy, investment evaluation, which

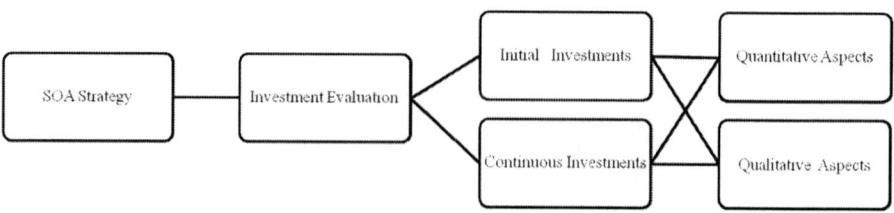

Fig. 1. Case Study Framework

---

[1] For an overview of the numerous models and valuation methods we refer to Frisk (2007) [20].

is split into initial investments and continuous investments to investigate on potential consideration of timing or lag effects. Examination on investment evaluation also reflects the distinction between quantitative and qualitative aspects influencing an IT manager's investment decision.

3 Case Analysis

Case 1 - Banking Group

The company is an Austrian banking group with a total of more than 2000 branches in Austria and the bank's strategy is focused on retail customers.

The touch point for SOA was the idea to integrate separate services from different technologies into a singular platform. Over the course of development the focus has shifted to business process management. Back then, cost savings were not an issue for the group because of its good financial performance, however, due to the financial crisis, cost pressure has become more severe and the need to reduce costs has suddenly emerged. Key business processes have been identified and categorized in their respective fields based on their automation potential, eventually it is planned to cover all the key processes in the bank with SOA, despite the fact that it was stated by the CIO that services reuse was very hard to achieve with SOA due to the high degree of services customization. Initial investments into the new SOA framework were primarily based on qualitative aspects and didn't require any quantitative measurements, whereas continuous investments needed a 5-year ROI calculation. Continuous investments were primarily based on automation potential, e.g. process execution time is measured "by standing behind the account manager with a stop-watch", Qualitative aspects are not measured for continuous investments; moreover no calculations for potential future benefits of the investment are required. Based on the interview, the company is satisfied with its measurement techniques, and the measurements they deliver are "rather precise".

Case 2 - Telecommunications Group

The company's core business includes all aspects of mobile communication ranging from telephony to data transmission. It holds a strong market position in the Austrian market and is present in 8 countries in Europe. Altogether, the group serves approximately 20 million customers and employs approximately 20,000 employees. First SOA attempts were carried out in 2004 to eliminate excessive utilization of resources, increase transparency and reduce time-to-market, which is crucial to success in the dynamic and fast-paced telecom sector. A major benefit which has been realized with the first SOA implementations was the enablement of enterprise-wide communication in terms of a bridging between IT and other departments. There is no overall strategy behind the company's SOA initiatives to transform the whole architecture service-oriented; it is primarily used as a connecting layer between different silos with standard software, e.g. ERP or CRM systems. In terms of initial SOA investment evaluation mostly qualitative aspects on the benefits side of the equation have been used. Investment decisions were solely driven by the IT department; therefore no business case had to be calculated. All continuous IT projects are usually initiated by a change request coming from the business side, which requires a static business case containing a 5-year total cost of ownership calculation. In cases like automating processes to lower the load of the call center, the amount of minutes saved by the investment, and therefore ROI, is relatively easy to calculate, however, it has been mentioned that is far more difficult to calculate revenue increases due the number of people who would for instance "switch their tariff plan based on the implementation of certain new services".

Case 3 – Federal Ministry of the Interior

The department within the ministry is responsible for managing Austria's central residence registry and the foundation of Austrian's e-government services. The company only uses open-source software; also the process engine is based on an open source workflow system. The SOA implementation started in 2001 with XML and there were no pilot projects in the beginning on its way to SOA. The main reasoning behind the SOA implementation was to become faster, reduce errors, enable flexibility and improve change management. As a result, the company managed to hold its costs at a constant level, while traffic and number of users and transactions more than tripled. It has been stated by the company that one of the major advantages gained by SOA is the reuse of separate SOA blocks; accordingly all the processes of the company are currently supported by SOA. The decision to pursue a SOA strategy was primarily taken on qualitative aspects as no calculations have been made prior to the investment. Ongoing projects are evaluated with an static ROI calculation and investment costs are simply compared to the costs saved. The company is satisfied with its investments in SOA and during the past 4 years, the company managed to run at the same budget, although the amount of traffic, users and transactions has more than tripled.

Case 4 – Publishing Company

The company is the leading publisher of newspapers and magazines in Austria. It publishes 3 daily newspapers, 6 weekly magazines and employs more than 2000 employees in Austria. The company operates in a very heterogeneous IT environment, therefore key challenges include the ability to build processes quicker, cost-efficiency and the possibility to implement open-source technology more easily. In contrast to the Telco case, the publishing business does not require very short time-to-market, nevertheless SOA investment decisions have also taken by the IT department in order to be better prepared for possible future developments.

Based on its strategic nature, initial SOA investments have not required any ROI calculation at all. Main aspects proposed by the IT department to evaluate potential investments have included cost and time savings in development, flexibility due to the implementation of open-source technologies as well as support of future business-to-business processes. In terms of continuous investments a 5-year ROI calculation is required. The IT department is responsible for calculating the costs, while business units are responsible for researching the benefits, either in cost savings or in revenue increase dimensions and subsequently both parts are compared over a 5-year period. No business project is approved if it doesn't break-even within the next 5 years, though cost savings had been identified very quickly during the first process automation projects, e.g. to reduce the workload of the call-center or implementation of an electronic billing system as the amount of minutes and postage expenses saved were easy to calculate and quick-wins. The company has recently started its company-wide SOA initiative and is satisfied with the results so far.

Case 5 - Data Service Provider

The company is the IT service provider for the Austrian public sector. Its two major customers include the Ministry of Finance and Ministry of Justice, whose IT infrastructure and architectural development are at very different stages.

The main drivers behind SOA initiatives were flexibility, quicker time to market, cost efficiency and synergy potential through services reuse. In the Ministry of Finance SOA is currently being implemented in the area of customs and taxes. The initial project started in 2007 and it was originally planned to transform the entire applications landscape service-oriented. In the Ministry of Justice SOA has been approached incrementally, focusing on optimization potential in the area of process development, currently services specification and development accounts for about 15% of the annual IT budget. For any initial SOA investments, similar to projects which are legally required to be implemented, no ROI calculations have to be completed. The main drivers that facilitated the decision to pursue SOA were qualitative aspects such as flexibility, transparency and faster time-to-market. Every continuous investment is evaluated based on several key criteria, e.g. IT strategy fit, investment costs, cost saved in operations and development. Current projects such as the electronic document exchange have been easy to justify as the calculation of time to be saved by eliminating manual process steps.

Case 6 - Insurance Group

The company is an international insurance group with its headquarters in Vienna. It is present in more than 20 countries in Europe and together with all its subsidiaries abroad, the company employs more than 20,000 employees and serves more than 7 million customers. The company is one of the SOA pioneers in Austria and has one of the most SOA-intensive architectures in the country. SOA currently accounts for approximately 1/3 of the total architecture of the group utilizing mostly web services, while large part of the group's architecture is comprised of old legacy systems. The main reasons behind the SOA implementation were expected efficiency increases, IT centralization as well as cost savings. After first steps in the late nineties, the scope of the first big SOA initiative included a business process management system, which was originally implemented as a pilot project by the company's own development team. However, the group decided to shift from in-house development to investing in ready-made solutions instead. The first large investment into SOA was not supported by any ROI calculation due to its strategic importance; still all continuous investments in the company are usually evaluated based on qualitative and quantitative aspects. Qualitative aspects include the alignment of IT and business processes overall efficiency improvements. Quantitative aspects consist of a 3-year Total Cost of Ownership calculation.

4 Research Results Analysis

After evaluating 6 cases of SOA implementations in large Austrian companies in various industries a distinct pattern of investment criteria has been examined which is depicted in Table 1. In almost all cases, initial investments into SOA were supported by primarily qualitative aspects. These included transparency, efficiency and flexibility, services reuse, faster time-to-market, strategic reasons or focus on key business processes. Some of the companies had conducted pilot projects prior to their investment decision, while others went straight to implementation. In none of the cases stakeholders were fully aware of all the costs and benefits of SOA implementations with regards specifically to their company. However, it must be stated, that all of the interview partners were satisfied with their investments into SOA. In spite of the numerous existing IT valuation methods, advanced techniques have hardly been used and managers interviewed in this study predominantly used methods they intuitively understood [21]. Some of the reasons put forward for the failure to monitor benefits of their investments are [22]:

- Assessment of benefits was not required due to business strategy guidelines
- Difficulty to asses benefits after project implementation

In contrast to initial SOA investments, the evaluation focus of continuous projects in SOA environments observed shift to more quantitative aspects, with ROI in various forms serving as primary evaluation metric, while qualitative aspects have been playing a secondary role at this stage. Interestingly, some interview partners have given contradictory statements about certain benefits of SOA, such as reuse. It became also transparent that still IT employees are responsible for decisions related to SOA budgeting and spending, which is highly questionable [23]. The main reasons for the inconsistency of SOA investment evaluation included the choice of incomprehensive evaluation methodologies, inadequate assessment of inputs and outputs and the complexity behind establishing the total value of SOA [24], therefore the most common and accepted measure to justify IT investments based on this study has been a static Return on Investment (ROI) calculation. Nevertheless, it is uncertain if such a financial measure and other metrics can reflect intangible benefits, long-term strategic advantages and also the risk that enfold SOA-related investments.

Table 1. Cross-Case Analysis

Investment Criteria	Case	Banking	Teleco	Ministry	Media	Data Service Provider	Insurance
	SOA Pilot				✓		✓
Qualitative Aspects	Transparency		✓				
	Efficiency	✓		✓		✓	✓
	Flexibility	✓	✓			✓	
	Re-use			✓	✓	✓	
	Time		✓			✓	
	Strategy		✓		✓		
	Process Focus				✓	✓	
Quantitative Aspects	Costs	✓	✓	✓	✓	✓	✓
	Revenue Increase		✓				
	ROI						✓

5 Conclusions and Further Research

It has been shown in this paper that it is still unclear how appropriate financial evaluation methods used in practice are to fully capture proposed SOA benefits and that a gap between science and practice is still undeniable. Based on the results presented we will concentrate our future research on (1) financial performance measures capable of evaluating all aspects that arise from SOA-related investments and (2) how to estimate payoffs from previous or future SOA investments. We are currently working on a real-options approach allowing to measure for risk and staged investments in a SOA context, which has been ignored in the past [11] and recommend to further investigate on this topic.

References

[1] Thomas, O., vom Brocke, J.: A value-driven approach to the design of service-oriented information systems - making use of conceptual models. Information Systems and e-Business Management (ISeB) 8(1), 67–97 (2009)

[2] Stewart, W., Coulson, S., Wilson, R.: Information Technology: When is it Worth the Investment? Communications of the IIMA 7(3), 119–122 (2007)

[3] Papazoglou, M.P., Traverso, P., Dustdar, S., Leymann, F.: Service-Oriented Computing: a Research Roadmap. Int. J. Cooperative Inf. Syst. 17(2), 223–255 (2008)

[4] Willcocks, L.: Evaluating Information Technology investments: research findings and reappraisal. Information Systems Journal 2, 243–268 (1992)

[5] Grembergen, W.: Information Technology Evaluation Methods and Management. John Wiley & Sons, Inc., New York (2001)

[6] DeLone, W.H., McLean, E.R.: The DeLone and McLean Model of Information Systems Success – A Ten-Year Update. Journal of Management Information Systems 19(4), 9–30 (2003)

[7] Lagerstrom, R., Ohrstrom, J.: A Framework for Assessing Business Value of Service Oriented Architectures. In: Services Computing, IEEE International Conference on Services Computing, pp. 670–671 (2007)

[8] Beimborn, D., Joachim, N., Weitzel, T.: Drivers and Inhibitors of SOA Business Value: Conceptualizing a Research Model. In: AMCIS 2008 Proceedings, Toronto (2008)

[9] Viering, G., Legner, C., Ahlemann, F.: The (Lacking) Business Perspective on SOA - Critical Themes in SOA Research. Wirtschaftsinformatik 1, 45–54 (2009)

[10] Becker, A., Buxmann, P., Widjaja, T.: Value Potential and Challenges of Service-Oriented Architectures - A User and Vendor Perspective. In: Proceedings of the 17th European Conference on Information Systems, Verona (2009)

[11] Dewan, S., Shi, C., Gurbaxani, V.: Investigating the risk-return relationship of information technology investment: firm-level empirical analysis. Management Science 53(12), 1829–1842 (2007)

[12] Yoon, T., Carter, P.: Investigating the Antecedents and Benefits of SOA Implementation: A Multi-Case Study Approach. In: AMCIS Proceedings (2007)

[13] Beimborn, D., Joachim, N., Münstermann, B.: Impact of Service-oriented Architectures (SOA) on Business Process Standardization - Proposing a Research Model. In: Proceedings of the 17th European Conference on Information Systems (ECIS), Verona (2009)

[14] Starke, G., Tilkov, S.: SOA Expertenwissen – Methoden, Konzepte und Praxis serviceorientierter Architekturen (2007)

[15] Schryen, G.: Preserving knowledge on IS business value: what literature reviews have done. In: Business & Information Systems Engineering (BISE), vol. 52(4), pp. 225–237 (2010)

[16] Glaser, B., Strauss, A.: The discovery of grounded theory. de Gruyter, New York (1967)

[17] Yin, R.: Case Study Research: Design & Methods, Thousand Oaks (2007)

[18] Kauffman, R.J., Weill, P.: An evaluative framework for research on the performance effects of information technology investment. In: Proc. 10th International Conference on Information Systems, Boston (1989)

[19] Miles, M.B., Huberman, A.M.: Qualitative data analysis: an expanded sourcebook, 2nd edn. Sage Publications, Thousand Oaks (2005)

[20] Frisk, E.: Categorization and overview of IT perspectives – A literature review. In: Proc. of the European Conference on Information Management and Evaluation (2007)

[21] Nijland, M.: Understanding the Use of IT Evaluation Methods in Organizations, London School of Economics, PhD Dissertation (2004)

[22] Lin, C., Pervan, G.: The practice of IS/IT benefits management in large Australian organizations. Inf. Manage. 41(1), 13–24 (2003)

[23] Ross, J., Weill, P.: Six IT decisions your IT people shouldn't make. Harvard Business Review 80(11), 5–11 (2002)

[24] Brynjolfsson, E.: The productivity paradox of information technology. Commun. ACM 36(12), 66–77 (1993)

A Metamodelling Approach for *i** Model Translations[*]

Carlos Cares[1,2] and Xavier Franch[1]

[1] Universitat Politècnica de Catalunya (UPC)
c/Jordi Girona, 1-3, E-08034 Barcelona, Spain
{ccares,franch}@essi.upc.edu
[2] Universidad de la Frontera (UFRO)
Fco. Salazar 01145 Temuco, Chile
carlos.cares@ceisufro.cl

Abstract. The *i** (i-star) framework has been widely adopted by the information systems community. Since the time it was proposed, different variations have arisen. Some of them just propose slight changes in the language definition, whilst others introduce constructs for particular usages. This flexibility is one of the reasons that makes *i** attractive, but it has as counterpart the impossibility of automatically porting *i** models from one context of use to another. This lack of interoperability makes difficult to build a repository of models, to adopt directly techniques defined for one variation, or to use *i** tools in a feature-oriented instead of a variant-oriented way. In this paper, we explore in more detail the interoperability problem from a metamodel perspective. We analyse the state of the art concerning variations of the *i** language, from these variations and following a proposal from Wachsmuth, we define a supermetamodel hosting identified variations, general enough so as to embrace others yet to exist. We present a translation algorithm oriented to semantic preservation and we use the XML-based iStarML interchange format to illustrate the interconnection of two tools.

Keywords: *i**, i-star, interoperability, semantic preservation, iStarML.

1 Introduction

The *i** (pronounced *i-star*) framework [1] is currently one of the most widespread goal- and agent-oriented modelling and reasoning frameworks. It has been applied for modelling organizations, business processes and system requirements, among others.

Throughout the years, different research groups have proposed variations to the modelling language proposed in the *i** framework (for the sake of brevity, we will name it "the *i** language"). There are basically two reasons behind this fact:

- The definition of the *i** language is loose in some parts, and some groups have opted by different solutions or proposed slight changes to the original definition. The absence of a universally agreed metamodel has accentuated this effect [2].

---

[*] This work has been partially supported by the Spanish project TIN2010-19130-c02-01

H. Mouratidis and C. Rolland (Eds.): CAiSE 2011, LNCS 6741, pp. 337–351, 2011.
© Springer-Verlag Berlin Heidelberg 2011

- Some groups have used the *i** framework with very different purposes thus different concepts have become necessary, from intentional ones like trust, delegation and compliance, to other more related with the modelling of things, like service or aspect (see [3] for an updated summary).

The adaptability of *i** to these different needs is part of its own nature, therefore these variations are not to be considered pernicious, on the contrary, flexibility may be considered one of the framework's key success features. However, there are some obvious implications that are not so desirable:

- It makes difficult to build a repository of *i** models shared and directly used by the whole community.
- It also hampers the possibility of interconnecting different *i** tools that are not compliant to the same *i** language variation.
- Finally, it makes techniques defined for one *i** variation not directly applicable into another variation.

The work presented here addresses these problems and specifically tries to answer the following research questions:

- What types of *i** variations are proposed and how can they be characterized?
- Which is an appropriate semantic framework for analysing *i** interoperability?
- Given two *i** variations *A* and *B*, to what extent is it possible to translate models built with *A* to *B* and the other way round according to this semantic framework?
- Given two *i** variations *A* and *B*, how can a model from *A* be translated to *B*, in the light of the limitations identified in the previous question?

The rest of the paper is structured as follows. Section 2 provides the background about *i** variations. Section 3 presents the metamodel framework for translation remarking why the concepts of supermetamodel and semantic-preservation can be used for dealing with interoperability among *i** variants. Section 4 proposes the supermetamodel for *i** variants, and Section 5 presents a translation algorithm to maximize semantic-preservation illustrated with an example of model interchange between two *i** tools. Finally, Section 6 states the conclusions and future work.

Basic knowledge of *i** is assumed, see [1] and the *i** wiki [4] for details.

2   The *i** Framework: Evolution and Existent Variations

The *i** framework was issued in the mid-nineties and the first full definition was included in the PhD thesis by Eric Yu [1]. Some of its concepts were previously proposed and used in KAOS [5] and in the NFR Framework [6]. This original work on *i** has been the most cited in the community. Recently, an updated version has been included as part of the *i** wiki [4], with minor differences with respect to the seminal one (e.g., richer types of contribution links).

From this major trunk, we may consider two main variations. On one hand, the Goal-oriented Requirement Language (GRL) which is part of the User Requirements Notation (URN) [7]. On the other hand, Tropos [8], an agent-oriented software methodology that adopts *i** as its modelling language. In both cases, the differences with respect to the seminal Yu's *i** are not that relevant to consider them as different

notations, but due to its adoption by the community we consider them as major variations. Thus, we may say that *i** has three main dialects: the seminal *i** currently represented by the wiki definition, GRL, and the language adopted by Tropos.

On top of these three main dialects, we may find many proposals for particular
purposes. Some of them are bound to a particular domain, e.g., security as in Secure-
Tropos [9], or norm compliance as in Nòmos [10]. Others propose very particular
concepts for a particular purpose, like the concept of module or constraint. Finally,
some others propose more fundamental variations affecting the way of modelling, as
the concepts of service, variation point or aspect.

Table 1 presents a comparative analysis of the proposals issued by the community
in the last 5 years. We have carried out a review in the following conferences and
journals for the period 2006-2010: CAiSE, REJ, DKE, IS Journal, RE, ER, RiGiM,
WER, *i** workshop, and also including the recent book on *i** [3]. Our goal has not
been carrying out a systematic review but to get a representative sample of the community proposals in this period as a way to know what the major trends concerning
language variability are. In total, we have found 146 papers about *i** in these sources
(without including papers talking about goal-modelling, since we are interested in
language-specific issues). From them, we have discarded 83 which are not really
relevant to our goals (i.e., papers not directly related with the constructs offered by the
language). For the remaining 63, the table shows how many of them propose addition,
removal or modification of concepts classified into six different types. It must be
taken into account that a single paper may propose more than one construct variation
and that similar changes are proposed or assumed in different papers. Also it is necessary to remark that most papers just focus on some specific part of the language, in
that case we assume that the other part remains unchanged.

Table 1. Variations proposed by the *i** community in the last 5 years (selected venues only).
Each paper increments at most in 1 each column.

	Actors	Actor links	Dependencies	Intentional elements (IE)	IE links	Diagrams
New	4	24	10	21	21	19
Removed	8	5	2	1	0	0
Changed	3	1	1	36	43	0

An analysis of this table follows:

- On actors. The most usual variation is getting rid of the distinction on types of
  actors, like remarkably GRL[1] does. Some special type (e.g., "team") may appear.
- On actor links. Most of the variants include *is_part_of* and *is_a* but some get rid
  of one (e.g., GRL just keeps *is_part_of*) or even both. Of course, having just a generic type of actor means not having the links bound to specific types like *plays*.
  Finally, some proposals use new actor links, like in Nòmos: A *embodies* B means
  the domain actor A has to be considered as the legal subject B in a law.
- On intentional elements. Although all virtually all variants keep the four standard
  types (*goal*, *softgoal*, *task* and *resource*), we may find a lot of proposals of new

---

[1] In the rest of the paper, we refer to the GRL implementation supported by the jUCMNav tool,
http://jucmnav.softwareengineering.ca/ucm/bin/view/ProjetSEG/.

intentional elements. To name a few, GRL adds *beliefs*, Nòmos adds *norms*, and even *aspects* appear as dependums. There are not many modification proposals, e.g., resources may be classified as physical or informational with consequences for class diagram generation in an MDD process.

– On intentional element links. Most of the variants keep the general idea of the three link types (*means-end*, *task decompositions* and *softgoal contributions*), some of them merge two of them, e.g., GRL defines a link *decomposition* that merges *means-end* and *task-decomposition*. Then we have lots of variations about types of decompositions (e.g, Tropos allows both *AND* and *OR* means-end links), contribution values (labels such as +,- vs. *make, help*, etc.), correctness conditions (e.g., whether a resource may be a mean for a goal), etc. Finally, some modifications usually occur in the form of labels, e.g., quantitative labels for contributions in GRL, multiplicity in some Tropos-based variants, etc. A special type of modification is relaxing some conditions, e.g., allowing links among intentional elements that belong to different actors, or contributions to goals.

– On dependencies. About modifications, we may find the addition of attributes which qualify the type of dependency, e.g., Secure-Tropos adds *trust* and *ownership* qualifiers. Then, we have new types of relationships that may be interpreted as dependencies, like Nòmos' legal relations. Also, a quite usual variation is to get rid of dependencies' strength, probably due to the difficulty of interpreting the concept in a reasoning framework. The type of depender and dependee also presents constraints sometimes, e.g., GRL forces them to be intentional elements, actors are not allowed in this context.

– On diagrams. The distinction among SD and SR diagrams is not always kept, some proposals just have a single model in which the actors may be gradually refined. One type of diagram that was depicted in Yu's thesis but not recognised as such was actor diagram, and some authors have promoted this third type of diagram as such. On addition, several proposals of types of diagrams exist, from the generic concept of module to specific proposals like interaction channel.

The result of this study shows the complexity of the model transformation problem. In fact, one may easily anticipate that it will be impossible to get an automatic transformation technique for any pair of existing proposals. It becomes necessary to investigate the limits of model transformation in *i** and provide a general customizable framework.

3   A Metamodel View of *i** Model Translation

Metamodels have been the traditional tool in Software Engineering to express valid models of a certain modelling language [11]. The language used to specify a metamodel is called metalanguage. Note that metamodels represent only what can be expressed in valid models but not what these expressions mean, i.e., a metamodel specifies the syntax of a modelling language but not its semantics.

In the case of *i** transformations, the different *i** variants mentioned in Section 2 correspond to their own metamodels which are expressed using different metalanguages (e.g., UML, EBNF, Telos). The problem of transforming a source model into a target model can be viewed as a particular case of applying general rules to transform the differences between the corresponding metamodels.

3.1 Wachsmuth's Proposal on Metamodel Adaptation

In 2007, Guido Wachsmuth presented a proposal [12] to deal with the problem of metamodel evolution and its implications for adapting its instance models according to this evolution (see Fig. 1, left). The basic hypothesis is that co-adaptation of models can be automatically derived from well-defined metamodel evolution steps. Wachsmuth defines different semantic-preserving categories and matches them with specific refactoring operations on metamodels. The way of handling semantic-preserving features respond to the concept of semantics already introduced, i.e., semantic preservation is not characterized by meaning but by structural changes on corresponding metamodels.

Here, we are proposing the adoption of this framework in the context of the problem of translating models among metamodels which have a common set of concepts (see Fig. 1, right). In other words, we change the perspective:

- **from:** given a model m_A created as instance of a metamodel M_A, translate it into another model m_B created as instance of a metamodel M_B via a metamodel correspondence,
- **into:** given a model m_A created as instance of a metamodel M_A, and given a metamodel evolution from M_A to M_B, co-adapt the model m_A into another model m_B created as instance of M_B via some metamodel refactoring operations.

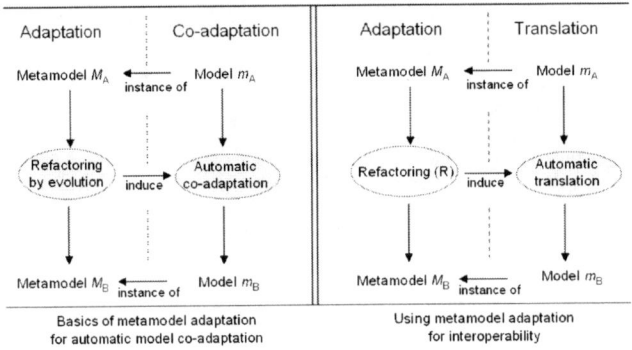

Fig. 1. Comparative between co-adaptation and interoperability via metamodel refactoring

3.2 Wachsmuth's Proposal: Relationships and Semantic Preservation

To characterize refactoring operations Wachsmuth proposes some basic concepts:

- $\mathcal{M}_M$ represents all the metamodels conforming to a specific metamodel formalism M, denoted by $\mathcal{M}_M := \{\mu \models M\}$. Although it is not really relevant, we may assume MOF 2.0 formalism in this paper.
- $C_M(\mu)$ represents the concepts defined by a particular metamodel μ. In our case, typical concepts would be *actor*, *intentional element*, etc.

- $I(\mu)$ represents the set of all metamodel instances conforming to a metamodel μ, denoted by $I(\mu):= \{ \iota \models \mu \}$. In our case, we focus on those μ which are a meta-model of some i* variation (e.g., i*-wiki metamodel, GRL metamodel, Tropos metamodel) and then for each μ, $I(\mu)$ are i* models built as instances of μ.
- $I_C(\mu)$ represents the set of instances $I(\mu)$ of restricted the specific set of concepts C, i.e., $I_C(\mu) \subseteq I(\mu)$. For instance, we may refer to the set of concepts C which are part of SD models, and then $I_C(\mu)$ would represent SD models built according to the metamodel μ.

Using these concepts, 5 types of generic relationships between metamodels are defined (see 1st and 2nd columns of Fig. 2.) which yield to 5 degrees of semantic preservation. The transformation from one metamodel to another implies a relationship R between the source and target metamodels, thus, the type of semantic preservation of R (if any) will depend of which of these generic relationships is subset (see 3rd column of Fig. 2). Besides, the different types of semantic preservation imply different types of instance preservation (see 4th column of Fig. 2).

Relation	Definition	Semantics-preserving (s-p)	Instance-preserving (i-p)
Equivalence	$\mu_1 \equiv \mu_2$ iff $I(\mu_1) = I(\mu_2)$	$R \subseteq \equiv$ is strictly s-p	is strictly i-p
Variant modulo φ	$\mu_1 \equiv_\varphi \mu_2$ iff $\varphi: I(\mu_1) \to I(\mu_2)$ is a bijective function	$R \subseteq \equiv_\varphi$ is s-p modulo variation	is i-p modulo variation
Submetamodel / Supermetamodel	$\mu_1 \leq \mu_2$ iff $C(\mu_1) \subseteq C(\mu_2)$, $I(\mu_1) = I_{C(\mu1)}(\mu_2)$	$R \subseteq \leq$ is introducing s-p $R \subseteq \leq^{-1}$ is eliminating s-p	is strictly i-p is partially i-p modulo variation
Enrichment	$\mu_1 \sqsubseteq \mu_2$ iff $C(\mu_1) = C(\mu_2)$, $I(\mu_1) \sqsubseteq I(\mu_2)$	$R \subseteq \sqsubseteq$ is increasing s-p $R \subseteq \sqsubseteq^{-1}$ is decreasing s-p	is strictly i-p is partially i-p modulo variation
Extension	$\mu_1 \sqsubseteq_\varphi \mu_2$ iff $C(\mu_1) = C(\mu_2)$, $\varphi: I(\mu_1) \to I(\mu_2)$ is an injective function	$R \subseteq \sqsubseteq_\varphi$ is increasing modulo variation $R \subseteq \sqsubseteq_{\varphi^{-1}}$ is decreasing modulo variation	is i-p modulo variation is partially i-p modulo variation

Fig. 2. Summary of semantic preserving relationships in Wachsmuth's framework [12]

3.3 Wachsmuth's Proposal: A Framework for *i** Interoperability

As we have already said, most of the i* variants have their own metamodel which conforms a different modelling formalism. However, this diversity of formalisms seems to be just a representational problem. We have assessed this opinion in earlier works by proposing an i* reference metamodel and proposing a set of refactoring operations to allow obtaining the different variants [13].

But this preliminary result that we obtained, although valuable as a first step, exhibits an important drawback that the set of common concepts was intentionally kept to a minimum (i.e., we wanted to represent the universally agreed concepts). Whilst providing a good ontological basis, this decision was damaging the model interoperability goal that we are targeting here. Wachsmuth allows stating the reason why: model translation was suffering from eliminating or decreasing semantic preser-vation. In this

work, we search for the fundamental property of instance preservation: given an *i**
model that is instance of a metamodel M_A that represents a source variation then, when
applying the mapping from M_A to M_B (the metamodel that represents the target variation) the model can also be considered an instance of M_B.

Let's assume that a model, named the *i** supermetamodel, exists, therefore any existing metamodel of *i** variation is a submetamodel of the *i** supermetamodel. Then,
if we could model refactoring operations from the *i** supermetamodel to the particular
variants, then we would have a feasible translation from each variant to another. This
hypothetical scenario would exhibit three advantages: (i) supporting at some extent
interoperability between models belonging to different metamodels; (ii) given *k* *i**
variants, providing a framework that offers translation from one variant to each other
with linear complexity in terms of transformation functions (*k* functions) instead of
quadratic (k^2-*k* pair-wise functions); (iii) the type of semantic preservation would be
characterized with a clear specification of preservation (strict, modulo varia-tion,
increasing or decreasing). In Figure 3 we illustrate this hypothetical assumption.

Although it may appear hard to sustain that such an *i** supermetamodel exists (due
to the continuous proposals that modify it), in the next section we will discuss
the conditions under which its existence appears reasonable to sustain and a first i*
supermetamodel approach will be presented.

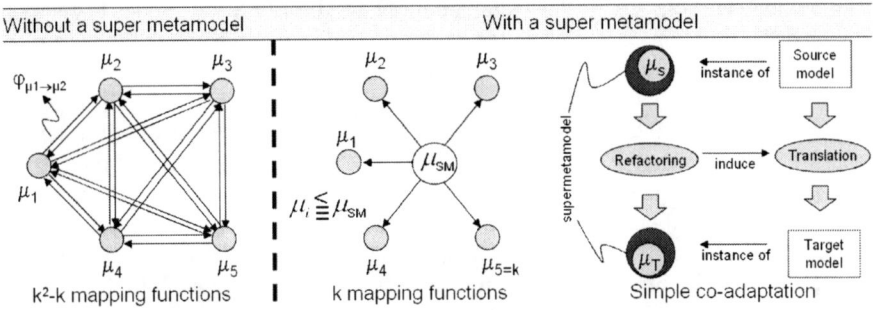

Fig. 3. Comparing absence and presence of an *i** supermetamodel for model translations

4   A Supermetamodel for *i**

From its definition, we can colloquially understand a supermetamodel as a metamodel
which contains the superset of language constructs existing on other metamodels. In
the case of the *i** framework, this means that if M is a supermetamodel for *i** then the
different values of softgoals contributions (some+, helps, makes, +, ++, - , --) should
be modelled in M. Besides, the same for intentional element types, actor types, etc.

Therefore, in the attempt of formulating a supermetamodel for the existing variants
and, ideally, upcoming ones, we need to answer two questions: (i) how to put under
the same metamodel a set of different language constructs coming from different *i**
variants?, (ii) how to make this supermetamodel stable enough in order to suffer
minimal modifications (if any) when a new *i** variant is proposed? To satisfactorily
answer both questions, the key concept is abstraction to allow putting different
concepts together. It is crucial to capture the right level of abstraction: if the

metamodel is too abstract (e.g., only differentiating nodes and links) it may fail in capturing the essence of *i**; if it is too detailed, the metamodel can result in a rigid structure which requires high effort to be refactorized. In the first case, an additional problem appears, because using a high abstraction level means adjusting basic syntax formations by means of textual (e.g., OCL) constraints, and textual constraints are not considered in the Wachsmuth's framework, therefore semantic preservation could not be qualified. Therefore, we are looking for a metamodel which allows representing different *i** variant structures and possible extensions whilst, at the same time, keeping the core *i** language constructions.

These two situations appear in the two most related works we may found in the literature. Amyot et al. have proposed a metamodel for GRL [14] that contains concepts such as metadata, links and groupings that enable the language to be extended and tailored, also using OCL constraints. So it may be classified as too abstract. In addition, it presents some peculiarities that forces its customization either in quite classical *i** contexts (e.g., types of actors are not defined, dependencies linking actors –not intentional elements– are not allowed; dependencies without dependums are allowed) or in non-classical contexts (e.g., types of boundaries are impossible to be set). On the other hand, the reference metamodel presented by Cares et al. [13] proposes the use of refactoring operations to map into other variants. However, there are specializations for representing specific *i** elements, therefore, adding a new language element would mean adding new classes to the metamodel. Thus, the reference model has a great value, but the problem comes when we want to use it in the context of model translation since it would imply alterations to classes.

The *i** supermetamodel proposal is based on the reference model but incorporates the concept of metadata appearing in the GRL metamodel. From the *i** reference model we obtain a more abstract metamodel using *i** related concepts and their extensions are handled with metainformation. We formalize this idea into UML stereotypes. The result is the metamodel that appears in Fig. 4. *Actor* and *IElement* are the central classes. Then *ActorLink* and *IElementLink* are recursive binary associations on them. *Boundary* is a binary association among *IElement* and *Actor* (note that an *IElement* may appear outside any boundary, e.g., dependums). Finally the concept of dependency is implemented with two associations: dependencies are divided into *DependencySegment* which is an easy way to allow different properties at each end, or even with just one end defined. Each *DependencySegment* connects an *Actor* (considered depender or dependee depending on the value of the *participatorType* attribute) and an *IElement* (the dependum) and then may (or not) be connected to a particular *IElement* that would be an internal element inside the corresponding *Actor*. We remark that this high-level model is providing stability since abstract concepts are shared in the different variants, and according to the historical track of the language, we may assume that future variants will still adhere to them.

The resulting UML stereotypes are: (a) <<XEnum>> which represents a special kind of enumeration class that may grow (i.e., may be assigned more values). We have included as class attributes only the most consolidated ones (i.e., *name* of *Actor* and *IElement*; *value* of *IElementLink* as optional for those links without values; *strengths* for *DependencySegment* among others). (b) <<XClass>> which allows having an additional list of attribute-value pairs. To take full profit of this definitions, plain associations are converted into association classes with stereotype <<XClass>>.

The *i** supermetamodel as presented is capable to represent as instances those *i** models built with any of the variations mentioned or referenced so far. In order to illustrate this expressive power we show, in Figure 5, an object diagram corresponding to the *i** supermetamodel. It represents a specific *i** model selected from [15]. We may observe different usual elements (types of *actors*, *goals*, *softgoals*, etc.) then some particular elements, more precisely costs in contribution links (both a label and a quantitative value). We have tested the *i** supermetamodel with additional representations including service-oriented *i** [16], *i** with norms [10] and the different secure-oriented *i** variants [9].

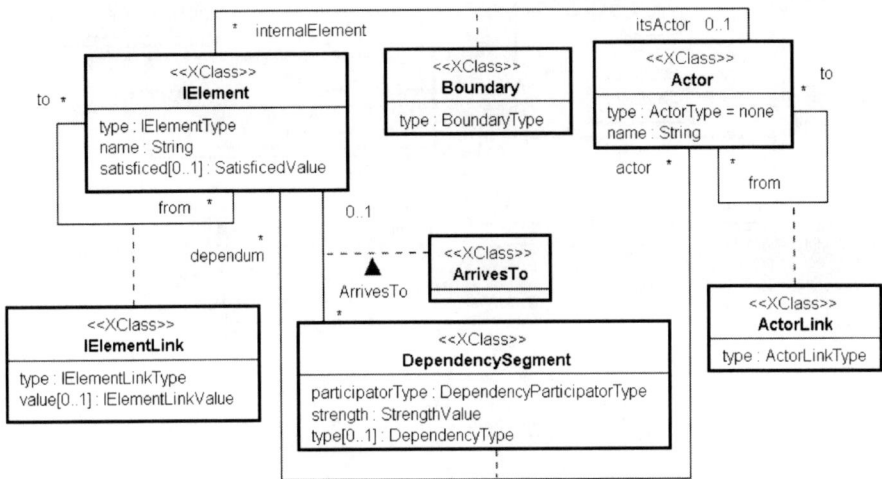

Fig. 4. The *i** supermetamodel

It is interesting to remark that, in spite of its expressive power, the *i** supermetamodel cannot be considered an *i** variant by itself. Although it is a metamodel, it just represents a wide set of possible *i** configurations but considered by itself, there are hundred of instances of the *i** supermetamodel that have not any sense into any *i** community, e.g., a *belief* decomposed into *resources*. Therefore, the *i** supermetamodel has to be considered just a reference framework for supporting model interoperability. Nevertheless, it must be mentioned that the *i** supermetamodel does impose basic syntactic validity conditions for models to be really considered an *i** variant. For instance, it is stated through multiplicities that an intentional element cannot belong to more than one actor. Other additional conditions are not shown graphically but exist in the form of OCL integrity constraints. Just to name one, in the case of *DependencySegments* that arrive to an *IElement*, it must hold that the *IElement* is inside the boundary of the *Actor* linked by the segment:

```
context DependencySegment::IElementInsideActor() inv:
   self.IElement->notEmpty() implies self.iElement.itsActor->notEmpty()
                and self.iElement.itsActor = self.Actor
```

It must be mentioned that the current *i** supermetamodel proposed here does not cover the complete range of constructs that appear in the state of the art, that remain

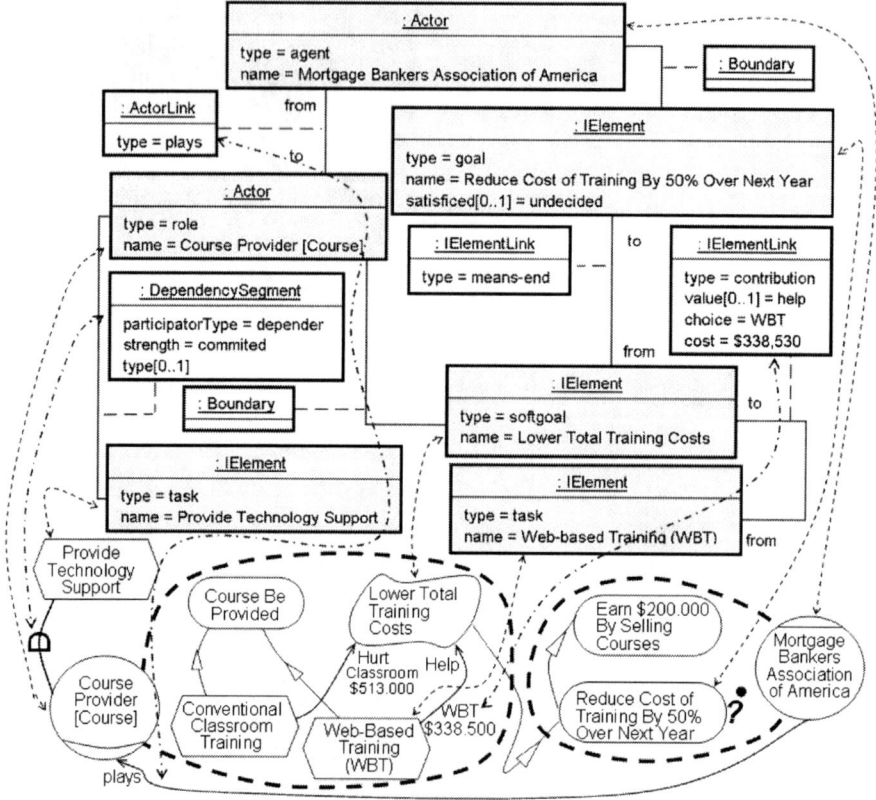

Fig. 5. An excerpt of a particular *i** model considered as an instance of the *i** supermetamodel

for the next version. The elements remarkably left are: links to external elements (i.e., from other conceptual models, e.g., UML classes), boundaries other than actors and some types of intentional links (e.g., GRL's correlations).

5   Implementing *i** Variants Translation

Now we face the ultimate goal of our work: given a model *m1* built as an instance of a metamodel *M1* that represents a particular *i** variant, how to proceed in order to obtain a model *m2* built as instance of a metamodel *M2* that represents a different *i** variant, so that the loss of information is kept to a minimum. To implement this translation, we need an algorithm and a computational representation of the *i** supermetamodel.

Let's start by the second point, which is simpler. As computational representation of the metamodel we use the iStarML interchange format [17]. It was designed with the reference model in mind but it may easily match the *i** supermetamodel as well. XML was chosen as interchange language, therefore we may use a broad set of

Table 2. Correspondence between the *i** supermetamodel and iStarML

*i** Supermetamodel Element	iStarML Construction
XClass Actor	\<actor\>
XClass Intentional Element	\<ielement\>
Association XClass Boundary	\<boundary\> nested under \<actor\>
Association XClass Dependency Segment	\<dependee\> or \<dependeder\> nested under \<dependency\>
XClass ActorLink	\<actorLink\>
XClass IElementLink	\<ielementLink\>
Association XClass ArrivesTo	\<dependency\> nested under \<ielement\>

technologies in order to parse and process iStarML files. The particular XML elements of iStarML correspond to supermetamodel concepts as we show in Table 2.

For the translation algorithm, it is important to start reminding from Section 3 that, since we are using the *i** supermetamodel, then the departing model *m1* is considered an instance of the *i** supermetamodel. Therefore the translation from the metamodel *M1* to *M2* should be considered in fact as a translation from the *i** supermetamodel to *M2*. Since the target variant corresponds to a restricted version of the very *i** supermetamodel, then the refactoring operations required for translation can be only to restrict attributes of the existing classes or to constraint the set of values of specific attributes. In our iStarML implementation, this means to omit some attribute of an existing XML element or to translate specific values of attributes to a different set. Both types of translations (if any) can imply different semantic-preserving situations.

In order to minimize information loss, an algorithm is proposed (Figure 6). It is presented as a UML activity diagram labelled with information about the semantic-preservation consequences considering Wachsmuth's framework. The activities are:

- *Copy known formations.* The part of *m1* that is also a valid instance of *M2* is directly considered as part of *m2*. In other words, the concepts which are shared by both metamodels *M1* and *M2* are kept. In case that the full model *m1* is a valid instance of *M2*, we finish and classify the translation as strictly semantic preserving. Example: a generic actor is always kept as a generic actor.
- *Translate using bijective mappings.* Let's name $m1_A$ the part of *m1* that has not been treated in the previous activity. The part of $m1_A$ for which we may establish a bijective mapping between its elements and corresponding elements, which are instance of *M2*, is translated using this bijective mapping. In other words, the concepts that can be expressed in both metamodels *M1* and *M2* but with different constructs, are just translated. In case that after this activity the full *m1* has been treated, then the translation is *semantic preserving modulo variation*. Example: a *task* can be translated into *plan* and a *plan* into a *task*.
- *Translate using injective mappings.* Let's name $m1_B$ the part of *m1* that has not been treated in the previous activity. The part of $m1_B$ for which we may establish an injective mapping from its elements to others which are instance of *M2*, is translated using this mapping. In case that after this activity the full *m1* has been treated, then the translation is *decreasing modulo variation* (the variation introduced by the mapping). Example: a *make* contribution from GRL can be translated into ++ contribution in seminal *i**, but not any ++ is a *make* contribution.

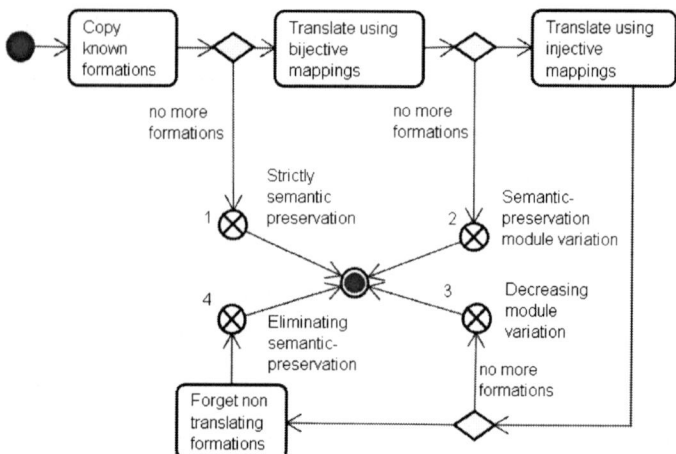

Fig. 6. Translation algorithm from the *i** supermetamodel to an *i** variant

- *Forget non translating formations.* Finally, those constructs in *m1* which have not been translated in the previous activities, are just removed. Example: a *belief* from GRL when translating into Aspectual *i**.

In order to illustrate the process, we have designed a proof-of-concept for translating models built with the OME tool [18] into the jUCMNav tool [19]. The metamodels involved are determined in this case by the implementation of the tool. Basically OME is offering the *i** variant available in the *i** wiki, whilst *jUCMNav* is implementing GRL's metamodel, although a closer look reveals some minor differences not relevant for the purposes of this paper. For the technical implementation of the algorithm we have used XSLT, a declarative language for transforming XML files [20]. The algorithm is implemented as a Java applet and currently available at http://www.essi.upc.edu/~gessi/iStarML/. Besides, jUCMNav has been modified in order to import and export iStarML files [21]. We prove the approach doing XSLT transformation from OME representations (*i**), as special case of supermetamodel, to jUCMNav representations (GRL). In Table 3 we show four submodels to illustrate the four different outputs of the translation algorithm. We explain below each row:

- Row 1: dependency from an intentional element into another. Strictly semantic preserving: all the model is translated without changes. Output 1 in Figure 6.
- Row 2: task decomposition with dependency to an intentional element. Semantic preservation modulo variation: the task decomposition in OME is translated into an AND-decomposition in jUCMMNav. Note that it would be possible to recreate the original model. Therefore, this is a bijective mapping. Output 2 in Figure 6.
- Row 3: dependency from an intentional element into an actor. Please note that jUCMNav does not admit dependencies with actors as dependers or dependees (i.e., the 0..1 multiplicity in *arrives-to* in the *i** supermetamodel of Figure 4 becomes 1 in jUCMNav). Decreasing modulo variation: it is possible to translate the dependency by creating an intentional element in the target actor and attaching dependency on it, but the original model can not be recreated, since it is not known if

the added intentional element is really new or not. In particular, note that this jUCMNav model is identical to the previous one, clearly showing the lack of bijection with respect to this particular point. Output 3 in Figure 6.

- Row 4: agent as instance of actor. Although the agent is converted into actor (decreasing modulo variation), the instance link is lost. Eliminating semantic preservation: the element can not be kept and is removed. Informational loss. Output 4 in Figure 6.

We remark that we are not proposing specific semantic equivalences from one variant to another, we are just showing a proof-of-concept of our approach by describing a general procedure to maximize semantic preservation reducing the complexity of

Table 3. Classification of specific model translations from OME to jUCMNav

the translation problem. The existence of many *i** variants implies the existence of different semantic-pragmatic communities and the equivalences or mappings among metamodels (in fact, from the *i** supermetamodel to variants' metamodels) should be a matter of a meaning-making process inside that specific community. Just to mention an example, row 3 and row 4 are proposing two different strategies for dealing with one specific construct (dependency with an actor as dependee) that is supported in the departing metamodel but not in the target metamodel. Choosing one or another depends on the target community.

6 Conclusions and Future Work

In this paper we have dealt with the problem of interoperability among *i** variants under a metamodel perspective. We organized the research into 4 questions which we think have been satisfactorily explored:

- We have surveyed 146 proposals presented by the community in the last 5 years, and we have classified them in terms of additions, removals and modifications of *i** constructs organized into six categories. Thus, we have obtained a quite complete characterization of the *i** variability to support interoperability goals.
- We have proposed a framework for the interoperability problem based on an approach that can be considered consolidated and widespread in the MDE community. We have customised this framework about model evolution into the *i** model interoperability problem. As cornerstone of this customization, we have defined a supermetamodel for *i** that eases interoperability by metamodel containment.
- Given the framework above, we have classified the surveyed *i** variation types in terms of the semantic impact of their translation, having then a general idea about what types of information loss may happen and to what extent the analyst may provide information (mappings) to minimize this loss.
- We have defined a process for translating a model compliant to one metamodel to another compliant to a different metamodel, and we have demonstrated how it works by exploring the translations of models built with the OME tool to the jUCMNav tool.

As a summary, we may say that we have provided a first consolidated step towards not just syntactic but also semantic interoperability in the *i** framework. Our approach may help creating a repository of *i** models (using the *i** supermetamodel as universal reference model), may favour the application of techniques that work over different metamodels, and may possibilitate the interchange of models between tools.

Our future work spreads along four different axes. First, improving the translation algorithm which is currently able to deal just with reductions, to tackle increasing modulo variations. Second, to offer a portfolio of tool interconnections in similar way to the one between OME to jUCMNav explained here (in fact, we have a more complete case of interconnection among the jUCMNav and H*i*ME [22] tools, described in [21]). Third, consider not just syntax and semantics but also ontological issues in the translation process. Forth, digging into more details of Wachsmuth's framework for proposing translation heuristics depending on the refactoring distance between the source and target metamodels, allowing thus having some default translation rules instead of a pure case-by-case analysis (although as remarked at the end of Section 5, translation will ultimately depend on the community ontological perception of *i**).

References

1. Yu, E.: Modelling Strategic Relationships for Process Reengineering. PhD. Computer Science, University of Toronto, Toronto (1995)
2. Franch, X.: Fostering the Adoption of *i** by Practitioners: Some Challenges and Research Directions. In: Intentional Perspectives on Information Systems Engineering. Springer, Berlin (2010)
3. Yu, E., Giorgini, P., Maiden, N., Mylopoulos, J. (eds.): Social Modeling for Requirements Engineering. The MIT Press, Cambridge (2011)
4. The i* Wiki, http://istar.rwth-aachen.de
5. Dardenne, A., Lamsweerde, A.v., Fickas, S.: v. and Fickas S.: Goal-directed Requirements Acquisition. Science of Computer Programming 20(1-2), 3–50 (1993)
6. Chung, L.K., Nixon, B.A., Yu, E., Mylopoulos, J.: Non-functional Requirements in Software Engineering. Kluwer Academic Publishing, Dordrecht (2000)
7. ITU-T Recommendation Z.151 (11/08), User Requirements Notation (URN) - Language Definition (2008), http://www.itu.int/rec/T-REC-Z.151/en
8. Bresciani, P., Perini, A., Giorgini, P., Giunchiglia, F., Mylopoulos, J.: Tropos: An Agent-Oriented Software Development Methodology. Autonomous Agents and Multi-Agent Systems 8(3), 203–236 (2004)
9. Mouratidis, H., Giorgini, P., Manson, G.: Integrating Security and Systems Engineering: Towards the Modelling of Secure Information Systems. In: Eder, J., Missikoff, M. (eds.) CAiSE 2003. LNCS, vol. 2681, pp. 63–78. Springer, Heidelberg (2003)
10. Siena, A.: Engineering Law-compliant Requirements. The Nòmos Framework. PhD. Thesis, University of Trento, Trento (2008)
11. Seidewitz, E.: What Models Mean. IEEE Software 20(5), 26–32 (2002)
12. Wachsmuth, G.: Metamodel Adaptation and Model Co-adaptation. In: Bateni, M. (ed.) ECOOP 2007. LNCS, vol. 4609, pp. 600–624. Springer, Heidelberg (2007)
13. Cares, C., Franch, X., Mayol, E., Quer, C.: A Reference Model for i*. In: Yu, E., Giorgini, P., Maiden, N., Mylopoulos, J. (eds.) Social Modeling for Requirements Engineering, pp. 573–606. The MIT Press, Cambridge (2011)
14. Amyot, D., Horkoff, J., Gross, D., Mussbacher, G.: A Lightweight GRL Profile for i* Modeling. In: Heuser, C.A., Pernul, G. (eds.) ER 2009. LNCS, vol. 5833, pp. 254–264. Springer, Heidelberg (2009)
15. Liu, L., Yu, E.: Designing Information Systems in Social Context: a Goal and Scenario Modelling Approach. Information Systems 29(2), 187–203 (2004)
16. Estrada, H., Martínez, A., Pastor, O., Mylopoulos, J., Giorgini, P.: Extending Organizational Modeling with Business Services Concepts: An Overview of the Proposed Architecture. In: Parsons, J., Saeki, M., Shoval, P., Woo, C., Wand, Y. (eds.) ER 2010. LNCS, vol. 6412, pp. 483–488. Springer, Heidelberg (2010)
17. Cares, C., Franch, X., Perini, A., Susi, A.: Towards *i** Interoperability using iStarML. Computer Standards and Interfaces 33, 69–79 (2010)
18. OME Tool, http://www.cs.toronto.edu/km/ome
19. jUCMNav Tool, http://jucmnav.softwareengineering.ca
20. XSL Transformations (XSLT) V1.0 W3C Consortium (1999), http://www.w3.org/TR/xslt (1999)
21. Colomer, D., Lopez, L., Cares, C., Franch, X.: Model Interchange and Tool Interoperability in the *i** Framework: A Proof of Concept. In: WER 2011 (2011)
22. López L., Franch X. and Marco J.: HiME: Hierarchical *i** Modeling Editor. *Revista de Informática Teórica e Aplicada* (RITA), 16, 2, (2009)

Automatic Generation of a Data-Centered View of Business Processes*

Cristina Cabanillas[1], Manuel Resinas[1],
Antonio Ruiz-Cortés[1], and Ahmed Awad[2]

[1] Universidad de Sevilla, Spain
{cristinacabanillas,resinas,aruiz}@us.es
[2] Hasso Plattner Institute at the University of Potsdam
ahmed.awad@hpi.uni-potsdam.de

Abstract. Most commonly used business process (BP) notations, such as BPMN, focus on defining the control flow of the activities of a BP, i.e., they are activity-centered. In these notations, data play a secondary role, just as inputs or outputs of the activities. However, there is an increasing interest in analysing the life cycle of the data objects that are handled in a BP because it helps understand how data is modified during the execution of the process, detect data anomalies such as checking whether an activity requires a data object in a state that is unreachable, and check data compliance rules such as checking whether only a certain role can change the state of a data object. To carry out such an analysis, it is very appealing to provide a mechanism to transform from the usual activity-centered model of a BP to the set of life cycles of all the data objects involved in the process (i.e., a data-centered model). Unfortunately, although some proposals describe such transformation, they do not deal with data anomalies in the original BP model nor include information about the activities of the BP that are executed in the state transitions of the data object, which limits the analysis capabilities of the life cycle models. In this paper, we describe a model-driven procedure to automatically transform from an activity-centered model to a data-centered model of a BP that solves the aforementioned limitations of other proposals.

Keywords: business process, data management, object life cycle, data anomalies, Petri net, reachability graph.

1 Introduction

It is widely known that business processes (BPs) involve different kinds of elements, to be named control flow, time, data and resources. However, most

* This work has been partially supported by the European Commission (FEDER), Spanish Government under the CICYT project SETI (TIN2009-07366); and projects THEOS (TIC-5906) and ISABEL (P07-TIC-2533) funded by the Andalusian Local Government.

H. Mouratidis and C. Rolland (Eds.): CAiSE 2011, LNCS 6741, pp. 352–366, 2011.

commonly used BP models and notations focus on the control flow and the timing of activities in the BP. As a consequence, in most BP models, data (e.g., documents, reports, invoices, emails and the like) play a secondary role, just as inputs or outputs of the activities of the process.

Nevertheless, understanding and analysing how data is modified during the execution of a BP is getting an increased interest from both industry and academy. For instance, BPMN, the de-facto standard for BP modelling, has incorporated more advanced constructs for data management in its last version [1]. In addition, there is an increasing number of research proposals to analyse the way data is used in a BP to detect anomalies [2,3,4] and to define data-aware compliance rules [5] for BPs. Therefore, providing a mechanism to transform from the usual activity-centered view of a BP to a data-centered view that focuses on the data handled during the process is very appealing to this goal of understanding and analysing how data is modified during the execution of a BP.

In this paper we describe a model-driven procedure based on Petri nets for carrying out this transformation automatically. In particular, the input of the procedure is a BP diagram expressed in BPMN 2.0 (cf. Figure 1). We use this notation because it is the de-facto standard for BP modelling. Such diagrams represent data objects connected to the BP activities that use them either to read them or write them, or for both things. A data object has a *type* and can have one or more *states* along the execution of a process. For instance, in the BP of opening a bank account, the data object *application* filled by the new customer could go through states *sent, accepted* and *stored*. The output of the procedure is a data-centered view composed of the set of object life cycles (OLCs) of all the data objects that are involved in a BP. They represent the allowed transitions between the states of the data object according to the BP diagram. In addition, these transitions also include information about the activities of the BP that are executed in the transition between states of the data object (cf. Figure 2). Furthermore our procedure also deals with some data anomalies that may appear in a BP model (cf. Section 4 for more details).

Our approach has the following advantages: (i) it is fully automated; (ii) it is based on Petri nets, which allows us to use efficient and well-tested Petri net algorithms; (iii) since it includes information about the activities that are executed in each transition, it provides the same full information required to understand BP execution as activity-centered process diagrams; and (iv) it is robust in the sense that it provides an accurate data-centered view despite having a BP with data anomalies as input. Moreover, it informs the user about these data anomalies.

The remaining of the paper is organised as follows. Section 2 introduces a use case used to exemplify the output produced by the procedure. Section 3 contains the description of the whole procedure for OLC generation. In Section 4 the detection and handling of data anomalies is introduced. Section 5 contains a summary of related work and in Section 6 we draw a set of conclusions and outline some future work.

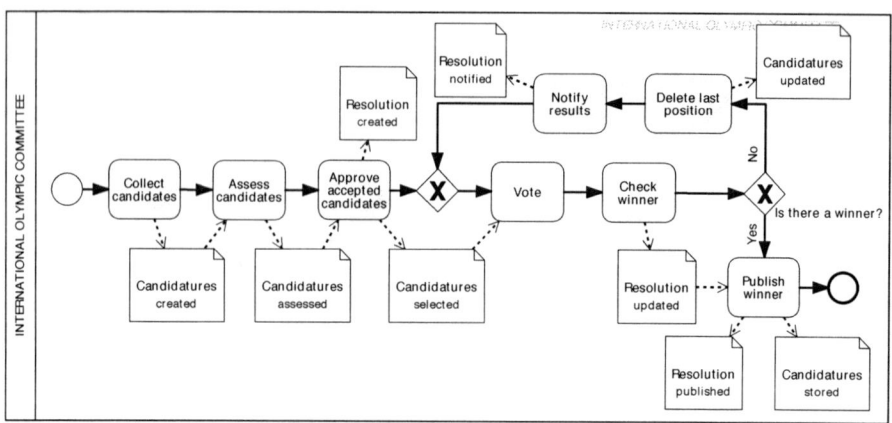

Fig. 1. Business process for assigning the venue for the Olympic Games

2   Use Case

To illustrate our approach we use the BP for assigning the venue for the Olympic Games (Figure 1) as use case in this paper[1]. The International Olympic Committee is in charge of this process. This committee first receives the applications of the cities that want to organize the Olympic Games. Each city is evaluated in order to keep only those which fulfill all the requirements. After this filter is applied, an approval of the final candidates is necessary. Once the list of candidates is ready, a secret voting is carried out. If there is consensus and only one city is selected, then the winner venue is published. Otherwise, the least voted city is eliminated from the list of candidates and a new voting is performed. This is repeated until there are only two cities left. Then, the city with a greatest number of votes wins.

There are two data objects in this BP model. Data object *Candidates* represents a document that contains a list of the cities that applied for the venue. The information of each candidate in the document includes the name of the city, its description, what it offers for each requirement needed, and the mark given by the committee to discern between accepted and rejected candidates. This document may be updated during the voting repetitive process. Data object *Resolution* represents the result of the voting and, thus, is a document with the same list of candidates and the number of votes each of them received. Again, this data object will be updated if more than one voting is performed. If there is no winner yet, the resolution is notified. Otherwise, the resolution is completed with the features of the final venue and published.

The output of the procedure presented in this paper is a set of finite-state machines (FSM) representing the life cycles of the data objects modelled in a

---

[1] Note that this process is used for illustration purposes only, so there may be differences with the actual process of the Olympic Games venue selection process.

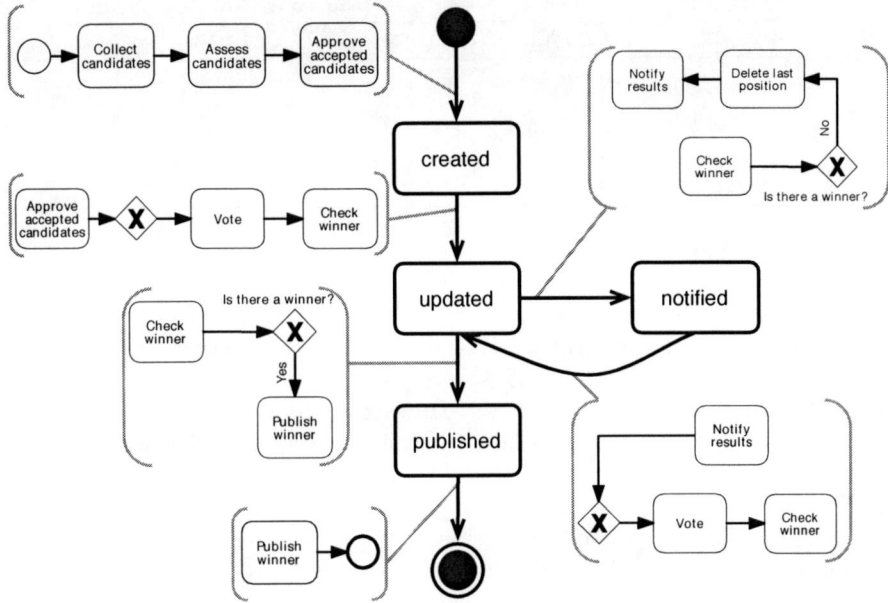

Fig. 2. Object life cycle of data object *Resolution* of the business process in Fig. 1

BP. Figure 2 depicts the life cycle of data object *Resolution* of our use case. The life cycles of a data object have one *start state* (represented with a filled circle), one *final state* (represented with a semi-filled circle), and one or more *intermediate states* (represented with a rectangle) that correspond with states of the data object in the BP model. Transitions (represented with directed arrows) connect two states and contain the parts of the BP that are executed in the transition between states of the data object.

3 BP2OLC Procedure

BP2OLC is our approach to automatically generate the OLCs of the data objects represented in a BPMN model[2]. As depicted in Figure 3, it is a three-step procedure based on model transformations which involves four different models. The procedure must be carried out for each *data object type* present in the BP model. We assume the source BP model has the following features:

1. As far as control flow is concerned, the BP model is sound, which basically means it has no control flow deadlocks and terminates properly [6].
2. There is only one copy of each data object in each instance of the process, e.g., there is only one data object *Resolution* in one instance of the process.

---

[2] All the terms referring to elements of a BP model are used in the same sense as in the BPMN 2.0 specification [1].

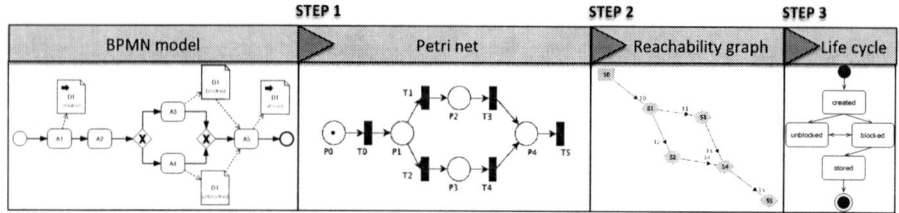

Fig. 3. Overview of the BP2OLC procedure

Besides, data objects are created within the BP instance that uses them (i.e. data objects created outside of the process are not considered).

3. Each data object has always a state. In case an appearance of a data object in the BP model is not associated with any state, this appearance will be ignored.

4. The BP model can contain data objects connected to any kind of activity (sub-processes are treated like task activities). Only XOR gateways can be used.

Assumption 1 is made because control-flow soundness is out of the scope of this paper. Assumptions 2 and 3 are reasonable and have also been made elsewhere [2]. The last assumption is related to the reach of the current approach.

3.1   Step 1. From BPMN Model to Petri Net

We believe that providing a semantic mapping [7] between a BPMN model and a target domain such as Petri nets, whose semantics has been formally defined, is a good approach because it allows one to use the techniques specific to the target semantic domain for analysing the source models. We chose Petri nets for two reasons: (i) plenty of processing algorithms on Petri nets have already been developed and can be useful for our purpose [6,8]; and (ii) the transformation of the control flow of a BP model into an equivalent Petri net has already been described in [6].

Definition 1. *A Petri net is a 3-tuple* $PN = (T_{PN}, P, F)$, *where:*

- $T_{PN} = \{t_1, t_2, ..., t_n\}$ *is the set of transitions of the Petri net, represented graphically as rectangles.*
- $P = \{p_1, p_2, ..., p_n\}$ *is the set of places of the Petri net, represented graphically as circles.*
- $F \subseteq (P \times T_{PN}) \bigcup (T_{PN} \times P)$ *is the set of arcs of the Petri net (flow relation), represented as arrows.*

A *marking (state)* or *markup* assigns a nonnegative integer to each place of a Petri net. If it assigns to place p a nonnegative integer k, we say that p is marked with k tokens. Pictorially, we place k black dots (tokens) in place p. A markup

Table 1. Mapping for data objects association with loop activities

BPMN	Description	Petri Net
a) Read-write loop activity with a condition on input state	Loop activity A has both a precondition and a postcondition on the state of the data object. The precondition will be considered only for the first execution of A. Then the object is assumed to be in state2.	
b) Read-write loop activity without a condition on input state	Loop activity A has no specific condition on the state of data object as input. We assume that after first completion of A the data object will always have state2.	

is denoted by M, an m-vector, where m is the total number of places. The pth component of M, denoted by $M(p)$, is the number of tokens in place p. The firing of an enabled transition will change the token distribution (marking) in a net [8].

We use the set of rules introduced by Awad et al. [2] to do the semantic mapping between elements of a BP model with data objects and elements of a Petri net. Let E_{BP} be the set of flow nodes of a BP (model), i.e. activities, gateways and events, D_{BP} the set of states of a data object of that BP, and $WRITERS_{BP} \subseteq E_{BP}$ be the set of activities of the BP that write that data object. The result of the semantic mapping is a Petri net with the following characteristics:

- The places of the Petri net are of two different kinds: control places P_C and data places P_D. Therefore $P = P_C \bigcup P_D$ and $P_C \bigcap P_D = \emptyset$.
  - $P_C = \{pc_1, pc_2, ..., pc_n\}$ corresponds to those places that represent sequence flow elements (arrows) of the business process. Each $pc_i = (ei_i, eo_i)$, where $ei_i, eo_i \in E_{BP}$ is a pair of values composed of the two flow nodes of the business process that the sequence flow element connects.
  - $P_D = \{pd_1, pd_2, ..., pd_n\} = D_{BP}$ corresponds to those places that represent states of the data object whose object life cycle we are generating. There is exactly one data place for each possible state of the data object.
- The transitions of the Petri net represent flow nodes of the business process model. It follows an $n : 1$ relationship, i.e., each transition represents only one flow node of the business process and a flow node may appear several times in a Petri net. Function $elem : T_{PN} \rightarrow E_{BP}$ represents such relation.

An example of the transformation rules is depicted in Table 1, which illustrates an extension of the catalogue of transformations proposed in [2] to deal with loop activities. As stated in [1], a loop activity executes the inner activity as long as a loop condition evaluates to true. An attribute can be set to specify a maximal number of iterations. An example of loop activity is an activity *Update order* that updates an order in a restaurant (by customer's command) until an event or a received message indicates no more updates are allowed. For more details about the other transformations we refer the reader to [2].

Finally, note that there is a small difference between this mapping and the one presented in [2] because in this paper we consider no data objects are supposed to exist before the execution of a BP in our BP2OLC procedure, whereas [2] considers data objects have an initial state when instantiating a BP. This difference causes the transformation in [2] referring to the writing of the data object has to be slightly changed for the first writing of the object in our BP2OLC procedure, in order to comply with our assumption 2. It means the first time the data object is written, the responsible transition of the Petri net does not have any input data places.

3.2   Step 2. Reachability Graph from Petri Net

Definition 2. *A reachability graph related to a Petri net is a 3-tuple $RG_{PN} = (N, M, T_{RG})$, where:*

- *$N = \{n_1, n_2, ..., n_n\}$ is the set of nodes of the reachability graph. $\forall n_i \in N, \bullet n_i$ and $n_i \bullet$ represent immediately previous and next nodes of n_i, respectively.*
- *$M : P \times N \to \mathbb{N}$ represents the markup of the net.*
- *$T_{RG} \subseteq (N \times N)$ are the transitions of the reachability graph.*

The reachability graph is obtained by analysing the Petri net by means of well-known algorithms. Each node of the reachability graph represents a reachable marking state of the net and each arc a possible change of state, i.e. the firing of a transition. However, due to the characteristics of our semantic mapping between BPMN and Petri net, in the reachability graph resulting from such Petri nets it holds that $M(p, n) \in [0, 1], \forall n \in N, \forall p \in P$. In addition, the information about the markup of the net contained in every node always corresponds with both a sequence flow of the BP model and a state of the data object, as illustrated in Figure 4. It means there is always one token in a control place of the Petri net and one in a data place, except in the beginning (until an activity writes the data object for the first time) and in the final nodes of the reachability graph (in which, on the contrary, all the tokens in control places have been consumed).

Given the previous definitions, the following functions can be defined:

- Function $map : T_{RG} \to T_{PN}$ is defined to map the transitions of a reachability graph into the transitions of a Petri net.
- Function $state : N \to P_D$ returns the state of the data object of the business process model contained in the current node of the reachability graph. $state(n) = \{p_d \in P_D : M(p_d, n) = 1\}$.

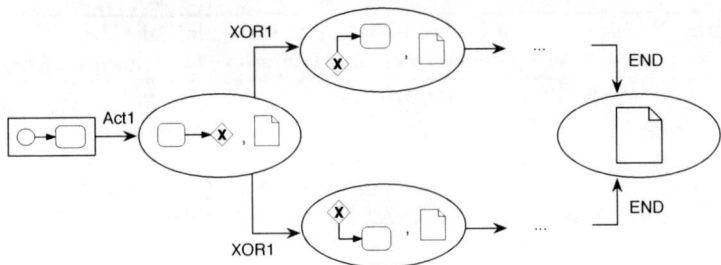

Fig. 4. Content of the arcs and nodes of a reachability graph

- Function $flow : \mathcal{P}(N) \rightarrow \mathcal{P}(P_C)$ returns the set of sequence flow elements of the business process model contained in a set of nodes of the reachability graph. $flow(N') = \{p_c \in P_c : \exists n \in N'(M(p_c, n) = 1))\}$.
- Function $activity : N \rightarrow E_{BP}$ returns the flow node of the business process model contained in the input arc of the current node of the reachability graph. $activity(n) = \{e_i \in E_{BP} : p_c = (e_i, e_o) \wedge M(p_c, n) = 1\}$.

The node of the reachability graph with no input arrows is called $firstNode \in N : \not\exists \bullet firstNode$ and it is the start node of a reachability graph. The nodes of the reachability graph with no output arrows, whose input is called END and with no tokens in a control place are *normal* final nodes of the reachability graph. We will describe *abnormal* final nodes in Section 3.3.

3.3   Step 3. Object Life Cycle from Reachability Graph

Definition 3. *An object life cycle of a data object of a business process is a 2-tuple $OLC = (S_{OLC}, T_{OLC})$, where:*

- $S_{OLC} = \{s_1, s_2, ..., s_n\}$ *is the set of states in which the data object can be.* $\forall s_i \in S_{OLC}, \bullet s_i$ *and* $s_i \bullet$ *represent immediately previous and next states of state s_i, respectively. Let start $\in S$ and end $\in S$ be the start and the final states of the OLC, respectively. Then, $S_{OLC} \setminus (start \bigcup end) = P_D = D_{BP}$*
- $T_{OLC} \subseteq S_{OLC} \times S_{OLC} \times \mathcal{P}(N)$ *is the set of transitions that appear in the object life cycle. Each transition contains a set of nodes of the reachability graph from which it has been generated. Function replace : $T_{OLC} \times N \times \mathcal{P}(N) \rightarrow T_{OLC}$ replaces the set of nodes before node N in the path of a transition for a specific set of nodes.*

We have defined Algorithms 1 and 2 to obtain an OLC from a reachability graph. Algorithm 1 receives the reachability graph resulting from the previous step and the list of activities of the BP that write the data object. Its output is the OLC together with a set of data anomalies found while creating it.

Algorithm 1. Algorithm to initialize an object life cycle, call Algorithm 2 from a reachability graph and post-process nodes already processed in Algorithm 2 (RG2OLC)

```
 1: IN: RG_DPN = (N, M, T_RG); WRITERS_BP
 2: OUT: S_OLC; T_OLC; WARN ⊆ N
 3: S_OLC ← {START_STATE}; T_OLC ← ∅
 4: INPUT ← (WRITERS, firstNode, START_STATE, ∅, ∅, ∅, ∅, S_OLC, T_OLC)
 5: (S_OLC, T_OLC, PNODES, PP, WARN) ← RG2OLC(INPUT)
 6: found ← 1 // Post-processing of nodes in PP
 7: while found ≠ 0 do
 8:    found ← 0
 9:    for all (node, assocPath) ∈ PP do
10:       for all (s_i, s_o, path) ∈ T_OLC do
11:          if node ∈ path then
12:             found ← found + 1; newT ← (s_i, s_o, path)
13:             T_OLC ← T_OLC ⋃ replace(newT, node, assocPath)
14:          end if
15:       end for
16:    end for
17: end while
18: return (S_OLC, T_OLC, WARN)
```

Its behaviour consists of calling Algorithm 2 with the appropriate parameters and post-processing the resulting reachability graph. Algorithm 2 is a recursive algorithm that builds an OLC by processing a reachability graph node by node from its start node. Its input set and steps are described below.

Input of Algorithm 2.
- $WRIT \subseteq E$ is the set of activities that write the data object.
- $cNode \in N$ is the node being processed.
- $cState \in D$ is the current state of the data object.
- $PNODES \subseteq N$ is the set of already processed nodes.
- $PATH \subseteq N$ contains a set of nodes of the reachability graph, which is the information required in the transitions of the object life cycle.
- $PP = \{pair_1, pair_2, ..., pair_n\}$, where $pair_i = (node, assocPath), node_i \in N, assocPath_i \subseteq N$ is a set of pairs containing a node of the reachability graph and a set of nodes associated to that node, which conceptually corresponds to the path contained in variable PATH when processing that node.
- $WARN \subseteq N$ is a set of nodes related to deadlocks in the Petri net.
- $S'_{OLC} \subseteq S_{OLC}$ is the set of states of the resulting object life cycle.
- $T'_{OLC} \subseteq T_{OLC}$ is the set of transitions of the resulting object life cycle.

Check for and add new transitions (lines 3-7). A new transition of one of the types shown in Figures 5a and 5b must be added to the OLC in case that a new state of the data object is found in the reachability graph. If, on the contrary, the node shows that the data object is still in the current state but

---

Algorithm 2. Algorithm to generate the life cycle of a data object from a reachability graph (RG2OLC)

1: **IN:** $WRIT, cNode, cState, PNODES, PATH, PP, WARN, S'_{OLC}, T'_{OLC}$
2: **OUT:** $S'_{OLC}, T'_{OLC}, PNODES, PP, WARN$
3: **if** $state(cNode) \neq \emptyset \quad \wedge \quad (cState \neq state(cNode) \quad \vee \quad activity(cNode) \in WRIT)$ **then**
4:     $S'_{OLC} \leftarrow S'_{OLC} \bigcup state(cNode)$
5:     $T'_{OLC} \leftarrow T'_{OLC} \bigcup (cState, state(cNode), PATH)$
6:     $cState \leftarrow state(cNode); PATH \leftarrow cNode$
7: **end if**
8: $PATH \leftarrow PATH \bigcup cNode$
9: **if** $cNode \notin PNODES$ **then**
10:     $PNODES \leftarrow PNODES \bigcup cNode$
11:     **if** $cNode\bullet = \emptyset$ **then**
12:         **if** $activity(cNode) = END$ **then**
13:             $S'_{OLC} \leftarrow S'_{OLC} \bigcup FINAL\_STATE$
14:             $T'_{OLC} \leftarrow T'_{OLC} \bigcup (cState, FINAL\_STATE, PATH)$
15:         **else**
16:             $WARN \leftarrow WARN \bigcup cNode$ // Deadlock detected
17:         **end if**
18:     **else**
19:         **for all** $next \in cNode\bullet$ **do**
20:             $IN \leftarrow (WRIT, next, cState, PNODES, PATH, PP, WARN, S'_{OLC}, T'_{OLC})$

21:             $(S''_{OLC}, T''_{OLC}, PNODES', PP', WARN') \leftarrow RG2OLC(IN)$
22:             $S'_{OLC} \leftarrow S'_{OLC} \bigcup S''_{OLC}; T'_{OLC} \leftarrow T'_{OLC} \bigcup T''_{OLC}; PP \leftarrow PP \bigcup PP'$
23:             $PNODES \leftarrow PNODES \bigcup PNODES'; WARN \leftarrow WARN \bigcup WARN'$
24:         **end for**
25:     **end if**
26: **else**
27:     **if** $activity(cNode) \notin WRIT$ **then**
28:         $PP \leftarrow PP \bigcup (cNode, PATH)$ // Save for post-processing
29:     **end if**
30: **end if**
31: **return** $(S'_{OLC}, T'_{OLC}, PNODES, PP, WARN)$

---

we find that the activity of the node is one of those that write the data object in the BP model, a self-transition will be added (Figure 5c). For instance: (i) in loops in a BP a data object may be written by an activity consecutively twice, giving rise to a self-transition (e.g. data object *Candidates* of Figure 1 has a self-transition in state *updated* due to a loop); (ii) loop activities also cause self-transitions, as can be inferred from Table 1.

Update variable $PATH$ **(line 8).** New nodes will be added to the path in order to collect the information contained in the transitions of the life cycle[3].

---

[3] Note that operator *union* ($\bigcup$) neither inserts duplicates nor null or empty values.

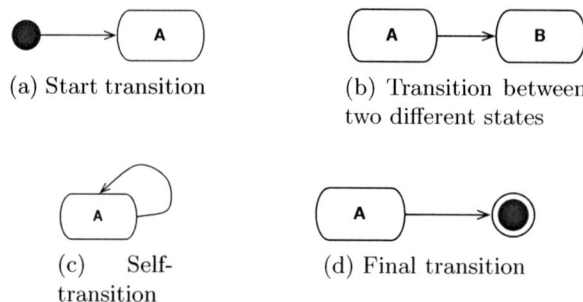

Fig. 5. Kinds of transitions of an object life cycle

Revise the last activity executed and act consistently (lines 9-31).
Algorithm 2 is returned either when a *normal* final node of the reachability
graph is reached, when an *abnormal* final node[4] is found, or when the current
node is in the list of processed nodes. In the last case, if, furthermore, the
activity represented in the node writes the data object, the node has been
properly processed in lines 3-7 and the rest of the reachability graph does not
have to be re-processed. Otherwise, the already processed node is saved in a
list of nodes that must be properly post-processed later. This way we avoid
processing nodes of the reachability graph more than once and we ensure that
Algorithm 2 always terminates. If none of the previous situations appears,
we must go on processing the reachability graph and update the variables
whose values must be propagated.

Post-process the necessary nodes (lines 6-17 of Algorithm 1). Some
scenarios represented in a BP model can give rise to the appearance of transitions between the same two states, which differ from each other in their
contents. This situation is detected in the reachability graph when reaching
a node that has already been processed and its corresponding activity of the
BP does not modify the data object being examined. In Algorithm 2 only
one of the transitions is added to the OLC. To add the new transition with
the right content, we must find the transition to which the node refers, add
a duplicate transition to the OLC and set its content to the proper value.

4 Detecting and Showing Data Anomalies

As aforementioned in this paper, we assume soundness (also called correctness)
in the control flow of BPs, but the process can be *unsound* regarding data
perspective. The data-related deadlocks that appear in a reachability graph (i.e.
abnormal final nodes) indicate data-related anomalous situations (known as *data*

---

[4] *Abnormal* final nodes are those with no output transitions and with no input transitions called END. They indicate there is a deadlock in the Petri net that stops the
execution. We collect them in a list of warnings that will have to be addressed later.

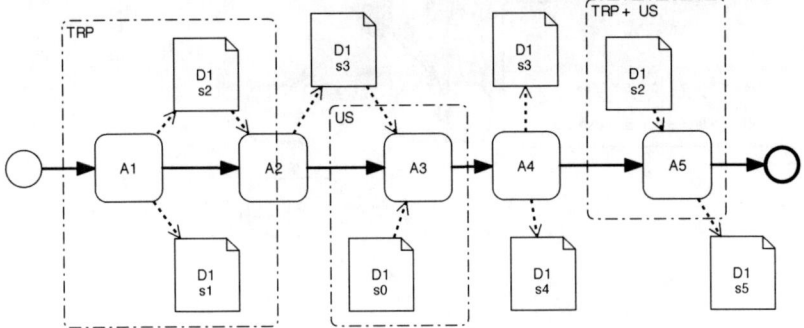

Fig. 6. Business process model with data anomalies

anomalies) in the represented BP model. All these data anomalies cause dead-locks in the Petri net, so the OLC generated is not complete. For example, in Figure 7 only the states and transitions outlined with black solid lines are generated when processing the actual reachability graph that represents the BP model in Figure 6[5]. States such as *s0* and *s5* are not detected without managing some data anomalies present in the BP model previously. There are two different groups of anomalous situations, which can be mapped into data-related problems defined in [2,3]. The resolution of the data anomalies in the BP is out of the scope of this paper. We explain how to modify the Petri net to solve the deadlocks and be able to simulate the whole execution of the BP modelled, with the aim of generating all the states and transitions there represented.

Too restrictive preconditions (TRP). This kind of problems appear when the data object specified as input of an activity can be in a different state at that moment, and so the activity may stay waiting indefinitely. For instance, in the BP model of Figure 6, activity *A2* will get blocked if *A1* writes data object *D1* in state *s1*, and *A5* will get blocked if *D1* is in either *s3* or *s4*. To fix this kind of problems we relax the precondition, assuming that the state of the data object that caused the deadlock is also a precondition of the blocked activity.

Unreachable states (US). In this case, the state specified for the input data object of an activity is unreachable (either because it does not exist yet or because the object is in another state at that moment), blocking the execution of the activity. It appears in activities *A3* and *A5* of the BP model in Figure 6. To fix it, we assume the data object can be at that state at that moment and so we continue processing the BP model from the blocked activity with that "unreachable" state.

There may be other alternatives for dealing with data anomalies, but their study is out of the scope of this paper. To apply the solutions described above, the Petri net has to be "re-constructed" in order to obtain a new reachability graph

---

[5] For the sake of clarity we are using a simplified BP to illustrate the data anomalies.

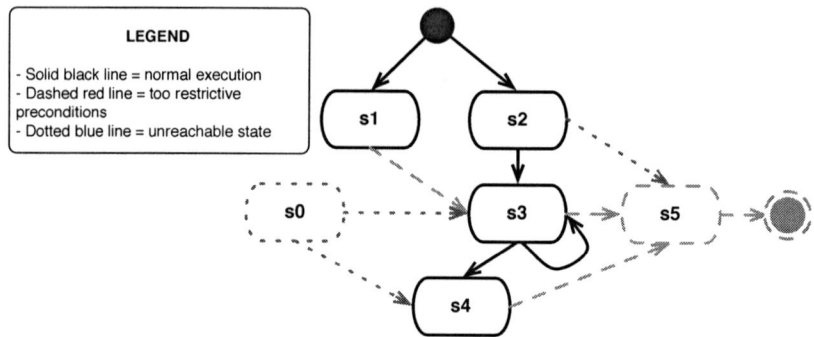

Fig. 7. Object life cycle of data object *D1* of the business process in Fig. 6

and then process it. For too restrictive preconditions, this re-construction has 4 steps. The resulting Petri net is kept while processing the rest of warnings.

1. Find the transition at which the deadlock takes place.
2. Identify the data place that is input of the blocked transition.
3. Identify the data place at which there is a token.
4. As we now know the transition that could not be triggered (step 1) and the actual current state of the data object (step 3), to re-construct the Petri net we have to duplicate the blocked transition and replace the arrow from the data place without token (step 2) by an arrow from the data place with token (step 3), leaving the rest of inputs and outputs like they are.

Unreachable states are reflected in a Petri net in the form of unfired transitions. To detect them and fix them, we have to find the transitions that were never triggered and set the markup of the net with a token in each of their input places. Then a new reachability graph with this configuration is obtained and we can examine the rest of the net from that point by processing it.

In order to warn the modeller/analyst about the presence of data anomalies in the BP model, new elements emerged from dealing with them are marked in a different way in the OLC. In Figure 7, transitions and states generated from too restrictive preconditions are shown with dashed red lines, and those corresponding to unreachable states have dotted blue lines.

5   Related Work

The importance of complementing the activity-centered view of BP models with an object-oriented view has been described by Snoeck et al. [9]. We are generating such a view from the data objects that appear in a BP model.

The work most related to our approach is the one of Ryndina et al. In [10] they present an ad-hoc approach for the automatic generation of OLCs from a BP model and propose some techniques to analyse the consistency of BPs and OLCs on the basis of the concepts of *conformance* and *coverage* between OLCs.

Their procedure is based on transformation rules that are applied directly to a BP model. However, no data anomalies are described nor detected in their approach. Besides, our use of well-known Petri nets algorithms makes it more unlikely to introduce errors while implementing the procedure. In [11] the opposite procedure is introduced, i.e. an approach for generating a BP model from OLCs of different data objects is described.

Data anomalies in BP models have been addressed by several researchers. Sadiq et al. [3] explain the importance of managing the data requirements in BPs and introduce some ideas related to the modelling and validation of data, such as the importance of considering the type of data and their structure. They also state some data anomalies that may appear in a BP model, which in turn are referenced by the authors in [4]. In that work, Sun et al. divide the same problems into three main groups with one or more scenarios, and then they explain the matching of every scenario with the data anomalies in [3]. Awad et al. describe three kinds of data anomalies that can also be mapped to anomalies defined in the previously mentioned work [2]. They have developed an approach for diagnosing and automatically repairing these three kinds of problems on the basis of Petri nets. A prototype has been implemented in Oryx [12]. Besides, they propose some validation algorithms targeted at fixing these data anomalies, which are being implemented to correct BPMN models. In our BP2OLC procedure, we use the transformations described in that work to carry out step 1 of the BP2OLC procedure. However, the mentioned work on data anomalies does not consider the generation of OLCs from a BP model.

Finally, Sakr et al. have developed a framework for querying both control flow and data flow perspectives of BPs [13]. Data perspective can be queried from OLCs. However, no automatic generation of OLCs is included and the framework is not targeted at the detection and management of data anomalies.

6 Conclusions and Future Work

In this paper we introduce a model-driven approach for the automatic generation of a data-centered view of a BP composed of the life cycles of the data objects the BP model has. It consists of mapping a BPMN model into a target semantic domain, Petri nets, which allows us to use techniques specific to that domain, in particular obtaining its reachability graph, for analysing the source model. Then, the reachability graph is mapped into an OLC model. An advantage of our procedure is that the resulting OLCs include information about the activities that are executed in each transition and, hence, it provides the same full information required to understand BP execution as activity-centered process diagrams.

Besides, our procedure is robust in the sense that it provides an accurate result despite having a BP with data anomalies as input. Furthermore, we detail how this procedure can be used to detect two kinds of data anomalies present in a BP model. For each group of data anomalies identified, the following questions have been answered: (i) what does the anomalous situation mean in terms of the

BP model and the resulting OLC?; (ii) how can it be detected in the reachability graph?; and (iii) how can the Petri net be re-constructed to fix the deadlock?

A prototype of the BP2OLC procedure has been implemented reusing the code of ProM[6], an open-source platform for process mining that counts on a number of plugins and components to work with Petri nets. The developed prototype corresponds to steps 2 and 3 of the BP2OLC procedure and also contains the detection and handling of the data anomalies described above. It receives a Petri net in format PNML 1.3.2. as input and returns the life cycle of the data object represented in that net. The software is available upon request.

As future work, we plan to extend the kinds of BP structures considered, to take data objects from repositories into account, and to study alternatives to manage and repair the detected data anomalies in the source BP model.

References

1. Bpmn 2.0, recommendation, OMG (2011)
2. Awad, A., Decker, G., Lohmann, N.: Diagnosing and repairing data anomalies in process models. In: BPM Workshops, pp. 5–16 (2009)
3. Sadiq, S., Orlowska, M.E., Sadiq, W., Foulger, C.: Data flow and validation in workflow modelling. In: Fifteenth Australasian Database Conference (ADC). CRPIT, vol. 27, pp. 207–214. ACS (2004)
4. Sun, S.X., Zhao, J.L., Nunamaker, J.F., Sheng, O.R.L.: Formulating the Data-Flow perspective for business process management. Info. Sys. Research 17(4), 374–391 (2006)
5. Awad, A., Weidlich, M., Weske, M.: Specification, verification and explanation of violation for data aware compliance rules. In: Baresi, L., Chi, C.-H., Suzuki, J. (eds.) ICSOC-ServiceWave 2009. LNCS, vol. 5900, pp. 500–515. Springer, Heidelberg (2009)
6. van der Aalst, W.M.P.: The application of petri nets to workflow management. Journal of Circuits, Systems, and Computers 8(1), 21–66 (1998)
7. Harel, D., Rumpe, B.: "Meaningful modeling: what's the semantics of "semantics"? Computer 37(10), 64–72 (2004)
8. Murata, T.: Petri nets: Properties, analysis and applications. Proceedings of the IEEE 77(4), 541–580 (1989)
9. Snoeck, M., Poelmans, S., Dedene, G.: An architecture for bridging oo and business process modelling. In: Proceedings of the 33rd International Conference on Technology of Object-Oriented Languages, TOOLS 33, pp. 132–143 (2000)
10. Ryndina, K., Kuster, J., Gall, H.: Consistency of business process models and object life cycles. Models in Software Engineering, 80–90 (2007)
11. Kuster, J., Ryndina, K., Gall, H.: Generation of business process models for object life cycle compliance. In: Business Process Management, pp. 165–181 (2007)
12. Decker, G., Overdick, H., Weske, M.: Oryx - an open modeling platform for the BPM community. In: Proceedings of the 6th International Conference on Business Process Management, pp. 382–385. Springer, Heidelberg (2008)
13. Sakr, S., Awad, A.: A framework for querying graph-based business process models. In: WWW, pp. 1297–1300 (2010)

---

[6] http://prom.win.tue.nl/tools/prom/

Connecting Security Requirements Analysis and Secure Design Using Patterns and UMLsec

Holger Schmidt[1] and Jan Jürjens[1,2]

[1] Software Engineering, Department of Computer Science, TU Dortmund, Germany
[2] Fraunhofer ISST, Germany
{holger.schmidt,jan.jurjens}@cs.tu-dortmund.de

Abstract. Existing approaches only provide informal guidelines for the transition from security requirements to secure design. Carrying out this transition is highly non-trivial and error-prone, leaving the risk of introducing vulnerabilities.

This paper presents a *pattern-oriented* approach to connect *security requirements analysis* and *secure architectural design*. Following the divide & conquer principle, a software development problem is divided into simpler subproblems based on *security requirements analysis patterns*. We complement each of these patterns with *architectural security patterns* tailored to solve classes of security subproblems. We use *UMLsec* together with the advanced modeling possibilities for software architectures of UML 2.3 to equip the architectural security patterns with security properties, and to allow *tool-supported* analysis and composition of instances of these patterns. We validate our approach using two case studies and illustrate its support for Common Criteria certifications.

Keywords: security requirement, secure design, architectural pattern.

1 Introduction

When building *secure systems*, it is instrumental to take *security requirements* into account right from the beginning of the development process to reach the best possible match between the expressed requirements and the developed software product, and to eliminate any source of error as early as possible. Knowing that building secure systems is a highly sensitive process, it is important to *reuse* the experience of commonly encountered challenges in this field. This idea of using *patterns* has proved to be of value in software engineering, and it is also a promising approach in *secure software engineering*. Moreover, *tool support* greatly increases the practical applicability of secure software engineering approaches. Tools not only guide software developers in their daily activities, they also help to make the construction of complex secure systems feasible and less error-prone.

In fact, there already exist a number of approaches to security requirements analysis and secure design. Although this can be considered a positive development, the different approaches are mostly not integrated with each other. In particular, relatively little work has been done on bridging the gap between security

H. Mouratidis and C. Rolland (Eds.): CAiSE 2011, LNCS 6741, pp. 367–382, 2011.

requirements analysis and design, and existing approaches only provide informal guidelines for the transition from security requirements to design. Carrying out the transition manually at the hand of these guidelines is highly non-trivial and error-prone, which leaves the risk of inadvertently introducing vulnerabilities in the process. Ultimately, this would lead to the security requirements not to be enforced in the system design (and later its implementation).

This paper presents an integrated and pattern-oriented approach connecting security requirements analysis and secure architectural design. We use a security requirement analysis method [14] that makes extensive use of different kinds of patterns for structuring, characterizing, analyzing, and finally realizing security requirements. We extend this approach by *architectural security patterns* to construct *platform-independent secure software architectures* that realize previously specified security requirements. We specify structural and behavioral views of these architectural security patterns using *UML*[1] (*Unified Modeling Language*) *class diagrams, composite structure diagrams,* and *sequence diagrams.* We annotate these diagrams based on an improved version of the security extension *UMLsec* [8] named *UMLsec4UML2* [15] to represent results from security requirements analysis in the architectural security patterns. More specifically, we apply the advanced modeling possibilities of UML2.3 and UMLsec4UML2 to architectural design to construct the architectural security patterns presented in this paper. Moreover, our approach allows the tool-supported analysis of instances of these patterns with respect to security.

The rest of the paper is organized as follows: we present background about the patterns for security requirements engineering in Sect. 2. In Sect. 3, we first give an overview of the UMLsec4UML2-profile that adopts UMLsec to support UML2.3. Then, we use this profile to specify security patterns for software components and architectures. Furthermore, we generally discuss the application of these patterns yielding global secure software architectures. In Sect. 4, we validate our approach using two case studies and illustrate its support for Common Criteria certifications. We consider related work in Sect. 5. In Sect. 6, we give a summary and directions for future research.

2   Pattern-Oriented Security Requirements Analysis

SEPP (*Security Engineering Process using Patterns*) (see [14] for a comprehensive overview) is a pattern-based approach to construct secure software systems that especially deals with the early software development phases. SEPP makes use of *security problem frames* (SPF) and *concretized security problem frames* (CSPF), which constitute *patterns* for security requirements analysis. (C)SPFs are inspired by *problem frames* invented by Jackson [7] for functional requirements. SPFs are patterns for structuring, characterizing, and analyzing problems that occur frequently in secure software engineering. Following the divide & conquer principle, SPFs are used to decompose an initially large software development problem into smaller subproblems. Then, for each instantiated SPF,

---
[1] http://www.omg.org/spec/UML/2.3/Superstructure/PDF/

a CSPF is selected and instantiated. CSPFs involve first solution approaches for the problems described by SPFs. For example, there exists an SPF for the problem class of confidential transmission of data over an insecure network, and a CSPF that represents the corresponding solution class of using cryptographic key-based symmetric encryption to protect such data transmissions.

Each CSPF contains a *machine domain*, which represents the software to be developed in order to fulfill the *requirement*. The environment, in which the software development problem is located, is described by *problem domains*. According to Jackson [7], we distinguish *causal* domains that comply with some physical laws, *lexical* domains that are data representations, and *biddable* domains that are usually people. Each domain has at least one *interface*. Interfaces consist of *shared phenomena*, which may be events, operation calls, messages, and the like. They are observable by at least two domains, but controlled by only one domain. Since requirements refer to the environment, *requirement references* between the domains and the requirement exist. At least one of these references is a *constraining* reference. That is, the domain this constraining references points to is influenced by the machine so that the requirement can be fulfilled. We developed a comprehensive set of SPFs for confidential and integrity-preserving data transmission and data storage, and authentication problems and the corresponding CSPFs that use symmetric and asymmetric encryption, keyed and non-keyed hashing, digital signatures, password-based and cryptographic key-based mechanisms (see [14] for details).

3   Pattern-Oriented Transition to Secure Architectural Design

This section contains the main scientific contributions of this paper. To proceed after security requirements analysis following SEPP to the development of secure software architectures that realize the security requirements, we develop in this section *architectural security patterns*. We specify these patterns using UML and an improved version of the security extension UMLsec, which is introduced in Sect. 3.1. We describe patterns for security components in Sect. 3.2, and we present patterns for secure software architectures related to CSPFs in Sect. 3.3. In Sect. 3.4, we briefly explain the process of instantiating GSAs. Then, we discuss the composition of different instances of GSAs yielding global secure software architectures in Sect. 3.5. Finally, we outline an approach to verify global secure software architectures based on the UMLsec4UML2-profile and the UMLsec tool suite in Sect 3.6.

3.1   UMLsec4UML2

In this section, we present an overview of a *notation* for the specification of structural as well as behavioral views of architectural security patterns based on UML. As explained in [11], UML includes special support for modeling software architectures since version 2.0. For example, the current UML version 2.3 supports typical architectural concepts such as *parts*, i.e., black-box components, *connectors*, and *required* and *provided interfaces* (see Sects. 3.2 and 3.3 for details). For this reason, we specify our architectural security patterns based on different kinds of UML2.3 diagram types, i.e., class diagrams, composite structure diagrams, and sequence diagrams. Moreover, we use *UMLsec* [8] to pick up results from security requirements analysis, and to annotate the different UML diagrams representing the structural and behavioral views of architectural security patterns accordingly. Since UMLsec is a profile for UML1.5[2], we developed a UML2.3-compatible profile called *UMLsec4UML2* that adopts the UML1.5-compatible profile UMLsec. The UMLsec4UML2-profile, all examples shown in this paper, as well as additional material are published in a technical report [15].

We constructed the UMLsec4UML2-profile using the *Papyrus UML*[3] editing tool. as a UML2.3 profile diagram. It defines several *stereotypes* and *tags*. Stereotypes give a specific meaning to the elements of a UML diagram they are attached to, and they are represented by labels surrounded by double angle brackets. A tag or tagged value is a name-value pair in curly brackets associating data with elements in a UML diagram.

The original version of UMLsec for UML1.5 is complemented by a tool suite[4] that supports static checks for stereotypes that restrict structural design models, a permission analyzer for access control mechanisms, and checks integrated with external verification tools to verify stereotypes that restrict behavioral design models. Basically, models created based on the UMLsec4UML2-profile can be verified using this tool suite. However, the UMLsec4UML2-profile introduces a novel way to verify models directly within the UML editing tool. For this purpose, the UMLsec4UML2-profile is enriched with constraints denoted in the *Object Constraint Language* (OCL)[5], which is part of UML2.3. OCL is a formal notation to describe *constraints* on object-oriented modeling artifacts. The static checks available in the tool suite of the original version of UMLsec are covered by the OCL constraints that are integrated into the UMLsec4UML2-profile.

We use the UMLsec4UML2-profile in the subsequent sections to specify structural as well as behavioral views of architectural security patterns. There, we also explain details about the profile where necessary.

3.2   Generic Security Components

The *generic security components* (GSC) discussed in this section constitute patterns for software components that realize concretized security requirements. We call them "generic", because they are a kind of conceptual pattern for concrete software components. They are *platform-independent*[6]. An example for a GSC is an encryption component defined neither referring to a specific encryption algorithm nor cryptographic keys with a certain structure and length. In addition

---

[2] http://www.omg.org/cgi-bin/doc?formal/03-03-01
[3] http://www.papyrusuml.org/
[4] http://www.umlsec.de/
[5] http://www.omg.org/docs/formal/06-05-01.pdf
[6] The term *platform-independent* is defined according to the Model-Driven Architecture (MDA) approach (http://www.omg.org/mda/).

to GSCs, *generic non-security components* (GNC) are necessary, which do not realize any security requirements. Instead, they represent auxiliary components for GSCs. Typical examples for GNCs are user interface, driver, and storage management components.

According to [16], the architecture of software is multifaceted: there exists a structural view, a process-oriented view, a function-oriented view, an object-oriented view with classes and relations, and a data flow view on a given software architecture. Hence, we specify each GSC and GNC based on a structural view using UML2.3 class and composite structure diagrams, and control and data flow views using UML2.3 sequence diagrams. We make required and provided interfaces of GSCs and GNCs explicit using sockets, lollipops, and interface classes. After GSCs are instantiated, the process-oriented and object-oriented views can be integrated seamlessly into the structural view. Semantic descriptions of the operations provided and used by the components' interfaces can be expressed as OCL pre- and postconditions.

We use GSCs and GNCs to structure the machine domain of a CSPF. The GSCs and GNCs describe the machine's interfaces to its environment and the machine-internal interfaces, i.e., the interfaces between the GSCs and GNCs. Each CSPF is related to a set of GSCs and GNCs.

Given a CSPF, the following procedure can be applied to construct GSCs and GNCs that help to realize the concretized security requirement of the CSPF:

1. Each interface of the machine with the environment must coincide with an interface of some GSCs and GNCs.
2. GSCs and GNCs that serve the same purpose can be represented by one such component, e.g., several storage management components can be represented by one storage management component.
3. For each interface between the machine and a biddable or display domain a user interface component should be used. If the same CSPF contains different interfaces between the machine and a biddable or display domain, user interface components represented by GNCs must be kept separate from user interface components represented by GSCs. For example, a generic non-security user interface component to edit some text should be kept separate from a generic security user interface component to enter a password.
4. For each interface from the machine to a lexical domain, a storage management component should be used. Symbolic phenomena correspond to return values of operations or to getter/setter operations.
5. For each interface of the machine domain with a causal domain, a driver component should be used. Causal phenomena correspond to operations provided by driver components.
6. GSCs adequate to realize the concretized security requirement should be used, such as components for symmetric / asymmetric encryption / decryption, cryptographic key handling, hash calculation, etc.

We enrich GSCs with UMLsec4UML2 language elements to express security properties based on the CSPFs the GSCs are related to. Since each CSPF considers at least one asset to be protected against the malicious environment, these

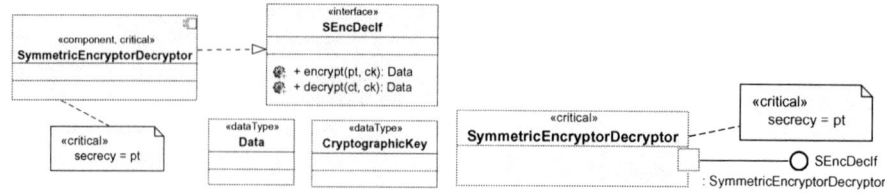

Fig. 1. Structural View of GSC SymmetricEncryptorDecryptor

assets should be considered by the GSCs associated to the CSPF. Consequently, we equip the GSCs dealing with the assets with the stereotype ≪critical≫, and we assign values (e.g., in terms of attributes, parameters, return values, etc.) to the tags of this stereotype accordingly. In case of an asset to be kept *confidential*, we assign this asset to the {secrecy} tag, and in case of preserving the *integrity* of an asset, we assign this asset to the {integrity} tag.

According to the described procedure, we have developed catalogs (see [14, pp. 150 ff.] for details) of GSCs and GNCs for each available CSPF. For instance, there exist GSCs for keyed and non-keyed hash processing, calculation of random numbers, digital signature processing, etc. In the following, we present the GSC SymmetricEncryptorDecryptor as an example.

GSC SymmetricEncryptorDecryptor. The SymmetricEncryptorDecryptor is a conceptual pattern for a component that provides symmetric encryption and decryption services (see [12, pp. 59 ff.] for details). Concrete implementations of symmetric encryption and decryption algorithms are, e.g., the *javax.crypto.Cipher* class provided by *SUN's Java 6 Standard Edition*[7], the *encryption.pbe.Standard-PBEStringEncryptor* class provided by *Jasypt*[8], and the *crypto.engines* class provided by *Bouncycastle*[9].

Figure 1 shows the structural view of this GSC using a class diagram and a composite structure diagram. The first diagram defines the type of the port used in the second diagram. Moreover, the first diagram explicates the provided interface of the GSC. For reasons of simplicity, we do not present the GSC's behavioral view here.

The GSC SymmetricEncryptorDecryptor abstracts from algorithm-specific details such as cryptographic key lengths, the stream and block modes of the algorithms, and so on. Instead, the SymmetricEncryptorDecryptor component is designed to represent the essence of symmetric encryption and decryption services, i.e., the usage of the same cryptographic key for encryption and decryption. The SymmetricEncryptorDecryptor provides the interface SEncDecIf, which contains an operation encrypt() to symmetrically encrypt a plaintext pt using a cryptographic key ck. The result is a ciphertext ct. Additionally, it provides an inverse

---

[7] http://java.sun.com/javase/6/docs/api/
[8] http://www.jasypt.org/
[9] http://www.bouncycastle.org/

operation decrypt() that calculates the plaintext pt given the ciphertext ct and the cryptographic key ck, which has to be equal to the cryptographic key used for the encryption process. This relation between the encryption and decryption functions can be formally expressed as follows: $\forall\, pt : Data;\ ck : CryptographicKey \mid decrypt(encrypt(pt, ck), ck) = pt$. We equip the GSC SymmetricEncryptorDecryptor with the stereotype ≪critical≫, and we assign the plaintext pt to the tag {secrecy}, since the goal of this GSC is to keep the plaintext confidential.

In the next section, we explain how different GSCs and GNCs are combined to obtain patterns for secure software architectures related to CSPFs.

3.3   Generic Security Architectures

We combine the GSCs and GNCs constructed or selected for a given CSPF to obtain *generic security architectures* (GSA). Such a GSA represents the structure of the machine domain of the CSPF. Since GSCs and GNCs are platform-independent, so are GSAs. Based on the connection of CSPFs and GSAs, traceability links are introduced. Hence, our approach allows to understand which security requirements are realized by the different parts of a software architecture. This improves the maintainability of the software. Similar to GSCs and GNCs, we specify GSAs based on structural views using UML2.3 composite structure diagrams, and control and data flow views using UML2.3 sequence diagrams. The structural as well as the behavioral views of GSAs comprise the composed views of the GSCs and GNCs they consist of. We construct GSAs according to the following procedure:

1. An adequate basic software architecture to connect the GSCs and GNCs has to be selected. The GSA presented in the following and the ones introduced in [14, pp. 160 ff.] organize the components in a layered architecture. For this purpose, each of the GSAs contains an *application* component, which coordinates the behavior of all other components and provides a simplified interface (compared to directly using the interfaces of the components it connects) to the environment.
2. If components can be connected directly, one connects these components.
3. If components cannot be connected directly (e.g., because a component produces incompatible output for another component), additional *adapter* components to connect them must be introduced.
4. Interfaces between the machine and its environment must be designed in the GSA according to the interfaces of the machine domain of the corresponding CSPF.

We enrich GSAs with UMLsec4UML2 language elements to express security properties based on the CSPFs the GSAs are related to. We apply the stereotype ≪secure dependency≫ to the specification of the structural views of GSAs according to the following procedure:

1. The structural view of a GSA should be organized in a package stereotyped ≪secure dependency≫, which contains a class diagram to define port types

for the composite structure diagram that is also contained in this package. Connections between components contained in the composite structure diagram are expressed using either simple connectors or lollipop notation.

2. As described in Sect. 3.2, GSCs refer to assets, and they are already equipped with the ≪critical≫ stereotype and corresponding tagged values. GSCs connected in the structural view's composite structure diagram with other GSCs or / and GNCs might allow the transmission of assets to these components. According to the ≪secure dependency≫ stereotype, these GSCs or / and GNCs should be stereotyped ≪critical≫, too. Moreover, the tagged values of these components should be equal to the tagged values of GSCs that are connected to them.

3. ≪use≫ dependencies between components stereotyped ≪critical≫ in the structural view's composite structure diagram should be stereotyped according to the tagged values of the ≪critical≫ stereotype. That is, if the tag {secrecy} is assigned a value, then the corresponding ≪use≫ dependency between the components should be stereotyped ≪secrecy≫, and if the tag {integrity} is assigned a value, then the corresponding ≪use≫ dependency between the components should be stereotyped ≪integrity≫. The dependencies stereotyped ≪use≫ between components and interfaces of components labeled ≪critical≫ in the structural view's class diagram should be stereotyped analogously.

Moreover, the behavioral views of GSAs are equipped with the ≪data security≫ stereotype. Given a package stereotyped ≪data security≫ containing a structure and a behavior diagram, the requirements defined in the structure diagram using the ≪critical≫ stereotype should be fulfilled with respect to the behavior diagram and environment description (especially the malicious environment and the value of the tag {adversary}). We apply the stereotype ≪data security≫ to the specification of the behavioral views of GSAs according to the following procedure:

1. The behavioral view should be organized in a package stereotyped ≪data security≫.
2. The structural view previously discussed should be reused by importing the corresponding package into the one of the behavioral view.
3. A specification in terms of a set of sequence diagrams should be included in the behavioral view to describe the collaboration between the different GSCs and GNCs contained in the GSA at hand.
4. The attacker model, i.e., especially the {adversary} tag, is not assigned a value on the level of patterns. Instead, the attacker model is fixed when instantiating GSAs (see Sect. 3.4 for details).

According to the described procedures, we have developed a catalog of GSAs (see [14, pp. 160 ff.] for details) for each available CSPF.

GSA for CSPF Confidential Data Transmission Using Cryptographic Key-Based Symmetric Encryption. In the following, we present as an

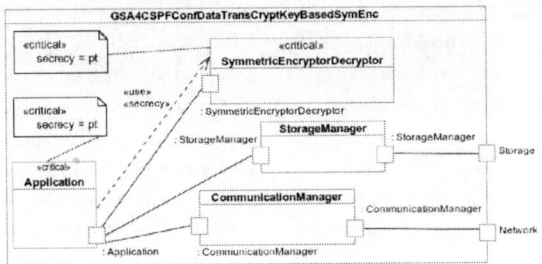

Fig. 2. Structural View of GSA for "CSPF Confidential Data Transmission Using Cryptographic Key-Based Symmetric Encryption"

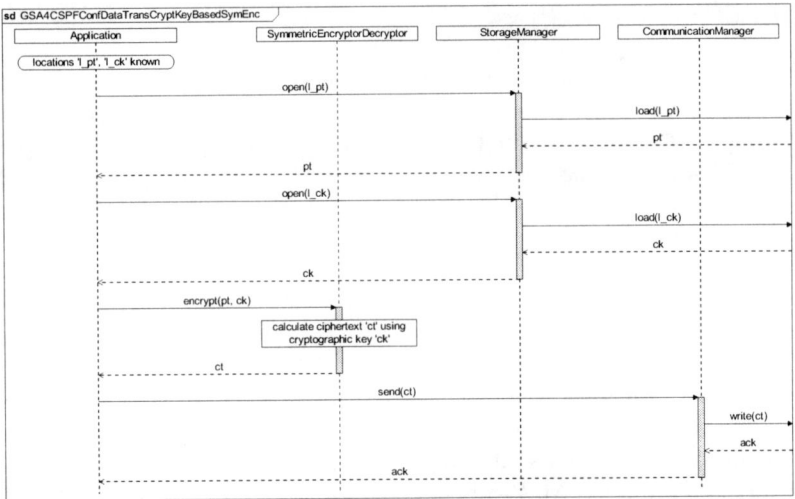

Fig. 3. Behavioral View Complementing Fig. 2

example a GSA for the machine domain of the CSPF confidential data transmission using cryptographic key-based symmetric encryption. Figure 2 shows the structural view of the GSA using a composite structure diagram. The Sender machine domain loads the Sent data and Cryptographic key₁ domains from a storage device. Hence, we introduce the GNC StorageManager to access a storage device. The Sent data domain is encrypted using the Cryptographic key₁ domain. For this reason, we introduce the GSC SymmetricEncryptorDecryptor presented in Sect. 3.2. Furthermore, the Sender machine domain sends the encrypted data to the Communication medium domain. Hence, we introduce the GNC CommunicationManager to access a network.

According to the CSPF this GSA is related to, the plaintext pt represented in the CSPF as lexical domain Sent data should be kept confidential. Hence, the GSC SymmetricEncryptorDecryptor and the GNC Application that makes use of this GSC are stereotyped ≪critical≫. Furthermore, the {secrecy} tag is assigned the plaintext pt, and the ≪use≫ dependency between the GSC SymmetricEncryptor-Decryptor and the GNC Application is stereotyped ≪secrecy≫.

The overall behavior of the GSA is depicted in Fig. 3. Initially, the locations of the plaintext pt (that corresponds to the Sent data domain) and the crypto-graphic key ck (that corresponds to the Cryptographic key₁ domain) are known, and they are retrieved from a storage device using the GNC StorageManager. Then, the ciphertext is constructed by the GSC SymmetricEncryptorDecryptor. Fi-nally, the GNC CommunicationManager sends the ciphertext to a network. Both the structural as well as the behavioral views are contained in a package (not explicitly depicted in this paper) stereotyped ≪data security≫.

In the next section, we discuss the instantiation of GSAs.

3.4   Instantiation of GSAs

GSAs are instantiated based on the corresponding CSPF instances. The selec-tion process for GSAs is heavily intertwined with the (security) requirements engineering phase and described in detail in [6]. An application layer compo-nent should have a name that equals the name of the machine domain of the corresponding CSPF instance. Moreover, the interfaces between a GSA and the environment should be named after the domains of the corresponding CSPF instance. The names of shared phenomena of the CSPFs should be re-used for the instantiation of the messages contained in the sequence diagrams of the be-havioral views of the corresponding GSAs. Thus, each component contained in a GSA is instantiated, too.

The CSPF instance that constitutes the basis for a GSA instance includes an environment description in terms of problem domains, interfaces in between, and domain knowledge. This information is used to construct an attacker model based on the UMLsec4UML2-stereotype ≪secure links≫. Given a package stereotyped ≪secure links≫ with the tagged value {adversary=default} containing a deployment diagram with a dependency stereotyped ≪secrecy≫ (or / and ≪integrity≫ or / and ≪high≫) between two nodes, these nodes should be either connected via a communication path stereotyped ≪LAN≫ or ≪encrypted≫ or ≪wire≫ (but not ≪Internet≫) or labeled ≪LAN≫. We now explain how we use the ≪secure links≫ stereotype to construct an at-tacker model based on the instantiated CSPFs:

1. An environment description in terms of a deployment diagram should be included in a package stereotyped ≪secure links≫. The diagram should represent the state of the environment before the envisaged software system is in operation. Hence, the constraint associated to the ≪secure links≫ stereotype is *not* fulfilled.
2. Each machine domain, lexical domain, and causal domain should be modeled as a node (3-D box) in the deployment diagram.
3. The physical connections between nodes should be modeled as communica-tion paths (solid line between two nodes) in the deployment diagram, and

each path should be stereotyped according to its type as either ≪LAN≫ or ≪encrypted≫ or ≪wire≫ or ≪Internet≫.

4. The tag {adversary} of the ≪secure links≫ stereotype should be assigned a value according to domain knowledge collected during requirements analysis. For example, if shared phenomena exist indicating that data items of a connection can be deleted, read, and inserted arbitrarily, then the ≪Internet≫ should be applied, and the tag {adversary} should be assigned the value default (see [8] for details).

5. The tag {adversary} of the ≪data security≫ stereotype of the behavioral views of the instantiated GSAs should be assigned the same value as the value of the equally named tag of the ≪secure links≫ stereotype.

In the next section, we show how different GSA instances are combined yielding global secure software architectures.

3.5   Composition of Different GSA Instances

Composing different GSA instances means that one must decide whether the components contained in more than one GSA instance should appear only once in the global architecture, i.e., whether they can be merged. Basically, three different categories of components must be considered: application layer components, GNC instances, and GSC instances.

Choppy et al. [2] developed for a set of *functional* subproblem classes a corresponding set of subarchitectures that solve these subproblems. Moreover, the subarchitectures are composed based on dependencies between the subproblems such as parallel, sequential, and alternative dependencies. We adopt the principles by Choppy et al. [2] to merge application layer components and GNC instances.

We now discuss the composition of GSA instances to obtain a global secure software architecture that still fulfills the security requirements realized by the corresponding GSA instances. Especially confidentiality requirements must be treated carefully, since the composition of incompatible components can lead to non-fulfillment of confidentiality requirements.

If two GSA instances contain GSC instances that serve the same purpose, then these components cannot be merged in general. The question to be answered is if the two GSC instances can use the same algorithm-specific configuration, e.g., a specific algorithm, key lengths, salt lengths, etc., to fulfill the different security requirements. For example, a specific encryption algorithm and specific key lengths might be sufficient to solve one security subproblem, while the same configuration would lead to a vulnerable system if applied to another subproblem. The level of abstraction of GSC instances might not allow to decide whether they can be merged or if they have to be kept separately. Then, it is necessary to refine the GSC instances to platform-specific security components to come to this decision. An approach to deal with the composition of GSC instances before their refinement to platform-specific security components is to merge GSC instances of the same type into one *configurable* GSC instance of this type. Here, configurable means that we equip such a component with a variable mode, i.e., the algorithm a component realizes as well as the used cryptographic key can be changed at runtime.

Choppy et al.[2] considered only the structural composition of subarchitectures. Since our GSA instances are additionally equipped with behavioral views, we also consider the composition of these descriptions. The resulting behavioral specification of the global secure architecture represents the life-cycle of the software to be constructed and its components. The behavior of GSA instances is given as a set of sequence diagrams, which should be composed based on the subproblem dependencies determined for the structural composition of the application layer components: in case of a sequential dependency, the sequence diagrams should be composed according to the order defined by this dependency. In case of a parallel dependency, the sequence diagrams should be composed in such a way that the effects and output realized by the different GSA instances are fulfilled jointly. If GSC instances are merged into one configurable GSC instance, then its re-configuration should be included in the corresponding sequence diagrams. The result is a platform-independent global secure software architecture described by structural and behavioral views.

We briefly discuss an approach to analyze global secure software architecture, i.e, to verify the constraints associated with UMLsec4UML2 stereotypes that are contained in such architectures.

3.6   Verification of Global Secure Software Architectures

We use the UMLsec4UML2-profile and the UMLsec tool suite to show that the GSC and GNC instances in a global secure software architecture work together in such a way that they fulfill the security requirements corresponding to the different subproblems. Based on the ≪secure dependency≫ stereotype and the OCL constraints contained in the UMLsec4UML2-profile, it is possible to check whether critical data items might be leaked ({secrecy} and {high}) or changed ({integrity}). These checks can be executed directly within compatible UML editing tools such as Papyrus UML and MagicDraw UML. Based on the stereotypes ≪data security≫ and ≪secure links≫ and the OCL constraints contained in the UMLsec4UML2-profile, it is possible to check whether behavior introduced by a GSA might compromise confidentiality ({secrecy} and {high}) or integrity ({integrity}). Such checks can be executed based on the UMLsec tool suite, which makes use of the SPASS theorem prover[10] for the verification of properties of behavioral models.

We now illustrate the previously presented approach in the next section.

4   Validation

We tested the approach presented in the previous sections using two case studies: a *secure text editor* and an *Internet-based password manager*. We performed

---

[10] http://www.spass-prover.org/

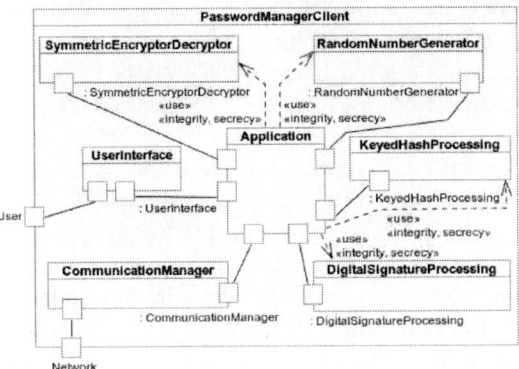

Fig. 4. Structural View of Password Manager Client

the complete development life-cycle for both case studies, i.e., from requirements engineering to architectural and fine-grained design to the implementation and testing of the programs. Moreover, we created all development artifacts as advocated by SEPP. While we present in this paper only a small part of the password manager case study, the complete results are contained in [14] and partly in a technical report about UMLsec4UML2 [15].

Security requirements analysis of the password manager following SEPP as outlined in Sect. 2 leads to the elicitation and analysis of 13 different security requirements, e.g., about the confidentiality and integrity of the different usernames and passwords. Due to partly overlapping security requirements, only 11 different CSPF instances are developed. Consequently, 11 different GSAs are instantiated and combined yielding a global secure software architecture. Figure 4 partly shows the structural view of global secure software architecture of the password manager client expressed using a composite structure diagram. There, instances of GSCs for encryption/ decryption, keyed hash processing, digital signature processing, generation of random numbers, as well as instances of GNCs for the user interface and network communication work together, coordinated by the application layer GNC instance, in order to ensure secure communication between the password manager client and server over the Internet. The global architecture consisting of instances of GSCs and GNCs significantly helped us to proceed with the development phases to follow. We identified adequate frameworks, off-the-shelf components, and API modules based on the generated artifacts, i.e., the components, the explicit interface descriptions, the protocol descriptions, and the UMLsec4UML2 security annotations. Consequently, the programming phase had to cover creating the glue code to connect the existing components and modules only. In summary, the case studies show that using patterns to bridge the gap between security requirements analysis and secure architectural design constitutes a feasible and promising contribution to the field of secure software engineering.

We also evaluated our approach with respect to ISO/IEC 15408:2005 aka Common Criteria (CC) certifications. The usage of the architectural artifacts generated following our approach for a CC certification is possible based on the TOE (Target Of Evaluation) Design Specification (TDS) of the class ADV Development. For instance, EAL (Evaluation Assurance Level) 5 requires a semi-formal modular design, i.e., a representation of the TOE's structure in terms of subsystems and a description of the parts the subsystem consists of in terms of modules. In addition to the TDS requirements for EAL 4, it is necessary to also describe those modules that represent SFR (Security Functional Requirement)-supporting modules in detail, i.e., by describing its SFR-related interfaces, return values from those interfaces, and called interfaces to other modules. These TDS requirements are met by the artifacts that describe the realizations of GSC and GNC instances. Moreover, a semiformal notation for the SFR-enforcing modules should be used. Since our approach makes use of UML2.3 diagrams, this requirement is fulfilled right away. The tool suite developed for the original version of UMLsec and checking the OCL constraints of the UMLsec4UML2-profile supports creating TDS documents. For instance, the stereotype «secure dependency» allows to track the occurence of assets in the complex TDS documents. Using an UML editing tool such as Papyrus UML the OCL constraints representing this stereotype can be verified to ensure that we covered all relevant occurences of an asset.

5   Related Work

Recently, an approach [10] to connect the security requirements analysis method *Secure Tropos* by Mouratidis et al. [3] and UMLsec [8] is published. Bryl et al. [1] extended the Secure Tropos variant by Massacci et al. [9] by an approach to automatically select design alternatives based on results from security requirements analysis. Compared to our work, these approaches are not based on patterns, and they rather focus on the transition to finer-grained secure design.

Choppy et al. [2] present architectural patterns for Jackson's basic problem frames [7]. The patterns constitute layered architectures described by UML composite structure diagrams. Similar to other approaches considering the connection between problem frames and software architectures such as [13, 4], the work by Choppy et al. does not consider security requirements, behavioral interface descriptions, and operation semantics. Furthermore, only a vague general procedure to derive components for a specific frame diagram is given in [2].

The vast body of patterns for secure software engineering (see [5] for an overview) can be used during the phase that follows the phase presented in this paper, i.e., these patterns are applied in fine-grained design of secure software. Hence, the existing security design patterns and our approach complement each other to such an extent that the existing patterns can be expressed in a unifying way based on SPFs, CSPFs, and GSAs.

6   Conclusions and Future Work

We presented in this paper a novel *pattern-oriented* and *tool-supported* approach to bridge the gap between security requirements analysis and secure architectural design. Its main benefit is that the construction of global secure software architectures based on results from security requirements engineering becomes more feasible, systematic, less error-prone, and a more routine engineering activity.

In the future, we plan to develop new UMLsec4UML2 stereotypes to specify assumptions and facts about the operational environment of the software. Moreover, we intend to develop patterns that support the systematic composition of different GSA instances thereby preserving the associated security requirements.

References

[1] Bryl, V., Massacci, F., Mylopoulos, J., Zannone, N.: Designing security requirements models through planning. In: Martinez, F.H., Pohl, K. (eds.) CAiSE 2006. LNCS, vol. 4001, pp. 33–47. Springer, Heidelberg (2006)

[2] Choppy, C., Hatebur, D., Heisel, M.: Component composition through architectural patterns for problem frames. In: Proceedings of the Asia Pacific Software Engineering Conference (APSEC), pp. 27–34. IEEE Computer Society, Washington, DC, USA (2006)

[3] Giorgini, P., Mouratidis, H.: Secure tropos: A security-oriented extension of the tropos methodology. International Journal of Software Engineering and Knowledge Engineering 17(2), 285–309 (2007)

[4] Hall, J.G., Jackson, M., Laney, R.C., Nuseibeh, B., Rapanotti, L.: Relating software requirements and architectures using problem frames. In: Proceedings of the IEEE International Requirements Engineering Conference (RE), pp. 137–144. IEEE Computer Society, Los Alamitos (2002)

[5] Heyman, T., Yskout, K., Scandariato, R., Joosen, W.: An analysis of the security patterns landscape. In: Proceedings of the International Workshop on Software Engineering for Secure Systems (SESS), pp. 3–10. IEEE Computer Society, Los Alamitos (2007)

[6] Heyman, T., Yskout, K., Scandariato, R., Schmidt, H., Yu, Y.: The security twin peaks. In: Erlingsson, Ú., Wieringa, R., Zannone, N. (eds.) ESSoS 2011. LNCS, vol. 6542, pp. 167–180. Springer, Heidelberg (2011)

[7] Jackson, M.: Problem Frames. In: Analyzing and structuring software development problems. Addison-Wesley, Reading (2001)

[8] Jürjens, J.: Principles for Secure Systems Design. PhD thesis, University of Oxford (2002)

[9] Massacci, F., Mylopoulos, J., Zannone, N.: An Ontology for Secure Socio-Technical Systems. Information Science Reference. In: Ontologies for Business Interaction, pp. 188–207 (2007)

[10] Mouratidis, H., Jürjens, J.: From goal-driven security requirements engineering to secure design. International Journal of Intelligent Systems – Special issue on Goal-Driven Requirements Engineering 25(8), 813–840 (2010)

[11] Pérez-Martínez, J.E., Sierra-Alonso, A.: UML 1.4 versus UML 2.0 as languages to describe software architectures. In: Oquendo, F., Warboys, B.C., Morrison, R. (eds.) EWSA 2004. LNCS, vol. 3047, pp. 88–102. Springer, Heidelberg (2004)

[12] Pfleeger, C.P., Pfleeger, S.L.: Security In Computing, 3rd edn. Prentice Hall PTR, Englewood Cliffs (2003)

[13] Rapanotti, L., Hall, J.G., Jackson, M., Nuseibeh, B.: Architecture-driven problem decomposition. In: Proceedings of the IEEE International Requirements Engineering Conference (RE), pp. 80–89. IEEE Computer Society, Los Alamitos (2004)

[14] Schmidt, H.: A Pattern- and Component-Based Method to Develop Secure Software. Deutscher Wissenschafts-Verlag (DWV) Baden-Baden (April 2010)

[15] Schmidt, H., Jürjens, J.: UMLsec4UML2 - adopting UMLsec to support UML2. Technical Report 838, Technical University of Dortmund (February 2011), http://hdl.handle.net/2003/27602

[16] Shaw, M., Garlan, D.: Software Architecture. Perspectives on an Emerging Discipline. Prentice Hall PTR, Englewood Cliffs (1996)

Transforming Enterprise Architecture Models: An Artificial Ontology View

Sandeep Purao[1], Richard Martin[2], and Edward Robertson[3]

[1] College of Information Sciences & Technology, Pennsylvania State University, PA 16802
spurao@ist.psu.edu
[2] Tinwisle Corp., Bloomington Indiana
richardm@tinwisle.com
[3] Indiana University and Persistent Systems Inc.
edrbtsn@indiana.edu

Abstract. Enterprise Architecture (EA) is, by definition, an artificial construct. It includes conceptual objects and attributes created for human purposes. EA models, therefore, require an ontological foundation that goes beyond the 'furniture of the world' metaphor. We develop an argument that supports this premise, and demonstrate how the perspective can help us understand operations on EA models. The paper demonstrates these operations with an example and briefly points to formalization efforts detailed elsewhere. The paper concludes with implications for research and practice.

Keywords: Enterprise Architecture, Artificial Ontology, Operations.

1 Introduction

Enterprise Architecture (EA) is the logic underlying a business (Ross et al. 2006). An EA outlines how the different technological, human and organizational elements within the business are structured, describes how they can be coordinated, and makes plain possibilities for manipulation for business improvements. Accurate yet mutable representations are, therefore, important for the practice of EA. These representations must also include not only the furniture of the world (Bunge 1977) but also conceptual constructs created for satisfying human goals (March and Allen 2007). Decision-makers can use these representations to chart the course for an enterprise; managers can use these to communicate with stakeholders (Schekkerman 2008).

A number of meta-models, frameworks and standards have been suggested for EA representations (e.g. Zachman 1987, DoD 2010, ISO Standards 42010, 15704). The manner in which practitioners use these meta-models and frameworks, however, remains unclear. Their use of these meta-models and frameworks cannot be the same as that for conventional software engineering where the designers' aim is to move from models to executable software. In contrast, EA efforts often include an archaeological expedition with the intention of charting a new course of action for the enterprise and facilitating stakeholder buy-in.

The research reported in this paper is presented against the above backdrop. We explore an alternative ontological basis for EA models and identify transformations

H. Mouratidis and C. Rolland (Eds.): CAiSE 2011, LNCS 6741, pp. 383–390, 2011.
© Springer-Verlag Berlin Heidelberg 2011

that capture how EA models are designed and used. The question we address is, therefore, the following: *What are the operations on EA models following an artificial ontology view?* The key contribution of our work is the development of these operations building on the ontology of the artificial (March and Allen 2007).

2   Conceptual Models and Modeling for Enterprise Architecture

Received wisdom for conceptual modeling for information systems (Wand and Weber 1990) builds on Bunge's (1977) ontology. It posits that conceptual models must show fidelity against the real-world they represent (Wand and Weber 2002). In contrast, EA practitioners use models for different purposes. Table 1 below summarizes the differences and Figure 1 outlines how EA models are used.

Table 1. Conceptual Models for Enterprise Architecture vs. Software

Concern	Enterprise Architecture	Software Applications
What	The underlying logic of a business	Structure and functionality of software
When	Often *post-facto*, after a business model is in place in the organization	Typically prior to the (detailed) design and implementation of software
How	A reflection of the architecture as it exists, and a vision of a desired state	Often based on the intended software implementation
Who	Provides a decision vehicle and what-if model for decision-makers	Translates user requirements into designs for software developers
Use	A communication mechanism to effect change	Produce a functional software that meets the requirements
Life cycle	Moving from extension to intension, effecting change in both	Moving from intension to extension in spite of round-trip engineering
Abstraction	Explicit use and instantiation from frameworks with multiple models	Implicit re-use of the ontology or domain models during design

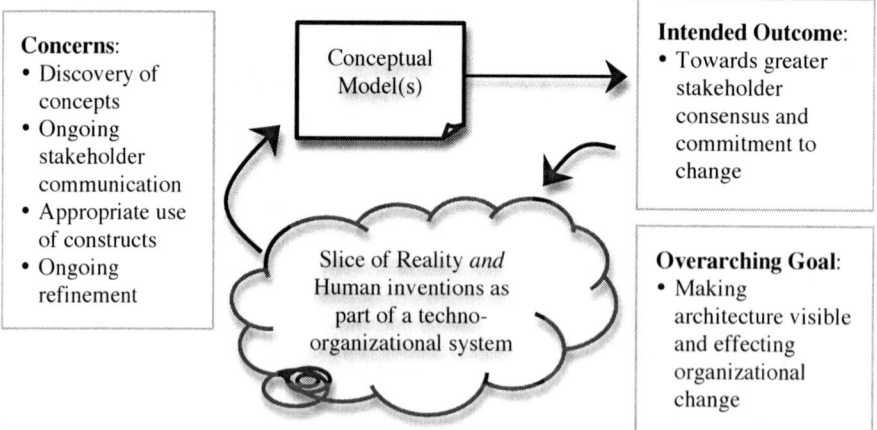

Concerns:
• Discovery of concepts
• Ongoing stakeholder communication
• Appropriate use of constructs
• Ongoing refinement

Conceptual Model(s)

Slice of Reality *and* Human inventions as part of a techno-organizational system

Intended Outcome:
• Towards greater stakeholder consensus and commitment to change

Overarching Goal:
• Making architecture visible and effecting organizational change

Fig. 1. Conceptual models and modeling for *Enterprise Architecture*

These differences provide inspiration for the work we present in this paper. We start by proposing an alternative to Bunge's ontology.

3 An Ontology of the Artificial

An ontology of the artificial begins with the premise that although business enterprises "use concrete objects such as people, machines, and buildings to accomplish their goals; they are concerned primarily with the meaning and purpose ascribed to concrete and conceptual objects and with invented rules ..." (March and Allen 2007). As a result, it "accommodate[s] occurrences of conceptual events that form the basis of social phenomena" (March and Allen 2007).

3.1 Fundamental Constructs

The constructs underlying the ontology of the artificial extend those suggested by Wand and Weber (1990). Table 2 outlines these (March and Allen 2007).

Table 2. Constructs in the Ontology

Construct	Elaboration
Objects: Concrete	Exist physically
Objects: Conceptual	Exist by human invention and social agreement (**new**)
Attributes: Substantial	Ascribed to concrete objects to represent human understanding of natural phenomena
Attributes: Invented	Ascribed to concrete and conceptual objects to enable social discourse (**new**). Can be used for individuation.
Types (classes)	Object grouped into Types based on ascription of one or more common attributes
Status (of an Object)	Is the set of values of its attributes at a point in time; the history of an object is the chronology of its status.
Events: Concrete	Effect changes to substantial properties of concrete objects; follow immutable, natural, discoverable laws.
Event: Conceptual	Effect changes to invented attributes of concrete or conceptual objects, and follow rules (designed by human agreement), which are mutable. (**new**)
Relationship	Objects affected by the same event are in a relationship with the event and hence, in relationship to each other through the event.
Decomposition	Objects as well as Events may compose and decompose to form other Objects or Events respectively.

March and Allen's (2007) proposal contains three new constructs. The *conceptual objects* and *invented attributes* allow addition of new objects and attributes as EA models are refined. The *conceptual events* allow negotiation of new modes of changes to objects and attributes, including a change in the stakeholders. Together, they allow possibilities for EA models. We use these to conceptualize operations on EA models.

4   Operations of EA Models

We identify six operations on EA models: Projection, Instantiation, Specialization, Refinement, Derivation, and Linking. Table 3 summarizes the operations.

Table 3. Operations on EA Models[*]

Operation	Description
Projection	Transforming a (set of) *objects* by selecting a subset of *attributes*
Instantiation	Creating a (set of) *objects/events* based on a construct in the meta-model
Refinement	Elaboration of a (set of) *conceptual objects/events* by addition of *invented attributes*; Decomposition of a (set of) *conceptual objects/events* by identifying *component objects*
Specialization	Adding variations to a (set of) *objects* by adding *invented attributes*
Derivation	Manipulation of *attributes* of a (set of) *objects* to derive values for related *objects;* Transformation of *objects* into other *objects*
Linking	Establishing a connection between a (set of) *objects/events*

* Note: The description uses constructs from the ontology (see Table 2), shown in *italics*.

4.1   Projection

Projection (P) refers to the transforming of an object into a form that consists only of those properties that a stakeholder requires. It is analogous to the existential quantification operation in predicate logic. The essential characteristic of Projection is discarding some information as a surjective and unary operation while preserving 'type'. Consider, for example, a shipping container manufacturer who uses a corrugating machine to produce a range of cardboard stock that must be cut, printed, folded, and glued on many machines to meet particular customer specifications. A Projection may involve extracting information related to machine capacities to structure a conveyance system that accumulates and forwards the stock to the appropriate printer and box machines.

4.2   Instantiation

Instantiation (I) is a constructive step. It adds the detail necessary for constructing a reification of a concept. For concrete entities, this can lead to tangible things in the world. For conceptual entities, instantiation can require assigning values to invented attributes. During the EA cycle, instantiation may *first* occur for the purpose of constructing an EA model from the framework or standard. A *second* level of instantiation may occur during the mapping and use of the EA model against specific instances of an Activity or Node. Instantiation of the first kind takes place during the design phase, whereas the second kind takes place during use of the EA models.

4.3   Refinement

Refinement (R) is an information adding operation. The essential characteristic of Refinement is the preservation of boundary. During an EA life cycle, Refinement is critical because it allows addition of information beyond the original creation of an object. It acknowledges that it is not feasible to fully specify the structure of an object upon its creation because of inputs needed from multiple stakeholders, and because of the artificial nature of the object itself. Refinement can occur via decomposition or elaboration. For example, one can refine the understanding of an automobile engine in two ways: adding facts about engine features, e.g. the displacement or horsepower (refinement); or exposing sub-components, e.g. block, head, and pistons (decomposition). Elaboration may require addition of Invented Attributes enhancing commitment from stakeholders; decomposition may result in addition of Conceptual Objects or Events providing a way to scale down commitment.

4.4 Specialization

Specialization (S) is the construction of variations on the basis of Concrete or Conceptual objects. It corresponds to the idea of creating sub-classes. It is qualitatively different from Elaboration in that Specialization involves spawning new objects that are still within the boundaries specified by the source object. Consider, for example, a global producer of manufactured products with customers in different countries. Each may have different trade rules regarding import tariffs. Here, Specialization may be applied to the object Country to distinguish between non-tariff countries and tariff countries by identifying and including different attributes that are part of each new object. During the EA life cycle, Specialization can help different stakeholders negotiate their spheres of responsibility.

4.5 Derivation

Derivation (D) refers to the changing of the form of one or more elements without changing its content. It is a critical operation during the EA life cycle because it goes beyond Specialization or Refinement. Unlike these two, Derivation allows manipulation of specific attributes, including attributes from multiple objects without the need to generate a new Object or a permanent elaboration of an Object. It allows stakeholders to combine attributes from a number of related Objects in response to their information needs. As an example, consider the derivation of elements necessary for an ISO 19439 (ISO 2006) model description from a Zachman Framework description. Derivation may also involve computation of aggregates based on values of attributes from component objects or attributes from other related objects.

4.6 Linking

Linking (L) is the idea that elements – Concrete and Conceptual Objects, Concrete and Conceptual Events – may be connected via arbitrary associations. The need for links arises because EA models can contain representations at different levels of abstraction. Consider, for example, the creation of a link within a meta-level or across meta-levels or between components in a part hierarchy. The transformations resulting from links are often implicit or unstated. The link between a process and the role that is going to be responsible for completion of that process transforms the process into a managed activity. The linking may also be the result of a business rule applied during a refinement in the architecture model.

The operations have been formalized. Further details of the formalization are available elsewhere (Martin et al. 2011).

5 Application

This section demonstrates their application to a case (adapted from Sessions 2007). The case describes an organization, MedAMore, that owns a regional chain of drugstores with a software package, MedAManage, consisting of three modules: M/Store, to run at the drug store; M/Warehouse, to run in a regional warehouse; and M/Home, to run at the home office. As the result of acquiring three regional chains, the software package has become an obstacle. M/Store now requires specializations such as regional insurance plans; M/Warehouse must reflect practices at regional warehouses; and the information-sharing approach cannot scale to the now 200 drugstores and regional offices. Upgrading is difficult because each module is large. The technical problems have created internal conflicts. The business side wants to acquire two more chains while IT was struggling to bring existing acquisitions online. EA is now being considered as a possible mechanism to build stronger partnerships across IT and business groups. Table 4 illustrates the operations on EA models that can help achieve these potentially stronger partnerships with the above case description.

Table 4. *Illustrating* operations with *MedAMore*

Operation	Described in the *MedAMore* Case
Projection	• Making intended consequences of IT investment decisions visible to the business stakeholders; making financial implications of IT investments visible to the Finance function.
Instantiation	• Constructing instances of elements in the framework, e.g. creating instances such as activity and object; and instantiating these, e.g. creating instances of Activity such as 'Data cleansing for pharmacy orders.'
Refinement	• Elaboration and Decomposition by adding new attributes to track the impact of IT investment decisions: investigation of financial impact and impact on IT infrastructure.
Specialization	• Generating regional or store-specific specialization for practices and modules. Distinguishing these based on criteria such as insurance plans or region resulting in changes to modules and reporting structure.
Derivation	• Transforming models specified by stakeholders into detailed requirements. Deriving information from these models about the impacts of IT investments for affects on different portions of the IT infrastructure.
Linking	• Linking operational views of business processes to design decisions, to organizational structure decisions, and IT investment decisions. Links traversed to understand impact of IT investments on business operations.

The example demonstrates how the operations can help understand design and use of EA models. We acknowledge that this does not constitute validation. We use the example as a demonstration, similar to the suggestion by Hevner et al (2004).

6 Discussion

Much prior research shows that EA models are realized through a series of actions by stakeholders, who not only design the models but also use them to transform the enterprise. The operations described in the paper emphasize this dual nature by capturing the evolutionary aspects of EA models. They build on the ontology of the artificial as a foundation, and allow explicit acknowledgement of the progression of EA models in practice (see, e.g. Grossman 2003; Ross et al 2006). Table 5 reconnects the arguments so far to the unique aspects of EA models.

Table 5. The Use of Operations for Conceptual Models of EA (see Table 1)

Dimension	Enterprise Architecture	Use of Operations
What	The underlying logic of a business	Difficult to specify up-front, must evolve; also stakeholders can have different views: **Refinement and Specialization**
When	Often *post-facto*, after a business model is in place in the organization	**Refinement and Linking** allows multiple stakeholders to take part in the definition and evolution of EA models
How	A reflection of the architecture as it currently exists, and with a vision of a desired state	**Decomposition and Specialization** allows fixing responsibility and allowing stakeholders to participate in modeling
Who	Provides a decision vehicle and what-if model for the decision-makers	**Decomposition and Derivation** can be useful to communicate with different stakeholders
Use	Used as a communication mechanism to effect change among stakeholders	**Projection** allows stakeholders to use and vary the different views of EA models and use Linking to ensure consistency
Life cycle	Moving from extension to intension, effecting change in both via iterations	Multiple levels of **Instantiation** including their reversal allows stakeholders to refine EA models
Abstraction	Explicit use and instantiation from frameworks with multiple models; for ensuring compliance	Use of frameworks along with **Instantiation and Derivation** allows consistency across EA models

We do not argue that the operations are complete. Arguing completeness will require greater scrutiny, and understanding of how the operations are carried out in practice. Our intent is to distinguish the different modes of action by stakeholders across the EA life cycle (see, e.g. Smolander et al. 2008).

This paper has argued for appropriateness of an ontology of the artificial for EA modeling. To the best of our knowledge, no prior work has unpacked the space of transformations that represent these changes in EA models. Our intent in describing these operations is not to provide a formal basis for their automation. Instead, we seek to identify these operations as a way to better understand the decisions and actions of stakeholders involved in the design and use of EA models. We hope that the operations we have identified can provide a foundation for further studies such as the locality of impact, traversing abstractions or construction of macros (e.g. Martin and Robertson 2008) and identification of evolution patterns, backed by empirical studies.

References

[1] Bunge, M.: Ontology: the furniture of the world. Springer, Heidelberg (1977)
[2] DoD Deputy Chief Information Officer (2010), DODAF Overview and Supporting Materials, http://cio-nii.defense.gov/sites/dodaf20/index.html (retrieved May 3, 2010)
[3] Grossman, I., Application of the NOAA Federated IT Enterprise Architecture Process. *Government Enterprise Architecture Conference*, June 2003.
[4] Hevner, A., et al.: Design Science Research in Information Systems. MIS Quarterly 28(1), 75–105 (2004)
[5] ISO 2000. International Organization for Standardization: Industrial Automation Systems - Requirements for Enterprise–Reference Architecture and Methodologies, ISO 15704:2000 (2000), http://www.iso.ch
[6] ISO 2006. International Organization for Standardization: Enterprise integration – framework for enterprise modeling, ISO 19439:2006 (2006), http://www.iso.ch
[7] ISO 2010 International Organization for Standardization: Systems and Software Engineering - Architecture Description (ISO/IEC CD1 42010) (2010), http://www.iso.ch (draft version of January 2010)
[8] March, S., Allen, G.: Challenges in Requirements Engineering: A Research Agenda for Conceptual Modeling. In: Design Requirements Workshop, Cleveland, OH, USA, June 3-6 (2007)
[9] Martin, R., Robertson, E.: Meta-matters. In: International Conference on Information Resources Management, Niagara Falls, Ontario (2008)
[10] Martin, R., Robertson, E., Purao, S.: Tracking Transformations in Enterprise Architecture Development. Working Paper (2011)
[11] McGann, T., Lyytinen, K.: How Information Systems Evolve by and for Use, Case Western Reserve University, USA. Sprouts: Working Papers on Information Systems 5(15) (2005), http://sprouts.aisnet.org/5-15
[12] Ross, J., Weill, P., Robertson, D.: Enterprise Architecture as Strategy. Harvard Business Press, Boston (2006)
[13] Schekkerman, J.: Enterprise Architecture Good Practices Guide: How to Manage the Enterprise Architecture Practice. Trafford Publishing (2008)
[14] Sessions, R., A comparison of top four enterprise integration methodologies (2007), http://msdn.microsoft.com/en-us/library/bb466232.aspx (retrieved 3 May 3, 2010)
[15] Smolander, K., Rossi, M., Purao, S.: Software architectures: Blueprint, literature, language or decision? European Journal of Information Systems, 114 (2008)
[16] Wand, Y., Weber, R.: An ontological model of an information system. IEEE Transactions on Software Engineering 16(11), 1282–1292 (1990)
[17] Wand, Y., Weber, R.: Research Commentary: Information Systems and Conceptual Modeling – A Research Agenda. Information Systems Research 13(4), 363–376 (2002)
[18] Zachman, J.: A framework for information systems architecture. IBM Systems Journal 26(3) (1987)

Handling Concept Drift in Process Mining

R.P. Jagadeesh Chandra Bose[1,2], Wil M.P. van der Aalst[1], Indrė Žliobaitė[1], and Mykola Pechenizkiy[1]

[1] Department of Mathematics and Computer Science, University of Technology, Eindhoven, The Netherlands
[2] Philips Healthcare, Veenpluis 5–6, Best, The Netherlands
{j.c.b.rantham.prabhakara,w.m.p.v.d.aalst,m.pechenizkiy}@tue.nl, zliobaite@gmail.com

Abstract. Operational processes need to change to adapt to changing circumstances, e.g., new legislation, extreme variations in supply and demand, seasonal effects, etc. While the topic of flexibility is well-researched in the BPM domain, contemporary *process mining* approaches assume the process to be in steady state. When discovering a process model from event logs, it is assumed that the process at the beginning of the recorded period is the same as the process at the end of the recorded period. Obviously, this is often not the case due to the phenomenon known as *concept drift*. While cases are being handled, the process itself may be changing. This paper presents an approach to analyze such *second-order dynamics*. The approach has been implemented in ProM[1] and evaluated by analyzing an evolving process.

Keywords: process mining, concept drift, flexibility, change patterns.

1 Introduction

In order to retain their competitive advantage in today's dynamic marketplace, it is increasingly necessary for enterprises to streamline their processes so as to reduce costs and to improve performance. Moreover, today's customers expect organizations to be flexible and adapt to changing circumstances. New legislation is also forcing organizations to change their processes. It is clear that the economic success of an organization is highly dependent on its ability to react to changes in its operating environment. Therefore, flexibility and change have been studied in-depth in the context of Business Process Management (BPM). For example, process-aware information systems have been extended to be able to flexibly adapt to changes in the process. State-of-the-art Workflow Management (WFM) and BPM systems provide flexibility. Moreover, in processes not driven by WFM/BPM systems there is even more flexibility as processes are controlled by people.

Although flexibility and change have been studied in-depth in the context of WFM and BPM systems, *existing process mining techniques assume processes*

---

[1] ProM is an extensible framework that provides a comprehensive set of tools/plugins for the discovery and analysis of process models from event logs. See http://www.processmining.org for more information and to download ProM.

H. Mouratidis and C. Rolland (Eds.): CAiSE 2011, LNCS 6741, pp. 391–405, 2011.

to be in steady state. Starting point for process mining is an event log containing a sequence of business events recorded by one or more information systems. Based on such an event log, processes can be discovered. Today's process discovery techniques are able to extract meaningful process models from event logs not containing any explicit process information. Using ProM, we have analyzed processes in more than 100 organizations. These practical experiences show that it is very unrealistic to assume that the process being studied is in steady state: while analyzing the process, changes can take place. For example, governmental and insurance organizations reduce the fraction of cases being checked when there is too much work in the pipeline. In case of a disaster, hospitals and banks change their operating procedures etc. Such changes are indirectly reflected in the event log. Moreover, analyzing such changes is of the utmost importance when supporting or improving operational processes.

In the data mining and machine learning communities, such second-order dynamics are referred to as *concept drift*, and has been studied in both supervised and unsupervised settings. Concept drift has been shown to be important in many applications and several successful stories have been reported in the literature [1,2,3]. However, existing work tends to focus on simple structures such as changing variables rather than changes to complex artifacts such as process models describing concurrency, choices, loops, cancelation, etc. In handling concept drifts in process mining, the following three main problems can be identified:

1. *Change (Point) Detection:* The first and most fundamental problem is to detect concept drift in processes, i.e., detect that a process change has taken place. If so, the next step is to identify the time periods at which changes have taken place.
2. *Change Localization and Characterization:* Once a point of change has been identified, the next step is to characterize the nature of change, and identify the region(s) of change (localization) in a process. Uncovering the nature of change is a challenging problem that involves both the identification of change perspective (for example, control-flow, data, resource, sudden, gradual etc.) and the exact change in itself.
3. *Unravel Process Evolution:* Having identified, localized and characterized the changes, it is necessary to put all of these in perspective. There is a need for techniques/tools that exploit and relate these discoveries. Unraveling the evolution of a process should result in the discovery of the change process (describing the second order dynamics).

In this paper, we focus on the first two problems. We propose features and techniques to detect changes (drifts), change points, and change localization in event logs from a control-flow perspective. The techniques proposed in this paper show significant promise in handling concept drifts. We further provide an outlook on some of the topics in concept drift and believe that this niche area, with its broad scope and relevance, evokes lots of interest in the research community.

The remainder of this paper is structured as follows. Related work is presented in Section 2. Section 3 describes the various aspects and nature of change. Section 4 introduces various features and techniques for detecting drifts in event logs. Section 5 describes the effectiveness of the features and techniques proposed in this paper in discovering change points and localization of changes through a case study. In Section 6, we project an outlook on some of the open research questions and directions in this area. The paper ends with some conclusions in Section 7.

2   Related Work

Over the last two decades many researchers have been working on process flexibility, e.g., making workflow systems adaptive. In [4,5] collections of typical change patterns are described. In [6,7] extensive taxonomies of the various flexibility approaches and mechanisms are provided. Ploesser et al. [8] have classified business process changes into three broad categories viz., sudden, anticipatory and evolutionary. This classification is used in this paper, but now in the context of event logs.

Despite that many publications on flexibility, most process mining techniques assume a steady state process. A notable exception is the approach by Günther et al. [9]. This approach uses process mining to provide an aggregated overview of all changes happened so far. However, this approach assumes that change logs are available, i.e., modifications of the workflow model are recorded. At this point in time very few information systems provide change logs. Therefore, this paper focuses on concept drift in process mining assuming only an event log as input. Concept drift refers to changes in the target variable(s)/concept induced by contextual shifts over time [10]. While the topic is well-studied in various branches of the data mining and machine learning community, the problem of concept drift has not been studied in the process mining community. While experiences from data mining and machine learning can be used to investigate concept drift in process mining, existing techniques cannot be used due to the complexity of process models and the nature of process change.

3   Aspects and Nature of Change in Business Processes

Three important perspectives in the context of business processes are the control-flow, data and resource perspective. One or more of these perspectives may be subjected to a change.

- *Control-flow/Behavioral Perspective:* This class of changes deals with the behavioral and structural changes in a process model. Just like the design patterns in software engineering, there exist *change patterns* capturing the common control-flow changes [4]. Control-flow changes can be classified into operations such as insertion, deletion, substitution and reordering of process fragments. For example, an organization which used to collect the fee after the processing and acceptance of an application can now change their process to enforce the payment of fee before the processing of an application. Here the *reordering* change pattern had been applied on the payment and application processing process fragments. As another example, with the addition of new product offerings, a *choice* construct is *inserted* into the product development process of an organization. In the context of PAIS, various control-flow change patterns have been proposed in [4,5]. Most of these control-flow change patterns are applicable to traditional information/workflow systems as well.

Sometimes, the control-flow structure of a process model can remain intact but the behavioral aspects of a model could have been changed. For example, consider an insurance agency that classifies claims as "high" or "low" depending on the amount claimed. An insurance claim of €1000 which would have been classified as high last year is categorized as a low insurance claim this year due to the organization's decision to increase the claim limit. The structure of the process remains intact but the routing of cases changes.

- *Data Perspective:* This class of changes refer to the changes in the requirement, usage, and generation of data in a process. Tasks may produce or require information/data. An example of change in a data perspective is enabling the execution of a task without the requirement of an otherwise needed data element d.
- *Resource Perspective:* This class deals with the changes in resources, their roles, and organizational structure, and their influence on the execution of a process. For example, there could have been a change pertaining to who executes an activity in what roles in a process. As another example, certain execution paths in a process could be enabled (disabled) upon the availability (non-availability) of resources. Furthermore, resources tend to work in a particular manner and this bias may change over time. For example, a resource can have a bias of executing a set of parallel activities in a specific sequential order. Such biases could be more prominent when a limited number of resources are available; the addition of new resources can remove this bias.

Based on the duration for which a change is active, one can classify changes into momentary and permanent. Momentary changes are short-lived and affect only a very few cases while permanent changes are persistent and stay for a while [6]. In this paper, we consider only permanent changes. Changes are perceived to induce a drift in the concept (process behavior). We identify four classes of drifts as depicted in Fig. 1 based on how they manifest.

- *Sudden Drift:* This corresponds to a substitution of an existing process M_1 with a new process M_2 as depicted in Fig. 1(a). M_1 ceases to exist from the moment of substitution. In other words, all cases (process instances) from the instant of substitution emanate from M_2. This class of drifts are typically seen in scenarios such as emergency response planning. As an example, airlines and airports changing their security processes due to a new regulation.
- *Recurring Drift:* This corresponds to the scenario where a set of processes reappear after some time (substituted back and forth) as depicted in Fig. 1(b). It is quite natural to see such a phenomenon with processes having a seasonal influence. For example, a travel agency might deploy a different process to attract customers during Christmas period. The recurrence of processes may be periodic or non-periodic. An example of a non-periodic recurrence is a deployment of a process subject to market conditions. The point of deployment and duration of deployment are both dependent on external factors (here, the market conditions).

- *Gradual Drift:* This refers to the scenario as depicted in Fig. 1(c) where a current process M_1 is replaced with a new process M_2. Unlike the sudden drift, here both processes coexist for some time with M_1 discontinued gradually. For example, a supply chain organization might introduce a new delivery process. However, this process is applicable only for orders taken henceforth. All previous orders still have to follow the older delivery process.
- *Incremental Drift:* This refers to the scenario where a substitution of process M_1 with M_N is done via smaller incremental changes as depicted in Fig. 1(d). This class of drifts is more pronounced in organizations adopting agile business process management methodology.

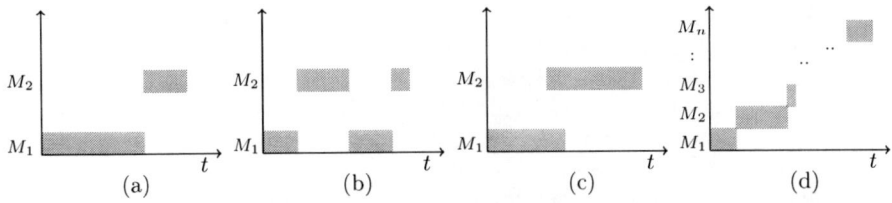

Fig. 1. Different types of drifts. (a) sudden drift (b) recurring drift (c) gradual drift and (d) incremental drift. X-axis indicate time and Y-axis indicate process variants. Shaded rectangles depict process instances.

4   Approaches to Detecting Drifts in Event Logs

We propose approaches to detect potential control-flow changes in a process manifested as sudden drifts over a period of time by analyzing its event log. Detecting drifts in data and resource perspectives and in the contexts of gradual, recurring and incremental drifts is beyond the scope of this paper.

4.1   Causal Footprints

Event logs are characterized by the relationships between activities. Dependencies between activities in an event log can be captured and expressed using the *follows* (or *precedes*) relationship. For any pair of activities, a and b $\in \Sigma$, one can determine whether they exhibit either *always*, *never*, or *sometimes* follows/precedes relationship. If b follows a in all the traces in an event log, then we say that b *always follows* a; if b follows a only in some subset of the traces or in none of the traces, then we say that b *sometimes follows* a, and b *never follows* a respectively. Consider an event log $\mathcal{L} = \{$acaebfh, ahijebd, aeghijk$\}$ containing three traces defined over $\Sigma =\{$a, b, c, d, e, f, g, h, i, j, k$\}$. The following relations hold in $\mathcal{L}$: e *always follows* a, e *never follows* b, and b *sometimes follows* a respectively. The variants of *precedes* relation can be defined on similar lines. The *follows/precedes* relationship is rich enough to reveal many control flow changes in a process. In the next section, we exploit this relationship and define various features for change detection.

4.2 Features Capturing the Manifestation of Activity Relationships

We distinguish between two classes of features (i) global features and (ii) local features. Global features are defined over an event log while local features can be defined at a trace level. Based on the follows (precedes) relation, we propose two global features viz., Relation Type Count and Relation Entropy, and two local features viz., Window Count and J-measure. These features are defined as follows:

- *Relation Type Count (RC):* The relation type count with respect to follows (precedes) relation is a function $f_{\mathrm{RC}} : \Sigma \to \mathbb{N}_0^3$ defined over the set of activities. f_{RC} of an activity, $\mathsf{b} \in \Sigma$ with respect to follows (precedes) relation over an event log $\mathcal{L}$ is a triple $\langle c_a, c_s, c_n \rangle$ where c_a, c_s, and c_n are the number of activities in Σ that always, sometimes, and never follow (precede) b in $\mathcal{L}$ respectively. For the event log $\mathcal{L}$ mentioned above, $f_{\mathrm{RC}}(\mathsf{a}) = \langle 2, 9, 0 \rangle$ since e and h always follows a while all other activities in $\Sigma \setminus \{\mathsf{e}, \mathsf{h}\}$ sometimes follows a. $f_{\mathrm{RC}}(\mathsf{i}) = \langle 1, 4, 6 \rangle$ since only j always follows i; b, d, e, and k sometimes follows i while a, c, f, g, h and i never follows i.
  For an event log containing $|\Sigma|$ activities, this results in a feature vector of dimension $3|\Sigma|$ (if either follows or precedes relation is considered) or $2 \times 3|\Sigma|$ (if both follows and precedes relation are considered).
- *Relation Entropy (RE):* The relation entropy with respect to follows (precedes) relation is a function $f_{\mathrm{RE}} : \Sigma \to \mathbb{R}^+$ defined over the set of activities. f_{RE} of an activity, $\mathsf{b} \in \Sigma$ with respect to follows (precedes) relation is the entropy of the relation type count metric. In other words, $f_{\mathrm{RE}}(\mathsf{b}) = -p_a \log p_a - p_s \log p_s - p_n \log p_n$ where $p_a = c_a/|\Sigma|, p_s = c_s/|\Sigma|$, and $p_n = c_n/|\Sigma|$.
  For the above example event log $\mathcal{L}$, $f_{\mathrm{RE}}(\mathsf{a}) = 0.68$ (corresponding to $f_{\mathrm{RC}}(\mathsf{a}) = \langle 2, 9, 0 \rangle$) and $f_{\mathrm{RE}}(\mathsf{i}) = 1.32$ (corresponding to $f_{\mathrm{RC}}(\mathsf{i}) = \langle 1, 4, 6 \rangle$). For an event log containing $|\Sigma|$ activities, this results in a feature vector of dimension $|\Sigma|$ or $2 \times |\Sigma|$ depending on whether either or both of follows/precedes relation is considered.
- *Window Count (WC):* The window count with respect to follows (precedes) relation is a function $f_{\mathrm{WC}} : \Sigma \times \Sigma \to \mathbb{N}_0$ defined over the set of activity pairs. Given a trace t and a window of size l, let S_l be the set of all subsequences $\mathsf{t}(i, i + l - 1)$, such that $\mathsf{t}(i) = \mathsf{a}$ and there exists a j such that $i < j < i + l$ and $\mathsf{t}(j) = \mathsf{b}$. The window count of the relation b *follows* a is defined as the number of sequences of length l in which b follows a. In other words, $f_{\mathrm{WC}}(\mathsf{a}, \mathsf{b}) = |S_l|$.
  For the above example event log $\mathcal{L}$, using a window of size $l = 4$, $f_{\mathrm{WC}}(\mathsf{a}, \mathsf{b}) = 1$ for trace `acaebfh` and 0 for traces `ahijebd` and `aeghijk`.
- *J-Measure:* Smyth and Goodman [11] have proposed a metric called J-measure based on [12] to quantify the information content (goodness) of a rule. We adopt this metric as a feature to characterize the significance of relationship between activities. The basis lies in the fact that one can consider the relation b follows a as a rule: "if activity a occurs, then activity b will probably occur". The J-measure with respect to follows (precedes) relation is a function $f_{\mathrm{J}} : \Sigma \times \Sigma \to \mathbb{R}^+$ defined over the set of activity pairs. Let $p(\mathsf{a})$ and $p(\mathsf{b})$ denote the probability of occurrence of activities a and b respectively in a trace t. Let $p_l(\mathsf{a}F\mathsf{b})$ denote the probability that b follows

a within a window of size l. Then the J-measure is defined as $f_J(\mathbf{a}, \mathbf{b}) = p(\mathbf{a})\mathrm{CE}_l(\mathbf{a}F\mathbf{b})$ where $\mathrm{CE}_l(\mathbf{a}F\mathbf{b})$ denotes the cross-entropy of **a** and **b** (**b** follows **a** within a window of size l) and is defined as

$$\mathrm{CE}_l(\mathbf{a}F\mathbf{b}) = p_l(\mathbf{a}F\mathbf{b}) \log \left(\frac{p_l(\mathbf{a}F\mathbf{b})}{p(\mathbf{b})} \right) + (1 - p_l(\mathbf{a}F\mathbf{b})) \log \left(\frac{1 - p_l(\mathbf{a}F\mathbf{b})}{1 - p(\mathbf{b})} \right)$$

The J-measure of **b** follows **a** for trace `acaebfh` using a window of size $l = 4$ is $f_J(\mathbf{a}, \mathbf{b}) = 0.147$.

Though local features are defined at a trace level, it is easy to lift them to the level of an entire event log.

4.3   Statistical Hypothesis Tests to Detect Drifts

One can consider an event log $\mathcal{L}$ as a time series of traces (traces ordered on their arrival time). Fig. 2 depicts such a perspective on an event log along with change points. An event log can be split into sub-logs of s traces each. We can consider either overlapping or non-overlapping windows when creating such sub-logs. Fig. 2 depicts the scenario where two subsequent sub-logs do not overlap. In this case, we have $k = \lceil \frac{n}{s} \rceil$ sub-logs for n traces. One can estimate the feature values for each trace separately (local features) or cumulatively over a subset of traces (local and global features) and generate a dataset defined by a matrix/vector of feature values over a sub-log/trace. For example, the relation count feature type will generate a dataset $\mathcal{D}$ of size $k \times 3|\Sigma|$ when either the follows/precedes relation counts of all activities are considered over $\mathcal{L}$. Instead, if the follows/precedes relation count of an individual activity is considered in isolation, it generates a dataset of size $k \times 3$ for $\mathcal{L}$. The J-measure generates a scalar value for each trace (sub-log) when an activity pair is considered thereby generating a vector of size $n \times 1$ or $k \times 1$ (depending on whether it is measured over traces or sub-logs) over $\mathcal{L}$. If all activity pairs are considered, then a dataset of size $n \times |\Sigma|^2$ or $k \times |\Sigma|^2$ is generated.

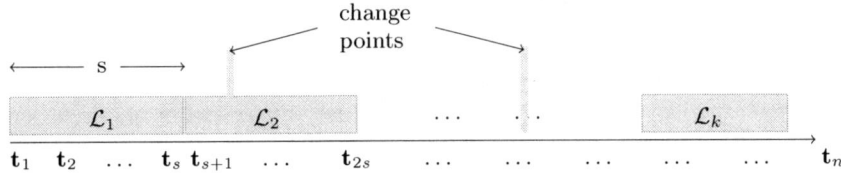

Fig. 2. An event log and change points

We believe that there should be a characteristic difference in the manifestation of feature values in the traces (sub-logs) before and after the change points with the difference being more pronounced at the boundaries. *The goal of concept drift in process mining is then to detect the change points and the nature of changes given an event log.* We propose the use of statistical hypothesis testing to discover these change points. Hypothesis testing is a procedure in which a *hypothesis* is evaluated on a sample data. One can distinguish between two classes of hypothesis tests (i) tests on a single population (single-sample tests) and (ii) tests on two populations (two-sample tests). Another classification of hypothesis tests is concerned with the dimensionality of each data element in a sample. Tests dealing with scalar data elements are called as *univariate* tests while those dealing with vector data elements are called as *multi-variate* tests. For our problem, *two-sample univariate and multi-variate tests are appropriate.* The dataset $\mathcal{D}$ of feature values can be considered as a time series as depicted in Fig. 3. Each $\mathbf{d}_i \in \mathcal{D}$ corresponds to a feature value for a trace (or sub-log) and can be a scalar or a vector. *The basic idea is to consider a series of successive populations of values (of size w) and investigate if there is a significant difference between the two populations.* The premise is that differences are expected to be perceived at change points provided appropriate characteristics of the change are captured as features. A moving window of size w is used to generate the populations. Fig. 3 depicts a scenario where two populations $P_1 = \langle \mathbf{d}_1, \mathbf{d}_2, \ldots, \mathbf{d}_w \rangle$ and $P_2 = \langle \mathbf{d}_{w+1}, \mathbf{d}_{w+2}, \ldots, \mathbf{d}_{2w} \rangle$ of size w are considered. In the next iteration, the populations correspond to $P_1 = \langle \mathbf{d}_2, \mathbf{d}_3, \ldots, \mathbf{d}_{w+1} \rangle$ and $P_2 = \langle \mathbf{d}_{w+2}, \mathbf{d}_{w+3}, \ldots, \mathbf{d}_{2w+1} \rangle$. Given a dataset of m values, the number of population pairs will be $m - 2w + 1$.

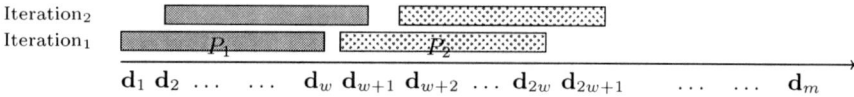

Fig. 3. Dataset of feature values considered as a time series for hypothesis tests. P_1 and P_2 are two populations of size w

We will use the univariate two sample *Kolmogorov-Smirnov* test (*KS* test) and *Mann-Whitney U* test (*MW* test) as hypothesis tests for univariate data, and the two sample *Hotelling T^2* test for multivariate data. The *KS* test evaluates the hypothesis "Do the two independent samples (populations P_1 and P_2) represent two different cumulative frequency distributions?" while the *MW* test evaluates the hypothesis "Do the two independent samples have different distributions with respect to the rank-ordering of the values?". The multi-variate Hotelling T^2 test is a generalization of the *t*-test and evaluates the hypothesis "Do the two samples have the same mean pattern?". All of these tests yield a *significance probability* assessing the validity of the hypothesis on the samples. We refer the reader to [13] for a classic introduction to various hypothesis tests.

5   Case Study and Discussion

We illustrate the concepts presented in this paper with an example process. The process corresponds to the handling of health insurance claims in a travel agency. Upon registration of a claim, a general questionnaire is sent to the claimant. In parallel, a registered claim is classified into a high or low claim. For low claims, two independent tasks, viz., check insurance and check medical history need to be executed. For high claims, three tasks need to be executed viz., check insurance, check medical history, and contact doctor/hospital for verification. If one of the checks shows that the claim is not valid, then the claim is rejected; otherwise, it is accepted. An insurance grant and acceptance decision letter is prepared in cases where a claim is accepted while a rejection decision letter is created for rejected claims. In both cases, a notification is sent to the claimant. Three modes of notification are supported viz., by email, by telephone (fax) and by postal mail. The case should be archived upon notifying the claimant. This can be done with or without the response for the questionnaire. However, the decision of ignoring the questionnaire can only be made after a notification is sent. The case is closed upon completion of archiving task.

Fig. 4 depicts five variants of this process represented in YAWL [14] notation. The dashed rectangles indicate regions where a change has been done in the process model with respect to its previous variant. The changes can have various reasons. For example, in Fig. 4(a), the different checks for high insurance claims are modeled using a parallel construct. However, a claim could be rejected if any one of the checks fail. In such cases, the time and resources spent on other checks go waste. To optimize this process, the agency can decide to enforce an order on these checks and proceed on checks only if the previous check results are positive. In other words, the process is modified with a *knockout* strategy adopted for high insurance checks as depicted in Fig. 4(b). As another example, the OR-construct pertaining to the sending of notification to claimants in Fig. 4(c) has been modified to an exclusive-or (XOR) construct in Fig. 4(d). The organization could have taken a decision to reduce their workforce as a cost-cutting measure. Due to availability of limited resources, they would like to minimize the redundancy of sending the notification through different modes of communication and restrict it to only one of the modes.

Let us denote these process variants as M_1, M_2, M_3, M_4 and M_5. We have modeled each of these process variants in CPN tools [15] and simulated 1200 traces for each model. We created an event log $\mathcal{L}$ of 6000 traces by juxtaposing each set of the 1200 traces. The event log contains 15 activities or event classes (i.e., $|\Sigma| = 15$) and 58953 events. Given this event log $\mathcal{L}$, *our first objective is to detect the four change points pertaining to these five process variants* as depicted in Fig. 5.

The ideas presented in this paper have been implemented as the *concept drift* plugin in ProM. We have considered global features (at sub-log level) and local features (both at trace and sub-log level) for our analysis. To facilitate this, we have split the log into 120 sub-logs using a split size of 50 traces. We have computed the relation type count (RC) of all 15 activities thereby generating a multi-variate vector of 45 features for each sub-log. We have applied the Hotelling T^2 hypothesis test on this multi-variate dataset using a moving window of size, $w = 8$. For this hypothesis test, we have randomly chosen 6 of the 45 features with a 10-fold cross validation. Fig. 6a depicts the average significance probability of the Hotelling T^2 test for the 10 folds on this feature set. The troughs in the plot signify that there is a change in the distribution of the feature values in the log. In other words, they indicate that there is drift (change) in the concept, which here corresponds to the process. It is interesting to see that the troughs

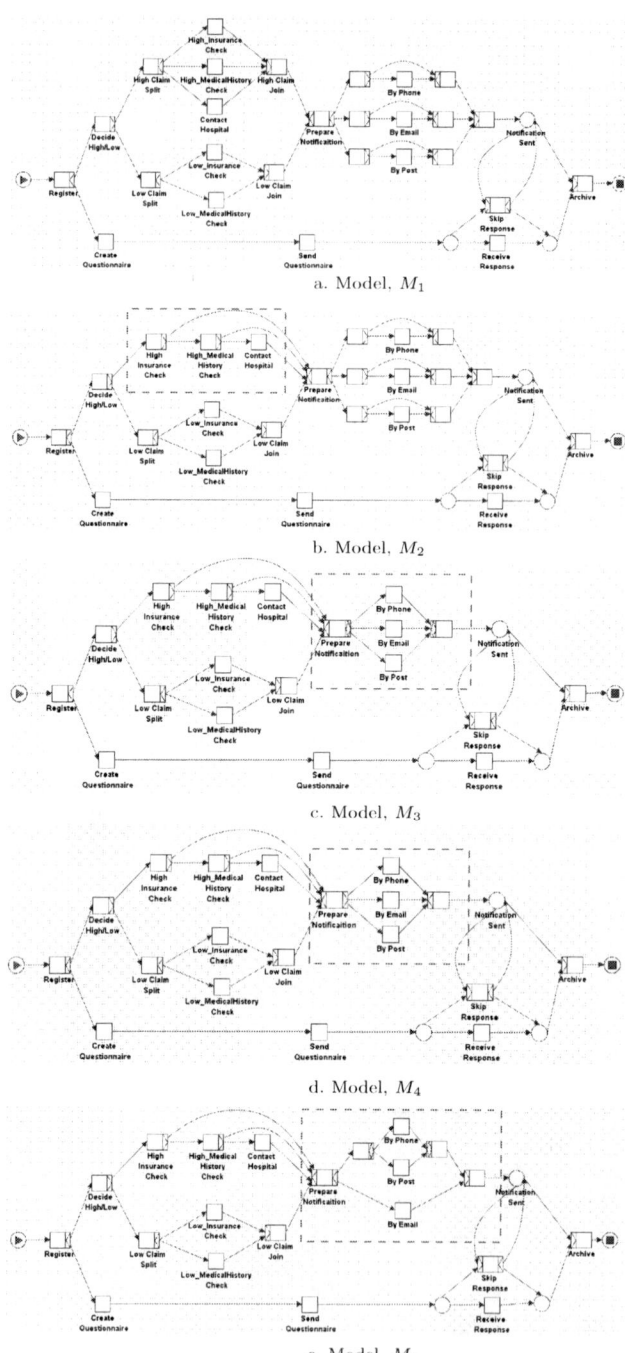

a. Model, M_1

b. Model, M_2

c. Model, M_3

d. Model, M_4

e. Model, M_5

Fig. 4. Five variants of an insurance claim process of a travel agency represented in YAWL notation. The dashed rectangles indicate the regions of change from its previous model.

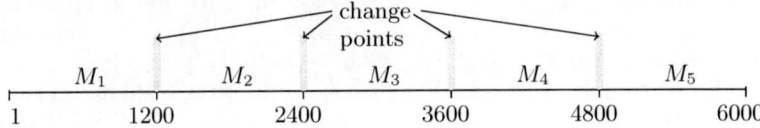

Fig. 5. Event log with traces from each of the five models juxtaposed. Also indicated are change points between models.

are observed around indices 24, 72 and 96 which are indeed the points of change (remember that we have split the log into 120 sub-logs with the change points at indices 24, 48, 72 and 96). The change at index 48 corresponding to the transition from M_2 to M_3 could not be uncovered using this feature set due to the fact that the relation type counts would be alike for logs generated from these two process variants.

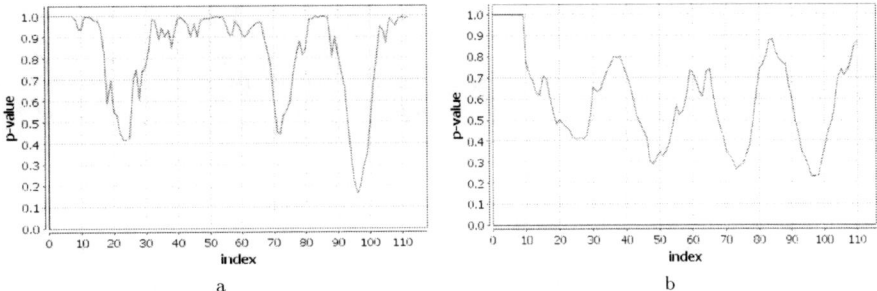

Fig. 6. (a) Significance probability of Hotelling T^2 test on relation counts (b) Average significance probability (over all activity pairs) of KS-test on J-measure. The event log is split into sub-logs of 50 traces each. X-axis represents the sub-log index. Y-axis represents the significance probability of the test. Troughs signify change points.

We have computed the J-measure for each sub-log and for every pair of activities, a, b in Σ (aFb, b follows a within a window of size 10). The univariate Kolmogorov-Smirnov test using a window size of $w = 10$ is applied on the J-measure of each activity pair. Fig. 6b depicts the average significance probability of KS-test on all activity pairs. It could be seen that significant troughs are formed at indices 24, 48, 72 and 96 which correspond to the actual change points. Unlike the relation type count feature, the J-measure feature is able to capture all the four changes in the models. This can be attributed to the fact that the J-measure uses the probability of occurrence of activities and their relations. In M_2, there could be cases where all the modes of notification are skipped (XOR construct). However in M_3 at least one of the modes need to be executed (OR construct). This results in a difference in the distribution of activity probabilities and their relationship probabilities which is elegantly captured in the J-measure.

We have considered the J-measure for each trace separately instead of at the sub-log level. Each activity pair generates a vector of dimension 6000 corresponding to the J-measure of that activity pair in each trace. The univariate

Kolmogorov-Smirnov test using a window size of $w = 400$ is applied to the vector corresponding to each activity pair in $\Sigma \times \Sigma$. Fig. 7 depicts the average significance probability of KS-test on all activity pairs. It could be seen that significant troughs are formed at indices 1200, 2400, 3600 and 4800. These are indeed the points where the models have been changed. Thus the features and approach proposed in this paper are shown to have significant promise in accurately identifying the points of change.

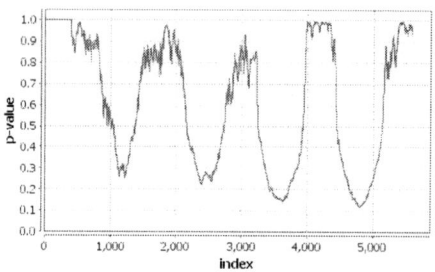

Fig. 7. Average significance probability (over all activity pairs) of KS-test on J-measure estimated for each trace. X-axis represents the trace index. Y-axis represents the significance probability of the test. Troughs signify change points.

The second objective in handling concept drift is that of *change localization*. In order to localize the changes (identify the regions of change), we need to consider each activity pair individually or a subset of activity pairs. For example, the change from M_1 to M_2 is localized in the region pertaining to high insurance claim checks. We expect characteristic changes in features pertaining to these activities and other activities related to these activities. For example, in M_1, the activities 'High Medical History Check' and 'Contact Hospital' always follow the activity 'Register' whenever a claim is classified as high. In contrast, in M_2, these activities need not always follow 'Register' due to the fact that both these activities are skipped if 'High Insurance Check' fails while 'Contact Hospital' is skipped if 'High Medical History Check' fails. During simulation, we have set the probability of success of a check to 90%. We have considered the window count (WC) feature for the activity relation 'Contact Hospital' follows 'Register' on a window size of 10 in each trace separately. Fig. 8a depicts the significance probability of the univariate KS-test using a window size of $w = 200$ on this feature. It could be seen that one dominant trough is formed at index 1200 indicating that there exists a change in the region between 'Register' and 'Contact Hospital'. No subsequent changes with respect to this activity pair is noticed which is indeed the case in the models.

As another example, we have considered the activity 'Prepare Notification' along with all the three 'Send Notification' activities. There exists a change pertaining to these activities between models M_2 and M_3, M_3 and M_4, and M_4 and M_5. More specifically, we have considered the window count feature on the activity relations 'Send Notification By Phone' follows 'Prepare Notification', 'Send Notification By email' follows 'Prepare Notification' and 'Send Notification

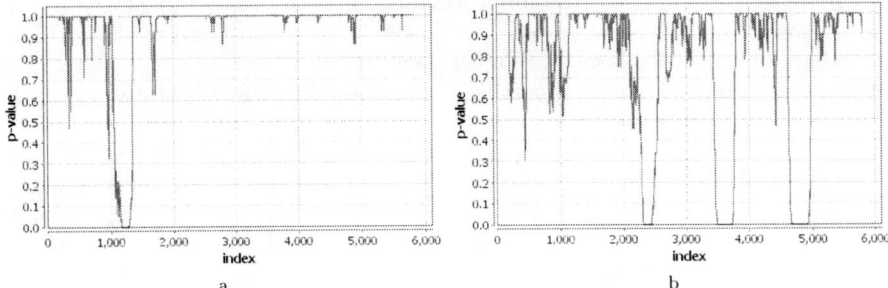

Fig. 8. (a) Significance probability of *KS*-test on *WC* feature estimated for the relation, 'Contact Hospital' follows 'Register'. Trough indicate change point w.r.t this feature. (b) Average significance probability (over activity pairs) of *KS*-test on *WC* feature estimated for the various modes of 'Send Notification' follows 'Prepare Notification' relation. Troughs indicate change point w.r.t these activities. *X*-axis represents the trace index. *Y*-axis represents the significance probability of the test.

By Post' follows 'Prepare Notification'. Fig. 8b depicts the average significance probability of the univariate *KS*-tests using a window size of $w = 200$ on the WC feature of these three activity pairs. We see three dominant troughs at around indices 2400, 3600 and 4800 signifying the changes in the models. Certain false alarms (minor troughs) can also be noticed in this plot. One means of alleviating this is to consider only those alarms with a significance probability less than a threshold, δ. In this fashion, by considering activities (and/or activity pairs) of interest, one can localize the regions of change. Furthermore, one can also get answers to diagnostic questions such as *"Is there a change with respect to activity a in the process at time period t"*?

6   Outlook

Dealing with concept drifts raises a number of scientific and practical challenges. In this section, we highlight some of these challenges.

- *Change-Pattern Specific Features:* In this paper, we presented very generic features (based on follows/precedes relation). These features are neither complete nor sufficient to detect all classes of changes. An important direction of research would be to define features catering to different classes of changes and investigate their effectiveness. A taxonomy/classification of change patterns and the appropriate features for detecting changes with respect to those patterns is needed. For example, if we would like to detect changes pertaining to a loop construct (insertion/removal/modification of loops as changes in process variants), tandem arrays [16] would be an appropriate feature to consider.
- *Feature Selection:* The feature sets presented in this paper result in a large number of features. For example, the activity relation count feature type generates $3|\Sigma|$ features whereas the window count and *J*-measure generate $|\Sigma|^2$ features (corresponding to all activity pairs). On the one hand, such high

dimensionality makes the computational complexity intractable for most real life logs. On the other hand, changes being typically concentrated in a small region of a process makes it unnecessary to consider all features. There is a need for dimensionality reduction techniques that can efficiently select the most appropriate features.

- *Holistic Approaches:* In this paper, we discussed ideas on change detection and localization in the context of sudden drifts and owing to the control-flow perspective of a process. However, as mentioned in Section 3, data and resource perspectives are also equally important. So are the contexts of gradual, recurring and incremental drifts. Features and techniques that can enable the detection of changes in these other perspectives need to be discovered. Furthermore, there could be instances where more than one perspective (e.g., both control and resource) change simultaneously. Hybrid approaches considering all aspects of change holistically need to be developed.

- *Techniques for Drift Detection:* In this paper, we explored just the Hotelling T^2 test to deal with multi-variate data. In addition, we have dealt with multiple features by considering univariate hypothesis tests on each feature separately and averaging the test results over all features. Further investigation needs to be done on hypothesis tests devised naturally for multi-variate data. Also, determining an appropriate size of the window for hypothesis tests is nontrivial; this mandates further study on understanding the influence of window size on the results. Alternatives to hypothesis testing that can uncover drifts and diagnose the changes are a welcome addition to the repertoire of techniques for handing concept drifts in process mining.

- *Sample Complexity:* Sample complexity refers to the number of traces (size of the event log) needed to detect, localize, and characterize changes within acceptable error bounds. This should be sensitive to the nature of changes, their influence and manifestation in traces, and the feature space and algorithms used for detecting drifts. On a broader note, the topic of sample complexity is relevant to all facets of process mining and is hardly addressed. For example, it would be interesting to know the lower bound on the number of traces required to discover a process model with a desired fitness.

7   Conclusions

This paper introduced the topic of concept drift in process mining, i.e., analyzing process changes based on event logs. We proposed feature sets and techniques to effectively detect the changes in event logs and identify the regions of change in a process. The approach has been implemented in ProM and evaluated using synthetic data. This is a first step in the direction of dealing with changes in any process monitoring and analysis efforts. We considered changes only with respect to the control-flow perspective manifested as sudden drifts. However, there is much to be done on various other perspectives mentioned in this paper. Moreover, to further validate the approach we plan to conduct extensive case studies based on real-life event logs.

Acknowledgments. R.P.J.C. Bose and W.M.P. van der Aalst are grateful to Philips Healthcare for funding the research in process mining.

References

1. Žliobaitė, I.: Learning under Concept Drift: an Overview. Technical report, Faculty of Mathematics and Informatics, Vilnius University, Vilnius, Lithuania (2009)
2. Pechenizkiy, M., Bakker, J., Žliobaitė, I., Ivannikov, A., Kärkkäinen, T.: Online Mass Flow Prediction in CFB Boilers with Explicit Detection of Sudden Concept Drift. SIGKDD Explorations 11(2), 109–116 (2009)
3. Tsymbal, A., Pechenizkiy, M., Cunningham, P., Puuronen, S.: Handling Local Concept Drift with Dynamic Integration of Classifiers: Domain of Antibiotic Resistance in Nosocomial Infections. In: CBMS, pp. 679–684 (2006)
4. Weber, B., Rinderle, S., Reichert, M.: Change Patterns and Change Support Features in Process-Aware Information Systems. In: Krogstie, J., Opdahl, A.L., Sindre, G. (eds.) CAiSE 2007 and WES 2007. LNCS, vol. 4495, pp. 574–588. Springer, Heidelberg (2007)
5. Mulyar, N.: Patterns for Process-Aware Information Systems: An Approach Based on Colored Petri Nets. PhD thesis, University of Technology, Eindhoven (2009)
6. Schonenberg, H., Mans, R., Russell, N., Mulyar, N., van der Aalst, W.M.P.: Process Flexibility: A Survey of Contemporary Approaches. In: Dietz, J., Albani, A., Barjis, J. (eds.) Advances in Enterprise Engineering I. LNBIP, vol. 10, pp. 16–30. Springer, Berlin (2008)
7. Regev, G., Soffer, P., Schmidt, R.: Taxonomy of Flexibility in Business Processes. In: Proceedings of the 7th Workshop on Business Process Modelling, Development and Support, BPMDS, Citeseer (2006)
8. Ploesser, K., Recker, J.C., Rosemann, M.: Towards a Classification and Lifecycle of Business Process Change. In: Proceedings of BPMDS, vol. 8 (2008)
9. Günther, C.W., Rinderle-Ma, S., Reichert, M., van der Aalst, W.M.P.: Using Process Mining to Learn from Process Changes in Evolutionary Systems. International Journal of Business Process Integration and Management 3(1), 61–78 (2008)
10. Widmer, G., Kubat, M.: Learning in the Presence of Concept Drift and Hidden Contexts. Machine learning 23(1), 69–101 (1996)
11. Smyth, P., Goodman, R.M.: Rule Induction Using Information Theory. In: Knowledge Discovery in Databases, pp. 159–176. AAAI Press, Menlo Park (1991)
12. Blachman, N.M.: The Amount of Information that y Gives About X. IEEE Transactions on Information Theory IT-14(1), 27–31 (1968)
13. Sheskin, D.: Handbook of Parametric and Nonparametric Statistical Procedures. Chapman & Hall/CRC (2004)
14. van der Aalst, W.M.P., ter Hofstede, A.H.M.: YAWL: Yet Another Workflow Language. Information Systems 30(4), 245–275 (2005)
15. Vinter Ratzer, A., Wells, L., Lassen, H.M., Laursen, M., Qvortrup, J.F., Stissing, M.S., Westergaard, M., Christensen, S., Jensen, K.: CPN Tools for Editing, Simulating, and Analysing Coloured Petri Nets. In: van der Aalst, W.M.P., Best, E. (eds.) ICATPN 2003. LNCS, vol. 2679, pp. 450–462. Springer, Heidelberg (2003)
16. Jagadeesh Chandra Bose, R.P., van der Aalst, W.M.P.: Abstractions in Process Mining: A Taxonomy of Patterns. In: Dayal, U., Eder, J., Koehler, J., Reijers, H.A. (eds.) BPM 2009. LNCS, vol. 5701, pp. 159–175. Springer, Heidelberg (2009)

An Iterative Approach for Business Process Template Synthesis from Compliance Rules

Ahmed Awad[1], Rajeev Goré[2], James Thomson[2], and Matthias Weidlich[1]

[1] Hasso Plattner Institute, University of Potsdam, Germany
{ahmed.awad,matthias.weidlich}@hpi.uni-potsdam.de
[2] School of Computer Science, The Australian National University, Australia
{Rajeev.Gore,jimmy.thomson}@anu.edu.au

Abstract. Companies have to adhere to compliance requirements. Typically, both, business experts and compliance experts, are involved in compliance analysis of business operations. Hence, these experts need a common understanding of the business processes for effective compliance management. In this paper, we argue that process templates generated out of compliance requirements can be used as a basis for negotiation among business and compliance experts. We introduce a semi automated approach to synthesize process templates out of compliance requirements expressed in Linear Temporal Logic (LTL). As part of that, we show how general constraints related to business process execution are incorporated. Building upon existing work on process mining algorithms, our approach to synthesize process templates considers not only control-flow, but also data-flow dependencies. Finally, we elaborate on the application of the derived process templates and present an implementation of our approach.

Keywords: Process synthesis, Analysis of business process compliance specification, Process mining.

1 Introduction

Recently, there has been a growing interest in compliance checking of business operations. Financial scandals in large companies led to legislative initiatives, such as SOX [1]. The purpose of these initiatives is to enforce controls on the business operations. Such controls relate to the execution order of business activities, the absence of activity execution in a dedicated data context, or restrictions on role resolution to realize separation of duty.

Driven by these trends, numerous approaches have been presented to address compliance management of business processes. In general, we can distinguish two types of approaches. First, compliance rules can guide the design of a business process [12,13]. These approaches ensure *compliance by design* by identifying compliance violations in the course of process model creation. Second, existing process models are *verified* against compliance rules [10,6]. Given compliance requirements and a process model as input, these approaches identify violations on the process model level.

Evidently, addressing compliance during the design of business operations has many advantages. Non-compliant processing is prevented at an early stage of process implementation and costly post-implementation compliance verification along with root

H. Mouratidis and C. Rolland (Eds.): CAiSE 2011, LNCS 6741, pp. 406–421, 2011.

cause analysis of non-compliance is not needed. In most cases, process models that are synthesized from compliance rules cannot be directly used for implementing a business process. Instead, they should be seen as a blueprint that is used as a basis for negotiation between business and compliance experts. Hence, we refer to these process models as *process templates* in order to emphasize that further refinements are needed to actually implement the business process. While this approach has been advocated by other authors, e.g., [12,11,25], existing approaches are limited when it comes to data-dependent compliance requirements.

In this paper, we present an approach to the synthesis of compliant process templates that avoids some of the pitfalls of existing approaches. We start with a set of compliance rules specified in LTL. Hence, we do not require the definition of explicit points in time as in [12,11], but focus on relative execution order dependencies. Further, we also consider data flow dependencies between activity executions, which is neglected in [25]. These rules are then enriched with general constraints related to business process execution to avoid phenomena such as vacuous satisfiability. Subsequently, a process template is generated automatically if the compliance requirements are satisfiable. We also illustrate how generated templates are applied during process design and how the template generation may identify inconsistencies and open questions. Hence, the template guides further refinements of the process model and the compliance requirements. To evaluate the applicability of our approach, we present a prototypical implementation. Our contribution is a complete approach to process design grounded in compliance rules.

Against this background, the remainder of this paper is structured as follows. The next section introduces preliminaries for our work, such as the applied formalism. Section 3 introduces our approach of synthesizing process templates from a given set of compliance rules. We also elaborate on how to use these templates as a basis for process design. A prototypical implementation of our approach is presented in Section 4. Finally, we discuss related work in Section 5 and conclude in Section 6.

2 Preliminaries

This section gives preliminaries for our work. Section 2.1 clarifies our notion of execution semantics. Section 2.2 presents LTL as the logic used in this paper. Section 2.3 summarizes existing work on generating a behavioral model from a given LTL formula.

2.1 Process Runs as Linear Sequences

In this paper, we rely on trace semantics for process models. An execution sequence σ of a process model is referred to as a *process run* or *trace* – a finite linear sequence of states $\sigma : s_0, s_1, \ldots, s_n$ with a start state s_0 and an end state s_n. Evidently, a process model as well as a set of compliance requirements allow for many conforming traces. Each state of a trace is labeled with propositions that refer to *actions* and *results*. Actions are the driving force of a trace and refer to the execution of business activities. This, in turn, may effect or be constrained by results, which relate to data values of the business process. As an example, think of an activity 'risk analysis' (ra) and a data object 'risk'. The action that represents the execution of this activity may have the result of setting the state of the data object to 'high' or 'low'. The execution of another activity, i.e., another action, may be allowed to happen solely if a certain result, e.g., the object has been set to 'high', occurred. Both, actions and results, are represented by Boolean propositions at each state. For instance, proposition ra being 'true' at a state s_i means that the action, i.e., execution of activity 'risk analysis', has happened at state s_i. In contrast, proposition ra being 'false' at state s_i means that the action did not happen at state s_i. Given a trace $\sigma : s_0, s_1, \ldots, s_n$, we write $p \in s_i$ to indicate that proposition p is true in state s_i, for $0 \le i \le n$ and $p \in \sigma$ if there is a state s_i in σ where $p \in s_i$, for some $0 \le i \le n$.

We represent an execution sequence as a linear sequence of states where states are labelled with both actions and results, and (unlabelled) edges between states represent the temporal ordering in the sequence. Hence, we rely on Linear Temporal Logic (LTL) in order to formulate statements about traces.

2.2   Linear Temporal Logic

Linear Temporal Logic (LTL) [20] is a logic specifically designed for expressing and reasoning about properties of linear sequences of states. The formulae of LTL are built from atomic propositions using the connectives of $\vee$ (or), $\wedge$ (and), $\neg$ (not) and $\Rightarrow$ (implication), and the following temporal connectives: **X** (next), **F** (eventually), **G** (always), **U** (until) and **B** (before). The latter are interpreted as follows:

X φ**:** in the ne**X**t state, φ holds
F φ**:** there is some state either now or in the **F**uture where φ holds
G φ**:** in every state **G**lobally from now on, φ holds
φ **U** ψ**:** there is some state, either now or in the future, where ψ holds, and φ holds in every state from now **U**ntil that state
φ **B** ψ**:** **B**efore ψ holds, if it ever does, φ must hold.

We apply LTL to encode compliance requirements. Hence, we obtain a set of formulae Γ expressing the constraints to which compliant traces have to conform.

2.3   Finding All LTL-Models of a Given LTL Formula

Given a collection of compliance requirements expressed as a set Γ of LTL-formulae, we seek to find a behavioral model that captures all *formula-models*, i.e., traces in our setting, which satisfy Γ. That is, such a model describes all linear sequences of states $s_0, s_1, \ldots, s_n$ such that Γ is true at s_0. Since Γ may contain *eventualities*, such as **X** φ or ψ_1 **U** ψ_2, ensuring that Γ is true at s_0 may require us to ensure that φ is true at s_1 or ψ_2 is true eventually at some state s_i with $0 \le i \le n$. In contrast to model checking [7] we are not given a single trace, but construct all traces satisfying the given constraints.

The first step is to determine whether the constraints are satisfiable. If not, the specification is erroneous since no trace can conform to the given constraints. The second step is the creation of the behavioral model that describes *all* traces.

For both steps, we use a tableaux-based method introduced in [24,23]. In essence, this approach works as follows. We start by creating a root node containing Γ and proceed in two phases. First, a finite (cyclic) graph of tableau nodes is created by applying tableau-expansion rules that capture the semantics of LTL and by pruning nodes containing local contradictions [24]. Second, once the graph is complete a reachability algorithm is used to determine which nodes do not satisfy their eventualities. These nodes are removed and the reachability algorithm is reapplied until no nodes may be removed. The set of formulae Γ is satisfiable, if and only if the root node has not been removed [24]. Further, the graph created by the tableau algorithm, referred to as the *pseudomodel*, describes all possible formula-models, i.e., possible traces [24]. We use this pseudomodel to extract possible traces during our synthesis approach.

3 Synthesis of Process Templates from Compliance Rules

In this section, we describe our approach to the synthesis of process models from a set of compliance rules expressed in temporal logic. First, Section 3.1 gives an overview of the approach and introduces an example set of compliance rules used to illustrate all subsequent steps. Section 3.2 describes the LTL encoding of the compliance rules and additional domain knowledge. Section 3.3 elaborates on the extraction of traces from a behavioral model, while Section 3.4 focuses on consistency of these traces. Synthesis of a process template from these traces is discussed in Section 3.5. Finally, we elaborate on the evaluation of synthesized templates in Section 3.6.

3.1 Overview

The process model in Fig. 1 visualizes the steps to synthesize a process template out of a set of compliance rules. First, a set of compliance rules is collected. In order to identify whether these requirements are consistent and thus a process template can be synthesized, related domain-specific knowledge is identified. In Section 3.2 we give details on the LTL encoding of both compliance rules and domain knowledge.

For the conjunction of these LTL formulae, we verify satisfiability as it has been summarized in Section 2.3. If the set is not satisfiable then no trace can be constructed to satisfy the given LTL formulae so the inconsistency is reported to the user. If the set is satisfiable then the satisfiability checker automatically returns the pseudomodel which is a behavioral model of all traces that obey the given constraints.

As a next step, finite traces are extracted from the pseudomodel by following all choice points and stopping when a trace becomes cyclic. We focus on this step in Section 3.3. Having a finite set of traces that satisfy the compliance rules, we check it for

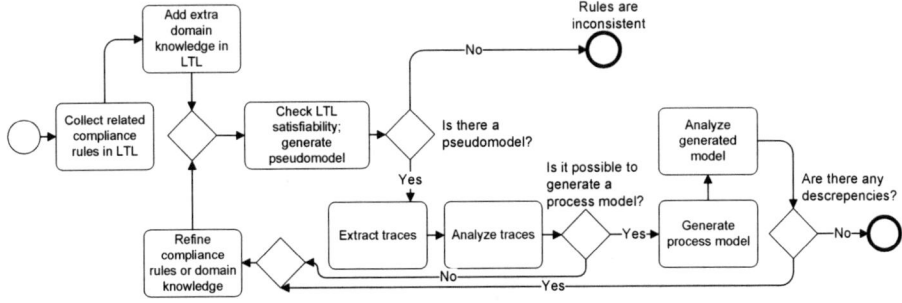

Fig. 1. Process Synthesis Approach

consistency. This check guarantees that a template can be generated. Inconsistent traces hint at issues in the specification, so that a new iteration of the synthesis may be started with refined compliance rules or adapted domain knowledge. We focus on the analysis of traces in Section 3.4. If the traces are consistent, we apply a process synthesis algorithm to extract a process template. Details on this step are given in Section 3.5. The synthesized template is then analyzed to identify discrepancies that stem, e.g., from underspecification. Depending on the result of this analysis, again, a new iteration of the synthesis may be started. We discuss the evaluation of process templates in Section 3.6.

Example. We illustrate our approach with an example from the financial domain. Anti money laundering guidelines [8] address financial institutes, e.g., banks, and define a set of checks to prevent money transfers with the purpose of financing criminal actions. We focus on the following guidelines for opening new bank accounts:

$R1$: A risk assessment has to be conducted for each 'open account' request.

$R2$: A due diligence evaluation has to be conducted for each 'open account' request.

$R3$: Before opening an account the risk associated with that account must be low. Otherwise, the account is not opened.

$R4$: If due diligence evaluation fails, the client has to be added to the bank's black list.

3.2   LTL Encoding

Once the compliance rules have been collected, a behavioral model that represents all traces conforming to these rules is created. In order to arrive at such a model, we need to collect extra domain-specific rules. Much of the domain-specific rules can be generated automatically from a higher level description. Such a description needs to be defined by a human expert in the first place and comprises the following information.

Actions and Goals. The set of all *actions* is denoted by A. The set of *goal* actions $G \subset A$ comprises activities that indicate the completion of a trace. Moreover, we capture contradicting actions that are not allowed to occur together in one trace in a relation $CA : A \times 2^A$.

Results and Initial Values. The set of all *results* is denoted by R. The initial values of data objects are defined by a set $IV \subset R$. Further, we define the set of negated results as $\overline{R} = \{\neg r | r \in R\}$. Similar to contradicting actions, we capture contradicting results in a relation $CR : R \times 2^R$.

Relation between Actions and Results. The mapping from actions to sets of results is given as a relation $AM : A \times 2^{R \cup \overline{R}}$. Mutually exclusive sets of results are captured in a relation $RE = \{S : \exists a \in A.(a, S) \in AM \wedge S \neq \emptyset\}$.

Based on this information and two additional actions *start* and *end* that represent the initial and final states of a trace (independent of any goal states), we derive LTL rules to represent the domain knowledge according to Table 1. Common process description languages, e.g., BPMN or EPCs, assume interleaving semantics, which is enforced by formula *interleaving* and *progress*. The information on exclusiveness constraints and on contradicting actions and results yields the formulae *mutex* and *contra*. The formula *causality* guarantees correct implementation of dependencies between actions and results. Finally, the formulae *once*, *final*, *goals*, and *initial* ensure

Table 1. The formulae making up the domain knowledge

Constraint Description	Formalization
To realize interleaving semantics, the formula *interleave* ensures that at most one action can be true, i.e., one activity can be executed, at any state. The formula *progress* guarantees that at least one action occurs at each state.	$interleave(a) = a \Rightarrow (\bigwedge_{b \in A \setminus \{a\}} \neg b))$ $interleave = \bigwedge_{a \in A} interleave(a)$ $progress = \bigvee_{a \in A} a$
The mutual exclusion constraints given in RE are enforced by the formula *mutex*, i.e., exclusive results cannot be true at the same time.	$mutex(S) = \bigwedge_{a,b \in S, \, a \neq b} \neg(a \wedge b)$ $mutex = \bigwedge_{S \in RE} mutex(S)$
Knowledge on contradicting actions or results is taken into account by the formulae, *con* and *conRes*.	$con(a) = a \Rightarrow \mathbf{G} \bigwedge_{b \in CA(a)} \neg b$ $conRes(r) = a \Rightarrow \mathbf{G} \bigwedge_{s \in CR(r)} \neg s$ $contra = \bigwedge_{a \in A \cup R} con(a) \wedge conRes(a)$
To implement the relation between actions and results, formula cau_1 states that for every entry $(a, S) \in AM$ the action a must cause at least one of the results in S. Formula cau_2 states that for every result r, that result can only be changed by one of the actions which can cause it.	$cau_1(a, S) = a \Rightarrow \bigvee_{r \in S} r$ $cau_2(r) =$ $r \Rightarrow (\mathbf{X} \bigvee_{(a,S) \in AM, \, \{r, \neg r\} \cap S \neq \emptyset} a) \, \mathbf{B} \, \neg r$ $causality =$ $\bigwedge_{(a,S) \in AM} cau_1(a, S) \wedge \bigwedge_{r \in R \cup \overline{R}} cau_2(r)$
The formula *once* enforces that all actions other than *end* occur at most once, in order to avoid infinite behavior. The formula *final* enforces that *end* persists forever to represent the process end.	$once(a) = a \Rightarrow \mathbf{X} \, \mathbf{G} \, \neg a$ $once = \bigwedge_{a \in A \setminus \{end\}} once(a)$ $final = end \Rightarrow \mathbf{G} \, end$
The formula *goals* is used to require that eventually the outcome of the process is determined, while *inital* ensures correct initial values for all objects.	$goals = \bigvee_{g \in G} g$ $initial = start \Rightarrow \bigwedge_{v \in IV} v$

correct initialization and successful termination of any trace. The combination of all these formulae yields the formula *domain*, which represents the domain knowledge.

$$domain = start \wedge \mathbf{G} \, initial \wedge \mathbf{F} \, goals \wedge \mathbf{F} \, end \wedge \mathbf{G} \, interleave \wedge \mathbf{G} \, progress$$
$$\wedge \, \mathbf{G} \, mutex \wedge \mathbf{G} \, causality \wedge \mathbf{G} \, once \wedge \mathbf{G} \, contra \wedge \mathbf{G} \, final$$

Example. For our example, an expert first identifies the following actions and results.

$Actions = \{ra, edd, og, od, bl\}$ $Results = \{ri, rh, rl, ei, ef, ep\}$

ra: conduct a risk assessment   *ri*: risk assessment is initial

edd: evaluate due-diligence   *rh*: risk was assessed as high

og: grant a request to open an account *rl*: risk was assessed as low

od: deny a request to open an account   *ei*: due-diligence evaluation is initial

bl: blacklist a client.   *ef*: due-diligence evaluation failed

                                          ep: due-diligence evaluation passed.

Note that the results are all descriptive statements, while the actions refer to activities. Moreover, we introduce *positive* representations for the states 'high' and 'low' of the risk object, even though both states are opposed. That is due to the three possible states of the risk object: high, low, or initial. The same holds true for the the due-diligence object.

Based on these actions and results, the compliance rules are encoded in LTL. As a process to open a bank account is considered, the process is assumed to start by receiving such a request. Therefore, rules 1 and 2 are interpreted as "A risk assessment has to be conducted" and "A due diligence evaluation has to be conducted", respectively. The third rule is interpreted to mean that the risk associated with opening an account must be low *at the time the request is granted*, rather than at some point in the past. Similarly is the case when denying the open request, the risk has to be high.

$R1$: A risk assessment has to be conducted.

    F ra "Eventually ra must hold"

$R2$: A due diligence evaluation has to be conducted.

    F edd "Eventually edd must hold"

$R3$: The risk associated with opening an account must be low when the request is granted.

    G $(og \Rightarrow rl) \land$ **G** $(od \Rightarrow rh)$ "Always, og only if rl, and always, od only if rh"

$R4$: If due diligence evaluation fails, the client has to be added to the bank's black list.

    G $(edd \land ef \Rightarrow$ **F** $bl)$ "Always, edd and ef imply eventually bl"

As a next step, the domain knowledge is defined in more detail. For instance, the action mapping defines $ra \mapsto \{rh, rl\}$ and $ra \mapsto \{\neg ri\}$. The former says that action ra causes the risk object to take a concrete value of 'high' or 'low'. The latter means that ra causes the risk to stop being 'initial' by forcing ri to not hold. Excluding results are defined, e.g., $\{ri, rl, rh\}$ states that at most one of the propositions ri, rh, rl can hold at a time. The goal of the process is defined as $\{og, od\}$ and the set of initial values $\{ri, ei\}$ signifies that initially, both risk and due-diligence objects, are put to an initial, unknown, value. There are also contradicting actions, $\{og \mapsto \{od\}, od \mapsto \{og\}\}$, ensuring that we cannot grant and deny a request within the same trace. Based on Table 1, this specification is converted into LTL. For example, this yields the formula $progress = ra \lor edd \lor og \lor od \lor bl \lor start \lor end$. The final set of LTL formulae is the union of the *domain* formula with all four formulae representing the compliance rules.

3.3 Extracting Traces

Given a set of LTL formulae, we apply the technique summarized in Section 2.3 to determine whether the constraints are satisfiable. If so, we obtain a pseudomodel that describes all traces that conform to the set of formulae. To create a process template, these traces are extracted. Any sequence $\sigma = s_0, \ldots, s_n$ of states, starting at the root node of the pseudomodel can be extended into a trace. As we are modeling finite sequences with an end state, we consider a trace to be complete if $end \in s_n$. Because of the *once* constraint introduced in the previous section, there will be no loops in the pseudomodel between the start and the end. Hence, the finite set of paths in the pseudomodel between the root state and a state labeled with end is the set of correct traces.

Table 2. Excerpt of the extracted traces

$\sigma_1 : start \wedge ei \wedge ri, edd \wedge ep \wedge ri, ra \wedge ep \wedge rh, bl \wedge ep \wedge rh, od \wedge ep \wedge rh, end \wedge ep \wedge rh$

$\sigma_2 : start \wedge ei \wedge ri, edd \wedge ep \wedge ri, ra \wedge ep \wedge rh, od \wedge ep \wedge rh, end \wedge ep \wedge rh$

$\ldots$

$\sigma_{37} : start \wedge ei \wedge ri, bl \wedge ei \wedge ri, edd \wedge ep \wedge ri, ra \wedge ep \wedge rh, od \wedge ep \wedge rh, end \wedge ep \wedge rh$

$\ldots$

$\sigma_{42} : start \wedge ei \wedge ri, bl \wedge ei \wedge ri, ra \wedge rl \wedge ei, og \wedge rl \wedge ei, edd \wedge ep \wedge rl, end \wedge ep \wedge rl$

Note that it is possible to extract traces that take repetition of activities into account by omitting the *once* constraint in the domain knowledge. Still, for our purpose, this does not seem to be appropriate. Compliance rules rarely forbid the repetition of activity execution, so that modeling all potential loops blurs up the structure of a generated process template. As this hinders discussions between business and compliance experts, we neglect potential repetition for our synthesis approach.

Example. Some of the traces extracted from the pseudomodel of our running example are illustrated in Table 2. Here, the states of a trace are characterized by the conjunction of propositions that hold true in the respective state.

3.4   Analysis of Extracted Traces

As stated earlier, the goal of synthesizing a process template out of compliance rules is to support experts in getting a better understanding of the compliance aspects and to discover missing or under-specified requirements. However, it is possible to detect such under-specification by analysis of extracted traces before proceeding to synthesizing a process template. Yet, not every semantical error in the specification can be detected, so that a human expert has to validate the synthesized process template. We address the issue of under-specified LTL specification by correctness criteria for the extracted traces.

Let $\mathcal{P}$ be a set of traces derived from a pseudomodel, cf., Section 3.3. We leverage the information whether an action $a \in A$ is optional for completing the process.

Definition 1 (Optional Actions). *Given a set of actions A and a set of traces $\mathcal{P}$, the set A_O of optional actions is defined as $A_O = \{a \in A : \exists \sigma \in \mathcal{P}.a \notin \sigma\}$.*

We argue that correctness of a specification where some activity is optional implies the existence of a specific data condition under which the optional activity is executed. For the traces in Table 2, for instance, og and od are optional activities. The condition under which og executes is $(rl \wedge ef) \vee (rl \wedge ep) \vee (rl \wedge ei)$, i.e., the risk object assumes the state 'low'. Action og is executed independently from the state of the due diligence evaluation object. For action od the condition is $(rh \wedge ef) \vee (rh \wedge ep) \vee (rh \wedge ei)$, i.e., the risk is 'high'. In contrast, action bl is executed under the condition $(ei \wedge ri) \vee (ei \wedge rh) \vee (ei \wedge rl) \vee (ef \wedge ri) \vee (ef \wedge rh) \vee (ef \wedge rl) \vee (ep \wedge rh) \vee (ep \wedge rl) \vee (ep \wedge ri)$. Hence, none of the objects influences the decision of executing bl, since bl appears with *all* combinations of data values. Yet, bl is optional. This indicates an under-specified LTL specification as conditions for executing optional activities are not stated explicitly.

Definition 2 (Optional Action Execution Condition). *Let A_O be the set of optional actions, $\mathcal{P}$ a set of traces, and RE the set of mutually exclusive results. For an action $a \in A_O$, the execution condition is defined as:*
$$cond_a = \{\{x_1, \ldots, x_n\} : \exists\, \sigma \in \mathcal{P}.\exists\, s \in \sigma.a \in s \wedge x_1 \in s \wedge x_1 \in S_1 \wedge S_1 \in RE \wedge \cdots \wedge x_n \in s \wedge x_n \in S_n \wedge S_n \in RE \wedge n = |RE|\}.$$

This definition describes the conditions under which an action executes by investigating for each observation of the action a the data effects that are true in the same state as a. If an optional activity a has an execution condition, which is a proper subset of the combination of non-exclusive results, then this indicates a well specified set of compliance rules. We formalize this trace correctness criterion as follows.

Definition 3 (Proper Execution of Optional Actions). *Let A_O be the set of optional actions with respect to a set of traces $\mathcal{P}$ and RE the set of mutually exclusive results. We define the set of all possible results interactions as $RI = \{\{x_1, \ldots, x_n\} : x_1 \in S_1 \wedge S_1 \in RE \wedge \cdots \wedge x_n \in S_n \wedge S_n \in RE \wedge n = |RE|\}$. An action $a \in A_O$ has a proper execution iff $cond_a \subset RI$.*

The *proper execution of actions* is the first correctness criterion to be investigated on traces before synthesizing a template. Referring to the set of traces in Table 2, we find that this criterion is not met for activity bl. This problem is reported to the user so that the compliance rules are refined and a new set of traces is extracted.

Another correctness criterion for a set of traces is *data-completeness*. A set of traces $\mathcal{P}$ is data-complete if for every possible combination of results resulting from the *mandatory* activities, there is a trace in which this combination occurs.

Definition 4 (Traces Data-Completeness). *Let $\mathcal{P}$ be a set of traces, AM be the set of action mappings, $A_M = A \setminus A_O$ be the set of mandatory actions and $RE_M = \{S : \exists\, a \in A_M.(a, S) \in AM \wedge S \neq \emptyset\}$ be the set of mutually exclusive results of mandatory actions. We define the set $CO = \{\{x_1, \ldots, x_n\} : x_1 \in S_1 \wedge S_1 \in RE_M \wedge \cdots \wedge x_n \in S_n \wedge S_n \in RE_M \wedge n = |RE_M|\}$. The set of traces $\mathcal{P}$ is data-complete iff $\forall\, C \in CO\, \exists\, \sigma \in \mathcal{P}\, \exists\, s_i \in \sigma : \forall\, x \in C\; x \in s_i$ where $i > 0$.*

Even if data incompleteness is detected for a set of traces, a process template may be generated. Nevertheless, the template would suffer from deadlocks as for some combinations of results, continuation of processing is not defined. Therefore, we proceed solely in case the set of traces shows data completeness.

Example. For our running example, we find that the set of traces lacks the proper execution condition for activity bl. To address this issue, a compliance expert might add an explicit condition to black list a client only if the evaluation fails. This is represented by an additional constraint **G** $(bl \Rightarrow ef)$. Repeating all steps from satisfiability checking to extracting traces yields a set of traces that satisfies the two correctness criteria above.

3.5   Generating Process Templates

Given a set of traces within which activities have proper execution conditions, *process mining* [4] is applied to generate a process template. Most mining algorithms neglect the difference between control flow dependencies and data flow dependencies when generating a process model. Therefore, we cannot apply an existing algorithm directly. Instead, we use the α-algorithm [4] and incorporate the respective data aspects.

Order of actions. As a first step, we extract the precedence of actions. To this end, we employ an adapted version of the order relations known from the α-algorithm [4].

Definition 5 (Order Relations). *Let $\mathcal{P}$ be a set of traces and A the sets of actions. We define the following order relations for two actions $a_1, a_2 \in A$ with $R_{a_2} = \bigcup_{(a_1,S)\in AM} S$ as the set of results of a_2.*

$a_1 > a_2$: *iff either*

$\quad R_{a_2} = \emptyset$ *(i.e. a_2 has no results), and there is a trace $\sigma : s_0, \ldots, s_n \in \mathcal{P}$, such that*
$\quad\quad a_1 \in s_i \wedge a_2 \in s_{i+1}$ *for some $0 \le i < n$, or*
$\quad R_{a_2} \neq \emptyset$ *(i.e. a_2 has results), and $\forall r \in R_{a_2}$ there is a trace $\sigma : s_0, \ldots, s_n \in \mathcal{P}$,*
$\quad\quad$ *such that $a_1 \in s_i \wedge a_2 \in s_{i+1} \wedge r \in s_{i+1}$ for some $0 \le i < n$.*

$a_1 \gg a_2$: *iff $a_1 > a_2$ and $a_2 \not> a_1$*
$a_1 \to a_2$: *iff $a_1 \gg a_2$ and $\nexists a_3 \in A : a_1 \gg^+ a_3 \wedge a_3 \gg a_2$ with $\gg^+$ as the transitive closure of $\gg$*

For two actions ordered by $>$, we know that the first action appears immediately before the second action. This notion of order is stronger than the one originally used in the α-algorithm [4]. If $a_1 > a_2$ and $a_2 \not> a_1$, then we conclude that a_1 precedes a_2, i.e., $a_1 \gg a_2$. However, it might be the case that $a_1 \gg a_2$, $a_2 \gg a_3$ and $a_1 \gg a_3$. In this case, we drop the dependency $a_1 \gg a_3$ as it unnecessarily complicates the template synthesis. Thus, we use the more strict precedence relation $\to$.

In contrast to common order relations known in process mining, the precedence dependencies in our approach may be guarded by conditions, captured as follows.

Definition 6 (Precedence Condition). *Let $\mathcal{P}$ be a set of traces, A the sets of actions, AM the mapping from actions to results, and $a_1, a_2 \in A$ two actions in precedence, $a_1 \to a_2$. Let $E = \{r | r \in \bigcup_{(a_1,S)\in AM} S \wedge (\exists \sigma = s_0, \ldots, s_i, s_{i+1}, \ldots, s_n \in \mathcal{P} : a_1 \in s_i \wedge r \in s_i \wedge a_2 \in s_{i+1})\}$ be the set of results of a_1 under which a_2 is observed. Then, we define the precedence condition $cond(a_1, a_2)$ as follows.*

$$cond(a_1, a_2) = \begin{cases} \bigvee_{r \in E} & \text{iff} \quad E \subset \bigcup_{(a_1,S)\in AM} S \wedge E \neq \emptyset. \\ true & \text{otherwise.} \end{cases}$$

According to this definition, we distinguish two types of precedence conditions. First, precedence holds for a proper subset of the results of the first action. Then, the precedence condition is the disjunction of results that can be caused by the first action. The second case captures unconditioned precedence, i.e., the precedence holds independent of any results. For our running example, we observe $ra \to og$. This precedence is guarded, as we observe this dependency solely in case of the result rl. In other words, only if action ra yields the result rl, we observe the action og subsequently.

Synthesis of process model. Based on the precedence among activities, the precedence conditions, along with the knowledge on optionality of activities, we proceed to build a process template. First, the overall structure of a process model is derived from the precedence relation. This step yields a graph with all nodes representing actions, while the precedence relation defines directed edges between them. Second, control nodes (split and join nodes) that realize the behavior routing in the process model have to be introduced whenever a node has more than one predecessor or successor.

Starting with split nodes, our approach inserts nodes that implement either AND-, XOR-, or OR-logic. The routing semantics depends on the precedence conditions for the edges to succeeding nodes. If all precedences originating at an action are unconditioned, an AND-split node is inserted. If all precedences are conditioned and those conditions do not overlap, an XOR-split node is inserted and each outgoing edge inherits the respective precedence condition. Similarly, an OR-split is applied if the conditions are overlapping.

The case of join nodes, nodes with multiple predecessors, is not straightforward. We distinguish the following cases.

- All precedences of an action a are conditioned, we use an AND-join to synchronize these conditions.
- Only a proper subset of precedences of an action a is conditioned, we use an OR-join to synchronize any subset of these conditions.
- All precedences of an action a are unconditioned. If a is mandatory, we apply either an OR-join or an AND-join. The former is used if at least one of the preceding actions is optional; the latter is used in all other cases. If a is optional, we proceed as follows. An AND-join is applied to synchronize all precedences. Moreover, for all combinations of results of preceding actions of a, we check for a state in which the execution of a is observed as well. In other words, we identify all combinations of results of preceding actions under which a can occur. The disjunction of these result combinations is then used as a precedence condition for the edge between the AND-join and action a.

Applying these steps yields a process template. Still, our approach to model synthesis is rather naive and may create OR-joins for which the synchronization behavior could be implemented using solely AND- and XOR-joins. However, existing methods for restructuring a process model are used to replace these OR-joins with a semantically corresponding structure of AND- and XOR-joins, see [22].

Example. After we adapted the set of constraints for our running example as discussed above, we derive the basic graph structure for the template based on the precedence relation. Fig. 2 visualizes this structure in a BPMN-like notation. Here, the start and end actions are represented by start and end events. Activities depicted with a dashed border are optional. After inserting control nodes (aka gateways in BPMN) into the graph, the complete process template is derived. The first ver-

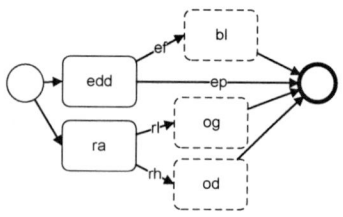

Fig. 2. Precedence among actions

sion of the generated process template is shown in Fig. 3a. Application of the restructuring according to [22] yields the process template shown in Fig. 3b.

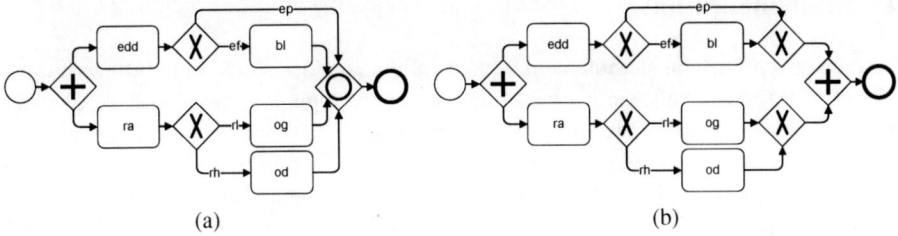

(a)   (b)

Fig. 3. (a) The process template for our example. (b) The restructured process template.

3.6 Evaluation of the Synthesized Process Template

Process templates aim at supporting experts in getting a better understanding of the compliance aspects and to discover missing or under-specified requirements. Such under-specification is manifested in the process template in terms of semantical problems. Those problems can only be detected by human experts. In this section, we will further elaborate on the running example to illustrate such problems. Using the process template in Fig. 3a as a basis of the discussion between compliance expert and business expert, they identify that the template allows for executing both black listing the client and granting to open the account in the same instance. This is an example of the aforementioned semantical problems caused by under-specified compliance rules. The compliance expert refines the set of constraints by indicating that black listing and granting open the account are contradicting, cf. the CA relation in Section 3.2, formalized as $\mathbf{G} \ (og \Rightarrow \mathbf{G} \ (\neg bl))$ and $\mathbf{G} \ (bl \Rightarrow \mathbf{G} \ (\neg og))$. Repeating the steps of our approach reveals that the adapted set of compliance rules yields a set of traces that is data incomplete. This is explained based on the two added constraints as follows. By forcing bl and og to be exclusive, we implicitly require bl to be executed only with the condition $ef \wedge rh$, while og is executed only with the condition $ep \wedge rl$. Other combinations of results are not considered. There is no trace that addresses the situation where $ef \wedge rl$ holds in some state. This contradicts with our requirement to execute either og or od in each run. Since the condition $ef \wedge rl$ enables neither of them, it is not observed in any of the generated traces.

As a consequence, another adaptation of our set of compliance requirements is needed. The missing interaction $ef \wedge rl$ has to be handled. One solution is to update the conditions under which og and od are executed, i.e., $\mathbf{G} \ (og \Rightarrow ep \wedge rl)$ and $\mathbf{G} \ (od \Rightarrow (ef \vee rh))$. With these updated constraints, another iteration of behavior synthesis is started. This time, the generated set of traces shows data completeness. The final generated process template is visualized in Fig. 4.

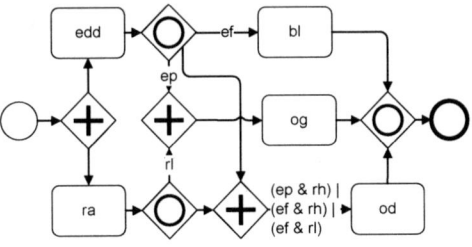

Fig. 4. A compliant process template where bl and og are exclusive and conditions adjusted

4   Implementation

We created a prototypical implementation to validate our approach. Fig. 5 shows a snapshot of it. It relies on a specification of domain knowledge, such as activity results and contradicting activities, which has to be defined once by a human expert. Given a set of compliance rules, our implementation adds extra rules to control the behavior synthesis and to enforce domain knowledge automatically. The satisfiability checking is done by an implementation of Wolper's method for checking LTL satisfiability [24] developed by the authors at the School of Computer Science of the Australian National University[1]. If the rules are satisfiable, the checker generates the pseudomodel of all possible traces. Next, our implementation extracts finite traces, analyzes them and synthesizes the process template, if the extracted traces pass quality tests, cf. Section 3.4. At that point, the resulting template is visualized using GraphViz [9]. In case that traces do not pass checks, the found problems are reported on the "Analysis result" tab.

There is the potential for a state space explosion, especially since the additional constraints of the process are unrestricted logical formulae. Even without pathological constraints, if there is a lot of freedom or non-local conditions then the satisfiability checking phase can take a considerable amount of time. The *once* constraint helps limit this, and too much freedom can often be a sign that other conditions have been omitted. We aim evaluating these issues in further case studies.

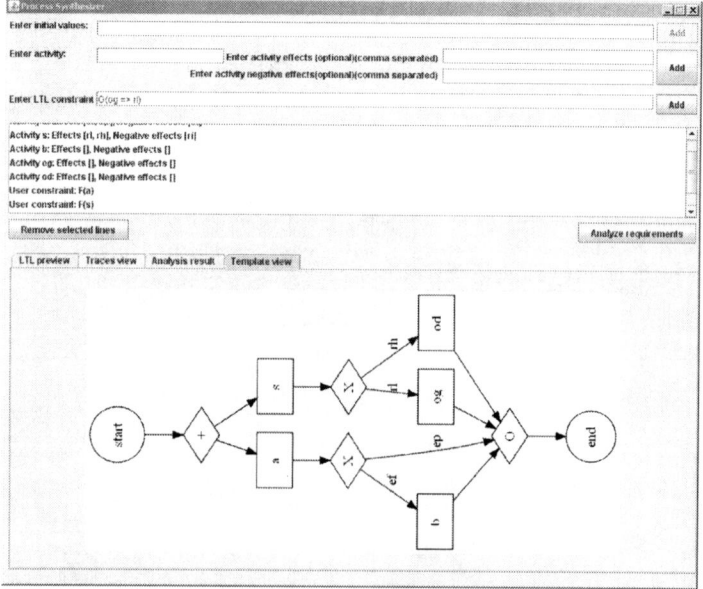

Fig. 5. A snapshot of the process synthesis tool

---

[1] Source code available at
http://users.cecs.anu.edu.au/~rpg/PLTLProvers/pltlmultipass.tar

5   Related Work

Compliance checking of business process models with a focus on execution order constraints has been approached from two angles: namely compliance by design and compliance checking of existing models. The latter has been tackled using model checking techniques [6,10,14]. Our work follows a compliance by design approach that has also been advocated in [11,12,13,15,16,25]. Close to our work, the authors of [12,11] employ temporal deontic assignments to specify what can or must be done at a certain point in time and synthesize a process template from these assignments. In contrast to our work, however, the approach is limited to temporal dependencies between activity executions and the underlying logic requires an encoding of these dependencies via explicit points in time. Another approach to synthesize compliant processes was introduced in [25]. The authors employ a set of compliance patterns expressed in Linear Temporal Logic (LTL). For each pattern a finite state automaton (FSA) is defined. To synthesize a process, the FSAs of the involved patterns are composed. Next, the user is required to select for each composition an execution path in order to synthesize the process. That approach is able to generate processes with sequence and choice only. Moreover, it does not consider data flow aspects in the synthesized process.

Related to our approach to process model synthesis is work on process mining, which aims at automatic construction of a process model from a set of logs [5,4,3]. We adapted the α-algorithm [4], a standard mining approach, for our purposes. Besides the commonalities, there are some important differences between process mining and process template synthesis. We consider control flow routing based on data values. This aspect is often neglect in process mining algorithms. Only recently, time information and data context have been considered when predicting the continuation of a trace based on its current state [21,2]. Further, process mining approaches have to be robust against incorrect data (log noise). As we derive a model from artificially generated traces, this is not an issue for our approach.

Work on declarative business process modeling is also related to our work. The authors of [17,19] propose to model processes by specifying a set of execution ordering constraints on a set of activities. These constraints are mapped onto LTL formulas; which are used to generate an automaton that is used to both guide the execution and monitor it. That is similar to our approach of generating a pseudomodel. Recently, the authors also showed how finite traces that respect interleaving semantics can be extracted from a set of LTL constraints [18]. The major difference from our work is that [18] does not model data constraints as we do. They also change the semantics of LTL rather than by using standard LTL as we do. Finally, we initially tried the approach of extracting Büchi automata from our LTL specifications for our example, but found that the automata approach required hours to return the automata whereas our LTL satisfiability checker returns a pseudomodel in less than a second.

6   Conclusion

In this paper, we introduced an approach to synthesize business process templates out of a set of compliance rules expressed in LTL. We also showed that extra domain-specific knowledge is required to decide about consistency of such requirements and introduced an LTL encoding for compliance rules and domain knowledge. This was used to generated traces, which are analyzed for inconsistencies. Finally, we proposed an approach to the synthesis of process templates that goes beyond existing work on process mining by focusing on data dependencies of activity execution. We also discussed the analysis of generated templates with respect to semantical errors.

In our approach, we addressed control- and data-flow aspects of compliance rules, in contrast to similar approaches that focus on control-flow aspects only. The consideration of data-flow aspects comes with new challenges which we addressed in this paper by introducing correctness criteria for the set of generated traces. We illustrate that data dependencies may show rather interactions that are hard to handle at the first place. As a consequence, our approach is iterative – the required knowledge is built incrementally each time constraints are under-specified. In future work, we want to consider constraints on role resolution for generating process templates.

References

1. Sarbanes-Oxley Act of 2002. US Public Law 107–204 (2002)
2. van der Aalst, W.M.P., Pesic, M., Song, M.: Beyond process mining: From the past to present and future. In: Pernici, B. (ed.) CAiSE 2010. LNCS, vol. 6051, pp. 38–52. Springer, Heidelberg (2010)
3. van der Aalst, W.M.P., Reijers, H.A., Weijters, A.J.M.M., van Dongen, B.F., de Medeiros, A.K.A., Song, M., Verbeek, H.M.W.E.: Business process mining: An industrial application. Inf. Syst. 32(5), 713–732 (2007)
4. van der Aalst, W.M.P., Weijters, T., Maruster, L.: Workflow mining: Discovering process models from event logs. IEEE Trans. Knowl. Data Eng. 16(9), 1128–1142 (2004)
5. Agrawal, R., Gunopulos, D., Leymann, F.: Mining process models from workflow logs. In: Schek, H.-J., Saltor, F., Ramos, I., Alonso, G. (eds.) EDBT 1998. LNCS, vol. 1377, pp. 469–483. Springer, Heidelberg (1998)
6. Awad, A., Weidlich, M., Weske, M.: Visually specifying compliance rules and explaining their violations for business processes. J. Vis. Lang. Comput. 22(1), 30–55 (2011)
7. Clarke, E.M., Grumberg, O., Peled, D.A.: Model Checking. MIT Press, Cambridge (1999)
8. Commission, F.S.: Guidelines on anti-money laundering & counter-financing of terrorism (2007)
9. Ellson, J., Gansner, E.R., Koutsofios, E., North, S.C., Woodhull, G.: Graphviz - open source graph drawing tools. In: Graph Drawing, pp. 483–484 (2001)
10. Förster, A., Engels, G., Schattkowsky, T., Van Der Straeten, R.: Verification of Business Process Quality Constraints Based on VisualProcess Patterns. In: TASE, pp. 197–208. IEEE Computer Society Press, Los Alamitos (2007)
11. Goedertier, S., Vanthienen, J.: Compliant and Flexible Business Processes with Business Rules. In: BPMDS. CEUR Workshop Proceedings. CEUR-WS.org, vol. 236 (2006)
12. Goedertier, S., Vanthienen, J.: Designing Compliant Business Processes with Obligations and Permissions. In: Eder, J., Dustdar, S. (eds.) BPM Workshops 2006. LNCS, vol. 4103, pp. 5–14. Springer, Heidelberg (2006)
13. Lu, R., Sadiq, S.K., Governatori, G.: Compliance Aware Business Process Design. In: ter Hofstede, A.H.M., Benatallah, B., Paik, H.-Y. (eds.) BPM Workshops 2007. LNCS, vol. 4928, pp. 120–131. Springer, Heidelberg (2008)

14. Lui, Y., Müller, S., Xu, K.: A Static Compliance-checking Framework for Business Process Models. IBM Systems Journal 46(2), 335–362 (2007)
15. Milosevic, Z., Sadiq, S., Orlowska, M.: Translating Business Contract into Compliant Business Processes. In: EDOC, pp. 211–220. IEEE Computer Society, Los Alamitos (2006)
16. Namiri, K., Stojanovic, N.: Pattern-Based Design and Validation of Business Process Compliance. In: Chung, S. (ed.) OTM 2007, Part I. LNCS, vol. 4803, pp. 59–76. Springer, Heidelberg (2007)
17. Pesic, M., van der Aalst, W.M.P.: A Declarative Approach for Flexible Business Processes Management. In: Eder, J., Dustdar, S. (eds.) BPM Workshops 2006. LNCS, vol. 4103, pp. 169–180. Springer, Heidelberg (2006)
18. Pesic, M., Bosnacki, D., van der Aalst, W.M.P.: Enacting declarative languages using ltl: Avoiding errors and improving performance. In: SPIN 2010. LNCS, vol. 6349, pp. 146–161. Springer (2010)
19. Pesic, M., Schonenberg, H., van der Aalst, W.M.P.: DECLARE: Full Support for Loosely-Structured Processes. In: EDOC, pp. 287–300. IEEE Computer Society, Los Alamitos (2007)
20. Pnueli, A.: The temporal logic of programs. In: SFCS, pp. 46–57. IEEE Computer Society, Washington, DC, USA (1977)
21. Schonenberg, H., Jian, J., Sidorova, N., van der Aalst, W.M.P.: Business trend analysis by simulation. In: Pernici, B. (ed.) CAiSE 2010. LNCS, vol. 6051, pp. 515–529. Springer, Heidelberg (2010)
22. Vanhatalo, J., Völzer, H., Leymann, F., Moser, S.: Automatic workflow graph refactoring and completion. In: Bouguettaya, A., Krueger, I., Margaria, T. (eds.) ICSOC 2008. LNCS, vol. 5364, pp. 100–115. Springer, Heidelberg (2008)
23. Wolper, P.: Temporal logic can be more expressive. Information and Control 56, 72–99 (1983)
24. Wolper, P.: The tableau method for temporal logic: an overview. Logique et Analyse 110-111, 119–136 (1985)
25. Yu, J., Han, Y., Han, J., Jin, Y., Falcarin, P., Morisio, M.: Synthesizing service composition models on the basis of temporal business rules. J. Comput. Sci. Technol. 23(6), 885–894 (2008)

A Design of Business-Technology Alignment Consulting Framework

Kecheng Liu[1], Lily Sun[2], Dian Jambari[2], Vaughan Michell[1], and Sam Chong[3]

[1] Informatics Research Centre, University of Reading, PO Box 241, Whiteknights,
Reading, RG6 6WB, UK
k.liu@henley.reading.ac.uk, v.a.michell@reading.ac.uk
[2] School of Systems Engineering, University of Reading, Whiteknights, Reading, Berkshire,
RG6 6AY, UK
{lily.sun,j.dianindrayani}@reading.ac.uk
[3] CTO Emerging Solution Group, Cisco APAC, Capital Tower,
168 Robinson Rd #26-01 to #29-01 Singapore
sachong@cisco.com

Abstract. Current work on applying scientific methods to capture the cultural values as requirements for business-IT alignment has been scarce, even though organisations acknowledge its significant impact. This paper introduces a Business-Technology Alignment Consulting Framework that adopts an Organisational Semiotics approach to capture cultural values from both formal norms and informal hidden social norms that can significantly impact the actual vs perceived alignment. A set of techniques in the framework are described for its use in conducting consulting analysis. Business Service Analysis is the core analysis that provides the holistic structure of the business services. Business Service Valuation calculates the service cultural values to complement the Business Service Analysis. Business Service Norms Analysis captures the business norms that govern the business service. A case study example is used to illustrate the analysis templates to holistically represent the business services. The significance of the consulting framework and future work are also discussed.

Keywords: business-technology alignment, consulting framework, socio-technical approach, consulting requirements analysis, norm analysis.

1 Introduction

Organisations have acknowledged the importance of a well aligned business and IT to ensure competitiveness. However, achieving alignment is difficult due to challenges such as poor shared knowledge management [1], rigid alignment strategy constricting the ability to adapt to changes [2] and miscommunication due to "language" differences between the business and IT domain [3]. The failure to capture and account for social views and influences can affect alignment problems [4]. IT spending is an organisation's major investments. Increasingly complex and expensive IT deliveries [5],[6]directly impacts an organisation's business performance and has reigned as one of their top concerns [7]. The alignment quality is measured by 1) the value added by

H. Mouratidis and C. Rolland (Eds.): CAiSE 2011, LNCS 6741, pp. 422–435, 2011.
© Springer-Verlag Berlin Heidelberg 2011

the business services to the enterprise to achieve their business goals, 2) the optimum "performance versus cost" of the IT capabilities to complement the value added business services. Therefore, a sensible alignment strategy between business and IT is essential to help organisations improve their financial efficiency particularly towards their IT investments. Organisations require a mechanism that can provide the mapping of the current business service with the "best fit" IT capabilities. They also need to be able to provide recommendations for the future IT strategy to assist in decision making for their IT investments, which inevitably improve the business service and IT applications alignment. However, before any alignment activities can be performed, a set of comprehensive and accurate requirements are needed to form a solid foundation to ensure well aligned business and IT. Business and technology consultants are facing challenges in articulating what IT applications are currently used for adding business value in an organisation. It is also difficult for consultants to recommend the future of those IT applications in relation to effective computing resources in the organisation. There are a number of factors contributing to these difficulties, 1) business environments have become complex and IT applications are deeply integrated with the business operations; 2) various IT applications serve different users for different purposes in their work; and 3) The socio-technical phenomenon impacts on business behavior towards IT applications. IT applications also add different benefits to the business performance and these benefits or values are normally perceived differently from a cultural perspective [8],[9]. Such multiple dimensional aspects can be described by complex business-IT alignment requirements which need to be captured by consultants. Well gathered and represented requirements enable the production of better analysis results for the current state of the business and IT alignment. It also helps the organisation in setting the future direction of the organisation to achieve their ultimate aims and objectives. Therefore, having the right requirements will ensure that the organisation will have the correct knowledge to establish a better alignment between their business and IT aspects [10],[11].

The Business-Technology Alignment Consulting Framework, therefore, has been developed to facilitate the analysis of business-IT alignment requirements in organisations. This framework enables consultants to establish a holistic view of the business situation and the IT applications supporting the business. The framework first defines whether a business operation is a core or supporting service. It then identifies and prioritises the future of the IT applications based on their support of these services and recommends, depending on the business service that the IT supports, if an IT application should be developed/upgraded with new functions to adequately serve a wider range of business operations, or be outsourced to reduce unnecessary incurred cost to the business. The framework techniques are developed and implemented in a consulting CASE tool. The paper is structured as follows; Section 2 discusses the issues of business and IT alignment vs. socio-technical aspects of Information Systems. This section also discusses the complexity of eliciting informal requirements to establish the business and IT alignment. Section 3 describes the adoption of an Organisational Semiotics approach in the consulting framework design to support this. Section 4 describes the Business-Technology Alignment Consulting Framework and three of its techniques, Business Service Analysis, Business Norms and Business Service Valuation for modeling the business landscape through articulating the cultural aspects in an organisation into an explicit form that reflects the business and IT alignment requirements and level. Section 5 draws conclusions and suggests future work.

2   Business and IT Alignment from a Socio-technical Viewpoint

The relationship between business and IT in organisations has a socio-technical aspect that can be viewed via socio-technical theory. Socio-technical theory is a set of explicit concepts that considers the complex interaction of the social aspects, which influence the usability and functionality of technology capabilities [12],[13],[14].The social aspects such as the behavioural patterns of the stakeholders involved are one of the important factors that impact the effectiveness of the technology that supports the business in an organisation [15]. However, the associated mapping and representation of the socio-technical aspects of the organisation is difficult in practice [16],[17], particularly to achieve business and IT alignment.

One approach to achieve business and IT alignment is through enterprise architecture frameworks (EAF) [18],[19] often using service oriented architecture (SOA). The Open Group Architecture Framework (TOGAF), aims to establish a proper alignment between the organisation's business strategy and IT capabilities in a well-structured, comprehensive and systematic manner [20]. Work done in [21] highlights how TOGAF can assist an enterprise to develop a new alignment between business and IT, or improve the existing alignment through the Architecture Development Model (ADM) life cycle. The emergence of SOA concepts has influenced a paradigm shift in business thinking where business components can be viewed as services [22],[23],[24]. Subsequently, these services can be assessed regarding the value they add to the profitability and sustainability of the business. The adoption of SOA in business architecture is beneficial as SOA concepts remove redundancies and align IT infrastructures [25]. The ability to deconstruct business components and organisational structures into sets of services behaving in a service-oriented manner, supports the enhancement of EAF [26]. Yet, the integration of EAF and the SOA concept is not simple as SOA is still considered as an immature technology with no specific foundational theory, and this indirectly complicates the process of defining granular and reusable business and IT services [22]. Work presented in [27],[28], illustrates the integration of EAF and SOA concepts by establishing the linkage of the business services with IT capabilities. However, these approaches focus more on business services than on IT services performance. The availability of information on the IT services performance is crucial to analyse the linkage between business and IT services.

The complexity of alignment is also raised by the fact that the business includes the organisation of people, which cultivates social and cultural informalities. However, accurate inclusion of the socio-cultural factors contributes to the success of the alignment as highlighted in [29],[16],[3],[17]. Work presented in [30] also recognizes the importance to missing non explicit stakeholder social information(s) in modeling enterprise architecture. Lagerstrom et al. [4] address such factors by focusing on stakeholders and their behaviour. However, there is no specific analysis for 1) how stakeholders gain value from their involvement in the business activities and 2) their important views of the value added, or otherwise, by various business services. Cultural aspects have been recognised to fill the informal requirements gap in the business and IT alignment. A study has shown that 30% of companies have failed in their attempt to achieve business and IT alignment [2]. The main problem highlighted is the miscommunication between business and IT caused by unclear specification of the organisation's business and IT requirements. Business and IT should have a well-defined understanding of their own domain within the organisation before proper alignment between them can be successfully achieved. It is often the case that the poor understanding between the business and IT requirements can critically affect the quality of the alignment [31]. Eliciting and representing the business requirements is important for accurate mapping of IT capabilities to obtain the optimal alignment between business service and IT capabilities [32]. Incomplete requirements were also found to be the top reason for failures in the Information System (IS) projects, whereas user involvement was found to have high influence on the IS project success factor [33]. The notion of "completeness" in requirements definition is problematic as there is no specific standard or easy procedure to determine the validity of the requirements information that is important and required by the consultant provided by the users (in our interest, the organisations) [34]. The missing requirements are found to arise from intangible information within the informal (e.g. socio-cultural) factors in the organisation that are difficult to elicit and represent [8],[35]. Incorporating the cultural aspects in the informal requirements is therefore important to fill in gaps in the requirements that forms the foundation for the business [36].

Techniques such as use case development, user centered design. structured interviews and informal modeling, which are among some of the widely practiced approaches by the IT consultant (acting as requirement engineers) [37], are available for eliciting requirements from stakeholders. However, very little attention is given to documentations providing clear understanding and management of the requirements. Well-recorded requirements ensure the identification of any incomplete requirements and enable the realisation of potential reuse of the requirements [38]. One problem identified in the requirement elicitation is the lack of ability to express and record the stakeholders' requirements in an understandable form, for not only the consultants to build the alignment, but also for the non-technical people in the organisation being analysed [39].

3   Articulation of Complex Business and IT Alignment Requirements

In an organisation context, business functions are performed within a social system, where people behave in a coordinated manner that corresponds to certain specified norms [40]. Organisational Semiotics (OS) [41],[8] provides a set of elaboration techniques for stakeholder identification and analysis. Treating any technology change, e.g. the introduction of a new IT system, as a course of action, the Stakeholder Identification method aids the analysis by placing the change as the focus of the analysis, which is surrounded by potential stakeholders. The roles, responsibilities and impact of the stakeholders in relation to each course of action can be articulated in a structured manner by careful application of semiotic based stakeholder theory. The application of OS concepts in modeling organisations is supported through a set of tools called Method for Eliciting, Analysing and Specifying Users' Requirements (MEASUR) [42], e.g. problem articulation method (PAM) and norm analysis method (NAM).

PAM provides a set of mechanisms to identify the main issues related to the organisation context, which enables an establishment of understanding of a complex problem situation faced by the business. Valuation Framing serves as a feedback in PAM to measure the satisfaction level of the user's needs to the target technical system and is performed iteratively to refine the requirements [41]. The diversity of cultural values among the users influences their perception of the business and IT being measured. Valuation Framing adapted from Hall's [43] ten cultural aspects can provide a basis for capturing and measuring these cultural values. The valuation framing process integrates Hall's metrics with traditional metrics covering quality, performance and providing the organisation with robust information (with added values from cultural aspects) to assist in their decision making process. An Organisational Onion technique facilitates a representation of the stakeholder's relationship to the business function and level of influence in the organisation. Through the definition of the formal, informal and technical factors within the Organisational Onion it addresses the relationships among the stakeholders. It supports the *"notion of viewing an organisation as a social system where the people involved internally and externally behaves in a structured patterns that are govern by a certain system of norms"* [8].

The norm analysis enables an organisation to study holistically the behaviours of their members and the active interactions between the members that are driven by norms [40]. The norms define the knowledge of the business processes in the structure of *<context>*followed by the conditions applied in *<state>* to the associated stake-holders affected or responsible *<agent>* and to the categorisation of the type of action *<deontic operator>*, and the *<action>* needed. According to Organisational Semiotics, norms can be categorised into formal and informal. Formal norms are a set of statements, such as business rules governing the business process, which define the expected or intended behavior of the business in an organisation. However, in practice, the stakeholders that are directly involved within the business may develop their own interpretation of the formal norms through adaptation from their cultural background. These informal norms may not be explicitly defined as the behavior of the business and they should not overwrite the formally defined norms. However, these informal norms provide knowledge of the actual social practices that relate to the formal norms of business. Using such knowledge, the organisation will have the capability to implement their strategy and ensure business processes that best fit the working culture, which will eventually improve business performance and productivity.

4   The Design of the Business-Technology Alignment Consulting Framework

The consulting framework, as it is developed in collaboration with Capgemini [44], is devised to aid a business and IT alignment. The methods in this framework are underpinned by SOA, TOGAF, PAM, and NAM. The framework consists of a requirements stage; valuation stage; and strategy formulation stage (see Fig. 1). In the first stage, the requirement elicitation for the overall business operations establishes a holistic view of the core and supporting business services. In the second stage, the requirements are analysed to assess the value of the IT applications that support the business services by applying the Valuation Framing techniques [41]. This assessment incorporates an assessment of the financial aspects, (in the IT Financial Analysis component), that influence the value of the IT applications towards the business services. The valuation phase outputs provide information on the state or performance of the IT applications in relation to the business services and enables recommendations to optimise the IS/IT performance in the third stage. The consulting framework guides the enterprise to formulate a strategic and flexible IT strategy for improving the business-IT alignment.

The verification of the design was obtained from a group of consultants in the Capgemini consulting team who worked on the actual client case studies. The design of the consulting methods with the techniques has been iteratively refined during the consulting activities.

4.1 Conceptual Model of the Consulting Framework

The consulting framework facilitates the analysis of business and IT alignment by a set of techniques as shown in Fig. 1 [44].

Business Domain Analysis analyses the organisations' high level background context, which includes the organisations' business goals and strategy, market and competitive conditions, internal structure, core business services, internal and external main stakeholders and finance flow. The component segments the enterprise's background information into five sections: organisation aspects, business structure, services, external stakeholders and finance. The segmentation is based on Osterwalder's business model [45].

Business Service is where details of the business service including the business processes, the association (if any) of the business services to other business services and the list of the IT applications assigned to support the business services are

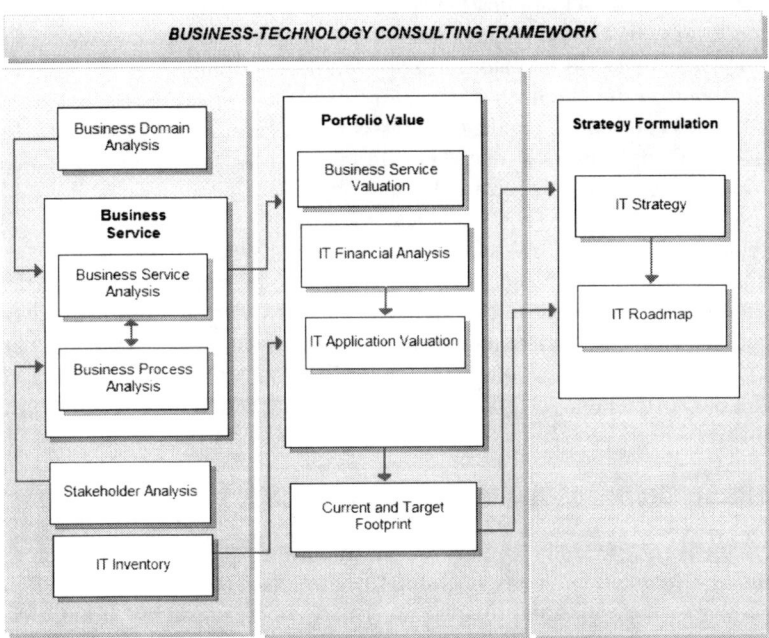

Fig. 1. The Business-Technology Consulting Framework

extracted. This Analysis captures the informal aspects through the *Business Service Norms Analysis,* eliciting the organisation's business norms within the business processes that support the business services. To complement the analysis in achieving the holistic view of the business service, its cultural value is evaluated in *Business Service Valuation*. The analysis will be presented and discussed in detail in the following section.

Business Process Analysis provides detailed descriptions of processes/sub-processes and activities involved in the business services. The business processes are influenced by, and are tightly integrated with, business norms. The business service norms analysis captures the behaviour patterns as requirements enabling us to derive the business rules from both the business and customer focused perspective. It also identifies the informal factors involved in the business service, such as strategies, cultures, unwritten conventions and common practices supporting the business capability. The analysis extends the details of the business rules and procedures with knowledge of any pre-conditions and post-condition from the Business Process Analysis.

Stakeholder Analysis is the component where the detailed analysis and elicitation of requirements of the associated stakeholders of the business service is performed. Stakeholder Analysis enables the identification and evaluation of highly influential stakeholders and their influence on business services in the enterprise, refining the stakeholder information identified in the business service analysis. Stakeholders perceive different values for the business services according to their perspectives. The stakeholder valuation of the business service is one of the important factors in the Business Service Analysis.

IT Inventory is the IT applications requirements analysis, where detailed information concerning the IT applications assigned to support the business services are elicited. The IT Application Inventory analyses all the IT applications in the enterprise to provide a foundation for further valuation assessment. The inventory component is divided into: 1) current IT applications (as-is state); and 2) recommendation for future changes to IT applications (to-be state). In the as-is state, each IT application available within the enterprise is recorded and described in detail: its capabilities, financial cost and technical value. The IT application value is evaluated in a separate component in the second stage, *IT Application Valuation* whereas, in the to-be state, the inventory includes recommendations for decisions to be made for the status of each IT applications (e.g. to be outsource, upgrade to improve performance, merge with other applications etc). The comprehensive requirements from both the business processes and services and IT applications form the basis for the components in the next stage, the Strategy Formulation to formulate future IT strategy that improves the business-IT alignment.

4.2   The Application of Business Services Analysis

Fig. 2 presents a case study to illustrate the framework. Techniotics™ is an advanced systems development company operating in three markets: advanced intelligent systems, defence electronics and robotics, alternative energy systems. It has global presence and market capitalisation and the following core competences: 1) highly skilled and rewarded, networked and virtual workforce; 2) patented processes for advanced electronics and autonomous systems manufacture; 3) extensive range of patents and intellectual property in the three sectors; 4) rapid design to manufacture virtual processes; and 5) innovative knowledge focused on new product development processes.

In the analysis process, we focus on a Techniotics™ example core competency: the innovative knowledge focused new product development processes. This is provided by Techniotics™'s unique range of engineers and scientists and marketing staff that form the core product concept teams, supported by global part-time problem solving consultants connected as virtual team members to the company and the extensive supporting IT systems. The key service is *Product Conceptualisation*. This internal service delivers an output: to identify three product concepts using the marketing departments' concept market specification that has been developed from an external client request for a new product/service. *Product Conceptualisation* is part of their *New Product Development* service that comprises several sub-services. The service operates according to a set of business rules or norms and is subject to constraints from business strategy, marketing and compliance. The set of business norms also have related social implications and social norms. For example a business norm specifies the internal company teams' tasks. This satisfies business norms that ensure for example, satisfaction with the task and support of the social contract that the key stakeholder value.

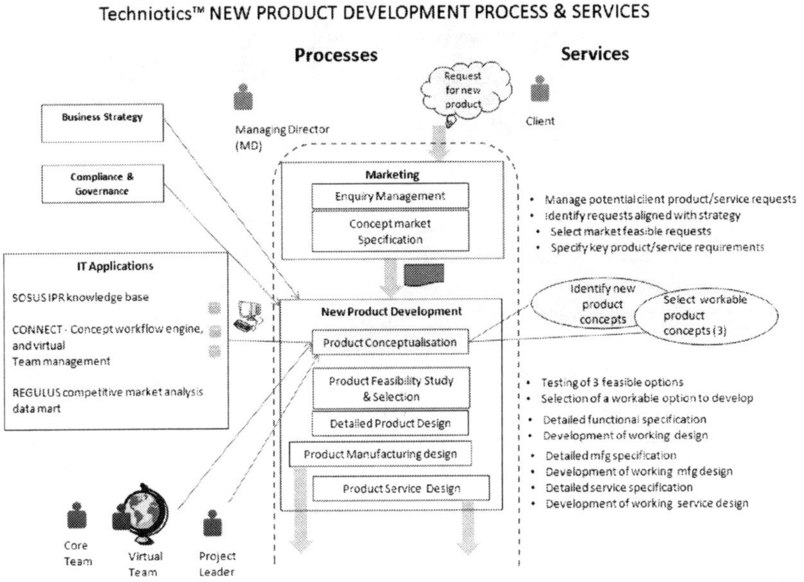

Fig. 2. Overview of Techniotics™ example business processes, services and IT applications

4.3 Business Service Analysis for the Alignment

Business Service Analysis in Fig. 3 assesses each business service which is defined by Business Domain Analysis with regard to their capabilities, stakeholder's participation in the business process, and IT applications support. In the business service analysis, a number of further techniques (see the shaded elements in Fig. 3), e.g. business cultural valuation, stakeholders analysis, definition of service norms, and business process modeling, IT Inventory, IT Application Valuation are applied and the corresponding outcomes recorded in this document.

BUSINESS SERVICE ANALYSIS					
Service name:	Product Conceptualisation	ID:	S2.1-NPD-PC	Type:	Core / ~~Support~~
		Business Cultural Value:	72%	Date:	29/04/2010
Description:	Identify three new product concepts based on product design norms that could proceed to feasibility study stage of new product development				
Stakeholders:	Marketing team, Core team, Virtual team, Project leader, Managing Director, Client				
Service Norms:	PCStart1001, PCIdentify1002, PCOptimize1003, PCApproval1004				
Business Process:	1) Project Leader received concept market specification from marketing team. 2) Core and Virtual team identify/generates new product concepts that fits the business strategy for approval by MD . 3) Virtual team assess workable product concepts to satisfy cost analysis and risk management criteria. 4) Core and Virtual team optimizes the best three concepts for approval by the project leader to proceed to the next stage: Product Feasibility Study& Selection				
Process Models:	P47: Product conceptualisation - as is v0.5				
Relationships					
part-of:	New product development				
shared with:	Marketing				
input/output:	inputs: concept market specification. Outputs: 3 approved product concept				
joint outcomes:	potential solution designs approved for Service-Product Feasibility Study & Selection (ID: S2.2)				
IT applications:	SOSUS IPR, CONNECT, REGULUS	IT Application value to Business Service:		73%	

Fig. 3. The analysis of business service requirements

The business service of *Product Conceptualisation* is a core service, which identifies new product concepts based on ideas from the clients. This business service can be described with service name, ID, description, date, and list of associated stakeholders, which are further analysed in detail by using the Stakeholder Analysis. The business behavior is described in P47 in conjunction with the service norms. *Product Conceptualisation* carries out its function in relation to other business services, e.g. it jointly produces the outcomes with Service-Product Feasibility Study & Selection.

Product Conceptualisation is supported by a list of IT applications, i.e. SOSUS, CONNECT, and REGULUS, each of which is detailed in the IT Inventory. Its value to this business service is calculated in the example as 73% by the IT Application Valuation. This value implies that the IT applications that the organisation has invested in, gives a good ROI in terms of its support of the functions in the business service. The requirements documented in this structure are used as the input for the subsequent analysis in the consulting process.

4.4 Norms for Governing the Business Behaviour

A norm is considered as a control mechanism for the business to deliver its value to customers through the business services. These business services are performed by the stakeholders whose behavior impacts on the business output. The consulting framework identifies two types of norms, i.e. service norms and social norms, which may govern the business services. Service norms as formal norms define a business context (i.e. business service) where the activities are conducted by the stakeholders through the business processes. Fig. 4 presents the service norms which govern the business behavior of *Product Conceptualisation*. Each norm focuses on the specific expected behavior of *Product Conceptualisation*. For example, *PCIdentify1002* defines that the proposed new product concepts can only be accepted by the MD if they satisfy the business strategy.

BUSINESS SERVICE NORM ANALYSIS					
Business Service ID:	S2.1-NPD-PC				
Rule ID	Whenever	if	then	is	to
	<context>	<state>	<agent>	<deontic operator>	<action>
PCStart1001	concept market specification is received	specification is complete and stable	Core Team and Virtual Team	obliged	confirmed acceptance by Project Leader.
PCIdentify1002	New product concepts are identified	each concepts fits business strategy	MD	obliged	proceed to optimization process.
PCAssess1003	Pre-selected concepts exist	concepts satisfy the cost analysis and risk management criteria	Virtual Team	obliged	optimized 3 product concepts.
PCApproval1004	3 product concepts has been optimized	product concept is to be approved	Project Leader	obliged	approve the product concepts report for Product Feasibility Study & Selection

Fig. 4. Description of the business norms that govern the business processes

Social norms as informal norms are implicit and not formally codified. These norms are used by individuals and groups to execute the actual behavior (vs the codified behaviour). Social norms need to be identified to appreciate the stakeholders' rights and interest in the business activities and the related technology. It is important to take these factors into consideration as stakeholders are the influential aspects in determining the value of the business activities and IT support. In the consulting framework, the social norms are derived from the Hall's ten cultural aspects which characterise some quantified measurements. The social norms in the consulting analysis are applied to evaluate *business cultural value* in a Business Service.

4.5 Evaluate the Cultural Values of Business Services

During the *Product Conceptualisation* analysis, the business cultural value needs to be assessed. Such cultural values should be taken into account in the business service analysis as this can provide a more truthful view of any lack of alignment, or forced perceived alignment due to political/social pressure affecting an individual's value judgment. In conducting a valuation, the Business Service Valuation is used in Fig. 5.

In the valuation matrix, the relevant stakeholders and criteria are also weighted accordingly to its degree of significance on the business service. The weighting schema or values for both are not fixed and can be adjusted to suit the objective of the consultancy exercise, provided the total weighting of all involved stakeholders in the business service and the overall criteria sums to 1 respectively. The stakeholders then rate each of the criteria based on the value range -3 to +3. In the valuation of *Product Conceptualisation*, six stakeholders have been identified and weighted. These stakeholders rated the business services and first, the overall value of the business service cultural value per stakeholder is calculated by VS_i. Then, the value is recalculated

BUSINESS SERVICE VALUATION							
Business Service ID:	S2.1 -NPD-PC						
Stakeholder (Si)		Marketing Team	Core Team	Virtual Team	Managing Director	Project Leader	Client
	Stakeholder weight	0.10	0.25	0.25	0.10	0.10	0.20
Criteria	Criteria weight						
Core competence	0.2	2	3	2	3	2	3
Knowledge/skill specialisation	0.05	2	2	2	2	3	-1
Learning capability	0.05	1	2	1	-1	2	-2
Task satisfaction	0.04	1	1	2	-1	2	-1
Social contract	0.01	1	1	1	1	1	-2
Adaptability	0.05	1	1	2	2	2	1
Interaction	0.1	2	2	2	3	3	3
Transparency	0.1	2	2	2	3	3	3
Operational risk	0.03	2	2	3	2	2	1
Environment risk	0.02	-1	1	2	2	2	1
Reputation risk	0.02	3	3	1	3	2	1
Security risk	0.03	1	2	1	3	2	1
Service productivity	0.02	1	2	2	3	3	3
Quality - timeliness	0	2	1	3	2	3	3
Quality - Output: General Product/service	0.1	2	3	2	3	3	3
Financial benefit	0.1	2	3	3	3	3	3
Reliability/consistency	0	-1	1	3	3	3	3
Cultural benefit	0	1	-2	-3	-3	-2	-1
Customer benefit	0.04	2	2	2	2	2	3
Sustainability	0	1	2	1	2	2	2
Information security	0.04	1	2	2	3	3	3
	VSi = Sum(Stakeholder rate * service criteria weight)	1.72	2.30	2.02	2.43	2.50	2.04
	V = VSi * Stakeholder weight	0.17	0.58	0.505	0.243	0.25	0.408
Total Business Service cultural value:	Sum(V)/3						72%

Fig. 5. The culture values perceived by the relevant stakeholders

with consideration of the stakeholders weight to provide the values according to the strength of their influences on the business service in V. The final value of 72% is then calculated by Sum(V) / 3 which provides a percentage value of the business cultural value according to the overall perceptions of the stakeholders involved. This cultural value indicates the positive perception of *Product Conceptualisation* by the stakeholders.

The business service analysis needs to be conducted on all business services in Techniotics™ to establish a holistic view of the business. As the framework requires a full spectrum of quantitative and qualitative analysis supporting alignment, the technical aspects (i.e. IT) in the organisation need to be thoroughly analysed by using Portfolio Value analysis which analyses the financial aspects and risk management. The outcomes of this analysis assists the formulation of IT strategies for Techniotics™, i.e. what IT applications should be invested in the future, what IT applications should be considered for outsourcing etc. A socially informed IT strategy can help organisations make the right decisions for IT applications based on the business and IT alignment.

5 Conclusion

The design of the consulting methodology has been discussed, focusing on the Business Service Analysis component and its close relationship with the Business Service Norms Analysis which provides a structured approach to elicit the social (cultural) aspect values embedded in the business. The articulation of the social values helps to provide a more complete and holistic representation of the business services in the organisation. We have, using the example, shown how to identify the business norms provided by the analysis methods and techniques. We have also addressed the need to include the cultural aspects that impact the business and IT alignment. Although the social aspect is considered as implicit knowledge, it can have a significant influence on the business service values and IT applications that support the business services.

The holistic representation from the business service analysis is complemented by the Business Service Valuation in the Portfolio Valuation where the emphasis is on determining the cultural values of the business service. The analysis is performed by capturing the organisation stakeholders' perception of the business services and IT applications. The outcome of the complete valuation phase enhances the holistic view of their business-IT alignment presented in the Business Service Analysis and contributes to the other analysis components, specifically the Strategy Formulation component, where a future IT strategy and a roadmap for an improved alignment can be developed for the organisation.

Future work will focus on the development of a complete consulting methodology. This will include a validation of the architecture using actual organisations. This is critical to test the framework against real business issues to ensure the validity of the methodology for practical use. Further development to complete the consulting methodology toolset is also under construction.

References

1. Chan, Y.E., Reich, B.H.: IT alignment: what have we learned? Journal of Information Technology 22, 297–315 (2007)
2. Tallon, P.: The Alignment Paradox. CIO Insight (2003),
   http://www.cioinsight.com/c/a/Past-News/Paul-Tallon-The-Alignment-Paradox/
3. Campbell, B.: Alignment: Resolving ambiguity within bounded choices. In: Pacific Asia Conference on Information Systems, Bangkok, Thailand (2005)
4. Lagerström, R., et al.: Enterprise Meta Modelling Methods - Combining a Stakeholder-Oriented and A Causality-Based Approach. In: 13th International Workshop on Exploring Modelling Methods in Systems Analysis and Design 2009, Amsterdam, The Netherlands (2009)
5. Gartner: Gartner Says Worldwide IT Spending On Pace to Surpass $3.4 Trillion in 2008 (2008), http://www.gartner.com/it/page.jsp?id=742913
6. Gartner: Gartner Says Worldwide IT Spending to Grow 5.3 Percent in 2010 (2010), http://www.gartner.com/it/page.jsp?id=1339013
7. Luftman, J., Kempaiah, R., Rigoni, E.H.: Key Issues for IT Executives 2008. MIS Quarterly Executive 8(3), 151–159 (2008)

8. Liu, K.: Semiotics in Information Systems Engineering. Cambridge University Press, Cambridge (2000)
9. Ulrich, W., McWhorter, N.: Defining Requirements for a Business Architecture Standard, B.A.S.I.G. Technical Report, The OMG (2010)
10. Versteeg, G., Bouwman, H.: Business architecture: A new paradigm to relate business strategy to ICT. Information Systems Frontiers 8(2), 91–102 (2006)
11. Luftman, J., Brier, T.: Achieving and Sustaining Business-IT Alignment. California Management Review 42(1), 109–122 (1999)
12. Trist, E.L., Bamforth, K.W.: Some Social and Psychological Consequences of the Longwall Method of Coal-Getting. Human Relations 4(1), 3–38 (1951)
13. Baxter, G., Sommerville, I.: Socio-technical systems: From design methods to systems engineering. Interacting with Computers, Corrected Proof (2008) (in press, Corrected Proof)
14. Walker, G.H., et al.: A review of sociotechnical systems theory: a classic concept for new command and control paradigms. Theoretical Issues in Ergonomics Science 9(6), 479–499 (2008)
15. Mumford, E.: A Socio-Technical Approach to Systems Design. Requirements Engineering 5(2), 125–133 (2000)
16. Chan, Y.E.: Why Haven't We Mastered Alignment? The Importance of the Informal Organization Structure. MIS Quarterly Executive 1(2), 97–112 (2002)
17. Zacarias, M., et al.: Adding a Human Perspective to Enterprise Architectures. In: Wagner, R., Revell, N., Pernul, G. (eds.) DEXA 2007. LNCS, vol. 4653, pp. 840–844. Springer, Heidelberg (2007)
18. Zachman, J.: A framework for information systems architecture. IBM Systems Journal 26(3), 276–292 (1987)
19. Jonkers, H., et al.: Concepts for Modeling Enterprise Architectures. International Journal of Cooperative Information Systems 13(3), 257–287 (2004)
20. TOGAF: TOGAF version 9 Enterprise Edition. The Open Group Architecture Framework, TOGAF (2009)
21. Buckl, S., et al.: Using Enterprise Architecture Management Patterns to Complement TOGAF. In: IEEE International on Enterprise Distributed Object Computing Conference, EDOC (2009)
22. Brahe, S.: BPM on Top of SOA: Experiences from the Financial Industry. Business Process Management 96–111 (2007)
23. Hagel Iii, J., Singer, M.: Unbundling the Corporation. Harvard Business Review 77(2), 133–141 (1999)
24. Iansiti, M., Levien, R.: The Keystone Advantage. Harvard Business School Press, Boston (2004)
25. Erl, T.: SOA: Principles of Service Design. The Prentice Hall Service-Oriented Computing Series from Thomas Erl. Prentice Hall/PearsonPTR (2008)
26. Bieberstein, et al.: Impact of service-oriented architecture on enterprise systems, organizational structures, and individuals. International Business Machines 44(18) (2005)
27. Orriens, B., Yang, J., Papazoglou, M.P.: A Rule Driven Approach for Developing Adaptive Service Oriented Business Collaboration. In: Benatallah, B., Casati, F., Traverso, P. (eds.) ICSOC 2005. LNCS, vol. 3826, pp. 61–72. Springer, Heidelberg (2005)
28. Cherbakov, L., et al.: Impact of service orientation at the business level. IBM Syst. J. 44(4), 653–668 (2005)
29. Reich, B.H., Benbasat, I.: Factors that Influence the Social Dimension of Alignment between Business and Information Technology Objectives. MIS Quarterly 24(1), 81–113 (2000)

30. Kilpeläinen, T.: From Genre-based Ontologies to Business Information Architecture Descriptions. In: 17th Australasian Conference on Information Systems, Adelaide, Australia (2006)
31. Luftman, J., Papp, R., Brier, T.: Enablers and inhibitors of business-IT alignment. Commun. AIS, 1(3es) (1999)
32. Grant, K., Hackney, R., Edgar, D.: Strategic Information Systems Management. Thomas Rennie (2010)
33. Hull, E., Jackson, K., Dick, J.: Requirements Engineering. Springer, Heidelberg (2005)
34. Kotonya, G., Sommerville, I.: Viewpoints for requirements definition. Software Engineering Journal 7(6), 375–387 (1992)
35. Liu, K., Sun, L., Tan, S.: Modelling complex systems for project planning: a semiotics motivated method. International Journal of General Systems 35(3), 313–327 (2006)
36. Coughlan, J., Macredie, R.D.: Effective Communication in Requirements Elicitation: A Comparison of Methodologies. Requirements Engineering 7(2), 47–60 (2002)
37. Neill, C.J., Laplante, P.A.: Requirements engineering: the state of the practice. IEEE Software 20(6), 40–45 (2003)
38. Samuel, R., et al.: A pattern-based method for building requirements documents in call-for-tender processes. Technical Report. Technomathematics Research Foundation (2009)
39. Toro, A.D., et al.: A Requirements Elicitation Approach Based in Templates and Patterns. In: WER 1999, pp. 17–29 (1999)
40. Stamper, R., et al.: Understanding the Roles of Signs and Norms in Organisations. Journal of Behaviour and Information Technology 19(1), 15–27 (2000)
41. Stamper, R.K.: Knowledge as action: a loci of social norms and individual affordances. Social Action and Artificial Intelligence (1985)
42. Stamper, R., et al.: Signs plus norms - one paradigm for organisation semiotics. In: The First International Workshop on Computational Semiotics, Paris, France (1997)
43. Hall, E.T.: The Silent Language. Doubleday and Company, New York (1959)
44. CEAR: Business Aligned IT Strategy (BAITS) - Methodology and User Guide. Capgemini Enterprise Architecture Research (2009)
45. Osterwalder, A.: The Business Model Ontology - a proposition in a design science approach. In: Institu d'Informatique et Organisation. University of Lausanne, Lausanne (2004)

ONTECTAS: Bridging the Gap between Collaborative Tagging Systems and Structured Data

Ali Moosavi, Tianyu Li, Laks V.S. Lakshmanan, and Rachel Pottinger

University of British Columbia, Vancouver, BC, Canada
{amoosavi,lty419,laks,rap}@cs.ubc.ca

Abstract. Ontologies define a set of terms and the relationships (e.g., IS-A and HAS-A) between them; they are the building block of the emerging semantic web. An ontology relating the tags in a collaborative tagging system (CTS) makes the CTS easier to understand. We propose an algorithm to automatically construct an ontology from CTS data and conduct a detailed empirical comparison with previous related work on four real data sets – Del.icio.us, LibraryThing, CiteULike, and IMDb. We also verify the effectiveness of our algorithm in detecting IS-A and HAS-A relationships.

Keywords: ontology, taxonomy, tag, collaborative tagging systems.

1 Introduction

Ontologies organize information in content management systems and are the core building blocks of the emerging Semantic Web. Substantial work has been done in extracting ontologies automatically from large repositories like text corpora, databases, and the web. This paper focuses on collaborative social tagging systems (CTSs) such as Del.icio.us (for tagging bookmarks), Flickr (for tagging photos), IMDb (for tagging movies), LibraryThing (for tagging books) and CiteULike (for tagging publications). These systems permit users to tag and share resources (documents, photos, videos, etc.). Our goal is to create a generic ontology of the tags from a CTS. By ontology, we mean a set of concepts from a domain, represented by the tags, and their (IS-A and HAS-A) relationships.

Learning an ontology from a CTS can help make the CTS more useful. For example, browsing an ontology of tags from a CTS can help users better refine their queries, either to find more items by using a more general term or to find fewer items by using a more specific term. This is especially important in a CTS since the resources are typically labeled by a small, sparse, set of tags — so discovering content in CTSs by simple keyword search is much harder than in document and web search. Another application of domain specific ontology builders is to enhance search engines with ontologies. E.g., the prototype Clever Search system [15] merges words and their word senses in the general ontology, WordNet[1], and returns more relevant result items to the user.

---

[1] http://wordnet.princeton.edu

H. Mouratidis and C. Rolland (Eds.): CAiSE 2011, LNCS 6741, pp. 436–451, 2011.

In principle, we could use a general purpose ontology such as WordNet to browse a CTS; there are two disadvantages. First, tags in CTSs are not based on a fixed vocabulary but constantly evolve. Thus, one cannot expect WordNet (or similar systems) to capture the vocabulary in a dynamic CTS, e.g., "Mac OS X". Secondly, as we demonstrate in Section 7, even when terms corresponding to tags in a CTS *are* present in WordNet, in many cases, valid IS-A relationships between them that are found by our algorithm are missing in WordNet. This mirrors a similar finding for the ontology extracted from Wikipedia using YAGO [25]; using a combination of WordNet and Wikipedia found significantly more ontological relationships (including IS-A) that were absent in WordNet.

This paper studies the following problem: given a collaborative tagging system consisting of users, resources (also called items), and tags assigned by users to items, extract an ontology consisting of tags in the CTS and IS-A and HAS-A relationships between the tags. We consider HAS-A relationships in addition to IS-A: indeed, IS-A and HAS-A relationships are among those most used in ontologies with rich relationships, such as WordNet.

Our algorithm for ontology extraction from CTSs is predicated on the hypothesis that tags assigned to a resource by a group of users tend to contain both child and parent tags. We have conducted experiments to validate this assumption in the full version of our paper[20]. A possible explanation for this phenomenon is that different users may use tags at different levels of abstraction (from an underlying ontology in their mind); thus tags for the same item may include more abstract or more specific terms as an aggregation effect of various tagging behaviors. We leverage this hypothesis using association rules [1] and lexico-syntactic patterns to find relationships between tags. Our approach accounts for bi-grams (which can affect the precision of detected relationships), multi-word tags, and also infer non-trivial IS-A relationships from detected ones. We make the following contributions:

- We propose (Sections 4 through 6) an algorithm for ontology extraction from a CTS, called ONTECTAS (for ONTology Extraction from Collaborative TAgging Systems). The highlights of the algorithm include:

  - Candidate IS-A relationships are mined using association rules, making use of both forward and reverse confidence (Section 4.1).
  - Invalid tuples are pruned based on discovering bi-grams (Section 4.3).
  - Headword detection is leveraged for discovering relationships between multi-word tags (Section 4.4).
  - Lexico-syntactic patterns are used for detecting IS-A and HAS-A relationships. To our knowledge, we are the first to explicitly extract HAS-A relationships from CTSs (Section 5).
  - Based on items in the ontology having a common (IS-A) child, additional IS-A relationships are inferred (Section 6).

- We demonstrate via a comprehensive set of experiments on four real datasets that our algorithm outperforms previous algorithms w.r.t. quality and richness of the extracted ontology (Section 7).

Section 2 discusses related work. Section 3 formalizes the problem studied in this paper. Section 8 concludes and discusses future work.

2   Related Work

Some other works have studied extracting ontologies from CTSs. Some approaches [16,18,2] match CTS tags to concepts in general purpose ontologies such as WordNet, resulting in a graph of tags. However, because CTSs are ad-hoc and use terms dynamically, general purpose ontologies miss many terms as well as edges (i.e., relationships). For example, our experiments show that WordNet misses more than 25% of correct edges between concepts extracted from Del.icio.us, *even when both parent and child concepts are in WordNet.*

Schmitz [23] constructs weighted graphs based on conditional probabilities between pairs of tags. His algorithm cannot identify the exact relationship (e.g., IS-A and HAS-A) between terms — it simply says they are related, not how. By contrast, our algorithm pinpoints IS-A and HAS-A relationships between terms.

Heymann and Garcia-Molina [10] create an ontology by vectorizing the tags and finding the cosine similarity between tags. However, their method puts every tag from the similarity matrix into the taxonomy which causes many erroneous edges. Their work lacks an evaluation.

Schmitz et al. [22] use association rule mining to build a tree of related tags from a CTS; however, they do not explain how the edges are built or what types of relationships they model. We explain this in depth and also use lexico-syntactic patterns and a search engine to detect accurate IS-A and HAS-A relationships. [24] extends [22] and [10] by considering the tag's context. Barla and Bieliková [3] consider tag context similarly to [24].

The DAG algorithm [5] distinguishes between subjective and objective tags. After calculating feature vectors for each objective tag, DAG places tags with higher entropy in higher levels of abstraction. Like many other previous works, DAG does not determine the type of relationship between concepts.

Lin et al. [19] build a subsumption graph from the folksonomy and use a random walk to sort tags by generality ranking. They put tags in the taxonomy based on support and confidence between candidate nodes from the graph. They only consider a single sense for each tag, which leads to missed relationships. The authors claim building transactions for tags associated to items by specific users will lead to the best taxonomy because it preserves most of the information. In contrast, we found that user information does not improve taxonomy quality.

Körner et al. [14] categorize users by the kind of tags they use. They show that excluding some users can reduce noise and improve precision. This improvement is orthogonal to the contribution we make in this paper and is applicable in our context as well. We leave adapting ONTECTAS to this as future work.

Hearst [9] defines a set of patterns that indicate IS-A relationships between words in text documents. [4,6] find patterns for detecting HAS-A relationships from text corpora. To our knowledge, our work is the first to extend the lexico-syntactic patterns to find relationships of any type between tags in CTSs.

In sum, in contrast to previous works on ontology extraction from CTSs, our method is capable of detecting both HAS-A and IS-A relationships and explicitly identifying each. Our multi-stage algorithm also extracts high quality relationships between multi-word tags.

3 Problem Statement

A *collaborative tagging system* [22] is a 4-tuple $C = (U, T, I, Y)$ where U is a set of users, T is the set of tags used by the users, I is the set of items (resources) to which tags are assigned by users, and Y, the set of tag assignments, is a ternary relation on tags, users, and items, i.e., $Y \subseteq U \times T \times I$.

Specific CTSs may vary in detail from our definition above, e.g., IMDb does not have user information. We can model such CTSs by dropping U and defining $Y \subseteq T \times I$ as a binary relation. CTSs such as [11] allow users to declare their own IS-A relationships. User-supplied IS-A relationships can augment those automatically extracted but cannot supplant them because of the scale.

This paper studies how to efficiently extract IS-A and HAS-A relationships between tags in a given CTS. The output ontology consists ⟨tag1, tag2, label⟩ tuples where tag1 is the super class and label is either IS-A or HAS-A.[2] E.g., the tuple ⟨OS, Windows, IS-A⟩ indicates that Windows a kind of OS.

4 Ontology Extraction from Collaborative TAgging Systems (ONTECTAS) Algorithm

---

Algorithm 1. ONTECTAS

Input: (D) A set of ⟨*item, tag*⟩ 2-tuples or ⟨*user, item, tag*⟩ 3-tuples
Output: (O) Ontology of tags with IS-A and HAS-A relationships
1: $D' \leftarrow$ Preprocess D.   /*D' is a set of ⟨*item, tag*⟩ tuples*/
2: ⟨T_{basic}, F⟩ $\leftarrow$ *Association_Rule_Tuple_Detection*(D') /*Algorithm 2*/
3: $T_{pruned} \leftarrow$ *Bigram_Filtering*(T_{basic})   /*Algorithm 3*/
4: ⟨$T_{headword}, O$⟩ $\leftarrow$ *Headword_Detection*(T_{pruned})   /*Algorithm 4*/
5: $O \leftarrow O \cup$ IS-A_*Relationship_Detection*($T_{headword}$, IS-A-patterns,
          IS-A $-$ *threshold*)   /*Algorithm 5*/
6: $O \leftarrow O \cup$ HAS-A_*Relationship_Detection*($T_{headword}$, HAS-A-patterns,
          HAS-A $-$ *threshold*)
7: $T_{co\_parent} \leftarrow$ *Co_Parent_Pruning*($T_{headword}, F$)   /*Algorithm 6*/
8: Return $O \cup$ IS-A_*Relationship_Detection*(T_{co_parent}, IS-A $-$ *patterns*,
          IS-A $-$ *threshold*)   /*Algorithm 5*/

Our ONTECTAS algorithm for ontology extraction (Algorithm 1) consists of six phases. First, data is preprocessed and cleaned. Next, we extract candidate tag tuples via association rule mining using forward and reverse confidence. We then

---

[2] In both relationships tag2 IS-A tag1 and tag1 HAS-A tag2, we refer to tag1 as the super class label or the parent label for convenience, by abusing terminology.

remove tuples corresponding to bigrams. Next, we detect headwords of multi-word tags and use this to infer additional IS-A relationships. We then use lexico-syntactic patterns to extract additional IS-A and HAS-A relationships. Finally, we leverage pairs of tags sharing a common child in the extracted ontology to infer additional IS-A relationships. The next three sections describe the phases.

4.1   Preprocessing

The preprocessing step is primarily a cleaning step. It takes as input a CTS and performs the following tasks: (1) Any user information is projected away; we found looking at transactions at the level of group of users was most effective in ontology extraction. (2) Words with non-English characters are removed from the input data using the same method as in [5]. This adequately removed non-English words from all of our datasets. (3) Basic stemming: singular nouns are substituted for their plural forms. (4) Since tags occurring very infrequently are not statistically reliable, we removed tags or items that occurred fewer than 5 times. This threshold was determined empirically. (5) Verbs and verb phrases are removed by applying the Stanford parser[3] to each tag. This prunes tags that are used for organizing but convey no meaning about the item being tagged [7].

4.2.   Detecting Potential Relationships Using Association Rules

Adapting tagged data to market basket analysis requires defining how to build transactions from tags, which in turn requires defining "co-occurrence". We explored three different definitions of co-occurrence. Empirically, we determined that the most effective co-occurrence definition is the following: Tags t and t' co-occur if both were used to tag the same item (by possibly different users). The frequency of $\{t, t'\}$ equals the number of distinct items which were assigned both tags t and t'. Our careful study of the best definition of co-occurrence [20] allows us to more optimally use association rules than previous approaches, e.g., [23].

We use the FP-tree association rule mining algorithm [8] to extract frequent tag sets[4] and interesting rules from the set of transactions. The *support* of a tag set X is the proportion of transactions containing tag set X and the *confidence* of a rule is defined as confidence$(X \Rightarrow Y) = $ support$(X \cup Y)/$support(X) — i.e., the proportion of transactions in which X and Y occur together among those in which X appears. In this paper, we refer to the well-known definition of confidence as forward confidence (FC). We also introduce a new notion, reverse confidence (RC) as follows: reverse_confidence$(X \Rightarrow Y) := $ support$(X \cup Y)/$support(Y).

We assume tags assigned by users tend to contain both a term in the ontology and another term that has relationship with it. Therefore, if two keywords co-occur frequently, they are likely to be related. We use *support* to filter sets of tags with a cardinality of two. However, popular unrelated terms may occur together frequently, so we use *confidence* to remove tuples containing unrelated tags. Because terms which co-occur with high confidence are sometimes synonyms

---

[3] http://nlp.stanford.edu/software/lex-parser.shtml
[4] Tag sets correspond to itemsets in the context of frequent itemset mining.

Algorithm 2. Association_Rule_Tuple_Detection

Input: (D) A set of 2-tuples in form of $\langle item, tag \rangle$
Output: (T) Preliminary tag tuples, (F) Set of frequent itemsets
1: Group D by item. /*create: $\langle item, \{tag_1, ..., tag_k\} \rangle$*/
2: $S \leftarrow$ Union of tags associated with each item (i.e., S is set of transactions)
3: $F \leftarrow$ Frequent itemsets of size two from S where support > $min\_support$
   /*FC_i and RC_i are forward and reverse confidence respectively*/
4: **for all** $F_i \in F$ **do**
5:    **if** (($FC_i \geq min\_conf.$) and ($RC_i \leq 1 - min\_conf.$)) OR (($RC_i \geq min\_conf.$)
      and ($FC_i \leq 1 - min\_conf.$)) **then**
6:        Add F_i to T
7:    **end if**
8: **end for**
9: Return $\langle T, F \rangle$

(e.g., "os" and "operating system"), we use confidence in the reverse direction to ensure that terms are related with IS-A or HAS-A relationships. Different values for $min\_support$ and $min\_conf.$ can drastically change the size of the ontology; in our experiments these values were chosen empirically. At the end of this step, we have not yet classified the relationships into IS-A and HAS-A.

4.3   Pruning Edges between Bi-gram Elements

In this phase, bi-gram tuples which are common phrases are automatically pruned using a search engine. Usually bi-grams are compound nouns in the form of "adjective + noun" (e.g., free software) or "noun + noun" (e.g., web browser). Bi-grams do not contain IS-A or HAS-A relationships but sometimes are incorrectly detected as edges of an ontology since they co-occur frequently.

Finding bigrams by using a search engine [26,12,17] has not previously been applied to extracting relationships between CTS tags. ONTECTAS sends two keyword queries to a search engine for each relationship tuple (Algorithm 3). The queries are the quoted permutations of the terms in the tuple. If the ratio of the number of results returned for the two queries is larger than a threshold, the terms in the relationship tuple are regarded as bi-grams. E.g., if the relationship tuple is $\langle software, free \rangle$, the queries are "free software" and "software free". Since the ratio is higher than the threshold for this tuple, it is detected as a bi-gram and pruned. We experimentally found that the optimal threshold for detecting bi-grams is between 50 and 100. Because words in text documents have Zipfian distribution, [12] suggests using a logarithmic transformation of returned result counts. We found that the logarithmic transformation is also more accurate in detecting bi-grams.

4.4   Detecting Headwords in Multi-word Tags

Since many CTS tags are multi-word tags in form of compound phrases such as "science-fiction" and "object-oriented-data-model", we use headword detection

---

Algorithm 3. Bi-gram_Filtering

Input: (T) A set of 2-tuples of the form $\langle tag1, tag2 \rangle$
Output: (T') A reduced set of 2-tuples
 1: $T' \leftarrow T$
 2: **for all** $T'_i \in T'$ **do**
 3:     $ratio1 \leftarrow$ # of hits of querying "$tag1\ tag2$" as a phrase
 4:     $ratio2 \leftarrow$ # of hits of querying "$tag2\ tag1$" as a phrase
 5:     $ratio \leftarrow \frac{\log(\max(ratio1, ratio2))}{\log(\min(ratio1, ratio2))}$
 6:     **if** $ratio \geq bi - gram\_threshold$ **then**
 7:         remove T'_i from T'
 8:     **end if**
 9: **end for**
10: Return T'

---

to extract additional IS-A relationships (Algorithm 4). First, the Stanford parser detects the *headwords* for each phrase. A headword is a phrase's grammatically most important word; it determines the phrase's syntactic type. We then extract an IS-A relationship for each multi-word tag by putting the headword as the parent of the whole phrase. E.g., we can infer "object-oriented data model" IS-A "model". In this phase, more candidate tuples are produced by using either whole phrases or their headwords as the tags in tuples.

Algorithm 4. Headword_Detection

Input: (T) A set of 2-tuples of the form $\langle tag1, tag2 \rangle$
Output: (T') A set of enhanced 2-tuples, (O) Ontology with IS-A relationships
 1: $T' \leftarrow T$
 2: **for all** $T_i \in T$ **do**
 3:     **if** T_i contains multi-tags **then**
 4:         $head1 \leftarrow$ headword in $tag1$
 5:         $head2 \leftarrow$ headword in $tag2$
 6:         $O \leftarrow O \cup \{\langle head1, tag1, \text{IS-A}\rangle\}$
 7:         $O \leftarrow O \cup \{\langle head2, tag2, \text{IS-A}\rangle\}$
 8:         $T' \leftarrow T' \cup \{\langle head1, tag2\rangle, \langle head2, tag1\rangle, \langle head1, head2\rangle\}$
 9:     **end if**
10: **end for**
11: Return $\langle T', O \rangle$

---

5   Using Lexico-Syntactic Patterns

Finally, we analyze occurrences of lexico-syntactic patterns to detecting IS-A and HAS-A relationships. Due to data sparsity, lexico-syntactic patterns do not occur frequently enough to accurately detect relationships between terms [21]. Hence, we build on [13] and query the web for more occurrences of the patterns.

The core of our lexico-syntactic search is shown in lines 3-6 of Algorithm 5: given two tags and a pattern, we generate two keyword queries by considering

---

Algorithm 5. IS-A_Relationship_Detection

Input: $(T, P, \text{threshold})$ Where T is a set of $\langle tag1, tag2 \rangle$ tuples , P is a set of patterns
Output: (I) A set of 2-tuples in form of $\langle parent\_tag, child\_tag \rangle$
1: **for all** $t_i \in T$ **do**
2:     **for all** $p_j \in P$ **do**
3:         $hits1 \leftarrow$ # of hits of querying "$t_i.tag1\ p_j\ t_i.tag2$" as a phrase
4:         $hits2 \leftarrow$ # of hits of querying "$t_i.tag2\ p_j\ t_i.tag1$" as a phrase
5:         $ratio_j.F \leftarrow \frac{hits1}{hits2}$
6:         $ratio_j.R \leftarrow \frac{hits2}{hits1}$
7:     **end for**
8:     $maximum_F \leftarrow \max(ratio_j.F)$ over all j
9:     $maximum_R \leftarrow \max(ratio_j.R)$ over all j
10:     $maximum \leftarrow \max(maximum_F, maximum_R)$
11:     **if** $((maximum = maximum_F)$ and $(maximum_F \geq threshold))$ **then**
12:         $I \leftarrow I \cup \{\langle tag1, tag2, \text{IS-A} \rangle\}$
13:     **else**
14:         **if** $((maximum = maximum_R)$ and $(maximum_R \geq threshold))$ **then**
15:             $I \leftarrow I \cup \{\langle tag2, tag1, \text{IS-A} \rangle\}$
16:         **end if**
17:     **end if**
18: **end for**
19: Return I

---

the two possible permutations of the tags in the pattern. E.g., given ("human", "body", "'s"), the two generated queries will be "human's body" and "body's human". Then, the ratios for both forward and reverse occurrences direction are calculated. It is clear that given any set of patterns for any relationship, this algorithm can be applied. We use the following patterns from [9] to identify IS-A relationships: (1) Pattern 1: NP_1 such as NP_2; (2) Pattern 2: NP_1 including NP_2; (3) Pattern 3: NP_1 especially NP_2.

Our HAS-A relationships are supersets of meronymy (part-of relationships), and are not limited to the physical perspective. We consider two noun phrases NP_1 and NP_2 to have a HAS-A relationship (with NP_1 as the parent) if one of the following statements is true: (1) NP_2 is a part of NP_1. E.g., "body" is a part of "human"; or (2) NP_1 has/have NP_2. E.g., "human" has "mind" and "google" has "googleMaps"; or (3) NP_1 may have NP_2. E.g., "human" may have "disease".

From the existing lexico-syntactic patterns mentioned in the literature such as [4,6], we use three following patterns to detect HAS-A relationships:(1) Pattern 1: NP_1's NP_2; (2) Pattern 2: NP_2 of the NP_1; (3) Pattern 3: NP_2 of NP_1.

While patterns 1 and 2 are among the most common English patterns [6], pattern 3 is not. However, pattern 3 can be used to detect HAS-A relationship between tags such as the tuple $\langle \text{Coffee}, \text{Caffeine} \rangle$.

All patterns for a relationship are fed into a search engine. If the largest ratio of a pattern is above a threshold, that tuple is labeled with the corresponding relationship and added to the ontology. Algorithm 5 shows the IS-A detection algorithm. The HAS-A algorithm is similar, but requires that pattern 1 *and* one of patterns 2 and 3 are above the threshold. Both thresholds were found experimentally. In our experiments, the IS-A threshold was 7 and HAS-A threshold ranged from 20 to 50.

6   Exploiting Co-parents to Find More IS-A Relationships

Examining the ontology built thus far reveals an interesting property when pairs of tags share the same child. Consider the following example: the ontology may contain "fiction → urban-fantasy" and "fantasy → urban-fantasy", where "fiction" and "fantasy" are both parents for "urban-fantasy" w.r.t. the IS-A relationship.[5] However, the IS-A relationship between "fiction" and "fantasy" may be missing. One possible reason for this is that people tend to use the more specific tags leading to "fiction → urban-fantasy" and "fantasy → urban-fantasy", so that "fiction → fantasy" does not occur above the relatively high threshold needed to avoid noise.

Hence we have the following hypothesis: in a co-parent structure it is more likely than usual that the two parents are in an IS-A relationship. Hence, we include the following additional step (Algorithm 6) to ONTECTAS: for such co-parent pairs, we re-examine the pair's confidences under a lower threshold and extract candidate tuples for an IS-A relationship.

Algorithm 6. Co_Parent_Pruning

Input: (T) A set of tuples with IS-A relationships in form $\langle parentTag, childTag \rangle$; (F) A set of frequent itemsets
Output: (T') An enhanced set of tuples with IS-A relationships
1: $T' \leftarrow T$
2: $G \leftarrow$ A graph where each tuple in T corresponds to an edge from $parentTag$ to $childTag$.
3: $S \leftarrow$ All tuples of tags $\langle parent_1, parent_2, child \rangle$
    s.t. (1) edge$(parent_1 \rightarrow child) \in G$ and (2) edge$(parent_2 \rightarrow child) \in G$ and
    (3) edge$(parent_1 \rightarrow parent_2) \notin G$ and (4) edge$(parent_2 \rightarrow parent_1) \notin G$.
4: **for all** $\langle parent_1, parent_2, child \rangle \in S$ **do**
5:    **if** $\{parent_1, parent_2\}$ is frequent and if it satisfies lower forward and reverse confidence thresholds **then**
6:       Add $\langle parent_1, parent_2 \rangle$ to T' with the more frequent tag as the parent.
7:    **end if**
8: **end for**
9: Return T'

As a final step of the ONTECTAS algorithm, following standard practice in ontology extraction algorithms, if the graph of relationships is disconnected, we add a generic "Entity" root node and make it the parent of all orphan nodes.

---

[5] Here, "fiction" → "urban-fantasy" means "urban-fantasy" IS-A "fiction".

7 Experiments

7.1 Datasets and Assumptions

Our experiments used four real datasets: Del.icio.us (a social bookmarking web service), IMDb (the Internet Movie Database), LibraryThing (for tagging books) and CiteULike (a service for storing, organizing, and sharing scholarly papers). Table 1 shows the characteristics of the datasets. User information is not available in the IMDb dataset, so competing algorithms were unable to create ontologies from it.

Table 1. Corpus Details in Some Collaborative Tagging Systems

	Del.icio.us (Dec. 2007)	CiteULike (Jan. 2010)	IMDb (Nov. 2009)	LibraryThing (corpus from Delft*)
Number of Tags	6,933,179	431,160	2,593,747	10,469
Number of Items	54,401,067	2,081,799	356,162	37,232
Number of Users	978,979	60,220	N/A	7,279
Number of Tag Assignments	450,113,886	7,922,454	2,625,237	2,415,517

* http://homepage.tudelft.nl/5q88p/LT

To show that general purpose ontologies are insufficient, we validated that WordNet misses many relationships between terms *even when it contains both terms*. To show this, we evaluated a sample ontology (from Del.icio.us) both manually and by using all parent-child senses (meanings) in WordNet. We limited our experiments to relationships where both parent and child term exist in WordNet. This gives WordNet an advantage since many tags do not appear in WordNet at all. In this case, we found WordNet is missing 26.9% of manually validated relationships discovered by ONTECTAS. For example, WordNet contains 3 senses for "python", but none of these senses is related to programming; as a result, "programming → python" is missing in WordNet.

Since our approach is successful, it is clear that our hypothesis that *a group of users tend to tag items with both parent and child tags* is validated. The full version of this paper [20] shows detailed experiments which validate this empirically. We discuss our results, beginning with HAS-A relationship detection.

7.2 Evaluation of ONTECTAS in Detecting HAS-A Relationships

Table 2 shows the precision of ONTECTAS in detecting HAS-A relationships. *None of the other competing algorithms address HAS-A relationships from CTSs.* Table 2 only reports precision for ONTECTAS, the first algorithm to detect HAS-A from CTS data.

One challenge in detecting HAS-A relationships was that pattern-based search engine queries such as "human's middle" and "middle of human" are frequently part of phrases such as "human's middle finger" and "middle of human history". Clearly, there is room for improvement in ONTECTAS' precision in HAS-A detection, which we plan to address in future work.

Table 2. Precision in detecting HAS-A relationships

	Del.icio.us	CiteULike	LibraryThing	IMDb
Precision	51.6%	61.9%	55.5%	33.3%

7.3   Evaluation of ONTECTAS in Detecting IS-A Relationships

In the following, we focus on IS-A relationships. All competing algorithms do not distinguish between IS-A and other relationships such as synonyms, whereas we clearly isolate IS-A relationships. We lump all other relationships into ANY and compare the performance of ONTECTAS on IS-A with that of other algorithms on IS-A and ANY, giving them an advantage, since in this evaluation, we do not give credit to ONTECTAS for correctly finding HAS-A relationships. We use the following standard performance measures: (1) Precision: We consider the precision of ONTECTAS on IS-A with that of other algorithms on IS-A+ ANY. Precision for both is the number of correct edges over the number of all edges. (2) Maximum depth and average depth of the IS-A taxonomy. (3) Average number of children. A higher value of the last two measures implies richer ontology is extracted. In addition, following [19], we compare all algorithms with a gold standard to see how they fare in trying to recreate manually-curated ontologies.

For depth and breadth metrics, we calculate these metrics on an ontology with only correct relationships to ensure algorithms cannot earn an artificially and unfairly high score on these by finding many incorrect relationships!

Absolute recall for ontology extraction from a large CTS is very hard to measure. Instead, we propose a new metric: *relative recall*. Relative recall for an algorithm is the number of valid IS-A relationships found by the algorithm divided by the total number of valid IS-A relationships found by all algorithms.

7.4   Comparing ONTECTAS to Other Algorithms

We compare ONTECTAS with the four algorithms from Section 2: 1) the algorithm from [19] (abbreviated "LFZ") 2) the DAG algorithm [5] ("DAG-ALG") 3) Schmitz's algorithm [23] ("Schmitz"), and 4) Barla and Bielikova's algorithm [3] ("BB"). Since these algorithms cannot process the IMDb dataset due to the lack of user information, we only compare them on Del.icio.us, LibraryThing, and CiteULike.

To have a fair comparison, we implemented the above algorithms as closely as possible to the way their authors had implemented them; we used the parameters that were described in the papers and contacted the authors for additional information about how to make their algorithms as competitive as possible.

Validating the edges manually required that each algorithm output a small number of edges. To do so, we put another threshold on the number of times a tag, an item, or a user must occur in order to be considered. To be fair, we used the same threshold to ensure that each algorithm output fewer than 150 edges.

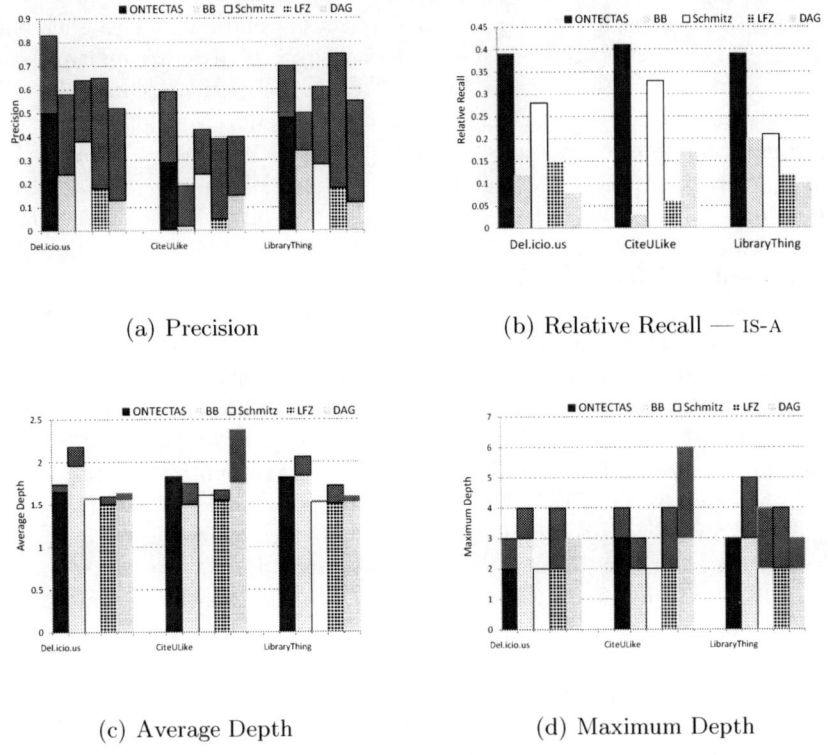

(a) Precision

(b) Relative Recall — IS-A

(c) Average Depth

(d) Maximum Depth

Fig. 1. Comparison of ONTECTAS to other algorithms for different metrics. Lower bars show IS-Arelationships and higher bars show "any" relationships.

Figure 1(a) shows the algorithms' precision for both IS-A relationships (the lower bars) and ANY relationships (the higher bars); for IS-A relationships, the precision of ONTECTAS is 0.50 for Del.icio.us, 0.48 for LibraryThing and 0.29 for CiteULike. ONTECTAS outperforms the precision of all other algorithms on all datasets. We also compare our precision on IS-A with that on IS-A+ ANY for the other algoritms since they do not distinguish IS-A from non-IS-A. Even then ONTECTAS outperforms the other algorithms in del.icio.us and CiteULike. On LibraryThing, the performance is close to the winner.

Figure 1(b) compares the algorithms' relative recall for IS-A relationships. ONTECTAS is the best performer for all three datasets. One reason for DAG-ALG's bad relative recall is that it detected many popular tags such as "web" and "software" as subjective tags, and pruned them before discovering the edges. BB had relatively low precision and recall in CiteULike because it detected many relationships with the tag "no-tag", which is a popular tag rather than an ontological tag. ONTECTAS performs the best for relative recall for ANY relationships [20].

Figures 1(c) and 1(d) measure the depth of the validated ontology detected by each algorithm for both IS-A (lower bars) and ANY relationships (higher bars). These measures quantify the richness of the ontology. If there are multiple paths from the root to a node n, the depth is the longest path. Because the other algorithms find just ANY relationship between elements in an ontology, rather determining the types of relationships, like ONTECTAS does, we measure both the IS-A relationships and ANY relationships found. We do not consider HAS-A since no other algorithms detect it. Notice that this gives an advantage to the competing algorithms. For the depth metrics, other algorithms usually find a long chain with combination of synonyms and IS-A relationships. Since ONTEC-TAS detects mostly IS-A or HAS-A (and not synonyms), maximum depth for ANY relationship in ONTECTAS is close to maximum depth of IS-A relationship because in general of chains containing IS-A and HAS-A are rare.

For IS-A relationships, ONTECTAS has the highest maximum depth for two out of three datasets. In the full version of the paper [20], we show that the average number of children is similar to the average depth. For the average number of children, ONTECTAS has the best performance for CiteULike, is roughly tied for Library thing, and is second best for Del.icio.us.

Even when competing algorithms are given credit for ANY relationships and ONTECTAS only for finding IS-A, ONTECTAS performs fairly well. This is because there are so many IS-A relationships detected as compared to the other relationship types.

For all of the depth/children metrics, we note that *all algorithms perform markedly better using our preprocessing step of removing verb phrases*. This step helped a lot in removing non-ontological tags such as "to-read" in the Del.icio.us dataset. By applying this to all algorithms, we have improved all algorithms' performance, not just ONTECTAS's. Figure 1 also shows that most of the algorithms performed better on most measures for the Deli.icio.us and Library-Thing datasets than on CiteULike. This validates the fact that the tags in these datasets are of better quality than the ones in CiteULike. This shows that we can compare different CTSs on the quality of tagging actions, using an ontology creation algorithm.

In summary, ONTECTAS outperforms the four other algorithms on precision and relative recall for IS-A relationships, and does well on the structural metrics of maximum depth, average depth and average number of children.

7.5   Comparing with a Gold Standard

Following [19], we compared how the algorithms extracted IS-A relationships against a "gold standard" ontology — the concept hierarchy from the Open Directory Project (ODP) [6]. To judge precision, recall, and F-measure, we use the lexical and taxonomic metrics from [19]. The lexical metrics measure how well the algorithms did in recreating the *concepts*, and the taxonomic metrics show how well the algorithms did in recreating the *structure*. Notice that comparing with a static ontology considered as gold standard has its problems since it

---

[6] http://dmoz.org

Table 3. Gold standard based lexical and taxonomic comparison

	Lexical					Taxonomic				
	ONT	LFZ	BB	DAG	Schmitz	ONT	LFZ	BB	DAG	Schmitz
Precision	0.261	0.743	0.183	**0.745**	0.128	**0.480**	0.077	0.434	0.123	0.329
Recall	0.240	0.006	**0.244**	0.025	0.007	0.723	0.023	0.711	**0.783**	0.256
F-Measure	0.044	0.011	0.043	**0.049**	0.014	**0.577**	0.035	0.539	0.212	0.288

may miss important concepts and relationships and a good algorithm that finds concepts and relationships manually verified to be correct may get penalized unfairly. We will return to this point. The full version of this paper [20] shows the formal definitions of the measures and the detailed results. Due to space limitations, we only cover the highlights in this paper.

We looked at the 25 highest-level concepts common across the five algorithms. Table 3 shows the results. Bolded entries represent the best performance.

ONTECTAS has the second highest overall lexical recall and f-measure, which shows that it did well at finding the desired concepts. While DAG had the highest lexical precision and f-measure, and BB had the highest lexical recall, they both did very poorly on taxonomic precision, leading to a low taxonomic f-measure.

LFZ had a very good lexical precision; however, this is achieved by reporting a very small number of correct concepts. ONTECTAS is superior to LFZ in terms of all three taxonomic measures.

Because the 25 highest level common concepts were very uneven in size, we performed an analysis of the 6 largest subtrees — otherwise algorithms would be testing against subtrees that were only one or two concepts large. When we considered only the 6 largest subtrees, ONTECTAS had the best lexical and taxonomic f-measure.

Comparing to a gold standard shows how well algorithms do against a manually created ontology. But since a gold standard ontology is static, this metric may unfairly penalize algorithms that genuinely find correct concepts and relationships. E.g., "dialect" and "software IS-A technology" is incorrect according to this standard. Thus, comparing algorithms should take into account other components discussed above as well.

8   Conclusion and Future Work

We proposed an algorithm (ONTECTAS) for building ontologies of keywords from collaborative tagging systems. ONTECTAS uses association rule mining, bi-gram pruning, exploiting pairs of tags with the same child, and lexico-syntactic patterns to detect relationships between tags. We also provided a thorough analysis of ONTECTAS and how it compares to other algorithms. Some of the important open problems include detecting spam users, improving accuracy of ontology extraction via supervised learning and by means of incorporation of part-of-speech detection. Our ongoing work addresses some of these.

References

1. Agrawal, R., Srikant, R.: Fast algorithms for mining association rules in large databases. In: VLDB, pp. 487–499 (1994)
2. An, Y.J., Geller, J., Wu, Y.-T., Chun, S.A.: Automatic generation of ontology from the deep web. In: Database and Expert Systems Applications (2007)
3. Barla, M., Bieliková, M.: On deriving tagsonomies: Keyword relations coming from crowd. In: Conference on Computational Collective Intelligence (2009)
4. Berland, M., Charniak, E.: Finding parts in very large corpora. In: Annual Meeting of the Association for Computational Linguistics, pp. 57–64 (1999)
5. Eda, T., Yoshikawa, M., Uchiyama, T.: The effectiveness of latent semantic analysis for building up a bottom-up taxonomy from folksonomy tags. World Wide Web 12(4), 421–440 (2009)
6. Girju, R., Badulescu, A., Moldovan, D.: Automatic discovery of part-whole relations. Comput. Linguist. 32(1), 83–135 (2006)
7. Golder, S., Huberman, B.A.: The structure of collaborative tagging systems. Journal of Information Science 32(2), 198–208 (2005)
8. Han, J., Pei, J., Yin, Y., Mao, R.: Mining frequent patterns without candidate generation: A frequent-pattern tree approach. Data Mining and Knowledge Discovery 8(1), 53–87 (2004)
9. Hearst, M.A.: Automatic acquisition of hyponyms from large text corpora. In: Conference on Computational linguistics, pp. 539–545 (1992)
10. Heymann, Garcia-Molina.: Collaborative creation of communal hierarchical taxonomies in social tagging systems. Technical Report 2006-10, Stanford (2006)
11. Hotho, A., Jäschke, R., Schmitz, C., Stumme, G.: BibSonomy: A social bookmark and publication sharing system. In: Conceptual Structures Tool Interoperability Workshop at the International Conference on Conceptual Structures (2006)
12. Keller, F., Lapata, M.: Using the web to obtain frequencies for unseen bigrams. Computational Linguistics 29(3), 459–484 (2003)
13. Keller, F., Lapata, M., Ourioupina, O.: Using the web to overcome data sparseness. In: ACL Conference on Empirical Methods in NLP, pp. 230–237 (2002)
14. Körner, C., Benz, D., Hotho, A., Strohmaier, M., Stumme, G.: Stop thinking, start tagging: tag semantics emerge from collaborative verbosity. In: WWW (2010)
15. Kruse, P.M., Naujoks, A., Rsner, D., Kunze, M.: Clever search: A wordnet based wrapper for internet search engines. In: Proceedings of the 2nd GermaNet Workshop (2005)
16. Laniado, D., Eynard, D., Colombetti, M.: Using wordnet to turn a folksonomy into a hierarchy of concepts. In: Semantic Web Application and Perspectives - Fourth Italian Semantic Web Workshop, pp. 192–201 (December 2007)
17. Lapata, M., Keller, F.: Web-based models for natural language processing. ACM Transactions on Speech and Language Processing 2, 1–31 (2005)
18. Lin, H., Davis, J., Zhou, Y.: An integrated approach to extracting ontological structures from folksonomies. In: Aroyo, L., Traverso, P., Ciravegna, F., Cimiano, P., Heath, T., Hyvönen, E., Mizoguchi, R., Oren, E., Sabou, M., Simperl, E. (eds.) ESWC 2009. LNCS, vol. 5554, pp. 654–668. Springer, Heidelberg (2009)
19. Liu, K., Fang, B., Zhang, W.: Ontology emergence from folksonomies. In: CIKM, pp. 1109–1118 (2010)
20. Moosavi, A., Li, T., Lakshmanan, L.V., Pottinger, R.: ONTECTAS: Bridging the gap between collaborative tagging systems and structured data (full version), http://www.cs.ubc.ca/~rap/ontectas.pdf

21. Sánchez, D., Moreno, A.: Learning non-taxonomic relationships from web documents for domain ontology construction. DKE 64(3), 600–623 (2008)
22. Schmitz, C., Hotho, A., Jäschke, R., Stumme, G.: Mining association rules in folksonomies. In: Classification, Data Analysis, and Knowledge Organization (2006)
23. Schmitz, P.: Inducing ontology from flickr tags. In: Collaborative Web Tagging Workshop at WWW (2006)
24. Schwarzkopf, E., Heckmann, D., Dengler, D., Kroner, E.: Mining the structure of tag spaces for user modeling. In: Wkshp. on Data Mining for User Model. (2007)
25. Suchanek, F.M., Kasneci, G., Weikum, G.: Yago: a core of semantic knowledge. In: WWW, pp. 697–706 (2007)
26. Zhu, X., Rosenfeld, R.: Improving trigram language modeling with the world wide web. In: Acoustics, Speech, and Signal Processing, pp. 533–536 (2001)

Cognitive Complexity in Business Process Modeling

Kathrin Figl[1] and Ralf Laue[2]

[1] Vienna University of Economics and Business Administration, Austria
kathrin.figl@wu.ac.at
[2] Computer Science Faculty, University of Leipzig, Germany
laue@ebus.informatik.uni-leipzig.de

Abstract. Although (business) process models are frequently used to promote human understanding of processes, practice shows that understanding complex models soon reach cognitive limits. The aim of this paper is to investigate the cognitive difficulty of understanding different relations between model elements. To allow for empirical assessment of this research question we systematically constructed model sets and comprehension questions. The results of an empirical study with 199 students tend to suggest that comprehension questions on order and concurrency are easier to answer than on repetition and exclusiveness. Additionally, results lend support to the hypothesis that interactivity of model elements influences cognitive difficulty. While our findings shed light on human comprehension of process models, they also contribute to the question on how to assure understandability of models in practice.

Keywords: Business Process Models, Understandability, Cognitive Complexity.

1 Introduction

Business process models (BPM) serve as a basis for communication between domain experts, business process analysts and software developers. To fulfill this purpose, such models have to be easy to understand and easy to maintain. Comprehension of process models is relevant for all tasks in which users interact with models, as for example in business process redesign or implementation of process-aware systems.

Many researchers have recently turned to investigate comprehensibility of process models and investigated various influence factors as modularity [1], domain knowledge [2] and notational aspects [3]. In addition, various complexity metrics have been proposed for BPM in the past years (see [4,5] for the discussion of relevant concepts and [6] for a comprehensive survey on related work). It has been shown that some of these metrics are significantly correlated with the number of control-flow errors in a BPM [5] and with the understandability of a BPM, measured in terms of correctly answered questions about the model [7,8]. [2] and [7] discuss how global measures (like the number of split nodes in a BPM) affect the understandability of a BPM.

H. Mouratidis and C. Rolland (Eds.): CAiSE 2011, LNCS 6741, pp. 452–466, 2011.

However, the scope of existing studies is limited, because the metrics used in these studies assign single (global) values to a BPM to describe its complexity. Ananda et al. [9] state: "Although studying the overall comprehensibility of a model is important, from a language evolution perspective it is even more relevant to discover *which* elements of a notation work well and which do not." With this paper, we want to give some first answers on the question *which* relations between model elements in a BPM are difficult to understand.

Despite increasing consciousness about the need to consider comprehensibility of process models, little research has been undertaken in order to improve and understand the relationships between modeling elements and comprehensibility. In this paper, we want to explore comprehensibility as a *local* property. This means that we measure the comprehensibility of a specific part of a BPM instead of the model as a whole. This way, we seek to investigate, which relations between elements in a graphical BPM are difficult to understand. In the research area of software complexity metrics, similar research has been published by Yang et al. [10]. Research results suggest that local complexity metrics could be a promising predictor for understandability. Therefore, we address the question as to when or under what circumstances similar relationships with local metrics will emerge in the context of BPM.

Our motivation is to complement the existing stream of work on improving the comprehensibility of BPM by examining local comprehensibility. In contrast to existing research such as [7,8] we assign own metrics to each comprehension question in our study, not to the models as a whole.

The remainder of this paper proceeds as follows: First, comprehensibility of process model elements and their relationships is placed in context with a review of relevant theoretical perspectives. Next, we articulate a research model. Then, we discuss design and findings of an empirical study. The final section discusses the limitations of our work and presents the implications of our research.

2 Measuring the Cognitive Load of Process Models

2.1 Comprehensibility and Cognitive Load Theory

For defining the term *comprehensibility*, we adapt the definition for understanding of computer programs given by Biggerstaff [11] by relating it to the modeling context and replacing the word "program" by "BPM":

"A person understands a BPM when they are able to explain the BPM, its structure, its behavior, its effects on its operational context, and its relationships to its application domain in terms that are qualitatively different from the tokens used to construct the BPM in a modeling language."

Further popular explanations of the term *comprehensibility* such as "the ease with which the ... model can be understood" [12] suggest that cognitive effort is an important factor determining model comprehensibility and should be as low as possible. Based on the complex relationships and control flow logic of organizational processes in practice, understanding of BPM is a task likely to demand high cognitive effort.

For conceptualizing model comprehensibility in greater detail we draw on the notion that understanding of a fact in a BPM becomes more difficult if the number of model elements that need to be attended to increases. This is backed by the work on Cognitive Load Theory. Cognitive Load Theory builds on the fact that the capacity of the working memory at a given point of time is limited [13]. If the amount of information to be processed exceeds this capacity, comprehension is affected negatively. It has been shown that an instructional design that avoids an overload of the working memory makes understanding of the instructional material easier [14]. Prior research on various visual languages like entity-relationship models [15] or UML class diagrams [16] suggests that reducing the cognitive load improves the understandability of visual models.

2.2   Influence Factors for Model Comprehensibility

To determine the relevant factors for the cognitive load involved in understanding elements and their relations in a model, we draw on work on BPM metrics.

Relations between Elements. Based on the similarity between structures in software code and process models, research results on code comprehensibility can serve as a profound basis for analyzing BPM comprehensibility. A large body of research exists on the cognitive complexity of different programming elements. Different control structures demand e.g. different levels of effort for understanding [17]. Little research has been undertaken to investigate the cognitive difficulty of different understanding tasks in process models. First efforts have been made by Melcher et al. [18]. In an experiment with 42 students reading a rather small BPM (containing 12 activities) they found that understandability values for questions on the four aspects order, concurrency, repetition and exclusiveness are different. As this is the only study in this context, further empirical research still needs to be done. Additionally, there is another strand of research stemming from the area of cognitive psychology, relating control flow elements and cognitive effort. Research on deductive reasoning has shown that systematic fallacies (so called 'illusory inferences') can occur when individuals construct or interpret mental models on premises concerning modeling-level connectives (like conjunctions or disjunctions) [19]. This situation may also be present for externalized visual BPM and may lead to higher error rates for understanding specific control flow elements. The current body of literature on error analysis of process models suggests for instance the existence of systematic reasoning fallacies concerning routing elements as inclusive OR gateways [20].

Element Interactivity. The cognitive load that a task imposes on a person is represented by the number of elements that have to be attended to. This number is determined by the level of interactivity between the elements. Elements interact if they are interrelated such that it is necessary to assimilate them simultaneously [21]. High interactivity leads to high cognitive load because each element has to be processed with references to other elements. On the other hand, cognitive load is low if the elements can be processed serially without referring to other elements.

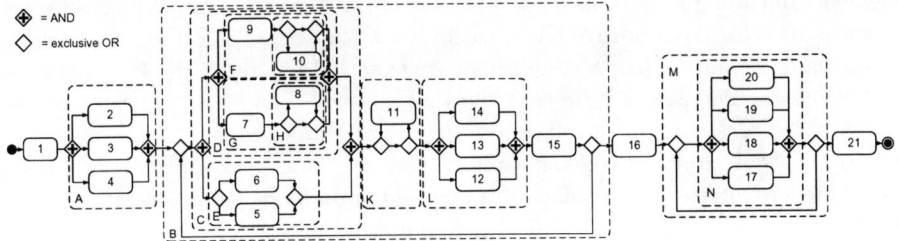

Fig. 1. Business process model, structured into regions

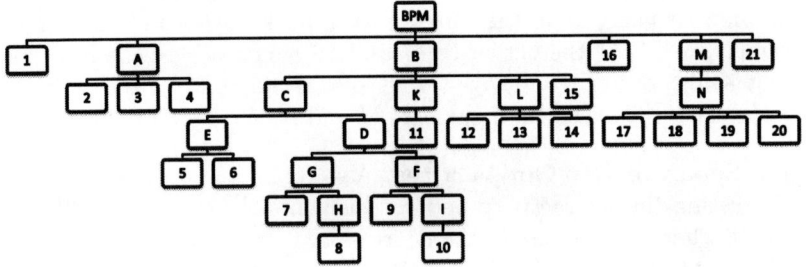

Fig. 2. PST for the model shown in Fig. 1

In order to define a measure for the cognitive load resulting from the effort to understand the relation between two elements in a BPM, we follow the idea of Vanhatalo et al. [22] to decompose the BPM into canonical fragments with a single entry and a single exit. These fragments can be arranged in a process-structure tree (PST) such that there is exactly one PST for each BPM. For details we refer to [22], but we introduce the concept of a PST by an example. Fig. 1 shows a BPM (similar to the ones used in our experiment) and its canonical fragments that form the PST. Additionally to the fragments that are marked with dotted boxes, all single activities and the model as a whole are canonical fragments in the PST. From the example, it can be seen that canonical fragments can be nested. For example, the fragments D and E are within a larger fragment C. The depth of the nesting shows how many routing constructs in the BPM have to be understood in order to reason about the execution of an activity. The PST of the model is shown in Fig. 2. For a better readability, the control nodes (called gateways in BPMN) are omitted in this graph.

We argue that the distance between two elements in the PST can serve as a measure for the interactivity between those elements. Each region in the PST represents one concept (for example the concept of an exclusive choice or the concept of parallel branching) that the reader of the model has to understand. If elements are located in deeply nested control-flow blocks, the reader has to understand a large number of concepts before being able to answer a question on the relation between those elements. In this case, the path between the two elements in the PST contains many arcs. On the other hand, if both elements are located into the same control block without additional nesting, they will also be in the same region of the PST, i.e. there are exactly two arcs in the PST between the elements. The assumption that the PST-distance can be an indicator of the difficulty to reason about a relation between two model elements is in line with the conceptual model of cognitive complexity by Cant et al. [23] that has been developed with respect to understanding software. Cant et al. discuss nesting within a a piece of software and argue that "the number of 'steps' [groups of control-flow statements; note from the authors] involved indicates the number of chunks which need to be considered" [23].

Formally, we define the *PST-distance* between two elements A and B of a BPM as the number of arcs between A and B in the PST minus one. This means that elements in a sequence or in the same control block have a PST-distance of 1. For example, in Fig. 1 the activities 17 and 18 which are executed in parallel inside the same control block have a PST-distance of 1 while the activities 16 and 17 (the latter is inside the fragments M and N) have a PST-distance of 3.

Element Separateness: Cut-Vertices. A second aspect we take into account when discussing the interactivity between elements A and B in a BPM is the special case where a single arc in the BPM separates the BPM into two disjoint parts P_1 and P_2 such that $A \in P_1$ and $B \in P_2$.

In terms of graph theory this means that the connected graph G that forms the BPM has a so-called cut-vertex on a path from A to B, i.e. a vertex that when removed causes that the remaining graph is not connected anymore. If such a cut-vertex between A and B exists, the mental model of the relationships between A and B becomes much easier, because A is located "before" and B is located "after" an easy-to-spot reference point (the cut-vertex). For example, in Fig. 1 it is easy to see that activity 7 cannot be executed after activity 17. Because of the cut-vertices before and after activity 16, this can be concluded without analyzing the control structures in which the activities 7 and 17 are embedded. The assumption that the presence of a cut-vertex makes it easier to understand a model is backed by results by Mendling and Strembeck [2] who found that a large number of cut-vertices in a model has a positive effect on its understandability.

3   Research Model

Having laid out the relevant theoretical factors related to local understandability of process models, we will now draw several propositions to suggest how these factors will influence cognitive difficulty in comprehension tasks. Prior research on process model comprehension has almost exclusively focused on global model understanding, a focus of study that we extend in this paper by looking at the understandability of relations between elements in a process model.

Fig. 3 shows our research model. The model proposes that the cognitive difficulty of understanding the relation between model elements is influenced by

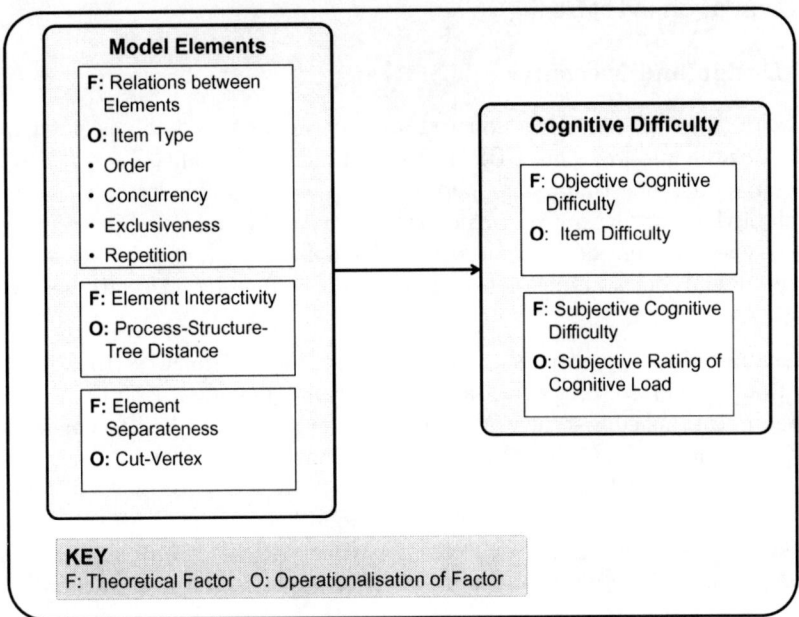

Fig. 3. Research Model

three factors: the type of relation between elements that has to be understood, the interactivity and the separateness of elements.

Following the research model, we now discuss three expected effects. As we anticipate similar effects on both objective as well as subjective side of the dependent variable 'cognitive difficulty', we formulate hypotheses for cognitive difficulty in general. First, we turn to different relations between elements. We state:

H1. The type of relation between elements that has to be understood (order, concurrency, repetition, exclusiveness) will have an influence on cognitive difficulty of understanding.

Second, we turn to the interactivity between elements. We expect that it is more difficult to understand relations between elements with a large PST-distance between them. Therefore, we have:

H 2. The interactivity between elements (high PST-distance) will be positively associated with the cognitive difficulty of understanding the relation between them.

Additionally we hypothesize if separateness of elements is low, understanding their relation gets easier:

H 3. High separateness between elements (existence of a cut-vertex between those elements) will be negatively associated with the cognitive difficulty of understanding the relations between them.

4   Research Method

4.1   Design and Measures

To test our hypotheses, we conducted an experiment in which the participants had to answer questions on a BPM. Model understandability (in terms of correctly answered questions) and perceived subjective difficulty were measured at each of the four levels *order, concurrency, repetition* and *exclusiveness* of the factor "type of comprehension question". To manipulate the main factor we constructed comprehension questions targeting the four different relations between activities.

Comprehension Questions. When selecting the questions, we took into consideration the work of Melcher et al. [18]. However, in comparison to [18] we formulated questions consistently, so that participants always had to consider two model elements (two activities) and their relationship for answering a question. Additionally we tried to use every-day-language in the questions. We used two different wordings to ask for the four relations between activities. To demonstrate the type of questions we refer to two activities with alphabetic names, although A and B were replaced with activity labels in the test material:

- Concurrency:
  - "A and B can be executed at the same point of time."
  - "A and B can be executed in parallel."
- Exclusiveness:
  - "In one process instance, A as well as B can be executed."
  - "The activities A and B are mutually exclusive."
- Order:
  - "If A as well as B are executed in a process instance, then A has to be finalized, before B can start."
  - "If A as well as B are executed in a process instance, then A is executed before B."
- Repetition:
  - "A can be executed more often than B."
  - "In each process instance A is executed exactly as often as B."

We took care that the wording in the questions is understandable, and we ran a pre-test in order to make sure that the participants understood the questions [24]. The comprehension questions, to which participants had to give a response of "right", "wrong" or "I don't know", were selected so that each activity was addressed approximately once in each diagram. The response option "I don't know" was included to lower guessing probability.

Questionnaire Construction. For each model in the questionnaire we posed the same eight types of comprehension questions. Despite the use of the same wording, it is obvious that there is a large number of possibilities how to ask these questions, because any two activities can be targeted with the same question. We identified two basic variations: 1) the statement given in the question is correct

	Version A	Version B
(process model diagram: A; B C D; G, I; E F, H J; K; L M N; O; P; Q R S; T; U)	• **Concurrency 1:** L and M can be executed at the same point of time. (correct, close)	• **Concurrency 1:** D and P can be executed at the same point of time. (wrong, distant)
	• **Concurrency 2:** G and S can be executed in parallel. (wrong, distant)	• **Concurrency 2:** G and H can be executed in parallel. (wrong, close)
	• **Exclusiveness 1:** In one process instance E as well as F can be executed. (correct, close)	• **Exclusiveness 1:** In one process instance K as well as S can be executed. (correct, distant)
	• **Exclusiveness 2:** The process steps C and R are mutually exclusive. (wrong, distant)	• **Exclusiveness 2:** The process steps E and F are mutually exclusive. (wrong, close)
	• **Sequence 1:** If T as well as J are executed in a process instance, then T has to be finalized, before J can start. (wrong, distant)	• **Sequence 1:** If A as well as C are executed in a process instance, then A has to be finalized, before C can start. (correct, close)
	• **Sequence 2:** If B as well as A are executed in a process instance, then B is executed before A. (wrong, close)	• **Sequence 2:** If B as well as T are executed in a process instance, then B is executed before T. (correct, distant)
	• **Repetition 1:** U can be executed more often than D. (correct, distant)	• **Repetition 1:** L can be executed more often than M. (wrong, close)
	• **Repetition 2:** In each process instance Q is executed exactly as often as P. (correct, close)	• **Repetition 2:** In each process instance O is executed exactly as often as I. (correct, distant)

Fig. 4. Example of Asking Model Comprehensability Questions

or wrong and 2) the location of the chosen activities. For varying the location of activities consistently, we decided to use pairs of activities, which are either close (≤ 1 activity between them) or distant (> 1 activity between them). As a consequence, we constructed the test material, such that each question was used once in each of four constellations (correct-close, correct-distant, wrong-close, wrong-distant), leading to 32 different question instances.

To ensure reliability of measurement we used a replication of the study design (questionnaire version A and B). In the replication, exactly the same models and comprehension questions were used, but the questions were asked for different activities in another constellation. Fig. 4 demonstrates how questions were asked for a specific process model.

Measured Variables. The outcome of our main dependent variable *comprehension* is cognitive per se, i.e. it is created in the viewer's cognition and not directly observable. Therefore, it can only be measured indirectly or via comprehension questions. According to Aranda et al. [9] there are four variables that can measure comprehensibility: *correctness* (did the participant give the right answer?), *confidence* (certainty of the participant in his answers), *perceived difficulty* (to answer the question, as subjective judgment by the participant) and *time* (required to give an answer). In our experiment, we chose to use the main objective and subjective measure of cognitive difficulty, viz. the percentage of correct answers (correctness) as objective measure and the user's rating of cognitive load as subjective measure (perceived difficulty). To measure the perceived difficulty, we asked the users to rate it on a 7-point Likert-scale (with the labels "very difficult", "difficult", "rather difficult", "neither difficult nor easy", "rather easy", "easy" and "very easy").

4.2 Materials

Questionnaire Parts. We used a pencil-and-paper questionnaire including three different sections in the experiment. The first section comprised items to obtain information about participants' demographic data, academic qualifications and modeling experience. Participants were asked about the number of years they had worked in the IT sector and the extent to which they had previously been involved with modeling in the context of education and work. After the first section, the questionnaire included a tutorial on process modeling, which covered all aspects the participants would need to know to perform the comprehension tasks. The third section included four different models with eight corresponding comprehension tasks per model. The amount of models used was determined by the selection of the comprehension questions during the experiment, as we wanted to ask 32 different instances of comprehension questions. To avoid order effects due to decreasing motivation or concentration of participants, we used two different scramblings. Models as well as comprehension questions were presented in different order, respectively.

Model Domain. The four models were selected from different domains such that we could expect that they are understandable for an average student with no special domain knowledge.

Model Language. Because it has been shown that the graphic style of a modeling language can influence the understandability of the model [25], we presented BPM modeled using different graphic styles. The models were modeled in different modeling directions and with three different routing symbol designs (UML Activity Diagrams, BPMN and Event-Driven Process Chains). These variations were included for allowing to generalize findings beyond specific layout and design restrictions. Additionally, they served as an experimental control to prevent a possible bias due to choosing only one modeling direction and routing symbols from a specific modeling language for all diagrams.

Model Layout. We took into account that a change in the graphical layout of a BPM can influence its comprehensibility [9]. For this reason, we took care that the graphical layout of the models did not impose additional challenges to the reader.

Model Size. Each of the four models used contained 21 activities. The model size was held constant for all models, because this variable is likely to have an influence on cognitive load of answering understandability questions.

4.3 Participants

A total of 199 business students participated in this study (125 males, 74 females), aged 23.5 years on average. Of all respondents, 36% were undergraduate students, 60% were master's level students and 4% had already completed their master's degrees. 67% had received training in modeling at university with 1.6 credit hours on average. About half of participants were familiar with Event-Driven Process Chains (60%) and UML Activity Diagrams (50%). 27% had work experience in the IT industry and 10% had already worked with BPMs.

5   Results

We first screened the data for its conformance with the assumptions of our statistical test. One assumption behind the use of ANCOVAs is that the variables are normally distributed. Kolmogorov-Smirnov tests confirmed that the dependent variables "percentage of correct answers" and "perceived difficulty" met this criterion ($p = 0.105$ and $p = 0.722$).

For each dependent variable, we ran a univariate ANCOVA with "relations between elements" and "element separateness" as independent factors and "element interactivity" as a covariate. According to the respective hypothesis, the percentage of correct answers and the perceived difficulty were the dependent variables. The ANCOVAs allow us to test the influence of the three predictor variables (two independent variables and a covariate) on the dependent variables as well as possible interaction effects. We use "element interactivity" as a covariate, because it is a continuous variable, which co-varies with the dependent variables percentage of correct answers ($r = -0.37$, $p = 0.002$) and perceived difficulty ($r = -0.61$, $p < 0.001$). Therefore, the covariate accounts for variance in the dependent variables and the inclusion of the covariate can increase the statistical power of the procedure.

5.1   Results for Hypothesis 1

ANCOVA results indicate that there is a effect of different types of relations between model elements (order, concurrency, repetition and exclusiveness) on cognitive difficulty of understanding their relation (H1). While there is only a trend for the percentage of correct answers ($F_{3,55} = 2.65$, $p = 0.058$), the effect on perceived difficulty is significant ($F_{3,55} = 4.20$, $p = 0.010$). According to our results, Hypothesis 1 is supported concerning subjective (perceived) difficulty, but only tentatively concerning objective difficulty (correct answers). Table 1 gives the percentage of correct answers and perceived difficulty for each type of relation. Order was the easiest relation (80% correct answers) with the lowest subjective difficulty (3.08), followed by concurrency (83%, 3.20). Exlusiveness was most the most difficult relation concerning correct answers (70%, 3.19) and repetition was rated as the most difficult by participants (71%, 3.58).

Table 1. Results for Hypothesis 1 (influence of types of relations between model elements), scale for perceived difficulty from 1="very easy" to 7="very difficult")

type of relation	percentage of correct answers		perceived difficulty	
	Mean	SD	Mean	SD
order	85%	6.16	3.08	0.35
concurrency	83%	10.09	3.20	0.28
repetition	71%	11.40	3.58	0.33
exclusiveness	70%	23.73	3.19	0.43

Table 2. Results for Hypothesis 2 (Influence of PST-Distance)

PST-distance	cases	percentage of correct answers		perceived difficulty	
		Mean	SD	Mean	SD
1	19	80.2%	20.62	2.89	0.31
2	13	79.5%	10.35	3.30	0.26
3	15	76.9%	15.06	3.42	0.27
4	5	76.4%	76.38	3.47	0.54
5	6	64.9%	64.85	3.55	0.29
6	4	74.8%	74.80	3.47	0.36
7	2	79.6%	79.64	3.54	0.28

5.2   Results for Hypothesis 2

As expected, the covariate PST-distance (element interactivity) has an influence on the percentage of correct answers ($F_{1,55} = 4.32$, $p = 0.042$). Additionally there is a highly significant effect on perceived difficulty ($F_{1,55} = 22.04$, $p < 0.001$). Table 2 shows the average percentages of correct answers and the average perceived difficulties across different PST-distances. Hypothesis 2 predicted that PST-distance will be positively associated with cognitive difficulty. Therefore, Hypothesis 2 is supported.

5.3   Results for Hypothesis 3

79.9% of the questions about two activities with a cut-vertex between them have been answered correctly, compared to 75.8% of the questions about two activities without a cut-vertex. Although the difference between means shows in the expected direction, the results of the ANCOVA indicate that this difference is not statistically significant and that there is also no significant influence on perceived difficulty. Moreover, there are also no interaction effects of "element separateness" (cut vertex) with "relations between elements".

Therefore, there is not enough evidence to support Hypothesis 3, which expected that the presence of a cut-vertex makes it easier to answer a question.

6   Discussion

This study provides empirical results on the influence of different relation types of elements, their interactivity and their separateness on cognitive difficulty of understanding the relation between elements.

In line with our predictions in Hypothesis 1, we found that different control structures in a BPM (like order or concurrency) differ according to their difficulty to be understood. Our results are in line with [18]. However, results are not directly comparable, as we used different wordings of possible understandability questions based on possible issues concerning ambiguousness (see [24] for details) and consistently addressed two model elements in the questions.

We further found that users perceive the relation between elements with a larger PST-distance as more difficult to understand. This effect has not been researched so far but is comparable to the discussion whether the nesting level in a BPM has an influence on its understandability. Mendling et al. [7,8] did not found a significant relationship between the nesting level and the understandability of a model. However, while Mendling et al. regarded the nesting level as a global attribute of a BPM, we related the PST-distance to the model elements we asked about.

While our results on Hypothesis 2 support the theory that the PST-distance is correlated with the difficulty of a task, the results are still not yet conclusive.

In particular, in our experiment there were too few cases with a PST-distance greater than 3 to come to reliable results about deeply nested model elements. Furthermore, in the models we used for our experiment, the presence of a cut-vertex was more likely between elements with a large PST-distance (like the activities 10 and 20 in Fig. 1). This can explain the fact that understanding for the elements with PST-distance 6 and 7 was better than for those elements with PST-distance 5 (see Tab. 2). Future research on this question will be necessary.

An interesting observation was that in some cases, a small PST-distance can even mislead to a wrong conclusion. For the model shown in Fig. 1, we asked the question whether both activity 5 as activity 6 can occur in the same process instance. Because of the exclusive OR-gateway before those activities, 76% of the participants answered "no". We assume that they did not bother to look at the parts of the model outside fragment E. Therefore they did not realize that this fragment is inside a loop and can be executed more than once.[1]

Our results did not confirm an influence of the existence of a cut vertex on the cognitive difficulty of an understanding task (H3). This relationship has also been discussed by other authors [7,8,2]. In contrast to our results, a similar experiment by Mendling and Strembeck [2] provided support for the hypothesis that a BPM with more cut-vertices is easier to be understood. Further studies [7,8] yielded inconsistent results on this topic, so further research will be necessary.

From a more general perspective, our findings highlight that reducing the cognitive load improves the understandability of process models as already demonstrated for other visual models [15,16]. Additionally our results provide support for the contention that the cognitive process for understanding a model depends on the actual task being carried out. This has already been substantiated in the research area of software comprehension by the work of Gilmore and Green [26].

7   Limitations

As with all studies, the research results presented here are subject to a number of limitations.

Model Size. We acknowledge that our models might not be representative for all kinds of BPM. Models from real projects are often much larger than the

---

[1] A very similar observation has been reported in [7].

ones used in our experiment. On the other hand, selecting rather simple models allowed us to keep the number of activities constant for all models and to avoid models that cannot be understood without the knowledge of a particular domain.

Questions. Right/wrong questions can introduce a measurement error, because on average 50% of the questions will be answered by guessing alone. For this reason, we left the possibility of checking "I don't know". Additionally, we acknowledge that for some questions users might have guessed the expected answer based on domain knowledge. Future research should collect similar data sets based on models with meaningless activity labels like "activity XY" as suggested in [27]. While we did not find understanding problems during the pre-test, in the analysis we realized that the statement "The activities A and B are mutually exclusive." gives room for misunderstandings ("A and B can not be processed either at the same time vs. both in the same process instance"). However, as those questions did not lead to more wrong answers as the alternative questions (see Sect. 4.1), we refrained from excluding these questions from our data.

Participants. The participants of our study were students who were familiar with the modeling languages, although they were not experts in this area. The results might differ if the experiment is replicated with experts in business process modeling [28]. We tried to select participants for the experiment so that there was a variation of little to medium experience with conceptual modeling, resembling potential users in practice. However, the results might not be generalizable to the entire population of BPM users.

Selection of Influence Factors. The factors for BPM understanding we have analyzed are not exhaustive. For example, we did not take into account the effect of the type of control structures (for example alternative or parallel branching) that are nested in the PST-tree. As related papers on this subject suggest that this factor should be considered as well [23], future research could examine this topic in detail.

8   Implications and Conclusion

This study is one of the first to investigate understandability as a local property in a model and denotes an important extension to the literature on influence factors for BPM understandability. Our main contribution is a first analysis of the cognitive difficulty of different relations between elements (order, concurrency, repetition, exclusiveness) in process models. Prior research has predominantly looked at global understandability of models and differences between models that influence understandability, in contrast we investigated local understandability of different items in a model.

Our results have implications for business process modeling practice and research. In terms of research, the results have an implication on the design of future experiments that measure understandability aspects of BPM. Our results demonstrate that several aspects of question selection (as the selection of the model elements and the type of the question) have an influence on cognitive difficulty. Implications of these results for researchers include exercising caution when aggregating answering rates of randomly chosen comprehension questions to total comprehension measures for models. As the choice of questions might significantly influence comprehension scores, balanced selection and construction of questions is highly relevant.

In addition, our work provides further evidence that high interactivity of elements may heighten cognitive load and lower comprehensibility of BPM. If possible, deep nesting of control-flow blocks should be avoided in order to make understanding easier and – in the end – to improve the quality of BPM and reduce modeling errors. Research on modularity of BPM [1] suggests that decomposing complex models into smaller submodels can improve model comprehensibility. Additionally syntax highlighting [29] can be used to heighten comprehensibility of deeply nested blocks.

Future research is needed to determine valid and reliable values for the cognitive difficulty of understanding specific relations between model elements. These values could make it possible to finally estimate understandability of models without the need of a user evaluation. Looking ahead, exact comprehension values could then be used to guide modeling tool developers to provide feedback on cognitive difficulty of models to users or to give hints on possible understandability problems in models.

References

1. Reijers, H., Mendling, J.: Modularity in process models: Review and effects. In: Proc. of the 6th Int. Conf. on Business Process Management, pp. 20–35. Springer, Heidelberg (2008)
2. Mendling, J., Strembeck, M.: Influence factors of understanding business process models. In: Schlender, B., Frielinghaus, W. (eds.) Business Information Systems. LNBIP, vol. 7, pp. 142–153. Springer, Heidelberg (1974)
3. Genon, N., Heymans, P., Moody, D., Amyot, D.: Improving the cognitive effectiveness of the bpmn 2.0 visual syntax (2010)
4. Gruhn, V., Laue, R.: Complexity metrics for business process models. In: 9th International Conference on Business Information, pp. 1–12. Springer, Heidelberg (2006)
5. Mendling, J.:Metrics for Process Models: Empirical Foundations of Verification, Error Prediction, and Guidelines for Correctness. LNBIP, vol. 6. Springer, Heidelberg (1974)
6. González, L.S., Rubio, F.G., González, F.R., Velthuis, M.P.: Measurement in business processes: a systematic review. Business Process Management Journal 16, 114–134 (2010)
7. Mendling, J., Reijers, H.A., Cardoso, J.: What makes process models understandable? In: Alonso, G., Dadam, P., Rosemann, M. (eds.) BPM 2007. LNCS, vol. 4714, pp. 48–63. Springer, Heidelberg (2007)
8. Reijers, H., Mendling, J.: A study into the factors that influence the understandability of business process models. IEEE Transactions on Systems, Man, and Cybernetics, Part A (2010)
9. Aranda, J., Ernst, N., Horkoff, J., Easterbrook, S.: A framework for empirical evaluation of model comprehensibility. In: MISE 2007: Proceedings of the International Workshop on Modeling in Software Engineering (2007)

10. Yang, J., Hendrix, T.D., Chang, K.H., Umphress, D.: An empirical validation of complexity profile graph. In: Proceedings of the 43rd Annual Southeast Regional Conference. ACM-SE 43, vol. 1, pp. 143–149. ACM, New York (2005)
11. Biggerstaff, T.J., Mitbander, B.G., Webster, D.: The concept assignment problem in program understanding. In: ICSE 1993: Proceedings of the 15th International Conference on Software Engineering, pp. 482–498 (1993)
12. Moody, D.L.: Metrics for evaluating the quality of entity relationship models. In: Ling, T.-W., Ram, S., Li Lee, M. (eds.) ER 1998. LNCS, vol. 1507, pp. 211–225. Springer, Heidelberg (1998)
13. Kirschner, P.A.: Cognitive load theory: implications of cognitive load theory on the design of learning. Learning and Instruction 12, 1–10 (2002)
14. Sweller, J.: Cognitive load during problem solving: Effects on learning. Cognitive Science 12, 257–285 (1988)
15. Moody, D.L.: Cognitive load effects on end user understanding of conceptual models: An experimental analysis. In: Benczúr, A.A., Demetrovics, J., Gottlob, G. (eds.) ADBIS 2004. LNCS, vol. 3255, pp. 129–143. Springer, Heidelberg (2004)
16. Genero, M., Manso, E., Visaggio, A., Canfora, G., Piattini, M.: Building measure-based prediction models for UML class diagram maintainability. Empirical Software Engineering 12, 517–549 (2007)
17. Cant, S.N., Jeffery, D.R.: A conceptual model of cognitive complexity of elements of the programming process. Information and Software Tech. 37, 351–362 (1995)
18. Melcher, J., Mendling, J., Reijers, H.A., Seese, D.: On measuring the understandability of process models. In: Rinderle-Ma, S., Sadiq, S., Leymann, F. (eds.) BPM 2009. Lecture Notes in Business Information Processing, vol. 43, pp. 465–476. Springer, Heidelberg (2010)
19. Khemlani, S., Johnson-Laird, P.N.: Disjunctive illusory inferences and how to eliminate them. Memory & Cognition 37, 615–623 (2009)
20. Mendling, J., Neumann, G., van der Aalst, W.M.P.: Understanding the occurrence of errors in process models based on metrics. In: Chung, S. (ed.) OTM 2007, Part I. LNCS, vol. 4803, pp. 113–130. Springer, Heidelberg (2007)
21. Sweller, J.: Cognitive load theory, learning difficulty, and instructional design. Learning and Instruction 4, 295–312 (1994)
22. Vanhatalo, J., Völzer, H., Koehler, J.: The refined process structure tree. Data & Knowledge Engineering 68, 793–818 (2009)
23. Cant, S., Jeffery, D., Henderson-Sellers, B.: A conceptual model of cognitive complexity of elements of the programming process. Information and Software Tech. 37, 351–362 (1995)
24. Laue, R., Gaddatsch, A.: Measuring the understandability of business process models - are we asking the right questions? In: 6th International Workshop on Business Process Design (2010)
25. Nordbotten, J.C., Crosby, M.E.: The effect of graphic style on data model interpretation. Inf. Syst. J. 9, 139–156 (1999)
26. Gilmore, D.J., Green, T.R.G.: Comprehension and recall of miniature programs. International Journal of Man-Machine Studies 21, 31–48 (1984)
27. Parsons, J., Cole, L.: What do the pictures mean?: guidelines for experimental evaluation of representation fidelity in diagrammatical conceptual modeling techniques. Data Knowl. Eng. 55, 327–342 (2005)
28. Petre, M.: Why looking isn't always seeing: readership skills and graphical programming. Commun. ACM 38, 33–44 (1995)
29. Reijers, H., Freytag, T., Mendling, J., Eckleder, A.: Syntax highlighting in business process models. Decision Support Systems (2011) (to appear)

Human-Centered Process Engineering Based on Content Analysis and Process View Aggregation

Sonja Kabicher and Stefanie Rinderle-Ma

University of Vienna, Faculty of Computer Science
Vienna, Austria
{sonja.kabicher,stefanie.rinderle-ma}@univie.ac.at

Abstract. In the context of business process modeling, the transformation of process information elicited from process participants into process models often remains a black box. This paper presents a method that supports the designer to extract a formal business process model from natural language captured in written form in a transparent and traceable way. To-do's of process participants are examined by means of the qualitative content analysis in order to identify essential process elements and to handle different levels of abstraction and labels used for describing operations. Results of the analysis serve as the basis for shaping process view logs which can be aggregated and merged into entire process models. The conduction of the method is described in a case study.

Keywords: Process design, Process views, Qualitative Content Analysis.

1 Introduction

Business process modeling is a widely used and accepted technique for capturing, understanding, and analyzing business processes in organizations. One main goal of business process modeling is to visualize processes (1) how they are actually performed (lived process), and (2) how they should be performed (intended process model). To strengthen organizations' effectiveness (doing the right things) and efficiency (doing things right), internal transparency and awareness of the organizations' business processes is needed. From the workflow perspective, business process modeling plays a key role in the phases design and diagnosis.

In the design phase, designers usually elicit information about the business process e.g., during workshops and interviews with selected employees (e.g. employees assigned to particular roles, middle and top management) [1]. At this stage, the processes are known and lived in the organization and often exist, at least process fragments if at all, simply as mental models in the heads of the member staff. The process designers face the challenge to transform implicit processes into explicit process models. In the diagnosis phase, which takes place some time after the enactment of the (new or improved) workflow system, the task of the designers is to find out how performers of the process work with the new system, as well as where and why they diverge from the intended process [2]. Common practices are, e.g., interviews, observations and process mining of event logs. Results are used to understand

H. Mouratidis and C. Rolland (Eds.): CAiSE 2011, LNCS 6741, pp. 467–481, 2011.
© Springer-Verlag Berlin Heidelberg 2011

work practice and to refine the process definition, or in other words, to improve the workflow system's support of employees in performing their business.

In this work, we present a technique of identifying lived business processes by means of To-do's of the performers in a process. Coming from a human-centric perspective of business processes and workflows, we argue that a bottom-up approach of capturing and documenting actually performed business processes leads to a more faithful process model to reality than a top-down approach. In a bottom-up approach, performed activities and tasks are collected from the performers of the process and are mapped to the process model. Thus, the process model is implicitly modeled by all process participants, whereas in a top-down approach the process model is mainly defined considering the knowledge about the company's processes available at various management levels, e.g. top-level, middle-, and lower management (depending on the company's organizational structure) and the collected data is optionally complemented with process information elicited from samples of further process participants. The top-down approach seems to be often used in practice when business processes need to be, e.g. defined and redesigned, as it is a question of time and costs to which extend all participants of the process are actively involved in the process elicitation phase.

We assume that personal To-do's implicate information about process fragments that are very close to reality and which include information about activities, tasks, roles, agents, decisions, delegations, time, data, and tools. By *personal To-do's* we understand activities and tasks listed by an individual organizational member in order to organize his or her work. These lists include activities and tasks (complemented with additional data if available) sorted in treated chronological order. We consider To-do lists as a resource of the individual used to perform a particular work task. We use the terms *activity* and *task* according to the definition of BPMN [3].

We use the *qualitative content analysis* as a technique to capture data of the To-do lists in a transparent, structured, uniform and comprehensive way. The content analysis supports the process designer to deal with (1) systematic identification of process elements, (2) homogenization of labels, and (3) granularity of activities. We show how the output of the analysis can be used to model or mine views of individual process participants and to aggregate and merge process fragments resulting from To-do's into an entire business process. By the term "process view" we understand a part of a process (process fragment) that is performed by a single organizational member. The method presented in the paper is evaluated in a case study.

The work is structured as described in the following. The next section presents the Business Process Model Extraction (BPME) Method that allows a transparent and traceable transformation of process information communicated in natural language to a formal process model. In Section 3 the qualitative content analysis and its potential to deal with challenges particularly arising in the context of business process modeling (level of abstraction, labeling and identification of essential process elements) are discussed. Section 4 focuses on the preparation of process view logs based on the output of the qualitative content analysis and the mining of the logs into one entire process model. The case study which was conducted for the teaching process in a real-world setting is presented in section 5. In Section 6 related work is reflected and Section 7 concludes our work.

2   Challenges and Overall Methodology

The elaboration of business process models is usually performed according to a common practice. In general, interviews and workshops are conducted in order to ascertain process information which is than interpreted by the designer by means of, e.g. visual and textual illustrations of the process [4]. So far, the step between information gathering and the presentation of the process models is handled like a black box. In this step, the designer's task is to transform mainly textual descriptions into natural language of process participants (humans) into a formal process model.

Our work concentrated on the analytical and designing step between data collection and the visualization of the process model. Our main research question was how to translate various process elements captured in To-do lists of individuals into a process model in a structured and traceable way. The following challenges are:

- How to extract process elements, like activities, tasks, roles, agents, time, data, tools, and decisions, from To-do lists of single organizational members in a structured and traceable way?
- How to handle different labeling of To-do's?
- How to deal with different granularity of To-do's?
- How to find reference points between the process views (connection, participation and delegation)?
- How to deal with special cases identified in To-do's?

Our approach included the following research instruments and techniques:

- Personal face-to-face interviews to elicit To-do lists from process participants.
- Qualitative content analysis to examine gathered To-do lists in order to identify key process elements and terms, and to transform the To-do's into process view logs.
- Mining of process view logs and the process model.

We propose a reusable semi-structured method that supports the designer to extract process models from text material (e.g. interviews, workshop protocols) in a traceable way. Traceability is supported by the use of the qualitative content analysis to code and categorize the gathered textual information and thus to identify and tag essential process elements. Results of the qualitative content analysis serve as the basis for the development of the process model which can be created by modeling or by transforming the data into process view logs.

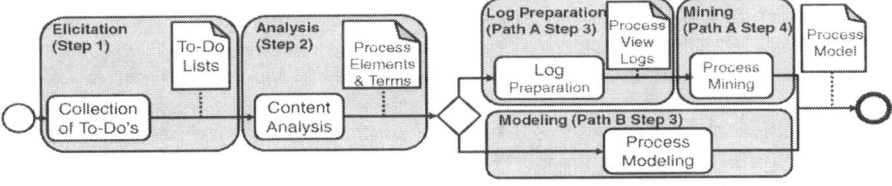

Fig. 1. Business Process Model Extraction (BPME) Method

Fig. 1 illustrates the steps of our method. In a first step, lived processes are elicit from process participants. Second, collected information is analyzed and prepared for further processing. There are two alternative ways of continuing: path A includes the capturing of collected process data in fictive logs (Step 3) which were then mined (Step 4). Choosing path B, the designer continues with manually designing process models. In this work we focus on the analysis (Step 2) and the steps of path A (log preparation and mining). In the following we describe the different steps at an abstraction level. In the Sections 3 and 4 the steps will be discussed in more detail.

Elicitation. The elicitation of process information is an important step as it determines the quality of the gathered material. In our view, an organizational member knows best what activities he or she actually performs in his or her job and how and when this work is done. To-do lists can be used to capture all the things that need to get done in a logical and also chronological order. To-do lists might also represent practical processes of work of single persons. For example, particularly new employees may notice their To-do's for particular scenarios (e.g. "What is usually to do?", and "What is to do in a special case?") during job instruction. Experienced employees often have their To-do lists in a particular context of work in their heads. There are several task management tools that support the organization of To-do's. If used by employees, such tools can be considered as a source of information.

Analysis. The To-do lists are examined with the qualitative content analysis. The method supports the designers to identify key process elements (e.g. activities, tasks, roles, agents, time, data, and tools) and terms (e.g. pre- and post conditions).

Log preparation. The process views are then transformed into process view logs. The logs illustrate personalized views of the process. Different views on the process can be generated by subsuming To-do's of process members with particular characteristics e.g. acting in the same roles, but also members of the whole process.

Mining. In contributions related to process mining, processes are commonly considered to be already performed. The resulting process model is automatically derived from generated logs. These logs include the IDs of each process instance and each event within the process instances. In a human-centric approach of identifying the business process model, as described in this work, processes must not be performed at the time of process mining. The process model is constructed on the basis of To-do lists of roles (humans) that are involved in the process.

3   Qualitative Content Analysis

In this section the qualitative content analysis and its potential in the context of business processes are described. Challenges of analyzing To-do's of process members are discussed, like unique labeling and the level of abstraction of To-do's.

3.1   Qualitative Content Analysis in General

The qualitative content analysis is a rule-guided technique to systematically analyze text [5]. Typically, the content analysis includes the steps: unitization, categorization, and coding [6]. Unitization means to divide the textual material into units of analysis (e.g. sections, thematic units, or syntactical units) [7]. In the categorization step a category scheme is created. The coding step includes the assignment of each unit of analysis to a category. Ideally, more than one coder performs the content analysis in order to assure inter-coder reliability of findings. There are tools that support the qualitative analysis of text material, e.g. the QDA Miner [8]. Such tools offer functionalities like coding retrieval, coding frequency, and coding co-occurrences. Existing experiments support automation of the analysis in order to reduce the high manual coding effort of large text material [9].

3.2   Qualitative Content Analysis in the Context of Business Processes

In the context of business process modeling, the qualitative context analysis offers a possibility to analyze textual material, e.g. interviews and workshop protocols collected during the phase of eliciting process information, in a structural and traceable way. Traceability is supported by coding text material, for example process views of individuals, or "process stories", and thus to identify, tag, and cluster essential process elements. Some questions that can be answered with the qualitative content analysis particularly in the context of business processes are:

- What are the tasks and activities of a particular process view?
- What tasks and activities can be subsumed according to their meaning?
- What persons and roles are involved in particular tasks and activities?
- What tools are used to perform particular tasks and activities?
- What kind of data is transferred, where, why and how?
- Are there decisions in the process view?
- Are there activities performed by several persons and roles?
- Are there activities in the process view that are delegated?

Results of the qualitative content analysis can be interpreted in several ways, but the final output of our method is always a business process model. The results may be manually modeled (resulting in process models, or process scenarios [1]), or further translated into logs. The latter alternative is described in more detail in Section 4.

3.3   Challenges of Qualitatively Analyzing To-Do's

In this section we address challenges the designers face when they examine To-do's of individual process members that belong to a particular process under investigation. We propose the elicitation of process information by means of To-do lists of individual process members. Several preconditions need to be considered when a qualitative content analysis of To-do's of individual process members is conducted.

Precondition 1: Clearly define the process under investigation.
Precondition 2: Identify at least one process member.
Precondition 3: Collect information: Ask process participants in individual exchange to list their To-do's in the process of investigation. Ask for the usual case and for special cases.
Precondition 4: Capture information. If necessary transform information into text.

The procedure of examining To-do's of individual process members based on the content analysis is described in the following. We discuss the steps specifically important for extracting business process models in more detail in Sections 3.4-3.6.

Step 1:   *Consider each To-do list of a process member as a case. For analysis use the entire text material.*

Step 2:   *If the roles of the process members are known, group the To-do lists of the same roles and analyze the cases considering their group affiliation. Cases of the same group are analyzed in sequence. Considering the roles during the analysis supports a faster identification of possibly identical To-do's and similar labeling. The designer develops a particular understanding of the context of work. If the roles are not known, the designer may identify other hints towards a particular ordering in analyzing the cases, e.g. data transfer or delegation to other persons mentioned in the To-do lists which allows a sequential analysis of the process flow views.*

Step 3:   *Define the unit of analysis. The unit of analysis may be, e.g. a set of semantically related To-do's, one To-do (thematic unit) or a sentence. To face the challenge of a uniform level of granularity of the To-do's, compare with Section 3.5.*

Step 4:   *Paraphrase the units of analysis. Reduce text to the key essence of the unit of analysis. In this phase, the identification of process elements takes place (compare with section 3.6).*

Step 5:   *Label or code To-do's. There are mainly two alternatives to perform labeling (compare with section 3.4): inductive or deductive labeling.*

Step 6:   *Evaluate results. For example, list all units of analysis with the same code, analyze the codes' frequency, the number of cases in which the codes were found, etc.*

3.4   Homogeneity of Activity Labels

The labeling of activities as recorded by process members might vary. Synonyms and homonyms need to be handled in order to design an understandable process model [10]. The content analysis supports two ways of determining unique labels [5]: the inductive and the deductive elaboration of labels.

Inductive elaboration of labels. Performing the inductive elaboration of labels means to derive labels from the text material. The labels are elaborated in a process of generalization. This procedure is used to attain a close reflection of the language and terms used in the material. Labels should reflect the content of the text material as good as possible. The inductive elaboration of labels can be complemented by the frequency measure of terms used in the material [11]. A visual representation of the terms' frequency offer word clouds. When the frequency measure is used to label operations misinterpretation may arise due to technical terms used by persons working in different roles and unequally long text segments. Therefore, designers always need to be aware of the content of the text material.

*Deductive determination of labels .*The deductive determination of labels implies that the labels are defined a priori. This means that the labels are specified before the text material is analyzed. In this case, terms are considered that are, e.g., commonly used in practice, in a particular sector or in a company. Formulations actually used by process members are not explicitly considered in the phase of labeling process steps.

3.5   Dealing with Different Granularity

The descriptions of the To-do's listed by the process members might vary in their granularity. *Summarization* is a technique that supports the structured and meaningful aggregation of activities according to their semantic relation [5].

The goal of the summarization is to reduce material towards a particular level of abstraction and thus supports the designer to find related tasks and activities in the text material. The technique includes a stepwise generalization of the text material. In a first step, the text segments are paraphrased which includes the omission of decorating text passages, the translation into a consistent level of language and the transformation of the text segments into a short form. The second step includes the generalization of the paraphrases towards a specified level of abstraction (aggregation of tasks). A second generalization can be done in order to subsume related activities to process fragments (aggregation of activities).

3.6 Identification of Process Specifications

The identification of process elements is appropriate in the paraphrasing step. In this step the unit of analysis is reduced to its key essence. The key essence of a To-do is the performed task or activity, and is captured in the code by a meaningful noun and a verb [12]. The labels serve as labels of the tasks and activities in the process model. Reference points important for process view aggregation are listed in the following. We suggest adding reference points as variables to the cases. Variables specify the properties associated with the case.

- Hints to data transfer (data_input and data_output) or events (e.g. eMail, Mail, PhoneCall) between two or more process participants (connection). *E.g. To-do of the AdministrativeCooperator: "Mail lecturing contract to lecturer". To-do of a Lecturer: "Receipt of the lecturing contract in my post office box".*
- Hints to group tasks and activities (participation). *E.g. To-do of the Lecturers and the TeachingCoordinators: "Participation in coordination meeting".*
- Hints to decisions. *E.g. To-do of a Lecturer: "Either I contact the Administrative-Cooperator if I want to book the lecture hall X or I contact the Secretary if I want to book the lab."*
- Hints to delegation. It can be assumed that the To-do's are performed by the process member who listed the To-do's. Otherwise a hint to another person is given. *E.g. To-do of a ModuleCoordinator: "A lecturer of our lecturer team books a lecture hall for all of us".*
- Hints to tools used. It might be the case that process participants rather mention the particular tool name (e.g. Fronter) than the general term (e.g. learning platform). *E.g. To-do of a Lecturer: "Then I enter the grades of the students into iswi."*
- Hints to reoccurring activities. *E.g. To-do of a Lecturer: "Reoccurring task: conduction of the units".*
- Time notes. We assume that To-do list already reflect a sequential order of the To-do's. The first To-do mentioned in the list is the first To-do that is performed by the person in the process under investigation. Explicitly mentioned time notes may be vague but support the designer to sequentially order tasks and activities when subsuming all tasks and activities of the various process members. *E.g. To-do of a Lecturer: "At the day of the unit I print the attendance list.", or "Before the semester starts I plan the course".*
- Name of the agent. The name of the process member is captured as well.

Furthermore, process members may mention comments that refer to activities performed by other process members or to the general process. *E.g. Mentioned by a Lecturer: "The AdministrativeCooperator books the lecture hall in the system".* Such information might be important for the designer when he or she composes the logs. We suggest to label such statements as *comments* to recall the data if necessary. In order to be able to transform results of the qualitative content analysis into process view logs, post conditions of the analysis are listed in the following.

Post condition 1:   *Each To-do of each person is considered as a case. A case is the unit of analysis.*

Post condition 2:   *Each case is labeled. The labels reflect the key essence of the To-do. The labels illustrate the labels of tasks and activities in the process model.*

Post condition 3:   *A precise coding guideline is elaborated that support comprehensibility and traceability.*

Post condition 4:   *Each case is complemented by variables. Variables necessary for process modeling or log preparation are: Agent (name of the process member), Agent Position (role of the process member), Tool (tool name), DataInput (data name), DataOutput (data name), OrganizationOfWork (if delegation name of process member, else (direct work) no entry), Connection (person name or role label), Participation (person name or role label), Decision (String), Recurrence (Boolean), Event (event type), TimeNote (String). Further variables might be, e.g. Granularity (GranularityLevel), FirstGeneralizationLabel (String), SecondGeneralizationLabel (String).*

Post condition 5:   *The beginning and the end of the entire process are identified (e.g. from time notes or from the semantic context).*

According to the BPME method, the designer can choose two alternative paths after analyzing the data. The process view models and the entire process model can be modeled either (1) manually, (2) using a tool that supports scenario-based process modeling like GRETA [1], or (3) the designer transforms the captured process view information into process view logs and entire process logs. In the following we focus on the latter. We argue that the extraction of the business process model out of process view logs offers a structured and traceable way of illustrating the identified process model. We present guidelines for preparing logs based on results of the qualitative content analysis of process member To-do's in the next section.

4 Preparation of Logs and Process Mining

We want to know how the processes of the individual process members as well as the whole process look like. In order to capture all possible variants of performing process paths, process view logs for each process participant are prepared for mining. The preparation of the logs is based on the results of the qualitative content analysis of To-do lists of process members. We suggest the designer to choose the level of abstraction that is included in the original To-do lists (results of the paraphrasing in the qualitative content analysis), as the To-do's reflect the "mental process steps" of each process participant. For each To-do list a process view log is elaborated. One output of the log preparation step are logs that reflect To-do's of a process member in a particular thematic process (e.g. sale process of a particular organization). The process view logs reflect process fragment scenarios. The most challenging task for the designer is now to transform reference points (e.g. operations that are connected to a message event between process participants (send and receive), operations that are performed by several process participants at the same time, like meetings, and time notes) into concrete (artificial) start- and end date and time stamps. The procedure of considering reference points in the process view logs is described in the following.

Step 1: *Highlight reference points that are bounded to time. Such reference points are the variables Connection, Participation, DataInput, DataOutput, Event, Recurrence, Tool and TimeNote. Connection and Participation: It may be the case that there are links identified to persons or roles whose To-do lists were not gathered in the elicitation phase (compare with Fig.2). This is an indication for the designer that not all process participants were considered in the ascertainment of the process.*

      DataInput, DataOutput and Events: A question that need to be answered by the designer is if events, such as "send document x to person B" and "receive document x from person A" are considered in the process models as operations. The overall question to be answered in the first step is for what purpose is the process model designed. If it is designed to illustrate the actual process as perceived by the process members, than events as illustrated in the example may be integrated as operations (particularly if the event of person B "receive document x from person A" triggers the process of person B). If the process model is designed as a basis for the implementation of a workflow system, then such events may not be considered in the process model, as in a workflow systems all data material is available for process participants.

Step 2: *Aggregation of tasks. If the level of abstraction considered by the designer is identical to the abstraction level of the original To-do's, then there might be detailed operations across process views (e.g."Writing on PhD thesis", "Exchange with mentor", "Enrolling in creative writing workshop") less detailed operations probably implicitly subsuming the detailed operations (e.g. "Elaboration of PhD thesis"). In such a case the designer has to decide if, e.g. such creative tasks should all be considered in the process model. An option for the designer is use by default the higher level task and to additionally consider these detailed operations that were (n/2+1)-times mentioned across process views (n=number of process participants).*

Step 3: *Set order of execution. Transfer explicit time notes mentioned in the To-do list and transform them into concrete date and time. Estimate the duration of the operation. If there is no explicit time note mentioned for a particular To-do, then first check time notes of reference points in To-do lists of other process members. Choose these time notes as support to set a concrete date and time and consider at the same time already set time specifications of other To-do's in the list. The sequential ordering of the To-do's in the list should be reflected in the time stamps. If necessary, the designer may also consider Comments for specifying time and date of operations, or directly contact the respective process member.*

      Connection and participation: Adjust time stamps according to connection and participation hints. A connection considers work- or data transfer to another process member. Participation considers operations that are collectively performed (e.g. meetings).

      Recurrence of tasks. If there are hints that a tasks is executed several times (loop) then the designer has to decide if the loop is considered in the log as loop or if each execution of the operation is captured as a particular step in the process model.

Step 4: *Check the agent of the task. If there are indications that an operation is delegated to another process participant, than the operations are still considered as operations of the delegator. In the field "Agent" the name of the delegatee is entered. The tasks of the delegatee remain as tasks of the process member in his or her process view log.*

Step 5: *Set number of process instances (cases). The number of process instances in a process view is determined by the number of the decision paths. If there are > 1 decisions, the product of the decision paths determines the maximum number of process instances. So, if there is decision n including m paths and decision d including e paths considered in one process view, than the designer needs to build a maximum of m x e process instances.*

Step 6: *Explicate conditions and rules. Particular decision paths must not or cannot be considered in the same process instance. Therefore, rules for joining decision paths (if decision ≥ 1) need to be defined. The designer keeps overview when the operations are organized into operations that are always executed and operations that are only executed if a particular condition occurs. Thus, the designer obtains "operation blocks" that can be merged into one process instance by considering merging rules and condition.*

Step 7: Assemble logs. Illustrate common and alternative process paths as process view instances. Consider thereby the entire process view (from start to the end) as one process instance. If there are decisions included, then transform the common and alternative process paths into different instances.

The next task is to prepare the log instances of the entire process including all process views. The output is a process model of the entire process that captures possible variations of process paths by considering the allowed decision combinations. For preparing such logs, the following preconditions need to be considered.

Precondition 1: The process that is analyzed is clearly defined and has a common thread (e.g. in a sales process, the common thread is the order on which the process participants work on throughout the whole process).

Precondition 2: The process views are already transformed into process view logs. The process view logs reflect common and alternative process paths.

The steps necessary to merge process view logs in such way that they reflect the whole process with its alternative paths are presented in the following. In addition to the mentioned steps below, exchange with the process participants during workshops might support the analyst to get an immediate "big picture" of the whole process and to reduce errors and misinterpretations in the whole process model.

Step 1: Determine the number of process instances (cases). The process instances illustrate possible entire process path. Elaborate per process instance which decision path of what process member is considered. Consider rules and conditions when assembling the process view instances.

Step 2: Bring the process fragment scenarios together and sort the operations according to the time stamps.

In this work our intension is not to present a new mining algorithm for process view mining. Our goal is rather to discuss the purpose and benefits of mining prepared logs. Mining of the process view logs and the entire process supports the designer in constructing and visualizing process models from data analyzed and prepared in a structured and traceable way. The mining of each single log of the individual process

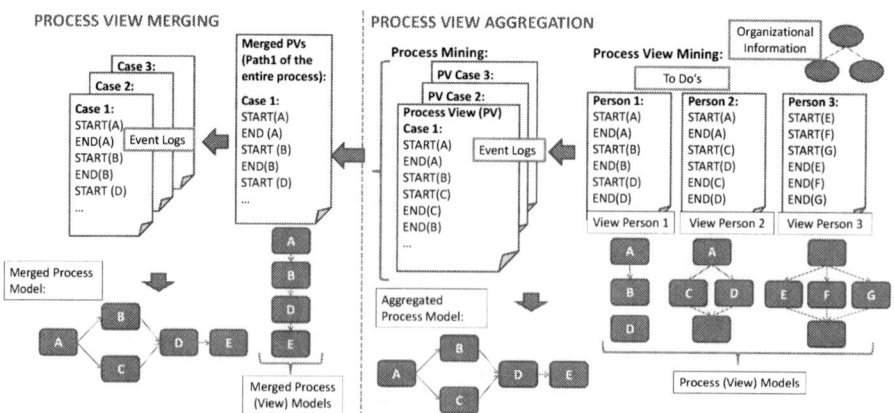

Fig. 2. Process view aggregation vs. merging

participants offers for each process participant a personalized process view. The designer obtains manageable process views. The process views can be aggregated or merged into one process model. Aggregation of process views means to consider each process view as an instance of the entire process in the logs, whereas merging of process views means to consider each process view as a part of an instance of the entire process in the logs (compare with Fig. 2).

5 Case Study

In this section we present a case study in which we applied the BPME method to extract the process model of teaching from To-do's of selected process participants at the Faculty of Computer Science, University of Vienna. For our investigation we considered both the administrative and creative process of teaching. A bottom-up approach was pursued to find out what activities the participants of the process perform to enable and support teaching at the faculty.

5.1 Collection of To-Do's in the Teaching Process

In face-to-face interviews selected persons working in key positions of the teaching process were asked to list their activities in the process under investigation similar to a scheduled To-do list. The interviews were conducted at the Faculty of Computer Science at the University of Vienna in October 2010. 12 persons participated in the survey. The selected persons were directly asked to participate in the study. The participation in the survey was voluntary. A question of the semi-structured interview was to specify the own role in the process of teaching. Some interviewees acted in several roles, e.g. the role of the module coordinator and the lecturer. The interviewer asked for the To-do's of each role successively and recorded the To-do list per role in separate protocols. Altogether, the activities appropriate to the following roles in the teaching process were collected (the cases illustrate persons): administrative cooperator (case 1), director of study program (case 2), lecturers (cases 3, 4, 5, 6, 7, 8, and 16), lecturer team coordinator (case 9), module coordinators (cases 10, 11), secretary (case 12), teaching coordinators (cases 13, 14), and technical staff (case 15).

5.2 Content Analysis of To-Do's, Log Preparation, and Mining

The protocols of the interviews were examined with the content analysis. Most of the activities were described in full sentences rather than by means of cues that are more common in personal To-do lists [13, p. 736]. The text material comprised 16 protocols and a total of 7,814 words. A case comprised between 65 and 1077 words. We conducted the content analysis with the demo version of the QDA Miner [8]. Analyzing functionalities, like coding frequency allow recognizing reference points among process participants. Fig. 3 illustrates an extract of the coding frequency analysis. "Inquire date and lecture hall", "Book date and lecture hall", and "Check scheduling conflict" are aggregations of more detailed tasks mentioned as To-do's. These aggregated tasks were subsumed to "reservation of lecture hall". "Inquire date and lecture hall" includes two reference points (connections) among process participants.

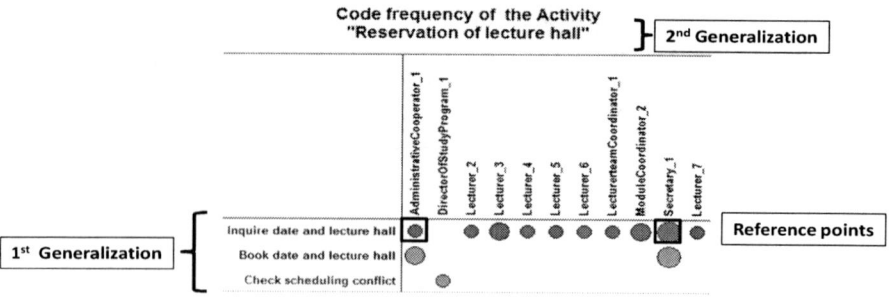

Fig. 3. Reference points among operations

Whereas the lecturers, module coordinators and lecturer team coordinators request for lecture halls, the administrative cooperator and the secretary receive the requests and book the halls.

Afterwards, each single To-do of the 399 To-do's was coded at the lowest possible level of abstraction already included in the To-do by paraphrasing. 195 labels reflecting the key essence of the To-do's were inductively derived. 77 hints to connections among process participants, 39 hints to operations with participation of several process participants (e.g. meetings), 49 hints to decisions (each operation of the alternative path was marked as a decision hint), 40 hints to events (e.g. eMail, phone calls, or messages), 12 hints to delegation, and 43 time notes could be identified. 8 tools could be extracted out of the To-do's as well as 30 different data material (e.g. "teaching scheduling sheet", "numbers of exams of the last semester", documented course schedule") could be identified.

In a next step we elaborated the process view logs per process participant (per role). The main challenges were (a) to define the artificial time stamps and (b) the inclusion of all possible combinations of process paths as these issues were the critical points for the quality of the mining result. Based on the number and conditions of the decisions we elaborated possible process paths per role and captured each possible path as a case (process instance) of each process view. The alpha mining algorithm was used for the transformation of the individual process view logs into process view models. We elaborated process logs in which each case illustrates an entire path of the entire process, and considered the course (with unique course ID) as the "common thread" throughout the process. To organize all the possible paths of the process per case the selected paths of the process views were documented. Log segments were compound to a possible way of the process. The heuristics mining algorithm was used to analyze the entire process logs. Fig. 4 illustrates a comparison of the result of applying the heuristic mining algorithm in ProM [18] of the collected, analyzed, and merged To-do's to a process model (left and middle part) and a manually modeled process segment in which BPMN notation was used (right part). The BPMN model segment was designed by the process-aware designer before the To-do's of the process participants were analyzed and mined. Both segment models gather the same activities (prepare, conduct, and post process unit).

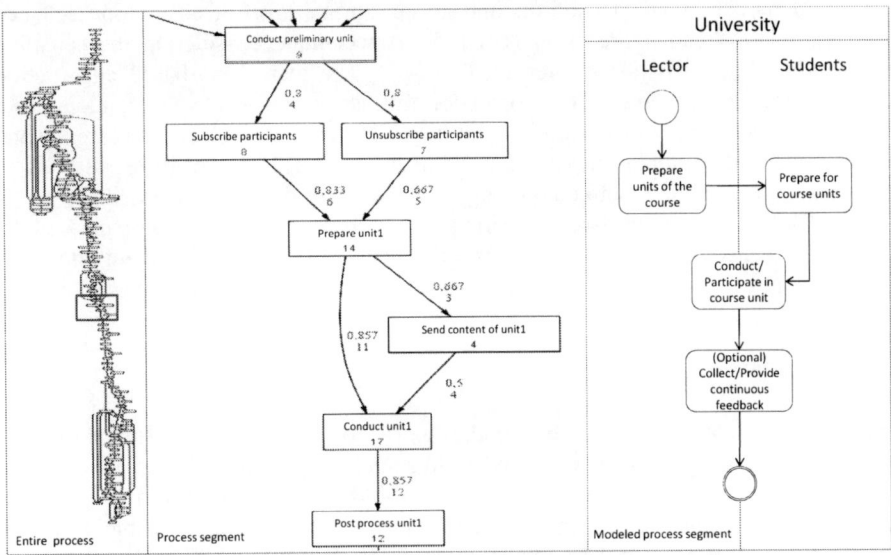

Fig. 4. Comparison of the merged process logs analyzed with the heuristics miner and a manually designed BPMN process model

Whereas the BPMN model provides more details about the participating roles due to its notation, the process segment of the mined To-do's offers information about how often the operations were considered in the logs (and thus in the To-do's of the process participants) and provides a more precise insight into the actual operations (e.g. several process participants explicitly mentioned the conduction of the preliminary unit, operations that were performed to fix the participants of the course, and operations among lecturers before a course unit was conducted).

6   Related Work

Process design is one of the main phases of the business process life cycle. There are basically two different approaches of process design that can be distinguished. First, business process (re-)engineering refers to the systematic process exploration based on methods such as interviews and the subsequent modeling of as-is processes, followed by an optimization of these processes (to-be processes). Much work has been spent on questions such as process modeling notations [17]. Only few approaches tackle process redesign [4]. Although there are known common practices to elicit process models, like questionnaires and workshops, no systematic approach for process exploration and design based on activities and tasks representing practical processes of work of single persons sorted in treated chronological order (To-Do's) has been presented yet. We utilize existing process mining algorithms [18] for our work not only to mine instances of the whole process but also operations that are performed by a single agent of the process (process view logs).

Commonly, the challenge of finding an appropriate level of abstraction is faced with elimination and aggregation [14] [15]. A technique considering the semantic relation between activities is presented in [16]. The meronymy-based aggregation uses business process domain ontologies for abstraction. To enable aggregation, process model activities are matched to an ontology. The technique offer transparency and traceability in determining a particular abstraction level. However, preconditions of this aggregation are a predefined ontology and the use of domain specific labels.

[1] proposes a scenario-based modeling paradigm which considers a process model as a set of scenarios. While [1] considers all process views and divide them into process fragments, we consider each process view of each process participant.

7 Conclusion

As so far the procedure of process models extraction from process information elicited from process participants was handled like a black box, our main research interest of this work was to find out how to translate various process elements captured in To-do lists of individuals into a process model in a structured and traceable way. We proposed the Business Process Model Extraction (BPME) method that supports a transparent way of transforming process information in natural language into formal process models. We discussed the potential of the qualitative content analysis in dealing with differing levels of abstraction, heterogeneous labeling and the identification of essential process elements. We presented guidelines for using the qualitative content analysis as a designer's tool to extract process information from To-do lists of process participants in a structured, documented and traceable way. In order to support transparency and traceability up to the final process model, we presented guidelines to transform results of the content analysis into view process logs and discussed two opportunities that lead to an entire process model by mining: the aggregation and the merging of process views. In the case study we showed that the BPME method could be successfully performed. The main challenges faced were the determination of artificial time stamps in the process view logs that were crucial for process mining and the elaboration of the possible process paths based on the decision paths in each process view. The use of the BPME method offered the following outputs: (a) a documented and traceable procedure of transforming To-do lists into the process model, (b) illustrative personalized process view models, and (c) an illustrative process model reflecting the To-do's of the process participants. In future work we will extend our work on process view mining. Furthermore, we will offer more details about the challenges of the elicitation step of the BPME method, and the validation of the resulting models that should support our approach.

References

1. Fahland, D., Weidlich, M.: Scenario-based process modeling with Greta. In: La Rosa, M. (ed.) Proceedings of BPM Demonstration Track. CEUR-WS.org, Hoboken, vol. 615 (2010)
2. Hammori, M., Herbst, J., Kleiner, N.: Interactive workflow mining - requirements, concepts and implementation. Data & Knowledge Engineering 56, 41–63 (2006)

3. OMG: Business Process Model and Notation (BPMN) Version 2.0 (2010)
4. Reijers, H.A.: Process Design and Redesign. In: Dumas, M., van der Aalst, W., ter Hofstede, A.H.M. (eds.) Process Aware Information Systems: Bridging People and Software Through Process Technology. Wiley-Interscience, Hoboken (2005)
5. Mayring, P.: Qualitative Inhaltsanalyse. Beltz, Weinheim (2003)
6. Srnka, K., Koeszegi, S.: From Words to Numbers: How to Transform Qualitative Data into Meaningful Quantitative Results. Schmalenbach Business Review 59, 29–57 (2007)
7. Strijbos, J.-W., Martens, R.L., Prins, F.J., Jochems, W.M.G.: Content analysis: What are they talking about? Computers & Education 46(1), 29–48 (2006)
8. Provalis Research, http://www.provalisresearch.com/
9. Nastase, V., Koeszegi, S., Szpakowicz, S.: Content Analysis Through the Machine Learning Mill. Group Decision and Negotiation 16, 335–346 (2007)
10. Rinderle-Ma, S., Reichert, M., Jurisch, M.: Equivalence of Web Services in Process-Aware Services. In: IEEE 7th International Conference on Web Services, pp. 501–508. IEEE, Los Angeles (2009)
11. Derntl, M., Neumann, S., Griffiths, D., Oberhuemer, P.: ISURE - Report on usage and recommendations of specification for instructional model, ICOPER Deliverable D3.2 (2010)
12. Mendling, J., Recker, J., Reijers, H.A.: On the Usage of Labels and Icons in Business Process Modeling. International Journal of Information System Modeling and Design 1, 40–58 (2009)
13. Bellotti, V., Dalal, B., Good, N., Flynn, P., Bobrow, D.G., Ducheneaut, N.: What a To-Do: Studies of Task Management Towards the Design of a Personal Task List Manager. In: Proceedings of the SIGCHI Conference on Human Factors in Computing Systems, pp. 735–742. ACM, New York (2004)
14. Polyvyanyy, A., Smirnov, S., Weske, M.: The Triconnected Abstraction of Process Models. In: Proceedings of the 7th International Conference on Business Process Management, pp. 229–244. Springer, Ulm (2009)
15. Bobrik, R., Reichert, M., Bauer, T.: View-based process visualization. In: Proceedings of the 5th International Conference on Business Process Management, pp. 88–95. Springer, Brisbane (2007)
16. Smirnov, S., Dijkman, R., Mendling, J., Weske, M.: Meronymy-Based Aggregation of Activities in Business Process Models. In: Parsons, J., Saeki, M., Shoval, P., Woo, C., Wand, Y. (eds.) ER 2010. LNCS, vol. 6412, pp. 1–14. Springer, Heidelberg (2010)
17. van der Aalst, W., ter Hofstede, A.H.M., Kiepuszewski, B., Barros, A.P.: Workflow patterns. Distributed and Parallel Databases 14, 5–51 (2003)
18. van der Aalst, W., Reijers, H.A., Weijters, A., Dongen, B.F., Alves de Medeiros, A.K., Song, M., Verbeek, H.M.W.: Business process mining: An industrial application. Information Systems 32, 713–732 (2007)

Process Model Generation from Natural Language Text

Fabian Friedrich[1], Jan Mendling[2], and Frank Puhlmann[1]

[1] inubit AG, Schöneberger Ufer 89-91, 10785 Berlin, Germany
{Fabian.Friedrich,Frank.Puhlmann}@inubit.com
[2] Humboldt-Universität zu Berlin, Unter den Linden 6, 10099 Berlin, Germany
jan.mendling@wiwi.hu-berlin.de

Abstract. Business process modeling has become an important tool for managing organizational change and for capturing requirements of software. A central problem in this area is the fact that the acquisition of as-is models consumes up to 60% of the time spent on process management projects. This is paradox as there are often extensive documentations available in companies, but not in a ready-to-use format. In this paper, we tackle this problem based on an automatic approach to generate BPMN models from natural language text. We combine existing tools from natural language processing in an innovative way and augmented them with a suitable anaphora resolution mechanism. The evaluation of our technique shows that for a set of 47 text-model pairs from industry and textbooks, we are able to generate on average 77% of the models correctly.

1 Introduction

Business process management is a discipline which seeks to increase the efficiency and effectiveness of companies by holistically analyzing and improving business processes across departmental boundaries. In order to be able to analyze a process, a thorough understanding of it is required first. The necessary level of insight can be obtained by creating a formal model for a given business process.

The required knowledge for constructing process models has to be made explicit by actors participating in the process [1]. However, these actors are usually not qualified to create formal models themselves [2]. For this reason, modeling experts are employed to iteratively formalize and validate process models in collaboration with the domain experts. This traditional procedure of extracting process models involves interviews, meetings, or workshops [3]. It entails considerable time and costs due to ambiguities or misunderstandings between the involved participants [4]. Therefore, the initial elicitation of conceptual models is considered to be a knowledge acquisition bottleneck [5]. According to Herbst [1] the acquisition of the as-is model in a workflow project requires 60% of the total time spent. Accordingly, substantial savings are possible by providing appropriate tool support to speed up the acquisition phase.

H. Mouratidis and C. Rolland (Eds.): CAiSE 2011, LNCS 6741, pp. 482–496, 2011.

In this context, it is a paradox that acquisition is costly although detailed information about processes is often already available in the form of informal textual specifications. Such textual documents can be policies, reports, forms, manuals, content of knowledge management systems, and e-mail messages. Content management professionals estimated that 85% of the information in companies is stored in such an unstructured format [6]. Moreover, the amount of unstructured text is growing at a much faster rate than structured data [7]. It seems reasonable to assume that these texts are relevant sources of information for the construction of conceptual models.

In this paper, we develop an approach to directly extract business process models from textual descriptions. Our contribution is a corresponding technique that does not make any assumptions about the structure of the provided text. We combine an extensive set of tools from natural language processing (NLP) in an innovative way and augment it with an anaphora resolution mechanism, which was particularly developed for our approach. The evaluation of our technique with a set of 47 text-model pairs from industry and textbooks reveals that on average 77% of the model is correctly generated. We furthermore discuss current limitations and directions of improvement.

The paper is structured as follows. Section 2 introduces the foundations of our approach, namely BPMN process models and natural language processing techniques. Section 3 identifies a set of language processing requirements, and illustrates how they are tackled in the various steps of our generation approach. Section 4 presents our evaluation results based on a sample of text-model pairs. Section 5 discusses related work before Section 6 concludes the paper.

2 Background

Generating models builds on understanding the essential concepts of BPMN process models and of state-of-the-art techniques for natural language processing. In this section, we introduce BPMN and then natural language processing tools.

The Business Process Model and Notation (BPMN) is a standard for process modeling that has been recently published in its version 2.0 [8]. It includes four categories of elements, namely Flow Objects (Activities, Events and Gateways), Swimlanes (Pools and Lanes), Artifacts (e.g. Data Objects, Text Annotations or Groups), and Connecting Objects (Sequence Flows, Message Flows and Associations). The first three are nodes, the latter ones are edges. Figure 1 shows a BPMN example of a claims handling process provided by QUT. The process is subdivided into three pools (one with two lanes) capturing the actors of the process. Activities are depicted as rounded boxes. Different events (round elements with icons) for sending and receiving messages affect the execution of the process. The diamond-shaped elements define specific routing behavior as gateways.

A BPMN process model is typically the result of analyzing textual descriptions of a process. A claims handling process provided by QUT is described as follows: "The process starts when a customer submits a claim by sending in relevant documentation. The Notification department at the car insurer checks

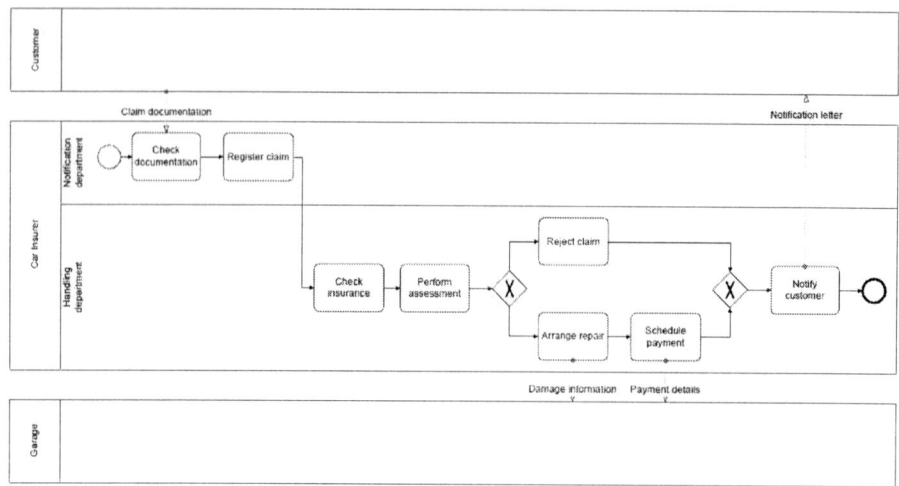

Fig. 1. Example of a claims handling process in BPMN

the documents upon completeness and registers the claim. Then, the Handling department picks up the claim and checks the insurance. Then, an assessment is performed. If the assessment is positive, a garage is phoned to authorise the repairs and the payment is scheduled (in this order). Otherwise, the claim is rejected. In any case (whether the outcome is positive or negative), a letter is sent to the customer and the process is considered to be complete." Such information is usually provided by people working in the process and then formalized as a model by system analysts [2].

For our model generation approach, we will employ methods from computational linguistics and natural language processing. This branch of artificial intelligence deals with analyzing and extracting useful information from natural language texts or speech. For our approach, three concepts are of vital importance: syntax parsing, which is the determination of a syntax tree and the grammatical relations between the parts of the sentence; semantic analysis, which is the extraction of the meaning of words or phrases; and anaphora resolution, which involves the identification of the concepts which are references using pronouns ("we","he","it") and certain articles ("this", "that"). For syntax parsing and semantic analysis, there are standard tools available.

The Stanford Parser is a syntax parsing tool for determining a syntax tree. This tree shows the dependencies between the words of the sentence through the tree structure [10]. Additionally, each word and phrase is labeled with an appropriate part-of-speech and phrase tag. The tags of the *Stanford Parser* are the same which can be found in the *Penn Tree Bank* [11]. The *Stanford Parser* also produces 55 different *Stanford Dependencies* [12]. These dependencies reflect the grammatical relationships between the words. Such grammatical relations provide an abstraction layer to the pure syntax tree. They also contain information about the syntactic role of all elements.

There are also tools available for semantic analysis. They provide semantic relations on different levels of detail. We use FrameNet [13] and the lexical database WordNet[14]. WordNet provides various links to synonyms, homonyms, and hypernyms for a particular class of meaning associated with a synonym-set. FrameNet defines semantic relations that are expected for specific words. These relations are useful, e.g., to recognize that a verb "send" would usually go with a particular object being sent. Syntax parsers and semantic analysis are used in our transformation approach, augmented with anaphora resolution.

3   Transformation Approach

The most important issue we are facing when trying to build a system for generating models is the complexity of natural language. We collected issues related to the structure of natural language texts from the scientific literature and analyzed the test data, which is described in section 4. Thereby, we were able to identify four broad categories of issues which we have to solve in order to analyze natural language process descriptions successfully (see Table 1). *Syntactic Leeway* relates to the fact that there is a mismatch between the semantic and syntactic layer of a text. *Atomicity* deals with the question of how to construct a proper phrase-activity mapping. *Relevance* has to check whether parts of the text might be irrelevant for the generated process model. Finally, *Referencing* addresses the question of how to resolve relative references between words and between sentences.

Table 1. References in the literature to the analyzed issues

Issue	Refs.	Issue	Refs.
1 Syntactic Leeway		3 Relevance	
1.1 Active-Passive	[15]	3.1 Relative Clause Importance	[16]
1.2 Rewording/Order	[17,18]	3.2 Example Sentences	[19]
1.3 Implicit Conditions	[20,21]	3.3 Meta-Sentences	[16]
2 Atomicity		4 Referencing	
2.1 Complex Sentences	[16,18]	4.1 Anaphora	[22,23]
2.2 Action Split over Sentences	[22]	4.2 Textual Links	[24]
2.3 Relative Clauses	[16]	4.3 End-of-block Recognition	[19,16]

Different solution strategies were applied in the works listed in Table 1 to overcome the stated problems, e.g. by constricting the format of the textual input [17], but no study considers all mentioned problems and offers a comprehensive solution strategy. Another interesting fact is that none of the works using a shallow parser shows how they deal with passive voice [15,22,23,17]. We solved this problem by using the grammatical relations of the Stanford Parser.

To obtain a structured representation of the knowledge we extract from the text, we decided to store it in a *World Model*, as opposed to a direct straight through model generation. This approach was also taken by most of the other works which built a similar system [17,22,18,23]. The data structure used by the approach of the University of Rio de Janeiro [23] was taken from the CREWS project [15]. The authors argue that it is suited well for this task as a scenario description corresponds to the description of a process model. Therefore, we also use the CREWS scenario metamodel as starting point. However, we modified several parts as, e.g., we explicitly represent connections between the elements using the class "Flow". Additionally, we explicitly considered traceability as a requirement. Thus, attributes relating an object to a sentence or a word are added to the World Model. The four main elements of our World Model are Actor, Resource, Action, and Flow. This World Model will be used throughout all phases of our transformation procedure to capture syntactic and semantic analysis results. Each phase is allowed to access, modify and add data.

The rest of this section is dedicated to analyzing and discussing the issues collected in Table 1. We will then seize the developed suggestions and reference these issues during the description of our transformation approach. Section 3.1 discusses sentence level analysis for finding actions. Section 3.2 investigates text level analysis for enriching the data stored in the world model. Finally, Section 3.3 describes the generation of a BPMN model. While we focus on the general procedure here, we documented details of all algorithms in [25].

3.1   Sentence Level Analysis

The first step of our transformation procedure is a sentence level analysis. The extraction procedure consists of the steps that are outlined as a BPMN model in Figure 2. This overview also shows the different components upon which our transformation procedure builds and their usage of Data Sources.

The text is processed in several stages. First, a tokenization splits up the text into individual sentences. The challenge here is to distinguish a period used for an abbreviation (e.g. M.Sc.) from a period marking the end of a sentence.

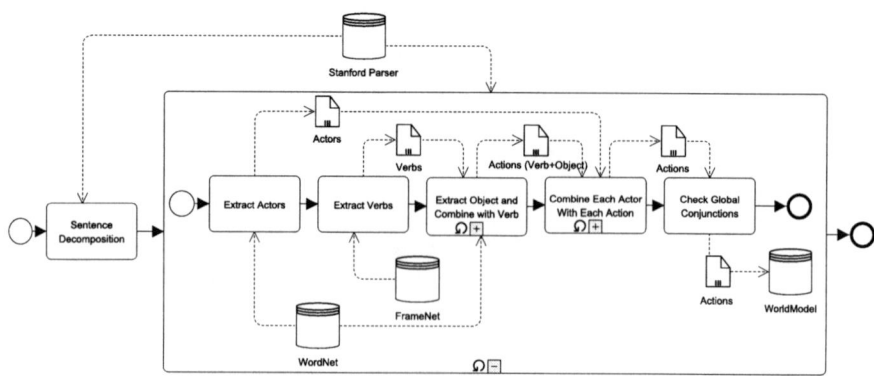

Fig. 2. Structural overview of the steps of the Sentence Level Analysis

Afterwards, each sentence is parsed by the *Stanford Parser* using the factored model for English [11]. We utilize the factored model and not the pure probabilistic context free grammar, because it provides better results in determining the dependencies between markers as "if" or "then", which are important for the process model generation. Next, complex sentences are split into individual phrases. This is accomplished by scanning for sentence tags on the top level of the Parse Tree and within nested prepositional, adverbial, and noun phrases.

Once the sentence is broken down into individual constituent phrases, actions can be extracted. First, we determine whether the *parsedSentence* is in active or passive voice by searching for the appropriate grammatical relations (Issue 1.1). Then, all Actors and Actions are extracted by analyzing the grammatical relations. To overcome the problem of example sentences mentioned earlier (Issue 3.2) the actions are also filtered. This filtering method simply checks whether the sentence contains a word of a stop word list called *example indicators*. Then, we extract all objects from the phrase and each Action is combined with each Object. The same is done with all Actors. This procedure is necessary as an Action is supposed to be atomic according to the BPMN specification [8] and Issue 2.1. Therefore, a new Action has to be created for each piece of information as illustrated in the following example sentences. In each sentence the conjunction relation which causes the extraction of several Actors, Actions or Resources is highlighted. As a last step, all extracted Actions are added to the World Model.

o "Likewise the old supplier **creates and sends** the final billing to the customer." (Action)
o "It is given either by **a sales representative or by a pre-sales employee** in case of a more technical presentation." (Actor)
o "At this point, the Assistant Registry Manager puts **the receipt and copied documents** into an envelope and posts it to the party." (Resource)

3.2   Text Level Analysis

This section describes the text level analysis. It analyzes the sentences taking their relationships into account. The structural overview of this phase is shown in Figure 3. We use the Stanford Parser and WordNet here, and also an anaphora resolution algorithm. During each of the five steps, the Actions previously added to the World Model are augmented with additional information.

An important part of the algorithm presented here is the determination heuristic for resolving relative references within the text (Issue 4.1). Existing libraries are not seamlessly integrateable with the output provided by the Stanford Parser. Therefore, we implemented a simple anaphora resolution technique for the resolution of determiner and pronouns. This procedure is described in detail in [25]. An experimental evaluation using our test data set showed that this approach achieved a good accuracy of 63.06%.

The second step in our analysis is the detection of conditional markers. These markers can either be a single word like "if", "then", "meanwhile" or "otherwise", or a short phrase like "in the meantime" or "in parallel". All of these

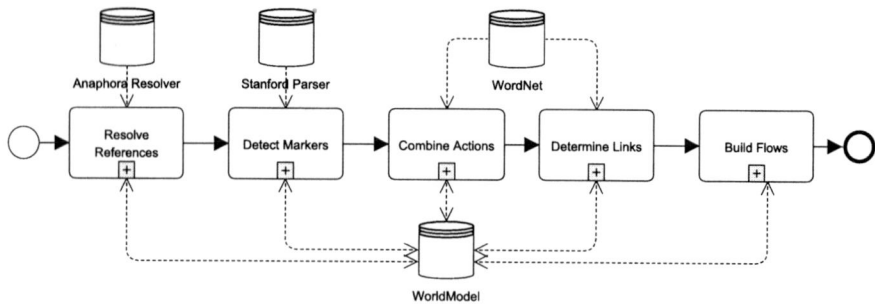

Fig. 3. Structural overview of the steps of the Text Level Analysis

markers have specific characteristics and can be mapped to different BPMN constructions. In order to capture this semantic information we compiled four lists, namely ConditionIndicators (exclusive gateway), ParallelIndicators (parallel gateway), ExceptionIndicators (for Error Intermediate Events), and SequenceIndicators (for the continuation of a branch of a gateway). These lists do not claim completeness and can be extended by the user, if necessary.

We can use the information gathered so far to combine the information contained in two different Actions. This procedure tackles the problem of Actions which are split up over several sentences (Issue 2.2). To consider two Actions as a candidate for a merger, a reference had to be established between them during the anaphora resolution phase. This reference can either directly point from the Actor or from the Object of this Action. But, for the case that the Object points to another Actor or Resource we also consider the Action which contains it as a possible candidate. Next, it is checked whether the objects can be merged by checking various characteristics of both Actions. If the actions truly complement each other, they can be merged and form one single action. When both Actions complement each other except for the negation modifier we can still enhance the information content of one action by copying information, as the initiating Actor, the Object, and/or the copula attribute. An example for such a case are these sentences: "Of course, asking the customer whether he is generally interested is also important." and "If this is not the case, we leave him alone, [...]"

For Issue 4.2, we defined three types of textual references: forward, backward, and jump references. In order to identify those links in the text automatically, we start by comparing all actions within our World Model to one another. It is then determined whether the selected actions can be linked or not. Within this method, we compare the following characteristics of both Actions: Copula Specifier, Negation Status, the initiating Actor (ActorFrom), the Object, the open clausal complement, and the Prepositional Specifiers, whose head word is "to" or "about". The elements are compared using their root form provided by WordNet. If the elements differ or an element is defined for one Action, but not for the other, the Actions cannot be merged. Otherwise, the Actions are considered equal and a link relationship can be established. Additionally, the type of the link relationship is determined and saved along with the link.

The last step of the text level analysis is the generation of Flows. A flow describes how activities are interacting with each other. Therefore, during the process model generation such Flows can be translated to BPMN connecting objects. When creating the Flows we build upon the assumption that a process is described sequentially and upon the information gathered in the previous steps. The word, which connected the items is important to determine how to proceed and what type of gateway we have to create. So far we support a distinction between "or", "and/or", and "and". Other conjunctions are skipped.

3.3   Process Model Generation

In the last phase of our approach the information contained in the World Model is transformed into its BPMN representation. We follow a nine step procedure, as depicted in Figure 3.3. The first 4 steps: creation of nodes, building of Sequence Flows, removal of dummy elements, the finishing of open ends, and the processing of meta activities are used to create an initial and complete model. Optionally, the model can be augmented by creating Black Box Pools and Data Objects. Finally, the model is laid out to achieve a human-readable representation.

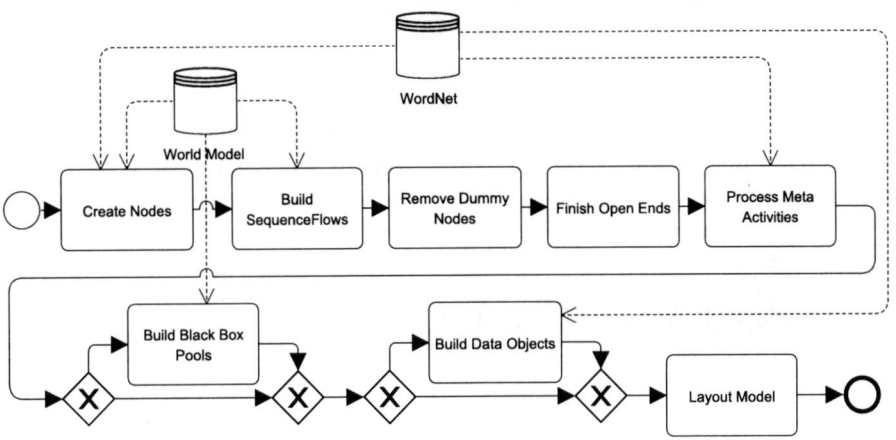

Fig. 4. Structural overview of the steps of the Process Model Generation phase

The first step required for the model creation is the construction of all nodes of the model. After the Flow Object was generated, we create a Lane Element representing the Actor initiating the Action. If no Lane was determined for an Action, it is added to the last Lane which was created successfully as we assume that the process is described in a sequential manner. The second step required during the model creation is the construction of all edges. Due to the definition of Flows within our World Model, this transformation is straight-forward. Whenever a Flow which is not of the type "Sequence" is encountered, a Gateway appropriate for the type of the flow is created. An exception to that is the type "Exception". If the World Model contains a flow of this type, an exception intermediate event is attached to the task which serves as a source and this Intermediate Event is connected to the target instead of the node itself. We then skip dummy actions, which were inserted between gateways directly following each other.

Step four is concerned with open ends. So far, no Start and End Events were created. This is accomplished in this step. The procedure is also straight forward. We create a preceding Start event to all Tasks which do not have any predecessors (in-flow = 0) and succeeding End Events to all Tasks which do not have any successors (out-flow = 0). Additionally, Gateways whose in- and out-flow is one receive an additional branch ending in an End Event.

The last step in the model creation phase handles Meta-Activities (Issue 3.3). We search and remove redundant nodes directly adjacent to Start or End Events. This is required as several texts contain sentences like "[...] the process flow at the customer also ends." or "The process of "winning" a new customer ends here." If such sentences are not filtered, we might find tasks labeled "process ends" right in front of an end event or "start workflow" following a start event. We remove nodes whose verb is contained in the hypernym tree of "end" or "start" in WordNet if they are adjacent to a Start or End Event.

The execution of these five steps yields a full BPMN model. As the elements of this model do not contain any position information yet, our generation procedure concludes with an automated layout algorithm. We utilize a simple grid layout approach similar to [26], enhanced with standard layout graph layout algorithms as Sugiyama [27] and the topology-shape-metric approach [28]. For the example text of the claims handling process from Section 2 we generated the model given in Figure 5. The question of how far this result can be considered to be accurate is discussed in the following section.

4   Evaluation of Generated Process Models

For the validation of our approach, we collected a test data set consisting of 47 of those text-model pairs, each including a textual process description and a corresponding BPMN models created by a human modeler. Different sources from research and practice were incorporated into our test data set: Academic (15 models), Industry (9 models), Textbook (9 models), and Public Sector (14 models), see Table 2. While the academic pairs were provided by our university partners, the industry models are taken from two main sources. First, we gathered four models from the websites of three BPM tool vendors, namely Active VOS, Oracle, and BizAgi. Four models stem from training material of inubit AG, another one from a German BPM practitioner, and further ones from two BPMN textbooks [29,9]. Finally, we included the definition of switch processes of the *Federal Network Agency* of Germany in its semi-structured tabular format and the corresponding model.

To avoid unintended effects while parsing, minor corrections were applied to the texts. Some models were translated, some were converted from other

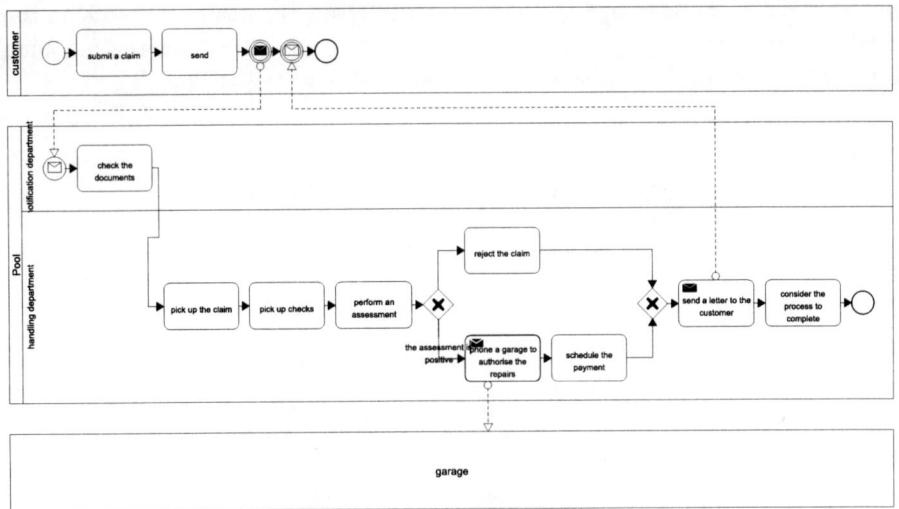

Fig. 5. The claims handling model as generated by our system

modeling languages to BPMN in order to compare them. Table 2 lists characteristics of the texts and models of our data set. The table captures the following
data: a unique ID, the number of models (M), the number of sentences (n), the
average length of sentences ($\varnothing l$), the size of the models in terms of nodes ($|N|$),
gateways ($|G|$), and edges ($|E|$). All our material is published in [25].

The evaluation results are based on the similarity (sim) between the manually
and automatically created models. We employ the metric of *Graph Edit Distance*.
To compute the Graph Edit Distance, the graph representation of the process
models is analyzed. The labels, attributes, the structural context, and behavior
are compared [32]. Afterwards a greedy graph matching heuristic [33] is employed

Table 2. Characteristics of the test data set by source (average values)

ID	Source	M	Type	n	$\varnothing l$	$\lvert N \rvert$	$\lvert G \rvert$	$\lvert E \rvert$
1	HU Berlin	4	academic	10.00	18.14	25.75	6.00	24.50
2	TU Berlin [30]	2	academic	34.00	21.17	71.00	9.50	79.50
3	QUT	8	academic	6.13	18.26	14.88	1.88	16.00
4	TU Eindhoven [31]	1	academic	40.00	18.45	38.00	8.00	37.00
5	Vendor Tutorials	4	industry	9.00	18.20	14.00	2.25	11.50
6	inubit AG	4	industry	11.50	18.38	24.00	4.25	21.25
7	BPM Practicioners	1	industry	7.00	9.71	13.00	1.00	9.00
8	BPMN Prac. Handbook [9]	3	textbook	4.67	17.03	13.00	1.33	14.67
9	BPMN M&R Guide [29]	6	textbook	7.00	20.77	23.83	3.00	23.67
10	FNA - Metrology Processes	14	public sector	6.43	13.95	24.43	3.14	25.93
	Total	**47**			**9.19**	**17.16**	**23.21**	**3.38**	**23.64**

to create pairs of nodes and edges. We use the greedy heuristic as it showed the best performance without considerable accuracy trade-offs. After the mapping is created, a Graph Edit Distance value can be calculated given:

- N_i - set of nodes in model i
- E_i - set of edges of model i
- $\overline{N_i}$ - the set of nodes in model i which were not mapped
- $\overline{E_i}$ - the set of edges in model i which were not mapped
- M - The mapping between the nodes of model 1 and 2

An indicator for the difference between the models can be calculated as:

$$m^* = \begin{cases} \sum_{i=1}^{|M|} 1 - sim(M_i) & \text{if} |M| > 0 \\ 1.0 & \text{otherwise} \end{cases} \quad (1)$$

As a last step weights for the importance of the differences (w_{map}), the unmapped Nodes (w_{uN}), and the unmapped Edges (w_{uE}) have to be defined. For our experiments we gave the difference a slightly higher importance and assigned $w_{\text{map}} = 0.4$ and $w_{\text{uN}} = w_{\text{uE}} = 0.3$. The overall graph edit distance then becomes:

$$sim(m_1, m_2) = 1 - (w_{\text{map}} * \frac{m^*}{|M|} + w_{\text{uN}} * \frac{|\overline{N_1}| + |\overline{N_2}|}{|N_1| + |N_2|} + w_{\text{uE}} * \frac{|\overline{E_1}| + |\overline{E_2}|}{|E_1| + |E_2|}) \quad (2)$$

This value ranges between 0 and 1. For the case that all nodes could be mapped with a similarity of 1.0 the terms will also become 1.0. If the mapping is not optimal, the term in parenthesis will grow steadily and the similarity decreases. If no nodes are mapped at all, the similarity will be 0.

For our evaluation, we generated the model for each text and calculated the similarity metric between it and the original BPMN model. The results are shown in Table 3. Columns 2-4 show that the concepts of meta sentences, relative references, and textual jumps are important for almost all elements within our test data. The following six columns show the average values of nodes, gateways,

Table 3. Result of the application of the evaluation metrics to the test data set

ID	m	r	j	$|N_{\text{gen}}|$	$\Delta|N_{\text{gen}}|$	$|G_{\text{gen}}|$	$\Delta|G_{\text{gen}}|$	$|E_{\text{gen}}|$	$\Delta|E_{\text{gen}}|$	sim
1	3	5,25	0	30,25	14,88%	5,50	-9,09%	28,75	14,78%	77,94%
2	7,50	7,50	2,50	91,50	22,40%	13,00	26,92%	94,00	15,43%	70,79%
3	0,50	1,38	0,00	20,25	26,54%	2,63	28,57%	20,13	20,50%	78,78%
4	8,00	4,00	1,00	63,00	39,68%	1,00	-700,00%	52,00	28,85%	41,54%
5	1,25	1,75	1,75	24,75	43,43%	4,25	47,06%	23,00	50,00%	63,63%
6	2,25	8,00	0,50	29,75	19,33%	2,75	-54,55%	25,25	15,84%	60,93%
7	0,00	5,00	0,00	14,00	7,14%	2,00	50,00%	11,00	18,18%	74,35%
8	0,00	5,00	0,33	13,33	2,50%	1,00	-33,33%	10,33	-41,94%	77,49%
9	0,83	1,50	0,33	22,33	-6,72%	3,33	10,00%	20,83	-13,60%	71,77%
10	0,00	0,21	0,36	25,29	3,39%	3,71	15,38%	27,29	4,97%	89,81%
Total	**1,23**	**2,60**	**0,49**	**27,43**	**15,36%**	**3,72**	**9,14%**	**26,77**	**11,69%**	**76,98%**

and edges within the generated models. We can see that the transformation procedure tends to produce models which are on average 9-15% larger in size then what a human would create. This can be partially explained by noise and meta sentences which were not filtered appropriately. On the other hand, humans tend to abstract during the process of modeling. Therefore, we often find more detail of the text also in the generated model. The results are highly encouraging as our approach is able to correctly recreate 77% of the model in average. On a model level up to 96% of similarity can be reached, which means that only minor corrections by a human modeler are required.

During the detailed analysis we determined different sources of failure, which resulted in a decreased metric value. These are noise, different levels of abstractions, and processing problems within our system. *Noise* includes sentences or phrases that are not part of the process description, as for instance "This object consists of data elements such as the customers name and address and the assigned power gauge." While such information can be important for the understanding of a process, it leads to unwanted Activities within the generated model. To tackle this problem, further filtering mechanisms are required. Low similarity also results from *difference in the level of granularity*. To solve this problem, we could apply automated abstraction techniques like [34] on the generated model. Finally, the employed *natural language processing components failed* during the analysis. At stages, the Stanford Parser failed at correctly classifying verbs. For instance, the parser classified "the second activity checks and configures" as a noun phrase, such that the verbs "check" and "configure" cannot be extracted into Actions. Furthermore, important verbs related to business processes are not contained in FrameNet, as "report". Therefore, no message flow is created between report activities and a Black Box Pool. We expect this problem to be solved in the future as the FrameNet database grows. With WordNet, for instance, there is a problem with times like "2:00 pm", where pm as an abbreviation for "Prime Minister" is classified as an Actor. To solve this problem a reliable sense disambiguation has to be conducted. Nevertheless, overall good results were achieved by using WordNet as a general purpose Ontology.

5   Related Work

Recently, there is an increasing interest in the derivation of conceptual models from text. This research is mainly conducted by six different groups.

Two approaches generate UML models. The Klagenfurt Conceptual Pre-design Model and a corresponding tool are used to parse German text and fill instances of a generic meta-model [35]. The stored information can be transformed to UML activity diagrams and class diagrams [18]. The transformation from text to the meta-model requires the user to make decisions about the relevant parts of a sentence. In contrast to that, the approach described in [36] is fully automated. It uses use-case descriptions in a format called RUCM to generate activity diagrams and class diagrams [17]. Yet, the system is not able to parse free-text. The RUCM input is required to be in a restricted format allowing only 26 types of sentence structures, which rely on keywords like "VALIDATES THAT" or "MEANWHILE". Therefore, it can hardly be used in the initial process definition phase as it would require rewriting of process-relevant documents.

The University of Rio de Janeiro focuses on the derivation of BPMN models from group stories provided in Portuguese [23]. The approach was tested with a course enrollment process modeled by students. The examples in their paper show that process models can be created successfully, but a couple of their exhibits show that syntactical problems can occur, e.g. implicit conditions, which we explicitly tackle with our approach. The *R-BPD* toolkit from the University of Wollongong uses a syntax parser to identify verb-object phrases [21]. It also identifies textual patterns like "If <condition/event>, [then] <action>" [20]. The result are rather BPMN snippets than fully connected models. Nevertheless, this toolkit is able to take existing models into account for cross validation.

A fifth approach is the one of Policy-Driven Process Mapping [37]. First, a procedure was developed which creates a BPMN diagram, given that data items, tasks, resources (actors), and constraints are identified in an input text document. Although the approach does not require a process description to be sequential, it does not support Pools, Data Objects, and Gateways other than an exclusive split. Furthermore, user-interaction is required at several stages.

The approach by Sinha et al. builds on a linguistic analysis pipeline [22,38]. First, text is preprocessed with a part-of-speech tagger. Next, words are annotated with dictionary concepts, which classify verbs using a domain ontology. Then, an anaphora resolution algorithm and a context annotator are applied. The resulting information is then transferred to a Use Case Description meta-model and later into a BPMN process model. The dictionary concepts, which are a vital part of their approach, rely on a domain ontology which has to be hand-crafted. This imposes a manual effort when transferring the system to other types of texts or languages. Instead, our approach builds on the free WordNet and FrameNet lexical databases, which are available for different languages.

6   Conclusion

In this paper, we presented an automatic approach to generate BPMN models from natural language text. We have combined existing tools from natural language processing in an innovative way and augmented them with a suitable anaphora resolution mechanism. The evaluation of our technique shows that for a set of 47 text-model pairs from industry and textbooks, we are able to generate on average 77% of the models correctly.

Despite these encouraging results, we still require empirical user studies. Such studies should investigate whether humans find the generated models useful and easy to adapt towards a fully accurate model. Furthermore, our system is able to read process descriptions consisting of full sentences. Furthermore, we assumed the description to be sequential and to contain no questions and little process-irrelevant information. Another prerequisite is that the text is grammatically correct and constituent. Thus, the parsing of structured input, like tables or texts making use of indentions, or texts which are of low quality is not possible at the moment and presents opportunities for further research.

While the evaluation conducted in this thesis evinced encouraging results different lines of research could be pursued in order to enhance the quality or scope of our process model generation procedure. As shown the occurrence of meta-sentences or noise in general is one of the severest problems affecting the generation results. Therefore, we could improve the quality of our results by adding further rules and heuristics to identify such noise. Another major source of problems was the syntax parser we employed. As an alternative, semantic parsers like [39] could be investigated.

References

1. Herbst, J., Karagiannis, D.: An inductive approach to the acquisition and adaptation of workflow models. In: Proceedings of the IJCAI, pp. 52–57 (1999)
2. Frederiks, P., Van der Weide, T.: Information modeling: the process and the required competencies of its participants. Data & Knowledge Engineering 58(1), 4–20 (2006)
3. Scheer, A.: ARIS-business process modeling. Springer, Heidelberg (2000)
4. Reijers, H., Limam, S., Van Der Aalst, W.: Product-based workflow design. Journal of Management Information Systems 20(1), 229–262 (2003)
5. Gruber, T.: Automated knowledge acquisition for strategic knowledge. Machine Learning 4(3), 293–336 (1989)
6. Blumberg, R., Atre, S.: The problem with unstructured data. DM Review 13, 42–49 (2003)
7. White, M.: Information overlook. EContent(26:7) (2003)
8. OMG, eds.: Business Process Model and Notation (BPMN) Version 2.0 (June 2010)
9. Freund, J., Rücker, B., Henninger, T.: Praxishandbuch BPMN. Hanser (2010)
10. Melčuk, I.: Dependency syntax: theory and practice, New York (1988)
11. Marcus, M., Marcinkiewicz, M., Santorini, B.: Building a large annotated corpus of English: The Penn Treebank. Computational Linguistics 19(2), 330 (1993)
12. de Marneffe, M., Manning, C.: The Stanford typed dependencies representation. In: Workshop on Cross-Framework and Cross-Domain Parser Evaluation, pp. 1–8 (2008)
13. Baker, C., Fillmore, C., Lowe, J.: The berkeley framenet project. In: 17th Int. Conf. on Computational Linguistics, pp. 86–90 (1998)
14. Miller, G.A.: Wordnet: A lexical database for english. CACM 38(11), 39–41 (1995)
15. Achour, C.B.: Guiding scenario authoring. In: 8th European-Japanese Conference on Information Modelling and Knowledge Bases, pp. 152–171. IOS Press, Amsterdam (1998)
16. Li, J., Wang, H., Zhang, Z., Zhao, J.: A policy-based process mining framework: mining business policy texts for discovering process models. ISEB 8(2), 169–188
17. Yue, T., Briand, L., Labiche, Y.: An Automated Approach to Transform Use Cases into Activity Diagrams. Modelling Foundations and Appl., 337–353 (2010)
18. Fliedl, G., Kop, C., Mayr, H., Salbrechter, A., Vöhringer, J., Weber, G., Winkler, C.: Deriving static and dynamic concepts from software requirements using sophisticated tagging. Data & Knowledge Engineering 61(3), 433–448 (2007)
19. Kop, C., Mayr, H.: Conceptual predesign–bridging the gap between requirements and conceptual design. In: 3rd Int. Conf. on Requirements Eng. p. 90 (1998)

20. Ghose, A., Koliadis, G., Chueng, A.: Process Discovery from Model and Text Artefacts. In: 2007 IEEE Congress on Services, pp. 167–174 (2007)
21. Ghose, A.K., Koliadis, G., Chueng, A.: Rapid business process discovery (R-BPD). In: Parent, C., Schewe, K.-D., Storey, V.C., Thalheim, B. (eds.) ER 2007. LNCS, vol. 4801, pp. 391–406. Springer, Heidelberg (2007)
22. Sinha, A., Paradkar, A., Kumanan, P., Boguraev, B.: An Analysis Engine for Dependable Elicitation on Natural Language Use Case Description and its Application to Industrial Use Cases. Technical report, IBM (2008)
23. de AR Gonçalves, J.C., Santoro, F.M., Baião, F.A.: A case study on designing processes based on collaborative and mining approaches. In: Int. Conf. on Computer Supported Cooperative Work in Design, Shanghai, China (2010)
24. Fliedl, G., Kop, C., Mayr, H.: From textual scenarios to a conceptual schema. Data & Knowledge Engineering 55(1), 20–37 (2005)
25. Friedrich, F.: Automated generation of business process models from natural language input. Master's thesis, Humboldt-Universität zu Berlin (November 2010)
26. Kitzmann, I., Konig, C., Lubke, D., Singer, L.: A Simple Algorithm for Automatic Layout of BPMN Processes. In: IEEE Conf. CEC, pp. 391–398 (2009)
27. Seemann, J.: Extending the sugiyama algorithm for drawing UML class diagrams: Towards automatic layout of object-oriented software diagrams. In: Graph Drawing, pp. 415–424. Springer, Heidelberg (1997)
28. Eiglsperger, M., Kaufmann, M., Siebenhaller, M.: A topology-shape-metrics approach for the automatic layout of UML class diagrams. In: Proceedings of the 2003 ACM Symposium on Software Visualization, p. 189. ACM, New York (2003)
29. White, S., Miers, D.: BPMN Modeling and Reference Guide: Understanding and Using BPMN. Future Strategies Inc. (2008)
30. Holschke, O.: Impact of granularity on adjustment behavior in adaptive reuse of business process models. In: Hull, R., Mendling, J., Tai, S. (eds.) BPM 2010. LNCS, vol. 6336, pp. 112–127. Springer, Heidelberg (2010)
31. Reijers, H.: Design and control of workflow processes: business process management for the service industry. Eindhoven University Press (2003)
32. Dijkman, R., Dumas, M., van Dongen, B., Käärik, R., Mendling, J.: Similarity of business process models: Metrics and evaluation. Inf. Sys. 36, 498–516 (2010)
33. Dijkman, R., Dumas, M., Garcia-Banuelos, L., Käärik, R.: Graph Matching Algorithms for Business Process Model Similarity Search. In: Dayal, U., Eder, J., Koehler, J., Reijers, H.A. (eds.) BPM 2009. LNCS, vol. 5701, pp. 48–63. Springer, Heidelberg (2009)
34. Polyvyanyy, A., Smirnov, S., Weske, M.: On application of structural decomposition for process model abstraction. In: 2nd Int. Conf. BPSC, pp. 110–122 (March 2009)
35. Kop, C., Vöhringer, J., Hölbling, M., Horn, T., Irrasch, C., Mayr, H.: Tool Supported Extraction of Behavior Models. In: Proc. 4th Int. Conf. ISTA (2005)
36. Yue, T., Briand, L., Labiche, Y.: Automatically Deriving a UML Analysis Model from a Use Case Model. Technical report, Carleton University (2009)
37. Wang, H.J., Zhao, J.L., Zhang, L.J.: Policy-Driven Process Mapping (PDPM): Discovering process models from business policies. DSS 48(1), 267–281 (2009)
38. Sinha, A., Paradkar, A.: Use Cases to Process Specifications in Business Process Modeling Notation. In: 2010 IEEE Int. Conf. on Web Services, pp. 473–480 (2010)
39. Shi, L., Mihalcea, R.: Putting Pieces Together: Combining FrameNet, VerbNet and WordNet for Robust Semantic Parsing. In: Proceedings of the 6th Int. Conf. on Computational Linguistics and Intelligent Text Processing, p. 100 (2005)

A Semantic Approach for Business Process Model Abstraction

Sergey Smirnov[1], Hajo A. Reijers[2], and Mathias Weske[1]

[1] Hasso Plattner Institute, University of Potsdam, Germany
{sergey.smirnov,mathias.weske}@hpi.uni-potsdam.de
[2] Eindhoven University of Technology, The Netherlands
h.a.reijers@tue.nl

Abstract. Models of business processes can easily become large and difficult to understand. Abstraction has proven to be an effective means to present a readable, high-level view of a business process model, by showing aggregated activities and leaving out irrelevant details. Yet, it is an open question how to combine activities into high-level tasks in a way that corresponds to such actions by experienced modelers. In this paper, an approach is presented that exploits *semantic* information within a process model, beyond *structural* information, to decide on which activities belong to one another. In an experimental validation, we used an industrial process model repository to compare this approach with actual modeling decisions, showing a strong correlation between the two. As such, this paper contributes to the development of modeling support for the application of effective process model abstraction, easing the use of business process models in practice.

Keywords: business process modeling, model management, business process model abstraction, activity clustering.

1 Introduction

Business process models are used within a range of organizational initiatives [19]. However, human readers are limited in their cognitive capabilities to make sense of large and complex business process models [2,33]. One well-known way to address this issue is by applying *abstraction*, the act of retaining essential properties of a process model on a particular level of analysis while hiding insignificant process details. Indeed, in a recent empirical investigation into the need for business process model abstraction [32], we found that its most prominent use case is the need for gaining a *quick overview* of the process. In such a situation, the user wants to familiarize herself with a business process but has only a large process model of many detailed activities at her disposal. To deal with such a demand, the process model can then be displayed as a partially ordered set of coarse-grained activities, each of which aggregates a number of lower-level activities. As an example, an abstraction of a process model that captures the creation of a forecasting report is shown in Fig. 1. In this figure, m is the initial model and m_a is the abstract model of the same process.

H. Mouratidis and C. Rolland (Eds.): CAiSE 2011, LNCS 6741, pp. 497–511, 2011.

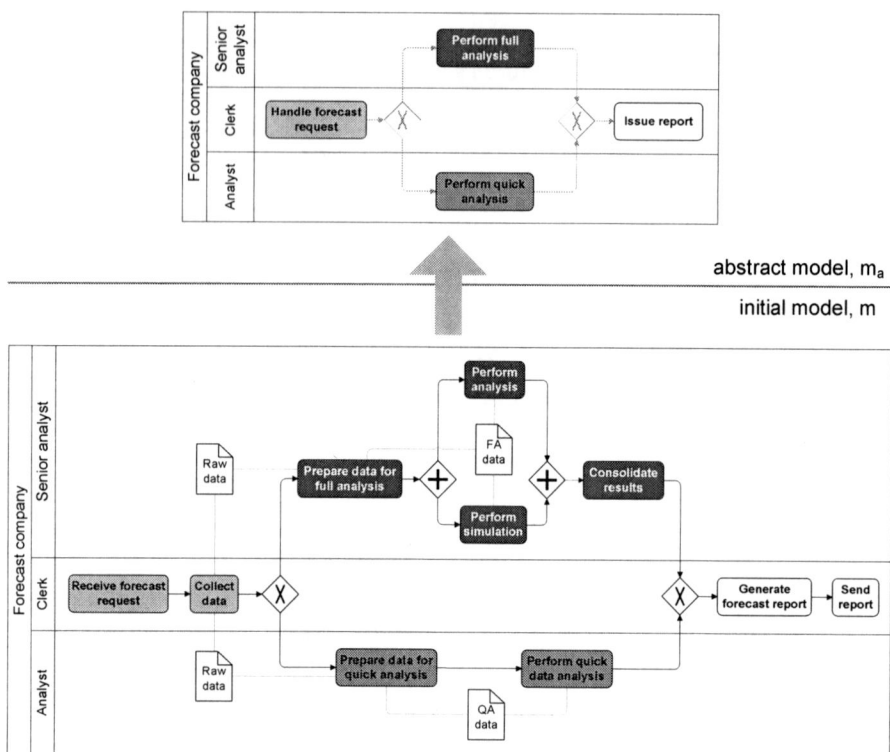

Fig. 1. Motivating example: initial model and its abstract counterpart

While it has been empirically shown that abstraction can significantly improve the sense-making of large process models [25], a limited insight exists into the criteria that experienced modelers use to decide on which activities to aggregate into new ones. A number of techniques has been proposed that exploit *structural* properties of a process model to arrive at abstract models [5,24]. It seems likely that experienced process modelers take a wider range of properties into account rather than just a model's control flow. For example, the fact that two activities use the same document and are executed by the same role may be used as relevant inputs in deciding to cluster these two into an aggregated activity. This situation applies, for example, to the activities *Prepare data for quick analysis* and *Perform quick data analysis* in Fig. 1.

In this paper, we complement the existing streams of work with respect to process model abstraction by proposing an abstraction technique that incorporates semantic aspects contained within a process model. We rely on the observation that industrial process models are often enriched with non-control flow model elements. Examples are: data that is being processed within an activity, IT systems invoked within particular activities, and roles assigned to activities. The central idea in this paper is that activities associated with the *same* non-control flow elements are semantically *related* and, therefore, more appropriate candidates to be aggregated into the same activity than activities without shared elements.

A number of recent contributions exist that consider *semantic* aspects for aggregation, e.g., [8,31]. However, their assumptions, e.g., the existence of an activity ontology [31], are too strict for generic use. Our approach is based on the application of the vector space model, an algebraic model popular in information retrieval [28]. As we will discuss in this paper, the use of vector spaces allows to determine the degree of similarity between activities according to several information types available in process models. We have validated the proposed technique applying it to a process model repository that is in use by a large European telecommunication organization. The repository incorporates hierarchical relations between high-level activities and the activities that they aggregate. Also, the process models contain various types of semantic information. The validation suggests that our approach closely approximates the decisions of the involved modelers to cluster activities.

The main contribution of this paper is a technique that may assist novice process modelers in the abstraction of complex process models by mimicking the abstraction decisions of more experienced modelers, as discovered from existing models. In this way, the technique allows to reuse activity aggregation principles for future aggregation decisions. Since the lack of experienced process modelers is a noted issue in many large modeling projects [26], this is a valuable asset to improve the process model quality. Meanwhile, the designed technique can also support experienced modelers enabling process model abstraction in conformance to their specific abstraction style. Hence, experts can accelerate their modeling routine configuring this technique, while staying in control over the modeling outcome. Finally, the technique can also be used to safeguard a particular "fingerprint" of a process model collection with respect to abstraction choices.

The paper is structured accordingly. We continue in Section 2 explaining the proposed algorithm, along with providing the required background knowledge. Section 3 empirically validates the proposed approach, using an industrial set of process models from the telecommunication sector. Finally, Section 4 contrasts our contribution with the related research, while Section 5 concludes the paper.

2   Activity Aggregation

This section elaborates on the proposed activity aggregation algorithm. After the introduction of the main concepts, we argue how activity aggregation can be interpreted as a clustering problem. We discuss a suitable clustering algorithm and alternative activity distance measures. The section focuses on one specific measure that enables the tuning of an activity aggregation. We explain how the aggregation setup is realized and show how the setup information can be mined from an existing process model collection.

2.1   Foundations

The designed aggregation algorithm inspects an activity environment, i.e., process model elements that are related to activities in a process model. Examples of such elements are data objects accessed by activities and roles supporting activity execution, e.g., see model in Fig. 1. The list of such model element types varies depending on the process modeling language, the tool at hand, modeling procedures taken into account, and the modeler's style. Each of the model element types can be considered as an activity property that has a specific value. Definition 1 formalizes the activity property concept.

Definition 1 (Activity Property Value and Activity Property Type). Let $\mathcal{P}$ be a finite nonempty set of activity property values. Alongside, $\mathcal{T}$ is a finite nonempty set of activity property types. Mapping $type : \mathcal{P} \to \mathcal{T}$ assigns a type to each value.

The process model in Fig. 1 illustrates Definition 1. *Raw data*, *FA data*, and *Analyst* are examples of activity property values. The process model presents two activity property types: *Role* and *Data object*. For instance, *type(Raw data)* = *Data object*, *type(FA data)* = *Data object*, and *type(Analyst)* = *Role*. Further, we define a process model as follows.

Definition 2 (Process Model). A tuple $m_i = (A_i, G_i, F_i, P_i, props_i)$ is a *process model*, where:

- A_i is a finite nonempty set of activities;
- G_i is a finite set of gateways;
- $N_i = A_i \dot\cup G_i$ is the set of nodes, where $\dot\cup$ denotes a disjoint union of sets;
- $F_i \subseteq N_i \times N_i$ is the flow relation;
- $P_i \subseteq \mathcal{P}$ is a set of activity property values;
- $props_i : A_i \to 2^{P_i}$ is a mapping that assigns property values to an activity.

Definition 2 does not make a distinction between different gateway types, since the future discussion does not make use of them. Mapping $props_i$ assigns activity property values to model activities. Referring to model m in the motivating example of Fig. 1, mapping $props_i$ can be illustrated as $props_i(Collect\ data)$ = $\{Clerk,\ Raw\ data\}$. Notice that Definitions 1 and 2 allow to manage the considered activity property types in a flexible fashion: it is enough to introduce a new activity property type to set $\mathcal{T}$, the values to $\mathcal{P}$, and respectively update mapping $type$. Thereafter, new activity properties can be easily considered within the activity aggregation. Finally, we postulate the concept of a process model collection.

Definition 3 (Process Model Collection). A tuple $c = (M, A, P, \sigma)$ is a *process model collection*, where:

- M is a nonempty finite set of n process models with elements $m_i = (A_i, G_i, F_i, P_i, props_i)$, where $i = 1, 2, \ldots, n$;
- $A = \dot\cup_{i=1,2,\ldots,n} A_i$ is a set of collection activities;
- $P = \cup_{i=1,2,\ldots,n} P_i$ is a set of collection activity property values;

- $\sigma \subseteq M \times M$ is a subprocess relation refining a process model with subprocess models, such that $\forall m_i, m_j \in M$, where $j = 1, 2, \ldots, n$ and $i \neq j$, if $(m_i, m_j) \in \sigma$ then $(m_j, m_i) \notin \sigma^+$, where σ^+ is a transitive reflexive closure of σ.

Definition 3 explicitly enumerates the model collection activities and property value types. The relation σ formalizes the subprocess relation that exists between models. Note that according to the definition, σ enables only a process model hierarchy without loops. Without loss of generality in the remainder of this paper we discuss abstraction of process models within a process model collection. Indeed, a process model m_i can be seen as a trivial process model collection $c = (\{m_i\}, A_i, P_i, \emptyset)$.

2.2   Activity Aggregation as Cluster Analysis Problem

In this paper we interpret activity aggregation as a problem of cluster analysis. Consider process model $m_i = (A_i, G_i, F_i, P_i, props_i)$ from process model collection $c = (M, A, P, \sigma)$. The set of objects to be clustered is the set of activities A_i. The objects are clustered according to a distance measure: objects that are "close" to each other according to this measure are put together. The distance between objects is evaluated through analysis of activity property values P. The cluster analysis outcome, activity clusters, correspond to coarse-grained activities of the abstract process model. While cluster analysis provides a large variety of algorithms, e.g., see [29], we focus on one algorithm that suits the business process model abstraction use case in focus.

In the considered scenario, the user demands control over the number of activities in the abstract process model. For example, a popular practical guideline is that five to seven activities are displayed on each level in the process model [30]. Provided a fixed number, e.g. 6, the clustering algorithm has to assure that the number of clusters equals the request by the user. We turn to the use of k-means clustering algorithm, as it is simple to implement and typically exhibits good performance [16]. K-means clustering partitions an activity set into k clusters. The algorithm assigns an activity to the cluster, which centroid is the closest to this activity. To evaluate an activity distance, we analyze activity property values P. We foresee a number of alternative activity distance measures and elaborate on them in the next section.

2.3   Activity Distance Measures

To introduce the distance measure among activities we represent activities as vectors in a vector space. Such an approach is inspired by the vector space model, an algebraic model widely used in information retrieval [28]. The space dimensions correspond to activity property values P and the vector space can be captured as vector $(p_1, \ldots, p_{|P|})$, where $p_j \in P$ for $j = 1, \ldots, |P|$. Consider an example set of property values $P' = \{FA\ data, QA\ data, Raw\ data\}$ and the corresponding vector space presented in Fig. 2. A vector v_a representing an activity $a \in A_i$ in process model $m_i = (A_i, G_i, F_i, P_i, props_i)$ is constructed as follows. If activity a

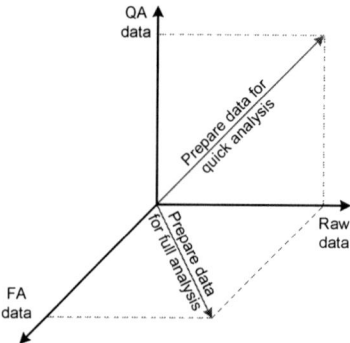

Fig. 2. Example of a vector space formed by dimensions *FA data, QA data, Raw data*

is associated with a property value $p_j \in P_i$, the corresponding vector dimension $\pi_j(\boldsymbol{v_a})$ has value 1; otherwise, the dimension $\pi_j(\boldsymbol{v_a})$ has value 0:

$$\pi_j(\boldsymbol{v_a}) = \begin{cases} 1, & \text{if } p_j \in props_i(a); \\ 0, & \text{otherwise.} \end{cases}$$

For process model m in Fig. 1, activities *Prepare data for quick analysis* and *Prepare data for full analysis* correspond, respectively, to vectors $\boldsymbol{v_1} = (0,1,1)$ and $\boldsymbol{v_2} = (1,0,1)$ in the vector space with dimensions *FA data, QA data, Raw data*, see Fig. 2.

Similarity of two vectors in the space is defined by the angle between these vectors: the larger the angle, the more distant the activities are. Typically, the cosine of the angle between two vectors is used as a vector similarity measure:

$$sim(a_1, a_2) = cos(v_{a_1}, v_{a_2}) = \frac{v_{a_1} \cdot v_{a_2}}{\|v_{a_1}\| \|v_{a_2}\|} \tag{1}$$

Then, the distance between two activities is:

$$dist(a_1, a_2) = 1 - sim(a_1, a_2) \tag{2}$$

By construction the vector dimension values are non-negative. Hence, the activity similarity and activity distance measures vary within the interval $[0, 1]$.

For a process model collection $c = (M, A, P, \sigma)$ we distinguish two types of vector spaces. On the one hand, a vector space can be formed by the dimensions corresponding to the activity property values disregard their type, i.e., all elements of P. We reference such spaces as *heterogeneous vector spaces*. An example of a heterogeneous vector space is a space with 6 dimensions *Analyst, Clerk, FA data, QA data, Raw data*, and *Senior analyst*. On the other hand, a vector space can be formed by the dimensions corresponding to the activity property values of a particular type. Given an activity property type t, such a space is formally defined by the set $P_t = \{\forall p \in P : type(p) = t\}$. We refer to such spaces as *homogeneous vector spaces*. Fig. 2 provides an example of a homogeneous vector space formed by activity properties of type *Data object*. We denote the activity distance in a heterogeneous space with $dist_h(a_1, a_2)$ and in a homogeneous vector space with $dist_t(a_1, a_2)$, where t is the respective activity property type. Both distance measures can be employed for activity aggregation. If the user wants to make use of one activity property type t only, the distance is defined by $dist_t$. To cluster activities according to several activity property types, $dist_h$ can be employed. In addition, we introduce an alternative distance measure $dist_{agg}$ that aggregates multiple homogeneous distance measures $dist_t$:

$$dist_{agg}(a_1, a_2) = \frac{1}{|T|} \sum_{\forall t \in T} w_t \cdot dist_t(a_1, a_2) \qquad (3)$$

In Equation 3, the set T corresponds to the activity property types that appear in process model collection c. Then, function $dist_{agg}$ is the weighted average value of distance measures in the vector spaces corresponding to the available activity property types. Coefficient w_t is the weight of $dist_t$ indicating the impact of the activity distance according to property type t. We reference all the weights in Equation 3 as $\boldsymbol{W} = (w_{t_1}, \ldots, w_{t_n})$, where $n = |T|$. In the remainder of this section we will explain the role of vector $\boldsymbol{W}$.

2.4 Process Model Collection Abstraction Fingerprint

The application of different abstraction operations to one process model leads to various abstract representations of the modeled business process. The differences between abstraction operations are explained by their pragmatics, i.e., various abstraction purposes. If the abstraction is realized by a human, the modeling habits of the designer are reflected in the abstraction operation as well. Hence, abstraction pragmatics and modeling habits of the designer are inherent properties of the abstraction operation and together form an *abstraction style*. We use vector $\boldsymbol{W}$ in Equation 3 to model an abstraction style.

From the user perspective vector $\boldsymbol{W}$ is the tool to express the desired abstraction style. We foresee two scenarios how vector $\boldsymbol{W}$ can be obtained. In the first scenario, the user explicitly specifies $\boldsymbol{W}$. This approach is useful if the user wants to introduce a new abstraction style. However, coming up with an appropriate value for $\boldsymbol{W}$ may be challenging. Hence, the second scenario implies that vector $\boldsymbol{W}$ is mined from a process model collection enriched with subprocess relation (formalized with σ in Definition 3). The discovered vector is a "fingerprint" of the process model collection with respect to the used abstraction style. We will now describe an approach how vector $\boldsymbol{W}$ can be discovered from such a process model collection.

The discovery process of a model collection's abstraction fingerprint is driven by the following argumentation. Activities of a process model collection are aggregated into aggregated activities, i.e. subprocess placeholders, by the model designer. We aim to achieve an activity clustering algorithm that approximates this aggregation behavior of a human. This is possible if an activity distance measure employed by the algorithm resembles the criteria that a human designer uses to aggregate activities into a subprocess. The exact criteria are unknown. Yet, for each pair of activities we can observe the outcome: Either the activities belong to different subprocesses or to the same one. For a process model collection $c = (M, A, P, \sigma)$ function *diff* formalizes this observation:

$$diff(a_k, a_l) = \begin{cases} 0, & \text{if } a_k, a_l \in A_i; \\ 1, & \text{otherwise.} \end{cases} \tag{4}$$

To mine the process model collection fingerprint $\boldsymbol{W}$ we select its value in such a way that the behavior of function $dist_{agg}$ approximates the behavior of *diff*. The discovery of vector $\boldsymbol{W}$ is realized by means of linear regression. In our setting, the values $dist_t$ are considered independent variables and the value of function *diff* the the dependent variable. Components of vector $\boldsymbol{W}$ are the regression coefficients. The standardized coefficients indicate the impact of each activity property type on the abstraction style. Hence, it is possible to reveal criteria employed by the human designer during abstraction. Furthermore, the regression's coefficient of determination R^2 allows to judge how well the obtained statistical model explains the observed behavior. For our purposes, R^2 suggests if the discovered statistical model can be used for business process model abstraction.

3   Empirical Validation

The proposed activity aggregation mechanism calls for validation. The goal of the validation is to learn how well the proposed operation approximates the abstraction style of human modelers. We performed an empirical validation of the approach by conducting an experiment with a real world business process model collection. This section provides a detailed discussion of the validation; it describes in detail the explored process model collection, explains the experiment design, and discusses the validation results.

3.1   Validation Setup

As a research object we choose a set of business process models from a large telecommunication service provider. This organization is currently in the process of setting up a repository with high-quality process models, which are brought together for the purpose of consultation and re-use by business users. The model set includes 30 elaborate models, enriched with activity properties of the following types: *roles, responsible roles, IT systems*, and *data objects*. It is noticeable that a special type of roles, i.e., *responsible roles*, is also distinguished in these models. In addition to these non-control flow types of information, we also study the impact that activity labels and activity neighboring control flow elements have on the decision to aggregate activities into the same subprocess. To compare activities with respect to their labels, the corresponding vector space is formed by the words that appear in the labels. Against this background, finding the distance between activities becomes an information retrieval task as labels can be treated as documents in information retrieval. The comparison of activities with respect to their neighbors shows whether the neighborhoods of the two

Table 1. Properties of business process models used in the validation

	Nodes	Activities	Role	Responsible role	IT system	Data object
Average	15.5	6.3	2.1	0.76	1.5	0.76
Minimum	5	1	0	0	0	0
Maximum	48	20	5	2	7	17

activities intersect, i.e., contain the same flow elements. Table 1 outlines the relevant properties of the process models. In the existing repository, the models are hierarchically organized using a subprocess relation. Within the model set, we have identified 8 subprocess hierarchies. Each hierarchy contains a root process model refined with subprocesses, allowing for several levels of refinement.

To formally validate how good the designed activity aggregation approximates the behavior of modelers clustering a set of activities into the same subprocess, we selected the following approach. For each pair of activities that belong to the same process hierarchy, we have evaluated two values in the process model collection: *diff* and *dist*. Here, *diff* describes the human abstraction style, which indicates whether the activities have been decided to be placed in the same subprocess or not. The value of *dist* represents the vector space distance between the two activities in accordance with our approach. To discover if the two approaches yield similar results, we study the correlation between the two variables. A strong correlation of two variables implies that *dist* is a good distance measure in the clustering algorithm. In this case, the inclusion of two activities within the same subprocess is mirrored by a close positioning of the corresponding vectors in the vector space. Given the nature of the observed variables, we employ Spearman's rank correlation coefficient.

In the following, we first investigate the human abstraction style in the model collection as a whole. Then, we verify the results organizing a K-fold cross validation. We partition the model sample into 4 subsamples, i.e., $k = 4$ and perform four tests. In each test, three subsamples are used to discover vector $\boldsymbol{W}$, while the fourth subsample is used to evaluate the correlation values between the *diff* and *dist* measures in different vector spaces. In this way, a more reliable insight is developed into the question whether the human abstraction style can be mimicked in contrast to using the whole process model collection for both the discovery and the evaluation of this correlation.

3.2 Validation Results

Table 2 outlines the validation's results. The columns in the table correspond to distance measures. While the first 6 columns correspond to distances in homogeneous spaces, the last three columns reflect the distance measure taking into account multiple activity properties. All three distance measures make use of the activity property types in columns 1–6. The distance $dist_h$ is measured in heterogeneous vector space, where dimensions are activity property values of types listed in columns 1–6. The distance measure $dist_{avg}$ is the average value of

Table 2. Correlation values observed in the K-fold cross validation

Experiment	$\rho(dist_t, diff)$						$\rho(dist_h, diff)$	$\rho(dist_{avg}, diff)$	$\rho(dist_{agg}, diff)$
	Role	Responsible role	IT system	Data object	Label	Neighbor			
All models	0.70	0.61	0.60	—	0.34	0.58	0.74	0.65	0.77
Test$_1$	0.79	0.76	0.75	—	0.42	0.60	0.79	0.69	0.79
Test$_2$	0.64	0.56	0.56	—	0.43	0.62	0.68	0.70	0.70
Test$_3$	0.68	0.58	0.58	—	0.53	0.64	0.68	0.72	0.71
Test$_4$	0.61	0.47	0.45	—	0.20	0.48	0.70	0.56	0.52
Average$_{1-4}$	0.68	0.59	0.58	—	0.39	0.59	0.71	0.67	0.68

distances in columns 1–6. The distance measure $dist_{agg}$ is evaluated according to Equation 3. Vector W used in $dist_{agg}$ is obtained using linear regression as described in the previous section. Rows of Table 2 correspond to experiments. The first row describes the study of the whole model collection. Rows 2–5 describe the results of 4 tests along the K-fold cross validation we explained earlier, while the last row provides the average correlations observed in the 4 separate tests.

The correlation values that are presented in Table 2 are all significant using a confidence level of 99%, i.e., all p values are lower than 0.01. However, no statistically significant results were obtained for the distance in the homogeneous vector space that corresponds to *Data objects*. Overall, the presented correlation values range around 0.7. This level is generally considered to indicate a strong correlation [11,12], particularly in situations where human decision making is involved. Therefore, we can speak of a strong relation between the *dist* and *diff* measures.

Among the distance measures in homogeneous spaces, one can point out the distance in the *Role* space that overall displays the highest correlation values for the different studies (0.61–0.79). In contrast, correlation values for *Label* are the lowest (0.20–0.53). Another observation is that distances taking into account multiple activity property types tend to have higher correlations. From these, $dist_{agg}$ outperforms all other distance measures with a value arriving at 0.77 when all models are considered. For the average values of the K-fold cross validations, however, $dist_h$, $dist_{avg}$, and $dist_{agg}$ demonstrate a similar performance, with correlation values of 0.71, 0.67, and 0.68 respectively. This observation can be explained by the fact that $dist_{agg}$ is parameterized by vector W—the abstraction fingerprint of a particular model set. Thus, the distance measure $dist_{agg}$ "trained" on one model set may never excel $dist_{avg}$, once the set of models is changed. Tests 1–4 support this argumentation. Note that this result does not restrict the applicability of the approach: in a real world setting, the goal is to transfer the abstraction style from one model set to another. The average values in the lower row should, therefore, be seen as most important from the ones displayed.

A careful inspection of the linear regression results associated with parameterizing vector W provides additional insights. In particular, we are interested in the observed R^2 values and the beta coefficients (also known as "standardized coefficients"). The R^2 for the whole model set, as well as the average value for the K-fold cross validation is 0.52. This value shows the explained level of variation in abstraction style as explained by the various distance measures under consideration and can be considered as moderately strong. The beta coefficients of the distance measures in various spaces reveal their impact on the activity aggregation. The beta coefficients for activity property types *Role* and *Responsible role* have average values of 0.55 and 0.37, respectively. At the same time, the standardized coefficients of *Neighbor* and *Label* property types fluctuate around 0. The average value for *IT systems* is in between, with a beta coefficient of 0.19. The provided numbers illustrate that the activity property types *Role* and *Responsible role* have a big impact on the abstraction style of the considered process model collection. *IT systems* also contributes to the activity aggregation, but the influence of activity labels and activity neighborhood is insignificant. Clearly, such insights may differ from one process model to the other.

The validation indicates that the suggested distance measures can be used in a close approximation of the abstraction style of human modelers. Among the introduced measures, $dist_{agg}$ is of great interest, as it takes into account the abstraction style of a particular process model collection. Furthermore, the validation revealed activity property types, *Role* and *Responsible role*, that have the highest impact on the abstraction style for this particular collection.

4   Related Work

The topic of business process model abstraction can be related to several research streams. We identify these streams looking both from the perspective of the disciplines of software engineering and business process management.

Model properties and relations are thoroughly investigated in the software engineering area. For instance, in [21,22] Kühne elaborates on the concepts of model, metamodel, model types, and model relations. These works systematically describe and organize relations, e.g., generalization and classification, which are seminal for the problem of model abstraction. Closely related are also the studies that cover model granularity. In [17], the authors investigate model and metamodel granularity. The authors compare several metamodels and come up with best practices with respect to granularity. One can observe that the relation between a coarse-grained activity in an abstract model and its counterparts in the initial model is the meronymy, or part-of, relation. Meronymy has been studied in depth in the software engineering domain [3,13]. Although the referenced papers do not provide concrete techniques for the implementation of abstraction within process models, they facilitate a better problem understanding and help to identify the main concepts in this domain.

Business process management is the discipline concerned with using methods, techniques, and software to design, enact, control, and analyze operational processes. A large body of knowledge corresponds to process model analysis based on model transformations. Model transformations can be reused in the context of the abstraction problem. An example of such a transformation consists of reduction rule sets for Petri nets, e.g., see [4,23,27]. Each reduction rule explicitly defines a structural fragment to be discovered in the model and a method of this fragment transformation. Hence, reduction rule sets enable process model abstraction through iterative rule application. As the transformed process fragments are explicitly defined, each reduction rule set handles only a particular model class. Thereby, each reduction rule set requires an argument about the model class reducible with the given rules. The model class limits the application of abstraction approaches based on reduction rules [5,10]. Process model decomposition approaches are free of this limitation: they seek for process fragments with particular properties. An example of such a decomposition is presented in [34], where single entry single exit fragments are discovered. The result of process model decomposition is the hierarchy of process fragments according to the containment relation, i.e., the process structure tree. Such a tree can be used for abstraction in process models [24]. Finally, one can distinguish model transformations that preserve process behavior properties. In [1], van der Aalst and Basten introduce three notions of behavioral inheritance for WF-nets and study inheritance properties. The paper suggests model transformations, such that the resulting model inherits the behavior of the initial model. An approach for process model abstraction can exploit such transformations as basic operations. While the outlined model transformations can support solving the general problem of process model abstraction, they all focus on structural and behavioral aspects of models and model transformations, leaving the semantic aspect out of scope.

Many tasks in the management of large process model collections can be traced back to the problem of *activity matching*, which is closely related to the problem of business process model abstraction. Examples of such management tasks are: the search for a particular process model over a process model set or ensuring the consistency of models capturing one and the same process from different perspectives. Activity matching is realized through analysis of activity properties: activity labels, referenced data objects and neighboring activities. In [9,35] the authors suggest activity matching algorithms and evaluate them. While the named works explore the existing process models and do not directly address the problem of process model abstraction, their results have a potential of being applied in business process model abstraction. Semantic aggregation of activities relates to research on semantic business process management. Notice that process models enriched with semantic information facilitate many process analysis tasks, see [18]. Along this line of research, several authors argue how to use activity ontologies to realize activity aggregation [6,7]. It should be noticed, however, that such works imply the existence of a semantic description for model elements and their relations, which is a restriction that rarely holds in real world settings.

Establishing an activity's granularity level is also a recurrent challenge in process mining, where logs contain records that are often very fine-granular. As such, the process models directly mined from the logs can be overloaded with information making them hard to comprehend. Activity clustering is an efficient means to raise the abstraction level for the mined models. In [14,15] Günther and van der Aalst propose activity aggregation mechanisms based on clustering algorithms. The mechanisms extensively use information present in process logs, but which are less common for process models, i.e., timestamps of activity starts and stops, activity frequencies, and transition probabilities. Thus, in contrast to the activity aggregation approach proposed in this paper, process mining considers other activity property types for clustering and utilize other clustering algorithms.

5   Conclusions and Future Work

Despite business process model abstraction has been addressed in a number of research endeavors, this paper proposes a novel approach in this area. Specifically, it exploits semantic aspects—beyond the control-flow perspective—to determine a similarity between different activities for the purpose of simplifying process model abstraction. Relevant levels of similarity can be determined on the basis of existing process models in which abstraction was already applied.

Our main contribution is a method to discover sets of related activities, where each set corresponds to a coarse-grained activity of an abstract process model. As a second contribution, we propose an approach to discover an abstraction style inherent to a given process model collection, which is reusable for abstraction of new process models. Both contributions are of practical interest, as they addresses model management issues recurrently appearing in process model projects. The experimental validation provides strong support for the applicability and effectiveness of the presented ideas.

Our approach is characterized by a number of limitations and assumptions. First of all, it builds on the assumption that all kinds of semantic information, such as data objects, roles, and resources, can be observed within the descriptions of process models in industrial collections. The process model collection we obtained through our cooperation with a large telecommunication company clearly confirms this idea, but this also applies to other industrial repositories, such as the SAP Reference Model [20]. Secondly, in our validation we have merely focused on the appearance or not of two activities being within the same subprocess or process model, although it can be imagined that a more fine-grained correspondence measure could yield even more useful results.

These and other limitations guide our future research plans. The direct next step for us is the use of advanced vector space models reflecting the relations between different activity property values. Such models enable activity clustering algorithms to consider the *structure* of organigrams and data object relations. Meanwhile, it can also be beneficial to consider other clustering algorithms and compare the outcome with the solution introduced in this paper. From a practical perspective, it is important to suggest names for coarse-grained activities that are products of activity aggregation. Finally, we would like to improve the validation method for activity aggregation. On the one hand, this implies replacing correlation with an alternative metric for activity aggregation quality. On the other hand, the validation will require an empirical study involving human modelers and stakeholders, who can evaluate the proposed activity aggregation.

References

1. van der Aalst, W.M.P., Basten, T.: Life-Cycle Inheritance: A Petri-Net-Based Approach. In: Azéma, P., Balbo, G. (eds.) ICATPN 1997. LNCS, vol. 1248, pp. 62–81. Springer, Heidelberg (1997)
2. Aguilar, E.R., Ruiz, F., García, F., Piattini, M.: Evaluation Measures for Business Process Models. In: SAC 2006, pp. 1567–1568. ACM, New York (2006)
3. Barbier, F., Henderson-Sellers, B., Le Parc-Lacayrelle, A., Bruel, J.-M.: Formalization of the Whole-Part Relationship in the Unified Modeling Language. IEEE TSE 29(5), 459–470 (2003)
4. Berthelot, G.: Transformations and Decompositions of Nets. In: Rozenberg, G. (ed.) APN 1987. LNCS, vol. 266, pp. 359–376. Springer, Heidelberg (1987)
5. Bobrik, R., Reichert, M., Bauer, T.: View-Based Process Visualization. In: Alonso, G., Dadam, P., Rosemann, M. (eds.) BPM 2007. LNCS, vol. 4714, pp. 88–95. Springer, Heidelberg (2007)
6. Casati, F., Shan, M.-C.: Semantic Analysis of Business Process Executions. In: EDBT 2002, pp. 287–296. Springer, Heidelberg (2002)
7. Alves de Medeiros, A.K., van der Aalst, W.M.P., Pedrinaci, C.: Semantic Process Mining Tools: Core Building Blocks. In: ECIS 2008, Galway, Ireland, pp. 1953–1964 (2008)
8. Di Francescomarino, C., Marchetto, A., Tonella, P.: Cluster-based Modularization of Processes Recovered from Web Applications. Journal of Software Maintenance and Evolution: Research and Practice (2010)
9. Dijkman, R., Dumas, M., García-Bañuelos, L.: Graph Matching Algorithms for Business Process Model Similarity Search. In: Dayal, U., Eder, J., Koehler, J., Reijers, H.A. (eds.) BPM 2009. LNCS, vol. 5701, pp. 48–63. Springer, Heidelberg (2009)
10. Eshuis, R., Grefen, P.: Constructing Customized Process Views. Data Knowl. Eng. 64(2), 419–438 (2008)
11. Franzblau, A.N.: A Primer of Statistics for Non-statisticians. Harcourt, Brace & World, New York (1958)
12. Gerstman, B.B.: StatPrimer – Version 6.4. Technical report, San Jose State University (2010), http://www.sjsu.edu/faculty/gerstman/StatPrimer/
13. Guizzardi, G.: Modal Aspects of Object Types and Part-Whole Relations and the de re/de dicto Distinction. In: Krogstie, J., Opdahl, A.L., Sindre, G. (eds.) CAiSE 2007 and WES 2007. LNCS, vol. 4495, pp. 5–20. Springer, Heidelberg (2007)
14. Günther, C.W., van der Aalst, W.M.P.: Mining Activity Clusters from Low-Level Event Logs. BETA Working Paper Series, WP, vol. 165 (2006)
15. Günther, C.W., van der Aalst, W.M.P.: Fuzzy Mining – Adaptive Process Simplification Based on Multi-perspective Metrics. In: Alonso, G., Dadam, P., Rosemann, M. (eds.) BPM 2007. LNCS, vol. 4714, pp. 328–343. Springer, Heidelberg (2007)
16. Hartigan, J.: Clustering Algorithms. John Wiley and Sons, New York (1975)

17. Henderson-Sellers, B., Gonzalez-Perez, C.: Granularity in Conceptual Modelling: Application to Metamodels. In: Parsons, J., Saeki, M., Shoval, P., Woo, C., Wand, Y. (eds.) ER 2010. LNCS, vol. 6412, pp. 219–232. Springer, Heidelberg (2010)

18. Hepp, M., Leymann, F., Domingue, J., Wahler, A., Fensel, D.: Semantic Business Process Management: A Vision Towards Using Semantic Web Services for Business Process Management. In: ICEBE, pp. 535–540. IEEE Computer Society, Los Alamitos (2005)

19. Indulska, M., Recker, J., Rosemann, M., Green, P.: Business Process Modeling: Current Issues and Future Challenges. In: van Eck, P., Gordijn, J., Wieringa, R. (eds.) CAiSE 2009. LNCS, vol. 5565, pp. 501–514. Springer, Heidelberg (2009)

20. Keller, G., Teufel, T.: SAP R/3 Process Oriented Implementation. Addison-Wesley Longman Publishing Co., Inc., Boston (1998)

21. Kühne, T.: Matters of (Meta-) Modeling. Software and Systems Modeling 5(4), 369–385 (2006)

22. Kühne, T.: Contrasting Classification with Generalisation. In: APCCM 2009. CRPIT, vol. 96 (January 2009)

23. Mendling, J., Verbeek, H., van Dongen, B., van der Aalst, W.M.P., Neumann, G.: Detection and Prediction of Errors in EPCs of the SAP Reference Model. Data Knowl. Eng. 64(1), 312–329 (2008)

24. Polyvyanyy, A., Smirnov, S., Weske, M.: The Triconnected Abstraction of Process Models. In: Dayal, U., Eder, J., Koehler, J., Reijers, H.A. (eds.) BPM 2009. LNCS, vol. 5701, pp. 229–244. Springer, Heidelberg (2009)

25. Reijers, H.A., Mendling, J.: Modularity in Process Models: Review and Effects. In: Dumas, M., Reichert, M., Shan, M.-C. (eds.) BPM 2008. LNCS, vol. 5240, pp. 20–35. Springer, Heidelberg (2008)

26. Rosemann, M.: Potential Pitfalls of Process Modeling: Part A. Business Process Management Journal 12(2), 249 (2006)

27. Sadiq, W., Orlowska, M.E.: Analyzing Process Models Using Graph Reduction Techniques. Information Systems 25(2), 117–134 (2000)

28. Salton, G., Wong, A., Yang, C.S.: A Vector Space Model for Automatic Indexing. Communications of the ACM 18(11), 613–620 (1975)

29. Schaeffer, S.E.: Graph Clustering. Computer Science Review 1(1), 27–64 (2007)

30. Sharp, A., McDermott, P.: Workflow Modeling: Tools for Process Improvement and Applications Development. Artech House Publishers, Boston (2008)

31. Smirnov, S., Dijkman, R., Mendling, J., Weske, M.: Meronymy-Based Aggregation of Activities in Business Process Models. In: Parsons, J., Saeki, M., Shoval, P., Woo, C., Wand, Y. (eds.) ER 2010. LNCS, vol. 6412, pp. 1–14. Springer, Heidelberg (2010)

32. Smirnov, S., Reijers, H.A., Nugteren, T., Weske, M.: Business Process Model Abstraction: Theory and Practice. Technical Report 35, Hasso Plattner Institute (2010), http://bpt.hpi.uni-potsdam.de/pub/Public/SergeySmirnov/abstractionUseCases.pdf

33. Vanderfeesten, I.T.P., Reijers, H.A., Mendling, J., van der Aalst, W.M.P., Cardoso, J.: On a Quest for Good Process Models: The Cross-Connectivity Metric. In: Bellahsène, Z., Léonard, M. (eds.) CAiSE 2008. LNCS, vol. 5074, pp. 480–494. Springer, Heidelberg (2008)

34. Vanhatalo, J., Völzer, H., Koehler, J.: The Refined Process Structure Tree. In: Dumas, M., Reichert, M., Shan, M.-C. (eds.) BPM 2008. LNCS, vol. 5240, pp. 100–115. Springer, Heidelberg (2008)

35. Weidlich, M., Dijkman, R.M., Mendling, J.: The ICoP Framework: Identification of Correspondences between Process Models. In: Pernici, B. (ed.) CAiSE 2010. LNCS, vol. 6051, pp. 483–498. Springer, Heidelberg (2010)

On the Automatic Labeling of Process Models

Henrik Leopold[1], Jan Mendling[1], and Hajo A. Reijers[2]

[1] Humboldt-Universität zu Berlin, Unter den Linden 6, 10099 Berlin, Germany
{henrik.leopold,jan.mendling}@wiwi.hu-berlin.de
[2] Eindhoven University of Technology
PO Box 513, 5600 MB Eindhoven, The Netherlands
h.a.reijers@tue.nl

Abstract. Process models are essential tools for managing, understanding and changing business processes. Yet, from a user perspective they can quickly become too complex to deal with. Abstraction – aggregating detailed fragments into more coarse-grained ones – has proven to be a valuable technique to simplify the view on a process model. Various techniques that automate the decision of which model fragments to aggregate have been defined and validated by recent research, but their application is hampered by the lack of abilities to generate meaningful names for such aggregated parts. In this paper, we address this problem by investigating naming strategies for individual model fragments and process models as a whole. Our contribution is an automatic naming approach that builds on the linguistic analysis of process models from industry.

1 Introduction

Business process management is a concept for enabling companies to cope with the increasing dynamics and challenges in a competitive business environment. A key element of process management is to map business processes to models in order to leverage understanding, analysis and improvement of processes. Today, many larger enterprises possess an extensive documentation of their business process in terms of several thousand models, often at a significant level of detail [1]. In order to make large and detailed models easier to understand, recent research has developed automatic abstraction techniques to generate coarse-grained model parts from more detailed ones [2,3].

The essential idea of abstraction is to identify fragments of a model that can be aggregated into a single activity. While this is valuable to reduce the structural complexity of a large model, existing techniques do not address how a suitable name for an aggregated part can be established. When using abstraction to render a high-level view of a process model for a human reader, which is the most popular use case for abstraction [4], this is troublesome. In this paper, we address the naming problem of aggregated model parts from the perspective of naming *a whole process model*. A complete process model is as much a collection of activities with mutual control-flow dependencies as an aggregated process

H. Mouratidis and C. Rolland (Eds.): CAiSE 2011, LNCS 6741, pp. 512–520, 2011.

model part is, although it is evidently not part of any higher-level model. Since in many industrial settings entire process models *themselves* carry names that convey indications of the business procedure that they capture, the underlying process model naming conventions are a valuable source of inspiration on how to name model parts. Our contribution is an automatic approach for generating name suggestions for a process model based on its events and activities, which is applicable to process model fragments as well. From a practical point of view, this approach paves the way to an integrated and automated abstraction of process models, in pursuit of the communication advantages associated with skilfully decomposed process models [5].

Against this background, the paper is structured as follows. Section 2 discusses the problem of assigning a meaningful name to a process and identifies a list of naming strategies in models from practice. Section 3 describes the different phases of our approach to generate process model names. Section 4 discusses our contribution in the light of related work. Finally, Section 5 summarizes the findings and provides an outlook on future research.

2   Naming of Process Models in Practice

Before considering an automatic approach for generating process model names, we have to understand how modelers assign names to process models. Guidelines exist on how activity names should be composed, e.g. [6] suggest a verb-object style putting the action first followed by the corresponding business object. While such guidelines advocate a certain *grammatical* structure of naming, they do not deal with its *content* by refraining to mention how to choose a particular verb or business object in the name of the model. In Section 2.1, we introduce Event-Driven Process Chains (EPCs), the process modeling language that we consider in this paper, and discuss directions for choosing a name for a model. In Section 2.2, we inspect three sets of process models from practice in order to identify strategies of naming. These strategies provide us with the foundation to automatically generate names for a fragment or the whole EPC.

2.1   Event-Driven Process Chains

An EPC is a graph-structured process model, which consists of different types of nodes: functions, events, and connectors. Functions define the business activities that have to be executed while events define the pre-conditions and post-conditions for starting a function. Figure 1 shows an example EPC from the SAP reference model with two functions (rounded boxes) and four events (hexagons). Functions and events are connected via control flow arcs in an alternating way. Complex routing is defined using connectors (circles). In the example, we observe an OR-split connector (symbol ∨) creating two end events. An EPC has at least one start event (no incoming arc) and at least one end event (no outgoing arc).

To illustrate the naming problem addressed in this paper, we re-consider the example EPC from Figure 1. One approach for naming the entire process would be to consider the different activities of the model. The two functions relate

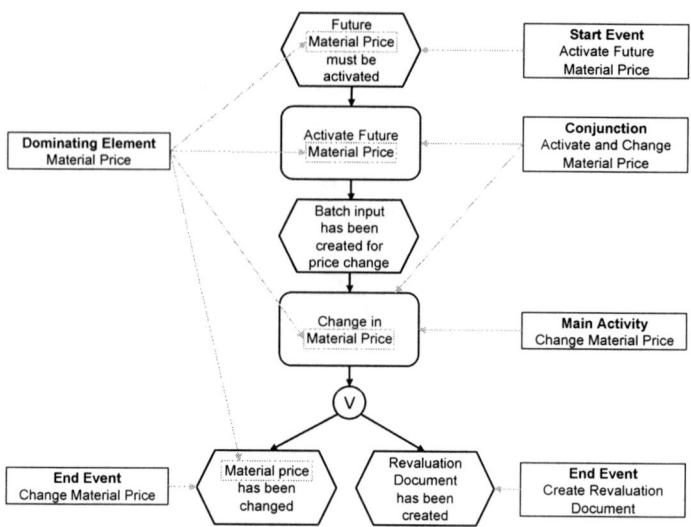

Fig. 1. Potential sources of process names for model *Change in Material Price*

to the business object *material price*, in respective connection with the actions *activate* and *change*, which therefore could be used as central elements of the overall name. Another option would be to look at the names of the end events since they refer to what is actually accomplished by executing the process. Hence, the *creation* of the *revaluation document* may represent another naming option. This discussion aims to emphasize that there are many options, and they are difficult to rank without prior domain knowledge. In the following part, we aim to systematically identify different strategies for choosing a name for a process.

2.2   Classification of Naming Strategies

We collected an extensive set of process models from practice in order to identify naming strategies. We used three different model collections, aiming in this respect for broad diversity in the underlying domains. First, we had access to the *SAP Reference Model* [7, pp. 145-164], a collection containing 604 EPCs organized in 29 functional branches of an enterprise such as sales, accounting and other areas, with a total number of 2,433 activity labels. Second, we have used a model collection from a *European utility vehicle manufacturer* consisting of one main procurement process with nine sub-processes and altogether 115 activity labels. Third, we inspected a model collection containing the incident management process from an *international IT service provider*. The process is captured as an EPC on three abstraction layers, containing 88 sub-processes and 293 activity labels.

We analyzed the names in the model sets by first identifying action and business objects in the name of the process model. Then, we used a self-developed tool that identified linguistic relations between the name parts and the activity and event labels. As a final step, we inspected each model manually. Based on this procedure, we were able to develop a classification of naming strategies. Five different approaches were observed including Dominating Element, Main Activity, Start or End Events, Conjunction of Activities, and Semantic Naming. Figure 1 illustrates some of these strategies, which we now aim to describe.

1. **Dominating Element:** If one particular business object or action is mentioned more often in activity labels than any other business object or action, this element is considered to be a dominating element. In the analyzed process model collections dominating elements were often used for naming the process model if such a dominating element existed. The example in Figure 1 has *Material Price* as such a dominating element.

2. **Main Activity:** Some of the analyzed processes contain one particular activity that is of central importance. The remaining activities have the character of side activities supporting, preparing or evaluating the result of this activity. Figure 1 is a good example since *Change in Material Price* has such characteristics. The process also contains an activity which is concerned with activating the future material price. However, from the choice of the modeler we can assume that this activity only plays a subordinate role, while the focus is on the activity of changing the material price.

3. **Start or End Events:** Especially when the state at the beginning or at the end of process define the overall goal of the process, the name of the whole model may be closely related to them. In Figure 1 this is visible in the start event *Future Material Price must be activated* and the end event *Material price has been changed.*

4. **Conjunction of Activities:** If the same action is performed on different business objects or different actions are applied on the same business object, these activities can be easily described in terms of a conjunction. Even whole activity labels may be connected if the resulting name is not too complex. For Figure 1, this would yield *Activate and Change Material Price.*

5. **Semantic Naming:** The previously introduced concepts always explicitly refer to the textual description of at least one element in the process model. By contrast, the concept of semantic naming does not refer to one or more model elements, but uses the broader context of the activities for naming the process model. This can be appropriate if there is a part-of relationship between the activities and the name of the process [8]. Hence, the process name, which is itself representing an activity, subsumes the given activities in the model. As an example, consider the SAP Process *Shipping*. It consists of five events and the two activities *Delivery for Returns* and *Goods Receipt Processing for Returns*. Apparently, the action shipping is not mentioned in any of the process elements. Nevertheless, *shipping* can be considered as a more general concept in semantic terms, which implies delivery and goods receipt. Clearly, the derivation of semantic names requires external knowledge, e.g. in terms of an ontology, and cannot be directly derived from the activity names.

3   An Automatic Approach to Generate Process Names

In this section we present an automatic approach for the generation of process model names, which builds on the strategies identified in the previous section. The main idea of the approach is to derive a set of potentially useful names for a given process model based on its activities and events. Subsequently, a modeler can select the most suitable name.

We organize our approach in three steps. Phase 1 serves as preparation for the main information extraction. In this phase all activities, start events and end events of the given process model are annotated with their respective action and business object. For this step, we use an algorithm for automatically identifying action and business objects from activity labels as presented in [9], and extended it with a capability to analyze different start and end event structures as defined in [10]. In Phase 3, the name candidates are transformed automatically to the verb-object style based on the techniques defined in [9]. In the following subsections, we introduce the specific techniques to generate the different proposals in Phase 2, as well as their interdependence. All of these assume that annotations of actions and business objects are available from Phase 1. The reader may wish to refer to [11] for the pseudo algorithms that abstract from our implementation in Java.

The *Dominating Element Extraction* technique investigates whether the given process model includes a dominating action and dominating business object. Therefore, for each element type, i.e. action or business object, the occurrence of the elements among all activities of the model is checked. If there exists an element that has a higher occurrence than all other elements among this type, it is saved as dominating element. If a dominating element has been identified, it can be used as input for the *Subordinate Element Extraction* technique. In case no dominating element could be detected, the further steps are limited to executing the *Event Extraction* and the *Main Activity Extraction* techniques, which do not require the input of a dominating element.

If one type of dominant element was detected, the *Subordinate Element Extraction* technique identifies those actions or business objects with which the dominant element is connected in the given process model. Therefore, all activities containing the dominating element are scanned and the complementing element is both extracted and saved to a list of subordinate elements. If, for instance, the dominating action *analyze* was derived from the two activities *Order Analysis* and *Program Analysis*, the subordinate elements are given by the business objects *order* and *program*. Hence, all activities containing the dominating element are selected and the subordinate elements are derived.

We introduce two techniques for constructing a process name based on the set of subordinate elements and the dominating element: *Lexical Conjunction* and *Logical Conjunction*. In case of the Lexical Conjunction the subordinate actions or business objects are replaced with a newly introduced element. In particular, the lexical database WordNet is consulted to detect common *holonyms* and *hypernyms* of the subordinate elements. If a proper holonym – a word that is more generic than a given word – or a hypernym is found – a word that is more specific than a given word – a name proposal is constructed using the dominating element and the according holonym or hypernym. In case of the Logical Conjunction, the subordinate elements are simply connected using the logical operators *and* or *or*.

Another technique that builds on the identified dominating elements is *Label Repository*. This technique uses the activities of other process models to build up a label repository. If a dominating element was identified with the *Dominating Element Extraction* technique, such a label repository can be consulted to find a corresponding element which is likely to be connected with the dominating element. As an example, consider the SAP process model *Capacity Planning* containing the dominating business object *Capacity*. As this business object in isolation is not a very comprehensive process name, the repository can be consulted. By browsing the repository for labels containing *Capacity*, we obtain, amongst others, the process name *Capacity Planning* which perfectly matches the original process name.

The *Event Extraction* technique derives potentially useful names from start and end events. Therefore, start and end events are inspected on their merit to provide information about the model content. That decision is based on the usage of particular signal terms given in the event label. For instance, it is not very probable that the term *was* in the start event *Asset was found* indicates what is *to be* performed in the process; rather, it represents a state that was required for triggering the first activity. By contrast, the term *is to be* in the start event *Asset is to be created* captures the necessity for the execution of a *subsequent* action within the process of consideration. Hence, (1) the identified start events are reduced to those where the signal term indicates that the event actually contains information about what is *going* to happen and (2) the end events are restricted to those that signal what *has* happened in the process. Based on an extensive classification of these terms from the investigated process model collections, this decision can be made in an automated fashion.

Referring to the main activity approach as briefly mentioned in Section 2.2, we further introduce the *Main Activity Extraction* technique. The objective of this technique is to automatically decide whether a considered activity represents a main activity for the given process model or not. In order to be able to make this decision for an individual activity, it is necessary to automatically derive the context of the process and subsequently decide about the role of the activity. This approach utilizes the insight of our analysis that approximately 85% of the main activities are found either at the beginning or at the end of the process. Accordingly, the main activity extraction presumes the existence of a main activity in the first or last position and selects the according activity labels as process name proposals.

In order to obtain an all-encompassing approach, we combined all techniques. To some extent the order of executing the techniques is fixed as some depend on the on the input of other, like the *Lexical Conjunction*. However, techniques such as *Main Activity Extraction* can be executed independently from other techniques and can be executed at any time.

4   Related Work

Several approaches have been proposed for process model abstraction. The work by Polyvyanyy et al. builds on an algorithm for aggregating activities based on a slider and specific abstraction criteria [3]. Abstraction criteria are discussed in [12,13]. A recent paper presents an abstraction approach based on behavioral profiles [14]. For a set of activities, this approach generates the control flow of the aggregated model. Both approaches do not generate names for aggregated activities, such that our work is complementary. A different approach based on meronymy relations is presented in [15]. This approach inspects meronymy relations between activity labels to find aggregation candidates. It integrates the problems of finding aggregation candidates and aggregation names. Our work is more general in that sense that it is able to derive names for arbitrary process model fragments, even if they do not share a meronymy relation.

The linguistic analysis of activity labels is also an import task in process model matching and similarity calculation [16,17,18]. Different approaches of matching process models are integrated in [19]. This area is also related to research on semantic annotation of business process models [20]. Recent research has also started using natural language processing techniques for generating process models from text. Gonçalves et al. generate process models from group stories [21]. The approach by the University of Klagenfurt combines linguistic analysis with user feedback [22]. The Rapid Business Process Discovery (R-BPD) framework uses natural language techniques for constructing BPMN models from corporate documentations or web-content [23]. Anaphora resolution is tackled in a recent approach to generate BPMN models [24].

5   Conclusion

In this paper, we have addressed the problem of automatically generating names for process models. Our work is motivated by the fact that existing works on process model abstraction require telling names for structurally aggregated process fragments. Our overall contribution is an automatic naming approach that builds on the linguistic analysis of the elements of process models from industry. The work presented in this paper has significant implications for research and practice. The automatic generation provides the basis not only for proposing names of whole processes, but also for process fragments. In this regard, our approach can be used for instance to dynamically generate abstractions of different granularity as the user is interacting with the modeling tool.

The main task for future research is the validation of the presented approach. This may include the comparison of the given with the generated names but also an applicability assessment by humans. In addition, we aim to further investigate the usability of different naming strategies. Currently, if a single name for an abstracted fragment is needed, a system can only make a random suggestion from the set of name proposals. We expect that the strategy itself, but also the length of the suggested name has a significant impact on the perceived usefulness. Based on such insight, we will be able to select the best name from a set of suggestions.

References

1. Rosemann, M.: Potential pitfalls of process modeling: part a. Business Process Management Journal 12(2), 249–254 (2006)
2. Eshuis, R., Grefen, P.: Constructing customized process views. Data Knowl. Eng. 64(2), 419–438 (2008)
3. Polyvyanyy, A., Smirnov, S., Weske, M.: Process model abstraction: A slider approach. In: Proceedings of EDOC (2008)
4. Smirnov, S., Reijers, H.A., Nugteren, T., Weske, M.: Business Process Model Abstraction: Theory and Practice. Technical Report 35 (2010)
5. Reijers, H.A., Mendling, J.: Modularity in process models: Review and effects. In: Dumas, M., Reichert, M., Shan, M.-C. (eds.) BPM 2008. LNCS, vol. 5240, pp. 20–35. Springer, Heidelberg (2008)
6. Mendling, J., Reijers, H.A., Recker, J.: Activity labeling in process modeling: Empirical insights and recommendations. Inf. Syst. 35(4), 467–482 (2010)
7. Keller, G., Teufel, T.: Sap R/3 Process Oriented Implementation. Addison-Wesley Longman Publishing Co., Inc., Boston (1998)
8. Reijers, H., Limam, S., van der Aalst, W.: Product-based workflow design. Journal of Management Information Systems 20(1), 229–262 (2003)
9. Leopold, H., Smirnov, S., Mendling, J.: Refactoring of process model activity labels. In: Hopfe, C.J., Rezgui, Y., Métais, E., Preece, A., Li, H. (eds.) NLDB 2010. LNCS, vol. 6177, pp. 268–276. Springer, Heidelberg (2010)
10. Decker, G., Mendling, J.: Process instantiation. Data Knowl. Eng. 68(9), 777–792 (2009)
11. Leopold, H.: Modularization of business process models using natural language techniques. Master's thesis, Humboldt-Universität zu Berlin (2010)
12. Eshuis, R., Grefen, P.: Constructing Customized Process Views. Data Knowl. Eng. 64(2), 419–438 (2008)
13. Günther, C.W., van der Aalst, W.M.P.: Fuzzy Mining – Adaptive Process Simplification Based on Multi-perspective Metrics. In: Alonso, G., Dadam, P., Rosemann, M. (eds.) BPM 2007. LNCS, vol. 4714, pp. 328–343. Springer, Heidelberg (2007)
14. Smirnov, S., Weidlich, M., Mendling, J.: Business process model abstraction based on behavioral profiles. In: Maglio, P.P., Weske, M., Yang, J., Fantinato, M. (eds.) ICSOC 2010. LNCS, vol. 6470, pp. 1–16. Springer, Heidelberg (2010)
15. Smirnov, S., Dijkman, R., Mendling, J., Weske, M.: Meronymy-based aggregation of activities in business process models. In: Parsons, J., Saeki, M., Shoval, P., Woo, C., Wand, Y. (eds.) ER 2010. LNCS, vol. 6412, pp. 1–14. Springer, Heidelberg (2010)
16. Ehrig, M., Koschmider, A., Oberweis, A.: Measuring similarity between semantic business process models. In: Proceedings of APCCM, pp. 71–80 (2007)
17. van Dongen, B.F., Dijkman, R., Mendling, J.: Measuring similarity between business process models. In: Bellahsène, Z., Léonard, M. (eds.) CAiSE 2008. LNCS, vol. 5074, pp. 450–464. Springer, Heidelberg (2008)
18. Dijkman, R., Dumas, M., van Dongen, B., Käärik, R., Mendling, J.: Similarity of business process models: Metrics and evaluation. Inf. Systems 36, 498–516 (2010)
19. Weidlich, M., Dijkman, R., Mendling, J.: The iCoP framework: Identification of correspondences between process models. In: Pernici, B. (ed.) CAiSE 2010. LNCS, vol. 6051, pp. 483–498. Springer, Heidelberg (2010)
20. Lin, Y., Ding, H.: Ontology-based semantic annotation for semantic interoperability of process models. In: Proc. of CIMCA/IAWTIC, pp. 162–167 (2005)

21. de AR Gonçalves, J.C., Santoro, F.M., Baiao, F.A.: A case study on designing processes based on collaborative and mining approaches. In: Int. Conf. CSCWD (2010)
22. Kop, C., Vöhringer, J., Hölbling, M., Horn, T., Mayr, H.C., Irrasch, C.: Tool supported extraction of behavior models. In: ISTA, pp. 114–123 (2005)
23. Ghose, A.K., Koliadis, G., Chueng, A.: Rapid business process discovery (R-BPD). In: Parent, C., Schewe, K.-D., Storey, V.C., Thalheim, B. (eds.) ER 2007. LNCS, vol. 4801, pp. 391–406. Springer, Heidelberg (2007)
24. Friedich, F., Mendling, J., Puhlmann, F.: Process model generation from natural language text. In: Proc. of CAISE. LNCS (2011)

Pattern-Based Modeling and Formalizing of Business Process Quality Constraints

Lial Khaluf, Christian Gerth, and Gregor Engels

Department of Computer Science
University of Paderborn
Paderborn, Germany
lial.khaluf@googlemail.com,
{gerth,engels}@uni-paderborn.de

Abstract. The quality of business processes can be checked by verifying their compliance with specific quality constraints. These constraints represent a set of required temporal and logical relationships between different steps of business processes. Quality constraints are usually formulated as informal texts, which makes them difficult to be verified, when business processes become complex. One way to solve this problem is by automating the verification of quality constraints on business processes by applying model checking. To apply model checking, both business processes and quality constraints have to be formalized. In this paper, we define a new visual language for modeling quality constraints and we provide a pattern-based translation for quality constraint models into Computation Tree Logic formulas.

Keywords: business process, quality constraint, visual pattern, CTL-formula.

1 Introduction

One of the most important factors of the success and reputation of any business is the quality of products and services it provides. For this reason, quality management has become an important competitive factor that must be considered on all levels including business processes. In this context, many standards were developed for total quality management, aiming at fulfilling the requirements of customers and improving the quality of products, as e.g. ISO 9001 regulations and constraints [1], which can be applied to any business. Quality constraints may be defined by producers or by customers. No matter who defines quality constraints, there must be a way to ensure that business processes satisfy them. However, since standard or user-defined quality constraints are usually documented as informal texts, it becomes difficult to prove their correctness, especially when business processes are complex. One solution for this problem is automating the verification of quality constraints on business processes by using the technique of model checking. To apply this technique, both business processes and quality constraints have to be formalized. To achieve this goal, many approaches were developed, where each one depends on a different temporal logic to formalize quality constraints, in order to enhance and increase their expressiveness. However, the major problem is still that no approach allows to formalize user-defined quality

H. Mouratidis and C. Rolland (Eds.): CAiSE 2011, LNCS 6741, pp. 521–535, 2011.
© Springer-Verlag Berlin Heidelberg 2011

constraints, which include a non-deterministic future, as e.g. formalizing a quality constraint or a part of it which must not hold for all control flows in a business process, but for at least one control flow.

Our goal in this paper is to overcome the expressiveness limitations concerning the branching logic, by allowing to formalize non-deterministic user-defined quality constraints, and to make this formalization intuitively understandable for business process users. For this reason, we have investigated the formalization approaches, which are based on UML Activity Diagrams [3], since these diagrams are a widely used standard and are familiar to business process users. One of these approaches is developed in [2]. It models business processes as UML 2.0 Activity Diagrams, and quality constraints as business process patterns using the Process Pattern Specification Language (PPSL) [4], which is a light weight extension of UML Activity Diagrams. Both business processes and business process patterns are then written in a formal way by transforming UML 2.0 Activity Diagrams into labeled transition systems, and translating business process patterns into LTL-formulas [5]. This enables the automated verification of quality constraints on business processes by using a model checker to verify LTL formulas on labeled transition systems. However, the Linear Temporal Logic (LTL) does not support the non-deterministic future. To support this kind of future, we replace LTL by the Computation Tree Logic (CTL) [6]. To achieve that, we extend PPSL to a new visual modeling language, Extended Process Pattern Specification Language (EPPSL), which has a formally defined semantics given by a translation into CTL-formulas. EPPSL is a heavy weight extension of UML Activity Diagrams. In other words, EPPSL uses the elements of UML Activity Diagrams, which semantics can serve to model quality constraints. It also defines new classes of elements to cover the semantics of quality constraints which are not defined by UML Activity Diagrams. We also provide a pattern-based translation for EPPSL models into CTL-formulas. In the following section, we give an overview of the related work. In Section 3, we provide a scenario for verifying business process quality constraints. In Section 4, we describe the modeling elements of EPPSL and how to use them for composing EPPSL models. In Section 5, we explain how to translate EPPSL models into CTL-formulas. In Section 6, we provide a conclusion and outlook for our approach.

2   Related Work

Modeling and formalizing business process constraints in order to verify their correctness are considered in several approaches. For example, PPSL is introduced in [4] and translated in [2] into LTL formulas. In [7], the graphical Business Property Specification Language is used to capture business process compliance rules, which are translated into LTL. In [8], DecSerFlow is mapped on LTL. DECLARE is defined in [9] and [10]. DECLARE models are also translated into LTL. BPMN-Q is developed in [11] to model requested compliance rules on business processes as queries. These queries are translated in [12] into PLTL [13][14] formulas. BPMN is used in [15] to check the semantic correctness of business process models by mapping them to Petri nets. In [16], the Object Constraint Language [17] expressions are used to refer to an integrated meta-model for different process models. In [18], process-independent compliance rules are specified using graph structures and formalized in terms of FOL [19]. All the previous approaches are not sufficient to express constraints, which contain a non-deterministic future and which have the degree of complexity required by users. But, in our approach, we define EPPSL which allows the users to build high complex shapes of quality constraints and we support the non-deterministic future by formalizing quality constraints as CTL-formulas. In [20], a lightweight, analyst-mediated approach introduces compliance patterns in terms of CTL as a heuristic basis for resolving the non compliance of process models, but it does not provide a way to map informal user-defined compliance rules into CTL-formulas. Additionally, expressing the non-deterministic future is limited to a set of structural and semantic patterns, which have a predefined shape of rules, but in our approach, we define EPPSL to enable modeling and formalizing quality constraints which can reach the complexity of CTL expressions.

3   Scenario

We assume that a company wants to hire new employees. The company has developed a business process to model the activities that must be carried out to accept or refuse applications. Fig. 1 shows the business process modeled as a UML Activity Diagram. To ensure the quality of this business process, some quality constraints may have to be checked on it, e.g. if a business analyst wants to know whether the lack of employees could result in accepting the application, then the correctness of QC_1 must be checked (We use an abbreviation QC to refer to a quality constraint).

QC_1: *"It is always the case, that when the number of required employees has not yet been reached, then there exists a possibility to accept the application"*.

By looking at the business process model, we see that it satisfies QC_1, since whenever we encounter the Guard "the number of required employees has not yet been reached", there exists at least one control flow, on which this Guard is followed by the Action "accept the application" where in between other Actions may occur and other Guards maybe be satisfied. However, verifying a textual quality constraint by looking at the business process model is prone to errors, especially when business processes become complex. In order to be able to verify QC_1 automatically on the business process model in Fig. 1 , we have to model QC_1 using specific patterns, which have a formally defined semantics given by a translation into temporal logic formulas.

The expression *"there exists a possibility"* in QC_1 states that the lack of employees enables the possibility of accepting the application, but does not enforce it. In other words, QC_1 states that accepting the application could be a non-deterministic future for the lack of employees. In order to have the ability to express this kind of future, it is not enough that we model QC_1 using specific patterns, which have a formally defined semantics given by a translation into temporal logic formulas, but also the temporal logic formulas should have the ability to express a non-deterministic future.

In this paper, we define a new modeling language EPPSL, which provides the patterns required by the previous scenario, since these patterns have a formally defined semantics given by a translation into CTL-formulas. We specify EPPSL by a meta model which we provide in [21]. In the following section, we introduce how to model quality constraints with EPPSL.

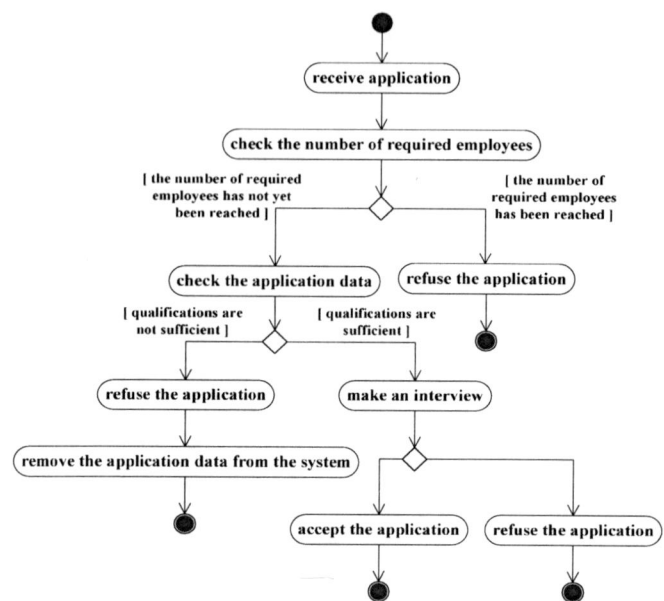

Fig. 1. Example of a business process model

4   Modeling Quality Constraints with EPPSL

We model a quality constraint by modeling its basic blocks and its temporal and logical relationships. The basic blocks could be actions, guards, anonymous steps, and partial quality constraints. The temporal and logical relationships are based on the semantics of the temporal and logical relationships of CTL [6] to provide the ability to express both deterministic and non-deterministic futures. In the following, we explain the semantics of both basic blocks and relationships and how to model them.

4.1   Modeling Basic Blocks

A quality constraint may include a mixture of the following basic blocks:

Actions: An action is an activity which is carried out by the system, the customer or any other entity in the business process. For example, QC_1 consists of the action "accept the application". The start of a business process and the end of a control flow in a business process are also considered to be actions. EPPSL uses the modeling elements "Action" (Fig. 2.a), "InitialNode" (Fig. 2.b), and "ActivityFinalNode" (Fig. 2.c) to model actions, the start of a business process, and the end of a control flow in a business process, respectively.

Guards: A guard is a condition which can be false or true. For example, QC_1 consists of the Guard "the number of required employees has not yet been reached". EPPSL uses the modeling element "Guard" (Fig. 2.d) to model guards.

Anonymous steps: A quality constraint may refer to an anonymous step which could be an unknown action or an unknown guard. For example, a quality constraint might be dedicated to ensure that the business process model in Fig. 1 includes a possibility to accept the application after 6 steps from starting the process without any need to know to which actions or guards these steps are referring. Quality constraints which counts anonymous steps are useful if the number of steps refers e.g. to the time consumed or the money paid to perform these steps, or if it plays a role in the satisfaction of the customer. For this reason, we introduce in EPPSL a modeling element called "AnonymousStep" (Fig. 2.e).

Partial quality constraints: A partial quality constraint is a quality constraint which is linked to other partial quality constraint(s) with a logical relationship to build another quality constraint. We use the concept of the partial quality constraint, since we need sometimes to model a quality constraint which consists of several quality constraints that are logically related, but temporally not related. For example, a quality constraint might be dedicated to ensure that if the business process model in Fig. 1 includes a possibility to make an interview, then it includes no possibility to accept the application online. The first possibility is not temporally related to the second one. However, they are logically related, since the first possibility implies the negation of the second one. We model partial quality constraints as separated units. For this reason, we provide in EPPSL a modeling element called "ConstraintContainer" (Fig. 2.f) which is dedicated to contain a partial quality constraint model separating it temporally from other partial quality constraint models.

The elements in Fig. 2.a, Fig. 2.b, Fig. 2.c, and Fig. 2.d are the same elements used by UML 2.0 Activity Diagrams to model Actions, InitialNodes, ActivityFinalNodes, and Guards. The elements in Fig. 2.e and Fig. 2.f are new classes defined by EPPSL.

a) Action   b) InitialNode   c) ActivityFinalNode   d) Guard   e) AnonymousStep   f) ConstraintContainer

Fig. 2. EPPSL modeling elements for the basic blocks

EPPSL models for quality constraints provide the ability to link the modeling elements of the basic blocks with temporal and logical relationships. In the following, we introduce how EPPSL can model these relationships.

4.2   Modeling Temporal Relationships

A temporal relationship determines the order of actions, guards, and anonymous steps. For example, QC_2, which we want to verify on the business process model in Fig. 1, consists of a temporal relationship "After":

QC_2: *"After checking the application data, the qualifications of the applicant are considered to be either sufficient or not sufficient"*.

The temporal relationship in QC_2 states that the action "check the application data" must be followed by one of the guards "qualifications are not sufficient", or "qualifications are sufficient".

EPPSL provides a set of modeling elements to express temporal relationships. Since EPPSL considers deterministic and non-deterministic futures, it provides for each temporal relationship two modeling elements. The first one represents the relationship when it holds for all control flows of a business process (deterministic). The second one represents the relationship when it holds for at least one control flow (non-deterministic). Fig. 3 shows the notation of EPPSL modeling elements for temporal relationships. These elements are new classes defined by EPPSL.

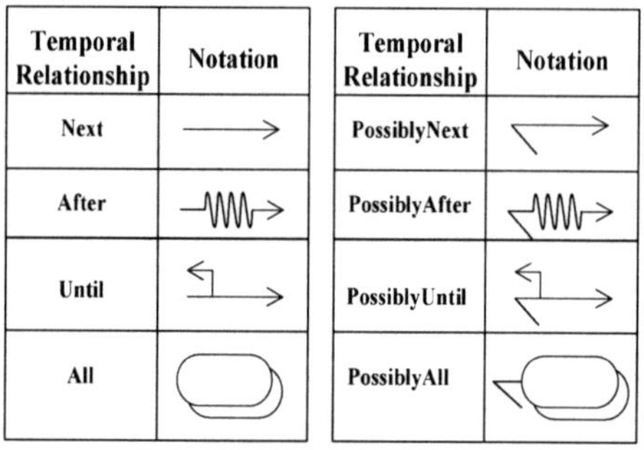

Temporal Relationship	Notation	Temporal Relationship	Notation
Next	⟶	PossiblyNext	⟶
After	⟶	PossiblyAfter	⟶
Until	⟵⟶	PossiblyUntil	⟵⟶
All	⬭	PossiblyAll	⬭

Fig. 3. EPPSL modeling elements for temporal relationships

Deterministic Temporal Relationships: The deterministic temporal relationships are "Next", "After", "Until", and "All". "Next" and "After" may link two basic blocks, which could be "Actions", "Guards" or "AnonymousSteps". "Next" states that the first block must be followed next by the second block on all control flows of a business process. "After" states that the first block must be followed by the second block on all control flows of the business process, no matter if other blocks occur between the first and the second block. "Until" connects two basic blocks, which could be two "Actions", or an "Action" and a "Guard". "Until" states that an action must be repeated on all control flows of a business process until another action takes place or a guard is satisfied. "All" refers to all instances of an action on all control flows of a business process. "All" is usually used to confirm that a quality constraint which includes an action, to which the temporal relationship "All" is applied, must hold for all instances of that action on all control flows of a business process.

For example, we want to verify QC_3 on a business process model for using a bank card to withdraw money.

QC_3: *"The pin number must always be entered repeatedly until the pin number is correct"*.

The EPPSL model in Fig. 4 models QC_3. It states that all instances of the action "enter the pin number" must be repeated on all control flows of the business process until the guard "pin number is correct" is satisfied. In this model, we have applied

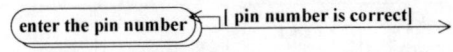

Fig. 4. EPPSL quality constraint model

an "All" temporal relationship on the action "enter the pin number", since the word "always" in QC_3 states that the temporal relationship "Until" must hold for all instances of the action "enter the pin number".

Non-Deterministic Temporal Relationships: The non-deterministic temporal relationships are "PossiblyNext", "PossiblyAfter", "PossiblyUntil", and "PossiblyAll". These relationships have the same semantics of "Next", "After", "Until", and "All" respectively with one different aspect that they must not hold for all control flows of a business process, but for at least one control flow. For example, we want to verify QC_4 on the business process model in Fig. 1.

QC_4: *"There exists a possibility after starting the process to check the application data which is possibly followed by making an interview"*.

The EPPSL model in Fig. 5 models QC_4. It states that there exists at least one control flow on which starting the process is followed by checking the application data which is followed on at least one control flow by making an interview.

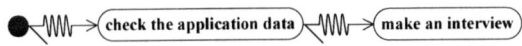

Fig. 5. EPPSL quality constraint model

4.3   Modeling Logical Relationships

A logical relationship may link actions, guards, and partial quality constraints. For example, QC_2 consists of a logical relationship "Or" which combines two guards and states that either the first guard "qualifications are sufficient" or the second one "qualifications are not sufficient" must follow the action "check the application data".
EPPSL provides the following set of modeling elements to express logical relationships, which may link Actions, Guards, and ConstraintContainers:

- **Join/ForkNode:** is used to model the logical relationship "And". "Join/ForkNodes" can be used to link Actions, Guards, and ConstraintContainers. We refer to Join/Fork Nodes as control nodes.
- **Decision/MergeNode:** is used to model the logical relationship "Or". "Decision/MergeNodes" can be used to link Actions, Guards, and ConstraintContainers. We refer to Decision/MergeNodes as control nodes.
- **Not:** is used to model the "Not" logical operator. "Not" can be applied to Actions, ConstraintContainers, and all EPPSL temporal relationship modeling elements.
- **Connector:** is used to model the "Imply" logical relationship. Connectors can link between two ConstraintContainers, or between ConstraintContainers and control nodes.

Logical Relationship	Notation	Logical Relationship	Notation
Join/ ForkNode	**|**	Not	✕
Decision/ MergeNode	◇	Connector	────•

Fig. 6. EPPSL modeling elements for logical relationships

Fig. 6 shows the notation of the EPPSL modeling elements for logical relationships. The "Join/ForkNodes" and "Decision/MergeNodes" are the same control nodes used by UML 2.0 Activity Diagrams. The "Not" and "Connector" elements are new classes defined by EPPSL.

To give an example for using the logical relationships modeling elements, we assume that we want to verify QC_5 on the business process model in Fig. 1.

QC_5: *"If there exists a possibility to make an interview, then there exists no possibility to accept the application online"*.

The EPPSL model in Fig. 7 models QC_5. It includes two EPPSL models for two partial quality constraints, which are temporally not related. This means that each one of them must be checked separately on the business process. For this reason, we separate each one in a ConstraintContainer. The first model states that there exists a possibility to make an interview after starting the process. The second model states that there exists no possibility to accept the application online after starting the process. The Constraint-Containers are linked with a Connector, which means that the partial quality constraint represented by the first model must imply the partial quality constraint represented by the second model.

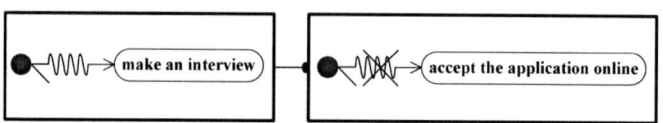

Fig. 7. EPPSL quality constraint model

5   Translation of EPPSL Models into CTL-Formulas

The Computation Tree Logic (CTL) [6] views the time as a tree. It considers all different paths, allowing the future to be non-deterministic. CTL-formulas are based on a set of atomic propositions (statements which truth value may change over time), logical connectives ($\neg$, $\wedge$, $\vee$, $\Rightarrow$), temporal operators (X: Next, F: Eventually, U: Until, G: Globally), and path quantifiers (A: On all paths, E: On at least one path). Whenever there is a temporal operator in a CTL-formula, a path quantifier must precede it. For example, if ψ is an atomic proposition, then $AX\psi$ is a CTL-formula, which states that, On all paths, ψ holds next.

In order to formalize quality constraints, we translate their EPPSL models into CTL-formulas. Here, we differentiate between two kinds of models depending on the number of control nodes they contain. Models which contain at most one control node are simple models. Models which contain more than one control node are complex models.

When we translate a simple or a complex model into a CTL-formula, we translate the basic blocks: Actions, Guards, InitialNodes, and ActivityFinalNodes into the atomic propositions: Actions labels, Guards texts, "Start", and "End" respectively. However, the strategy for constructing the CTL-formula for a simple model differs from the strategy for constructing it for a complex model. In the following, we explain the different strategies.

5.1   The Translation Strategy for Simple Models

The translation of simple models depends on analyzing them to specific EPPSL patterns, for which we provide a translation into CTL-formulas [21]. These patterns enable all required combinations of EPPSL modeling elements, in order to cover all possible semantics of quality constraints when the model contains at most one control node. For example, we want to formalize QC_6 which has to be verified on the business process model in Fig. 1:

QC_6: *"It is always true that checking the application data might be followed by refusing the application, which is followed next by removing the application data from the system"*.

First, we model QC_6 with EPPSL (see Fig. 8), then we translate the model into a CTL-formula by comparing it to the EPPSL patterns given by the translation tables, which we provide in [21].

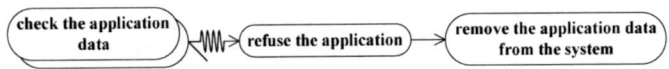

Fig. 8. EPPSL quality constraint model

Due to lack of space, we show in Fig. 9 only a part of the translation tables dedicated for simple models, where S represents an EPPSL pattern, which is attached to the element preceding S, and S^* refers to the CTL-formula, to which S is translated. C represents a partial quality constraint model, and C^* refers to the CTL-formula, to which C is translated. M represents an EPPSL modeling element, and M^* refers to the CTL-formula, to which M is translated.

We can translate simple models from right to left or left to right. Here e.g., we translate the model of QC_6 from right to left as explained by Fig. 10. In Step 1, we translate the Action "remove the application data from the system" to an atomic proposition represented by the Action label. In Steps 2, 3, 4, and 5, we divide the model into EPPSL patterns provided by the translation table in Fig. 9. This enables us to translate each pattern separately, until we reach the CTL-formula, which represents the whole model in Step 5.

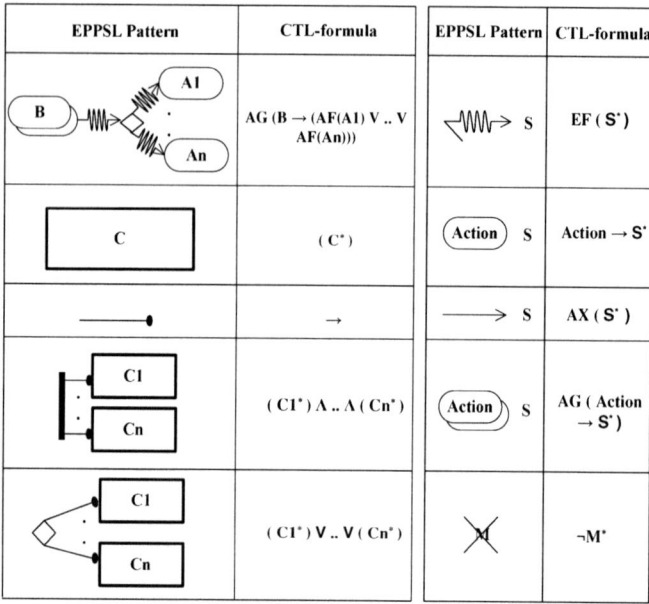

Fig. 9. Translation of EPPSL patterns into CTL-formulas

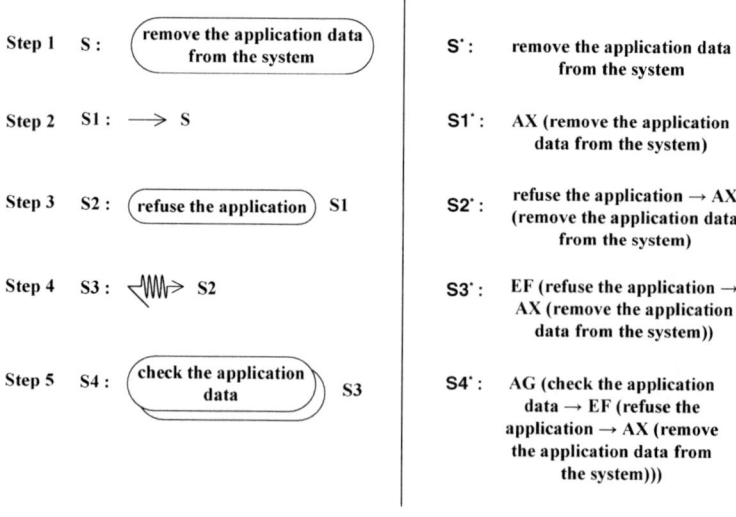

Fig. 10. Example of the translation strategy for a simple EPPSL model into a CTL-formula

5.2   The Translation Strategies for Complex Models

Logical relationships may link actions and guards, or may link partial quality constraints with respect to the basic assumptions defined in [21]. For this purpose, we define two different translation strategies for complex models. The first one is used when more than one control node link Actions and Guards (e.g. the model in Fig. 11), and the second one is used when more than one control node link ConstraintContainers (e.g. the model in Fig. 13).

Strategy 1: Given an EPPSL complex model, e.g. the complex model of QC_7 in Fig. 11, which we want to verify on the business process model in Fig. 1:

QC_7: *"Receiving an application is always followed by making an interview or refusing the application and remove the application data from the system".*

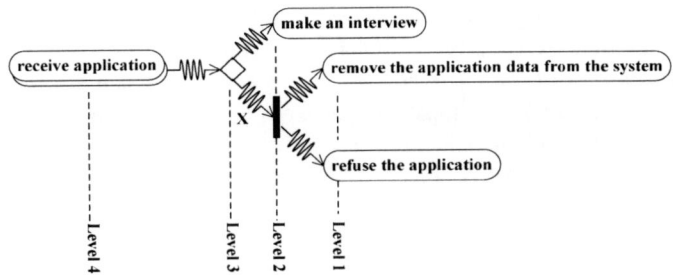

Fig. 11. Example of a complex model

We translate the model in Fig. 11 by applying the following steps:

1. We divide the model into levels numbering them in the opposite direction of the temporal relationships. In Fig. 11, we have 4 levels.
2. We assign a variable name to each temporal relationship followed and preceded directly by a control node. In Fig. 11, we have one variable X.
3. When a control node is preceded directly only by a variable with no assigned value and followed directly by temporal relationships attached only to Actions or Guards, then the node has to be translated according to the translation table dedicated for control nodes preceded by variables [21] (due to lack of space, we present only a part of it in Fig. 12) and the resulting CTL-formula is assigned to the variable. According to the table in Fig. 12, we translate the ForkNode on Level 2 in Fig. 11, and we assign its formula to X:

   X: $(\mathbf{AF}(\text{refuse the application}) \wedge \mathbf{AF}(\text{remove the application data from the system}))$
4. When a control node is followed directly only by a variable with no assigned value and preceded directly by temporal relationships attached only to Actions or Guards, then the node has to be translated according to the translation table of control nodes followed by variables [21], and the resulting CTL-formula is assigned to the variable. Fig. 11 does not encounter this case.

Control Nodes	The Value of X
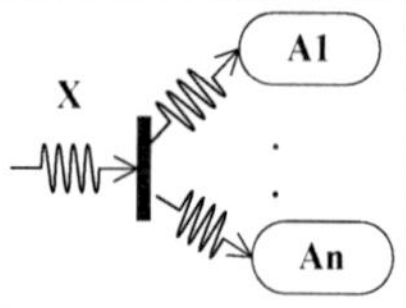 X ... A1 ... An	$(AF (A1) \wedge .. \wedge AF (An))$

Fig. 12. Example of translating control nodes preceded by a variable into CTL-formulas

5. We translate the control nodes of the model if they are yet not translated starting at the second level (the first level always includes Actions or Guards which are translated into atomic propositions). The variables which are yet not assigned a CTL-formula and followed directly by control nodes, which are in turn directly followed by Actions, Guards or variables already assigned a CTL-formula in terms of Actions or Guards, are assigned the CTL-formula to which the control node is translated according to the translation table of control nodes preceded by variables [21]. Fig. 11 does not encounter this case.

   If we reach a control node, which all incoming and outgoing temporal relationships are attached to Actions, Guards, or variables with assigned values, we translate it according to the translation tables of control nodes dedicated for simple models [21]. For example, we translate the DecisionNode on Level 3 in Fig. 11 depending on the first EPPSL Pattern in Fig. 9:
   $AG(receive\ application \rightarrow (AF(X) \vee AF(make\ an\ interview)))$
   We substitute X by its value:
   $AG(receive\ application \rightarrow (AF((AF(refuse\ the\ application) \wedge AF(remove\ the\ application\ data\ from\ the\ system))) \vee AF(make\ an\ interview)))$
   The previous CTL-formula represents the whole complex model in Fig. 11.

Strategy 2: Given an EPPSL complex model, e.g. the complex model of QC_8 in Fig. 13, which we want to verify on the business process model in Fig. 1:

   QC_8: *"If there exists a possibility to make an interview, then there exists no possibility to accept the application online and there exists a possibility to make an interview per phone or there exists a possibility to make an interview per Internet".*

   We translate the model in Fig. 13 by applying the following steps:

1. Each control node may have either multiple incoming connectors or multiple outgoing connectors. In both cases, we combine the control node with its multiple connectors and their attached ConstraintContainers in one ConstraintContainer. For example, in Fig. 13, we apply this rule on the DecisionNode and the ForkNode resulting in two new ConstraintContainers as shown in Fig. 14.
2. We start to translate the partial quality constraint models which do not include ConstraintContainers. For example, in Fig. 14, we translate (1), (2), (3), and (4) depending on the EPPSL patterns in Fig. 9:
   (1): $Start \rightarrow EF(make\ an\ interview)$
   (2): $Start \rightarrow \neg EF(accept\ the\ application\ online)$

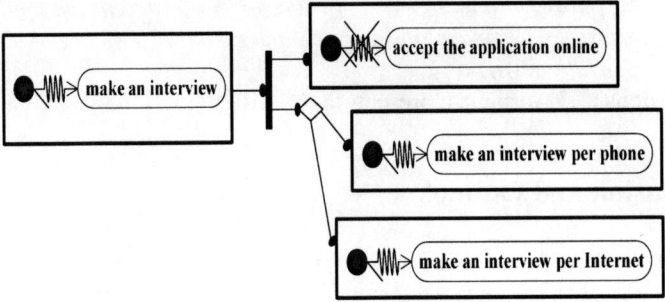

Fig. 13. Example of a complex model

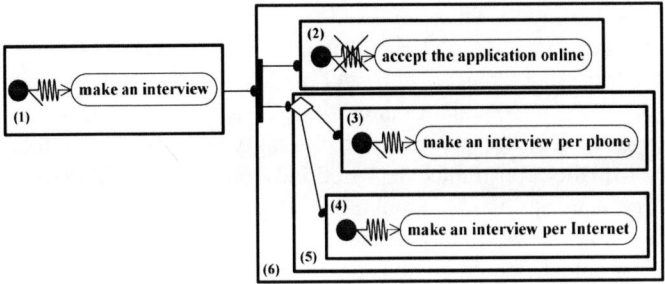

Fig. 14. Translation strategy of complex models, where control nodes link ConstraintContainers

(3): *Start → **EF**(make an interview per phone)*
(4): *Start → **EF**(make an interview per Internet)*

3. We translate the partial quality constraint models which contain ConstraintContainers, which in turn include partial quality constraint models already translated into CTL-formulas. For example, since (3) and (4) in Fig. 14 are already translated, we translate (5) depending on the EPPSL patterns in Fig. 9:

(5): *(Start → **EF**(make an interview per phone)) ∨ (Start → **EF**(make an interview per Internet))*

Then we translate (6) since (2) and (5) are already translated, depending on the EPPSL patterns in Fig. 9:

(6): *(Start → ¬ **EF**(accept the application online)) ∧ ((Start → **EF**(make an interview per phone)) ∨ (Start → **EF**(make an interview per Internet)))*

4. We always reach a state, where we have two ConstraintContainers, including two translated partial quality constraint models, and a connector, which connects the first to the second one. We translate this state depending on the EPPSL patterns for simple models [21] by stating that the CTL-formula representing the first ConstraintContainer implies the CTL-formula representing the second one. In Fig. 14, we reach a state where (1) implies (6). We translate it depending on the EPPSL patterns in Fig. 9:

*(Start → **EF**(make an interview)) → ((Start → ¬ **EF**(accept the application online)) ∧ ((Start → **EF**(make an interview per phone)) ∨ (Start → **EF**(make an interview per Internet))))*

The previous CTL-formula represents the whole complex model in Fig. 13.

6 Conclusion and Outlook

In this paper, we have introduced a new approach for modeling and formalizing business process quality constraints aiming at providing the users with more flexibility by allowing them to construct constraints with either deterministic or non-deterministic future and by that to enhance the expressiveness ability. Our approach introduces the Extended Process Pattern Specification Language (EPPSL), which is a heavy weight extension of UML Activity Diagrams, and could be easily transformed to be based on any other business process modeling language. EPPSL provides a set of intuitively understandable modeling elements to model quality constraints in terms of branching temporal logic. We also provide a pattern-based translation for EPPSL models into CTL-formulas to achieve the formalization of quality constraints. In our approach, the basic blocks of quality constraints are based only on actions and guards, since these blocks could be actions, guards, anonymous steps (refer to unknown actions or unknown guards), and partial quality constraints (based on the previous three blocks). Later, other basic blocks could be considered, e.g. data objects. In our approach, we only consider future temporal relationships. Past temporal relationships and real-time relationships are open for future work. In our approach, we do not consider the identification of conflicting constraints, which may contain contradicting semantics. Later, this aspect could be considered.

References

1. ISO 9001:2000:Quality Management Systems - Requirements. ISO International Organization for Standardization (2000)
2. Förster, A., Engels, G., Schattkowsky, T., Van Der Straeten, R.: Verification of Business Process Quality Constraints Based on Visual Process Patterns. In: The First Joint IEEE/IFIP Symposium on Theoretical Aspects of Software Engineering (TASE 2007), pp. 197–208. IEEE Computer Society, Shanghai (2007)
3. Object Management Group:UML 2.0 Superstructure. Version 2.0 (2005), http://www.omg.org/spec/UML/2.0/Superstructure/PDF/ (last visited 2.12.2010)
4. Förster, A., Engels, G., Schattkowsky, T.: Activity Diagram Patterns for Modeling Quality Constraints in Business Processes. In: Briand, L.C., Williams, C. (eds.) MoDELS 2005. LNCS, vol. 3713, pp. 2–16. Springer, Heidelberg (2005)
5. Pnueli, A.: The temporal logic of programs. In: Proceedings of the 18th Annual Symposium on Foundations of Computer Science (FOCS 1977), pp. 46–57 (1977)
6. Emerson, E.A.: Temporal and Modal Logic. In: van Leeuwen, J. (ed.) Handbook of Theoretical Computer Science, vol. B, pp. 955–1072. MIT Press, Cambridge (1990)
7. Liu, Y., Müller, S., Xu, K.: A Static Compliance-Checking Framework for Business Process Models. IBM Systems Journal 46, 335–361 (2007)

8. van der Aalst, W.M.P., Pesic, M.: DecSerFlow: Towards a Truly Declarative Service Flow Language. In: Bravetti, M., Núñez, M., Tennenholtz, M. (eds.) WS-FM 2006. LNCS, vol. 4184, pp. 1–23. Springer, Heidelberg (2006)
9. Pesic, M., Schonenberg, H., van der Aalst, W.M.P.: DECLARE: Full Support for Loosely-Structured Processes. In: 11th IEEE International Enterprise Distributed Object Computing Conference (EDOC 2007), Annapolis, Maryland, USA, pp. 287–300 (October 2007)
10. Pesic, M.: Constraint-Based Workflow Management Systems: Shifting Control to Users. Dissertation. TU Eindhoven (2008)
11. Awad, A.:BPMN-Q: A Language to Query Business Processes. In: EMISA 2007. LNI, vol. P-119, pp.115-128. GI (2007)
12. Awad, A., Decker, G., Weske, M.: Efficient Compliance Checking Using BPMN-Q and Temporal Logic. In: Dumas, M., Reichert, M., Shan, M.-C. (eds.) BPM 2008. LNCS, vol. 5240, pp. 326–341. Springer, Heidelberg (2008)
13. Laroussinie, F., Schnoebelen, P.: A Hierarchy of Temporal Logics with Past. Theoretical Computer Science 148, 303–324 (1995)
14. Zuck, L.: Past Temporal Logic. PhD thesis. Weizmann Intitute, Rehovet, Israel (1986)
15. Dijkman, R.M., Dumas, M., Ouyang, C.: Semantics and analysis of business process models in BPMN. Information and Software Technology 50, 1281–1294 (2008)
16. Wörzberger, R., Kurpick, T., Heer, T.: Checking Correctness and Compliance of Integrated Process Models. In: Proceedings of the 10th International Symposium on Symbolic and Numeric Algorithms for Scientific Computing (SYNASC 2008), pp. 576–583. IEEE Computer Society, Los Alamitos (2008)
17. Object Management Group: Object Constraint Language (OCL) Specification - Version 2.0 (May 2006), http://www.omg.org/cgi-bin/doc?formal/2006-05-01 (last visited 2.12.2010)
18. Ly, L.T., Rinderle-Ma, S., Dadam, P.: Design and Verification of Instantiable Compliance Rule Graphs in Process-Aware Information Systems. In: Pernici, B. (ed.) CAiSE 2010. LNCS, vol. 6051, pp. 9–23. Springer, Heidelberg (2010)
19. Hodges, W.: Classical Logic I: First Order Logic. In: Goble, L. (ed.) The Blackwell Guide to Philosophical Logic. Blackwell, Malden (2001)
20. Ghose, A.K., Koliadis, G.: Auditing Business Process Compliance. In: Krämer, B.J., Lin, K.-J., Narasimhan, P. (eds.) ICSOC 2007. LNCS, vol. 4749, pp. 169–180. Springer, Heidelberg (2007)
21. Khaluf, L.: Business Process Quality Assurance. Master thesis. University of Paderborn, Paderborn, Germany (May 2010)

Quality Evaluation and Improvement Framework for Database Schemas - Using Defect Taxonomies

Jonathan Lemaitre and Jean-Luc Hainaut

Laboratory of Database Application Engineering - PReCISE research Center
Faculty of Computer Science, University of Namur
Rue Grandgagnage 21 - B-5000 Namur, Belgium
{jle,jlh}@info.fundp.ac.be
http://www.fundp.ac.be/precise

Abstract. Just like any software artefact, database schemas can (or should) be evaluated against quality criteria such as understandability, expressiveness, maintainability and evolvability. Most quality evaluation approaches rely on global metrics counting simple pattern instances in schemas. Recently, we have developed a new approach based on the identification of semantic classes of definite patterns. The members of a class are proved to be semantically equivalent (through the use of semantics preserving transformations) but are assigned different quality scores according to each criteria. In this paper, we explore in more detail the concept of bad pattern by proposing an intuitive taxonomy of defective patterns together with, for each of them, a better alternative. We identify four main classes of defects, namely *complex constructs*, *redundant constructs*, *foreign constructs* and *irregular constructs*. For each of them, we develop some representative examples and we discuss ways of improvement against three quality criteria: *simplicity*, *expressiveness* and *evolvability*. This taxonomy makes it possible to apply the framework to quality assessment and improvement in a simple and intuitive way.

Keywords: Conceptual data schema, quality, schema improvement, schema evaluation, schema transformation.

1 Introduction

Modern engineering approaches to system development lead to methods in which modeling activities have become prominent, notably through the so-called *model driven engineering (MDE)* initiative. According to these methods, the design of a complex software system appears as a hierarchy of models, starting from the goal model down to the source code of the concrete artifacts. Models derive from each other through transformations that preserve some of their intrinsic properties, such as correctness, information capacity or performance. In addition, most models use components of other models. Through these derivation and use dependencies, a defect in a source model potentially propagates to many dependent models. In such an interconnected model network, the quality of the whole system critically depends on the quality of each of its models.

H. Mouratidis and C. Rolland (Eds.): CAiSE 2011, LNCS 6741, pp. 536–550, 2011.
© Springer-Verlag Berlin Heidelberg 2011

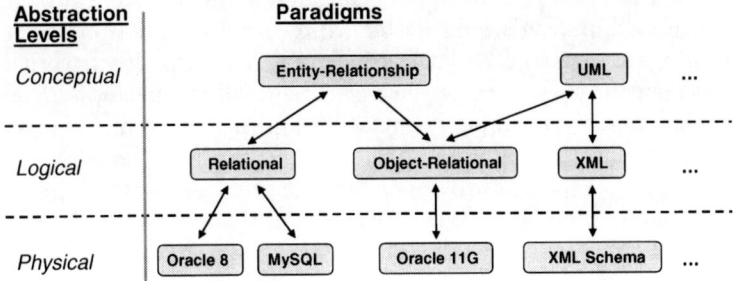

Fig. 1. A representation of the schema abstraction levels and some paradigms

When dealing with the quality of information systems, one has to pay particular attention to the database component because of the significant role of this
component in the whole system. Typically, a database can be the data provider
of thousands of programs. Any flaw in the schema of this database may cause
data inconsistencies, program malfunction or, at best, program code complexity
to compensate for this defect. The total cost to pay will be even higher: database
and program evolution will prove to be more complex and risky since both sane,
compensating and flawed components will need to evolve.

Although the MDE is often seen as a new approach in the software engineering community, it has been the standard way of developing databases since
the seventies, where the three-level methodologies were designed and progressively applied. They are based on a hierarchy of *schemas* (the database name
for *models*[1]), namely the conceptual, logical, physical schemas, the latter being
translated into DDL code. Those three levels are called *abstraction levels*. In
addition, a schema is expressed in a specification language, based on a definite
paradigm (figure 1). Since logical and physical schemas mostly derive from the
conceptual schema through semantics preserving transformations, ensuring the
highest quality of conceptual schemas is particularly important.

In [1], we proposed a database quality evaluation and improvement framework based on transformation techniques. Instead of computing global metrics
for the whole schema, it first tries to identify semantically significant constructs
that represent possible defects in schemas. Let us call *construct* an instance of a
definite pattern comprising data structures and constraints[2]. An n-ary relationship type, an entity type without attributes, a series of attributes with similar
names are all examples of constructs. A defect is a construct that is considered
sub-optimal to translate the intention of the modeler. A *relationship entity type*
(an entity type whose instances are used to connect instances of two other entity

---

[1] From this point, we will use the database terminology, i.e. a schema is the representation of the application domain and is expressed in a specification language called
model (e.g. entity-relationship model, relational model).

[2] In the following discussion, for simplicity reason and where no ambiguity may arise,
we will sometimes use the name *construct* to denote such a pattern as well as one of
its instances.

types) may be, in some circumstances, considered a defect since a mere many-to-many relationship type would better express the intention of the designer. Considering a set of about 20 basic conceptual patterns, we have defined as many equivalence classes, each of them gathering all the patterns that express the same semantics. For example, the relationship entity type and many-to-many relationship type patterns appear in the same equivalence class. In each class, we can identify its representative member, that is, the pattern that best expresses the common intention of the members of this class (and for this, called *best practice*). For example, the semantic pattern *many-to-many association* will be described by a class that includes, among a dozen equivalent patterns, the many-to-many relationship type, the relationship entity type, the multi-valued foreign key, the multi-valued embedded component. Clearly, the first pattern will be the representative member of this class. We will see in the section 3, that the best practice of an equivalence class depends on the quality criteria for the evaluation of which this class is used.

The qualification *defective* of a construct is not absolute[3] but depends on three factors, namely the abstraction level, the modeling paradigm and the quality criterion. For example, at the logical level, the *foreign key*, as the expression of a many-to-one relationship type, is optimal in a class of logical constructs but sub-optimal in a class of conceptual constructs. It is optimal in the SQL paradigm but not in the ADO Microsoft interface, based on a simple Entity-relationship model. It may be considered sub-optimal in an XML schema where element embedding may be preferred for performance reason.

In this paper, we will deepen the framework by exploring the space of conceptual defects and by attempting to classify them into an ontology of natural defect types. These reference defect types contribute to a better understanding of the third factor mentioned above: quality criterion. This classification will be used to improve our quality evaluation framework, but it has also been used in database design education [2] in the perspective of building high quality schemas.

Since most, if not all, database schemas include a certain amount of defects and considering that database design mainly is a creative task, we can expect the catalog of schema defect types being very large. In the following sections, we will concentrate on defects that degrade otherwise correct schemas. For example, a relational table that is not in 3NF is not intrinsically incorrect but it leads, among others, to expressiveness (two fact types are represented in the same table) and performance (space and update time) problems. The process of identifying these defects and improving their *structural* quality is generally known as *Conceptual normalization*.

The paper will be structured as follows. Section 2 presents a short state of the art in the role of defects in database schema quality. We recall the main concepts of the framework in section 3. Section 4 describes the bases of quality analysis for conceptual schemas. In section 5, we present a taxonomy of conceptual data schema defects and discuss their improvement. The use of the framework extended by this taxonomy is presented in section 6. Section 7 concludes the paper.

---

[3] For this reason, we have avoided the term *anti-pattern*.

2   State of the Art

Transformations are usually related to the functional requirements aspects of database schemas. Transforming a source schema must (should) preserve its information capacity[4] in such a way that the eventual DDL code completely translates the semantics of the conceptual schema. The use of transformations in the context of schema quality mainly concerns non-functional requirements, and, in this context, it has been rather limited. However, a few authors have already considered processes in which a local set of objects in a schema is replaced by another one in a way that improves some quality properties of the schema.

A first major (historical) proposal is the relational schema normalization process [4], based on functional dependencies mainly in order to remove redundancies at the data level. Though the term *transformation* was not used at that time, normalization decomposition actually makes use of semantics-preserving transformations[5]. These transformations can also be used to influence the performance of the database. Leaving the semantics of data unaltered but improving its redundancy or performance state, relational normalization clearly contribute to make the schema meet non-functional requirements. In [5], the authors studied the impact of relationships types attributes on the clarity of ER schemas. In a similar way in [6], Gemino and Wand have analysed the difference between the use of the mandatory and optional properties, also in ER schemas. Though these papers naturally called for substitution techniques to improve the readability of schemas, the authors did not push their analysis to this point.

Only a few authors have explicitly used semantics-preserving transformations for improving the schema quality. Among them, we can underline the framework of Assenova and Johanesson [7] for dealing with understandability of conceptual schemas. They assign qualitative quality scores to a set of transformations and propose to use them in order to improve schema quality. Rauh and Stickel [8] also use transformations in the context of conceptual schemas in order to normalize them and therefore to improve their quality.

The framework we propose is close to the work of Assenova and Johanesson [7]. Yet, we paid particular attention to genericity, referring to the possibility to use the framework on different abstraction levels, different paradigms and considering different quality criteria. Also, we did not associate quality preferences to the transformations themselves but to the structures.

3   Framework Reminder

In section 1, we introduced the main principles of our framework. In this section, we present some detail of the framework, based on reference [1], where the reader can find an extended description.

---

[4] A discussion on semantics preserving transformation and information capacity can be found in [3].

[5] The concept of semantics preservation is a bit more complex in this context since data preservation and functional dependency preservation may conflict.

The framework is based on the use of semantics-preserving transformations and on the identification of specific structures in schemas. It relies on the fact that there are generally different ways to express a set of facts of the application domain, and that some of them are better than others according to definite criteria. In order to make it generic enough to deal with different data model, we use the GER model [9], a wide spectrum model that encompasses the main data models (ER, EER, Relational, UML, etc.) and allows to use object types that belong to different abstraction levels and paradigms in a single schema. The framework relies on four concepts: the **equivalence classes**, the **contexts**, the **ratings**, the **quality methods**.

An **equivalence class** EC_i is a set of constructs, i.e., $EC_i = \{C_{i1}, \ldots, C_{in}\}$, that all represent the same modeling intention. Moreover for any couple of distinct constructs C_{ij}, C_{ik} of EC_i, there exists a transformation sequence T composed of semantics preserving transformations such that $T(C_{ij}) = C_{ik}$. An equivalence class represents a common modeling intention which refers to a specific type of facts of the application domain. Similar constructs are defined as instances of a generic pattern, so that equivalence classes can be reduced to about 20 useful classes made up of patterns.

We illustrate the *non-functional*[6] *binary association* equivalence class in figure 2. It collects popular generic data structures intended to represent many-to-many associations between the members of two object classes. *(a)* is a many-to-many relationship type. *(b)* is a relationship entity type accompanied by two one-to-many relationship types. *(c)* and *(d)* represent associations through multivalued (c) and single-valued (d) foreign keys. Finally, *(e)* and *(f)* use two-way and one-way object references, borrowed from the object oriented data model.

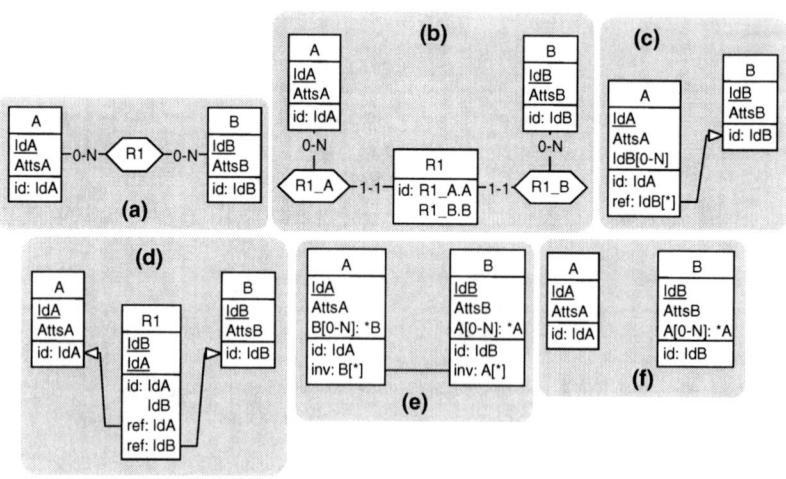

Fig. 2. *Non-functional association* equivalence class

---

[6] In the sense that it supports no functional dependencies.

A **context** describes the requirements against which the schema is evaluated. It is defined by a triple *(A,P,D)*, where *A* is an abstraction level, *P* is a data modeling paradigm and *D* is a set of quality requirements. The abstraction level *A* and the paradigm *P* define the use of a specific model, e.g., UML class diagrams at the logical level. According to level *A*, some constructs of *P* will become undesired. For example, a conceptual ER-like schema should not include constructs that explicitly or implicitly define a foreign key.

A **rating** is an ordering of the members of an equivalence class for a specific context. It states the extent to which each member meets the quality criteria *D* of the context. Different methods have been proposed to define these scores. Collecting expert estimation is the preferred technique since it requires less effort than standard empirical studies based on schema global evaluation. Ratings can be used in several ways, notably (1) to compute metrics for the schema under investigation and (2) to suggest schema improvement. A rating also allows to identify in the equivalence class a *best practice* as the construct that has the highest score for the context of the rating.

A **quality method** comprises analysis, evaluation and improvement methods. An evaluation method provides global and detailed quality scores for the schema. An improvement method is based on the replacement of constructs with a low rating for a context by a better construct in its equivalence class, for instance the best practice of the context.

4   Quality Requirements

Quality has become a major research field in software engineering, though its scope, its objectives and its evaluation techniques have not gained sufficient consensus so far to consider it a mature domain. For example, similar but still significantly different definitions of the very basic concept of *understandability* can be found in [10], [11], [12] and [13]. Definitions have evolved with time and standards have been proposed such as the ISO quality standard [12,14]. Unfortunately definitions available in a standard often appear too general and not intuitive enough when addressing the quality of a specific software product. This lack of precision also makes it uneasy to develop convincing operational methodologies and to build supporting tools.

In this paper, we consider three essential qualities of conceptual schema constructs, namely simplicity, expressiveness and evolvability. In the following, we provide definitions and interpretations of these non-functional requirements.

- **Simplicity:** Simplicity is a sub-characteristic of the understandability quality requirement in the ISO/IEC 9126 standard [12]. In [15], *a schema is said to be simple if it is constructed upon simple concepts.* The notion of *simple concept* relies on a measure of the complexity, which is itself related to the number of some specific elements, such as relationship types. Instead we define the simplicity as the property of types of facts of the application domain being represented *as simply as possible.* This definition encompasses the notion of minimality and low complexity. A practical definition could be based on the number of objects and the nature of objects. A construct has a better minimality score than another one if it uses fewer objects to express the same fact type. We also consider the structural and cognitive complexity of the objects. For example, to represent a definite fact type, an elementary attribute is clearly less complex than a compound attribute or an entity type.

– **Expressiveness:** Like simplicity, expressiveness has a high impact on understandability in the ISO standard [12]. An early definition of expressiveness was suggested by Batini et al. in [16], where the authors defined it as *the richness of a schema*. In this paper, we follow the definition given in [15] inspired from the previous early definition, and more specifically the subconcepts of *concept expressiveness*, that measures whether *the concept* [the constructs] *of the schema are expressive enough to capture the main aspects of the reality*. We relate the expressiveness to the fact that a type of facts of the application domain is represented by a construct that naturally and clearly refers to its nature. For instance, the fact that *domestic appliances* form a variety of *products* should be represented by a subtype/supertype relation, provided the data model includes this type of constructs.

– **Evolvability:** We define evolvability as the ability of a construct to support possible changes in the application domain and to trigger as little impact as possible on the system artefacts that include (e.g., the schemas) and use (e.g., programs and HCI) this construct. While previous quality requirements address the understandability of the conceptual schema, this definition is adapted from the *changeability* definition of the ISO Quality standard [12]. Being able to adapt the schemas following new application domain changes is important, especially in the context of the MDE methods.

These three qualities synthesize some of the most important requirements for a database schema. In addition, they can be formally defined through our framework, in which they form the D component of a context. We will apply them to evaluate and illustrate the taxonomy of section 5. Besides, these qualities are not independent: a construct that increases the simplicity of a schema may lower its expressiveness. They can be considered separately or combined in order to reach a trade-off.

5   Defect Taxonomy

Teaching, modeling experience and schema analysis eventually allow to come up with a set of good modeling practices. The latter are the structures an experienced designer would most probably use to represent specific fact types of the application domain. We observe that many designers make other, sometimes unfortunate, choices, so that design flaws may often be found in database schemas. In addition, the fact that a definite construct should be used is obviously context-dependent. Besides, a sound construct may not always meet all the requirements stated for some databases.

Based on experience and schema study, we progressively elaborated a set of constructs that appear to be poor design choices or defects in most situations. Among them, some can be replaced without altering the information capacity of the schema, that is, through the use of semantics preserving transformations. In such cases, there is a strong bond between the taxonomy and the framework since each couple of constructs (the *good* and the *bad*) belong to the same equivalence class. We gather these defective constructs into four categories, namely *complex*, *redundant*, *foreign* and *irregular* constructs.

We also analyse the impact of these defects on the three quality requirements defined in section 4 by comparing defects with the recommended alternative. Due to space limit, we discuss some defects only but we mention other popular constructs in each of the four categories. Additional description of defects can be found in chapter 17 of [2].

5.1 Complex Constructs

We qualify a construct *complex* when it is not the most straighforward way to represent the fact type of the application domain under consideration. So far, we have identified about a dozen complex construct classes. We detail three of them:

- **Attribute entity type** The construct represents a *property* of a concept of the application domain. It is made up of an entity type E_1 with one attribute A and that plays a *[1-1]* or *[1-N]* role in a binary relationship type R. The identifier of E_1 includes this attribute. It appears that the only goal of E_1 is to define a property of the concept represented by the other entity type in R. We call it E_0. E_0 and E_1 correspond respectively to the entity types *BOOK* and *KEYWORD* in figure 3 *(1.a)*. This goal would have been better achieved by a mere attribute. The suggested improvement would be to migrate A from E_1 to E_0 as illustrated in the figure 3 *(1.b)*. Considering the quality requirements we have defined, this transformation increases the simplicity and the expressiveness of the schema. Indeed, the new construct contains fewer objects and is closer to the application domain structure. On the contrary, the original construct favours evolvability. Indeed, should E_1 represent an important concept in the future, it could receive new attributes and play new roles without painful restructuring.
- **N-ary relationship type with a $[0-1]$ or $[1-1]$ role** As an identifier of the relationship type, this role is the functional determinant of each of the other roles. Therefore, the relationship type is decomposable into two binary functional relationship types as illustrated in figure 3 *(2)*. N-ary relationship types are intrinsically more complex and less intuitive than binary ones[7]. In general, the decomposed pattern is simpler, more expressive and easier to evolve than the source construct.

---

[7] As testified by the endless discussions on the semantics of N-ary associations in UML class diagrams!

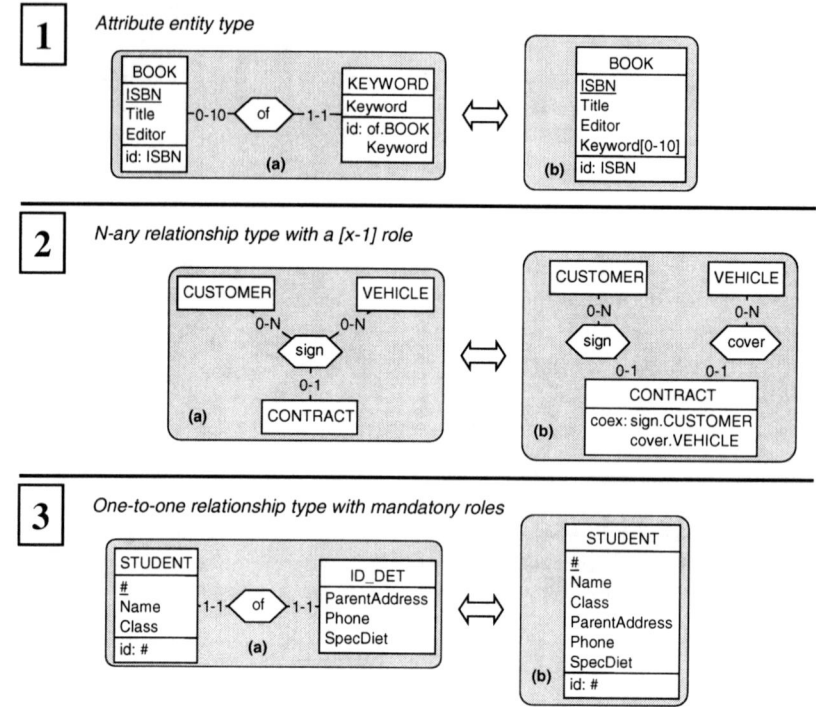

Fig. 3. Three complex constructs and their suggested alternatives.

- **One-to-one relationship type with mandatory roles** This pattern expresses a strong link between two concepts (no instance of a concept exists independently of an instance of the other one) (figure 3 *(3.a)*). In most situations, the latter just represent two aspects, or two complementary fragments, of the same concept. A better alternative (figure 3 *3.b*) can be to merge the entity types into a single entity type. This improves the simplicity and the expressiveness of the schema. Normally, evolvability should stay unchanged though, in some cases, it may decrease, depending on the semantics of each fragment.

The other complex constructs are classified in table 1, with the qualitative score difference when they are replaced by the alternative construct we suggest.

5.2 Redundant Contructs

A construct A is redundant with construct B when, in any database state, the instances of one of them can be computed from the instances of the other one. More complex situations may occcur (for instance where A and B are mutually dependent), but they will be ignored in the framework inasmuch as their solving requires expert knowledge. The correction of such defects consists in removing

Table 1. Complex constructs: qualitative evaluation

	Simplicity	Expressiveness	Evolvability
Entity type			
ET with a weakly specified subtype → *supertype attribute*	+	-	-
ET with an empty subtype → *supertype attribute*	+	+	≈
Relationship entity types → *relationship type*	?	+	≈+
Implicit Is-A rel.: materialization → *explicit is-a*	+	+	≈+
Implicit Is-A rel.: upward inheritance → *explicit is-a*	-	+	≈+
Implicit Is-A rel.: downward inheritance → *explicit is-a*	≈	+	+
Relationship type			
Inter-attribute functional dependencies → *decomposition*	-	+	+
Attribute			
Complex attributes → *entity type*	-	+	+
One-component compound attribute → *desagregation*	+	+	?
Inter-ET functional dependencies → *decomposition*	-	+	+
Constraint			
Decomposed existence constraints → *merging*	+	≈	+

Legend: (-) Quality decrease   (+) Quality improvement
(≈) Equivalent quality (?) Indeterminate quality change
(≈+) Nearly equivalent quality with slight improvement

one of the source constructs (normally the lowest quality construct according to its ranking in its equivalence class), which is (trivially) a semantics-preserving transformation.

We classify the redundant constructs in two categories. The first category includes patterns in which some fact types of the application domain are expressed more than once. The **relationship/foreign key redundancy** (figure 4 *(a)*) pattern is an example of this category. This pattern comprises an attribute that references entities of another type (therefore acting as a foreign key) while a relationship type already expresses such relationships explicitly. These constructs are redundant and one of them must be removed. Since the foreign key suffers from another problem (it appears as a *foreign construct* at the conceptual level - see below), we remove it from the schema. A second example, not illustrated, is that of attribute *Amount* of entity type *ORDER*, the values of which can be computed from the values of attributes *OrderedQuantity* of *DETAIL* and *UnitPrice* of *PRODUCT*. The derived objects should be removed from the schema in order to increase its simplicity and evolvability. The evaluation of expressiveness is somehow less intuitive. As we removed the less expressive objects, we consider that the expressiveness of the whole construct has increased too.

The second category includes constraints that can be formally inferred from other constructs on the schema. We describe two examples.

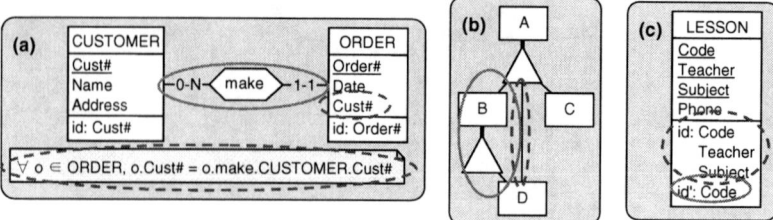

Fig. 4. Examples of redundant constructs

- **Transitive is-a relation** (figure 4 *(b)*). Based on the set-theoretic inclusion relation, is-a relations are transitive: if entity type D is a subtype of B, which itself is a subtype of A, then D is also a subtype of A. Specifying the latter property in the schema is needless. Removing it improves simplicity, expressiveness and evolvability.
- **Non-minimal identifiers** (figure 4 *(c)*). Non-minimal identifier I includes components that can be discarded without I loosing its uniqueness property. This pattern can be detected if the minimal subset of I has been declared an identifier of the entity type, as shown in the figure. According to the Armstrong inference rules, the largest identifier can be derived from the smaller one, and therefore can be discarded, which will improve the three quality requirements.

5.3   Foreign Constructs

Foreign constructs are groups of objects that technically comply with the model but that are highly influenced by the modeling practices of another model. Such constructs may appear due to cultural habits of the database designer, to missunderstanding of the *philosophy* (way to perceive the world) of the model or as left-over of a too straightforward migration process. Because of the large variety of models practically used, many different foreign constructs can be found. Let us mention two classical cases:

- **Referencing attribute.** Such attribute expresses relationships between concepts of the application domain. The attribute is accompanied by an informal constraint describing its intention. Such pattern is illustrated in figure 4 *(a)*. In order to improve the schema quality, one transforms the attribute into a many-to-one binary relationship type (unless it is, in addition, redundant). This substitution increases the expressiveness of the schema without having a major impact on the other requirements.
- **IMS style.** As shown in figure 5 *(a)*, a many-to-many relationship type is expressed in the the legacy IMS style, using three intermediate entity types following the hierarchical database modeling practices [17]. Such construct has obviously a harmful impact on the schema quality. Through semantics-preserving transformations, the source flawed pattern can be replaced by a many-to-many relationship type (figure 5 *(b)*) which significantly increases the three quality requirements.

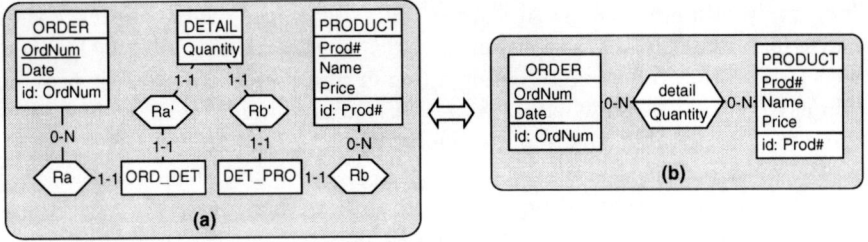

Fig. 5. Example of IMS style construct and its best practice alternative

5.4   Irregular Constructs

The last category of defects we identified are irregular constructs. They appear in large schemas when similar types of facts are expressed by different types of constructs. This anomaly does not affect individual constructs but the schema as a whole, which appears inconsistent. If these constructs are correct, they belong to the same equivalence class, so that each non-optimal construct can be replaced by a better one from this class. This substitution do not decrease the quality requirements but strengthen the evolvability of the schema.

6   Framework Application

In the previous section, we defined a taxonomy of constructs that can easily be related to the equivalence classes of the framework. This taxonomy suggests a way of using and applying the framework as it gives defect detection criteria and it provides a method for the evaluation and the improvement of schemas. Indeed, the taxonomy (1) brings a well-focused study of defects, (2) is developed in a specific context and (3) provides examples that suggest improvement techniques. In this section, we will discuss the integration of the taxonomy in our framework.

An equivalence class theoretically comprises all the constructs that can be used for representing a specific type of facts, independently of the context. Given a definite context, each construct of a class will receive a score, that defines its level of quality. The taxonomy allows to identify more precisely the constructs that are considered as defects and therefore to specialize the framework to quality evaluation and improvement in a particular situation.

An important aspect of the framework is the development of ratings, i.e., the definition of quality scales and the application of these scales to the scoring of the constructs. Obviously the definition of ratings for all constructs of each equivalence class, for all possible contexts should be a huge task. Again, the taxonomy can be used to specialize the framework. Simple ratings can be produced using the quality differences between the elements of each couple *(problem, solution)* of the *complex* and the *foreign* construct categories (the *irregular* and *redundant* construct categories will not be used here as they do not provide such evaluations). Example of ratings and a deeper discussion about the evaluation of schema quality using the framework can be found in [1].

The quality difference in a couple is an indicator on the relative scores of these constructs. Obviously a coarse evaluation such as the one provided in the previous section will not allow us to define a fine-grained scale. However, it provides a primitive but usable scale with 2 values *[0,1]*. Through a Condorcet-like voting technique, it is possible to designate the best practice as the construct that has been the most preferred in all the defect fixing suggestions. Then, the 1 score is assigned to the best practice and the 0 score to the other constructs of the class.

Because of the limited information (constructs and quality indicators) available in the taxonomy, the production of more precise ratings should require more investigation. Other scales are discussed in [1], in which we propose for example to use an ordinal scale based on five grades (e.g., *very bad, bad, neutral, good, very good*).

Improving the schema following a single quality requirements (e.g., simplicity, expressiveness or evolvability) becomes an easy task. However in practice, quality requirements are often combined and may lead to conflicting suggestions. When combining criteria, two situations appear. In the first one, all the criteria come to the same conclusion, i.e., there exists one best alternative construct that improves all the requirements. In the other situation, there are conflicts between different possibilities and we have to rely on trade-off techniques. For example, we can assign a weight to the requirements and compute an average score if the rating are properly defined (their scale is composed of a sufficient number of values).

7   Conclusion

The principle of taxonomy of defective constructs presented in this paper allows us to refine the quality evaluation and improvement framework proposed in [1], notably since it contributes to populating the equivalence classes. The taxonomy is semi-empirical. It derives from good practices published in the litterature and from modeling experience. The identified defects are probably representative of the common practical defects of this last decade. It also provides designers with guidelines to identify potential problems in database schemas and to apply solutions according to quality criteria such as *simplicity, expressiveness* and *evolvability*. It is important to note that this approach, based on the evaluation of *semantically significant constructs*, does not oppose classical metrics approaches counting atomic objects in the target schema. On the contrary, once defects violating definite quality criteria have been identified, they can be counted and weighted (according to their severity) in order to produce detailed and global metrics.

Though the illustrations (taxonomy and example schemas) of this paper concern the conceptual abstraction level only, the principles we have developed are valid for all abstraction levels and all data modeling paradigms. A demand exists for *relational schema* evaluation, inasmuch as software quality evaluation mainly addresses software metrics at the code level (high level model evaluation still is emerging). At this level, the quality criteria and the taxonomy are specific. For example, *time* and *space performance* as well as *DDL portability* criteria may become important. On the taxonomy side, such defects as *implicit foreign key, concatenated columns* and *missing primary key* will appear.

At the present time, we have defined about 20 equivalence classes with their rankings, as well as an extended taxonomy of conceptual defects for Entity-relationship schemas. We are also developing a suite of tools to identify instances of schema patterns (based on a declarative pattern description language), to compute various metrics of these instances and to apply improvement transformations.

The future work will address (1) the validation of the framework[8] and of the tools with the collaboration of experts in database engineering and (2) the extension of the framework to relational database evaluation and improvement.

References

1. Lemaitre, J., Hainaut, J.L.: Transformation-based framework for the evaluation and improvement of database schemas. In: Pernici, B. (ed.) CAiSE 2010. LNCS, vol. 6051, pp. 317–331. Springer, Heidelberg (2010)
2. Hainaut, J.L.: Bases de données: Concepts, utilisation et développement. Dunod (2009)
3. McBrien, P., Poulovassilis, A.: A formal framework for er schema transformation. In: Embley, D.W. (ed.) ER 1997. LNCS, vol. 1331, pp. 408–421. Springer, Heidelberg (1997)
4. Codd, E.F.: Normalized data structure: A brief tutorial. In: SIGFIDET Workshop, pp. 1–17. ACM, New York (1971)
5. Burton-Jones, A., Weber, R.: Understanding relationships with attributes in entity-relationship diagrams. In: ICIS 1999: Proc. of the 20th International Conference on Information Systems, Atlanta, GA, USA. Association for Information Systems, pp. 214–228 (1999)
6. Gemino, A., Wand, Y.: Complexity and clarity in conceptual modeling: comparison of mandatory and optional properties. Data Knowl. Eng. 55(3), 301–326 (2005)
7. Assenova, P., Johannesson, P.: Improving quality in conceptual modelling by the use of schema transformations. In: Thalheim, B. (ed.) ER 1996. LNCS, vol. 1157, pp. 277–291. Springer, Heidelberg (1996)
8. Rauh, O., Stickel, E.: Standard transformations for the normalization of er schemata. In: CAiSE, pp. 313–326. Springer, Heidelberg (1995)
9. Hainaut, J.L.: The transformational approach to database engineering. In: Lämmel, R., Saraiva, J., Visser, J. (eds.) GTTSE 2005. LNCS, vol. 4143, pp. 95–143. Springer, Heidelberg (2006)
10. Davis, A., Overmyer, S., Jordan, K., Caruso, J., Dandashi, F., Dinh, A., Kincaid, G., Ledeboer, G., Reynolds, P., Sitaram, P., Ta, A., Theofanos, M.: Identifying and measuring quality in a software requirements specification. In: Proceedings of the First International Software Metrics Symposium, pp. 141–152 (1993)
11. Moody, D.L., Shanks, G.G.: Improving the quality of data models: empirical validation of a quality management framework. Inf. Syst. 28(6), 619–650 (2003)
12. ISO/IEC: ISO 9126-1:2001, Software engineering - Product quality, Part 1: Quality model. ISO/IEC (2001)

---

[8] As suggested by a reviewer, the use of social network could help in the scoring of constructs.

13. Bansiya, J., Davis, C.G.: A hierarchical model for object-oriented design quality assessment. IEEE Trans. Software Eng. 28(1), 4–17 (2002)
14. ISO/IEC: Software Engineering – Software product Quality Requirements and Evaluation (SQuaRE) – Guide to SQuaRE. ISO/IEC (2005)
15. Si-Said Cherfi, S., Akoka, J., Comyn-Wattiau, I.: Perceived vs. measured quality of conceptual schemas: An experimental comparison. In: ER (Tutorials, Posters, Panels & Industrial Contributions), Australian Computer Society, pp. 185–190 (2007)
16. Batini, C., Ceri, S., Navathe, S.B.: Conceptual database design: An Entity-relationship approach. Benjamin-Cummings Publishing Co., Inc., Redwood City (1992)
17. Hainaut, J.-L.: Hierarchical data model. In: Encyclopedia of Database Systems, pp. 1294–1300 (2009)

Validation of Families of Business Processes

Gerd Gröner[1], Christian Wende[2], Marko Bošković[3], Fernando Silva Parreiras[1],
Tobias Walter[1], Florian Heidenreich[2], Dragan Gašević[3], and Steffen Staab[1]

[1] WeST Institute, University of Koblenz-Landau, Germany
{groener,parreiras,walter,staab}@uni-koblenz.de
[2] Technische Universität Dresden, Germany
{c.wende,florian.heidenreich}@tu-dresden.de
[3] Athabasca University, Canada
{marko.boskovic,dragang}@athabascau.ca

Abstract. A Software Product Line (SPL) is a set of programs that
are developed as a whole and share a set of common features. Product
line's variability is typically specified using problem space models (i.e.,
feature models), solution space models that specify the realization of
functionality and mapping models that link problem and solution space
artifacts. In this paper, we consider this concept in the scope of families
of business processes, whose specificity is that the solution space is defined with business process models. Solution space models are typically
specified as model templates, and thus in the rest of the paper we will
refer to business process model templates. While the previous research
tackled the concepts of families of business processes, there have been
very limited research on their validation.

Keywords: business process families, well-formedness constraints,
validation, process model variability, configuration.

1 Introduction

The increasing number of software systems with similar required functionality
has led software engineers to move from development of single software systems
to the development of Software Product Lines (SPLs). A SPL is a set of software
systems that share most of the features [1]. Because of the shared commonalities,
development of families improves reusability and is more cost effective [2].

A SPL[1] is typically specified with three kinds of models: *problem space models*, *solution space models* and *mapping models* [3]. *Problem space models* define
available features of the members of the SPL, as well as their interdependencies. They are typically used by stakeholders for selection of desired features of
the product. The set of selected features is called *configuration*. *Solution space
models* are comprehensive models that specify the realization of complete SPLs.
In this paper, we focus on business process families, i.e., families whose solution space models are business process model templates. Business process model

---

[1] In this paper, we will use product line and software family interchangeably, even
though one can easily argue that they can not be considered synonymous.

H. Mouratidis and C. Rolland (Eds.): CAiSE 2011, LNCS 6741, pp. 551–565, 2011.
© Springer-Verlag Berlin Heidelberg 2011

templates are specified by business process modeling languages and composed of business process patterns. From solution space models, a particular product, i.e., a business process model is derived by removing or adding parts of it. Finally, *mapping models* define mapping relations between problem space and solution space models. They regulate which parts might be removed from the business process model according to the selected features from the problem space model.

Given such a representation of business process families, we have to guarantee that each process model, that is built according to a feature configuration, does not violate any well-formedness constraints of the business process model template. Due to the size of contemporary business process families, it is time consuming, costly and error-prone to manually validate that each configuration has a well-formed corresponding business process model. For these reasons, an automated approach for the constraint validation is necessary.

To address this problem, in this paper, we propose a classification of interrelationships between elements of business process models and demonstrate how this classification can be used for the validation. The classification is based on an analysis of basic workflow patterns, a set of conceptual basis for process languages. This classification is specified in Description Logics (DL), which we use as means of validating business process templates.

2   Application Context of Business Process Families

A typical SPL consists of three kinds of artifacts, representing its problem space, solution space and mappings between the problem and solution space [3]. We introduce one such SPL, a part of the Electronic Store (e-store) SPL [4].

Fig. 1 depicts a snippet of the business process family of the e-store case study. The representation contains a feature model to represent commonality and variability of the business process family, a business process model template, specified in the Business Process Modeling Notation (BPMN) and mappings between features and elements of the business process model template.

Interdependencies in the feature model are specified with *mandatory* and *optional* parent-child relationships and *alternative* and *or feature groups*. A *mandatory* parent-child relationship specifies that if a parent feature is selected in a certain configuration, its mandatory child feature has to be too (e.g., E-Shop and StoreFront). An *optional* parent-child relationship specifies a possibility of the selection, e.g., StoreFront and WishList. An *alternative feature group*, or *xor feature group*, (e.g., Basic and Advanced), specifies that when their parent feature is selected, *exactly one* of the members of the group can be selected. Finally, an *or group* (e.g., Emails, ProductFlagging and AssignmentToPageTypesForDisplay) defines a set of features from which at least one has to be selected.

Feature models also contain interdependencies between features that are not captured by the tree structure of feature diagrams, called cross-tree constraints, namely includes and excludes. Includes means that if an including feature is in a configuration, the included feature has to be as well (e.g., EmailWishList and Registration). Excludes is the opposite to includes.

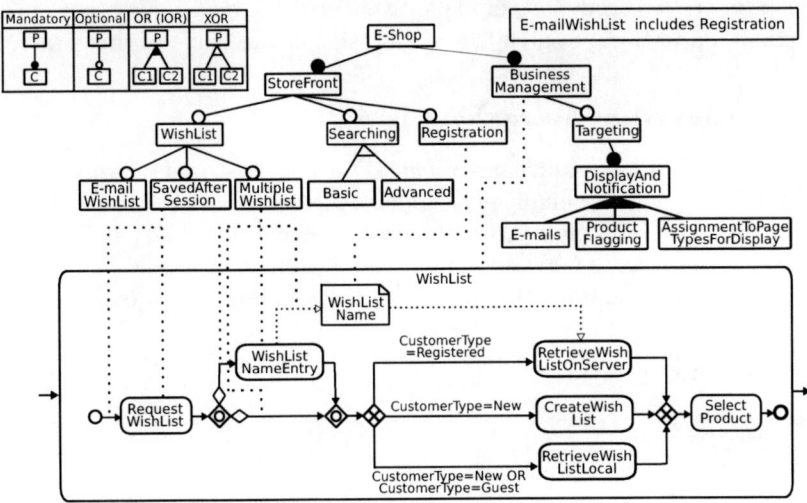

Fig. 1. A Part of the E-Store [4] Software Product Line

Business process model templates are specified in a business process modeling language. In Fig. 1, we use BPMN to specify a business process model template. Such a template typically consists of process patterns like subprocesses (WishList), activities (e.g., WishListNameEntry), gateways (diamonds), conditional sequence flows and data objects (WishListName). These process patterns impose well-formedness constraints like grouping of activities.

Finally, mappings connect the features with elements of the solution space model that implement the business logic of particular configurations. For example, every configuration that contains the Registration feature, contains the WishListName data object. On the contrary, every configuration that does not contain the feature MultipleWishList leads to a business process model where the corresponding mapped activity WishListNameEntry is missing. All elements of the business process model template that are not mapped to any feature are contained in every business process model.

The problem in this context is given by the constraints of both modeling spaces in combination with mappings between the feature model and the business process model template. Given a particular feature configuration and a mapping, we have to ensure, that the corresponding business process model satisfies the well-formedness constraints of the template.

3 Solution Space Model Dependencies and Validation

In this paper, Description Logic (DL) [5] is used to formalize the constraints of interests in models of interests and enable validation services. We could have used some other formalism, but we opted for DL as it is precise and expressive enough to serve our purpose - formalize constraints that need to hold between our models of interest. Discussions of expansiveness of DL over some other options, although an important research topic, is outside of the scope of this paper.

3.1   Modeling with Description Logics

DL is a decidable subset of first-order logic (FOL). A DL-based knowledge base is established by a set of terminological axioms (TBox) and assertions (ABox). The TBox is used to specify classes, which denote sets of individuals and properties defining binary relations between individuals. The main syntactic constructs are depicted in Table 1, supplemented by the corresponding (FOL) expressions.

Table 1. Constructs and Notations in DL and FOL Syntax

Construct Name	DL Syntax	FOL Syntax
atomic class, atomic object property	C, R	$C(x), R(x,y)$
subclass relation	$C \sqsubseteq D$	$\forall x.C(x) \rightarrow D(x)$
union class expression	$C_1 \sqcup \ldots \sqcup C_n$	$C_1(x) \vee \ldots \vee C_n(x)$
intersection class expression	$C_1 \sqcap \ldots \sqcap C_n$	$C_1(x) \wedge \ldots \wedge C_n(x)$
complement class expression	$\neg C$	$\neg C(x)$
universal quantification	$\forall P.C$	$\forall y.(P(x,y) \rightarrow C(y))$
existential quantification	$\exists P.C$	$\exists y.(P(x,y) \wedge C(y))$
object subproperty	$R \sqsubseteq S$	$\forall x,y.R(x,y) \rightarrow S(x,y)$

3.2   Representing Models of Business Process Families

In this section, we provide formal definitions of the three considered modeling spaces: (i) problem space, (ii) solution space and (iii) mapping space that combines features from the problem space with entities in the solution space.

Definition 1. *A Feature Model* $\Phi = \langle \mathcal{F}, \mathcal{F}_P, \mathcal{F}_M, \mathcal{F}_{IOR}, \mathcal{F}_{XOR}, \mathcal{F}_{cr} \rangle$ *is a tree structure that consists of features* $\mathcal{F}$. $\mathcal{F}_p \subset \mathcal{F} \times \mathcal{F}$ *is a set of parent and child feature pairs,* $\mathcal{F}_M \subset \mathcal{F}$ *is a set of mandatory features with their parents. All other features are optional.* $\mathcal{F}_{IOR} \subset \mathcal{P}(\mathcal{F}) \times \mathcal{F}$ *and* $\mathcal{F}_{XOR} \subset \mathcal{P}(\mathcal{F}) \times \mathcal{F}$ *are sets of pairs of child features and its common parent feature. The child features are either inclusive or exclusive features. Finally,* $\mathcal{F}_{cr} \subset \mathcal{F} \times \mathcal{F}$ *is a set of cross-tree constrained feature pairs that are either in an* includes *or* excludes *relation.*

$$E - Shop \equiv \exists\, hasFeature.StoreFront \sqcap \exists hasFeature.BusinessManagement \quad (1)$$
$$StoreFront \sqsubseteq \exists\, parent.E - Shop \quad (2)$$
$$Searching \equiv (\exists\, hasFeature.Basic \sqcup \exists\, hasFeature.Advanced) \sqcap$$
$$\neg(\exists\, hasFeature.Basic \sqcap \exists\, hasFeature.Advanced) \quad (3)$$
$$EmailWishList \sqsubseteq \exists includes.Registration \quad (4)$$

Axiom 1 defines StoreFront and BusinessManagement as mandatory child features of E-Shop. There is no such axiom for optional child features. Except of the root feature, each feature has a parent feature. Axiom 2 defines E-Shop as the parent feature of StoreFront. For inclusive and exclusive child features, a class union is used. For instance, Axiom 3 defines Basic and Advanced as exclusive child features of the parent feature Searching. The axiom ensures that only one feature is selected (cf. [6]). Axiom 4 depicts a cross-tree constraint where the feature EmailWishList includes Registration.

The solution space is defined by a business process model template. In Def. 2, we give generic definitions for solution space models. We focus on elements that are mapped to features, i.e., these elements realize a certain feature. Obviously, the granularity of mapping solution space models (e.g., classes, attributes) depends on the applications. In the reset of this paper, all elements that are subclasses of the BPMN class Element can be mapped to features.

Definition 2. *A Solution Space Model $\Omega = \langle \mathcal{S}, \mathcal{T} \rangle$ consists of entities $\mathcal{S}$ that could be mapped to features and entities that are not mapped and do not directly realize features ($\mathcal{T}$). The sets $\mathcal{S}$ and $\mathcal{T}$ are disjoint.*

Concrete process models found in the solution space are automatically transformed to a knowledge base Σ_Ω. The transformation creates classes for constructs of the BPMN metamodel, like the classes Activity and SequenceEdge, independently of the concrete BPMN model template. All elements of the BPMN model template are modeled as subclasses of Element (i.e., elements of $\mathcal{S}$).

$$Vertex,\ Activity,\ SequenceEdge \sqsubseteq Element \tag{5}$$
$$RequestWishList \sqsubseteq Activity \tag{6}$$
$$E1 \sqsubseteq SequenceEdge \tag{7}$$
$$RequestWishList \sqsubseteq \exists outgoingEdge.E1 \tag{8}$$

Vertex, Activity and SequenceEdge are subclasses of Element (Axiom 5). RequestWishList is a concrete activity, defined by a subclass axiom too (Axiom 6). Likewise, an edge E1 in the process model is represented by a subclass axiom (Axiom 7). The relations from activities to edges are described using existential restrictions on the object property outgoingEdge (Axiom 8).

Mappings (Def. 3) connect features $\mathcal{F}$ and the elements $\mathcal{S}$ of the solution space model Ω. A feature can be mapped to multiple elements and likewise an element in the solution space might realize multiple features.

Definition 3. *For a feature model $\Phi = \langle \mathcal{F}, \mathcal{F}_P, \mathcal{F}_M, \mathcal{F}_{IOR}, \mathcal{F}_{XOR}, \mathcal{F}_{cr} \rangle$ and a solution space model $\Omega = \langle \mathcal{S}, \mathcal{T} \rangle$, a Mapping Model M is a relation $M \subseteq \mathcal{F} \times \mathcal{S}$, that is defined as $\mathcal{F} \times \mathcal{S} := \{(f, s) : f \in \mathcal{F} \wedge s \in \mathcal{S}\}$.*

Finally, we transform the mapping models into a TBox Σ_M. For each mapped element $E \in \mathcal{S}$ in Ω, we introduce a class Map_E that is a subclass of Map. The object property feature is used to describe the mappings from elements to features. A mapping from an element E to a feature F is represented by an axiom $Map_E \sqsubseteq \exists feature.F$ (Axiom 9). The mapping classes Map_E are introduced to separate the feature mapping from the solution space. In order to ease the validation later on, we define the property $hasFeature$ as a subproperty of the property $feature$ form the mapping model (Axiom 10).

$$Map_{RequestWishList} \sqsubseteq \exists \, feature \, . \, SavedAfterSession \tag{9}$$

$$hasFeature \sqsubseteq feature \tag{10}$$

3.3   Well-Formedness in the Business Process Model Templates

Besides solution space models, we have to represent syntactic and structural well-formedness constraints of the process models. They are imposed by the BPMN language and by basic workflow patterns, as described in [7].

We focus in our work on structural well-formedness for two reasons. Firstly, for the class of structural models, structural constraints coincide with behavioral constraints (see the work on behavioral profiles that are derived from process structure trees [8]). Secondly, there are techniques to derive structured models for a broad class of unstructured models [9].

Elements of the business process model template that are not mapped to any feature occur in each process model. In contrast, the appearance of a mapped element in a process model depends on the feature configuration. For the sake of a more compact representation, we introduce auxiliary constructs. In Axiom 11, each element with at least one mapping to a mapping class is defined by the class $MappedElement$. Axioms 12, 13 and 14 define (composite) properties.

$$MappedElement \sqsubseteq Element \sqcap \exists mapped.Map \tag{11}$$

$$incomingEdge \, \circ \, source \sqsubseteq predecessor \tag{12}$$

$$outgoingEdge \, \circ \, target \sqsubseteq successor \tag{13}$$

$$successor \, \circ \, predecessor \sqsubseteq sibling \tag{14}$$

We identify three types of constraints. (i) An element might *require* another element. (ii) An element always appear in *conjunction* with another element. (iii) Elements of a process model might *exclude* each other. There is no influence on elements in inclusive or-branches by feature mappings, since well-formedness violations only occur due to missing elements in the process model.

We introduce three classes *Required, Conjunct* and *Exclude* to capture elements according to the different constraints. For instance, if element E is classified as a required element, we expect that E is subsumed by the (super-) class *Required*. Additionally, a property *isRequired* is introduced to get for an element its required element. The corresponding element E' that requires E is obtained by a derivable axiom like $E \sqsubseteq \exists isRequired.E'$. Likewise, we can use the property *sibling* (Axiom 14) to get the sibling activities.

Sequence. A sequence describes a series of activities that are connected by sequence edges. In a BPMN model, there are no dangling edges allowed, i.e., if there is a mapping to an activity, a violation might occur, if the activity is missing. Let us consider the mapping of the optional feature $MultipleWishList$ to the activity $WishListNameEntry$. In this case, we have to guarantee that sequence edges require their corresponding source and target vertex. Axiom 15 defines each mapped activity with an incoming or outgoing sequence edge as *required*. The corresponding properties *incomingEdge* and *outgoingEdge* are defined as subproperties of the introduced property *isRequired* (Axiom 16).

$$MappedElement \ \sqcap \ (\exists \ incomingEdge.SequenceEdge$$
$$\sqcup \ \exists \ outgoingEdge.SequenceEdge) \sqsubseteq Required \qquad (15)$$
$$incomingEdge, \ outgoingEdge \sqsubseteq isRequired \qquad (16)$$

Container Element. BPMN models facilitate grouping and decomposition of activities. Axiom 17 describes that container elements such as groups and subprocesses are supposed to be required by their elements. Groups in a BPMN model use the property *activities* to define which activities are members of this group, while subprocesses use the properties *vertices* and *sequenceEdges*. These properties are defined as subproperties of *isRequired* in order to have a connection to the element's counterparts (Axiom 18).

$$MappedElement \ \sqcap \ (\exists \ activites.Element \ \sqcup \ \exists \ vertices.Element$$
$$\sqcup \ \exists \ sequenceEdges.Element) \sqsubseteq Required \qquad (17)$$
$$activities, \ vertices, \ sequenceEdges \sqsubseteq isRequired \qquad (18)$$

Conditional Flow. Activities with outgoing conditional sequence edges impose the existence of at least one unconditional outgoing sequence edge. This constraint might be violated in the case an unconditional sequence edge is mapped and could be possibly removed in a certain configuration. For instance, consider the mapping of the optional feature *MultipleWishList* to the conditional outgoing edges of the activity *RequestWishList* in Fig. 1. In this case, a feature configuration could result in a process model without any outgoing unconditional edge. In order to avoid this, we describe in Axiom 19 that a mapped unconditional edge which is the only outgoing edge is required.

$$MappedElement \sqcap UnConditionalEdge \sqcap \exists \ source. \ (Activity \ \sqcap$$
$$\exists_{\leq 1} outgoingEdge.UnConditionalEdge) \sqsubseteq Required \qquad (19)$$
$$source \circ outgoingEdge \sqsubseteq isRequired \qquad (20)$$

Opening Gateway. An opening gateway is used to express the divergence of a control flow. While the number of outgoing branches is arbitrary in general, it is required that there are at least two outgoing sequence edges. Otherwise the gateway does not represent any divergence. Axiom 21 defines a mapped sequence edge that is one of at most two outgoing edges as a required element. Axiom 22 defines the property composition of *source* and *outgoingEdge* as subproperty of *isRequired* in order to find the counterparts.

$$MappedElement \sqcap SequenceEdge \sqcap \exists source.(OpeningGateway$$
$$\sqcap \exists_{\leq 2} outgoingEdge.SequenceEdge) \sqsubseteq Required \qquad (21)$$
$$source \circ outgoingEdge \sqsubseteq isRequired \qquad (22)$$

Closing Gateway. The convergence of a control flow in a process is described by closing gateways. Like for opening gateways, a closing gateway needs at least two incoming sequence edges. Axioms 23 and 24 are similar to the previous Axioms just for closing gateway.

$$MappedElement \sqcap SequenceEdge \sqcap \exists target.(ClosingGateway$$
$$\sqcap \exists_{\leq 2} incomingEdge.SequenceEdge) \sqsubseteq Required \quad (23)$$
$$target \circ incomingEdge \sqsubseteq isRequired \quad (24)$$

Exclusive Branching. Activities that appear in exclusive branches are supposed to be exclusive, i.e., it is not allowed to execute activities from alternative branches in a process execution. Axiom 25 defines each mapped element between XOR-gateways as a subclass of Exclude.

$$MappedElement \sqcap \exists successor.XORgateway \sqcap \exists predecessor.XORgateway \sqsubseteq Exclude \quad (25)$$

Parallel Branching. If activities occur in parallel branches, they have to be executed commonly, i.e., if an activity of the first branch is executed, then also the activity of the second branch is executed. In Axiom 26, we define elements between AND-gateways as subclasses of the class *Conjunct*.

$$MappedElement \sqcap \exists successor.ANDgateway \sqcap \exists predecessor.ANDgateway \sqsubseteq Conjunct \quad (26)$$

4   Validation Using Description Logics

Our aim is to ensure that the well-formedness constraints are satisfied in all process models that can be derived from a process model template. The constraints are represented by logical formulas, DL expressions in our case. We use the expression f_Φ to represent a constraint on features and f_Ω to represent a constraint on elements of the business process. From a logical point of view, checking whether the constraint given by f_Φ ensures that the constraints represented by f_Ω hold, can be realized by checking whether the implication $f_\Phi \Rightarrow f_\Omega$ holds for each interpretation, i.e., $f_\Phi \Rightarrow f_\Omega$ is a tautology.

In order to test the implication $f_\Phi \Rightarrow f_\Omega$ for all mapped elements, we have to tackle the following problems. (i) While the constraints in the problem space are directly represented by the axioms, the constraints of the process model template are implicitly given by the element's dependencies (cf. Sect. 3). Hence, we have to *classify and derive* constraints of the mapped elements. (ii) We have to guarantee that the same vocabulary and the same expression structure is used in f_Φ and f_Ω. This is realized by *building constraint expressions*. (iii) Finally, we have to check in DL whether the *implication* holds.

In Sect. 3.3, we specify implicit constraints of business processes and define axioms in order to allow a dynamic classification of mapped and constrained elements. These elements are categorized as subclasses of the classes Required, Conjunct and Exclude. Defined properties (sibling and isRequired) as well as their inverse properties help to get their counterparts. E.g., for the conjunct element E, the counterpart (E') can be found by using the subsumption $E \sqsubseteq \exists sibling.E'$.

Building Constraint Expressions. For the mapped elements, we build class expressions f_Φ and f_Ω. To check the implication (subsumption in DL $(f_\Phi \sqsubseteq f_\Omega)$), we describe f_Φ and f_Ω by complex class expressions in DL. Additionally, we introduce a class expression f_Ψ which will be used in the next step (Def. 4).

Depending on the constraint type (*Conjunct*, *Required* or *Exclude*) of the element E, the expression f_Ω is built either as intersection, implication or exclusive class expression. However, instead of the element E, we use the corresponding mapping class Map_E (cf. Axiom 9) of the mapping model and the property *feature* that is a superproperty of *hasFeature* (cf. Axiom 10). This guarantees the alignment of classes and properties between Σ_Φ and Σ_Ω.

For the feature model, f_Φ is the intersection of parents and cross-tree constraints of all mapped features ($\mathcal{F}$) of E. The absence of optional child features in the parent definition of the feature model (Axioms 1-3) directly meets the need of the feature representation in f_Φ, since for an optional mapped feature, we can not guarantee the appearance of the corresponding element in each business process model. The set $\mathcal{F}$ of mapped features F is obtained from the mapping knowledge base Σ_M by axioms like $Map_E \sqsubseteq \exists feature.F$. Cross-tree constraints are captured by the expression $cr(F)$. To allow a subsumption checking of the expressions in $cr(F)$, we define the properties *includes* and *excludes* as subproperties of *feature*. The functions *elements* and *element* are abbreviations. The element(s) is/are either the element that require another element or sibling elements, they can be found by using the introduced properties *isRequired* and *sibling*.

Definition 4. *The final knowledge base Σ is constructed from the problem, solution and mapping space knowledge bases, i.e., $\Sigma := \Sigma_\Phi \cup \Sigma_\Omega \cup \Sigma_M$. Moreover, for each mapped and constrained element, an axiom $f_\Psi \equiv \neg f_\Phi \sqcup f_\Omega$ is added to Σ, where f_Φ and f_Ω are defined as follows:*

- *for each element $E \sqsubseteq Conjunct$ and $\mathcal{S} := elements(sibling, E)$:*
  $f_\Omega \equiv \bigsqcap_{E' \in \mathcal{S}} Map_{E'}$ and $f_\Phi \equiv \bigsqcap_{F \in \mathcal{F}} Parent(F) \sqcap cr(F)$
- *for each element $E \sqsubseteq Required$ and $E\prime := element(isRequired, E)$:*
  $f_\Omega \equiv \neg Map_{E'} \sqcup Map_E$ and $f_\Phi \equiv \bigsqcap_{F \in \mathcal{F}} Parent(F) \sqcap cr(F)$
- *for each element $E \sqsubseteq Exclude$ and $\mathcal{S} := elements(sibling, E)$:*
  $f_\Omega \equiv \bigsqcup_{E' \in \mathcal{S}} Map_{E'} \sqcap \neg \bigsqcup (Map_{E''} \sqcap Map_{E'''})$ for $(E'', E''' \in \mathcal{S})$
  and $f_\Phi \equiv \bigsqcap_{F \in \mathcal{F}} Parent(F) \sqcap cr(F)$

Implication Checking. We reduce subsumption checking $f_\Phi \sqsubseteq f_\Omega$ to a classification problem by introducing f_Ψ. According to Def. 4, we add for each subsumption checking problem the corresponding axiom $f_\Psi \equiv \neg f_\Phi \sqcup f_\Omega$. Finally, we check for each class expression f_Ψ whether it is equivalent with the top class ($f_\Psi \equiv \top$). In this case, the solution space constraints (f_Ω) are satisfied, otherwise there is a violation. Moreover, f_Ω is a subclass of the corresponding mapping class Map_E. This directly indicates the element that violates the constraint.

The validation effort is determined by the number of mappings and the number of elements that are involved in one of the constraints (Required, Conjunct and Exclusive). There are two steps where reasoning is applied. (i) We classify the knowledge base in order to find the constrained elements. To build the class expressions f_Ω (for each mapping one expression), we use class subsumption to get the counterparts. For this purpose, we have to iterate over each element that is a subclass of Required, Conjunct and Exclusive, but without any further classification. (ii) The second step is a further knowledge base classification, in order to find those expressions f_Ψ that are not equivalent to the top class $\top$. Class subsumption and classification are both standard reasoning services that are quite tractable in practice. The DL expressivity is $\mathcal{SHOIN}$.

5   Correctness of the Validation

This section demonstrates the correct capturing of the constraints in DL by the implication $f_\Phi \Rightarrow f_\Omega$. We start with an consideration of the constraint coverage by these expressions. Afterwards, we show that well-formedness of the business process model template can be concluded from the implication checking.

Constraint Coverage. In our case, we know that both models are correct on its own. Hence, a violation of the well-formedness constraints can only be caused by the mappings. Our aim is to guarantee the well-formedness of each process model from the business process model template, for each valid feature configuration. We consider different cases how an element E might be involved in a feature mapping. In case the element is not mapped, there might be no violation, since the constraint types only contains mapped elements. E remains in each process model and is not involved in any constraint with another element.

If E is mapped to at least one feature F, it depends whether E is involved in a well-formedness constraint. In case E is not involved, there cannot be any violation of a well-formedness constraint, due to the same reasons as for unmapped elements. More difficult is the case when E is involved in one of the constraints. The constraint expression f_Ω (Def. 4) encodes the corresponding constraint of E with its counterpart elements, e.g., sibling elements. The intention of f_Ω is, that the encoded constraint has to be satisfied in all business process models. Hence, we have to check whether f_Ω holds for each feature configuration.

Again, we know that without any mapping there is no violation in the business process model template. Therefore, we know that only the constraints of the mapped feature F might lead to a violation of f_Ω. The expression f_Φ captures these constraints of all mapped features F of E. We build f_Φ as a conjunction on all these features (cf. Def. 4) to capture the case that there are multiple mappings of one element E. The constraints of the features are directly given by the definition of the parent features and the cross-tree constraints of F.

The alignment is solved by a design decision of the mapping model (cf. Sect. 3.3). The property $feature$ maps an element E to a feature F, by an axiom $Map_E \sqsubseteq \exists feature.F$. The property $hasFeature$ is defined as a subproperty of $feature$ from the mapping space (Axiom 10). Hence, all class expressions from the problem space using the property $hasFeature$ are subsumed by expressions where this property is replaced by its superproperty $feature$. In Def. 4, we use Map_E instead of elements E in the expression f_Ω. Hence, f_Φ and also f_Ω

only contain classes of the feature model and *hasFeature* and *feature* are in a subproperty relation. The expression f_Ω is composed of the mapped elements (Map_E of Σ_M) according to the logical meaning of their constraint classification (*Required, Conjunct* or *Exclude*).

Formula Representation in DL. In the validation, we check whether f_Ψ ($f_\Psi \equiv \neg f_\Phi \sqcup f_\Omega$) is equivalent with $\top$, which means to test whether the subsumption $f_\Phi \sqsubseteq f_\Omega$ holds for each interpretation (tautology). Due to the alignment, we can compare f_Φ and f_Ω by DL reasoning. Finally, we have to demonstrate that this subsumption ensures that the solution space constraints are satisfied for each allowed feature configuration (Lemma 1).

Lemma 1 (Correctness of the Validation). *For mappings from an element E to a set of features $\mathcal{F}$, f_Φ are the constraints of $\mathcal{F}$ and f_Ω the constraints of E. If f_Φ is subsumed by f_Ω then the well-formedness constraints of all elements E hold.*

Proof. Looking to the different types of constraints in both spaces, we basically deal with *implication, and, or* and *xor*. Hence, we have to consider all possible combinations in both spaces and check whether $f_\Phi \Rightarrow f_\Omega$ is a tautology. This kind of logical problem is in the nature of propositional logic. Hence in Def. 4, we define the DL expressions f_Φ and f_Ω in a propositional style. The term connectors are the DL counterparts, e.g., the intersection ($\sqcap$) for an and ($\wedge$). Instead of propositional variables, there are class expressions like $\exists hasFeature.F$ containing features and the properties *feature* and *hasFeature* from Σ_Φ and Σ_M. It is easy to see in Def. 4 that f_Ω is built as a DL expression representing either a conjunctive, exclusive or implicative combination. In f_Φ, we conjunctively connect the parent features and the cross-tree constraints that are already represented in this modeling style.

6 Proof-of-Concept and Discussion

The evaluation of our approach has been conducted by providing a proof-of-concept which has been developed by integrating the FeatureMapper [10] and the transformation of the control flow parts of BPMN to DL, as described in [11].

Setting. We applied the validation to the case study that was introduced in Sect. 2 and is part of the e-store SPL [4]. The feature model consists of 287 features, 2 top features, 192 of the features are leaf features and all others are parent features. There are 21 cross-tree references, including mandatory and optional as well as OR-grouped features. The process model contains 84 activities.

In the settings, we validated feature models with 154 features and with the entire feature model (287 features). In both cases, we build either 22 or 48 mappings. The average validation time using the Pellet reasoner is 2970 ms for 154 features with 22 mappings and 4430 ms for 287 features with 48 mappings. The time for the transformation to DL is less that the validation time. This is based on the fact that we use the DL-oriented feature model of [6] and we only transform the relevant control flow informations of BPMN to DL.

Validation Exemplified.   We demonstrate the validation of one mapping for an easy example from the case study excerpt of Fig. 1. We assume a mapping from the subprocess $WishList$ to the mandatory feature $BusinessManagement$. The mapping is represented in Σ by an axiom $Map_{WishList} \sqsubseteq \exists feature.BusinessManagement$. We expect no constraint violations since the feature $BusinessManagement$ is mandatory. Concerning the validation, f_Φ is build using the parent definition, i.e., $f_\Phi \equiv \exists hasFeature.StoreFront \sqcap \exists hasFeature.BusinessManagement$. The expression f_Ω is build using the class $Map_{WishList}$ from the mapping model ($f_\Omega \equiv \exists feature.BusinessManagement$). For the subsumption checking of f_Φ by f_Ω, we can replace the property $feature$ by $hasFeature$. It is easy to see that due to the negation of f_Φ, f_Ψ is equivalent to the top class $\top$: $f_\Psi \equiv \forall hasFeature.\neg StoreFront \sqcup \forall hasFeature.\neg BusinessManagement \sqcup \exists hasFeature.BusinessManagement$. In case a particular mapping causes a violation, the user finds the corresponding constraint expression f_Ψ classified as not equal to the top class ($\top$).

Lessons Learned.   In Sect. 3.3, we already distinguished the focus of our work on well-formedness constraints to the work on behavioral profiles of [12]. After a deeper comparison of both formalisms, we find two interesting aspects that directly impose further research challenges. Firstly, behavioral profiles are efficient to compare process behavior and behavior consistency. It might be a promising step to extend our well-formedness constraint validation towards a behavior constraint validation, while we still offer the same feature-oriented configuration view in combination with the mappings. Secondly, in this context, we see potential on using DL for the validation. The main challenge from the DL modeling perspective is to handle the possibility of concurrent executions of activities. This problem seems to be in line with the descriptive modeling style of DL to capture this kind of execution potentiality.

7   Related Work

Due to the increasing need of business processes customization, several approaches for the development of families of business processes have been introduced like Schnieders et al. [13], Boffoli et al. [14], La Rosa et al. [15] and van der Aalst et al. [16]. Schnieders et al. [13] model families of business process models as a variant-rich business process model. A configuration of such a family is performed by directly selecting business process elements of variant-rich processes. In order to support such an approach, Schnieders et al. extend BPMN with concepts for modeling variation. However, in order to perform it, such an approach requires from a customer knowledge of business process modeling.

Boffoli et al. [14] and La Rosa et al. [15] also distinguish between business process models and problem space models. Boffoli et al. model problem space as variability table, while La Rosa et al. provide variability by questionnaires. They provide guidance to derive valid configurations, while our aim is to guarantee that for each possible and valid feature configuration there is a corresponding valid process model that satisfies the well-formedness constraints.

More similar to our objective is the approach for process configuration from van der Aalst et al. [16]. Their framework ensures correctness-preserving configuration of (reference) process models. In contrast to our work, they capture the variability directly in the workflow net by variation points of transitions. Accordingly, a configuration is built by assigning a value to the transitions, while our approach uses feature selections.

Weidlich et al. [8,12] derive behavior profiles to describe the essential behavior in terms of activity relations like exclusivity, interleaving and ordering of activities. Weber et al. [17] extend process models by semantic annotations and use them for the validation of process behavior correctness that captures control-flow interaction and behavior of activities. In contrast to our work, their focus is on behavioral constraints, while we consider structural well-formedness constraints. Moreover, our particular emphasis is on the feature-oriented process family representation.

In the context of SPLs several approaches have been introduced, in order to ensure the well-formedness of solution space models. Czarnecki et al. [18] specify constraints on solution space model configurations using OCL constraints. Problem space models, solution space models with OCL constraints, and mappings between them are transformed to Binary-Decision Diagrams.

Thaker et al. [19] introduce an approach for the verification of type safety, i.e., the absence of references to undefined classes, methods, and variables, in solution space models w.r.t. all possible problem space configurations. They specify the models and their relations as propositional formulas and use SAT solvers to detect inconsistencies. Janota et al. [20] and van der Storm [21] introduce approaches to validate the correctness of mappings between feature and component models. They use propositional logics too.

8 Conclusion

As shown in the related work section, our contribution is primarily related to the validation of families of business processes. While the concept of business process families was previously introduced and even covered in our own work [22], there have been very limited (if any) attempts to propose a validation of such families.

Our proposal validates business process models w.r.t. their well-formedness constraints; mappings to problem space models; and dependencies in the problem space models. Hence, unlike other approaches on validation of (model-driven) software product lines, our approach also considers the very nature of business process models through the set of business process practices encoded in control flow patterns. Even though, in this paper, we used BPMN for defining the solution space of business process families, our approach is easily generalizable to other types of business process modeling languages. This can be deduced from control flow patterns used in this paper and control flow support analyzes presented in the relevant literature [23].

We evaluated our work with the largest publicly-available case study, for which we were able to find all the three types of models. While this case study has a realistic size, we would like to have a benchmarking framework which will allow for simulating larger solution space and mapping models. This is similar to what has been already proposed for feature models [24], but now to be enriched for the generation of business process model templates of different characteristics. We plan to organize a user study where the proposed approach will be evaluated by asking software modelers to complete some tasks by applying our tooling. A further plan is to extend our validation formalism towards behavioral constraints in the business process model template.

Acknowledgements. This work has been supported by the EU Project MOST (ICT-FP7-2008 216691).

References

1. Pohl, K., Böckle, G., van der Linden, F.J.: Software Product Line Engineering: Foundations, Principles and Techniques. Springer, Heidelberg (2005)
2. McGregor, J., Muthig, D., Yoshimura, K., Jensen, P.: Successful Software Product Line Practices. IEEE Software 27(3), 16–21 (2010)
3. Czarnecki, K., Eisenecker, U.W.: Generative Programming: Methods, Tools, and Applications. ACM Press, New York (2000)
4. Lau, S.Q.: Domain Analysis of E-Commerce Systems Using Feature-Based Model Templates. Master's thesis, University of Waterloo, Waterloo (2006)
5. Baader, F., Calvanese, D., McGuinness, D., Nardi, D., Patel-Schneider, P.: The Description Logic Handbook. Cambridge University Press, Cambridge (2007)
6. Wang, H., Li, Y., Sun, J., Zhang, H., Pan, J.: Verifying Feature Models using OWL. J. of Web Semantics 5(2), 117–129 (2007)
7. van der Aalst, W., ter Hofstede, A., Kiepuszewski, B., Barros, A.: Workflow Patterns. In: Distributed and Parallel Databases (2003)
8. Weidlich, M., Polyvyanyy, A., Mendling, J., Weske, M.: Efficient Computation of Causal Behavioural Profiles Using Structural Decomposition. In: Lilius, J., Penczek, W. (eds.) PETRI NETS 2010. LNCS, vol. 6128, pp. 63–83. Springer, Heidelberg (2010)
9. Polyvyanyy, A., García-Bañuelos, L., Dumas, M.: Structuring Acyclic Process Models. In: Hull, R., Mendling, J., Tai, S. (eds.) BPM 2010. LNCS, vol. 6336, pp. 276–293. Springer, Heidelberg (2010)
10. Heidenreich, F., Kopcsek, J., Wende, C.: FeatureMapper: Mapping Features to Models. In: ICSE 2008 Companion, pp. 943–944. ACM, New York (2008)
11. Ren, Y., Gröner, G., Lemcke, J., Rahmani, T., Friesen, A., Zhao, Y., Pan, J.Z., Staab, S.: Validating Process Refinement with Ontologies. In: Description Logics. CEUR Workshop Proceedings, CEUR-WS.org, vol. 477 (2009)
12. Weidlich, M., Mendling, J., Weske, M.: Efficient Consistency Measurement Based on Behavioural Profiles of Process Models. IEEE Transactions on Software Engineering 99 (2010)
13. Schnieders, A., Puhlmann, F.: Variability Mechanisms in E-Business Process Families. In: BIS 2006: 9th Int Conf. on Business Information Systems, pp. 583–601 (2006)

14. Boffoli, N., Cimitile, M., Maggi, F.M.: Managing Business Process Flexibility and Reuse through Business Process Lines. In: Cordeiro, J., Ranchordas, A., Shishkov, B. (eds.) ICSOFT 2009. Communications in Computer and Information Science, vol. 50, pp. 61–68. Springer, Heidelberg (2011)
15. Rosa, M.L., van der Aalst, W.M.P., Dumas, M., ter Hofstede, A.H.M.: Questionnaire-based Variability Modeling for System Configuration. SoSyM 8(2), 251–274 (2009)
16. van der Aalst, W.M.P., Dumas, M., Gottschalk, F., ter Hofstede, A.H.M., Rosa, M.L., Mendling, J.: Preserving Correctness during Business Process Model Configuration. Formal Asp. Comput. 22(3-4), 459–482 (2010)
17. Weber, I., Hoffmann, J., Mendling, J.: Beyond Soundness: On the Verification of Semantic Business Process Models. Distributed and Parallel Databases 27(3), 271–343 (2010)
18. Czarnecki, K., Pietroszek, K.: Verifying Feature-Based Model Templates Against Well-Formedness OCL Constraints. In: GPCE 2006, pp. 211–220. ACM, New York (2006)
19. Thaker, S., Batory, D., Kitchin, D., Cook, W.: Safe Composition of Product Lines. In: GPCE 2007, pp. 95–104. ACM, New York (2007)
20. Janota, M., Botterweck, G.: Formal Approach to Integrating Feature and Architecture Models. In: Fiadeiro, J.L., Inverardi, P. (eds.) FASE 2008. LNCS, vol. 4961, pp. 31–45. Springer, Heidelberg (2008)
21. van der Storm, T.: Generic Feature-Based Software Composition. In: Lumpe, M., Vanderperren, W. (eds.) SC 2007. LNCS, vol. 4829, pp. 66–80. Springer, Heidelberg (2007)
22. Mohabbati, B., Hatala, M., Gašević, D., Asadi, M., Bošković, M.: Development and Configuration of Service-Oriented Systems Families. In: Proceedings of the 26th ACM Symposium on Applied Computing (2011)
23. Wohed, P., van der Aalst, W.M.P., Dumas, M., ter Hofstede, A.H.M., Russell, N.: On the suitability of BPMN for business process modelling. In: Dustdar, S., Fiadeiro, J.L., Sheth, A.P. (eds.) BPM 2006. LNCS, vol. 4102, pp. 161–176. Springer, Heidelberg (2006)
24. White, J., Schmidt, D., Benavides, D., Trinidad, P., Ruiz-Cortés, A.: Automated Diagnosis of Product-Line Configuration Errors in Feature Models. In: SPLC 2008, pp. 225–234. IEEE Computer Society, Los Alamitos (2008)

Using SOA Governance Design Methodologies to Augment Enterprise Service Descriptions

Marcus Roy[1,2], Basem Suleiman[1], Dennis Schmidt[1],
Ingo Weber[2], and Boualem Benatallah[2]

[1] SAP Research, Sydney NSW 2060, Australia
[2] School of Computer Science and Engineering, UNSW, Sydney NSW 2052, Australia
{m.roy,basem.suleiman,dennis.schmidt}@sap.com,
{m.roy,ingo.weber,boualem}@cse.unsw.edu.au

Abstract. In large-scale SOA development projects, organizations utilize Enterprise Services to implement new composite applications. Such Enterprise Services are commonly developed based on service design methodologies of a SOA Governance process to feasibly deal with a large set of Enterprise Services. However, this usually reduces their understandability and affects the discovery by potential service consumers. In this paper, we first present a way to derive concepts and their relationships from such a service design methodology. Second, we automatically annotate Enterprise Services with these concepts that can be used to facilitate the discovery of Enterprise Services. Based on our prototypical implementation, we evaluated the approach on a set of real Enterprise Service operations provided by SAP. Our evaluation shows a high degree of annotation completeness, accuracy and correctness.

Keywords: SOA Governance, Enterprise Services, Annotation.

1 Introduction

Service-oriented Architectures (SOA) allow developers to create flexible and agile composite applications by reusing existing and loosely coupled Web services [5]. Such Web services can be roughly grouped into two categories: public and corporate Web services. On the one hand, there is a large body of public Web services currently available on the Web (cf. seekda[1]). These Web services are typically scattered across various domains (e.g. finance, weather etc.) with heterogeneous service definitions solely based on the service provider's preferences. On the other hand, modern Enterprise Applications, e.g. Enterprise Resource Planing (ERP), are often based on the SOA paradigm enabling organizations using these applications to expose internal data and functionality as (mainly proprietary) Web Services. In this context, Web services are referred to as *Enterprise Service* (ES) [11]. In contrast to public Web services, Enterprise Services are mostly offered through internal[2] and centralized UDDI-like repositories, e.g.,

---

[1] http://webservices.seekda.com (28.579 Web Services 11/2010).
[2] Some ESs may also be exposed publicly; both categories are not necessarily mutually exclusive.

H. Mouratidis and C. Rolland (Eds.): CAiSE 2011, LNCS 6741, pp. 566–581, 2011.

SAP's Enterprise Service Repository[3]. Their service design process is typically standardized and aligned to some kind of business modeling [21], endorsing a close cooperation and coordination of an organization's business and IT departments [27]. In order to manage a great number of ESs, which can easily be in the thousands for real-world enterprise applications [13], some application providers apply governance processes to manage their SOA – summarized under *SOA Governance* – to cover various area of service design, implementation, and provisioning. We herein focus on the design aspect, where SOA Governance often defines best practices to foster high reuse and avoid duplication of Enterprise Services. This includes guidelines to unambiguous naming of interfaces and operations for an increasing number of Enterprise Services. However, the gains in manageability for organizations developing services comes at the price of decreased understandability for service consumers. To demonstrate this point, we refer the reader to the examples of Enterprise Service operation names as shown in Table 1. These have been returned as part of a keyword search for ESs related to "Sales Order" on SAP's ESR. From the perspective of less professional and non SAP-educated developers, these examples of Enterprise Services interfaces may be perceived as (i) long, (ii) technical and (iii) similar sounding. The authors in [3] demonstrated that these characteristics – among others – impose additional difficulties to cognitively understand and locate Enterprise Services.

Table 1. Examples of Enterprise Services taken from SAP's ESR

#	Enterprise Services
1	SalesOrderItemScheduleLineChangeRequestConfirmation_In
2	SalesOrderERPCreditManagementApproveRequestConfirmation_In
3	SalesOrderERPItemConditionPropertyByIDQueryResponse_In

In order to facilitate techniques, e.g. discovery, that can be built on top of Enterprise Services, it is beneficial to augment the description of Enterprise Services as described hereinafter. In most cases, Enterprise Services are defined in terms of a standardized description language, e.g. WSDL[4], that defines what operations can be invoked using which parameters. Description languages like WSDL are, however, not well suited to sufficiently and formally describe the meaning of Enterprise Services. In such cases, semantic languages, e.g. RDF[5], allow to add and relate semantic concepts (e.g. defined by means of ontologies) to parts of the service description. Although semantic approaches have been shown to effectively improve e.g. service discovery and composition, they still require mostly manual effort to create required ontologies and generate related annotations.

---

[3] ESR: http://www.sdn.sap.com/irj/sdn/nw-esr (>4000 Enterprise Services 11/2010).

[4] Web Service Description Language (WSDL) 1.1: http://www.w3.org/TR/wsdl

[5] Resource Description Framework:
http://www.w3.org/TR/1999/REC-rdf-syntax-19990222/

In this work, we address the latter point by utilizing service design methodologies from a SOA Governance process to automatically annotate Enterprise
Service descriptions. Our approach can be regarded as deriving semantic annotations from ESs to make their discovery easier. We therefore propose a solution
to automatically augment the description of Enterprise Services with concepts
stemming from such a design methodology. Using the example of an SAP service design process, we first identified this service design methodology as well
as naming conventions used as part of a SOA Governance process (cf. Figure 1
- Conceptual Layer). Second, we used RDF/S[6] to describe and represent this

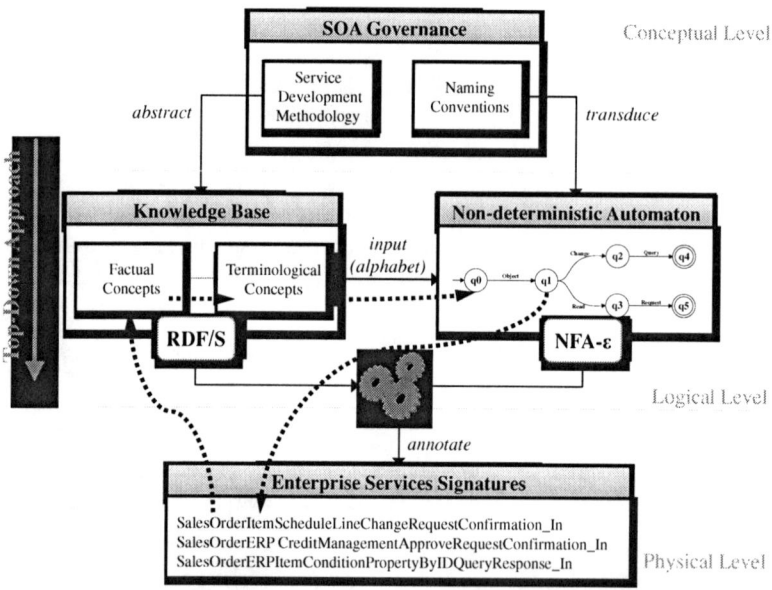

Fig. 1. Architectural overview of our approach to utilize SOA governance

service development methodology, abstractly. We formally defined this representational schema as a hierarchy of terminological and factual concepts, referred to
as service knowledge base (cf. Figure 1 - Logical Layer). Concepts in that knowledge base – among others – are used to define signatures of Enterprise Services.
We then use non-deterministic automata to formally define a language of Enterprise Service signatures derived from available naming conventions (cf. Figure 1
- Logical Layer). We refer to the knowledge base and automaton as service annotation framework. Third, we applied our framework to the example of SAP
to automatically annotate Enterprise Services (cf. Figure 1 - Physical Layer).
That is, an Enterprise Service signature that is accepted by the automaton is a
valid concatenation of words (i.e. factual concepts) that relate to terminological

---

[6] RDF Schema: http://www.w3.org/TR/rdf-schema

concepts in the service knowledge base. The set of detected concepts is then used to annotate the respective Enterprise Service. The corresponding steps are illustrated as dashed lines in Figure 1. Our prototypical implementation shows that a majority of available Enterprise Services can be annotated automatically. An evaluation of generated annotations in fact demonstrates a high degree of completeness, accuracy and correctness.

In the remainder, we explain an example of SOA Governance in Section 2. We then describe the service annotation framework in Section 3. In Section 4, we present a solution and evaluate generated annotations in Section 5. Finally, we refer to related work in Section 6 and provide a conclusion in Section 7.

2   SOA Governance

Organizations use SOA Governance to better manage their SOA development. This has been ascertained by [16], predicting a considerably high risk of failures for midsize to large SOA projects of more than 50 Web Services without any applied SOA governance mechanism. Large companies such as SAP, IBM and Oracle employ SOA Governance mechanisms to ensure a consistent, effective and business aligned development of SOA-based applications. We refer to SOA Governance as policies to make consistent decisions on how to build usable and long-living services. SOA Governance, similar to IT Governance, typically covers multiple phases of a service life-cycle, e.g. service planning, design, definition, implementation etc. [10]. In terms of service design, developers are guided in their task to create interfaces of future Enterprise Services. Governance applied during the design phase typically encompasses guidelines and best practices to effectively create services that are (ideally) mutually exclusive and exhaustive regarding coverage of functionality. It also creates a common agreement and semantic alignment of concepts used during the service development. For instance "Sales Order" is a business entity defining a contractual order that is commonly understood by developers, customers and partners across corporate boundaries.

2.1   SOA Governance - A Service Design Example

As a motivating example, we describe one possible way to abstract and utilize information used as part of SAP's service development methodology for creating Enterprise Services. Specifically the definition of Enterprise Service signatures can be quite versatile, encompassing the use of multiple (i) concepts arranged by some kind of (ii) naming convention [2,21]. We herein refer to concepts as *terminological concepts* and to its instance data as *factual concepts*. The schema in Figure 2 shows an example of such concepts as part of our definition of a *knowledge base*. Particularly, the schema illustrates – but is not restricted to – the application of two main terminological concepts: a domain-specific *data model* and service development *pattern*. We used RDF/S to describe the hierarchy of both types of concepts in a single schema focusing only on a minimal example of terminological concepts, i.e. data model and pattern, and naming conventions used to describe Enterprise Services. We refer to the Enterprise Service (S1) as an end-to-end example to illustrate used design principles as well as our approach:

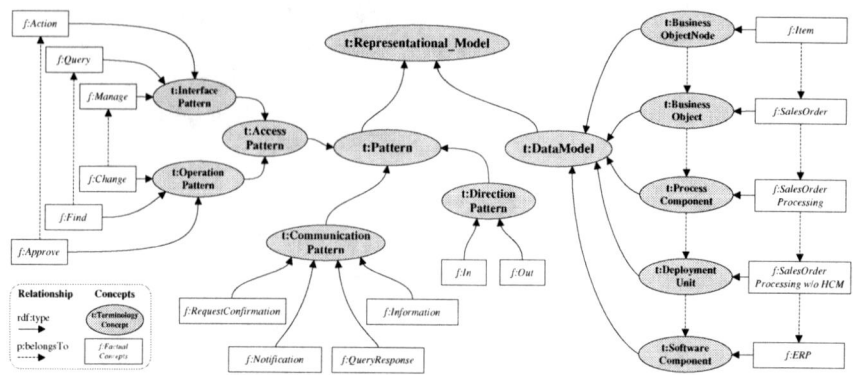

Fig. 2. RDF Schema of the knowledge base showing terminological and factual concepts

SalesOrderItemScheduleLineChangeRequestConfirmation_In   (S1)

Signature (S1) defines an incoming request operation (invoked by a seller) to change the schedule line of an item contained in an existing sales order of a specific customer. From this signature, we can recognize factual concepts that have been modeled as part of our schema. On the one hand, the terms "Sales Order", "Item" and "Schedule Line" refer to factual concepts belonging to terminological concepts listed under data model (cf. Fig. 2). On the other hand, the terms "Change", "Request" and "In" represent factual concepts of a particular service development pattern, also a terminological concept. In a final step, these terms are arranged according to a specific order that is defined by a set of naming conventions. Next, we explain terminological concepts of the data model and pattern and show how they are utilized to create Enterprise Service signatures using naming conventions.

A domain-specific data model. Typically, Enterprise Services are built on key entities of a domain-specific data model. For instance, signature (S1) is specified according to the factual concept "Sales Order", "Item" and "Schedule Line" as inferred from the data model defined as part of the knowledge base (cf. Fig. 2). This model represents an abstract and trimmed-down version of an existing information model used by SAP. Such an information model describes, in more detail, the relationship of business-related entities used by an organization to realize internal operations. Each terminological concepts has factual concepts associated with it. Applying this to our example, the terminological concept "Business Object Node" (BO Node) belongs to "Business Object" (BO), which implies that the factual concept "Item" belongs to "Sales Order": <u>SalesOrderItem</u>ScheduleLineChangeRequestConfirmation_In

Service development patterns. Apart from the data model, the service knowledge base also describes development patterns recurrently used by developers to uniformly define Enterprise Services. These patterns can further be separated into access, communication and direction pattern (cf. Fig. 2). Firstly, the access pattern specifies predefined ways on how to access objects in the data model, e.g. "create" or "change". Secondly, the communication pattern describes the type of interaction to be e.g. "request-confirmation" to define a request operation that causes some action and a confirmation message being returned. Thirdly, the direction pattern describes a service operation to be either "inbound" (e.g. incoming request to create a sales order) or "outbound" (e.g. outgoing credit card authorization request). We derived these patterns directly from available SOA Governance information. Referring to (S1), the signature employs factual concepts "change", "request-confirmation" and "in" belonging to terminological concepts access, communication and direction pattern respectively: `SalesOrderItemScheduleLineChangeRequestConfirmation_In`

Naming Conventions. The examples (N1)-(N3) below illustrate that naming conventions already come in a pre-structured way (cf. [21]) and are used to guide the creation of Enterprise Service signatures. The showcased syntax is defined on the basis of non-terminal symbols, i.e. terminological concepts, and terminal symbols, i.e. factual concepts.

$$NC_1 = \texttt{<BO>(<BO Node>)}^* \tag{N1}$$

$$NC_2 = \texttt{("Change"|"Create"|"Cancel"|"Read")} \tag{N2}$$

$$NC_3 = \texttt{<CommunicationPattern>} \tag{N3}$$

In terms of signature definition, (N1) for instance describes that any factual concept related to a BO Node has to be positioned after the factual concept of a particular BO. Finally, to receive a single (possibly incomplete) building rule, we consolidated naming conventions (N1), (N2) and (N3) based on documentation that outlines how to combine them. We further substituted the access pattern for the terminal words in (N2). As a result, the building rule (B1) represents one possible way of (partially) describing our Enterprise Service signature (S1):

BR = `<BO><BO Node><AccessPattern><CommunicationPattern>`

= `SalesOrderItemScheduleLineChangeRequestConfirmation_In` (B1)

3   Automated Annotation Framework

In this section, we formally describe the two components used in our framework to automatically annotate Enterprise Services, i.e. the service knowledge base and the non-deterministic automaton. These components are based on our modeling of SOA Governance as described in the previous Section. Therefore, we first define the service knowledge base as an abstract model and data that represents a given service design methodology (cf. data model and pattern in Section 2.1). Second, we define a non-deterministic automaton describing a formal language of Enterprise Service signatures. To illustrate this procedure, we will use the naming conventions from Section 2.1.

3.1   Service Knowledge Base

In the following, we formally define the service knowledge base as $\mathcal{KB} = \langle D, F \rangle$. We refer to D as a set of terminological concepts describing an abstract representation of a service development methodology. With F, we describe a set of factual concepts that are mapped to related terminological concepts in D.

Definition 1 (Methodology Representation D). *We define the abstract representation D as a directed graph $D = (T, R, E)$ with T representing a set of terminological concepts $T = \{t_1, \ldots, t_n\}$, R denoting a set of relationships $R = \{r_1, \ldots, r_m\}$ and E a set of directed edges between two terminological concepts belonging to a specific relationship such that $E = \{e_1, \ldots, e_k\}$ with $e_i = (t_o, r_y, t_p)$, $0 \le i \le k$, $0 \le o, p \le n$, $0 \le y \le m$.*

Example 1 (Methodology Representation D). We use the example of terminological concepts as described in Section 2.1 to define the conceptual part $D = (T, R, E)$ of the service knowledge base $\mathcal{KB} = \langle D, F \rangle$. As such, the representation D describes a child-relationship of the concept "Business Object Node" to the concept "Business Object".

$T := \{t_1, t_2\} = \{\texttt{Business Object}, \texttt{Business Object Node}\}$,
$R := \{r_1\} = \{\texttt{containsBON}\}$, $E := \{e_1\}$ with
$e_1 := (t_1, r_1, t_2) = (\texttt{Business Object}, \texttt{containsBON}, \texttt{Business Object Node})$

Definition 2 (Factual Concepts F). *We define F as a set of factual concepts $F = \{f_1, f_2, \ldots, f_m\}$. We further define a mapping $\Phi : F \to T$, $\Phi(f) = t$ for $f \in F, t \in T$, such that $\forall f \in F : \exists t \in T : \Phi(f) = t$. Furthermore, for each $t \in T$ we denote the (possibly empty) subset $F_t \subset F$ such that $\forall f \in F_t : \Phi(f) = t$. Obviously these subsets are distinct for different t, i.e. $\forall t_i, t_j \in T : t_i \ne t_j \to F_{t_i} \cap F_{t_j} = \emptyset$*

Example 2 (Factual Concepts F). Referring to the examples of factual concepts as shown in Section 2.1, we use F to represent a set of factual concepts, i.e. "Sales Order", "Purchase Order" and "Item". We further have distinct subsets F_{t_1} and F_{t_2} of F, whereas "Sales Order" and "Purchase Order" represent F_{t_1} and "Item" forms F_{t_2}. The mapping Φ describes the relationship of factual concepts in F_{t_1} and F_{t_2} to T, which practically relates "Sales Order" and "Purchase Order" to "Business Object" and "Item" to "Business Object Node".

$F := \{f_1, f_2, f_3\} = \{\texttt{Sales Order}, \texttt{Purchase Order}, \texttt{Item}\}$,
$\Phi(f_1) := t_1 \to \Phi(\texttt{Sales Order}) = \texttt{Business Object}$
$\Phi(f_2) := t_1 \to \Phi(\texttt{Purchase Order}) = \texttt{Business Object}$
$\Phi(f_3) := t_2 \to \Phi(\texttt{Item}) = \texttt{Business Object Node}$
$F_{t_1} := \{f_1, f_2\} = \{\texttt{Sales Order}, \texttt{Purchase Order}\}$, $F_{t_2} := \{f_3\} = \{\texttt{Item}\}$

3.2   Service Signature Automaton

In this section, we use the notation of a non-deterministic finite automaton (NFA) with ε-moves to formally define a language of accepted Enterprise Services signatures. We use previously defined terminological concepts T (Def. 1) as the set of input symbols (i.e. alphabet) to initiate state changes. A path through this automaton, i.e. a finite sequence of connected states, ending in a final state represents a concatenation of terminological concepts that defines a language of Enterprise Service signatures.

Furthermore, we decided for an NFA-ε and against a DFA (deterministic finite automaton) for the following reason. Although we expect SOA governance-compliant Enterprise Services to correctly employ naming conventions, we cannot assume them to be exhaustive to completely describe any Enterprise Services signature. This means that the automaton should be able to ignore parts of the signature that are unknown, i.e. not defined by a naming rule or where concepts are not recognized. For this, we included empty-word transitions (ε-moves).

Definition 3 (Automaton A). *We define the NFA-ε as $A = (Q, T, M, q_0, Z)$ with Q denoting a finite set of states, T used as input symbols, $M : Q \times (T \cup \{\varepsilon\}) \to P(Q)$ as the transition function (including ε-moves) to a powerset of Q, $q_0 \in Q$ representing the start state and $Z \subseteq Q$ denoting a (possibly empty) set of final states. We further define the powerset of a particular state $P(\{q\})$, $q \in Q$ as the set of states that can be reached from q with input $t \in T$ and ε such that $P(\{q\}) = \{p \in Q : q \xrightarrow{t,\varepsilon} p\}$ (ε-closure). The powerset of all states is defined as a union $P(Q) = \bigcup_{q \in Q} P(\{q\})$.*

Example 3 (NFA-ε). In Figure 3, we depicted an example of an automaton consisting of nine states $Q = \{q_0, \ldots, q_8\}$, an alphabet of nine symbols $T = \{t_1, \ldots, t_8, \varepsilon\}$, two accepting states $Z = \{q_7, q_8\}$ and a set of transitions M represented as edges in Figure 3. We further refer to the following examples of Enterprise Service signatures (S1) and (S2) that are accepted by this automaton using the set of transitions M_{S1} and M_{S2}. Moreover, Enterprise Service signature (S2) illustrates the need for ε-moves.

(S1) `SalesOrderItemScheduleLineChangeRequestConfirmation_In`

$$M_{S1} = \{q_0 \xrightarrow{\Phi(\text{Sales Order})} q_1, q_1 \xrightarrow{\Phi(\text{Item})} q_2, q_2 \xrightarrow{\Phi(\text{Schedule Line})} q_2,$$
$$q_2 \xrightarrow{\Phi(\text{Change})} q_3, q_3 \xrightarrow{\Phi(\text{Request Confirmation})} q_5, q_5 \xrightarrow{\Phi(\text{In})} q_7\}$$

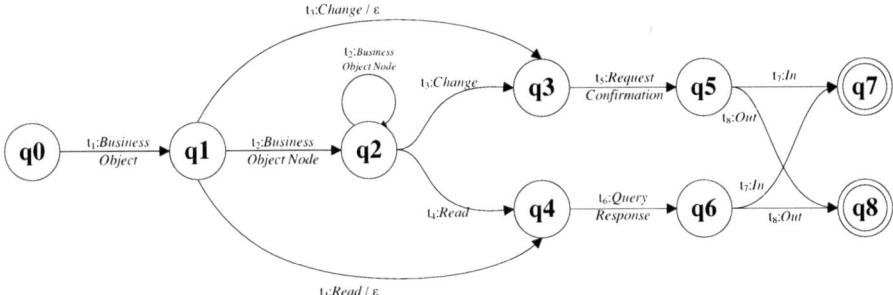

Fig. 3. Example of a NFA-ε accepting e.g. Enterprise Service signature (S1) and (S2)

(S2) `SalesOrderCRMItemSimpleByIDQueryResponse_In`

$$M_{S2} = \{q_0 \xrightarrow{\varPhi(\text{Sales Order})} q_1, q_1 \xrightarrow{\varepsilon} q_4, q_4 \xrightarrow{\varPhi(\text{Query Response})} q_6, q_6 \xrightarrow{\varPhi(\text{In})} q_7\}$$

4   Automated Annotation Solution

In this section, we briefly describe the construction of automata and an accepting algorithm used to detect concepts for the service annotation.

4.1   Automaton Construction

Automata, as defined in the previous section (cf. Def. 3), are generated from a set of rules that represent naming conventions. As referred to in Section 2.1, these rules consist of the following two parts. Firstly, there are non-terminal and terminal elements, e.g., `<BO>` and `SalesOrder` respectively, representing conditions for automata transitions as shown in (N1) - (N3) in Section 2.1. Each rule is transformed into a regular expression, which is abstractly defined using terminological terms $t \in T$, e.g. `/<BO>/`. Subsequently we replace the terminological concepts with the respective set of factual concepts $f \in F$ from the knowledge base – for instance `/<BO>/` → `/(SalesOrder|PurchaseOrder|...)/`.

Secondly, concatenation instructions define the automaton's structure as shown in (B1) in Section 2.1. These instructions specify the actual states. For each of these states, we assign potential input transitions, from the just created transition set, that need to be satisfied to enter this state. Furthermore, these rules define the states that are subsequently reachable.

Based on these constructed automata, Enterprise Service signatures are received as input to start the annotation procedure as described hereinafter.

4.2   Annotation Procedure

The actual annotation of Enterprise Services is realized by Algorithm 1, which accepts Enterprise Service signatures by detecting used concept. This accepting algorithm works recursively, starting with state q_0. Each state checks its incoming state transition, i.e. compares its own regular expression against the input provided. If the input transition is satisfied (line 1), i.e. a valid concept was found, the matched part is stored (line 2) and removed from the beginning of the input and passed on to its subsequent states (line 5). In case of an empty-word transition, obviously the same input is passed on to its subsequent states. When the base case has been reached in form of an accepting state (line 3), it returns its matched concept to its predecessors (line 13). For each of its preceding states, a set of matched concepts, i.e. their own plus previously matched concepts (line 10), is returned to their predecessors and so forth. The choice on what match results are returned depends on the quality of matched concepts. Therefore, the result with the largest number of different concepts covering most of the input is chosen (line 6 and 7). Eventually, the recursive algorithm terminates, detecting a maximum number of concepts used to annotate Enterprise Services.

We implemented this algorithm in Java with roughly 2000 lines of code. The knowledge base and generated annotations are stored as RDF triples. The

Algorithm 1. Annotate Enterprise Service: $annotate(q, i)$

Input: $q \in Q; i \in F^*$
Output: $A \subseteq TF; TF = \{(t, f)\}, t \in T, f \in F$

1: **if** $\exists w, w', q' : i = ww' \wedge w, w' \in (F \cup \{\epsilon\})^* \wedge q \xrightarrow{\Phi(w)} P(\{q\}) \wedge q' \in P(\{q\})$ **then**
2:    $A \leftarrow \{(\Phi(w), w)\}$
3:    **if** $q' \notin Z$ **then**
4:      **for** $p \in P(\{q\})$ **do**
5:        $S' \leftarrow annotate(p, w')$
6:        **if** $|S'| > |S|$ **or** S is undefined **then** {choose only the best path}
7:          $S \leftarrow S'$
8:        **end if**
9:      **end for**
10:     $A \leftarrow A \cup S$ {S undefined $\rightarrow$ A undefined}
11:    **end if**
12: **end if**
13: **return** A {undefined A indicates a failed path}

Call: annotate(q_0, SalesOrderItemScheduleLineChangeRequestConfirmation_In)

Sesame framework in version 2.3.2[7] has been used for RDF storage, querying, and inferencing.

5 Automated Annotation Evaluation

In this section, we evaluate our automated annotation approach using our prototypical implementation (cf. Section 4) based on a set of Enterprise Services. We then analyze the annotation results in terms of *completeness*, *accuracy* and *correctness*, which we define throughout the section. These criteria allow a direct evaluation of our annotation approach; a more exhaustive user study will be conducted once we devised a fully-fledged search.

5.1 Evaluation Environment

We conducted our evaluation using 1654 Enterprise Service signatures taken from SAP's ARIS Designer from various SAP applications, such as ERP, CRM, etc. These services are from the group of so-called A2X (Application-to-Unknown) ESs, as they have a coherent naming scheme. Based on our representational model, i.e. terminological concepts and their relationships, we automatically extracted the corresponding instance data, i.e. the factual concepts, from semistructured sources to populate the service knowledge base. Examples of these sources are Web pages from SAP's Enterprise Service Workplace that provide documentation for each Enterprise Service as well as internal documents (Excel sheets and the like) exported from SAP's ARIS platform. To extract and store this information as RDF triples, we developed extractors for each source of information (i.e. HTML/XML/Excel extractor). As a next step, we used documents

---

[7] http://www.openrdf.org/

describing naming conventions [21] to define the corresponding automaton as described in Definition 3. Finally, we used the 1654 Enterprise Services as the input to the algorithm as described in the previous section. The resulting set of detected concepts has been stored as annotations in form of RDF triples referencing the original Enterprise Service. This is the basis of the analysis below.

5.2   Annotation Completeness

The annotation completeness represents the number of Enterprise Services that have been partially or fully annotated. For this, we calculated the expected maximal number of annotations for each Enterprise Service operation by taking advantage of their Camel Case notation. We refer to the annotation accuracy as the ratio of actual number of generated annotations compared to the expected number of annotations. To determine the annotation completeness, we only considered Enterprise Services with an accuracy greater than null. As a result, we achieved an overall annotation completeness of 1583 out of 1654 Enterprise Services, which is equivalent to 95.7%. The missing 4.3% stem from Enterprise Service signatures that did not comply with existing naming conventions.

5.3   Annotation Accuracy

In this part of the evaluation, we only considered the 1583 fully or partially annotated Enterprise Services from above. To determine the accuracy of annotations, we grouped them into categories of 100% to 40% annotation accuracy (using a 10% scale). In terms of annotation accuracy, we refer to the ratio of actual vs. expected annotations from the previous section. We set the lower margin to 40% based on the lowest accuracy of all 1583 annotated Enterprise Services. For that level of accuracy, we only found four Enterprise Services. In fact, less than 1% of Enterprise Services have been annotated with less than 50% accuracy. On the other hand, the majority of Enterprise Services, i.e. 73.0%, have been fully annotated as illustrated in Figure 4. For an annotation accuracy of 80% or more, the percentage of annotated Enterprise Services increases to 91.4%. The whole procedure on the entire data set took less than 5 minutes on an Intel(R) Core(TM)2 Duo Processor T7300 machine @ 2GHz CPU and 3GB of RAM. These numbers lead to two observations: (i) the naming conventions were largely followed in the tested sample of ESs, and (ii) our approach delivered an effective annotation.

5.4   Annotation Correctness

To the best of our knowledge, there is no obvious solution to automatically validate the correctness of any generated annotation; the baseline therefore is manual verification. Therefore, we first selected a 10% sample of the completely annotated Enterprise Services to evaluate their correctness. Half of these services were strategically selected by a domain expert to cover various applications as well as a variety of design concepts and naming conventions. The other half was randomly selected to avoid any biased decision regarding the selection of concepts. In a second step, an independent expert in SOA Governance has been

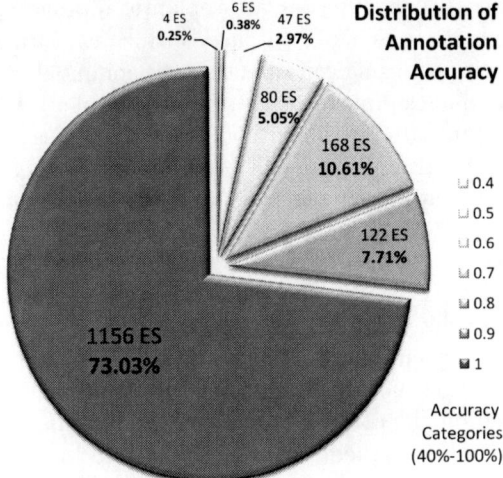

Fig. 4. Result Categories of the Annotation Accuracy Experiment

briefed on the workings of the service annotation approach. He then manually performed the annotation based on the technical names of these 10% sample Enterprise Services. In a final step, the expert's manual annotations have been analyzed and compared to the automatically generated annotations by one of our developers. As a result, we confirmed that approximately 94% of automated annotations match the manual annotations in terms of number of annotations and annotated concepts. We further investigated the 6% range of annotation mismatches, which we separated into two categories: governance violation and missing record. The former category describes Enterprise Services that, for whatever reason, did not fully comply with the SOA Governance design methodology. The latter denotes that the specific concept is missing in the service knowledge base and has been wrongly annotated as a different concept. While mismatches resulting from governance violations are out of our control, mismatches stemming from missing records can be fixed by extending the knowledge base. Either way, we believe that mismatches of less than 6% can be considered acceptable.

6   Related Work

Before we examine related work on proposed annotation techniques, we first outline some fundamental approaches that have been proposed to add additional knowledge (metadata) to the technical description of services, e.g WSDL. For instance, linguistic approaches using clustering techniques – as done in the Woogle search engine [8] – aim to derive meaningful concepts from the technical description of Web Services using WSDL. Although it effectively improves the search, the approach lacks certainty about generated concepts making an automatic solution less feasible. In contrast, semantic approaches [4,25,23,26,1,18] are based on an explicit modeling of ontologies which capture a specific domain knowledge in a generic way using concepts that are commonly agreed and understood [9]. As a consequence, machines become more able to interpret data and documents that are annotated with concepts from ontologies requiring less or no human intervention. In general, research communities distinguish between two types of metadata, i.e. ontological concepts and annotations encoding references to concepts.

In terms of ontological concepts, we described what can be seen as a rather closed ontology derived from a given methodology. Such an ontology can be considered nearly complete with respect to the universe of discussion. However, we did not focus on the semi-automatic creation of ontologies [15,6,12] such as generating ontologies from natural language text or automatic reuse of existing ontologies, e.g. based on the context of a document [22]. In fact, reusing public ontologies, e.g. TAP [7] or others, are less applicable in context of our work as design methodologies in an enterprise context rather describe proprietary concepts (e.g. data model) that cannot be necessarily expressed with public or community-contributed [24] ontologies.

In terms of annotations, existing approaches can be first categorized into degrees of automation, i.e. fully-automated, semi-automated [9,15] or manual [14,19]. Second, the annotation level of detail depends on the type of representation ranging from e.g. entire service descriptions, operation or input/ouptut parameters, non-functional requirements etc. Third, annotations can be used for different fields of applications, e.g. human or automatic service discovery [17], invocation, composition [22] etc. In this work, we proposed a fully-automated approach to annotate a large set of Enterprise Services based on their signatures, i.e. interface and operation names. Ultimately, we currently see the main purpose of the service annotation approach in improving the service discovery specifically for business users and non-professional developers. In this context, using automata can more precisely resolve disambiguities by determining correct concepts based on their expected position within the respective signature. As a result, we can increase the accuracy of generated annotations compared to other approaches, e.g. using similarity functions [7] (SemTag), natural phrase processing [9] (via SMES) or detecting association of concepts by confidence level [15]. Note, this is particularly feasible as we only focus on a small portion of explicitly defined text, i.e. Enterprise Service signatures, rather than a large body of text, i.e. Web pages.

For reasons of simplicity, we used RDF(/S) over OWL-S or WSMO to store our ontology and annotations. We consider RDF(/S) sufficient for our purpose to represent additional knowledge as we do not require sophisticated logical reasoning. The results of the annotation could, e.g., be stored in SAWSDL [25,23] format, where annotations are directly added to WSDL. However, at this point we are not entitled to change Enterprise Service descriptions, and decided to keep annotations independent from any specific Web Service standard.

7 Conclusion and Future Work

In this paper we presented an approach for the automatic annotation of Enterprise Services, based on a SOA Governance design methodology. We described a concrete methodology used at SAP, but presented a generic and formal model for capturing the structure of SOA Governance design methodologies. The model consists of terminological concepts and factual concepts, and automata for capturing naming conventions built from these concepts. Naming rules are specified using a (typically very small) set of terminological concepts; from those we construct a consolidated automaton and populate it with the respective factual concepts. Using the detailed automaton, we can automatically annotate service names that (at least partially) adhere to the naming conventions.

We evaluated the work on a set of more than 1500 Enterprise Services from SAP, and obtained highly encouraging results: more than 90% of the services could be annotated with more than 80% correctness. This was largely verified in a small experiment with an independent expert. We observed that some of the mismatches came from ESs that did not adhere to naming conventions. As such, our approach can also be used to check adherence to naming conventions, and thus, improve the management of SOA Governance. In terms of the annotation procedure, this is an one-off operation that only needs to be executed when the concepts or the service names change. All the above services were annotated in a matter of minutes. Hence, performance is unlikely to be a problem.

In an earlier instance of this work [20], we used a strongly simplified model, yielding significantly lower accuracy: only parts of the data model and patterns could be annotated. However, we there also showed how these annotations can be used in discovery of Enterprise Services for business users and developers who are unfamiliar with the set of services. As such, we can use our approach to facilitate an effective search by further reaching out into areas of automatic query extension and query suggestion.

Future work will build on the presented approach as follows. Firstly, we will attempt to evaluate the applicability of the approach on services from other sources; there is, however, a high risk that we might not be able to obtain detailed access to naming rules. Secondly, we will investigate improved search algorithms and other application scenarios making use of the produced annotations.

References

1. Akkiraju, R., Farell, J., Miller, J.A., Nagarajan, M., Sheth, A., Verma, K.: Web service semantics – WSDL-S. In: W3C Workshop on Frameworks for Semantic in Web Services (2005)
2. Artus, D.J.: SOA Realization: Service Design Principles (February 2006), http://www.ibm.com/developerworks/webservices/library/ws-soa-design/
3. Beaton, J., Jeong, S.Y., Xie, Y., Stylos, J., Myers, B.A.: Usability Challenges for Enterprise Service-oriented Architecture apis. In: VLHCC 2008: Proceedings of the 2008 IEEE Symposium on Visual Languages and Human-Centric Computing, pp. 193–196. IEEE Computer Society, Washington, DC, USA (2008)

4. Berners-Lee, T., Hendler, J., Lassila, O.: The Semantic Web. Scientific American 284(5), 34–43 (2001)
5. Curbera, F., Khalaf, R., Mukhi, N., Tai, S., Weerawarana, S.: The Next Step in Web Services. Commun. ACM 46, 29–34 (2003)
6. De Silva, L., Jayaratne, L.: Wikionto: A System for Semi-automatic Extraction and Modeling of Ontologies Using Wikipedia XML Corpus. In: IEEE International Conference on Semantic Computing, ICSC 2009, pp. 571–576 (2009)
7. Dill, S., Eiron, N., Gibson, D., Gruhl, D., Guha, R., Jhingran, A., Kanungo, T., Mccurley, K.S., Rajagopalan, S., Tomkins, A., Tomlin, J.A., Zien, J.Y.: A Case for Automated Large Scale Semantic Annotations. Journal of Web Semantics (2003)
8. Dong, X., Halevy, A., Madhavan, J., Nemes, E., Zhang, J.: Similarity Search for Web Services. In: VLDB 2004: Proceedings of the Thirtieth international conference on Very large data bases. VLDB Endowment, pp. 372–383 (2004)
9. Erdmann, M., Maedche, A., Schnurr, H.-P., Staab, S.: From Manual to Semi-automatic Semantic Annotation About Ontology-based Text Annotation Tools. In: Proc. of the COLING 2000 Workshop on Semantic Annotation and Intelligent Content, Luxembourg (August 2000)
10. Erl, T.: Service-Oriented Architecture: Concepts, Technology, and Design. Prentice Hall Professional Technical Reference (2005)
11. Haentjes, V.: SOA Made Easy with SAP (February 2010), http://www.sdn.sap.com/irj/sdn/go/portal/prtroot/docs/library/uuid/903aa937-03f2-2c10-968e-8e7d649cd352
12. Harmelen, F.V., Fensel, D.: Practical Knowledge Representation for the Web. In: In Proc. of the 2000 Description Logic Workshop DL 2000, pp. 89–97 (1999)
13. Hoffmann, J., Weber, I., Kraft, F.M.: SAP Speaks PDDL. In: AAAI (2010)
14. Kungas, P., Dumas, M.: Cost-Effective Semantic Annotation of XML Schemas and Web Service interfaces. In: SCC 2009. IEEE International Conference on Services Computing, pp. 372–379 (September 2009)
15. Maedche, A., Staab, S.: Semi-automatic Engineering of Ontologies from Text. In: Proc. of 12th Int. Conf. on Software and Knowledge Eng., Chicago, IL (2000)
16. Malinverno, P.: Service-Oriented Architecture Craves Governance (October 2006), http://www.gartner.com/DisplayDocument?id=488180
17. Paolucci, M., Kawamura, T., Payne, T.R., Sycara, K.: Semantic matching of web services capabilities. In: Horrocks, I., Hendler, J. (eds.) ISWC 2002. LNCS, vol. 2342, p. 333. Springer, Heidelberg (2002)
18. Patil, A.A., Oundhakar, S.A., Sheth, A.P., Verma, K.: Meteor-S Web Service Annotation Framework. In: Proceedings of the 13th International Conference on World Wide Web, WWW 2004, pp. 553–562. ACM, New York (2004)
19. Rao, J., Dimitrov, D., Hofmann, P., Sadeh, N.: A Mixed Initiative Approach to Semantic Web Service Discovery and Composition: SAP's Guided Procedures Framework. In: ICWS 2006: Proc. of the IEEE Int. Conf. on Web Services, pp. 401–410. IEEE Computer Society, Washington, DC, USA (2006)
20. Roy, M., Suleiman, B., Weber, I.: Facilitating enterprise service discovery for non-technical business users. In: Maximilien, E.M., Rossi, G., Yuan, S.-T., Ludwig, H., Fantinato, M. (eds.) ICSOC 2010. LNCS, vol. 6568, pp. 100–110. Springer, Heidelberg (2011)
21. SAP AG. Governance for Modeling and Implementing Enterprise Services at SAP (April 2007), http://www.sdn.sap.com/irj/sdn/go/portal/prtroot/docs/library/uuid/f0763dbc-abd3-2910-4686-ab7adfc8ed92
22. Segev, A., Toch, E.: Context-Based Matching and Ranking of Web Services for Composition. IEEE Transactions on Services Computing (2009)

23. Sivashanmujgam, K., Verma, K., Sheth, A., Miller, J.: Adding Semantics to Web Services Standards. In: Int. Conference on Web Services ICWS 2003 (June 2003)
24. Staab, S., Angele, J., Decker, S., Erdmann, M., Hotho, A., Maedche, A., Schnurr, H.-P., Studer, R., Sure, Y.: Semantic Community Web Portals. In: Proc. of the 9th Int. WWW Conference on Computer Networks, pp. 473–491. North-Holland Publishing Co., Amsterdam (2000)
25. Verma, K., Sheth, A.: Semantically Annotating a Web Service. IEEE Internet Computing 11(2), 83–85 (2007)
26. Vitvar, T., Mocan, A., Kerrigan, M., Zaremba, M., Zaremba, M., Moran, M., Cimpian, E., Haselwanter, T., Fensel, D.: Semantically-enabled Service Oriented Architecture: Concepts, Technology and Application. Service Oriented Computing and Applications 1(2), 129–154 (2007)
27. Woolf, B.: Introduction to SOA Governance (July 2007), http://www.ibm.com/developerworks/library/ar-servgov/

Management Services – A Framework for Design

Hans Weigand[1], Paul Johannesson[2], Birger Andersson[2],
Jeewanie Jayasinghe Arachchige[1], and Maria Bergholtz[2]

[1] Tilburg University, P.O.Box 90153,
5000 LE Tilburg, The Netherlands
{H.Weigand,J.JayasingheArachchig}@uvt.nl
[2] Royal Institute of Technology
Department of Computer and Systems Sciences, Sweden
{pajo,ba,maria}@dsv.su.se

Abstract. The Service-Oriented Architecture has rapidly become the de facto standard for modern information systems. Although recently considerable research attention has been paid to the management of services, several gaps can still be observed. Service management as far as it is automated is either mixed up with the operational service logic itself, or handled in a separate not service-oriented system, such as a BAM platform. In addition, there is a growing business demand for value-driven service management. In this paper, a general framework for management service design is presented that covers both business services and software services and is rooted in the business ontology REA, extended with a REA management ontology. The framework is applied to two different case studies, one in the Italian wine industry and one related to a robot cleaner.

Keywords: service design, REA, autonomic computing, management control.

1 Introduction

The Service-Oriented Architecture (SOA) has rapidly become the de facto standard for modern information systems. Having started with a focus on service description and discovery, SOA research shifted its attention to service composition in the second phase. The next phase, according to [13] in 2007, would focus on service management defined as "the control and monitoring of SOA-based applications throughout their lifecycle". The most prominent functions of service management include SLA management, auditing, monitoring and troubleshooting, dynamic resource provisioning, service lifecycle management (e.g. versioning) and scalability/extensibility. However, although considerable work has been done on these topics in the last few years, results so far are fragmented and limited. Early standards related to service management (MUWS/MOWS; see oasis.org) have become obsolete.

Several research gaps can be observed. Service management as far as it is automated is often mixed up with the operational service logic itself, or it is handled in a separate not service-oriented system, such as a BAM (Business Activity Monitoring) platform. In addition, service management, including business process management, is still mainly focused on execution correctness [1], whereas there is a growing

H. Mouratidis and C. Rolland (Eds.): CAiSE 2011, LNCS 6741, pp. 582–596, 2011.

business demand for value-driven service management. This also requires a better integration of software services and business services [10] – which still is the vision of service science anyway.

Along with SOA, the last decade has witnessed increased research efforts on self-adaptive or autonomic software. We refer to [20, 4] for recent overviews. Self-adaptive software embodies a closed-loop control mechanism that includes sensors and effectors, linked through processes of monitoring, detecting, deciding and acting. Despite considerable progress in sub areas, there are also still many challenges, including the question on how to integrate self-adaptation functionality in the SOA architecture. Is it possible to provide Management as a Service?

The control loop is taken by Forrester analyst Bartels as the backbone for Smart Computing [1]. Smart Computing is supposed to be the challenge of the coming decade and integrates the following technologies: Awareness (sensors, RFID chips, video cameras), Analysis (business intelligence, process mining), Alternatives (rule engines, workflow systems) and Actions (leveraging existing products), with Auditability as an overall concern. The big business challenge lies in optimizing the value of and the return on assets and minimizing the costs and risks from liabilities by far better *real-time* awareness of their status. Assets include both physical resources such as buildings and trucks, but also intangible ones such as software, or brand.

The research objective of this paper is to develop a general framework for management service design that covers both business services (as in Smart Computing) and software services (as in Autonomic Computing). The framework must be value-driven and truly service-oriented. To arrive at rigorous and relevant research results, we use Peffers' design science phases [14]. The *problem identification and motivation* is stated in section 1 and 2. Our *solution objective* is to develop an integrated framework for value-driven service management. In section 3 we lay a formal foundation by extending the REA business ontology with a REA management ontology and introduce a general framework of services that is applied recursively to management services (*design and development*). The framework is used in two case studies (*demonstration*). The first case was developed in the context of the S-Cube project [17] and concerns an Italian wine industry. The advantage of using this case study is the comparison it allows with other approaches. To explore the applicability range of our approach, the second case is about a robot cleaner.

2 Related Research

Given limited space, we can only present a brief overview of all the relevant fields.

The vision of Autonomic Computing was presented by Kephart and Chess in 2003 [7], and recently evaluated in [4]. The evaluation observes that the vision has been broadened to "the application of advanced technology to the management of advanced technology" and as such is still highly relevant. A notable omission in the original vision was the communication component, and research has been devoted to developing a so-called *knowledge plane*. What is also still lacking is an understanding of the broader software engineering aspects of autonomic system development. This includes such basic questions as when a system can be said to be "correct" if its behavior is expected to change over time. [4] also pleads for a comprehensive systems theory for adaptive systems that allows to reason not only about the "what" of monitoring and adaptation but also about the "how". Finally, it claims that current solutions are too much focused on isolated problems, and that we need an integrated autonomic systems engineering approach to avoid undesirable feature interaction.

Within the software engineering domain, runtime adaptation has become a highly relevant research topic. In this context, Model-Driven development and Software Product Lines modeling techniques are used and extended to Dynamic Software Product Lines that include variability transformations [3, 26, 8]. This work confirms the research gap identified by [4] that the "how" of the adaptation is no longer something that can be abstracted from.

Another line of research is considering modeling and monitoring requirements for adaptive systems. Goal-based approaches, such as Tropos [2] have been applied to the specification of requirements of self-adaptive systems. Goal-based models are well suited to exploring high-level requirements and it is natural to use goal models to represent alternative behaviors that are possible when the environment changes. Goal-based models have a natural fit with agent-oriented implementation platforms [11].

In the management literature, system theory and control theory have been around for a long time. In the standard management control textbook of Simons [19], four types of "control systems" are distinguished: diagnostic control systems, interactive control systems, belief systems and boundary systems. *Diagnostic control* systems correspond to the traditional cybernetic approach and are aimed at controlling results using a closed control cycle. This mode of control is important but it also has its limitations, according to Simons. *Belief systems* express norms and core values in the organization that are aimed at controlling (or influencing) the value systems of the people involved. *Boundary systems* constrain the behavior of the organization in the face of risks to be avoided. *Interactive control* is focused on handling uncertainties and on "doing the right thing", rather than doing the things right, as in diagnostic control, and is typically realized in the form of interaction.

3   A Framework for Management Service Design

3.1   The REA Business Ontology

The Resource-Event-Agent (REA) ontology was first formulated in [9] and has been developed further, e.g. in [21,5,6]. The following is a short overview of the core concepts of the REA ontology based on [23].

A *resource* is any object that is under the control of an agent and regarded as valuable by some agent. This includes goods and services. Resources are modified or exchanged in processes. A *conversion process* uses some input resources to produce new or modify existing resources, like in manufacturing. An *exchange process* occurs as two agents exchange (provide, receive) resources. To acquire a resource an agent has to give up some other resource. An *agent* is an individual or organization capable of having control over economic resources, and transferring or receiving the control to or from other agents. The constituents of processes are called *economic events*. REA recognizes two kinds of duality axioms between events: conversion duality and exchange duality.

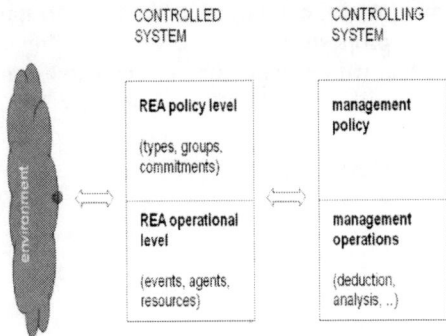

Fig. 1. REA levels of controlled enterprise system, derived from [22]

The event records give an answer to the question "what has happened?", but not to the question "what is planned or scheduled – what *should* happen?" In REA this is modeled at the policy level. This level allows talking about types and groups as well as commitments. Although no specific details for the structure of these types and commitments are given, examples in [5] show how REA does handle schedules, plans and integrity constraints in this way. [22] gives an enterprise architecture based on REA that makes a distinction between controlled system and controlling system (Fig. 1). The controlling system interacts with both levels of the controlled system, monitoring the facts at the operational level and changing the plans, standards, schedules etc at the policy level. As explained by [22], the controlling system is also a subsystem of the enterprise, so it does also have an operational and policy level.

3.2 REA Management Ontology

The business ontology of REA is strong in formalizing the operational level of the enterprise on an abstract economic level. However, the later extensions on the policy level have not been worked out as thoroughly. The occurrence of general abstractions (types, groups) on the one hand and commitments on the other hand in the same basket "policy level" is not really satisfactory. The dynamics of this policy level, e.g. how commitments are created and resolved, are also not covered, although this has a clear business impact and cannot be left out of scope. The work of [22] adds a useful distinction between controlled system and controlling system, but it does not spell out what exactly is processed in the controlling system when it talks about deduction, analysis, etc. Therefore we have developed an extension of REA ("management ontology") that we will use as a basis for our design framework (Fig.2).

The lower left corner summarizes the REA concepts of resource, event and agent (at operational level) as well as types and groups (of resources, events and agents). We group them together under the category *REA referent*. REA referents are "referred" to in *intentional resources*. Three categories of intentional resources are distinguished on the basis of what Searle [18] has called direction-of-fit: *assertives*, *directives* and *evaluatives*. An (basic) assertive says that some REA event has occurred, whereas a directive says that some REA event should occur. Directives are defined as having a world-to-word fit and as such are a generalization of commitments that were already in the REA business ontology, as part of the policy level.

However, unlike types and groups, commitments do not correspond to an abstraction; like assertives, they have a propositional structure with references to REA referents. We have extended REA also with a value dimension. Values qualify REA referents, in terms of good, bad, satisfactory, fast etc. *Evaluations* are intentional resources that are produced using other intentional resources and assign values; they do not have a direction-of-fit, like the expressive in [18].

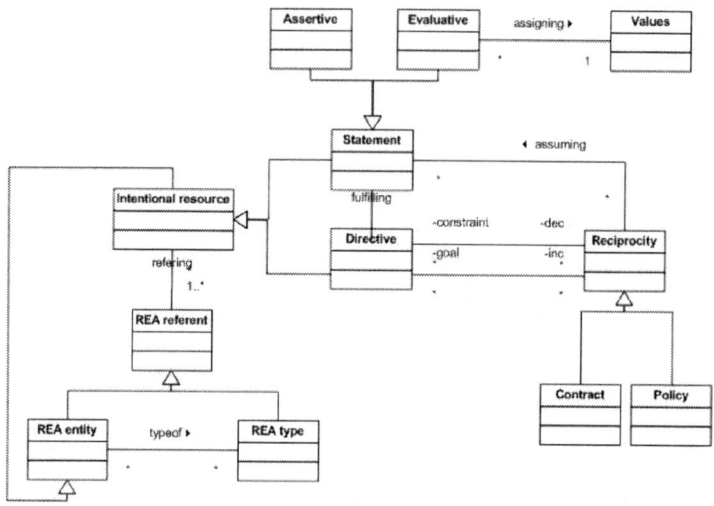

Fig. 2. REA management ontology (UML style)

As any resources, intentional resources are processed by events that we call *intentional events*. For instance, commitments are created or removed by commit and decommit events [24]. Intentional events are typically realized by means of communicative acts to be represented on the process level.

In REA, commitments are grouped together in contracts. A *contract* contains reciprocal commitments (what the agent will give versus what he aims to receive) and may contain additional terms. From a management perspective, *policies* are important concepts as well. According to the BusinessDictionary.com, a policy is "*a set of basic principles and associated guidelines, formulated and enforced by the governing body of an organization, to direct and limit its actions in pursuit of long-term goals*". We have formalized them in analogy to contracts as a group of intentional resources obeying the reciprocity principle: what the agent gives in versus what he aims to achieve. *Constraints* are what the agent gives in (directives that limit the actions of the controlled system) and *goals* are what the agent wants to gain in return (directives to be fulfilled by monitored assertives or evaluations). In addition, the policy may contain *assumptions* in the form of testable assertives.

The definition of policy can be seen as a generalization of the policy pattern in [6], which defines a policy entity as something that encapsulates constraints on economic exchanges and conversions and that "applies" to a group. In our definition, constraints are part of the policy, but always linked via the policy to goals. In our view, the "apply" relationship is a special kind of "referring".

Policies can be defined on two levels, the "what" and the "how". This abstraction is important, since it allows for adaptation of the controlled system, within given boundaries. The way the abstract policy is enforced may depend on the (monitored) situation and on design choices made by the controlling system [25]. In the following, we will use the term "specification" for the enforced (and executable) policy. How far the policy abstracts from the specification is a design choice that will differ from one case to the other.

A *management process* is like a REA economic process containing both conversion events and exchange events. For instance, an inference service *uses* certain assertives and some rule base (formally, these are also assertive) in order to *produce* other assertives. Management discourse is based on an exchange duality balancing incoming assertives ("subscribe") with outgoing ones ("publish").

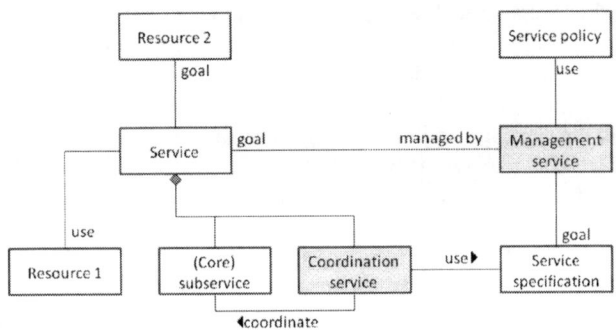

Fig. 3. Generic service model

3.3 Generic Service Model

Our generic service model is based on [23] and illustrated informally in Fig. 3. A service *adds value* to some *resource*, called its goal, consuming or making use of other resources. Including the resources in the service model is in line with REA and has been argued for independently by [16]. A service can be regarded as a small service system and decomposed into *subservices*. *Core subservices* contribute to the service goal, whereas *coordination services* are subservices that coordinate – manage the dependencies between – core subservices [24], using a *specification* derived from the *policy*. We model subservices with a part-of relationship, but in terms of the REA ontology, they are related to the super service by a use relationship just like resources. *Management services* have been defined in [23] as a kind of enhancing services that adds value to a service. They are not subservices of the service in question, with an active role in the operational process, but manipulate the (operational) service by enforcing one or more service policies.

The service model is intended to be generic. It can be applied to business services, such as a transport, or a hair-dressing, but also to a software service. Let us illustrate the transport case: the resources that are used or consumed include the lorry, the petrol and the driver's labor hours, each of them having a certain value. The transport adds value to the transported goods. This value is reflected in the price that the customer is willing to pay for the transport service. Transport *management* is an extensive task that includes capacity planning and scheduling as well as actual route planning. The transport itself can be decomposed into subservices such as driving, unloading and a coordination service for selecting a route. A policy may contain the constraint that the lorry should select the cheapest route in terms of fuel consumption.

The software case can be illustrated as follows. The same transport company may have a tracking & tracing service for customers. The resources used by this software service are informational in nature: in particular, they include the GPS data about the lorries. The service adds value to the transport (the transport service has a higher value for the customer when he can track his cargo – a case of "information enrichment" [12]). For its implementation this web service makes use of utility or infrastructural services based on resources such as CPU time, data storage and the network. Once the service is in place, there may be a management service to support it. For instance, the management service may monitor and adapt the service interface in order to maintain interoperability. Note that management services may be automated or semi-automated independently of the automation of the service itself.

The *value of a service* in a certain period of time is the difference between the total value increase on the one side and the total decrease at the other side. In order to realize value-based service management, it is therefore necessary that each resource is valued. There is a long tradition of cost accounting that we can rely on to implement this requirement, the details of which are not in the scope of this paper. What is important for keeping the valuation consistent is to integrate all services into the *value cycle* of the company, denoted as "cash wheel" in [19]. As traditional audit theory teaches, each company has a value cycle consisting of sales processes that generate money, consuming resources, and purchasing processes that acquire resources spending money. Depending on the type of company – sales, manufacturing, etc. – the value cycle can be drawn a bit more precise, but the structure is always the same. The *value of the value cycle* is the profit that is generated over a certain period of time minus the investments that have been added. Each service is directly or indirectly included in the value cycle, akin to Porter's distinction between primary and support activities of the value chain [15]. Directly included are services such as manufacturing, sales and purchasing. Indirectly included are for instance management services that contribute to primary services or other enhancing services. At the end of the day, the valuations of all services should be consistent, that is, the sum of these valuations should be equal to the value of the value cycle (for a certain period). We regard this principle not only important from an accounting point of view, but also as a useful constraint for business modeling.

3.4 Management as a Service

We have conceptualized management as an enhancing service. In Software Engineering, the idea of separating operational and management concerns is not new. In the field of self-adaptive software, an equivalent distinction is made between internal and external adaptation [20]. Internal approaches intertwine application and adaptation logic. This has certain drawbacks. External approaches use an external adaptation engine or manager that contains the adaptation logic, the other part being called the "adaptable software". By conceptualizing management as a service, we follow an external approach. Note that this is not a formal necessity but an architectural choice.

We propose a *fractal* design approach in the sense that the same generic service model is applied to management services. So a management service is itself something that uses resources. These include operational resources, e.g. labour hours, and intentional resources. The management service may have subservices, such as a monitoring service, and a management subservice may have a manager service. This service-oriented approach increases reusability. The advantage of the fractal approach is that it makes the design completely service-oriented, not only its operational part. This is in contrast to other approaches that for instance conceptualise BDI agents as management services, or monolithic BAM software.

What does a management service actually do? Most current approaches in the field of self-adaptive software follow classical control theory and posit a control loop, also called MAPE cycle (Monitor-Analyze-Plan-Execute). The same cycle underlies the deliberation cycle that is used in multi-agent systems (with Beliefs, Desires and Intentions). Our approach is required to be business-driven and service-oriented. Following the management control literature ([19]) we call it the diagnostic control cycle.

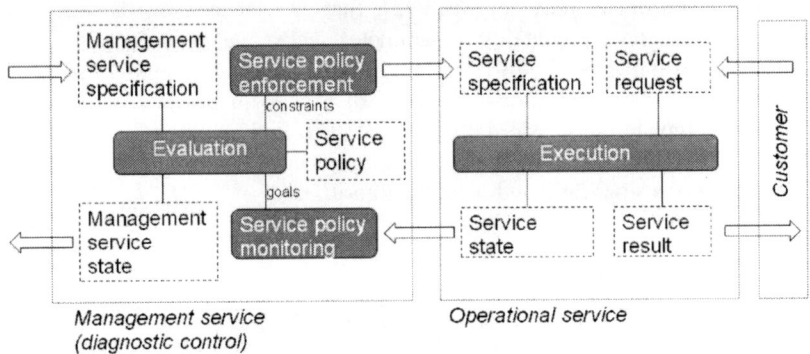

Fig. 4. Diagnostic control cycle together with the service interaction cycle. Dashed rectangles indicate intentional resources, coloured boxes indicate services

Fig. 4 depicts the generic service-oriented management architecture for the diagnostic control cycle. On the right hand side, we see the traditional service interaction cycle: the customer sends a request to the service provider. The execution produces, perhaps iteratively, a certain state that corresponds to and so fulfills the request, and this result is returned to the customer for evaluation. However, the execution does more than that. From a management perspective, the execution is the realization of the service specification. So there is another interaction cycle, between management service and operational service: the manager enforces a *service policy* on the operational service. In the case of software services, the *service specification* may take the form of a BPEL specification, or a set of business rules (cf. [8, 26] for how to implement policy enforcement based on such models). The execution produces a certain state (set of *assertive* – this reporting is also governed by policy constraints). The state information is returned to the manager, where it is typically aggregated by monitoring services and then evaluated. If the evaluation is not satisfactory (does not match the policy *goals*), the service specification is adapted. The policy will usually contain conditional constraints that become effective in the case of contingencies, akin to terms in a contract or a mitigation plan.

It is possible that the operational service policy has to evolve itself. However, this is not the responsibility of the management service. If we want this type of self-adaptation, a second management level has to be introduced, in accordance with our fractal design principle. In that case, the Management Service specification in Fig. 4 is not fixed but itself the result of a Management Policy enforcement process.

Three kinds of management subservices can be distinguished. A *monitoring* service uses and produces assertives. An *evaluation* service uses assertives and produces evaluatives. An *enforcement* service enforces policies, using evaluations and possibly assertives and producing directives. Further specializations are, for instance, sensor services, aggregation services, inference services and data transformation services as specific monitoring services.

The flexibility within the service policy that enables varying enforcements can be realized in several ways that go beyond the scope of this paper. We just want to mention two options. The policy may consist of a fixed set of alternatives, as in the variability transformations approach [3], that are selected on the basis of the evaluation. Or the policy contains a parameter whose value is dependent on the evaluation. An example is "credit level" in an order processing service. If the credit level is too low, the company losses because of non-payments. If the credit level is too high, the company misses sales opportunities. To find an optimal credit level and adapt if when the circumstances change the manager can (re-)calculate the parameter value by means of a stochastic optimization algorithm.

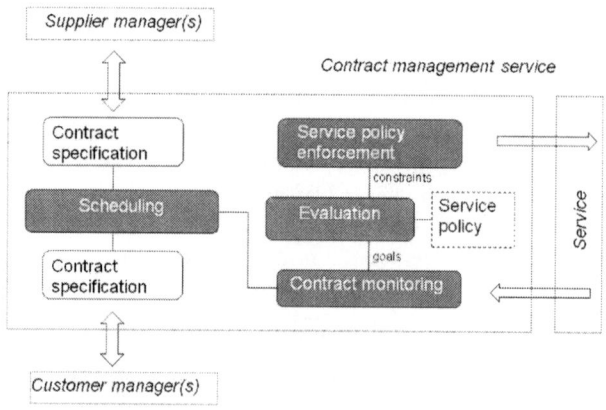

Fig. 5. Contract management cycle

According to [19], diagnostic control is the "automated pilot" that allows the human manager to spend his time on other things, in particular interactive control. The way Simons presents this it is not a homogeneous group. An important subclass concerns interactions with service stakeholders about the service requirements. These requirements can be diverse, but include requirements on future capacity. These requirements are passed through the value chain in reverse direction, from customer (market demands) to sales and further on to production and procurement. To account for this kind of interactive control, we need another management cycle (Fig. 5).

The manager interacts with other managers in the value chain at customer and supplier side. The customer manager's requests do not concern a particular service instance, but a certain state or quality of the service as such. For instance, that the service has a certain capacity at a specified time. This leads to certain *commitments* that are part of a REA *contract*. The synchronization of contracts is what is called scheduling. The purpose of a schedule is to make sure that for all services the needed resources are identified, as well as when they will be needed [6:108]. A schedule is a collection of increment and decrement commitments, as well as mitigation plans. The increment commitments indicate the availability of the service at some future time, or the availability of the resource produced by the service. Decrement commitments concern the resources (subservices or resources produced by subservices) needed to fulfill the increment commitments. These decrement commitments must be gained from the managers of the supplying subservices. In our conceptualization, the schedule is not a separate entity but the combination of these contractual commitments. Note that the scheduling usually runs independently from the operational service. It only prevents the operational service to break down when the actual service requests come. However, the scheduling may influence the operational service. For instance, if the subservice providers are not able to commit to the required resource capacity, this is forwarded as such via the contract monitoring, so that the service policy enforcement can pro-actively find and bind other suppliers.

A second important subclass of interactive control distinguished by Simons is the ongoing conversations on probing the assumptions underlying the diagnostic control settings. One of the manager's interactive control tasks is to adapt the service policy when its assumptions do not hold anymore or to anticipate such a break-down. This can be realized by a *discourse* between managers, akin to the above-mentioned knowledge plane [4].

As mentioned earlier, more control systems could be distinguished – boundary systems and belief systems. Whether these can be realized as special cases of the other ones, or deserve to be identified independently, is a question for future research.

3.5 Design Method

Following the fractal approach, a comprehensive design method for management services (we ignore other aspects here) looks roughly as follows:

Step 1	Identify core business services
Step 2	Identify coordination and management services per core service
Step 3	Identify management subservices per management service
Step 4	Identify software services that may support any core service, coordination service, or management (sub) service
Step 5	Identify software services that manage the software services from Step 4 as well as subservices of these management services

The first step is general. To identify core business services, it is recommendable to use the value cycle of the enterprise as a reference. In step 2, management services are identified. Whether this is feasible for all services in the enterprise depends on their *maturity level*. Some enterprises will require this maturity only for services that are of strategic importance. For the selected management services, a decomposition step is made in which subservices are modeled, related to the three different management control types distinguished above. Step 4 is making the match between required services and available IT support. This IT support can range from traditional MIS support software to business intelligence tools and to ubiquitous computing tools such as smart sensors. Each of the tools must get a service interface. At this point, the method can be complemented by existing industrial service engineering methods, as long as the architectural distinction between operational and management service is maintained. In step 5, we repeat step 2-4 of determining management services, but now for the software services. Some of these management services will be fully automated (autonomic computing), other semi-automated.

4 Demonstration

4.1 BSRM Modeling Notation

We use a simple modeling notation to model the services called BSRM (Business Service and Resource Modeling – not published yet). For the clear differentiation between services and the other resources we use different symbols. As far as terminology is concerned, we avoid adding the word "service" to each service name. A summary of the modeling notation is as follows:

- Services are denoted as rounded rectangle
    - exchange services are denoted as rounded rectangles with "exchange" label
    - conversion services are denoted as colored rounded rectangles
- Physical resources are denoted as rectangles
- Intentional resource are denoted as dashed rectangles
- PartOf relationship is denoted as a line with diamond end
- Management relationship is denoted as solid arrow with "+" label
- Stockflow relationship :
    - Inflow : arrow pointing to the resource/service
    - Outflow : arrow pointing from the resource/service

4.2 Italian Wine Producing

The proposed management service model has been evaluated in a real world case study of wine production [17]. According to the case description, the goal of the Wine Producer is to maximize his production in order to adapt the monitored market needs. During the wine producing process quality assurance plays a major role. The Quality Manager, the Agronomist who is an expert of a branch of agriculture which deals with field-crop production and soil management, and the Oenologist who is an expert in wine and wine production involved in this process. They have to observe the vineyard parameters and to react to critical conditions that may happen during the cultivation phase. The wine production case has major phases namely vineyard cultivation, harvesting, fermentation and wine distribution and selling.

Following the management service modeling approach the first step is modeling core business services at the operational level (Fig.6).

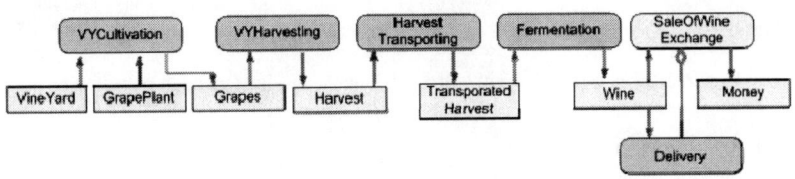

Fig. 6. BSRM model for wine production (core services operational layer)

The core services of the wine production are vineyard cultivation, vineyard harvesting, harvest transporting, fermentation, and sales of wine. There is one exchange service which is sales of wine and all the others are conversion services. Vineyard cultivation is the first step of the wine production process. It is a conversion service and it uses resources Grape plants and Vineyard and produces Grapes. The next conversion service VYHarvesting uses the Grapes and produces Harvest. The rest of the core services use and produce resources are as depicted in Fig. 6. Sales of wine which is an exchange service generates money in return for selling wine. To close the value cycle, the money derived from the sales of wine is spent on different activities in the wine producing process, for example to purchase grape plants, but we did not include it here because of space limitations.

Next, we take the step 2 and 3 from 3.5 together, focusing on the vineyard cultivation core service (Fig. 7). We look for management services corresponding to the three control cycles. Vineyard cultivation management service *VYCultivationMgt* is a *contract management service* responsible for the cultivation process and the VY activity planning and labor allocation are subservices to this management service. VY activity planning builds on contracts set up between *VYCultivation* and Sales, indirectly based on market information. VY quality management (*VYQualityMgt.*) service is a *diagnostic control service*. It is possible to identify two subservices to VYQualityMgt, namely VY activity monitoring and recovery management. It uses a number of intentional resources as input and produces a service *policy* in the form of a recovery action list. Climatic data is an example of assertives that are defined as *policy assumptions*. To acquire these assertives, the management service presumably relies on a *discourse* (not included in the model). VY parameters are assertives for *monitoring* and the critical condition list represents *values* that support the *evaluation*. It turns out that the three control cycles and their subservices provide a very good framework to structure and integrate the VY management phenomena.

The BSRM model aims to provide a first graphical overview of the management services. For each management subservice identified, a more precise definition is to be made in a next step. This may be done using data models and data flow diagrams.

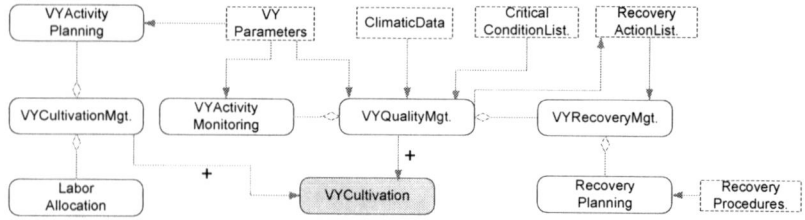

Fig. 7. BSRM model for vineyard cultivation service (management layer)

4.3 Robot Vacuum Cleaner

The next illustration is a fully automated vacuum cleaner [11]. The main goal of this vacuum cleaner is searching dust and keeping the room clean. To perform this task, it has sensors to detect dust, mop, and broom and dust box. The dustbin and the battery charging station are located in the building. Once the dust box is full the cleaner has to move to dustbin to empty it and as s the strength of the battery is low, it has to move to the battery station to recharge. Fig. 8a shows the core services of the vacuum cleaner.

Cleaning the room is the central service and it has two subservices namely *Move* and *CollectDust* – a service which collects dust by brushing and mopping. The cleaning room service involves at least two resources: room and the motor. This service converts the room into a cleaned room, so room is the service goal. The motor is defined as a resource to the main service (*CleaningRoom*), ensuring that it inherits to the subservices as well. There is another service related to the motor which is power supply (with battery). Finally the resources broom, mop and dust box are used by the *CollectDust* subservice.

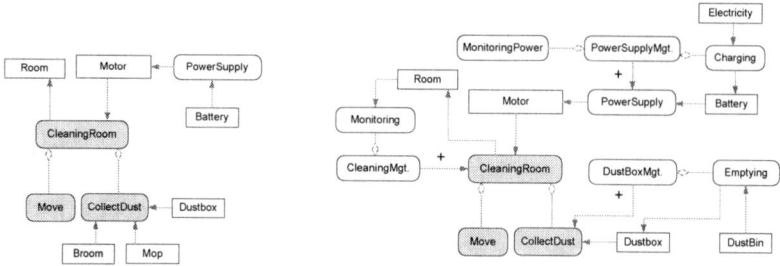

Fig. 8. (a) Core services (b) core + management services for automated vacuum cleaner

The next two steps are identifying management services. The service model for these steps is depicted in Fig. 8b. Since the robot is supposed to work autonomously, self-management is essential and only diagnostic control is relevant. It follows that the cleaning service has a cleaning management service which controls the cleaning process. Hence "monitoring" becomes a subservice of *CleaningMgt*. The next important management service is dust box management which intends to empty the dust box when it detects that dust box is full. *Emptying* works as a subservice.

PowerSupplyMgt monitors the supply of power to the battery and if it finds out that the battery strength is not on the required level, charging is executed.

A comparison of our approach with [11] reveals interesting information to the service designer. The modeling approach in [11] starts from an extended version of the Tropos [2] goal model. Then independently, it identifies non-intentional entities: external resources (room, dustbin and battery charger) and internal ones (dust box and battery). Entities are then related to the goals via the concept of *condition*, corresponding to the extended REA notion of policy constraint.

The management service modeling approach starts with the core services and identifies internal and some external resources needed or influenced by them. It then derives management services and management subservices. For example, *PowerSupplyMgt* has Monitoring and Charging subservices. The services have service policies. These include goals, but note that the goals are not defined globally, as in the previous approach, but per service. In the end, the entities identified are the same in both approaches, but the service-service and service-resource relationships in our model are not considered in the [11] approach.

5 Conclusion

In this paper, we have developed a design framework for management services, based on an established business ontology and applied to two case studies from the literature. The evaluation suggests that the framework provides useful and specific modeling support and that it is widely applicable. The theoretical relevance of the paper consists in the extended REA management ontology, as well as the three management control cycles that we have distinguished and described in SOA terms and that go beyond current work in adaptive systems considering a diagnostic control cycle only. The paper may also have practical relevance to service engineers interested in aligning business services and software services and to whom current service design methods do not give much specific support when it comes to service management design.

Although a design framework is not easy to evaluate in practice, we intend to strengthen the validation by applying it to real-world cases and extending the comparison with related work in management information systems, software engineering, multi-agent systems and the field of adaptive systems.

An interesting topic for future research is the design and development of generic management service software, e.g. to deploy a diagnostic service on the basis of a policy definition and a given environment (available services) only.

References

[1] Bartels, A.: Smart Computing Drives The New Era of IT Growth. Forrester (2009)
[2] Bresciani, P., Giorgini, P., Giunchiglia, F., Mylopoulos, J., Perini, A.: Tropos: An Agent-Oriented Software Development Methodology. Proc. AAMAS 8(3), 203–236 (2004)
[3] Cetina, C., Haugen, O., Zhang, X., Fleurey, F., Pelechano, V.: Strategies for Variability Transformation at Run-time. In: Proc. 13th Int. Software Product Lines Conf., SPLC (2009)
[4] Dobson, S., Sterritt, R., Nixon, P., Hinchey, M.: Fulfilling the Vision of Autonomic Computing. IEEE Computer 43, 35–41 (2010)

[5] Geerts, G., McCarthy, W.: Policy-Level Specifications in REA Enterprise Information Systems. Journal of Information Systems 20(2), 37–63 (2006)

[6] Hruby, P.: Model-Driven Design of Software Applications with Business Patterns. Springer, Heidelberg (2006)

[7] Kephart, J., Chess, D.: The Vision of Autonomic Computing. Computer, 41–50 (2003)

[8] Moscinat, A., Binder, W., Jazayeri, M.: Runtime Adaptability through Automated Model Evolution. In: Proc. 14th IEEE Enterprise Distributed Object Computing Conference, pp. 217–226 (2010)

[9] McCarthy W.E., The REA Accounting Model: A Generalized Framework for Accounting Systems in a Shared Data Environment. The Accounting Review, 544-577 (1982)

[10] Mueller, I., Han, J., Schneider, J.-G., Versteeg, S.: A Conceptual Framework for Unified and Comprehensive SOA Management. In: Feuerlicht, G., Lamersdorf, W. (eds.) ICSOC 2008. LNCS, vol. 5472, pp. 28–40. Springer, Heidelberg (2009)

[11] Morandini. M., Penserini. L., Perini. A.: Towards goal-oriented development of self-adaptive systems. In: Proc. Int. Workshop on Software Engineering for Adaptive and Self-Managing Systems, SEAMS (2008)

[12] Mason-Jones, R., Towill, D.R.: Information enrichment: designing the supply chain for competitive advantage. Supply Chain Management 2(4), 137–148 (1997)

[13] Papazoglou, M.P., van den Heuvel, W.J.: Service oriented architectures: approaches, technologies and research issues. VLDB Journal 16(3), 389–415 (2007)

[14] Peffers, K., Tuunanen, T., Rothenberger, M., Chatterjee, S.: A Design Science Research Methodology for Information Systems Research. Journal of Management Information Systems 24(3), 45–77 (2008)

[15] Porter, M.: Competitive Advantage. Free Press, New York (1985)

[16] La Rosa, M., Dumas, M., ter Hofstede, A.H.M., Mendling, J., Gottschalk, F.: Beyond Control-Flow: Extending Business Process Configuration to Roles and Objects. In: Li, Q., Spaccapietra, S., Yu, E., Olivé, A. (eds.) ER 2008. LNCS, vol. 5231, pp. 199–215. Springer, Heidelberg (2008)

[17] http://www.s-cube-network.eu/results/deliverables/wp-ia-2.2/CD-IA-2.2.2_Collection_of_industrial best_practices_scenarios_and_business_cases.pdf/view

[18] Searle, J.: A Classification of Illocutionary Acts. Language in Society 5, 1–24 (1976)

[19] Simons, R.: Performance Measurement and Control Systems for Implementing Strategy. Prentice Hall, Englewood Cliffs (2000)

[20] Salehie, M., Tahvildari, L.: Self-adaptive Software: Landscape and Research Challenges. ACM Transactions on Autonomous and Adaptive Systems 4(2), 1–42 (2009)

[21] UN/CEFACT Modelling Methodology (UMM) User Guide (2003), http://www.unece.org/cefact/umm/UMM_userguide_220606.pdf

[22] Vymetal, D., Hunka, F., Hucka, M., Kasik, J.: Enterprise modeling: process and REA value chain perspective (2010), http://mpra.ub.uni-muenchen.de/24617/

[23] Weigand, H., Johannesson, P., Andersson, B., Bergholtz, M.: Value-Based Service Modeling and Design: Toward a Unified View of Services. In: van Eck, P., Gordijn, J., Wieringa, R. (eds.) CAiSE 2009. LNCS, vol. 5565, pp. 410–424. Springer, Heidelberg (2009)

[24] Weigand, H., Johannesson, P., Andersson, B., Bergholtz, M., Jayasinghe Arachchige, J.: Closing the User-Centric Coordination Cycle. In: Proc. CAiSE 2010 Forum. LNBIP, vol. 72, pp. 267–282. Springer, Heidelberg (2010)

[25] Weigand, H., Heuvel, W.J., van den Hiel, M.: Business Policy Compliance in Service-Oriented Systems. Information Systems 36, 791–807 (2011)

[26] Yu, J., Sheng, Q.Z., Swee, J.K.Y.: Model-Driven Development of Adaptive Service-Based Systems with Aspects and Rules. In: Chen, L., Triantafillou, P., Suel, T. (eds.) WISE 2010. LNCS, vol. 6488, pp. 548–563. Springer, Heidelberg (2010)

Bottom-Up Fault Management
in Composite Web Services

Brahim Medjahed[1] and Zaki Malik[2]

[1] Department of Computer and Information Science,
University of Michigan – Dearborn,
brahim@umich.edu
[2] Department of Computer Science,
Wayne State University
zaki@wayne.edu

Abstract. We propose an approach for managing bottom-up faults in composite Web services. We define bottom-up faults as abnormal conditions/defects or changes in component services that may lead to run-time failures in composite services. The proposed approach uses soft-state signaling to propagate faults from components to composite services. Soft-state denotes a class of protocols where state (e.g., whether a service is alive) is constantly refreshed by periodic messages. Its advantages include implicit error recovery and easier fault management, resulting in high availability. We introduce a bottom-up fault model for composite services. Then, we propose a soft-state protocol for bottom-up fault propagation in composite services. Finally, we present experiments to assess the performance of our approach.

Keywords: Service Composition – Fault Management – Bottom-up Fault - Soft-State – Fault Coordinator.

1 Introduction

Service-oriented architecture (SOA) has recently emerged as a promising approach for application integration [1,14]. It utilizes services (commonly Web services) as the building blocks for developing software systems distributed within and across organizations. The primary value of SOA is the ability to (1) reuse pre-developed, autonomous, and independently provided resources (e.g., legacy applications, sensors, databases, storage devices, COTS products) as Web services and (2) combine pre-existing services, called *participants*, into higher level services, called *composite services*, which perform more complex functions [9,15].

Because of the dynamic and volatile nature of SOAs, Web services are subject to unavoidable faults during their lifetime. The relationship between a fault and failure is given in ISO/CD 10303-226 document, where a fault is defined as an abnormal condition or defect at the component, equipment, or sub-system level which may lead to a failure. In their seminal work, Avizienis et al. [2] state that faults cycle between dormant and active states, and a failure occurs when a fault becomes active. In this

H. Mouratidis and C. Rolland (Eds.): CAiSE 2011, LNCS 6741, pp. 597–611, 2011.
© Springer-Verlag Berlin Heidelberg 2011

paper, we focus on faults that propagate from participants to composite services. We refer to these faults as *bottom-up*. Two examples of bottom-up faults are (1) a shutdown scheduled by a participant's provider for maintenance; and (2) an update to a participant's policy (e.g., new message parameters added to a WSDL specification) that may affect the way that participant is consumed. Hence, composite services must rapidly detect and handle faults in their participants to avoid run-time failures and maintain consistency.

Web services generally use HTTP as the underlying message transport. Hence, they are either guaranteed message delivery or notified if a message was not delivered (e.g., because of a server unavailability). In the latter case, composite services become aware of a fault only at the time they interact with their participants *not* at the time that fault occurred. This may decrease the availability of composite services. Besides, users' requests are pending as long as the composite service did not recover from the fault (e.g., by replacing the faulty participant with an equivalent one). This calls for a framework in which composite services are able to detect and handle bottom-up faults as soon as those faults occur in their participants.

In this paper, we introduce a framework for managing bottom-up faults in composite services. The proposed framework uses *soft-state* signaling to propagate faults from participants to composite services. Soft state denotes a type of protocols where state (e.g., whether a server is alive) is constantly refreshed by periodic messages; state which is not refreshed in time expires [8]. This is in contrast to hard-state where installed state remains installed unless explicitly removed by the receipt of a state-teardown message. Advantages of the soft-state approach include implicit error recovery and easier fault management, resulting in high availability [16]. Soft state was introduced in the late 1980s and has been widely used in various Internet protocols (e.g., RSVP). However, to the best of our knowledge, this work is the first to use soft-state for fault management in composite services. The major contributions of this paper are summarized below:

- We introduce a bottom-up fault model for composite services. The model includes a taxonomy of bottom-up faults, a definition of (composite) service, and peer-peer topology for fault management.
- We propose a soft-state protocol for bottom-up fault propagation.
- We conduct experiments to assess the performance of the proposed framework.

The rest of this paper is organized as follows. In Section 2, we describe the bottom-up fault model. In Section 3, we propose the soft-state protocol for bottom-up fault propagation. In Section 4, we present experiments to assess the performance of the proposed approach. In Section 5, we give a brief survey of related work. We finally provide concluding remarks in Section 6.

2   Fault Model

In this section, we describe our model for bottom-up fault management. We first provide a categorization of bottom-up faults. Then, we define the notion of participant's state. Finally, we introduce a peer-to-peer topology for bottom-up fault management.

2.1 Bottom-up Fault Taxonomy

A fault management approach must refer to a taxonomy that describes the different types of faults that composite services are expected to be able to manage. We identify two types of bottom-up faults: *physical* and *logical* (Fig. 1). *Physical faults* are related to the infrastructure that supports Web services. In this paper, *the underlying communication system is assumed to be failure-free*: there is no creation, alteration, loss, or duplication of messages. However, *node faults* are still possible. A *node fault* occurs if the servers (e.g., application server, Web server) hosting a participant are out of action. *Logical faults* are initiated by service providers; this is in contrast to physical faults which are out of service providers' control. We categorize logical faults as *status change, participation refusal*, and *policy change*.

Status change occurs if the service provider explicitly modifies the availability status of its service. The status may be changed through freeze or stop. In the *freeze* fault, providers shut down their services for limited time periods (e.g., for maintenance, unavailability of a product in a supply chain's provider). In the *stop* fault, providers make their services permanently unavailable (e.g., a company going out-of-business).

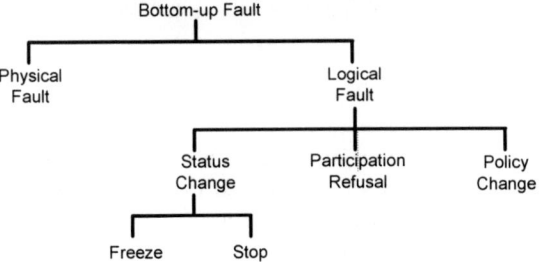

Fig. 1. Bottom-up Fault Taxonomy

Participation refusal occurs if a service is not willing to participate in a given composition. The way participation decision is made varies from a Web service to another. We give below four techniques on how such decisions could be made:

- A service load balancer may check that the server workload will not exceed a given threshold if the service participates in a new composition. The threshold could, for instance, be defined to maintain a minimum quality of service.
- A policy compatibility checker may also verify the compliance of the service policies (e.g. privacy policies) with the composite service policies.
- A service reputation manager may verify that the reputation (e.g., defined by users' ratings) of the composite service is higher than a threshold set by the participant.
- A notification may be sent to the participant's provider who will decide whether the service should participate in the composition or not.

In the rest, we assume the existence of a pre-defined function *Agreed2Join()* used by services to decide whether they are willing to participate in compositions. Each service may provide its own definition and implementation of the *Agreed2Join()* function. However, the way is *Agreed2Join()* defined is out of the scope of this paper.

Policy change occurs if the provider updates one of its service policies. We adopt a broad definition of policy, encompassing all requirements under which a service may be consumed. We adopt the categorization proposed in [11] considering policies as either vertical or horizontal. The vertical category refers to application domain-dependent policies (e.g., shipping policy in business-to-business e-commerce). The horizontal category refers to policies that are applicable across domains. It is composed of three sub-categories: functional, non-functional, and valued-added. Functional category describes the operational features of a Web service (e.g., in WSDL [1,14], OWL-S [10]). Non-functional category relates to Quality of Service metrics. The value-added category brings "better" environments for Web service interactions. It refers to a set of specifications for supporting optional (but important) requirements for the service (e.g., security, privacy, negotiation, conversation). Changes in the policies of a participant WS_i may impact the way a composite service CS_j interacts with WS_i. Hence, they should be considered as logical faults. For instance, CS_j invocations to WS_i may lead to run-time failures if WS_i provider changed the input message required by WS_i (e.g., new message parameters added, changes made to data types).

2.2 State of a Service

Soft-state signaling enables the propagation of bottom-up faults from participants to composite services. The main idea of this class of signaling is that the *state* of each participant is periodically sent to the composite service. The composite service will then use the received state to determine whether there was any physical or logical fault in the participant. Several questions need to be tackled when designing a soft-state protocol: what is the definition of a state? And how is the state computed? We will give answers to these questions in the rest of this section and paper.

The proposed framework must deal with all types of faults depicted in Fig 2. Physical faults and status changes are detected by composite services in an *implicit* manner; if a node fault occurs or a participant is stopped/frozen, then the composite service will not receive a state from that participant. The participation refusal fault is *explicitly* communicated by participants if they are not willing to be part of a composition.

Policy change faults are transmitted as part of the *participant's state*. To keep track of policy changes, each participant WS_i maintains a data structure called $State_i$ (Fig. 2). $State_i$ is defined by two attributes: *ChangeStatus* and *ChangeDetails*. *ChangeStatus* is equal to True if policy changes have been made to WS_i. Several changes may occur in WS_i during a time period; details about these changes are stored in the *ChangeDetails* set. Each element of this set represents a policy change; it is defined by a couple (C,S) where C is the *category* of the policy and S is the *scope* of the change. The initial values of ChangeStatus and ChangeDetails are False and $\emptyset$, respectively. If a change (C,S) is detected on WS_i, $State_i$.ChangeStatus is set to True and (C,S) is added to $State_i$.ChangeDetails. The content of $State_i$ is periodically communicated to composite services. If a composite service does not receive $State_i$ after a certain period of time, then it assumes a physical fault or status change in WS_i. Otherwise, the composite service reads $State_i$ to find out about the changes made to WS_i.

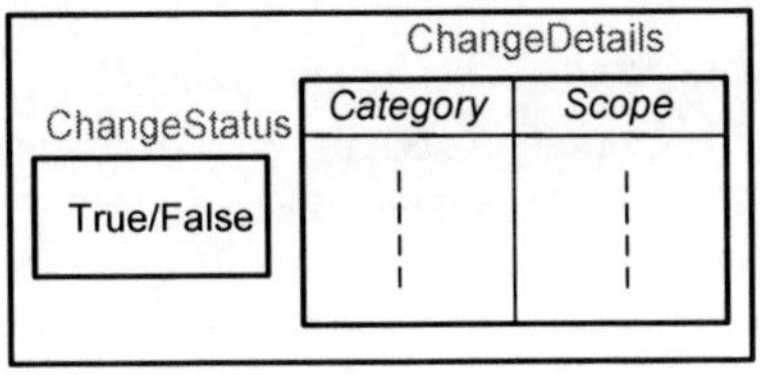

Fig 2. State of a Participant Service

A policy *category* refers to the type of requirements specified by a policy. As mentioned in Section 2.1, it refers to functional, non-functional, value-added, or domain (i.e., vertical) categories. Policies are specified in XML-based Web service languages/standards (e.g., WSDL [1], WS-Security [14]). The *scope* of a change defines the subject to which that change was applied. It includes details about (i) the location of the modified policy specification and (ii) the element that has been updated within that specification. The specification location is given by the URI of the XML file that stores the specification. The updated element is identified by the XPath query of that element within the specification. For instance, let us consider the following WSDL file located in "http://www.ws.com/sq.wsdl":

```
<definitions>
    <types>  <!-- XML Schema -->  </types>
    <message name="getQuote_In"> ....
    <message name="getQuote_Out"> ...
    <portType name="StockQuoteServiceInterface">
      <operation name="getQuote">
        <input message="getQuote_In" />
        <output message="getQuote_Out" />
      </operation>
    </portType>
    ...
```

Let us assume that the name of the operation "getQuote" has been modified in the WSDL document. The category and scope of the change are defined as follows:

- Category = (Functional,WSDL).
- Scope = (URL,Q) where:
  - URL = "http://www.ws.com/stockquote.wsdl"
  - Q = "definitions/portType/operation/@name"

The following definition summarizes the properties of $State_i$ maintained by a participant WS_i.

Definition: The state, denoted State$_i$, of a participant WS$_i$ is defined by (ChangeStatus,ChangeDetails) where:

- ChangeStatus = True $\Leftrightarrow$ changes have been made to WS$_i$.
- ChangeDetails = {(C,S) / C and S are the category and scope of a change in WS$_i$}.
- Initially do: State$_i$.ChangeStatus = False and State$_i$.ChangeDetails = $\varnothing$.
- At the occurrence of a change (C,S) in WS$_i$ do: State$_i$.ChangeStatus = True, State$_i$.ChangeDetails = State$_i$.ChangeDetails $\cup$ {(C,S)}. $\square$

2.3 Fault Coordinators

In the proposed framework, fault management is a collaborative process between architectural modules called *fault coordinators*. Each Web service (participant or composite) has a coordinator associated to it. This peer-to-peer topology distributes control and externalizes fault management, hence creating a clear separation between the business logic of the services and fault management tasks.

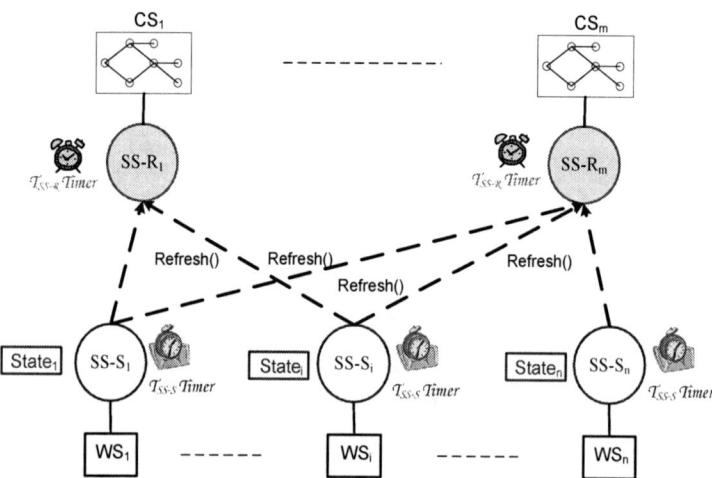

Fig. 3. Fault Coordinators

We define two types of coordinators (Fig. 3): *soft-state senders* (SS-S) and *soft-state receivers* (SS-R). Each participant (resp., composite service) has a sender (resp., receiver) attached to it. A sender SS-S$_i$ maintains the State$_i$ data structure. To keep track of its receivers, SS-S$_i$ maintains a Receivers(SS-S$_i$) data structure. If WS$_i$ (attached to SS-S$_i$) participates in CS$_j$ (attached to SS-R$_j$) then SS-R$_j$ $\in$ Receivers(SS-S$_i$). SS-S$_i$ periodically sends State$_i$ to its receivers via Refresh() messages. The refresh period is determined by the τ_{SSS} timer maintained by SS-S$_i$. A receiver SS-R$_j$ maintains two data structures: Senders(SS-R$_j$) and τ_{SSR}. Senders(SS-R$_j$) is the set of senders from which SS-R$_j$ expects to receive Refresh(). If WS$_i$ participates in CS$_j$ then

SS-S$_i$ $\in$ Senders(SS-R$_j$). τ_{SSR} is a timer used by SS-R$_j$ to process Refresh() messages received from its senders.

Bottom-up fault management involves three major tasks: *fault detection, fault propagation*, and *fault reaction*. The sequence diagram in Fig. 4 depicts the relationship between these tasks:

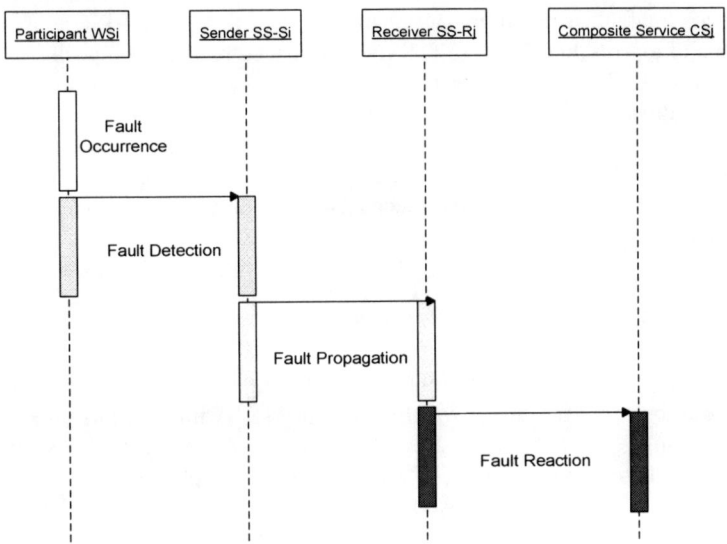

Fig. 4. Sequence Diagram for Bottom-up Fault Management

- *Fault detection*: SS-S$_i$ first detects faults that occurred in the attached WS$_i$. SS-S$_i$ does not need to detect physical and status change faults in WS$_i$ as these are implicitly propagated to receivers if the latter do not receive Refresh() messages from faulty senders. Participation refusal faults are communicated to SS-S$_i$ using one of the techniques mentioned in Section 2.1. Policy changes in WS$_i$ may be detected by SS-S$_i$ using various techniques. For instance, SS-S$_i$ may use XML version control algorithms to detect changes in WS$_i$ policy specifications; another solution is to provide an interface in SS-S$_i$ through which WS$_i$ provider submits all policy changes; a third solution is to use the publish/subscribe model [3] where SS-S$_i$ subscribes with WS$_i$ on policy changes. Due to space limitations, details about the detection of participation refusals and policy changes are out of the scope of this paper.
- *Fault propagation*: SS-S$_i$ then propagates the fault detected in the previous phase to SS-R$_j$. We propose a soft-state protocol for fault propagation. Details about this protocol are given in Section 3.
- *Fault Reaction*: Finally, SS-Rj and/or CSj execute appropriate measures to react to the fault. The techniques proposed in [12,7] can be extended for that purpose. In [12], we use ECA (Event Condition Actions) to react to changes: the event part of an ECA rule refers to change notifications; the action part allows for the specification of change control policies; the use of conditions allows the specialization of

these policies depending on pre-defined parameters. In [7], we define a Petri Net model for the specification of high-level recovery policies in composite services. These recovery policies are generic directives that model exceptions at design time together with a set of primitive operations used at run time to handle the occurrence of exceptions. The proposed model identifies a set of recovery policies that are useful and commonly needed in many practical situations. Details about the techniques for fault reaction are out of the scope of this paper. We assume the existence of a procedure React (FT, SS) implemented by each composite service to react to faults. The FT parameter is the fault type and takes one of the following values: "Refusal" to refer to a participation refusal fault, "No Refresh" to refer to the non-reception of a Refresh() message (i.e., physical fault or status change), and "policy" to refer to a policy change fault. The SS parameter refers to the sender of faulty participant; the reaction mechanism may in fact vary from a participant to another.

3   The Fault Propagation Protocol

In this section, we describe the algorithms executed by senders and receivers for propagating bottom-up faults. We assume that WS_i (with SS-S_i as attached sender) participates in CS_j (with SS-R_j as attached receiver). As mentioned in Section 2.1, we assume that there is no creation, alteration, loss, or duplication of messages. The propagation protocol adapts the well-know soft-state signaling described in [8,16] to service-oriented environments. It enables the propagation of participation refusal and policy changes faults to receivers. Physical faults and status changes are implicitly propagated to receivers if the latter do not get Refresh() messages from faulty senders.

3.1   Soft-State Sender Algorithm

Table 1 gives the algorithm executed by SS-S_i. SS-S_i receives two types of messages from SS-R_j: Join(SS-R_j) and Leave(). Join(SS-R_j) is the first message that SS-S_i receives from SS-R_j; it invites WS_i to participate in CS_j (lines 1-10). SS-S_i calls the *Agreed2Join()* function to figure out whether WS_i is willing to participate in CS_j (see Section 2.1). If *Agree2Join(SS-R_j)* returns False, SS-S_i sends the Decision2Join(SS-S_i,False) message to SS-R_j. Otherwise, SS-S_i adds SS-R_j to its receivers. If SS-R_j is the first receiver of SS-S_i, SS-S_i initializes $State_i$ and starts its τ_{sss} timer. Finally, SS-S_i sends its decision to SS-R_j through the Decision2Join(SS-S_i,True) message. At any time, SS-S_i may receive a Leave() message from SS-R_j (lines 11-13). This message indicates that CS_j is no longer using WS_i as a participant. In this case, SS-S_i removes SS-R_j from its receivers.

At the detection of a policy change (with a category C and scope S) in WS_i (lines 14-17), SS-S_i sets $State_i$.ChangeStatus to True. SS-S_i keeps track of that change by inserting (C,S) in $State_i$.ChangeDetails. In this way, the state of SS-S_i to be sent to receivers at the end of τ_{sss} cycle includes all changes that have occurred during that cycle. At the end each period (denoted by τ_{sss} timer), SS-S_i sends a Refresh()

message to each one of its receivers (lines 18-23). This message includes $State_i$ as a parameter, hence notifying SS-R_j about all policy changes that occurred in WS$_i$ during the last τ_{SSS} period. SS-S_i then reinitializes $State_i$ and restarts its τ_{SSS} timer.

Table 1. Sender's Propagation Algorithm

```
(01) At Reception of Join(SS-R_j) Do
(02)   If Agreed2Join(SS-R_j) = True Then  Receivers_i = Receivers_i ∪ {SS-R_j};
(03)            If | Receivers_i | = 1 Then  State_i.ChangeStatus = False;
(04)                            State_i.ChangeDetails = ∅;
(05)                            Start τ_SSS timer of SS-S_i;
(06)            EndIf
(07)            Send Decision2Join(SS-S_i,True) to SS-R_j
(08)   Else  Send Decision2Join(SS-S_i,False) to SS-R_j
(09)   EndIf
(10) End

(11) At Reception of Leave(SS-R_j) Do
(12)    Receivers_i = Receivers_i − {SS-R_j};
(13) End

(14) At the detection of Change(C,S) in WS_i Do
(15)    State_i.ChangeStatus = True;
(16)    State_i.ChangeDetails = State_i.ChangeDetails ∪ {(C,S)};
(17) End

(18) At the end of τ_SSS timer of SS-S_i Do
(19)    For each SS-R_j / SS-R_j ∈ Receivers_i Do Send Refresh(State_i) to SS-R_j;  EndFor
(20)    State_i.ChangeStatus = False;
(21)    State_i.ChangeDetails = ∅;
(22)    Re-start τ_SSS timer of SS-S_i;
(23) End
```

3.2 Soft-State Receiver Algorithm

The aim of SS-R_j protocol is to detect faults in senders. For that purpose, SS-R_j maintains a local table called *SR-Table$_j$*. SR-Table$_j$ allows SS-R_j to keep track of Refresh() messages transmitted by senders. It contains an entry for each SS-S_i that belongs to Senders(SS-R_j). Each entry contains two columns:

- *Refreshed: SR-Table$_j$[SS-S$_i$,Refreshed]* equals True iff SS-R_j received a Refresh() from SS-S_i in the current τ_{SSR} cycle.
- *Retry: SR-Table$_j$[SS-S$_i$,Retry]* contains the number of consecutive cycles during which SS-R_j did not receive Refresh() from SS-S_i.

A temporary node failure in SS-S_i may prevent SS-S_i from sending Refresh() to SS-R_j during a τ_{SSR} cycle. In this case, SS-R_j may want to give SS-S_i a second chance for sending Refresh() during the next τ_{SSR} cycle. For that purpose, SS-R_j maintains a variable (positive integer) *Max-Retry$_j$*. If SS-R_j does not receive Refresh() from SS-S_i during Max-Retry$_j$ consecutive τ_{SSR} cycles, it considers WS$_i$ as faulty. The value of Max-Retry$_j$ is set by CS$_j$ composer and may vary from a composite service to another. The smaller is Max-Retry$_j$, the more pessimistic is CS$_j$ composer about the occurrence of faults in participants.

Table 2. Receiver's Propagation Algorithm

```
(01) At addition of WSᵢ to CSⱼ Do
(02)     Send Join(SS-Rⱼ) to SS-Sᵢ;
(03) End

(04) At deletion of WSᵢ from CSⱼ Do
(05)     Send Leave(SS-Rⱼ) to SS-Sᵢ;
(06)     Delete SS-Sᵢ entry from SR-Tableⱼ;
(07) End

(08) At Reception of Decision2Join(SS-Sᵢ,decision) Do
(09)     If decision = True Then Sendersⱼ = Sendersⱼ ∪ {SS-Sᵢ};
(10)                             Create an entry for SS-Sᵢ in SR-Tableⱼ;
(11)                             SR-Tableᵢ[SS-Sᵢ,Refreshed] = False;
(12)                             SR-Tableᵢ[SS-Sᵢ,Retry] = 0;
(13)                             If | Sendersⱼ| = 1 Then Start τₛₛᵣ timer of SS-Rⱼ EndIf
(14)     Else React("Refusal",SS-Sᵢ);
(15)     EndIf
(16) End

(17) At Reception of Refresh(Stateᵢ) From SS-Sᵢ Do
(18)     SR-Tableᵢ[SS-Sᵢ, Refreshed] = True;
(19)     If Stateᵢ.ChangeStatus = True Then React("policy", SS-Sᵢ, Stateᵢ.ChangeDetails) EndIf
(20) End

(21) At the end of τₛₛᵣ timer of SS-Rⱼ Do
(22)     For each SS-Sᵢ / SS-Sᵢ ∈ Sendersⱼ Do
(23)         If SR-Tableⱼ[SS-Sᵢ, Refreshed] = True Then SR-Tableⱼ[SS-Sᵢ, Refreshed] = False;
(24)                                                     SR-Tableⱼ[SS-Sᵢ, Retry] = 0;
(25)         Else  SR-Tableⱼ[SS-Sᵢ, Retry]++
(26)             If SR-Tableⱼ[SS-Sᵢ,Retry] = Max-Retryⱼ Then
(27)                 React("No Refresh", SS-Sᵢ);
(28)             EndIf
(29)         EndIf
(30)     EndFor
(31)     Re-start τₛₛᵣ timer of SS-Rⱼ;
(32) End
```

$SS-R_j$ submits two types of messages to $SS-S_i$: Join() and Leave(). It also receives two types of messages from $SS-S_i$: Decision2Join() and Refresh(). Table 2 gives the algorithm executed by $SS-R_j$. Whenever a new participant WS_i is added to CS_j, $SS-R_j$ sends a Join($SS-R_j$) message to $SS-S_i$ (lines 1-3). At the deletion of WS_i from CS_j, $SS-R_j$ sends a Leave($SS-R_j$) message to $SS-S_i$ and removes $SS-S_i$ entry from SR-Table$_j$ (lines 4-7). At the reception of Decision2Join($SS-S_i$,True), $SS-R_j$ adds $SS-S_i$ to the list of senders (lines 8-16). It also creates a new entry for $SS-S_i$ in SR-Table$_j$ and initializes the Refreshed and Retry columns of that entry to False and 0, respectively. If $SS-S_i$ is the first sender of $SS-R_j$, $SS-R_j$ starts its τ_{SSR} timer. At the reception of Decision2Join($SS-S_i$,False), $SS-R_j$ calls the *React()* procedure to process the participation refusal fault issued by $SS-S_i$ (see Section 2.3).

At the reception of Refresh(State$_i$), $SS-R_j$ sets SR-Table$_j$[$SS-S_i$,Refreshed] to True (lines 17-20). If State$_i$.ChangeStatus is True, $SS-R_j$ calls the *React()* procedure to process all changes that occurred in $SS-S_i$ during the last τ_{SSR} cycle. At the end of τ_{SSR} timer (lines 21-32), $SS-R_j$ checks if it received Refresh() from each of its senders. If $SS-R_j$ received Refresh() from $SS-S_i$, it re-initializes the Refreshed and Retry columns of SS-S$_i$ entry in SR-Table$_j$ to False and 0, respectively. Otherwise, SS-R$_j$ increments SR-Table$_j$[SS-R$_j$,Retry] If SR-Table$_j$[SS-S$_i$,Retry] equals Max-Retry$_j$ (i.e., SS-R$_j$ did not receive Refresh() from SS-S$_i$ during Max-Retry$_j$ consecutive τ_{SSR} cycles), SS-R$_j$ assumes a physical (node) fault in SS-S$_i$ and hence, calls the *React()* procedure to process that fault. SS-R$_j$ finally restarts its τ_{SSR} timer.

3.3 Example

Let us consider a Web service WS$_3$ (with a soft-state sender SS-S$_3$) that participates in two composite services CS$_1$ and CS$_2$ (with soft-state receivers SS-R$_1$ and SS-R$_2$, respectively). We assume that $\tau_{SSR1} = \tau_{SSR2} = 2 \times \tau_{SSS3}$. Fig 5 depicts the interactions between SS-S$_3$ and SS-R$_1$/SS-R$_2$.

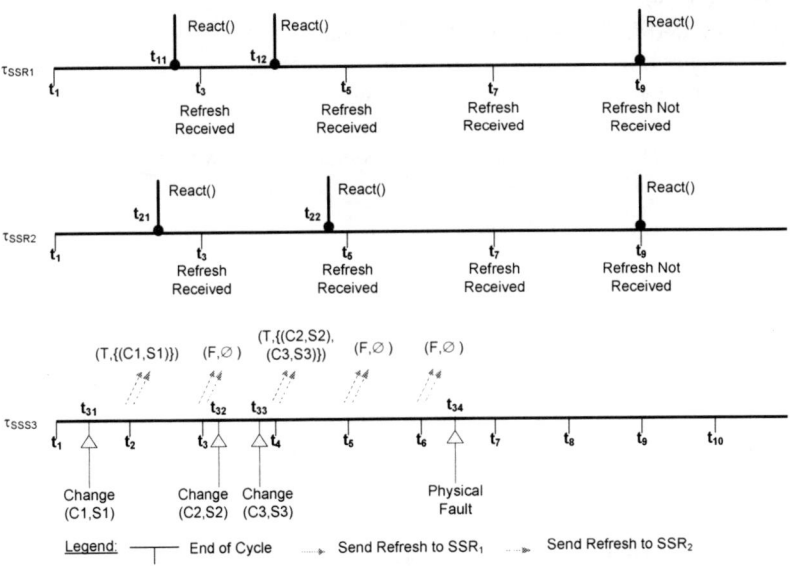

Fig. 5. Example of Fault Propagation

At time t$_{31}$, SS-S$_3$ detects a change (with category C$_1$ and scope S$_1$) in WS$_3$. SS-S$_3$ assigns True to State$_3$.ChangeStatus and inserts (C$_1$,S$_1$) in State$_3$.ChangeDetails. At time t$_2$, SS-S$_3$ sends Refresh(True,{(C$_1$,S$_1$)}) to SS-R$_1$ and SS-R$_2$, and reinitializes State$_3$. SS-R$_1$ and SS-R$_2$ process those changes by calling their *React()* procedure at times t$_{11}$ and t$_{21}$, respectively. At t$_3$, SS-R$_1$ and SS-R$_2$ note the reception of the Refresh() sent by SS-S$_3$. At this same time, SS-S$_3$ sends Refresh() to both receivers with the parameters (False,∅) since no changes have been detected in the second SS-S$_3$ cycle. SS-S$_3$ detects two changes (C$_2$,S$_2$) and (C$_3$,S$_3$) in WS$_3$ at t$_{32}$ and t$_{33}$, respectively. At t$_{33}$, State$_3$.ChangeStatus equals True and State$_3$.ChangeDetails contains {(C$_2$,S$_2$),(C$_3$,S$_3$)}. At t$_4$, SS-S$_3$ sends Refresh(True,{(C$_2$,S$_2$,),(C$_3$,S$_3$)}) to SS-R$_1$ and SS-R$_2$. SS-R$_1$ and SS-R$_2$ process those changes at t$_{12}$ and t$_{22}$, respectively. At t$_5$, SS-R$_1$ and SS-R$_2$ note the reception of the Refresh() sent by SS-S$_3$. At times t$_5$ and t$_6$,

SS-S$_3$ sends Refresh() to SS-R$_1$ and SS-R$_2$ with the parameters (False,∅) since no changes have been detected in the corresponding SS-S$_3$ cycle. At t$_7$, SS-R$_1$ and SS-R$_2$ note the reception of the Refresh() sent by SS-S$_3$. Let us now assume a server failure in WS$_3$ (and hence SS-S$_3$) at t$_{34}$. At t$_9$, SS-R$_1$ and SS-R$_2$ find out that they did not receive Refresh() from SS-S$_3$ during the last SS-R$_2$ cycle. If Max-Retry$_2$ is equal to 1, SS-R$_1$ and SS-R$_2$ conclude that SS-S$_3$ failed and hence call the *React()* function.

4 Performance Evaluation

We conducted experiments to assess the different parameters that may impact the performance of the proposed protocol. We used Microsoft Windows Server 2003 (operating system), Microsoft Visual Studio 8 (development kit), UDDI Server, IIS Server, and SQL Server. We ran our experiments on Intel(R) processor (1500MHz) and 512MB of RAM. Soft-state senders and receivers have been developed in C#. We created twenty (20) receivers and fifty (50) senders, and registered them in UDDI. Each receiver has ten (10) senders randomly selected among the existing senders. In the rest of this section, we analyze the relationship between τ_{SSR}/τ_{SSS} values and the following two parameters: *fault propagation time* and *false faults*.

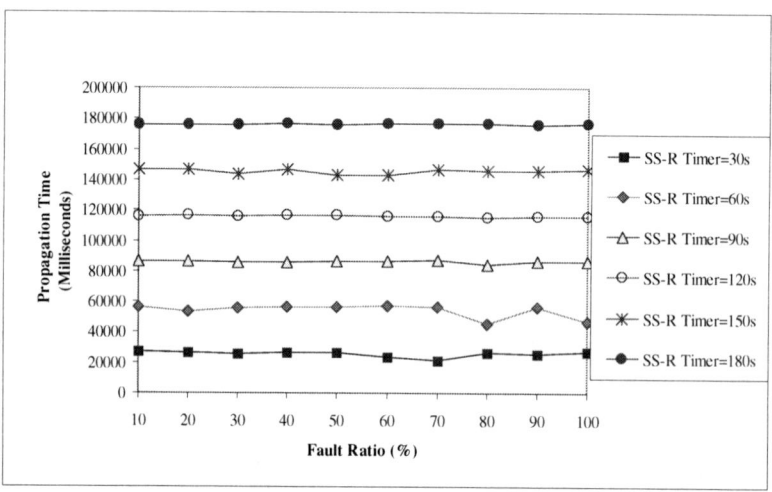

Fig. 6. Impact of τ_{SSR} on Fault Propagation Time

Fault propagation time is the first performance parameter we analyze in our study (Fig. 6). Let us assume that a fault occurred in a sender at time t$_1$ and has been detected by a receiver at time t$_2$. The fault propagation time is equal to t$_2$-t$_1$ (i.e., the time it took to the receiver to detect a fault in its senders). Fig. 6 compares the average fault propagation time for various τ_{SSR} timer values. We consider different fault ratios for each τ_{SSR} timer value. For instance, a fault ratio of 10 means that 10% (1 out of 10) of participants within a composite service failed. We focused on physical node faults; these are created by physically stopping the services corresponding to faulty senders (selected randomly). Fig. 6 shows that the τ_{SSR} timer value has an impact on the fault propagation time. The smaller is τ_{SSR}, the shorter is the fault propagation time.

False faults refer to the situation where receivers assume faults that did not occur in their senders. Fig. 7 depicts the relationship between false faults and timer difference (i.e., τ_{SSS}-τ_{SSR}). We set τ_{SSR} to 20s and vary τ_{SSS} from 20s to 25s, 30s, 35s, 40s, etc. Fig. 7 shows that false faults occur if τ_{SSS}-$\tau_{SSR}\geq0$ (i.e., $\tau_{SSS}\geq\tau_{SSR}$). In addition, the bigger is τ_{SSS} (compared to τ_{SSR}), the larger is the number of false faults. These faults correspond to cases where Refresh() messages are sent after the end of the corresponding τ_{SSR} cycles.

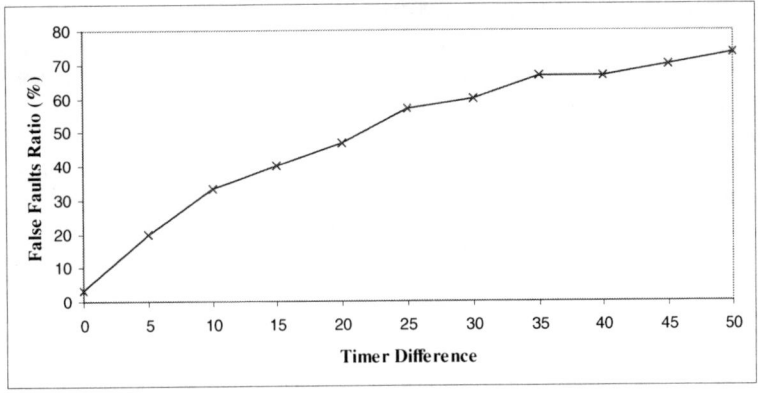

Fig. 7. Relationship between τ_{SSS} and τ_{SSR} and impact on False Faults ratio

5   Related Work

Mechanisms that support fault management in software systems have been around for a long time. Workflow has traditionally been used to deal with faults in business processes. Faults in workflows have usually been modeled as exceptions. SOAs consider faults as the rule, and any solution to fault management would need to treat them as such. Traditional software engineering solutions for fault management (e.g., exceptions and run-time assertion checking) have hard-coded, internal, and application-specific capabilities that limit their generalization and reuse. They disperse the adaptation logic throughout the application, making it costly to modify and maintain [4]. These factors have actuated research dealing with the concept of self-healing systems. A comprehensive survey of major self-healing software engineering approaches is presented in [5]. However, such approaches focus on "traditional" applications *not* service-oriented. The peculiarity of faults, interaction models, and architectural style in SOAs as well as the autonomy, distribution, and heterogeneity of services make these approaches difficult to apply in SOAs.

Current techniques for coping with faults in SOAs allow developers to include constructs in their service specifications (e.g., fault elements in SOAP and WSDL, exception handling in BPEL). Such techniques are static, ad hoc, and make service design complex [13]. Efforts have recently been made to add self-healing capabilities to SOAs (e.g., [6]). However, these efforts mostly focus on monitoring service level agreements and quality of service.

6 Conclusion

In this paper, we proposed a soft-state approach for managing bottom-up faults in composite services. The approach includes techniques for fault detection, propagation, and reaction. We then focused on describing a fault propagation protocol. The protocol inherits the advantages of the soft-state concept: implicit error recovery and easier fault management resulting in high availability. A future extension of our protocol is to combine soft-state with hard-state signaling. For instance, if WS_i provider is scheduling a shut-down, $SS-S_i$ sends an explicit Shutdown() message to $SS-R_j$; this allows $SS-R_j$ to differentiate between physical faults and status changes and hence, detect status changes as soon as they occur. Another possible extension is to transmit Refresh() reliably (e.g., via the use of ACK timers) and integrate a notification mechanism through which receivers inform senders about their view (in terms of failure) on those senders.

References

1. Alonso, G., Casati, F., Kuno, H., Machiraju, V.: Web Services: Concepts, Architecture, and Applications. Springer, Heidelberg (2003) ISBN: 3540440089
2. Avizienis, A., Laprie, J.C., Randell, B., Landwehr, C.: Basic Concepts and Taxonomy of Dependable and Secure Computing. ACM Trans. on Dependable and Secure Computing 1(1), 11–33 (2004)
3. Eugster, P.T., Felber, P.A., Guerraoui, R., Kermarrec, A.-M.: The Many Faces of Publish/Subscribe. ACM Computing Surveys 35(2), 114–131 (2003)
4. Garlan, D., Schmerl, B.R.: Model-based Adaptation for Self-healing Systems. In: WOSS Workshop (November 2002)
5. Ghosh, D., Sharman, R., Rao, H.R., Upadhyaya, S.: Self-Healing Systems - Survey and Synthesis. Decision Support Systems 42 (2007)
6. Guinea, S.: Self-healing Web Service Compositions. In: ICSE Conference (May 2005)
7. Hamadi, R., Medjahed, B., Benatallah, B.: Self-Adapting Recovery Nets for Policy-Driven Exception Handling in Business Processes. Distributed and Parallel Databases (DAPD), An International Journal 22(1) (February 2008)
8. Ji, P., Ge, Z., Kurose, J., Towsley, D.: A Comparison of Hard-state and Soft-state Signaling Protocols. In: SIGCOMM Conference (August 2003)
9. Khalaf, R., Keller, A., Leymann, F.: Business Processes for Web Services: Principles and Applications. IBM Systems Journal 45(2) (2006)
10. Martin, D., Paolucci, M., McIlraith, S., Burstein, M., McDermott, D., McGuinness, D., Parsia, B., Payne, T., Sabou, M., Solanki, M., Srinivasan, N., Sycara, K.: Bringing Semantics to Web Services: The OWL-S Approach. In: First International Workshop on Semantic Web Services and Web Process Composition (July 2004)
11. Medjahed, B., Atif, Y.: Context-Based Matching for Web Service Composition. Distributed And Parallel Databases 21(1) (February 2007)

12. Medjahed, B., Benatallah, B., Bouguettaya, A., Elmagarmid, A.: WebBIS: An Infrastructure for Agile Integration of Web Services. International Journal on Cooperative Information Systems (IJCIS) 13(2) (June 2004)
13. Verma, K., Sheth, A.P.: Autonomic Web Processes. In: ICSOC Conference (December 2005)
14. Papazoglou, M.P.: Web Services: Principles and Technology. Prentice Hall, Englewood Cliffs (2007) ISBN: 9780321155559
15. Qi, Y., Liu, X., Bouguettaya, A., Medjahed, B.: Deploying Web Services on the Semantic Web. VLDB Journal 17(3) (March 2008)
16. Raman, S., McCanne, S.: A Model, Analysis, and Protocol Framework for Soft State-Based Communication. In: ACM SIGCOMM 1999 Conference on Applications, Technologies, Architectures, and Protocols for Computer Communication, pp. 15–25 (September 1999)

Understanding the Diversity of Services Based on Users' Identities

Junjun Sun[1], Feng Liu[1], He Zhang[1], Lin Liu[1,*], and Eric Yu[2]

Key Laboratory for Information System Security, Ministry of Education
Tsinghua National Laboratory for Information Science and Technology (TNList)
[1]School of Software, Tsinghua University, Beijing, China
linliu@tsinghua.edu.cn
[2]Faculty of Information, Toronto University, Toronto, Canada

Abstract. Internet services involve complex networks of relationships among users and providers - human and automated - acting in many different capacities under interconnected and dynamic contexts. Due to the vast amount of information and services available, it is not easy for novice users to identify the right services that fit his purposes and preferences best. At the same time, it is not easy for service providers to build a service with a customizable set of features that satisfies the most people. This paper proposes to further extend the strategic actors modeling framework *i** to analyze the diverse needs of users by modeling explicitly the personal characteristics, organizational positions, and service related roles. We assume that service users' needs and preferences are determined by their personal background, organizational roles, and the immediate operational context in combination. In this way, the origin of the diversity of service needs, quality preferences, and usage constraints, can be ascribed and used as a basis to make rationale selection from currently available types of services, and to reconfigure service interfaces and structures. Example usage scenarios ofweb services are used to illustrate the proposed approach.

Keywords: Personalization, Context, Service adaptation, i* modeling.

1 Introduction

To provide better web services for large scale social events, such as the Olympic Games, or the World Expo, it is important to understand the needs of all involved parties better. User populations of today's web services are becoming more diversified, and it becomes difficult to identify a prototypical user. Diversity in this sense refers to the variety in user needs which involves accommodating users with different skills, knowledge, age, gender, disabilities, disabling conditions (mobility, sunlight, noise), literacy, culture, income, and so on [12]. For example, designingweb site interface needs to consider the users'computing skills and knowledge. Design of search engines can include basic and advanced searches for different users. Web

---

* Corresponding Author.

H. Mouratidis and C. Rolland (Eds.): CAiSE 2011, LNCS 6741, pp. 612–626, 2011.
© Springer-Verlag Berlin Heidelberg 2011

pages can accommodate users from all over the world:e-commerce sitesprovide multi-language product catalogue, and description tuned to regional requirements. Cognitively impaired users with learning disabilities and poor memory can be accommodated with modest changes in layout and controlled vocabulary. Expert and frequent users could as well benefit fromcustomizations that speeds high-volume users and macros to support repeated operations.

Literally, diversity can be defined as the characteristics that differentiate people as individuals, as well as the characteristics which make them alike. It is considered to indicate variety. In traditional product-oriented design, diversity assumes at least two roles: build assurances of variety and choice into its processes and products, and it can also be the source or catalyst for change. Most of the existing research in service adaptation focuses on the physical aspect of the service's environmentand context, e.g., time, location, as an origin of how users' needs changes. However, in addition to the physical settings, human factors also play an important role in the formation, change and evolution of the users' need.

In existing context-aware computing, context information is often pre-assumed to include a limited number of variables, e.g., time, location, and run-time status of the platform. Although a few context service frameworks [10] also consider the preferences of users, there is a pressing need for frameworks and models to support the analysis and design of complex social relationships and identities. We need to understand where the diversities of users come from, how they determine user needs, how they affect each other under different contexts, and how such information can be used when making service selection, adaptation, and reconfiguration decisions.

In this paper, we propose an approach to further extend the strategic actors modeling language $i*$ to facilitate the analysis of user's identity information and the underlining social context for Internet Services. Using this modeling approach, we are able to represent different types of identities, social dependencies between identity users and owners, service users and providers, and third party mediators. A reasoning process linking the steps of analyzing user's identity, deriving service needs and preferences, drilling down into service constraints, and making service selection and adaptation is introduced. This modeling approach will help service vendors to provide customizable solutions, user organizations to form integrated identity information management solution, system operators and administrators to accommodate changes. Typical scenarios of a map web service for users with diversified needs are used to illustrate the proposed approach.

This paper is structured as follows. In Section 2, we extend the actor concept of the $i*$ modelling framework and introduce an $i*$-Context to explain the social and identity origin of the users' diverse needs in the internet service environment. In Section 3, two scenarios are given to illustrate how the human factors of $i*$-Context will impact the preference-based internet service selection and customization. Section 4 introduces our design for a web service selection and adaptation system. An Olympic map service case study will also be presented to illustrate the reasoning process. Finally, in Section 5 and 6, we review the related work, and conclude the paper with a discussion of limitations and future directions.

2   Modeling User Diversity in Service Environment

There are many dimensions in human diversity that beyond obvious differences such as race, gender, age, physical disabilities, and marital status. Less obvious dimensions include: education, lifestyle, nationality, religion, political affiliation, culture and skills. To cope with these diversities, we need a comprehensive modeling framework to capture and understand theseuser diversities and to predict users' needs and preferences accordingly.

Strategic actors modeling framework $i*$ could be used to understand the user's preferences towards certain kinds of web-services. However, current $i*$ modeling lacks of an explanation for the origin of the users' diverse needs and preferences.We extend the actor concept in the $i*$ modelling framework to express the user diversity in the IT Service environment. $i*$ will also be introduced to capture modeling these services related user diversity.

2.1 Actors in Service Environment

In the strategic actor modeling framework, $i*$, system players are modeled on three different abstraction levels, namely, role, position, and agent [16]. Putting the actor concept into the web service environment, the meaning of three different levels of actors in the i* modeling framework can be further clarified to model various service users on different abstraction levels.

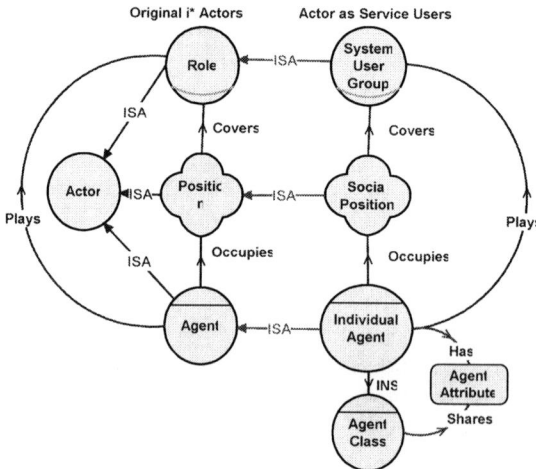

Fig. 1. Mapping service actors

- *Service role*: In the definition of $i*$ modeling language, a role is defined as "an abstract characterization of the behavior of a social actor within some specialized context or domain of endeavor." [16] Particularly, a role in the web service context could be interpreted as the behavior of a specific group of users towards a certain functionalities of service. For example:

$$\text{Role}[map\ serviceuser], \text{Role}[transportation\ serviceuser]$$

- *Social Position*: Normally, the term social position is used to represent the social status of an individual. In the service setting, the concept of social position extends the positionconcept in $i*$by referring to aggregations of social roles[7]. It covers not only the organizational perspectives, but also other community-relatedsocial roles and social relations among a group of people. For instance, in the Olympic Games setting, there are:

Position[*coach*],**Position**[*athlete*]

One could have more than one social position at a time, and social positions all come together with concrete background settings, such as affiliations, certain events or related groups. Different social positions may lead to different user needs and preferences. When there are conflicts between social positions, the actor could compare to find the most important position along with its related events, and select the services accordingly. The importance ofeach particular social positioncould vary as physical context changes.

- *Individual Agent*: In the original $i*$ modeling framework where agent is defined as "an actor with concrete, physical manifestations, such as human individual." [16].In the service environment, individual agent emphasizes on users, as executers and decision makers of pre-defined services.

Agent[*Jim*],**Agent**[*John*]

Agent Attributes: Agent Attributes represent Individual Agent's personal attributes, e.g. age, gender, religion, education level, etc. These agent attributes are comparatively stable and do not change very often.Once the participation of an agent in a social relation relinquishes, the obtained knowledge and history will be transformed into individual attributes.

Attribute[Jim, age]= *20*,**Attribute**[John, education]=*PhD*

We believeindividual agent's preferences most likely depend on his agent attributes.Although some of these relationships could be clearly defined, most of the agent attribute-preference relationships still remain latent or unknown.

- *Agent Class:* We use Agent Class to represent an abstract group of people who has the same value for a certain agent attribute which could lead to similar service needs or preferences. Such as:

John INS **Agent Class**[*vegetarian*]: **Attribute**[John, eating habit] = *vegetarian*

For instance, John is an athlete attending the Paralympics, being a vegetarian, and wants to use the map service to find a nearest restaurant. In this scenario, he is a user of the map service(**Role**[user [*map service*]]); his social position is athlete (**Position**[*athlete*]); as individual agent, his name is John (**Agent**[*John*]), who has a group of concerned attributes: he is a vegetarian (belong to **Agent Class**[*vegetarian*]); he is handicapped and uses a wheel chair (belong to **Agent Class**[*wheel chair user*]).

2.2 *i**-Context for Web Service

Incorporating results from current context-aware system practices, where location, identity, activity and time related information are used to capture the service context[3]. We propose to extend the current i* concept by putting the position occupation and role playing relation under context. In other words, the agent-position and position-rolerelationship may vary according to physical context such as time, location, etc. Thus, we use service user's social positions and his agent classesas *i**-Context to capture and model the diversity of service settings.

Table 1. Categories of Social Positions and Agent Classes in *i**-Context

	Category	Examples
Social Position	Occupation	Position[*athlete*],
		Position[*coach*]
	Family	Position[*father*]
		Position[*daughter*]
	Hobby	Position[*golf club member*]
		Position[*classic music fan*]
Agent Class	Demographical	Agent Class[*male*]
		Agent Class[*teenager*]
	Cultural	Agent Class[*Buddhist*]
	Physical	Agent Class[*wheel chair user*]
	Education & Skill	Agent Class[*C++ expert*]
	Habit	Agent Class[*vegetarian*]

Table 1 shows some example categories of social positions and agent classes used in *i**-Context.As individual agent may occupy more than one social positions at a time, the importance of each position under certain physical conditions is also modeled in the *i**-Context.

Back to our case Study, Table 2 shows the *i**-Context of a Paralympics athlete, John. Being an athlete and father of a six year old girl, Johnis also avegetarian, Chinese, practicing Buddhism and using a wheel chair.This information brings John the social positions of **Position[*Father*]**, **Position[*Athlete*]**, and also reveals the agent classes John belongs to, such as **Agent Class[*vegetarian*]**, **Agent Class[*Chinese*]**, **Agent Class[*Buddhist*]** and **Agent Class[*wheel chair user*]**. Along with these social positions and agent classes, *i**-Context also provides importance ratings for social positions in various physical contexts. To keep the problem clear, we use a simplified importance value to represent the importance of social positions.

Based on our observation, although individual agents may have more than one social positions at a time, there is usually one position that surpasses all others in importance, and decides what kind of action the agent should performina givenphysical context. Hereby, we introduce the concept of "foreground position".

- **Foreground Position**: Foreground position is the social position that overweighs all other social positions in importance in a certain physical context for a particular agent.Define***Positions*(Agent[*a*], *pc*)** asthe set of all social positions Agent[*a*] plays in certain physical context*pc*.

Foreground Position(Agent[a], pc):$\exists d \in$ *Positions(*Agent[a], *pc)/*

$\forall p \in Positions($**Agent**[a], *pc), p \neq d \Rightarrow$Importance (**Agent**[a], **Position**[p], *pc*)
$$<= \text{Importance } (\textbf{Agent}[a], \textbf{Position}[d], pc)$$

Table 2. Paralympics Athlete John's *i**-Context

Category	Value(s)	
Social Positions	**Position**[*Athlete*] **Position**[*Father*]	
Agent Class	**Agent Class**[*vegetarian*] **Agent Class**[*wheel chair user*] **Agent Class**[*Buddhist*] **Agent Class**[*Chinese*]	
Position	Importance Factor	Physical Context(Time, Location)
Position[*Athlete*]	3 (Very Important) (+)	**Time**: {9am~11am, Aug 30th, 2012} &**Location**: at Venue
Position[*Athlete*]	2 (Important)	**Location**: at Gym
Position[*Athlete*]	0 (Not Important)	**Time**: {Aug 31st ~ Sept 4th, 2012}
Position[*Father*]	3 (Very Important) (*)	**Time**: {June 30th }
Position[*Father*]	2 (Important)	**Location**: at home
Position[*Father*]	1 (Less Important)	**Time**: {Aug 25st ~ Sept 4th, 2012}

(+) John's competition is scheduled on the morning of Aug 30th, 2012
(*)June 30th is the birthday of John's daughter

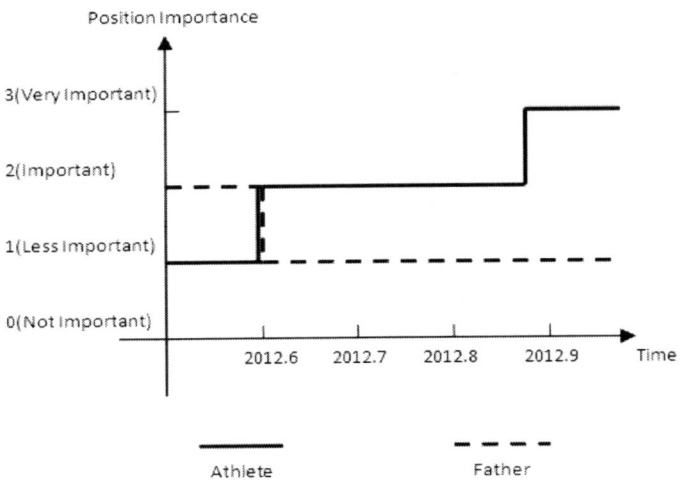

Fig. 2. Foreground Social Positionof John

For example, as shown in Fig. 2, the importance of John's social position changes according to time.Before 2012.6, John was with his daughter and had not started his intensive Paralympicstraining yet. In this case, during that period of time, **Position**[*father*] surpasses **Position**[*athlete*] to be the foreground position of John's. However, after 2012.6,John leaves home to take part in a national Paralympics training camp in Beijing, and finally join the wheel fencingcompetition. During these three months, John's social position as **Position**[*athlete*] is more important, and hereby it replaces **Position**[*father*]to become his foreground position.

3   Bridging User Diversity and Service Diversity

In this section, we will explore how the service diversity will comply with the user diversity, and will propose a set of*i**-modeling scenarios to illustrate the propagation of the users' diverse needs under the web service environment. We are in the belief that the users' diverse requirements towards web services originate from the users' social positions and their individual characters defined in its unique *i**-*Context*. This diversity will then propagate through the service selection and customization processes, and will eventually lead to the variations in the choice of services and their settings.

3.1   From User Diversity to Service Diversity

Fig. 3 shows how the user's own diversity will affect the service selection and customization process. User's needs aremainly decided by the social positions that the user occupies at the time, and it will then become the major concerns for service selection; while the user's preferences are mostly beingaffected by the user's agent attributes, and will further be used to customize the alternative services.

In the following two parts, we will use two particular scenarios to explain how the user's social position will influence the service selection, and how the user's agent attributes will affect the service customization process.

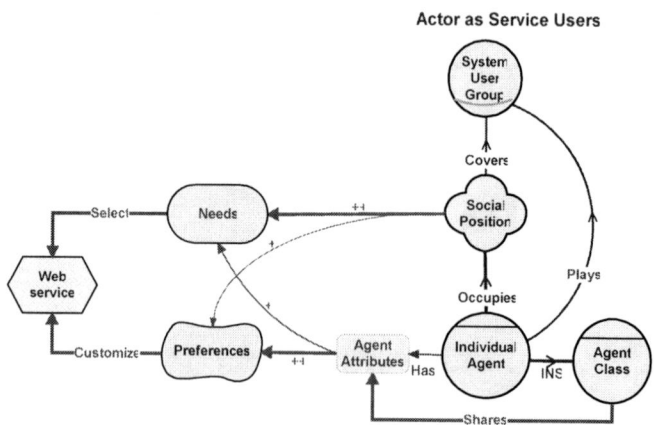

Fig. 3. Propagation Path from User Diversity to Service Diversity

3.2 Social Position's Influence over Service Selection

In this scenario, we will explore how the service user's social position can decide the user's needs and eventually influence over service selection. To make the relationship between social position and services-selections clearer, we choose the same individual agent occupying different positions in the same physical context.

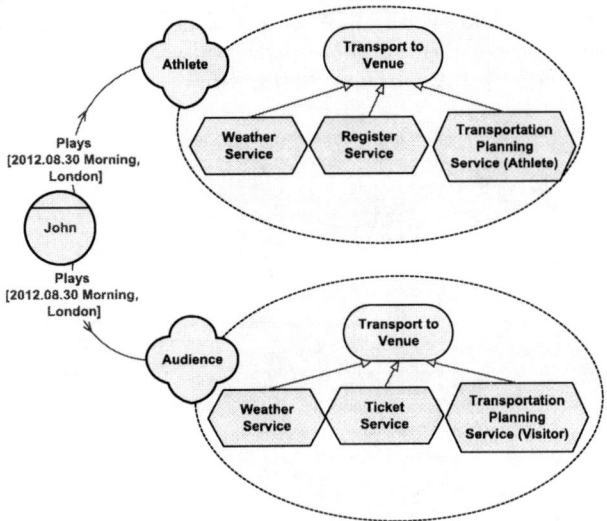

Fig. 4. Social Position's Influence over Service Selection

As depicted in the Fig. 4, during London Olympics, John is aspectator as well as anathlete.Service discovery follows a similar process as the *i** goal-task iterative refinementprocess. The top goals in the goal decomposition structure stands for the high level abstract user needs. As these goals are refined into more detailed levels, they will finally reach a level of abstraction that refers to concrete tasks which could matchgiven web services.As shown in Fig. 4, even if the social position spectator and athleteshare the same top goal "Transport to Venue", they could also lead to different service selections. This scenario shows, there are basically three possible results for the same user to pursue the same top -level goal using different positions:

1. The derived services are exactly the same, such as the "weather service", where social positions make no difference.
2. The selected services are completely different, such as the "register service" and "ticket service", which implicates social position leads to completely different procedures for achieving the same goal.
3. The selected services are basically the same type, but have different variations, such as the "transportation planning service" has different entries for athletes or spectators.

As we can see throughthis analysis, it is important to capture as much information about user's social identity as possible to make rationale service decisions. At the same time, users should be aware of the implications of his social identities when requesting services.

3.3 Agent Attributes' Influence over Service Customization

In the second scenario physical context and the social position are the same, but there are two different individual agents. Two individual agents Alan and Steve have the same social position as the **Position**[*athlete*]. Their diverse service preferences could be derived from theirindividual attributes respectively.

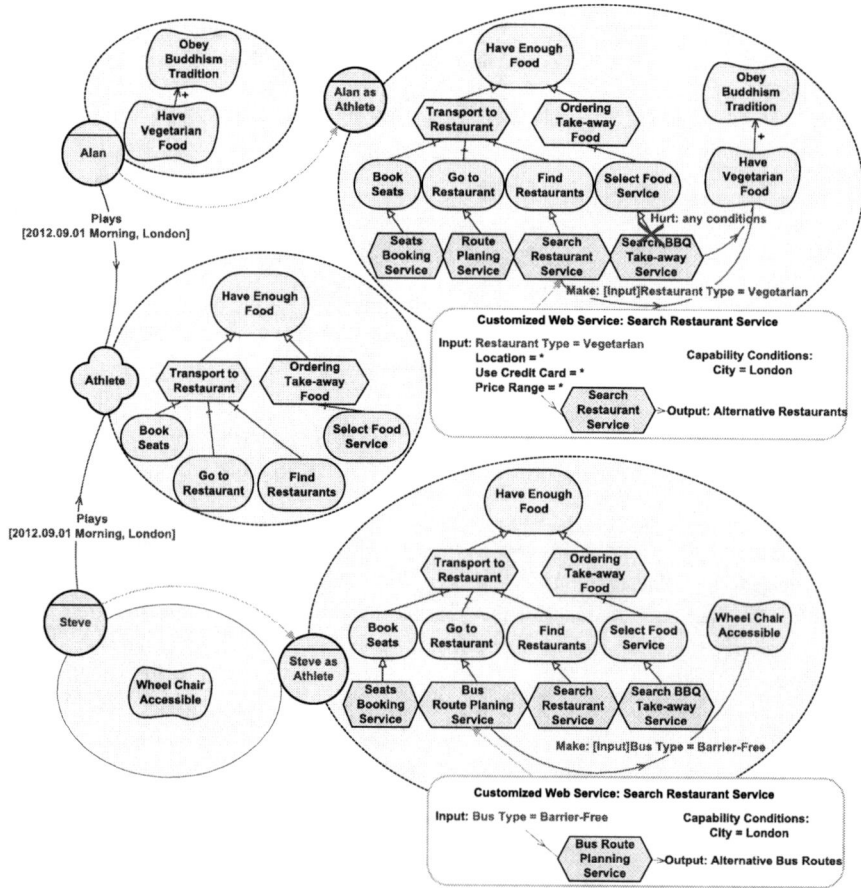

Fig. 5. Individual User Attributes' Influence over Service selection and Customization

In Fig.5, Alan and Steve's agent attributes will propagate to their social position athlete. In *i** model, the user preferences deduced from the user's Softgoals could be derived from individual attributes. Now we will use the Softgoal to select and configure web service.

Fig. 5 shows different individuals with the same social position and the same physical context may lead to different web service choices. In the selection case, take the example of contribution link from **Service**[*Search BBQ Take-away Service*] to **Softgoal**[*Vegetarian*], a service may not be selected as it hurts the user's softgoal. In the configuration case, such as contribution link between **Service**[*Search Restaurant Service*] to **Softgoal**[*Vegetarian*], the vegetarian setting should be add to the service input of **Service**[*Search Restaurant Service*].

4   Web Service Selection and Customization using i*-Context

We have discussed how the user diversity captured by the *i*-Context* would influence the serviceselection and customization. In this section, we propose a systematic process to select and customize services according to the users'personal needs and preferences.

4.1   Web Service Selection and Customization Process

Fig.6 shows the main process of our i*-Context based web service selection and customization approach. This process consists of four different steps, namely: Obtaining *i*-Context*; Creating Goal-Task Decomposition Tree; Generating Softgoal Decomposition Structure; and Selecting and Customizing Services. In the following part of this section, we will use a case study to illustrate this process.

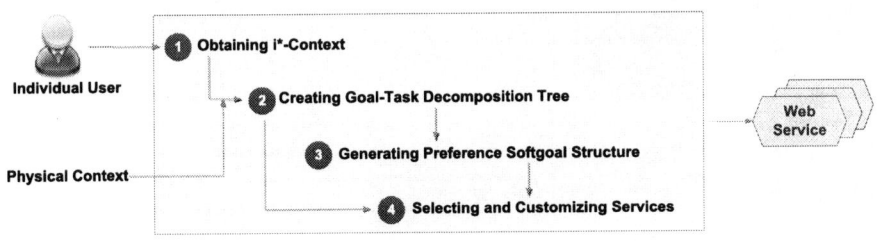

Fig. 6. Service Selection and Customization Process

4.2   Case Study

In the previous sections, we have described how the users' diverse needs propagate through the web service selection and customization operations. In this section, we will propose a systematic process to help the service users choose and personalize web services according to their diverse needs. A case study as well as a demo system will be given to illustrate this process.

In our case takesJohn as anexample, according to John's i*-Context provided in Table 2:John is a Chinese athlete in the 2012 Paralympics Games;heis also a father to a six year old girl. As an individual agent, Johnis vegetarian, Chinese, using a wheel chair andpracticing Buddhism.

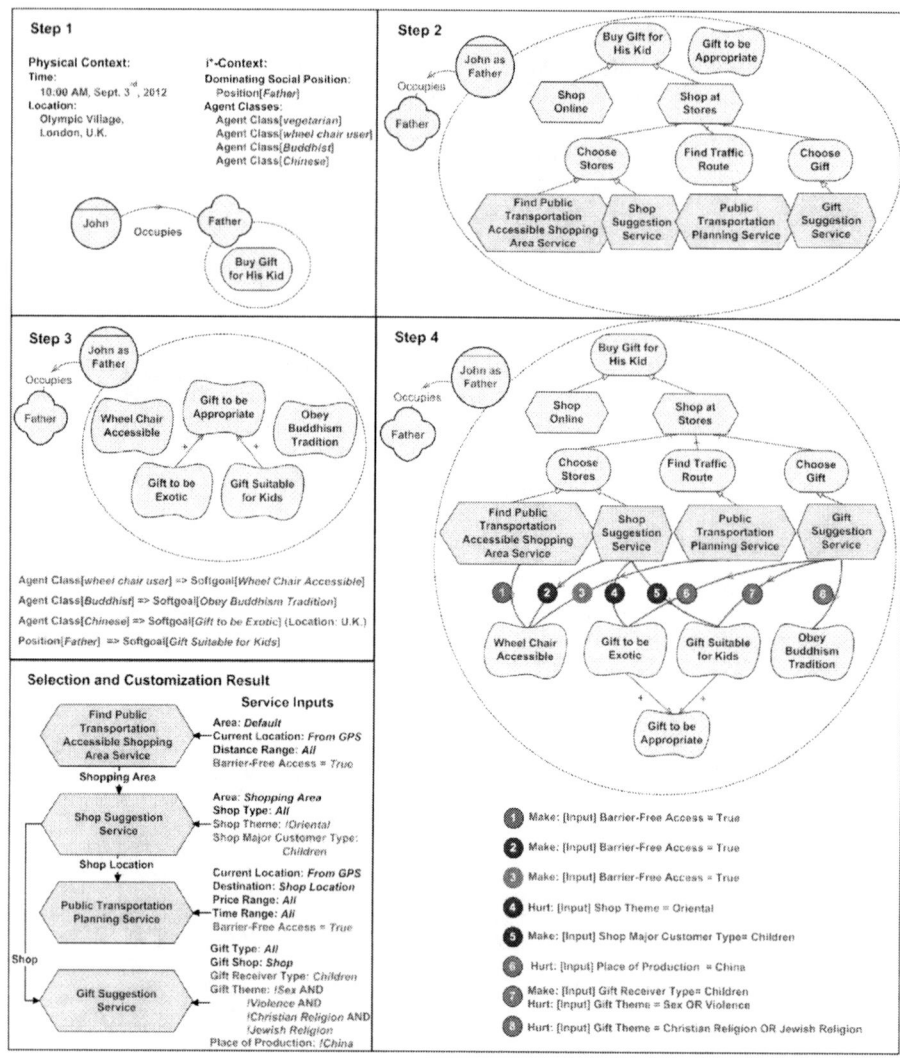

Fig.7. Service Selection and Customization Case Study

As shown in Fig. 7, with the process introduced in the previous part, services selection and customization for our case study will be carried out in the following four steps:

Step 1. Obtaining i*-Context:

In this step, i*-Context will be obtained from the user's profile. Meanwhile, the user's foreground social position as well as hiscurrent top goal will also be derived or collected.

In John's case study, when John uses the system, the physical context is 10:00AM, Sept 3rd, 2012, in the Olympic Village in London according to the sensors. Based on John's i*-Context (Table 2), his foreground social position is **Position**[*Father*]. This is because he has already finished his race, and now doing something to compensatefor being away from his little daughter. According to his to-do list as **Position**[*Father*], buying a gift for his daughter is on the top of the list.

Step 2. Creating Goal-Task Decomposition Tree:
The second step of the process focuses on the user's needs, and uses iterate refinement technique to create a goal-task decomposition tree based on the top user's goal.

In John's case study, **Goal**[*Buy Gift for His Kid*] could be satisfied either online or at a gift shop. When we look into **Task**[*Shop at Stores*], the task could then be further decomposed into three separate sub-goals, namely, **Goal**[*Choose Stores*], **Goal**[*Find Traffic Route*] and **Goal**[*Choose Gift*]. After that, these sub-goals are being operationalized by a set of different web services (modeled as the bottom-level yellow hexagons). In this step, softgoals derived from goal and task decomposition will also be modeled (see **Softgoal**[*Gift to be Appropriate*]).

Step 3. Generating Preference Softgoal Structure:
In the third step of the process, preference softgoals are being deduced from user's agent classes, his foreground position and the physical context.

As shown in Fig. 7, according to the pre-defined preference deduction rules, John's preference softgoal structure can be generated. For instance, **Softgoal**[*Wheel Chair Accessible*] is introduced since hebelongs to **Agent Class**[*wheel chair user*]. While as **Softgoal**[*Gift to be Exotic*] is derived from the existing **Softgoal**[*Gift to be Appropriate*], because John is from China and he is currently in U.K. Two other preference softgoals can be acquired in the same way.

Step 4. Selecting & Customizing Services:
In the last step of the process, the alternative services obtained from step 2 will be further selected and customized to better comply with the user's preference softgoals.

As the bottom-level service tasks and the preference softgoal being identified, the relationships between these services and softgoals could be used for service selection and customization. Assuming all the web services are stateless services, the relationships can be represented with the existing conditions of the contribution links in between. Take relationship number 3 as an example, **Service**[*Public Transportation Planning*] could make the **Softgoal**[*Wheel Chair Accessible*] possible, only if the information service provide a barrier-free access route. Thefinal service selection and customization result after this process is shown in the lower left corner of Fig. 7.

5   Related Work

We discuss related work in three areas: agent intentional modelling in requirements engineering, context-aware web services research and preference-based service selection and customization practices.

In information systems and software engineering research, organizational modeling has been of interest, often in connection with requirements engineering. Goal and agent oriented approaches have been used in this context, and agents or actors are often part of the modeling ontology [13][6]. However, the*i**-Context approach is distinctive in its treatment of agents/actors as being strategic and context-sensitive[5], and thus readily adaptable to the service requirements adaptation based on identity illustrated in this paper.

Context-aware system and web service are also a very important issue in recent years. Regarding the definition of context for web services, OASIS's WS-Context Standard defines service context as the kind of information that explains "what an activity is and what services it will require in order to perform that work, will depend upon the execution environment and application in which it is used[10]."Dey and Abowd surveyed the existing work in context-aware computing, and have defined context to be any information that can be used to characterize thesituation of an entity. Location, identity, time and activity are the four primary context types that they have summarized.[3]. In Baldauf, Dustdar and Rosenberg's survey, they summarized the different design principles and context models for context-aware systems. According to their survey, the context models used by existing context-aware systems involves key-value models, markup scheme models, graphical models, object oriented models, logic based models and ontology based models [2]. Comparing to the existing context-awareweb service practices covered by these surveys, our approach focuses more on exploring the origin of the users diverse needs and preferences rather than summarizing the directly related context of some specific kinds of services.

There are also a research and practiceson preference-based web service selection and customization. By utilizing decision tree algorithm, Hong, Suh, Kim and Kim have proposed an agent based framework for providing the personalized services using context history [4]. While in Lamparteret. al.'s [14] and Medjahed et. al.'s [8] research, ontology based modelling and reasoning approaches are being introduced to configure the service and system according to user's preferences.

Finally, there is a body of literature on personal and contextual requirements engineering. Sutcliffeet. al. proposed a framework for requirements analysis that accounts for individual and personal goals, and the effect of time and context on personal requirements[1]. The implications of the framework on system architecture are considered as three implementation pathways: functional specifications, development of customizable features and automatic adaptation by the system.Salifuet. al. presents a problem-orientedapproach to represent and reason about contextualvariability and assess its impact on requirements[9]. Ali et. al. proposes a goal-orientedRE modeling and reasoning framework for systems operatingin varying contexts[11]. Liaskoset. al. proposes a variability intensive approach to goal decomposition supporting requirements identification for highly customizable software [15]. Contextualgoal models are introduced to relate goals and contexts;reasoning techniques to derive requirements reflectingthe context and users priorities at runtime; and finally,design time reasoning techniques to derive requirementsfor a system. While these works are all along the same line with our approach, we emphasis on the user's social and personal identities to rationalize the diverse needs they may have when using web services.

6 Conclusion and Discussion

We have outlined an approach for modeling and analyzing diversities of web service users. The approach is based on social and intentional analysiscentered on actor's social and personal identities under different contexts.It allows us to go beyond mechanistic behavior, to deal with the opportunistic and rationaledecision making of strategic actors. Interdependencies among actors' identitybring opportunities as well as place constraints on their service privileges. Strategic actors seek to achieve goals (hard and soft) by obtaining new identities from service providers, taking into account the opportunities and disadvantages arising from various casual relationships, as illustrated in the examples.

Our approach is complementary to existing frameworks and techniques in service personalization and context-aware service provision. Weemphasizethat the systematic analysis of relationships among social rolesand personal background for a given individual actor may play an essential role in eliciting needs for service users. It supports the exploration and management of service alternatives, based on a balanced consideration of all competing requirements, thus complementing the various solutions of recent service selection and adaptation techniques.

While this paper has outlined some basic modeling concepts, much remains to be done.To evaluate the effectiveness of the proposed approach, we are currently developing an application for integrating online banking information services based on user's uniquebackground and settings.

There are also several options to further extend our initial approach. Identity management is increasingly connected with other activities in enterprise management. The proposed diversity rationalization approach provides a way of linking identity related analysis to serviceneeds analysis and technology configuration analysis. The conceptual modeling approach can thus provide a unifying framework for service broker systems, supporting decision making and the management of changes across technical services development, business services model development, and identity management.

Meanwhile, there is also much potential in the synergy between social position modeling and the foundational principles in context modeling. For example, in analyzing the implications of an identity, one would like to model the inter-relatedness among their subject matters. The interaction between social concepts and relationships (actors, goals, preferences) and physical ones (e.g., processes, information assets, time, etc.) need to be detailed. Libraries of domain knowledge, service design knowledge with regard to identity management and context management would be very helpful during modeling and analysis. These are topics of ongoing and future research.

References

1. Sutcliffe, A., Fickas, S., Sohlberg, M.M.: Personal and ContextualRequirements Engineering. Proceedings. In: 13th IEEE International Conference on Requirements Engineering, pp. 19–30 (2005)
2. Baldauf, M., Dustdar, S., Rosenberg, F.: A survey on context-aware systems. International Journal of Ad Hoc and Ubiquitous Computing 2(4), 263–277 (2007)

3. Gregory, D., Abowd, A.K.: Towards a better understanding of context and context-awareness. In: Gellersen, H.-W. (ed.) HUC 1999. LNCS, vol. 1707, pp. 304–307. Springer, Heidelberg (1999)
4. Hong, J., Suh, E.-H., Kim, J., Kim, S.: Context-aware system for proactive personalized service based on context history. Expert Systems with Applications 36(4), 7448–7457 (2009)
5. Keidl, M., Kemper, A.: Towards context-aware adaptable web services, p. 55. ACM Press, New York (2004)
6. Khalid, H.: Embracing diversity in user needs for affective design. Applied Ergonomics (2006)
7. Masolo, C., Vieu, L., Bottazzi, E.: Social roles and their descriptions. In: Proceedings of Principles of Knowledge Representation and Reasoning (2004)
8. Medjahed, B., Atif, Y.: Context-based matching for Web service composition. Distributed and Parallel Databases 21(1), 5–37 (2006)
9. Salifu, M., Yu, Y., Nuseibeh, B.: Specifying Monitoring and Switching Problems in Context. In: RE 2007, 211–220 (2007)
10. OASIS, Web services context (WS-Context) Standard. OASIS Standards (2003)
11. Ali, R., Dalpiaz, F., Giorgini, P.: A goal-based framework for contextual requirements modeling and analysis. Requir. Eng. 15(4), 439–458 (2010)
12. Schneiderman, B.: Universal usability. Commun. ACM 43(5), 85–91 (2000)
13. Shivakumar, N., Jannink, J., Widom, J.: Per-user profile replication in mobile environments: Algorithms, analysis, and simulation results. Mobile Networks and Applications 2 (1997)
14. Lamparter, S., Ankolekar, A., Studer, R., Grimm, S.: Preference-based selection of highly configurable web services. In: Proceedings of the 16th International Conference on World Wide Web - WWW 2007. ACM Press, New York (2007)
15. Liaskos, S., Lapouchnian, A., Yu, Y., Yu, E., Mylopoulos, J.: On Goal-based Variability Acquisition and Analysis. In: Proceedings of the 14th IEEE International Conference on Requirements Engineering (RE 2006), pp. 76–85 (2006)
16. Yu, E.: Modelling strategic relationships for process reengineering. PhD Thesis, University of Toronto, Department of Computer Science (1995)

Request/Response Aspects for Web Services

Ernst Juhnke[1], Dominik Seiler[2], Ralph Ewerth[1],
Matthew Smith[3], and Bernd Freisleben[1]

[1] Department of Mathematics & Computer Science, University of Marburg
Hans-Meerwein-Str. 3, D-35032 Marburg, Germany
{ejuhnke,ewerth,freisleb}@informatik.uni-marburg.de
[2] Information Systems Institute, University of Siegen
Hölderlinstr. 3, D-57068 Siegen, Germany
d.seiler@fb5.uni-siegen.de
[3] RRZN, University of Hannover
Schloßwender Straße 5, D-30159 Hannover, Germany
smith@rvs.uni-hannover.de

Abstract. Web services rely on standardized interface descriptions and
communication protocols to realize loosely-coupled distributed applications that are executed on several interconnected hosts. However, the
extension of a web service with non-functional requirements, such as efficient data transfer or security, is a tedious task that also requires access to
the web service implementations. In this paper, we present *request/re-
sponse aspects* for web services to allow software developers to easily
and transparently change the data exchange between web services *without* modifying their implementations or their interfaces. A framework
supporting request/response aspects for web services is presented, and
implementation issues are discussed. The usefulness of request/response
aspects is illustrated by three use cases.

Keywords: Aspect-oriented Programming, Web Service, Service-oriented Architecture, SOAP.

1 Introduction

With the advent of service-oriented architectures (SOA) and web services as
their most widely used implementation technology, applications can be composed
of existing web services, promising higher reusability, faster development, and
consequently, reduced costs. Web services are identified by their interfaces that in
turn are defined using the Web Services Description Language (WSDL, [23]). A
WSDL document contains a set of operations and defines input/output messages,
faults and bindings for transport protocols. Typically, web services communicate
via SOAP [22], relying on a request/response message exchange pattern based
on XML documents. Given a set of web services, composition languages such
as the Business Process Execution Language for Web Services (BPEL, [2]) can
be used to compose them into a more complex service. The original services act
as the basic activities in the newly constructed service; hence, this paradigm is

H. Mouratidis and C. Rolland (Eds.): CAiSE 2011, LNCS 6741, pp. 627–641, 2011.

often referred to as "Programming in the Large". The composite web service orchestrates the control flow between the original web services, stating the order in which their methods are called. To keep services decoupled, their interaction is managed by a client (i.e., a composer service like a workflow engine), and the services do not have any information about the service to be called next.

While powerful in terms of compositionality, the downside of the "Programming-in-the-Large" paradigm is that the control flow dictates its structure onto data-flow related concerns, which may not always be appropriate for them. We have experienced problems related to this dominance of the control flow structure during the development of a large-scale service-oriented platform for content-based search in image and video databases [10,11]. Following the service-oriented paradigm enables us to reuse multimedia algorithms, encapsulated as services, in different workflow configurations. However, it also forces data transfers between two subsequent multimedia services in a workflow to travel via the composition client in the middle, which is neither necessary nor desirable: The available network bandwidth of the machine hosting the BPEL engine can very quickly become a bottleneck and the runtime performance will decrease significantly.

One way to solve this issue is to avoid the transfer of huge data via the BPEL engine by using the Flex-SwA framework [12]. Instead of huge binary data, only a small reference is transferred from the data-producing service via the composition client to the data-consuming service. The data consuming service resolves the reference and gets the binary data. However, this technique requires that the involved services have reference data types in their method signatures and in their return values, respectively. Furthermore, modularity and decoupling of web services would suffer from scattered code for data handling.

It is preferable to have a solution that allows efficient data transmission in data-intensive service workflows without noticeable additional development efforts and without losing modularity and decoupling of services. One would like to be able to superimpose a structure over the service control flow such that the data transfer cuts across the control flow. Aspect-oriented programming (AOP) is a paradigm that is aimed at increasing the modularity of software [14]. Cross-cutting concerns such as the efficient transmission of data can be modularized in aspects. AOP enables a developer to integrate aspects into existing applications via join points, i.e., particular points in the control flow that specify when such modularized code (called "advice") should be executed; the description of a set of join points is called pointcut.

In this paper, we present *request/response aspects* for web services to address this problem. The proposed request/response aspects allow to add non-functional requirements at the web service communication level, and to execute aspect code on a remote host where a web service is running. They are independent of a web service's implementation language and do not assume that a web service implementation itself provides related functionality. For example, in the multimedia workflow mentioned above, the components for efficient data transmission can be woven dynamically into the communication infrastructure on top of which the web services run *without* changing their implementations or their interfaces.

We have implemented three use cases concerned with data transmission, data compression, and data encryption to demonstrate the feasibility of the proposed request/response aspects.

The paper is organized as follows. Section 2 reviews related work. The design of the proposed aspect framework for web services is presented in Section 3. In Section 4, implementation issues are discussed. The three use cases are presented in Section 5. Section 6 concludes the paper and outlines areas of future work.

2   Related Work

Several efforts have already been made to modularize cross-cutting concerns in a web service environment. Related approaches can roughly be grouped into three categories: (a) they operate on the composition of services; (b) they introduce an intermediary layer that encapsulates services; or (c) they work at the remote service, i.e. not a physically co-located service.

AO4BPEL [6] is an extension of BPEL to improve modularity and to support dynamic adaptation of the composition logic. In this setting, every BPEL activity is a possible candidate for a join point. AO4BPEL facilitates the modularization at the level of the BPEL engine. Despite the fact that due to its similarity to traditional programs this seems natural when weaving aspects into a service-oriented architecture, it is not possible to realize a data transfer aspect like the one described in our video analysis workflow, because an adaptation of the service is necessary.

Courbis and Finkelstein [8] present an adaptable BPEL infrastructure that is extensible both statically and dynamically. The extension is not limited to the engine itself, but also includes a BPEL process. While this is an interesting concept for developing a BPEL engine, this approach is – again – clearly limited to the adaptation of the BPEL engine and the BPEL process.

Other approaches [7,13] have also identified the necessity of modularizing crosscutting concerns in a service-oriented environment. To achieve this objective, they modularize scattered concerns at the composition level. This might be sufficient, e.g., for expressing authentication against remote services. However, the mentioned approaches cannot be used when crosscutting concerns affect the functionality of a remote service.

The Web Services Management Layer (WSML, [21]) introduces an intermediate layer between client and web service. The layer is used for the just-in-time integration of web services into a client. Thus, the scope of WSML is limited to the client-side. The thereby induced scattered code (by using multiple services) can be faced by utilizing Aspect Beans and Connectors of JAsCo [18]. JAsCO can work with client requests, but not with the requests on the service side.

Binder et al. [5] have introduced Service Invocation Triggers, a lightweight infrastructure based on a proxy layer that routes messages and thereby optimizes the data transfer when workflows are orchestrated. If the proxies are located on the same host as the service the proxy communicates with, a comparable situation to Flex-SwA is achieved in terms of optimizing the data transfer. However, this is achieved at the expense of transferring the control flow to the proxy architecture that in this way represents a hidden orchestration layer, whereas in our approach it remains at the orchestration engine.

DJCutter [16] is a framework that provides remote pointcuts as a new language construct for distributed aspect-oriented programming. While an aspect is distributed to remote machines, it only notifies a central server when a pointcut becomes active. In turn, this server executes the corresponding advice. Referring to the data transfer example again, the data must first be transferred to the server that processes it and afterwards transferred back. This will certainly undo the desired performance gain.

Baligand and Monfort [4] present a framework to separate crosscutting concerns within the service implementation, but do not deal with crosscutting concerns over multiple services or hosts. Their focus is on the weaving of aspects into the (Java) byte code of the service implementation. Thus, they pin the presented framework to web services implemented in Java, whereas request/response aspects operate on the message level. This allows us to be independent of the concrete implementation language of a service.

AWED [15] is an aspect-oriented programming language with explicit support for the distribution of aspects and advice. In contrast to request/response aspects, it operates on Java. However, web services can be implemented in any programming language, and only their XML-based SOAP-interface is known. Again, it cannot be assumed that Java is the implementation language of web services, such that AWED cannot be used in web service environments where Java is not used as the implementation language.

3   Request/Response Aspects

In this section, the pointcut description for defining request/response aspects in web service environments and a supporting framework are presented. The pointcut description is independent of any implementation language, whereas the framework itself is located at the web service middleware (i.e., a software stack that provides support for SOAP communication, like Apache Axis [3]) and provides the capabilities of interpreting pointcuts and applying the corresponding advice. The framework consists of: 1.) an extension of the web service middleware that enables the weaving of aspects into web services without changing their implementation or their interface, including an aspect configurator service, a request/response aspect weaver, and a security manager; 2.) tools that operate at the client side, in particular for the seamless integration in BPEL workflows. This framework is an essential prerequisite for writing and weaving request/response aspects. By remaining independent of the service implementation, the framework respects the black-box idea of service-oriented architectures.

3.1   Framework for Request/Response Aspects

The execution of web services is typically performed by an installed middleware, such as a web service container and a SOAP processor. The underlying idea of the proposed framework is to weave aspects into the SOAP message processing chain. The weaving of aspects at this point enables us to be independent of the concrete service implementation: the processing chain typically operates on XML documents and weaving into it does not interfere with the service implementation. In advance, an aspect configurator enhances the existing middleware by allowing to weave aspects and, e.g., to check whether it is allowed or possible to weave an aspect into a service based on the aspect definition and the enforced security constraints. The framework is based on the following design considerations:

Dynamic weaving: Since the implementation language of a web service is typically unknown, and the client's and the service's administrative domain differ, aspect weaving must be performed dynamically. As mentioned before, the middleware container hosting web services should expose web service methods for adding and removing aspects during runtime.

SOAP message chain: Due to the contract-first conception of web services, aspects are not allowed to change the previously negotiated interface (i.e., the WSDL description) of a service. Ideally, the aspects have to be placed in the SOAP processing chain to circumvent a modification of the interface. They must be able to deal with SOAP request and response messages such that a modification of them does not change their syntactical representation.

Caller/callee interaction: On the client side, the framework must have the ability to distinguish between different web service invocations and communication partners who are possibly affected by an aspect. An aspect affecting two subsequent services can only be added to a service if the corresponding aspect on the subsequent service can also be applied. On failure, the aspect has to be removed. Consequently, an atomic behavior of aspect weaving must be ensured. This includes the possibility for a caller to check whether an aspect is available that can be enabled, and to determine which aspects are currently active for this particular caller.

Scope: Aspects must have a certain scope [19]. A scope specifies whether the aspect is only valid for the client that has woven the aspect or whether the aspect is valid for all callers.

Security: A security manager must ensure that only authorized clients and aspects make use of the framework. The security manager can enforce that only aspects that meet security requirements can be used or deployed. Using established authentication mechanisms (e.g., SSL/TLS with client verification [20]), only authorized users are permitted to use the aspect configurator and to deploy signed aspects.

State: Since web services are stateless, aspects and advice should not maintain any internal state, besides the information stored in the external key value store (see 3.2) and in the message itself, respectively.

In Figure 1, the general design of the proposed framework for request/response aspects is shown. Our framework operates on the server side as well as on the client side. At the server side, there is a configuration interface (to be called by the client), a configuration manager and a component that performs the actual

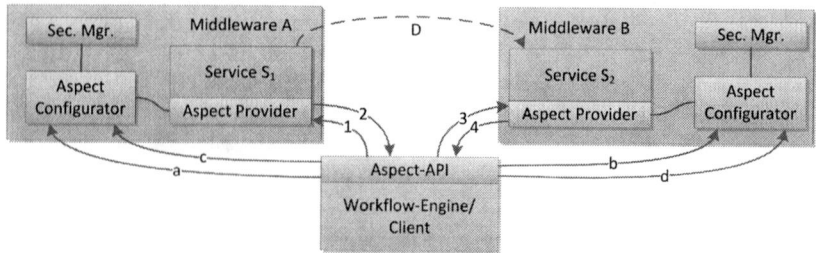

Fig. 1. Web service invocations with aspects

weaving of aspects. At the client side, mainly an API is provided that offers methods for interacting with the server component.

The interaction between a client and the framework is as follows. First, the client's Aspect-API asks the aspect configurators at the web services S_1 and S_2 whether they support the needed aspect (Figure 1: steps a and b). If both sites support the aspect and respond accordingly, then the aspect is added by the client (steps c and d) into the response of S_1 and the request of S_2. When the actual SOAP request is sent to Service S_1 (step 1), it is answered by the response message (step 2). Then, the client forwards the request (step 3) to Service S_2 that due to the request aspect can process the request and eventually responds (step 4). In our use case for data transfer, an aspect that supports direct and efficient data transmission from Service S_1 to Service S_2 will be used. Due to the use of request/response aspects, the response message of Service S_1 does not include the data D in this scenario, since the aspects enable the direct data transmission of data D to Service S_2.

3.2   Pointcut Description

A request/response aspect for web services is defined by the following tuple:

$$\mathcal{A} = \{\mathcal{P}, \mathcal{O}, \mathcal{F}, \mathcal{M}, \mathcal{I}, \mathcal{K}, \mathcal{D}, \mathcal{C}\}.$$

$\mathcal{P}$ (*porttype*) defines the qualified name (QName) of the web service porttype the aspect should be applied to. This element is required to cope with a wild-card operator, meaning that all porttypes within a service are relevant for this particular aspect. $\mathcal{O}$ (*operation*) specifies the name of the operation of the given porttype. A support for a wildcard operator is also needed. $\mathcal{F}$ (*field*) determines the actual WSDL message part the aspect should be applied to. Since XML schema elements can be nested complex data types, this field is realized as an XPath expression[24] , in case the aspect should be applied only to a particular part of a complex type. The XPath language also allows the use of wildcard operators and thus enables the application of the aspect to multiple elements or to multiple parts of (multiple) elements. $\mathcal{M}$ (*mode*) is an element of the enumeration containing the values {request, response, both}. The first one means that the aspect is applied to the request message, the same holds for response. When the mode is set to both, the aspect is applied to the request as well as to the response message. $\mathcal{I}$ (*ID*) determines a reference to the actual advice that has to be applied by providing an ID containing a reference to the advice.

The preceding elements are mandatory, whereas the following elements are optional: $\mathcal{K}$ (*key value store*) provides configuration data in the form of key-value pairs optionally needed by the aspect. $\mathcal{D}$ (*depends*) and $\mathcal{C}$ (*conflicts*) contain a list of aspects that refer to woven aspects. The first one enforces the listed aspects to be woven, whereas the latter one forbids them to be applied. This enables the weaving operation to respect (in-)compatibilities between different aspects.

The pointcut description is represented by the XML schema type shown in Listing 1.1. The first six elements of the tuple, namely $\mathcal{P}$, $\mathcal{O}$, $\mathcal{F}$, $\mathcal{M}$, $\mathcal{I}$, $\mathcal{K}$, are mapped to the schema type. The fields $\mathcal{C}$ (conflicts) and $\mathcal{D}$ (depends) are not present in the schema type, because they are expressed by the implementation of the advice itself. Hence, they are omitted in the schema declaration.

```
<complexType name="Aspect">
  <sequence>
    <element name="portType" type="xsd:QName" />
    <element name="operationName" type="xsd:string" />
    <element name="field" type="xsd:string" />
    <element name="mode" type="xsd:string" />
    <element name="aspectPlugIn" type="xsd:string" />
    <element name="aspectData" type="tns1:HashMap" />
  </sequence>
</complexType>
```

Listing 1.1. XML Schema type of an aspect

4   Implementation

Our Java-based, prototypical implementation[1] of the aspect framework rests upon Apache Tomcat 6 as the application container in combination with Apache Axis 1.4 as the SOAP processing engine and web service execution environment. On the caller's side, the Axis client libraries are used. The orchestration engine for composing web services is ActiveBPEL [1], an open-source implementation of the BPEL standard.

Adding and removing of request/response aspects and listing of available advice are performed by a single web service that offers the corresponding methods. Further methods provided by this aspect configuration service are (1) a method to check whether an advice of an aspect is supported, and (2) a method to determine whether an aspect is woven into a service. These two methods allow the client-side component to ensure the atomic behavior of the overall framework.

---

[1] The source code is available on request.

4.1   Advice Interface

An advice of a request/response aspect is referenced by its ID (element $\mathcal{I}$ of the pointcut description). In our implementation, this ID represents the base name of the Java class that actually implements the advice. A class representing an advice has to implement a specific Java interface. Its methods are called by the aspect framework when a join point shadow becomes a join point (i.e, a defined join point is triggered) for the specific advice. The argument passed to these two methods is resolved by the expression in the field element $\mathcal{F}$ (see pointcut description) of the aspect. Their return values substitute the original value. If a wildcard operator is in the field element $\mathcal{F}$, multiple attributes match this expression. Multiple attributes represent multiple fields in a complex data type. The corresponding advice is applied iteratively to each of these attributes.

The concrete value of the scope of an aspect is either local or global with respect to communication. The value local means that the join point of the aspect respects the client, meaning that it is only active for the client that has woven it in, whereas global indicates that an aspect is active for all clients. The global scope implicates that no other service is affected by this aspect, because the affection depends on the actual client in order to determine the succeeding service. Since web services are loosely-coupled, such a succeeding service cannot be identified in general. Advice must furthermore define whether there is a subsequent service that is affected by this advice and to which an aspect has to be applied. Furthermore, this value (none, direct, dataflow, or controlflow) controls how such a service is detected. For example, an aspect for profiling the communication returns none, because there is no other partner. The value direct might be used for reliable messaging since it affects both partners that are communicating directly.

4.2   Aspect Configurator

The *aspect configurator* is responsible for validating and weaving aspects and is realized as a dedicated web service. To decouple the aspect framework from the conventional web services, the aspect configurator is implemented as a distinct service, i.e. a remote interface for weaving aspects. It operates on a registry that contains all necessary information about woven request/response aspects. Since it is realized as a hashtable, the number of deployed aspects has only a marginal impact on the overall aspect-weaving runtime.

4.3   Aspect Provider

The activation of (web service) join points is realized as an AspectJ aspect, the *Aspect Provider*. It is woven into the global handler chain of Apache Axis. If a web service is called or sends its response, the around advice is executed. It checks the registry to find out whether a request/response aspect has been woven into this concrete web service. If yes, the scope and the caller are identified, and in case of a complete match, the advice of the request/response aspect gets executed. For this purpose, the pointcut of the AspectJ aspect matches the invokeMethod() method of the RPCHandler of Apache Axis, which in turn is in charge of invoking the concrete implementation of the web service. In Listing 1.2, the pointcut (line 1 – 5) and the advice (line 7 – 13) of the AspectProvider is shown. The AspectProvider is integrated into the handler chain of Apache Axis. Request/response aspects are applied in the order in which they were woven. A more sophisticated management strategy is subject to future work. The implementation as an AspectJ aspect circumvents the modification of the configuration of Axis (e.g., deploying a specific aspect deployment handler into each service) and potentially allows us to use this code within other middlewares supporting web services, such as JBoss or Spring.

```
  pointcut invokeMethod( MessageContext  msgC ,Method  method ,
2   Object obj ,Object[] args)  :  call(
   protected Object RPCProvider.invokeMethod(
4    MessageContext ,Method ,Object ,Object[]) throws Exception)
   && args(msgContext ,  method ,  obj ,  argValues);

6
  Object around(MessageContext msgC ,Method method ,Object obj ,
8   Object[] args) throws Exception  :
   invokeMethod(msgC ,method ,obj ,args) {
10      handleRequest(argValues ,method ,msgC);
       Object response = (method.invoke(obj ,  args));
12      return handleResponse(response ,method ,msgC);
   }
```

Listing 1.2. AspectProvider

4.4 Security Manager

The security manager mentioned in Section 3.1 uses a public key infrastructure based on the X.509[9] standard in order to authenticate users who call the aspect configurator. If an authorized user tries to deploy an aspect, the signature of the aspect is validated, and it is only deployed into the system if the validation is successful. Otherwise, the call fails and no aspect is deployed.

4.5 Aspect Invocation Handler

To facilitate the use of request/response aspects during service orchestration, a custom invocation handler of the ActiveBPEL engine is called each time a web service is called within a BPEL process. The weaving of aspects is initiated by the Aspect-InvokeHandler (AIH). The AIH includes all the client-side functionality described in Section 3 and is called every time the workflow engine performs a web service invocation. In order to register and use the AIH, an extension mechanism provided by the workflow engine itself is utilized. Thus, a modification of the implementation of the workflow engine is not necessary. When the AIH is called, it checks whether the actual web service should use an aspect. This information is provided by the workflow developer during the design of the workflow. If this is the case, the AIH deploys the aspect to the actual web service and – depending on the defined relevance of the (request/response) advice – deploys it also to subsequent web services.

5   Evaluation

In this section, the realization of three use cases for request/response aspects is discussed. First, the use case of efficient data transmission described above is considered. The goal is to realize efficient data transmission between web services without additional implementation efforts (except for developing the aspects themselves once), without changing the web services, and without losing modularity and decoupling of web services. The second use case is also motivated by the multimedia workflow and realizes data compression. The third use case provides a cryptographic data transfer via request/response aspects. This use case is motivated by a workflow that performs medical analysis.

5.1   Use Case 1: Data Transfer

For the first use case, consider the general design shown in Figure 1 – before the request of Service S_2 can be handled by an aspect, the response of Service S_1 must have been handled by an aspect, too. In this way, the aspect woven into Service S_1 implements the IResponseAdvice. This implementation first takes the binary data returned by the service and then creates a new Flex-SwA reference that points to the data. This reference is encoded into the response to fit syntactically into the data structure the service returns. After this (response) message has been passed to Service S_2, the aspect woven into this service implements the IRequestAdvice. This implementation expects a Flex-SwA reference encoded in the data structure the service receives. Then, the encoded reference is resolved, and the data is transferred.

To test the Flex-SwA advice on a broader variety of data types, three different service implementations were investigated. The first echo service implementation works on a byte array, the second service on strings, and the third service uses SOAP with Attachments. The performance is evaluated using these three web services and the Flex-SwA advice mentioned before. This specific advice handles the request as well as the response message of the services and also deals with their (different) message formats. A client that utilizes the presented framework only needs to call the aspect configurator service with an aspect (see Listing 1.3) as an argument. This is all a client has to do in order to perform the weaving of the aspect for efficient data transmission.

Obviously, it can be expected that the weaving of the Flex-SwA aspect improves the workflow runtime significantly. The tests were performed on three machines running Fedora Linux. Each machine has the same hardware, namely an Intel Core 2 Duo E8600, 4 GB of RAM and a 100 MBit/s Ethernet network. One of these machines operated as the client, whereas the other two hosted the echo services. Different transmission types and different files sizes were used, but not all tested transmission types support arbitrary data sizes. Due to the XML encoding – especially of arrays – the test of a byte array was limited by the available heap space for the Java virtual machine. The heap space was set to 2 GB and the Java Garbage Collector was allowed to run concurrently. Only with this setup we were able to test the transmission of byte arrays containing up to 800, 000 single byte values.

```
Aspect serviceaAspect = new Aspect(
    new QName("http://fb12.de/AosStringTestService",
      "AosStringTestService"), "echoStringA", "/data",
    Aspect.AOP_RESPONSE_MODE, "FlexSwAPlugIn")
```

Listing 1.3. Java bean constructor of an aspect

We have compared the workflow execution times for the three different echo service scenarios (each measurement was repeated 100 times) to show their relative speedups. The less effective the data transfer mechanism is, the higher the (relative) runtime improvement is, if request/response aspects are used. In case of the byte array, a large improvement of up to 50% could be achieved, i.e., the transmission of a byte array in the size of 800 kbytes took about 95824 sec using plain SOAP communication and about 50190 sec using request/response aspects in combination with Flex-SwA. The runtime improvement of up to 50 % can be explained as follows: As indicated in Figure 1, each SOAP transmission first requires a serialization, then the actual transmission over a network and finally a deserialization – repeated in each step 1 – 4. Thus, the overall runtime time can be determined as $T_{\text{SOAP}} = 4 * t(n) + \varepsilon$, where $t(n)$ is the network transmission time, including the serialization and deserialization time, and n is the amount of transmitted data (ε represents negligible processing times). In case of the Flex-SwA request/response aspect, step 2 and step 3 now transport the woven reference instead of the actual payload. Since the size of a reference is independent of the referenced amount of data, the runtime can be expressed as $T_{\text{R/R}} = 2 * t(n) + 2 * t' + t_D(n) + \varepsilon$, where t' is the corresponding time for transferring a reference and $t_D(n)$ is the time needed to transfer the payload via Flex-SwA. For large n, we obtain $t' \ll t(n)$, $t_D(n) \leq t(n)$ and thus we get $T_{\text{R/R}}^{n \to \infty} = 2 * t(n) + t_D(n)$. The comparison of T_{SOAP} and $T_{\text{R/R}}^{n \to \infty}$ indicates that the theoretical runtime improvement is (slightly) below 50 %. Our measurements show that the proposed framework can come close to this theoretical limit. On the other hand, the overhead introduced by the request/response aspect is noticeable for the small data sizes when using strings or SOAP with Attachments (cf., Figure 2), but for larger data sizes (> 40 KBytes) the achieved runtime improvement outweighs this overhead.

It is worth mentioning that not only the execution time of such a workflow can be accelerated by a significant factor, but also the development time is shortened considerably. This reduction in development time results from the fact that simple data types can be used for the development of the services and that

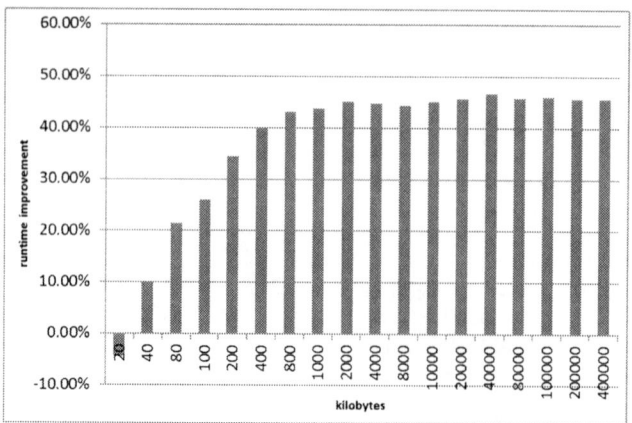

Fig. 2. SOAP with Attachments (SwA) – relative runtime improvement using Flex-SwA aspects

the more complex reference component for data transfer can be introduced using the proposed aspect framework. While the measurements only show the benefits for synthetical echo services, applying it to the concrete multimedia workflow also leads to a significant speedup. The plain workflow execution times without aspects are about 4422 seconds (where the SOAP communication requires 94% of the execution time). After weaving aspects into the services, the overall runtime only takes about 245 seconds (where now only 8% of this execution time is needed for communication). The overhead for the aspect handling (i.e., the time needed for executing the AspectProvider) is about 11 seconds.

5.2   Use Case 2: Data Compression

Text detection in videos is another workflow of our motivating multimedia analysis scenario: here, large text documents might be generated, depending on a given input video. These documents can be transported by Flex-SwA again, but it is also desirable to compress the text data. We implemented an advice to compress data and encode it afterwards with BASE64 in order to embed it again in an XML document. This aspect can reduce the message size to 60% of the original size. Another possibility would be to use the message-based compression offered by the web service container. Aspect-based compression allows us to use field-specific compression algorithms for different parts of the message, which is possible by the field operator of the pointcut description. This is reasonable since different types of data (text, images, videos etc.) may require different compression techniques. If a better compression algorithm becomes available, it is much easier to replace the applied compression using the proposed aspect-oriented solution. Otherwise, a new compression approach would have to be realized in all related web services' implementations.

This example indicates that the incorporation of new non-functional requirements via an advice is straightforward, since it effectively represents a filter operation applied to basic data types. Such advice only need to be applied to the primitive data types to be employed by the proposed framework.

5.3 Use Case 3: Data Encryption

Secure messaging is another crosscutting concern that can also be handled by request/response aspects. The secure messaging problem is illustrated by a workflow that we have developed during a cooperation with medical researchers. When performing patient studies, it is important that personal data of patients is kept private by either making it anonymous or by encrypting it. For this reason, we have developed an advice that performs an encryption at the data source, such that all subsequent services are not able to read sensitive data. Eventually, the decryption aspect is located at the service that merges the results into a patient's record. By using such an aspect, services can be encrypted without changing the service implementation or configuring the middleware.

The workflow that motivates this use case originates from the area of sleep research and basically performs an ECG (electrocardiogram) analysis and, based on the obtained results, conducts apnoea detection. The implementation uses the Physio Toolkit [17], a common set of open source tools in the biomedical sciences. Since the data format of the recorded vital signs is different from the format required by the Physio Toolkit, a data conversion is needed. Afterwards, the ECG records are processed to detect medically relevant peaks in the signal. The results are passed to an annotation reader service that in turn decodes the input and passes the results to a beat detection service that detects particular waves within the signal. In parallel, the output is passed to the apnoea detection service that analyzes the signal and detects respiration dropouts to diagnose the sleep apnoea syndrome.

The data exchanged by the services contain the actual ECG measurements and also some identification attributes. To prevent the misuse of these attributes, we have developed a privacy advice that uses public key cryptography. The support of the wildcard operator allows us to encrypt (and decrypt) all of the patient related data. These data are encrypted when they are initially retrieved from a database. During the processing by the mentioned services, the personal information is encrypted (while the ECG data remain unencrypted). Finally, when the analysis is finished and the result is stored in the database, the corresponding advice decrypts the personal data.

6   Conclusions

In this paper, we have proposed request/response aspects for web services that allow developers of service-oriented applications to easily enrich web services with additional non-functional requirements, such as efficient data transmission, data compression, or other crosscutting concerns. They can be woven dynamically into remote web services without changing their implementations or their interfaces. The presented framework supporting request/response aspects includes a pointcut description for SOAP-based web service environments.

Request/response aspects offer several advantages: They allow adding non-functional requirements at the web service communication level to offer the possibility of executing aspect code on the remote host where a called web service is running. Furthermore, they are independent of a web service's implementation language and do not assume that the web service implementation provides related functionality. By using the aspect framework, the development of web services is simplified. This is demonstrated by adding the non-functional requirement of efficiently transmitting large amounts of data in a web service workflow and thus circumventing the bottleneck at the client or workflow engine, respectively. Runtime measurements for a multimedia application that requires efficient transmission of large amounts of data have been presented. Two further use cases for data compression and encryption have demonstrated the benefits of the proposed approach.

There are several areas for future work. For example, instead of transferring the aspect-ID ($\mathcal{I}$), the whole aspect could be copied either as (Java) binary code or as an interpretable description to the remote service. Ideas like sequence pointcuts (sophisticated management of multiple advice for the same pointcut) and shared states between aspects executed on different hosts [15] are other interesting enhancements of our approach. Finally, to prevent a congestion of a service with aspects over time, investigations for sophisticated life cycle management (e.g., according to wall-clock time or communication patterns) are areas of further research.

Acknowledgements

This work is supported by the German Ministry of Education and Research (BMBF, D-Grid) and by the German Research Foundation (DFG, PAK 509).

References

1. ActiveEndpoints: ActiveBPEL Business Process Execution Engine,
   http://www.activebpel.org
2. Andrews, T., Curbera, F., Dholakia, H., Goland, Y., Klein, J., Leymann, F., Liu, K., Roller, D., Smith, D., Thatte, S., Trickovic, I., Weerawarana, S.: Business Process Execution Language for Web Services Version 1.1. Microsoft, IBM, Siebel, BEA und SAP, 1.1 edn. (May 2003)
3. Apache Foundation: Apache Axis., http://ws.apache.org/axis/
4. Baligand, F., Monfort, V.: A Concrete Solution for Web Services Adaptability Using Policies and Aspects. In: Proc. of the 2nd Intl. Conf. on Service Oriented Computing, pp. 134–142. ACM, New York (2004)
5. Binder, W., Constantinescu, I., Faltings, B.: Service Invocation Triggers: A Lightweight Routing Infrastructure for Decentralized Workflow Orchestration. In: Intl. Conf. on Advanced Information Networking and Applications, vol. 2, pp. 917–921 (2006)
6. Charfi, A., Mezini, M.: Aspect-oriented Web Service Composition with AO4BPEL. In: Proc. of the European Conf. on Web Services, pp. 168–182. Springer, Heidelberg (2004)

7. Cibrán, M., Verheecke, B.: Dynamic Business Rules for Web Service Composition. In: 2nd Dynamic Aspects Workshop (DAW 2005), pp. 13–18 (2005)
8. Courbis, C., Finkelstein, A.: Towards an Aspect Weaving BPEL Engine. In: The Third AOSD Workshop on Aspects, Components, and Patterns for Infrastructure Software (ACP4IS), Lancaster, UK, pp. 1–5 (2004)
9. Cooper, D., Farrell, S., Boeyen, S., Housley, R., Polk, W.: Internet X.509 Public Key Infrastructure Certificate and Certificate Revocation List (CRL) Profile, http://tools.ietf.org/html/rfc5280
10. Ewerth, R., Freisleben, B.: Semi-Supervised Learning for Semantic Video Retrieval. In: Proc. of the 6th ACM Intl. Conf. on Image and Video Retrieval, pp. 154–161. ACM, New York (2007)
11. Ewerth, R., Mühling, M., Freisleben, B.: Self-Supervised Learning of Face Appearances in TV Casts and Movies. In: Proc. of the Eighth IEEE Intl. Symposium on Multimedia, pp. 78–85. IEEE Computer Society, Los Alamitos (2006)
12. Heinzl, S., Mathes, M., Friese, T., Smith, M., Freisleben, B.: Flex-SwA: Flexible Exchange of Binary Data Based on SOAP Messages with Attachments (2006)
13. Joncheere, N., Deridder, D., Straeten, R., Jonckers, V.: A Framework for Advanced Modularization and Data Flow in Workflow Systems. In: Proc. of the 6th Intl. Conf. on Service-Oriented Computing, pp. 592–598. Springer, Heidelberg (2008)
14. Kiczales, G., Hilsdale, E., Hugunin, J., Kersten, M., Palm, J., Griswold, W.: An Overview of AspectJ. In: Proc. of the 15th European Conf. on Object-Oriented Programming, pp. 327–353 (2001)
15. Navarro, L., Südholt, M., Vanderperren, W., De Fraine, B., Suvée, D.: Explicitly Distributed AOP using AWED. In: Proc. of the 5th Intl. Conf. on Aspect-Oriented Software Development, pp. 51–62. ACM, New York (2006)
16. Nishizawa, M., Chiba, S., Tatsubori, M.: Remote Pointcut: A Language Construct for Distributed AOP. In: Proc. of the 3rd Intl. Conf. on Aspect-Oriented Software Development, pp. 7–15. ACM, New York (2004)
17. PhysioNet: PhysioToolkit, http://www.physionet.org/physiotools/
18. Suvee, D., Vanderperren, W., Jonckers, V.: JAsCo: An Aspect-Oriented Approach Tailored for Component Based Software Development. In: Proc. of the 2nd Intl. Conf. on Aspect-Oriented Software Development, pp. 21–29. ACM, New York (2003)
19. Tanter, É.: Expressive Scoping of Dynamically-Deployed Aspects. In: Proc. of the 7th Intl. Conf. on Aspect-Oriented Software Development, pp. 168–179. ACM, New York (2008)
20. Transport Layer Security, http://datatracker.ietf.org/wg/tls/charter/
21. Verheecke, B., Cibran, M., Vanderperren, W., Suvee, D., Jonckers, V.: AOP for Dynamic Configuration and Management of Web Services. Intl. Journal of Web Services Research 1(3), 25–41 (2004)
22. World Wide Web Consortium (W3C): W3C SOAP Specification, http://www.w3.org/TR/soap/
23. World Wide Web Consortium (W3C): Web Services Definition Language (WSDL) 1.1, http://www.w3.org/TR/wsdl
24. World Wide Web Consortium (W3C): XML Path Language (XPath), Version 1.0, http://www.w3.org/TR/xpath, http://www.w3.org/TR/xpath

Using Graph Aggregation for Service Interaction Message Correlation

Adnene Guabtni[1,2], Hamid Reza Motahari-Nezhad[3], and Boualem Benatallah[1]

[1] The University of New South Wales, Sydney, Australia
{aguabtni, boualem}@cse.unsw.edu.au
[2] National ICT Australia (NICTA), Sydney, Australia
aguabtni@nicta.com.au
[3] HP Labs, Palo Alto, CA, USA
hamid.motahari@hp.com

Abstract. Discovering the behavior of services and their interactions in an enterprise requires the ability to correlate service interaction messages into process instances. The service interaction logic (or process model) is then discovered from the set of process instances that are the result of a given way of correlating messages. However, sometimes, the Correlation Conditions (CC) allowing to identify correlations of messages from a service interaction log are not known. In such cases, and with a large number of message's correlator attributes, we are facing a large space of possible ways messages may be correlated which makes identifying process instances difficult. In this paper, we propose an approach based on message indexation and aggregation to generate a size-efficient Aggregated Correlation Graph (ACG) that exhibits all the ways messages correlate in a service interaction log not only for disparate pairs of messages but also for sequences of messages corresponding to process instances. Adapted filtering techniques based on user defined heuristics are then applied on such a graph to help the analysts efficiently identify the most frequently executed processes from their sequences of CCs. The approach has been implemented and experiments show its effectiveness to identify relevant sequences of CCs from large service interaction logs.

Keywords: SOA, Process mining, Correlation, Aggregation.

1 Introduction

As the number of services in organizations are growing and service interactions are getting more dynamic, there is a significant interest in understanding the relationships and behavior of services in the enterprise. Approaches for discovering the behavior of systems and services (also known as process or workflow discovery [10, 4]) take process instances as input. A process instance is a sequence of service messages corresponding to one execution of a process model. In the context of services, a process instance consists of a sequence of messages that are exchanged by services. However, identifying process instances, i.e. correlating messages so that we know which service messages belong to the same process instance, is not

H. Mouratidis and C. Rolland (Eds.): CAiSE 2011, LNCS 6741, pp. 642–656, 2011.

always straightforward. This is because information about correlator attributes may not be known to a monitoring service which overseas and captures service interactions in a log. Although each service internally knows how it correlates its messages with those of its immediate interacting partners, this information may not be well documented, or may be buried in the code of the service which is sometimes outsourced or the documentation may become obsolete. Therefore, it is often necessary to perform an automated message correlation in order to identify process instances.

We consider two messages in a service interaction log as correlated if a *Correlation Condition* (CC) is verified. A typical CC may be the equality of two messages' attributes [8]. When messages generated by the same process instance are correlated, the sequence of their CCs can be considered as the "fingerprint" of such process instance. Therefore, discovering sequences of CCs allows to step ahead towards the identification of process instances.

Motivated by the goal of providing a light-weight approach that can help an analyst to quickly identify relevant sequences of CCs, we propose in this paper an approach using message indexation and aggregations to generate an Aggregated Correlation Graph (ACG) that exhibits all the sequences of CCs identified in a service interaction log. In an ACG graph, each node corresponds to an aggregation of messages, and each edge represents a CC between all pairs of messages in the two nodes. Therefore, the ACG graph represents all correlations of messages in an aggregated representation allowing to quickly identify, using weighted nodes and edges, the most frequent sequences of CCs revealing frequent process executions. The approach has been implemented and we offer an interactive filtering/browsing of the ACG graph helpful for analysts to better understand the way messages are correlated and the potential process models those correlations may reveal. In particular, we make the following contributions:

(1) We propose an approach for service interaction message aggregation based on their CCs. The resulting ACG graph is size-efficient and exhibits all the sequences of CCs identified in the log.

(2) We provide a method, based on graph filtering techniques and user-defined criteria, to efficiently identify relevant sequences of CCs and visually browse them using an interactive ACG graph visualization tool.

(3) We have implemented the approach in a tool available online, and performed experiments on a number of service interaction logs. The experiments show that the generated ACG has a stable size regardless of the size of the log being processed. Furthermore, the interactive ACG visualizer allows analysts to quickly find relevant sequences of CCs and identify process instances.

The rest of the paper is organized as follows: In section 2, assumptions and notations used in this paper are presented. We present the service interaction message log format, and define the notations used in this paper. In section 3, we propose our approach and discuss its strengths and limitations. In section 4, we describe the implementation and discuss its results and applications. In section 5, we discuss related work. Finally, in section 6, we summarize the contribution of this paper and present future work.

2   Assumptions and Notations

2.1   Event Log Format and Sample

Process discovery techniques usually assume that service interaction logs have certain format that is useful for analysis. The most common format consists of mono-valued attributes describing service interaction messages. A message is generated by a service and represents an explicit or implicit transition in a process execution. Each message has a set of attributes and their associated values. We consider the set of attributes in a log as $\mathcal{A} = \{a_1, a_2, ..., a_i\}$ and the set of messages in a log as $\mathcal{M} = \{m_1, m_2, ..., m_j\}$. While it is difficult to ensure a global clock with infinite precision to have a total or partial order for messages generated by disparate services, we propose to consider the order that the messages have been inserted in the event log. Therefore, we assume in this work that one centralized log file is used, having exclusive write access to insure that we will always have messages inserted one by one as they are captured. Thus, we define a total order function $\prec$ for messages which corresponds to the order they have been inserted into the event log. The following notations are also used in this paper:

- $\forall\, m\, \in\, \mathcal{M}$, $\mathcal{A}(m) \subset \mathcal{A}$ is the set of attributes represented in the message m.
- $\forall a \in \mathcal{A}(m)$, $\mathcal{V}(a, m)$ is the value of the attribute a in the message m.
- $\mathcal{V} = \{v_1, ..., v_m\}$ is the set of values assigned to attributes of messages in $\mathcal{M}$.

2.2   Message Correlation

Two messages $m_i, m_j \in \mathcal{M}$ are considered as correlated if a Correlation Condition (CC) is verified. We follow our previous work [4] for the definition of a CC. We consider a CC as a relation between attribute's values of messages based on a correlation function $cf : \mathcal{V} \longmapsto \mathcal{V}$ such that $\mathcal{V}(a_i, m_i) = cf(\mathcal{V}(a_j, m_j))$. For simplicity reasons, we assume in this paper that $cf(x) = x$.

The correlation of two messages m_i and $m_j \in \mathcal{M}$ is denoted $m_i \lll m_j$, in which $\lll$ is the correlation relation, iff $m_i \prec m_j$ and $\exists\, a_k \in \mathcal{A}(m_i)$ and $a_l \in \mathcal{A}(m_j)$ having $\mathcal{V}(a_k, m_i) = \mathcal{V}(a_l, m_j)$.

The Correlation Condition (CC) of two correlated messages can be atomic or composite depending on the number of couples of attributes of the two messages having equal values. The representation of an atomic CC is denoted as the equality of two attributes: $a_i = a_j$ and we note $\mathcal{ACC}$ the set of all possible unique atomic CCs for a given set of attributes. A composite CC is a set of atomic CCs verified for the same couple of correlated messages. We note a composite CC as the conjunction of atomic correlation conditions: $a_i = a_j \wedge a_k = a_l \wedge \,...$ For both atomic and composite correlation conditions, we generalize the definition of a correlation condition as follows:

Definition 1 (Correlation Condition). *We note the CC of two correlated messages as a function $CCond : (\mathcal{M}, \mathcal{M}) \longmapsto P(\mathcal{ACC})$ such that $\forall m_i, m_j \in \mathcal{M} \,/\, m_i \lll m_j$, $CCond(m_i, m_j)$ is the set of **all** atomic correlation conditions verified by the correlated events m_i and m_j.*

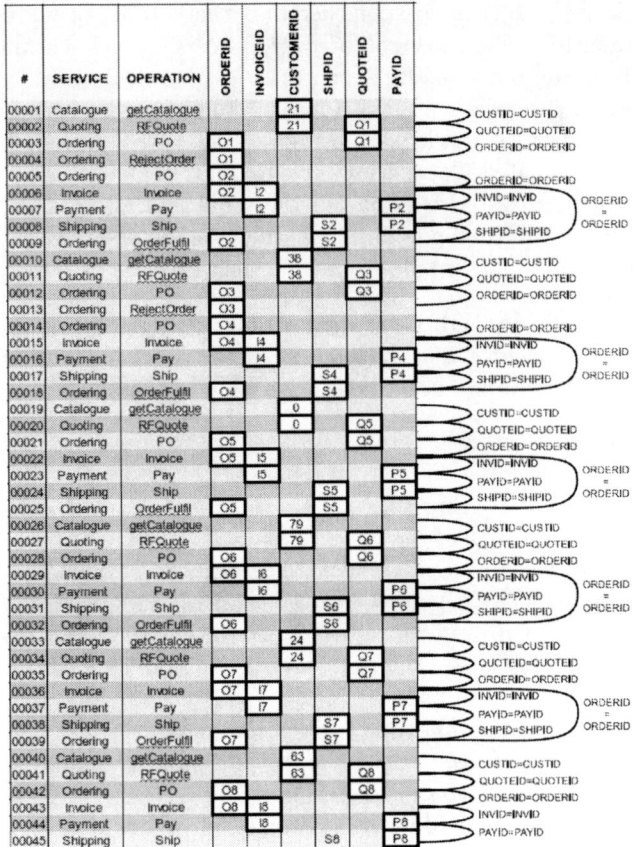

Fig. 1. Example of a real world service interaction log with correlated messages

The example of figure 1 illustrates the message correlations and their corresponding correlation conditions in a small portion of a real world service interaction log related to the Retailer services, respectively "Catalogue", "Quoting", "Ordering", "Invoice", "Payment" and "Shipping" services. For example, the two messages "0001" and "0002" describing respectively the invocation of the catalog and the quoting services, are correlated with the CC CUSTID=CUSTID.

3   Proposed Approach

3.1   Philosophy, Definitions and Properties

Correlation Conditions (CCs) are meant to describe the correlation of two messages. Having a set of correlated messages, it is possible to build a graph in which nodes are messages and edges are correlations. We call such a graph 'Correlation Graph'. Each edge is described using the CC correlating its source and destination nodes. Building such a graph for the entire event log generated a large graph when large event logs are considered as the number of nodes is equal to the number of messages in the log. Moreover, processing such a graph for process discovery can be computationally expensive.

We propose in this paper a novel approach to efficiently discover and evaluate all CCs from large event logs by using message indexation and aggregation and build an Aggregated Correlation Graph (ACG). The objective is that the ACG graph aggregates and exhibits all the sequences of message correlations (described using their corresponding CCs) identified in a service interaction log using a size-efficient single graph. We define such ACG as follows:

Definition 2 (Aggregated Correlation Graph (ACG)). *We define the ACG graph as a weighted and oriented graph in which nodes and edges have the following properties and notations:*

- **Nodes properties:** *Every node is associated with a set of messages. The set of nodes of the ACG is noted $\mathcal{N}(ACG)$.*
- **Edges properties:** *Every edge is associated with a set of message correlations. The set of edges of the ACG is noted $\mathcal{E}d(ACG)$. An edge is oriented and links two nodes. Two nodes can be linked with at most one edge. If two nodes $n_1, n_2 \in \mathcal{N}(ACG)$ are linked with an edge $e \in \mathcal{E}d(ACG)$, we note $ed(n_1, n_2) = n_1 \curvearrowright n_2$ and we say n_1 is correlated to n_2. The weight $\mathcal{W}(e)$ of an edge $e \in \mathcal{E}d(ACG)$ is equal to the number of event correlations associated to it.*
- **Constraints on message-node association:** *Every message in the log is associated with one single node in the ACG. Two correlated messages cannot be associated with the same node in the ACG.*
- **Constraints on message correlation-edge association:** *All message correlations associated to the same edge share the same correlation condition. Such a correlation condition is noted $CCond(ed)$ for an edge $ed \in \mathcal{E}d(ACG)$. Every correlation of two messages is associated with one single edge in the ACG.*
- **Root node:** *The ACG contains necessary one default node called root node where any message m_i is placed if $\nexists \ m_j \prec m_i \ / \ m_j \lll m_i$.*

In the Aggregated Correlation Graph, each node corresponds to a set of messages and each edge represents correlations between all pairs of messages in the two sets of messages[1]. The edge's weight represents the number of pairs of correlated messages in the two sets of messages contained in its source and destination nodes. The edge's type corresponds to their CC, ensuring that all correlations represented by the same edge have the same CC.

The correlations of messages are represented in the ACG graph using oriented edges mainly for two reasons. Firstly, time is important in correlation discovery as if a message m_1 is correlated with a message m_2 and $m_1 \prec m_2$, the edge representing such a correlation would be oriented in consequence (from the node containing m_1 to the node containing m_2). The second reason is that the ACG

---

[1] Only binary correlations are represented (correlations of pairs of messages).

graph is the result of an aggregation of messages into sets of messages (nodes of the ACG) and an edge in the ACG represents correlations between all pairs of messages in the two sets of messages. Without oriented edges, we won't be able to decide which messages would be aggregated. For example, let assume $m_1 \lll m_2$ are correlated with a CC c_1 and $m_3 \lll m_4$ are correlated with the same CC c_1. Should we aggregate m_1 with m_3 and m_2 with m_4 or should we aggregate m_1 with m_4 and m_2 with m_3? The orientation of the edges makes the decision easier as the sources of edges are aggregated together and the destinations of edges are aggregated also together.

Role of the Root Node in ACG. When processing each message sequentially, some messages may not be correlated to messages previously placed into the ACG. Such messages are associated by default with the root node which corresponds to all non correlated messages (singles) and all messages being the starting messages of process instances (starters). Without using such a root node, we would create a separate node for every message not correlated to previous messages which can be very frequent and thus can lead to a larger ACG. By aggregating those events in the root node, we do not loose any information and we reduce drastically the number of nodes in ACG.

Additionally, the weight of an edge corresponds to the number of couples of messages associated with, on one hand its source node and on the other hand its destination node, being correlated with the CC of the edge. Therefore, the more a CC is verified for couples of messages, the bigger is the weight of the corresponding edge(s)[2].

3.2   Step By Step Scenario of Building the ACG

In this section we describe a step by step scenario of building the ACG from a real world service interaction log as illustrated in figure 2. The ACG is built gradually by processing every message sequentially. At each step, an overview of the messages being processed as well as the resulting ACG under construction. Each node of the ACG is labeled using the service operation of its members (messages). The root node has an additional label corresponding to the service operation of its source message (within the root node).

Let's start with the first message in the service interaction log. In the example below, message 0001 is placed in the root node as this is the default node for any message that is not correlated with previously placed messages. As this is the first message to place in the ACG, it is obviously placed in the root node.

A second message, 0002, is correlated with message 0001 with a CC "curtomerid=customerid". However, there is no node in the ACG linked to the root node with correlation condition "curtomerid=customerid". The message 0002 is then placed in a newly created node and that node is linked to the root node using an edge labeled with the CC "curtomerid=customerid". The new

---

[2] Many edges within the ACG could have the same CC. This is made possible if such CC is involved in many different processes.

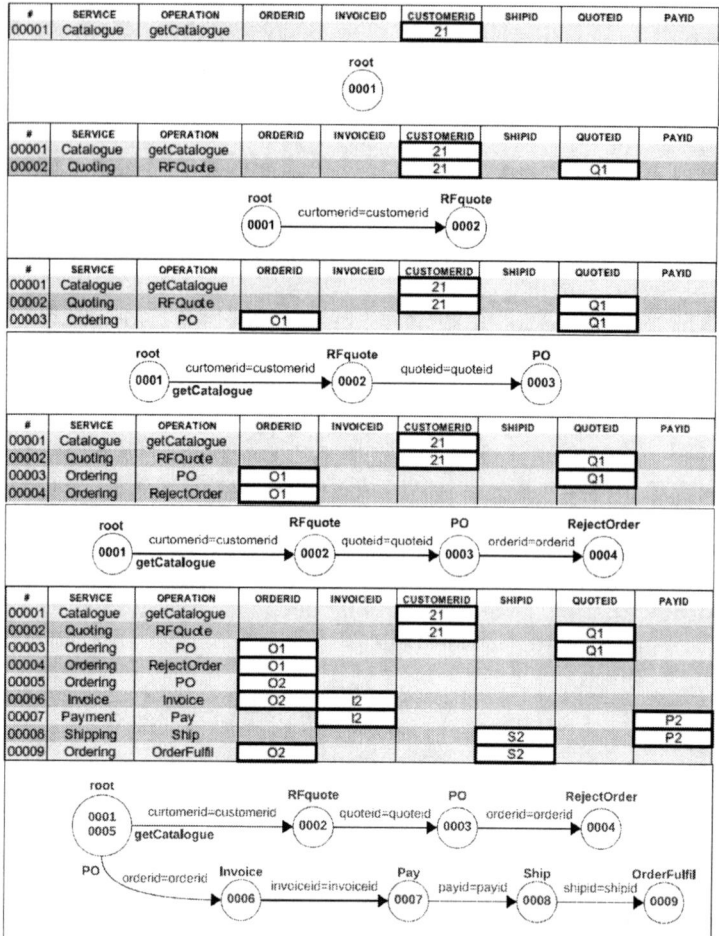

Fig. 2. Step by step scenario of building the ACG

node is labeled with the service operation generating message 0002, which is "RFquote". A similar scenario concerns the next messages 0003 and 0004 as described in the illustration below. Message 0005 is not correlated to any previous message. Therefore, it is placed in the root node. Then, messages 0006 to 0009 are placed in newly created nodes for each of them. Each time a message is placed in an existing node, its size is incremented and its inner and outer edges have their weight incremented. The heavier the edges are, the more frequent are their associated correlations.

3.3   Algorithm for Building the ACG

Building the ACG requires a method to aggregate the correlated messages and generate the expected graph without necessary generating the correlation graph ("non aggregated"). Assuming that the ACG is built by processing messages one by one, sorted by their total order from a service interaction log, the proposed approach follows some rules when processing a message m:

- If m is correlated to previous messages $m_1, m_2, ..., m_p$ which are necessary already associated with existing nodes $n_1, n_2, ..., n_q$ in the ACG then two cases are possible:

  - *CASE 1:* Every message in $\{m_1, m_2, ..., m_p\}$ is associated to a distinct node in $\{n_1, n_2, ..., n_p\}$. In that case, m is associated with a node n such that $\forall 1 < i < p$, $CCond(ed(n_i, n)) = CCond(m_i, m)$ and the label of n is the name of the service generating m. If such a node does not exist in the ACG, then it is inserted and its correlations to the adequate other nodes are also added.

  - *CASE 2:* $\exists\, n_i \in \{n_1, n_2, ..., n_q\}$ such that two or more messages from $\{m_1, m_2, ..., m_p\}$ are associated with it. If those messages are correlated to m using the same CC and n_i has the service generating m as label, the first case applies. If those messages are correlated to m using different CCs, each of them is then moved from its existing node to a new node inheriting all incoming edges of the original node. This is done for every node verified until the first case can be applied.

- If m is not correlated to any of the previous messages then it is associated with a predefined node in the ACG called *root node*.

Following those rules, it is possible to gradually build the ACG graph by processing the service interaction log message by message. At any stage of the ACG construction, each message to be processed is matched with the nodes of the ACG (sets of messages) to apply the above rules.

The algorithm described below associates messages with their corresponding nodes using a single parsing of the event log. Additional notations concerning the relation between a message and nodes sharing some of its attribute's values are used as follows:

$\forall m \in \mathcal{M}$, if $\exists m' \in \mathcal{M}$ associated with n such that $m' \lll m$, we note $n \lll m$ as an order relation between a node and a message meaning that the node has to be correlated to the node associated with the message. In that case, we also note $CCond(n, m) = CCond(m', m)$ the correlation condition between the node n and the node associated to a message m.

For each message m in the message log, the proposed algorithm allows to identify an existing node n in the ACG or add a new node to the ACG which can be associated to m with respect to the ACG properties previously defined in section 3.1 of this paper.

Algorithm 1. Building the ACG graph

Require: $\mathcal{M}$ as the event stream sorted by time
Ensure: Building the Aggregated Correlation Graph of all messages in $\mathcal{M}$
1: **for all** $m \in \mathcal{M}$ **do**
2:    **if** $\exists n \in \mathcal{N}(ACG)$ such that

        $\forall n' \in \mathcal{N}(ACG) \; / \; \exists ed \in \mathcal{E}d(ACG), \; ed = n' \frown n,$

          $n' \lll m$ and $CCond(ed) = CCond(n', m)$
        and

        $\forall n' \in \mathcal{N}(ACG) \; / \; n' \lll m,$

          $\exists ed \in \mathcal{E}d(ACG) \; / \; ed = n' \frown n$ and $CCond(n', m) = CCond(ed)$ **then**
3:        m is associated with n
4:        **for all** $n' \in \mathcal{N}(ACG) \; / \; n' \lll m$ **do**
5:            $W(n' \frown n)) + +$ {The weight of the edge between the two nodes is increased as a new correlation has been associated to it.}
6:        **end for**
7:    **else if** $\exists cn \in \mathcal{N}(ACG) \; / \; cn \lll m$ **then**
8:        Create a new node n in $\mathcal{N}(ACG)$
9:        **for all** $n' \in \mathcal{N}(ACG) \; / \; n' \lll m$ **do**
10:          Add a new edge $n' \frown n$ to $\mathcal{E}d(ACG)$
11:          Initialize the weight of $n' \frown n$ to 1 {This is because only one correlation is associated to it so far.}
12:        **end for**
13:    **else**
14:        m is associated with *root* which is the root node of the ACG. {The event is not correlated to any existing node}
15:    **end if**
16: **end for**

3.4 Using Inverted Indexes for Efficient Message-Node Association

Inverted indexes are widely used in database systems to efficiently locate information [12]. For example, an inverted index for a collection of documents is a data structure that stores, for each term (word) occurring in the collection, information about the locations where it occurs. Such inverted indexes allow to make the location of items more efficient. In the following, we justify the need and we describe the use of inverted indexes in the proposed approach to ensure an efficient identification of nodes associated to a given message.

Correlating messages is based on the equality of their attribute's values. Therefore, creating an inverted index of attribute's values of all messages in the log is an obvious solution to make correlation identification more efficient. Having such an inverted index of values, every value refers to all its couples of message/attribute having the same value. The inverted index can be formalized as a function *InvInd* mapping values to couples of attributes and messages:

$$InvInd \; : \; \mathcal{V} \longmapsto P(\mathcal{A}, \mathcal{M})$$

such that $\forall v \in \mathcal{V}, \forall m_i \in \mathcal{M}, \forall a_j \in \mathcal{A}$, if $V(a_j, m_i) = v$, then $(a_j, m_i) \subset InvInd(v)$.

However, building the inverted index *InvInd* concerns all couples of messages and attributes of the log and makes parsing/updating the index inefficient when large logs are used. Moreover, such an inverted index is helpful to build a correlation graph of the entire event log instead of an aggregated correlation graph.

We propose in this section an inverted index handling correlation of message aggregations (nodes in the ACG) instead of the original messages themselves. We formalize such an index as a function $AggInvInd$ mapping values to couples of attributes and ACG nodes:

$$AggInvInd : \mathcal{V} \longmapsto P(\mathcal{A}, \mathcal{N}(ACG))$$
such that $\forall v \in \mathcal{V}, \forall n \in \mathcal{N}(ACG), \forall a \in \mathcal{A}$, if $\exists m \in \mathcal{M}(n)$ and $V(a, m) = v$, then
$$(a, n) \subset SumInvInd(v).$$

The proposed algorithm is building the SCG and while processing each message sequentially, the index $AggInvInd$ is incrementally populated and used. For every message $m \in \mathcal{M}$, its attribute's values are used to access the inverted index $AggInvInd$, identify nodes correlated to the event and check if an existing node in the ACG is suitable to be associated with the message.

3.5   Using the ACG for Identifying Process Instances

In this section we discuss how the ACG can be used for identifying process instances and the proper use of graph filtering.

Starting from the root node, the ACG is read as different branches, each of them corresponding to a different sequence of correlation conditions revealing a bunch of similarly correlated process instances. The following example of ACG, illustrated in figure 3 (left) corresponds to an ACG generated from a log of 4000 messages. Such graph contains all sorts of correlations discovered from the log and is not easy to read as many of those correlations are non frequent (small node sizes). When the ACG contains a high number of nodes and edges, we propose to apply filtering techniques allowing to reduce the number of edges and nodes and make it clearer to read and easier to interpret.

Assuming that the relevance of a correlation condition depends on the weight of its associated edges in the ACG, applying graph filtering techniques to the ACG allows to discover relevant correlation conditions. Graph filtering techniques can consist of removing low weighted edges, and consequently the potential resulting orphan nodes, those removed edges are non frequent cases. After applying the filtering, it is then possible to read the filtered ACG and clearly identify relevant correlations.

The relevance of a correlation conditions depends on the number of process instances actually verifying that correlation condition in the log. Therefore, graph filtering can play an important role in highlighting potentially relevant correlation conditions from the ACG. An example of filtered ACG compared to an unfiltered ACG is illustrated in figure 3. The size of a node in the graph corresponds to the number of messages associated to it in the ACG. The thickness of an edge corresponds to its weight in the ACG.

3.6   Discussion

In some cases, correlator attributes in the log could have the same value for a large number of messages. This has a major consequence on the resulting ACG

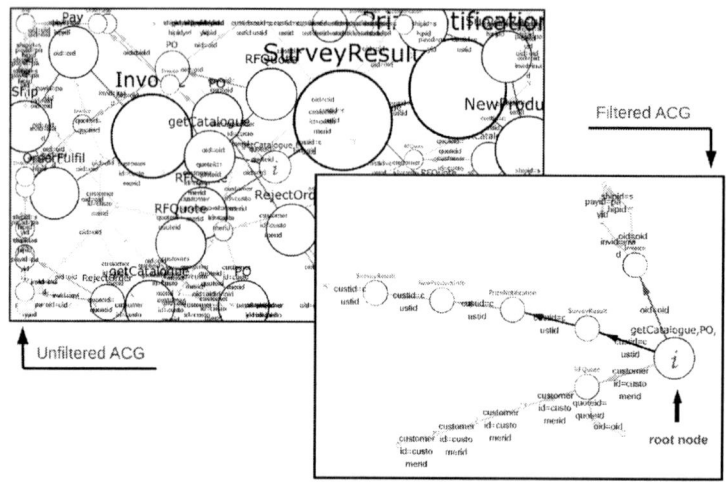

Fig. 3. Example of unfiltered ACG (left side) and filtered ACG (right side)

as two correlated messages cannot be associated with the same node in the ACG and therefore, if a large number of messages are correlated with each others, the ACG would have a large number of nodes and that makes its size inefficient. The relevance of an attribute depends on the diversity of its values within the service interaction log. A threshold is set to ignore attributes having low diversity meaning that most of the messages have the same value for the same attribute.

In some other cases, messages of iterative service calls (loops in a process) can occur in the log. The sequences of correlation conditions referring to a same process model may have various lengths as each execution introduces a variable number of loops, and thus includes a variable number of repetitive subsequences in the ACG graph. Therefore, such sequences are not aggregated together and lead to the identification of several processes.

Also, the approach works well if the log contains correlator attributes and every two consecutive messages belonging to a same process instance are actually correlated using some of those correlator attributes. Therefore, if the process is larger, it is more likely to have broken correlation chains. This is the case when two consecutive messages belonging to a same process instance are not correlated using any pair of correlator attributes. In such cases, a large process will be fragmented in the ACG as multiple smaller processes.

Finally, noise in service logs affects the result of correlation discovery as widely observed [5] [6]. Real-world logs are imperfect, i.e., they are incomplete (correspond to a subset of possible execution) and noisy (e.g., do not record some messages). A known approach to deal with noise in logs is to use a frequency threshold to filter noisy data [11]. In previous work [5], we have presented a quantitative approach for estimating a noise threshold used to filter noise from service logs. In this paper, we assume that the log is free of noise, or has been cleaned from noise in a pre-processing step.

4   Implementations and Experiments

A prototype has been developed to implement the proposed approach and offers
4 user driven steps:

Step 1 - Uploading Event Log Data
The format considered in this paper to represent service interaction log files is
the Comma Separated Values (CSV). The user can upload a CSV file using a
Web interface. For our experiments, we used a CSV log file generated using HP
SOA Manager (SOAM) which is a monitoring tool for Web services. SOAM cap-
tures all the messages that are exchanged to/from the set of registered services.
The resulting log represents a scenario generated based on WS-I (Web Service
Interoperability Organization) for a set of services in the supply chain (Retailer
services). Figure 1 illustrates a short sample of data generated by SOAM.

Step 2 - Indexing Messages and Selection of Correlator Attributes
This step allows to build indexes for each attribute in the service interaction
log. Such indexes allow to speedup the algorithm for building the ACG. We use
information about the size of each index to suggest which attributes are to be
considered as relevant for message correlation.

Step 3 - Aggregating Messages to Build the ACG Graph
This step allows to build the ACG by executing algorithm 1. The inverted index described in section 3.4 is also built incrementally during this step. Once
the ACG and the inverted index are created, they are stored on the back-end
database.

Step 4 - Filtering and Visualizing the ACG Graph
This is the final step and it allows, optionally, the filtering of the ACG, and the
visualization of the ACG graph highlighting the relevant sequences of correlation
conditions.

An evaluation of the tool has been conducted using service interaction logs
of various sizes, based on the Retailer services. Four logs containing respectively
1000, 2000, 3000 and 4000 messages have been processed to generate their associated ACG. A first analysis concerns the impact of the number of messages in
the log on the size of the filtered ACG as illustrated in figure 4 (a). It shows the
number of nodes in the resulting filtered ACG (with 50% threshold) for service
interaction logs of various sizes. This experiment shows that the size of the filtered ACG is relatively the same regardless of the size of the service interaction
log. This is due to the fact that filtered ACG shows the most frequent processes
and therefore do not include all the ways messages are correlated.

A second analysis concerns the impact of the number of messages in the log on
the size of the ACG (unfiltered) as illustrated in figure 4 (b). It shows the number
of nodes in the resulting unfiltered ACG for service interaction logs of various
sizes. The number of nodes in the unfiltered ACG starts stabilizing to a certain
level at around 3000 messages in the log. This is due to the fact that process
instances are usually repetitive in the log and the proposed algorithm aggregates
similar sequences of correlation conditions. Therefore, when a large number of
messages are placed in the ACG, most of the sequences of correlation conditions

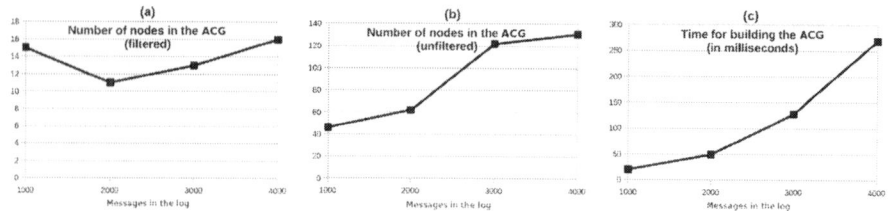

Fig. 4. Impact of log size on filtered ACG size (a), on unfiltered ACG size (b), and on processing time for generating the ACG (c)

become represented in the ACG. Thus, processing more messages would lead to increasing the size of existing nodes and the weight of existing edges but not increasing the number of nodes or edges. The stability in the number of nodes of the ACG makes the proposed approach effective in generating a useful ACG regardless of the size of the message interaction log.

The third and last analysis concerns the processing time needed to generate the ACG. Figure 4 (c) shows the time needed to build the ACG from service interaction logs of various sizes. Although the size of the ACG is stabilized over a certain number of messages, it is required to process all available messages from the log, just in case additional unexpected sequences of correlation conditions may occur. Therefore, it takes longer to process a larger log as illustrated in figure 4 (c).

5   Related Work

The need for automated approaches to message correlation in Web services has first been reported in [4] where a real situation on how to correlate service messages is presented. A categorization of various correlation methods in Web service workflows are presented in [1]. However, no automated support for message correlation is reported. The need for automated approaches for correlation of service messages in composite business applications is also raised in IBM Websphere platform [9]. This work presents an approach for the discovery of correlation identifiers in messages from the log of service interactions. This approach identifies the correlation between message types (e.g., PurchaseOrder and Invoice message types). This approach only considers atomic conditions, while we also consider composite correlation conditions. Also, this approach only concerns correlations between pairs of message types which does not allow to reason at the process instance level.

In related work also Web usage mining investigates the problem of session reconstruction [10]. A session represents all the activities of a user on a Web site during a single visit. Identification of users is usually achieved using cookies and IP addresses, if available, or through heuristics on the duration and behavior of the user. By contrast, when correlating messages, we assume that the correlation information is in the content of events (their attributes), and the problem is that of discovering which attributes or combination thereof are correlators. In contrast, in time-based session reconstruction, the issue of how to decide on the boundaries of the session is often related to the specification of the time delays between user activities.

In [2], an approach for constructing process instances from sender-receiver information has been proposed in the context of Web services. However, the used correlation criteria does not look at the content of exchanged messages for understanding the correlation between messages.

As a complementary work to our paper, a probabilistic approach based on iterative model convergence is used in [3] for discovering process models from unlabeled event logs. In our approach, we consider the message payload for the purpose of correlation and building the set of process instances.

Finally, in our previous work [8,7] various types of CCs are identified and used to discover interesting CCs and build process instances. The interestingness of CCs are defined based on heuristic-based criteria and user input. The aim of that approach is to prune the space of all possible CCs to only visit the relevant ones by reducing the number of considered CCs depending on their evaluated interestingness. Such approach is favoring the exploration of space of possible CCs. However, the notion of interestingness of CCs is based on heuristics and assumptions leading to ignoring many non-interesting correlations. However, interestingness of correlations is subjective. For example, one can aim to discover rare or exceptional correlations for identifying how a process in an organization executes in exceptional circumstances. Heuristics may be adapted to do so, however, it implies re-executing CC's interestingness evaluation according to every new heuristic. In contrast, our approach allows to try heuristic-based filtering on the ACG without the need to re-generate it.

6   Conclusion and Perspectives

In this paper, we proposed an efficient approach that quickly discovers potentially relevant sequences of correlation conditions from large service interaction logs. The approach reads a service interaction log, and builds an Aggregated Correlation Graph (ACG), which is size-efficient and exhibits all the sequences of correlation conditions identified in the log in a single graph. Such ACG is visualized and refined by the user through filtering parameters to identify relevant sequences of correlation conditions which reveal process instances.

Existing process discovery algorithms require as input the list of all process instances. An important step in future work would be the discovery of process models directly from the ACG, taking advantage of its efficient size and properties. This step is currently under experiments and is based on a mapping between the ACG and a BPMN representation of the process model. The major challenge here is to identify control flow operators (AND join, AND split, etc.). Another future work would be to offer an open interface for process-aware information systems to log in real-time messages directly into the proposed system for immediate inclusion in the ACG. Finally, further experiments with much larger service interaction logs would unveil the need of taking advantage of the inverted index compression techniques to further enhance the performance.

References

1. Barros, A., Decker, G., Dumas, M., Weber, F.: Correlation patterns in service-oriented architectures. In: Proceedings of FASE Conference, pp. 245–259. Springer, Heidelberg (2007)
2. De Pauw, W., Lei, M., Pring, E., Villard, L., Arnold, M., Morar, J.F.: Web services navigator: visualizing the execution of web services. IBM Syst. J. 44, 821–845 (2005)
3. Ferreira, D.R., Gillblad, D.: Discovering Process Models from Unlabelled Event Logs. In: Dayal, U., Eder, J., Koehler, J., Reijers, H.A. (eds.) BPM 2009. LNCS, vol. 5701, pp. 143–158. Springer, Heidelberg (2009)
4. Motahari, H.R., Benatallah, B., Saint-Paul, R.: Protocol discovery from imperfect service interaction data. In: Proceedings of VLDB Ph.D. Workshop (2006)
5. Motahari-Nezhad, H.R., Saint-Paul, R., Benatallah, B., Casati, F.: Deriving protocol models from imperfect service conversation logs. TKDE 20, 1683–1698 (2008)
6. Măruşter, L., Weijters, A.J., Aalst, W.M., Bosch, A.: A rule-based approach for process discovery: Dealing with noise and imbalance in process logs. Data Min. Knowl. Discov. 13(1), 67–87 (2006)
7. Motahari Nezhad, H.R., Benatallah, B., Saint-Paul, R., Casati, F., Andritsos, P.: Peocess spaceship: Discovering process views in process spaces. Technical Report UNSW-CSE-TR-0721. The University of New South Wales, Australia (2007)
8. Nezhad, H.R.M., Benatallah, B., Saint-Paul, R., Casati, F., Andritsos, P.: Process spaceship: discovering and exploring process views from event logs in data spaces. PVLDB 1(2), 1412–1415 (2008)
9. Pauw, W.D., Hoch, R., Huang, Y.: Discovering conversations in web services using semantic correlation analysis. In: Proceedings of ICWS Conference, 639–646 (2007)
10. Spiliopoulou, M., Mobasher, B., Berendt, B., Nakagawa, M.: A framework for the evaluation of session reconstruction heuristics in web-usage analysis. Informs J. on Computing 15(2), 171–190 (2003)
11. van der Aalst, W.M.P., van Dongen, B.F., Herbst, J., Maruster, L., Schimm, G., Weijters, A.J.M.M.: Workflow mining: a survey of issues and approaches. Data Knowl. Eng. 47(2), 237–267 (2003)
12. Zhang, J., Long, X., Suel, T.: Performance of compressed inverted list caching in search engines. In: Proceeding of WWW Conference, pp. 387–396. ACM, New York (2008)

Supporting Dynamic, People-Driven Processes through Self-learning of Message Flows

Christoph Dorn[1,2] and Schahram Dustdar[2]

[1] Institute for Software Research, University of California, Irvine, CA 92697-3455
cdorn@uci.edu
[2] Distributed Systems Group, Vienna University of Technology, 1040 Vienna, Austria
lastname@infosys.tuwien.ac.at

Abstract. Flexibility and automatic learning are key aspects to support users in dynamic business environments such as value chains across SMEs or when organizing a large event. Process centric information systems need to adapt to changing environmental constraints as reflected in the user's behavior in order to provide suitable activity recommendations. This paper addresses the problem of automatically detecting and managing message flows in evolving people-driven processes. We introduce a probabilistic process model and message state model to learn message-activity dependencies, predict message occurrence, and keep the process model in line with real world user behavior. Our probabilistic process engine demonstrates rapid learning of message flow evolution while maintaining the quality of activity recommendations.

Keywords: message prediction, process log mining, people-driven processes, process evolution, message activity dependencies.

1 Introduction

Modern information systems need to enable flexibility and automatic adaptation capabilities in order to cope with continuously evolving environments where a-priori fixed requirements are rarely applicable. Organization of multi-national events such as the Olympic Games or management of value chains across a large set of Small and Medium-sized Enterprises (SMEs) are just two examples where exact work practices cannot be precisely defined and executed. In such environments, users engage in knowledge and coordination intensive workflows that are subject to continuous change. Processes evolve as participants tune their work practice to increase efficiency and effectiveness. Thus, users want to focus on their tasks rather than managing and updating their workflow. Instead the involved information systems should learn from and adapt to the users automatically.

In this paper, we address the case of people-driven dynamic processes. There exists a process model that describes the activities carried out by humans representing their expertise. Users are, however, completely free to deviate from the underlying model which cannot foresee all possible situations. These processes heavily rely on exchanging unstructured or semi-structured messages such as

H. Mouratidis and C. Rolland (Eds.): CAiSE 2011, LNCS 6741, pp. 657–671, 2011.

emails. When organizing a large event, these messages are the main artifacts to coordinate between participants. While the necessary activities are relatively clear, the type of messages, their occurrence, and evolution will remain dynamic. We address three major problems (i) learning of message-activity dependencies, (ii) prediction of which messages will arrive, and (iii) automatically reflecting work evolution in the process model. The subsequent challenges are then to distinguish between accidental one-time deviations and desired process improvements, managing the interleaving of messages, and the inability to observe the complete set of user actions.

Our contributions in this paper are (i) a probabilistic process model and execution engine that does not rely on a precise occurrence or absence of messages; (ii) a self-learning (thus unsupervised) message flow algorithm to detect message-activity dependencies and updates thereof; and (iii) a message recommendation mechanism to support the analysis of unstructured and semi-structured messages. The main applied approaches are log sequence mining, and providing — respectively extracting — message-activity correlation information.

Our contributions bring benefit to both process designers and process users. Process designers need not capture all possible deviations nor the exact mapping of input and output messages to activities as this is automatically learned from process users. These in turn see the immediate effect of their applied expertise in form of process model changes without having to include a dedicated process designer.

The remainder of this paper is structured as follows: A motivating scenario sets the scene for our self-learning message flow algorithm (Section 1.1). We discuss related work in Section 2. Section 3 introduces our approach, followed by the probabilistic models in Section 4. Section 5 describes the recommendation and learning mechanism, which we evaluate in Section 6. We provide a short conclusion and outlook on future work in Section 7.

1.1   Motivating Scenario

The example in Figure 1 depicts a flexible people-driven order process. The individual work steps describe a general order of user activities to successfully complete a process instance. The outlined flow, however, does not enforce the exact order of activities, which is up to the user, and covers no exceptions or process adaptations that might arise due to specific customer request, incomplete information, or user specific expertise. Consequently, the listed document types that characterize the exchanged messages specify merely an initial set of expected documents. The process visualization in Figure 1 does not apply any particular process modeling language but rather presents an intuitive view on the involved documents that represent input and output of activities. In this scenario, we encounter various forms of message flow evolution:

Missing Documents : When the *Replenish* step (C) is updated to make use of an automatic restocking system, only user confirmation is required and the *Quote* message (3) no longer occurs.

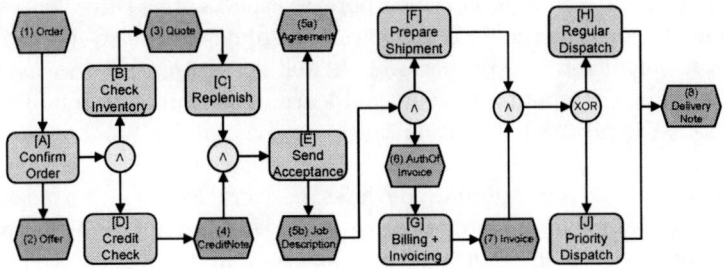

Fig. 1. Generic order process and associated document types

Delayed Documents : A company potentially decides to delay the *Agreement*
(5a) message until the *PrepareShipment* (F) activity signals the completion
of the production and packaging sub process. Such a company might depend
on just-in-time delivery of subcontracted parts and thus cannot guarantee
order completion any earlier.

Premature Documents : For premium customers, shipping becomes independent of billing thus the *DeliveryNote* message (8) potentially occurs before
Invoice (7).

Shifted Documents : To guarantee that the *Invoice* (7) always exactly reflects
the packaged goods, the ERP system no longer issues the *Authenticationof
Invoice* message (6). Instead the *PrepareShipment* (F) step triggers this
message.

2   Related Work

In the last decades considerable research effort was spent on systems for supporting flexible processes. In dynamic environments where business requirements
continuously change, processes need to be able to adapt to fluctuating constraints. Processes cannot remain rigidly structured but need to support ad-hoc
human control and evolve along alternative execution paths. Depending on the
supported level of flexibility, we can distinguish between roughly three types of
processes (including some example works):

Ad-hoc processes provide no constraints on the order of process activities and
provide the user complete freedom of choice [9,6,20,3].

Semi-structured processes (or case-based processes [17]) contain some structure and capture best practices (e.g., from previous process instances) but
are still too complex to be fully specified for automatic execution [18,2,15,19].

Well-structured processes are rigidly configured and determine for each condition the exact flow of control, data, and the involved resources and actors.
User involvement is limited to human tasks (if at all) while process management and execution control is fully automated [16,1,14].

On the one hand, our approach incorporates aspects of all three process types. Our notion of people-driven process allows users to determine ad-hoc the process execution order. There exists, however, a well structured process model as a baseline guidance. Recommendations and learning is similar to semi-structured systems as we apply past execution traces to dynamically update the process model.

On the other hand, our approach deviates in several aspects from existing flexible process recommendation systems. We relax the assumption of complete observability of user actions and instead rely on a combination of observed messages and actions. The primary focus in this paper, however, lies on the self-adaptation of message flows, leaving a reordering and adjustment of process activities aside. We addressed this aspect of automatic people-driven process adaptation in our previous work [5]. In addition, we expect the message structure and thus the recognized types to change over time. Subsequently, we determine a dependency between message types and activities dynamically through analysis of execution traces. In a similar effort, Lakshmanan et al. [12] describe how an ant colony optimization algorithm learns dependencies of document contents (e.g., the impact of certain values within a message) to predict the flow and outcome of a process. They also apply exponential aging to keep the decision probabilities up to date. The underlying process model, however, remains unchanged.

Related work that explicitly applies autonomic computing principles [10] for adaptive workflow and process support systems react to system internal events such as workload fluctuations [8], goal changes [7], or service replacements [21] when executing well-structured processes. These approaches, however, target only system elements and offer no support for having the user dynamically adapt the process. In contrast we aim to apply those autonomic principles primarily for achieving unsupervised learning and user recommendation.

3 Approach

Successful activity recommendations in people-driven processes (i.e., the user actually carries out the proposed activity) depend on correct classification of incoming and outgoing messages and their associated activities. Supporting users in dynamically evolving processes consists of two aspects: (i) run-time tracing of process progress including probabilistic message prediction, and (ii) automatic refinement of process model and message probabilities upon process termination.

The main phases of the run-time recommendation support is depicted in Figure 2. The core of the *Probabilistic Process Engine* consists of the process model that describes the general structure and current interdependencies of steps and messages (see Sec. 4.1). The Message State Model captures the likelihood for a message to occur in that process on time, early, late, or repeatedly (see Sec. 4.2).

For each intercepted message and observed user action, the process engine extracts the correlation of messages and activities to determine which activity produced a particular message, and which message served as input to a given activity (1). Next, the engine analyzes whether the observed messages and actions

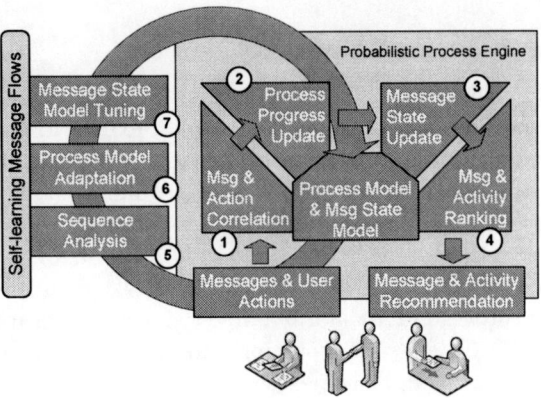

Fig. 2. Supporting message prediction and activity recommendation through self-learning message flows

advance the progress of the process (2). Any completed activity potentially results in an update of one or more messages, respectively their expectation states: messages are activated, become missing, or should no longer occur (3). Finally, the engine ranks messages according to their probability to occur, and activities to be carried out next (4). Activity recommendations support the user in carrying out his/her work. Message recommendations support the classification of unstructured and semi-structured messages that are exchanged between process participants.

After successful completion of a process instance, the mechanism we present in the following sections analyzes the sequence and timing of activities and messages (5) to update the process model (6) and message state model (7) to accurately reflect the changes in the real world.

Our approach specifically targets dynamic business environments that mostly rely on unstructured or semi-structured messaging to communicate and coordinate processes. SMEs and event organization usually coordinate and collaborate via email. Our approach relies on an infrastructure for intercepting those messages, extracting relevant information, and mapping those messages to document types. In our case, the EU FP7 Project Commius provides the necessary framework to extract information from emails and conduct a basic, process-unaware email content analysis. Details on the actual process orchestration as well as message interception, extraction, analysis, and user interfaces are outside the scope of this paper. The interested reader is referred to previous project-related publications [4,13,11]. Our prototype implements the algorithms and techniques introduced in this paper to support the email content analysis by determining the expected document types. The activity recommendations allow the annotation of emails with process-relevant information (as defined in the activity description) before forwarding them to the actual recipient.

4   Models

4.1   Probabilistic Process Model

In an environment where most communication happens via unstructured and semi-structured messages such as email, we cannot expect to have an integrated IT infrastructure (such as traditional process engines require) that allows a unified observation of all user actions. Likewise, we cannot assume a tight coupling of process activities and corresponding message types.

Our process model (see also Figure 3 left side) consists of *ActivityNodes* and *Connectors* which link multiple activities together via *FlowDirection* arcs. Activity nodes, connectors, and arcs form a directed acyclical graph where activity nodes always exhibit exactly one incoming and one outgoing arc. Only connectors potentially have in- and out-degree above 1. The *FlowData* annotations of the *FlowDirection* arcs describe the corresponding activity input and output messages. The *Occurence* attribute describes in the interval $[0, 1]$ the likelihood a message of the stated type will arrive. A connector describes how multiple activities are interlinked applying the traditional flow constructs of *AND*, *OR*, *XOR* splits, respectively joins. An activity consists of one or more *Actions* the user needs to carry out in order to consider that activity as completed. Alternatively, we consider an activity successfully performed when all specified output *Messages* have been transmitted.

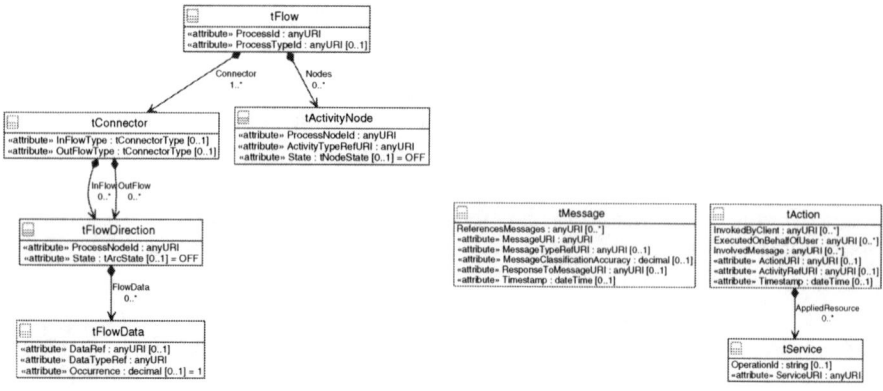

Fig. 3. Flow Model supporting the annotation of transitions with probabilistic message types; Action model and Message meta data model for correlating messages and activities

NodeStates (Off, Active, Completed, Skipped) and *ArcStates (Off, Active, Selected)* track the process progress and determine which subsequent arcs and nodes describe upcoming messages to expect and activities to carry out. The flow model describes the involved messages types and their probability to occur on a particular place in the process. The overall probability that a message will arrive at a given point in time, however, is hard to establish. We, therefore, pair the flow model with a *Probabilistic Message State Model* for each involved message type.

4.2   Probabilistic Message State Model

The message state model describes the possible message states (such as scheduled, expected, arrived) and the likelihood for transitions between those states during the process. We track only messages that are defined in the process model, each in a separate state model instance that is valid only for that particular process type and message type.

The state model is defined as a directed graph $G_{State}(V, E)$ where states are represented by vertices ($v \in V$) and labeled edges ($e \in E$) represent transitions. A labeled transition between neighboring states describes the probability that this transition will occur ($w(v_i, v_j) = [0, 1]$). The sum of outgoing transition probabilities is always 1. The state model can be interpreted as a Markov chain that describes the probability of reaching a particular state when a transition occurs. We neglect self-transitions as the duration a message type remains in a certain state is irrelevant for our purpose. We are only interested in the probabilities of reaching each of the subsequent states.

State transitions from *Start* to *Scheduled, Expected, Missing,* and *NotExpected* are driven by the process progress. Transitions to any *Received* state and to *Repeated* are message driven. The process termination finally triggers the transition to *Occurred* and *NotOccurred*. The initial transition probabilities (depicted in Figure 4) are optimistic: we assume a message to occur exactly once when its *Expected* or *Missing*, and when *Not Expected* to remain absent.

Scheduled. Initially every message is scheduled to take place at some time during the process lifetime.

Expected. A message becomes expected once it is needed to activate an activity or an activity is activated and is expected to produce the message as output.

Missing. Whenever an activity is completed without the required input, we mark its input message as missing. Likewise, we also mark any expected output message of an active preceding activity as missing.

Not Expected. When a user explicitly skips an activity, or an alternative XOR branch is completed — thereby skipping the involved activities — any message type to or from such activity is no longer expected.

Received Early. A message arriving while in state *Scheduled* indicates a potential change in the process model or just a one-time deviation.

Received On Time. A message arrives on time according to the process model.

Received Late. The arrival of a missing message indicates a potential change in the process model or just a one-time deviation.

Received Unexpected. Indicates that a skipped activity needed to be carried out anyway or a parallel branch is incorrectly specified as XOR.

Repeated. A message is received multiple times.

Occurred. The message has been sent or received at process end.

Never Occurred. The message did not occur for whatever reasons.

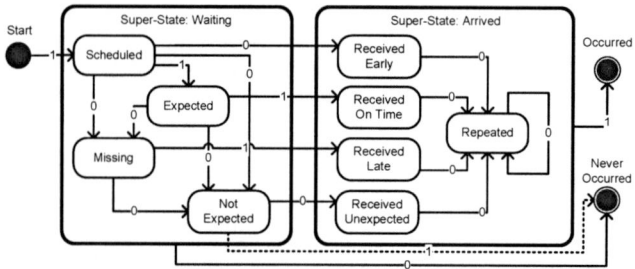

Fig. 4. State Transition model for a single message type. Each model instance provides for a message type and process model the specific transition probabilities

5   Prediction and Self-adjusting Mechanisms

5.1   Probabilistic Process Management

Messages in form of *FlowData* annotations play a central role in the progress tracking and process step activation (and ultimately recommendation). Messages often define the trigger condition for an activity. For example in the scenario: *Quote* (3) is required to continue with *Replenish* (C) but does not arrive for whatever reasons (unobservable communication channel, delayed arrival, or simply no longer necessary) it will halt the process progress. The recommendation mechanism would not continue to suggest to execute inactive process step (C). We, therefore, need a mechanism to decide when the process engine should wait for a message to arrive, and when to continue with analyzing and activating the subsequent *FlowDirection* arcs and process steps.

Whenever the probabilistic flow engine checks a *FlowDirection* arc for activation, it compiles a list of available M_a and missing M_m messages for that arc. For all messages, we calculate the sum of probabilities $p_{total} = \sum p_{occur}(msg \in M_m \vee M_a)$ and additionally the likelihood $p_{miss} = \sum p_{occur}(msg \in M_m)$ that at least one of the missing messages will arrive. We then apply following rules to determine whether the message conditions on an arc can be considered satisfied (*GO*) or not (*NOGO*):

1. *NOGO* if $p_{total} == 1 \bigvee |M_a| < 1$: the first condition describes a set of alternative messages and if non of those have yet arrive (second condition) we continue to wait.
2. *NOGO* if at least one missing message ($msg \in M_m$) always arrives ($p_{occur}(msg) == 1$) we continue to wait.
3. *GO* if $|M_m| == \emptyset$: no message is missing.
4. *GO* if $p_{total} == |M_a|$: its sufficient when the arrived messages cover the probability of all documents.
5. *GO* if $Random(0,1) > p_{miss}$ else *NOGO*: if the set of messages neither constitute XOR alternatives, nor a compulsory AND set, but one (or several) out of many we apply a random value from the interval $[0,1]$. If that random value is larger than the probability that any missing message will

arrive p_{miss}, then we consider the *FlowData* as satisfied and activate the corresponding arc.

Let is consider the incoming arc of *Billing and Invoicing* [G] from the scenario: messages (5b) and (6) always occur thus their probability is 1. Assume (5b) arrives then rule 2 applies, as we are still waiting for (6). Once message (6) arrives, rule 3 would consider the arc conditions satisfied and subsequently enable activity [G].

When the observed messages suffice to activate the underlying arc, we iterate through all non-observed alternative messages (that still reside in state *Scheduled, Expected,* or *Missing*) and switch them to *NotExpected* as we no longer require their appearance.

5.2   Message Prediction

Message prediction is purely based on the transition labels between message states. In any of the *Waiting* states the probability is determined based on the transition to the respective received state. From any of the *Arrived* states the probability is determined by the transition weight to the *Repeated* state.

After any change to any of the messages' state model, the messages are ranked according to their probability to occur in their current state. The top ranked message(s) constitute the prediction and are applied during the classification of newly intercepted messages. Suppose we have two messages and their corresponding message state models depicted in Figure 5(a) and (b): the first one currently in state *Expected*, the second in state *Missing*. Hence, we would predict the occurrence of message (b) with $p_{occur}(b) = 0.9$ rather than message (a) with $p_{occur}(a) = 0.7$. We are thus able to address and manage messages that over multiple process instances no longer adhere to the process model but have become delayed.

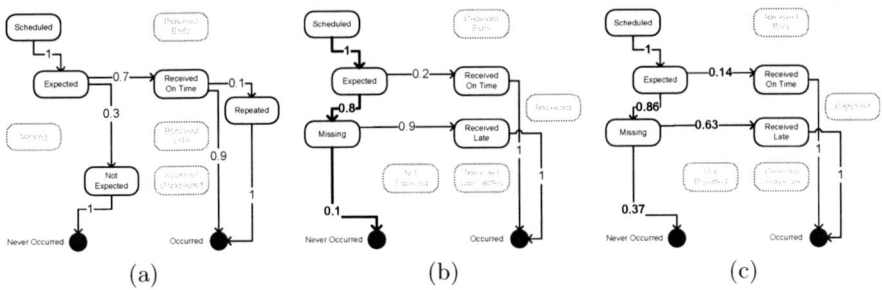

(a)   (b)   (c)

Fig. 5. Examples of message state model instances: only non-zero transitions are included for sake of clarity. Subfigure (a) displays a message that when expected occurs with 70% probability and is repeated in 10% of process cases. Subfigure (b) describes a message that occurs only 20% on time, but still arrives in 90% of all cases when missing. The effect on the transition probabilities of that message never occurring (thick lines and labels) in a single process instance is depicted in subfigure (c).

5.3   Message State Adjustment

After successful process termination, the message state model of each message in the scope of that process is adjusted. During the process execution, each model stores the sequence of message state transition (Seq_{ST}). All transition leaving any state listed in the sequence set are processed so their transition weights reflect the latest process instance. We apply an exponentially weighted moving average (EWMA) to update the transition probabilities.

$$w_{t+1}(v_i, v_j) = \begin{cases} 1 * \alpha + (1 - \alpha) * w_t(v_i, v_j) & \text{if } v_i, v_j \in Seq_{ST}, \\ 0 * \alpha + (1 - \alpha) * w_t(v_i, v_j) & \text{if } v_i, v_j \notin Seq_{ST}. \end{cases}$$

With EWMA, old process sequences have exponentially lower impact on the new probability value than more recent sequences. The coefficient α determines the significance of older values. With α close to 0, new values have next to no impact on the new probability and vice versa for α close to 1. As the updated probability depends only on the previous probability and the current state sequences, little memory is required to store the transition labels. Figure 5 (c) display the transition update result for the underlying message state after the state sequences in subfigure (b) are evaluated.

5.4   Self-learning Message Flows

After each process, we analyze the order in which activities and messages have occurred and compare it to the underlying process model (PM). Any message deviation needs to become reflected in the process model, e.g., a new message type emerged, an activity did not produce the expected message, a message did not serve as input as expected. During process execution, we captured the sequence of all message-activity dependency tuples ($tup(msg, act) \in Seq_{MA}$) by extracting message references from user actions, and activity references from messages. Algorithm 1 describes the technique for updating a process model's *FlowData*. For each tuple, we step through the process model in breadth-first style (lines 4-7, 20-21) and locate the corresponding *FlowDirection* arc and *FlowData* annotation (lines 8-11, 19). Once found, we increase the *FlowData* occurrence value ($occ(fd, mt)$) using exponentially weighted moving average (EWMA) (lines 12-15). At the end, *FlowData* annotations that are not covered by sequence tuples ($tup(msg, act)$) receive a lower occurrence value (lines 22-23):

$$occ_{t+1}(fd, mt) = \begin{cases} 1 * \beta + (1 - \beta) * occ_t(fd, mt) & \text{if } \exists tup(msg, act) \equiv tup(fd, mt) \\ & : tup(msg, act) \in Seq_{MA}, \\ 0 * \beta + (1 - \beta) * occ_t(fd, mt) & \text{otherwise} \end{cases}$$

where the tuple $tup(fd, mt)$ describes a *FlowDirection* fd with a *FlowData* annotation referencing message type mt. Coefficient β determines again how quickly a new message emerges or disappears on an arc.

Message types that show up for the first time, or messages types that arrive early result in a new *FlowData* annotation (lines 16-18). In the latter case, we reduce the original annotation occurrence value in addition. Message types that take place at a later stage of the process than specified receive the same treatment. We reduce their initial annotation value, and create a new *FlowData* annotation where we actually observed the message. When we have not observed a message type at all, we reduce all instances of the respective *FlowData* annotation. At the end, we clean the process model from rarely used *FlowData* annotations that would otherwise interfere with the message state model as message would be activated too early or remain expected too long (line 24). Experiments have identified a suitable cutoff value of 0.1. Arcs, however, that are inactive at the end of the process (due to explicit user skipping or non executed XOR branches) need no refreshing and are ignored. We insure this in Algorithm 1 by providing only the set of active *FlowDirections* in FD_{active} in the first place.

Algorithm 1. Self-learning Dependencies Algorithm $\mathcal{A}(PM, Seq_{AM}, FD_{active})$.

1: **for all** $tup(msg, act) \in Seq_{AM}$ **do**   ▷ For each activity-message tuple.
2:     $N \leftarrow PM.startNode$ ▷ List of nodes that we haven't checked yet
3:     $found \leftarrow false$
4:     **while** $!found \lor !N.empty$ **do**   ▷ While not found and not at the process end
5:         **for** $i = N.size \rightarrow 0$ **do**
6:             $Node\ n \leftarrow N_i$
7:             $N \leftarrow N - n$
8:             **for all** $FlowDirection\ df \in n.outFlow()$ **do**
9:                 **if** $fd.getActivity == act$ **then**
10:                     $actOk \leftarrow true$ ▷ Found the correct activity
11:                 $FlowData\ data == fd.get(msg)$
12:                 **if** $data \neq null \lor actOk \lor fd.dir == tup(msg, act).dir$ **then**
13:                     $increaseByEWMA(data)$
14:                     $FD_{active} \leftarrow FD_{active} - data$
15:                     $msgOk \leftarrow true$
16:                 **if** $actOk \lor !msgOk \lor fd.dir == tup(msg, act).dir$ **then**
17:                     $fd \leftarrow fd + FlowData(msg)$ ▷ Add a new *FlowData* to the arc
18:                     $msgOk \leftarrow true$
19:                 $found \leftarrow actOk \lor msgOk$
20:             **if** $!found$ **then**   ▷ Adding the next set of nodes to the search list
21:                 $N \leftarrow n.getSuccessorNodes()$
22:     **for all** $FlowData\ data \in FD_{active}$ **do**
23:         $reduceByEWMA(data)$
24: $removeFlowData(cutoffValue)$ ▷ remove rarely occurring *FlowData*.

---

6   Evaluation

We evaluate our approach based on the motivating scenario in Section 1.1. Specifically, we are interested in the time it takes to learn an evolved message flow given a fixed process model. In addition, we observe how the updated message states and *FlowData* annotation affect activity recommendations. We simulate user behavior though prepared log sequences that consist of activities and messages. The sequences serve as input to the Probabilistic Process Engine which is initiated with the original scenario process.

6.1   Experiment Setup and Success Metrics

Two quality metrics measure the success of our learning and adaptation mechanism. The Message Classification Error (MCE in range $[0, 1]$) determines for each incoming unknown message during the experiments how close our message prediction algorithm gets. $MCE = 0$ when the actual message type and highest ranked predicted message type are identical, otherwise we extract the matching message type from the ranking list and take the inverse of its expected occurrence probability $1 - p_{occur}$) (i.e., the lower its probability, the higher the MCE). Suppose for an incoming message the algorithm produces following ranking result $[(7) : 0.5], [(8) : 0.4], [(2) : 0.1]$: MCE would yield 0.6 when the message is actually of type (8). For measuring the effect towards the user, we apply the Activity Recommendation Error (ARE in range $[0, 1]$) which is analogue to the MCE: when the next element in the log sequence is an activity, we retrieve an activity recommendation and locate the matching activity. We keep the activity recommendation mechanism intentionally simple to focus only on the effect of the message state model and probabilistic flow model. Active activities are ranked based on their time since activation, thus the longer an activity remains unfinished, the higher it will score.

Two activity message log sequences describe an evolution of the initial process model. The *Quote* (3) message is no longer used, the *Agreement* (5a) message is delayed until completion of the *PrepareShipment* (F) activity which also triggers the *Authorization Of Invoice* message (6). The *Invoice* (7) is produced when executing *Priority Dispatch* (J). The *Delivery Note* (8) only applies to *Regular Dispatch* (H). During all experiments, we set EWMA coefficients $\alpha = 0.3$ and $\beta = 0.4$ which is a trade-off between rapid uptake of novel user behavior and robustness against one-time deviations. We set $\beta > \alpha$ as the arcs need to learn and forget message changes quicker than the state model (which tracks the message probabilities across the whole process and not just a single, local arc occurrence). Log Sequence A: 1, A, 2, B, C, D, 4, E, 5b, F, 6, 5a, G, J, 7 Log Sequence B: 1, A, 2, D, 4, B, C, E, 5b, F, 6, 5a, G, H, 8

6.2   Results

In experiment 1 (Figure 6 a and c), we play each of the log sequences 30 times against the scenario process (dotted lines, green +, and blue x) — as well as alternating each sequence (full lines, red circles) — and determine overall MCE_p and ARE_p (sum of all recommendation errors within one process instance).

MCE_p rapidly drops to zero within a few iterations for individual log sequences A and B. The interleaving sequences take longer to produce a stable process model as sequences A and B display an opposing user behavior (to the extend allowed by the process model). ARE_p shows similar behavior and settles to normal error rates once the probabilistic message states and flow model updates have settled. The consistent error rates above zero are due to the simple activity recommendation algorithm. Activities at branching points receive almost equal probabilities, and the log sequences happen to prefer the second highest choice.

In experiment 2 (Figure 6 b and d), we apply the same sequences but take an empty process model i.e., all *FlowData* annotations are on the first arc towards activity (A). As expected, MCE_p and ARE_p are high during the first few iterations, but quickly decrease to low error rates and then settle to the same rates as the evolved process model in experiment 1.

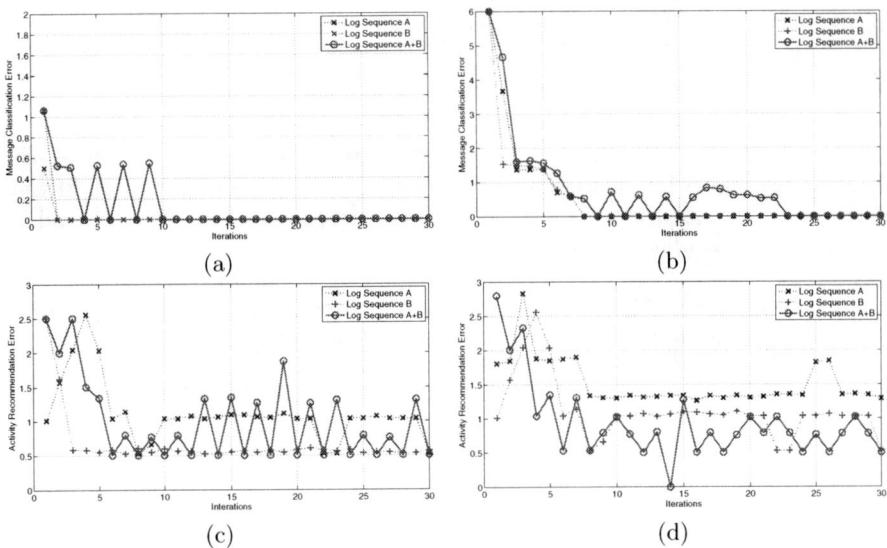

Fig. 6. Overall Message Classification Error MCE_p (a,b) and Activity Classification Error ARE_p (c,d) over 30 process iterations for fixed and alternating log sequences; for existing (a,c) and empty process model (b,d)

6.3   Discussion

Two important characteristics describe our self-learning approach. First, the process model and state management model allow a quick stabilization which reflects in the low prediction and recommendation error rates. As a positive side-effect, the user is hardly affected by incorrect activity recommendation. Second, the described techniques apply not only to cases of process evolution but also efficiently support a grass-roots approach to processes learning. We observe similarly swift adaptation when no message-activity dependencies are a-priori known. The applied 30 iterations are sufficient to demonstrate the learning behavior as our algorithm considers all changes simultaneously. The presented results are still valid for larger processes where individual segments undergo evolution one at a time. In that case our results describe the behavior for a single segment.

7   Conclusion

Users in collaboration and coordination intensive people-driven processes require self-learning mechanisms to reflect the evolution of message types in the underlying process model. In this paper we presented an approach based on a probabilistic process model and message state model to predict when and which messages will arrive to ultimately give suitable activity recommendations. Analysis of sequence logs containing both messages and activities allows updating the process model automatically. Evaluation based on a motivating scenario demonstrated successfully that our mechanisms work correctly and efficiently.

Future work consists of two main tasks: on the one hand we aim at improving the message prediction algorithm by extracting message patterns such as request-reply from interaction logs, including temporal aspects, and integrating the process learning techniques introduced in our previous work. Pairing the probabilistic algorithms with semantic message analysis is expected to further improve the classification success rate. On the other hand, we plan to conduct additional user studies based on a larger process and log sequences set.

Acknowledgment. This work has been partially supported by the EU STREP project Commius (FP7-213876) and Austrian Science Fund (FWF) J3068-N23.

References

1. Adams, M., Edmond, D., ter Hofstede, A.H.M.: The application of activity theory to dynamic workflow adaptation issues. In: 7th Pacific Asia Conference on Information Systems, pp. 1836–1852 (2003)
2. Adams, M., Hofstede, A., Edmond, D., van der Aalst, W.: Facilitating flexibility and dynamic exception handling in workflows through worklets. In: Proceedings of the CAiSE 2005 Forum, FEUP, pp. 45–50 (2005)
3. Burkhart, T., Loos, P.: Flexible business processes - evaluation of current approaches. Proceedings Multikonferenz Wirtschaftsinformatik, MKWI-2010 (February 2010)
4. Burkhart, T., Werth, D., Loos, P.: Commius – An Email Based Interoperability Solution Tailored For SMEs. Journal Of Digital Information Management 6 (2008)
5. Dorn, C., Burkhart, T., Werth, D., Dustdar, S.: Self-adjusting recommendations for people-driven ad-hoc processes. In: Proceedings of International Conference on Business Process Modelling. Springer, Heidelberg (September 2010)
6. Dustdar, S.: Caramba Process-Aware Collaboration System Supporting Ad hoc and Collaborative Processes in Virtual Teams. Distributed Parallel Databases 15(1), 45–66 (2004)
7. Greenwood, D., Rimassa, G.: Autonomic goal-oriented business process management. In: ICAS 2007: Proceedings of the Third International Conference on Autonomic and Autonomous Systems, p. 43. IEEE Computer Society, Washington, DC, USA (2007)

8. Heinis, T., Pautasso, C., Alonso, G.: Design and evaluation of an autonomic work-flow engine. In: Proceedings of the Second International Conference on Automatic Computing, pp. 27–38. IEEE Computer Society, Washington, DC, USA (2005), http://portal.acm.org/citation.cfm?id=1078027.1078524
9. Huth, C., Erdmann, I., Nastansky, L.: Groupprocess: Using process knowledge from the participative design and practical operation of ad hoc processes for the design of structured workflows. In: HICSS (2001)
10. Kephart, J.O., Chess, D.M.: The vision of autonomic computing. Computer 36(1), 41–50 (2003), http://dx.doi.org/10.1109/MC.2003.1160055
11. Laclavik, M., Dlugolinsky, S., Seleng, M., Kvassay, M., Gatial, E., Balogh, Z., Hluchy, L.: Email analysis and information extraction for enterprise benefit. Computing and Informatics Journal, Special Issue on Business Collaboration Support for Micro, Small, and Medium-Sized Enterprises 30(1), 57–78 (2011)
12. Lakshmanan, G.T., Duan, S., Keyser, P.T., Khalaf, R., Curbera, F.: A heuristic approach for making predictions for semi- structured case oriented business processes. In: Proceedings of First Workshop on Traceability and Compliance of Semi-Structured Processes @BPM 2010. Springer, Heidelberg (2010)
13. Marín, C.A., Stalker, I.D., Mehandjiev, N.: Engineering business ecosystems using environment-mediated interactions. In: Weyns, D., Brueckner, S.A., Demazeau, Y. (eds.) EEMMAS 2007. LNCS (LNAI), vol. 5049, pp. 240–258. Springer, Heidelberg (2008)
14. Müller, R., Greiner, U., Rahm, E.: $A_{gent}w_{ork}$: a workflow system supporting rule-based workflow adaptation. Data Knowl. Eng. 51(2), 223–256 (2004)
15. Pesic, M., van der Aalst, W.M.P.: A declarative approach for flexible business processes management. In: Business Process Management Workshops, pp. 169–180 (2006)
16. Reichert, M., Rinderle, S., Dadam, P.: Adept workflow management system: flexible support for enterprise-wide business processes. In: van der Aalst, W.M.P., ter Hofstede, A.H.M., Weske, M. (eds.) BPM 2003. LNCS, vol. 2678, pp. 370–379. Springer, Heidelberg (2003)
17. Reijers, H., Rigter, J., Aalst, W.V.D.: The case handling case. International Journal of Cooperative Information Systems 12, 365–391 (2003)
18. Sadiq, S., Sadiq, W., Orlowska, M.: Pockets of flexibility in workflow specifications. In: Proc. ER 2001 Conf., pp. 513–526 (2001)
19. Schonenberg, H., Weber, B., van Dongen, B., van der Aalst, W.: Supporting flexible processes through recommendations based on history. In: Dumas, M., Reichert, M., Shan, M.-C. (eds.) BPM 2008. LNCS, vol. 5240, pp. 51–66. Springer, Heidelberg (2008)
20. Stoitsev, T., Scheidl, S., Spahn, M.: A framework for light-weight composition and management of ad-hoc business processes. In: Winckler, M., Johnson, H. (eds.) TAMODIA 2007. LNCS, vol. 4849, pp. 213–226. Springer, Heidelberg (2007)
21. Yu, T., Lin, K.J.: Adaptive algorithms for finding replacement services in autonomic distributed business processes. In: Proceedings of the Autonomous Decentralized Systems, ISADS 2005, pp. 427–434 (April 2005)

Business Process Service Oriented Methodology (BPSOM) with Service Generation in SoaML

Andrea Delgado[1], Francisco Ruiz[2], Ignacio García-Rodríguez de Guzmán[2], and Mario Piattini[2]

[1] Computer Science Institute, Faculty of Engineering, University of the Republica
Julio Herrera y Reissig 565
CP 11300, Montevideo, Uruguay
[2] Alarcos Research Group, Information Tech & Systems Dep., University of
Castilla-La Mancha
Paseo de la Universidad No.4
CP 13071, Ciudad Real, España
adelgado@fing.edu.uy,
{francisco.ruizg,ignacio.grodriguez,mario.piattini}@uclm.es

Abstract. Carrying out business processes by means of software services helps to close the business–systems gap, by introducing an intermediate layer between business process definition and software systems, thus permitting not only better independence, but also more traceability between them. Despite the fact that technologies have matured to support this new reality, there is a lack of methodologies and notations, although some have been proposed to guide service development with different visions of service design and implementation. Service modeling is the basis for, among other things, the automation of several development steps by means of the model-driven development paradigm. The SoaML standard is a major step towards service modeling in UML. In this paper we extend our Business Process Service Oriented Methodology (BPSOM) for service development from business processes by integrating two main aspects: service modeling using SoaML and QVT transformations to obtain SoaML service models from BPMN BP models.

Keywords: Business Process Management (BPSOM), Service Oriented Computing (SOC), Model Driven Development (MDD), BPMN, SoaML.

1 Introduction

The modeling of business process as the means to show explicitly how organizations carry out their business has gained importance in recent years. Although the business area has several mature techniques with which to manage its business processes, based on the Business Process Management (BPM) [1][2] paradigm, the software area has recently been integrating this vision into software development, supported by the Service Oriented Computing (SOC) [3] paradigm. Carrying out business processes by means of software services based on a Service Oriented Architecture (SOA) [4][5] style, helps to close the business-system gap which has come about as a result of the differences between business and software area visions of the organization. The

H. Mouratidis and C. Rolland (Eds.): CAiSE 2011, LNCS 6741, pp. 672–680, 2011.

Model Driven Development (MDD) [6] paradigm, along with Model Driven Architecture (MDA) [7] have an important role to play. They allow correspondences between models to be defined, since they are key development artifacts, permitting the generation of code in different technologies. Although technologies have matured to support this new reality, few methodologies have been proposed to guide the service development process. The Service Oriented Architecture Modeling Language (SoaML) [8], recently defined by OMG, is a major step towards the modeling of services using UML and specific service stereotypes.

The standardized framework MINERVA [9] we have defined aims to support the Business Process (BP) lifecycle [1] by applying service-oriented and model-driven paradigms to business processes; it can be viewed on-line in [10]. The Business Process Service Oriented Methodology (BPSOM) [11] integrated in MINERVA provides the methodological guide with which to develop services from business processes. This paper extends the definition of BPSOM shown in [11] by integrating two new key aspects: the use of the SoaML standard for service modeling, and transformations using the Query/Views/ Transformations (QVT) [12] language, to generate SoaML service models, when possible, from business process models in Business Process Modeling Notation (BPMN) [13].

The remainder of the paper is organized as follows: BPSOM is presented in Section 2, along with the use of BPMN for BP modeling and SoaML for service modeling. In Section 3 service generation from business process is presented, related work is described in Section 4, and conclusions and future work are in Section 5.

2 BPSOM Definition

BPSOM has been defined for integration into the existing software development process used in the organization, with the aim of reusing existing knowledge, by adding only specific elements for service oriented development from business processes. Fig. 1 shows the definition of BPSOM and its use within the base process.

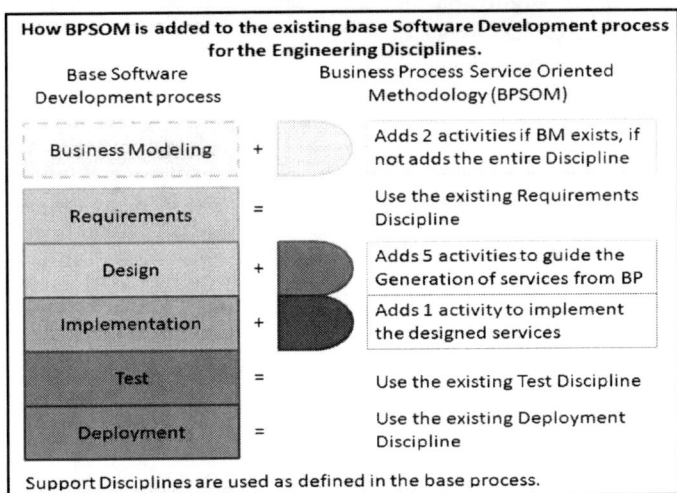

Fig. 1. How BPSOM is added to the existing software development process

We have based our work on the broad idea of methodology as outlined in [14]: a set of methods or techniques related, along with a process model and a set of deliverables, metrics, tools and management guidelines (including roles and organization team work). The definition of BPSOM began in 2005 and its disciplines, activities, roles and artifacts are detailed in [11]. In this paper we focus on the extension of BPSOM by adding service modeling in SoaML, along with the automatic generation of SoaML service models from BPMN BP models through QVT transformations, first described in [15]. The key aspect is to show where and how to use SoaML diagrams in BPSOM, as the standard provides their description, but no guide for using them. Transformations help obtaining some of the SoaML models automatically, providing support to activities, although human intervention is needed.

2.1 BPMN Use in the BPSOM Business Modeling Discipline

There is a great variety of notations for business process modeling [16], although in recent years, BPMN has emerged as the one preferred. Business people can use it to model business processes by themselves and then pass it to the software area.

BM1 – Assess the target organization. This activity aims to involve the project team in the organization for which the development is being carried out. The participating roles in this activity are the Business Analyst (from the business area), the Analyst and the Architect (from the software project team). The OMG Business Motivation Model (BMM) [17] can also be used for modeling goals and information which can be linked to SoaML services.

BM2 – Identify Business Processes. This is one of the key activities in the development of services from BP, since it is the main input needed to understand and describe BP in the organization. We use BPMN to specify them, which provides elements such as swimlanes (pool, lane), flow objects (activities, gateways), connecting objects (sequence, message) and artifacts (group, data) to model BP. Fig. 2 shows the "Patient Admission and Registration for Major Ambulatory Surgery (MAS)" BP from the Ciudad Real General Hospital project on which we are working, adapted to be used as example.

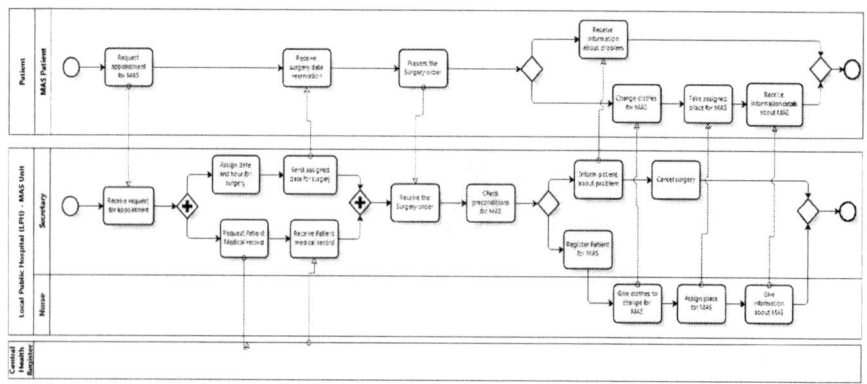

Fig. 2. "Patient Admission and Registration for MAS" Business Process in BPMN

2.2 SoaML Use in BPSOM Design Discipline

There is also a wide choice of notations for service modeling [16], UML being the one preferred. The SoaML profile extends UML by adding specific elements for service modeling, and will therefore soon be adopted by the community. It provides several stereotypes with which to specify services (contract, interfaces, operations, parameters) and the service architecture for the business process.

D1 – Identify and categorize services. This activity aims to identify the services needed to perform the business process under development and it is a key one in our approach. One of the main inputs of this activity is the BP model specified previously. The use of SoaML implies defining the Service Architecture (SA) which specifies the participants, contracts for the services and the roles they play as provider or consumer. Fig. 3 shows the SoaML SA for the example.

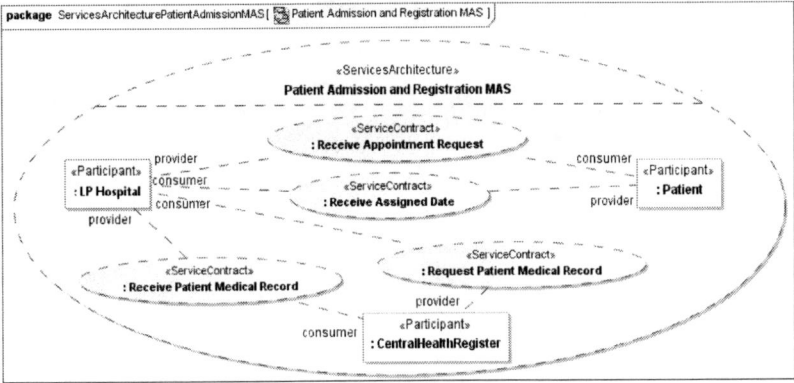

Fig. 3. SoaML ServicesArchitecture diagram for business process in Fig. 2

Services that the organization needs to provide to other parties and services that the organization has to consume from other parties are identified, based on the messages exchanged, each party being defined by a pool. To identify the services to support the business process, we look at each message exchanged between the pools (participants), setting the activity type to "ServiceTask" when we define it as a service. The ones that present incoming messages will be providers and those with outgoing

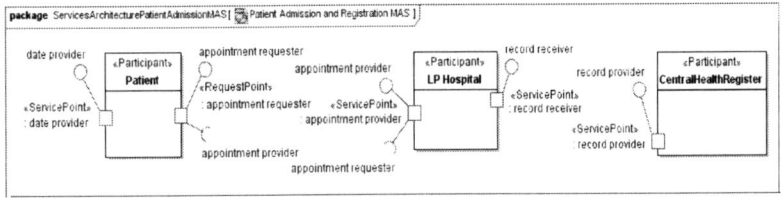

Fig. 4. Service and Request Ports for Participants

messages to a service task will be consumers. Services will be assigned to the participant corresponding to the pool containing each activity, and the associated Service Contract will hold all service information. Fig. 4 displays some Service and Request Ports (formerly Points) for the services defined, showing the bidirectional and unidirectional pattern of communication that can be defined.

D2 – Specify services. The specification of services corresponds to the definition of all the information needed, including the associated Service Contract with interfaces, operations, input and output parameters, among others. The information related to the in and out messages must be specified, indicating the parameters and data to be exchanged between the parties. The choreography defined by the Service Contract must also be specified, based on the interaction between participants. Once all this information has been considered, the most important parts of the ServiceContract can be generated and this can then be completed by the Architect or developers, who will also have to give the implementation details. Fig. 5 presents the ServiceContract definition and its choreography for the "ReceiveAppointmentRequest" service.

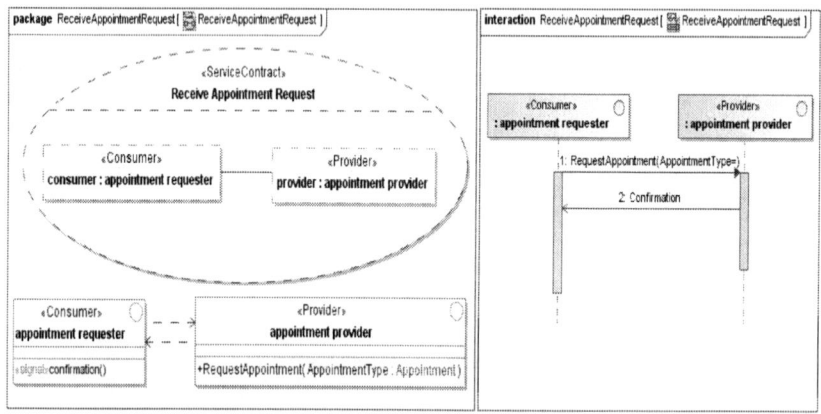

Fig. 5. ServiceContract and Choreography for the "ReceiveAppointmentRequest" service

D3 – Investigate existing services. The principal goal of this activity is to reuse the organization's existing services, as far as possible. To do so, a central Service Catalogue is defined, which has to be searched in each service development project. In the SoaML component diagram an adapter or wrapper has to be defined to relate the design service to its existing implementation, linking them in activity D4.

D4 – Assign components to services. The components that will implement the services generated must be defined and shown in the components diagram. For each service, a component with which to implement it has to be defined. SoaML provides the participant component with which to define the implementation of participants and services, defining new components to be generated and, if one exists, defining adapters or wrappers to use it.

D5 – Define services interaction. Service interaction can be defined as the orchestration or choreography of services, [4][5], as is done for business processes. This insight is provided by a sequence diagram showing all the services, or by various diagrams showing subsets of services for different sub-processes in the BP. That activity has no corresponding diagram in SoaML, so it is shown by a UML sequence diagram.

3   From BPMN Models to SoaML Models and Beyond

The BPSOM methodological and automated guide is used to derive and generate services from BP models, thus constituting the basis for its implementation. BPSOM defines how to derive services from BP in a conceptual manner. It identifies the participants involved and the services they provide and request, along with the associated contracts and interfaces, parameters, and the messages exchanged, using the SoaML standard. The automation in BPSOM focuses on the generation of services from BP by means of QVT transformations defined between the SoaML and the BPMN meta-models. We follow the MDA approach based completely on the use of OMG standards. The BPMN BP model constitutes the CIM, and the SoaML service model the PIM, which can then be used to generate code, using MDA engines. The QVT transformations are based on a defined ontology [18] which relates BP models to service models, conceptualizing their elements and relationships. Fig. 6 shows an overview of the relationship between the BP in Fig. 2 and ServicesArchitecture in Fig. 3. The QVT transformations code itself is not shown here, as it can be seen in [15].

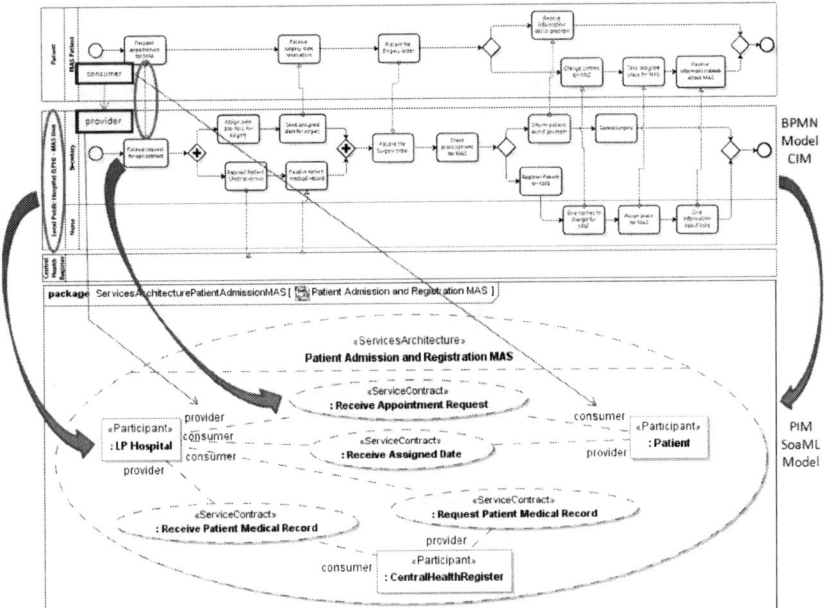

Fig. 6. BPMN to SoaML QVT transformations definitions for the example

We use the Eclipse environment in MINERVA, integrating several plug-ins to support BPSOM definitions, including MediniQVT as the QVT engine. The BPMN BP model is created by business and software people in a BPMN modeler which exports the model. It is then loaded into Eclipse and marked by the Architect with information to apply the QVT transformations, obtaining participants and its ports. To generate the code, the MDA engine needs all the SoaML diagrams, completed later by developers. For the example, we have integrated the MagicDraw Cameo SOA+ and ModelPro Eclipse plug-ins, which can be downloaded from [19] with the BPMN BP and SoaML services models, QVT transformations and input and output XMI files.

6 Related Work

We carried out a systematic review regarding the application of SOC and MDD para-digms to BP, presented in [16]. To the best of our knowledge, there is no other work that relates BPMN models directly to SoaML models the way we do. Regarding the methodological approach, BPSOM has been defined over the same period as other proposals shown in [11][16]. Nevertheless, it is worth mentioning [20], which defines a methodology for service development focusing on WS, the survey of methodologies presented in [21], as well as a consolidated methodology for defining business and software services, the SOMA plug-in for the RUP [22], which, as ours does, adds activities, but to RUP, and Shape [23] ,which also uses SoaML, but with different guides and no generation. For the model driven approach it is worth mentioning [24], which defines guidelines and transformations from one model to other, [25] proposing a method for service composition with a process to model generation, metamodels and artifacts to be obtained, adding in [26] a value model for deriving services using ATL [27].This is also used in [28], in which models, metamodels and transformations are defined, moving from collaborative BP to a SOA model, generating BPEL. Our proposal differs from these in several ways: firstly, BPSOM can be added to any ex-isting base software development process, thus promoting reuse and making it easier to adopt. Secondly, QVT transformations are integrated in the development environ-ment, obtaining the models from which to generate code. Thirdly, the conceptual and automatic guide is fully integrated in BPSOM. Finally, MINERVA framework inte-grates existing standards, promoting standardization of development.

7 Conclusions and Future Work

BPSOM has been defined to guide service development from business processes, integrated into MINERVA framework for continuous BP improvement. Its contribu-tions are as follows: it allows the reuse of existing knowledge in the developing or-ganization, by using the base software development process, adding specific elements for service development. The use of the SoaML standard to model services supports the definition of meaningful elements in specifying services from BP, in both a con-ceptual and an automatic way. Finally, we have defined QVT transformations from the BPMN metamodel to the SoaML metamodel that can be executed in the Eclipse environment, obtaining an initial definition of service models. These QVT transfor-mations were defined for previous versions of BPMN and SoaML, so we are updating and completing them using the BPMN 2.0 and SoaML beta2 standards recently released by OMG. There are few implementations of SoaML, so we are developing our own to show the service models graphically. From these diagrams, code can be generated using existing MDA engines. We are working on case studies at the Ciudad Real General Hospital to validate the proposal.

Acknowledgments. This work has been partially funded by the Agencia Nacional de Investigación e Innovación (ANII,Uruguay), ALTAMIRA project (Junta de Comunidades de Castilla-La Mancha, Spain, F. Soc. Europeo, PII2I09-0106-2463), PEGASO/MAGO project (Ministerio Ciencia e Innovacion MICINN, Spain, FEDER, TIN2009-13718-C02-01) and INGENIOSO project (Junta de Comunidades de Castilla-La Mancha, Spain, PEII11-0025-9533).

References

1. Weske, M.: BPM Concepts, Languages, Architectures. Springer, Heidelberg (2007)
2. Smith, H., Fingar, P.: Business Process Management:The third wave. Meghan-Kieffer, Tampa (2003)
3. Papazoglou, M., Traverso, P., Dustdar, S., Leymann, F.: Service-Oriented Computing: State of the Art and Research Challenge. IEEE Computer Society, Los Alamitos (2003)
4. Krafzig, D., Banke, K., Slama, D.: Enterprise, SOA, Best Practices. Prentice-Hall, Englewood Cliffs (2005)
5. Erl, T.: SOA: Concepts, Technology, and Design. Prentice-Hall, Englewood Cliffs (2005)
6. Mellor, S., Clark, A., Futagami, T.: Model Driven Development. IEEE Comp.Society, Los Alamitos (2003)
7. Object Management Group (OMG), Model Driven Architecture, MDA (2003)
8. Object Management Group (OMG), SOA Modeling Language, SoaML (2009)
9. Delgado, A., Ruiz, F., García-Rodríguez de Guzmán, I., Piattini, M.: MINERVA: Model drIveN and sErvice oRiented framework for the continuous business process improVement and relAted tools. In: Dan, A., Gittler, F., Toumani, F. (eds.) ICSOC/ServiceWave 2009. LNCS, vol. 6275, pp. 456–466. Springer, Heidelberg (2010)
10. Delgado, A.: MINERVA framework (2010), http://alarcos.esi.uclm.es/MINERVA/
11. Delgado, A., Ruiz, F., García - Rodríguez de Guzmán, I., and Piattini, M.: Towards a Service-Oriented and Model-Driven framework with business processes as first-class citizens. In: 2nd International Conference on BP and Services Computing, BPSC 2009 (2009)
12. Object Management Group (OMG), Query/Views/Transformations, QVT (2008)
13. Object Management Group (OMG), Business Process Modeling Notation, BPMN (2009)
14. Graham, I., Henderson-Sellers, B., Younessi, H.: The OPEN Process Specification. ACM Press, Addison-Wesley(1997)
15. Delgado, A., García - Rodríguez de Guzmán, I., Ruiz, F., Piattini, M.: From BPMN business process models to SoaML service models: a transformation-driven approach. In: 2nd Int.Conf. on Software Tech. and Engineering (ICSTE 2010), San Juan de Puerto Rico (October 2010)
16. Delgado, A., Ruiz, F., García-Rodríguez de Guzmán, I., Piattini, M.: Application of service-oriented computing and model-driven development paradigms to BP: a systematic review. In: 5th Int. Conf. on SW and Data Technologies (ICSOFT 2010), Athens (2010)
17. Object Management Group (OMG), Business Motivation Model, BMM (2010)
18. Delgado, A., Ruiz, F., García-Rodríguez de Guzmán, I., Piattini, M.: Towards an ontology for SO modeling supporting BP. In: 4th. Int. Conf. on Research Challenges IS, RCIS 2010 (2010)

19. Delgado, A.: BPSOM methodology example (2010),
    http://alarcos.esi.uclm.es/MINERVA/BPSOM/BPSOMexample.zip
20. Papazoglou, M., van den Heuvel, W.: Service-oriented design and development methodology. Int. J. Web Engineering and Technology 2(4), 412–462 (2006)
21. Kohlborn, T., Korthaus, A., Chan, T., Rosemann, M.: Identification and Analysis of Business and SE Services- A Consolidated Approach. IEEE Transactions on Services Comp. (2009)
22. IBM-SOMA, http://www.ibm.com/developerworks/rational/
    downloads/06/rmc_soma/
23. Stollberg, M., et al.: A Customizable Methodology for the MDE of Service-based System Landscapes. In: 4th Workshop on Modeling, Design, and Analysis for the Service Cloud (MDA4ServiceCloud 2010), with ECMFA 2010, Paris (June 2010)
24. Herold, S., Rausch, A., Bosl, A., Ebell, J., Linsmeier, C., Peters, D.: A Seamless Modeling Approach for service-oriented IS. In: 5th Int. Conf. on IT:New Generations, ITNG 2008 (2008)
25. de Castro, V., Marcos, E., López Sanz, M.: A model driven method for service composition modelling: a case study. Int. J. Web Engineering and Technology 2(4) (2006)
26. de Castro, V., Vara Mesa, J.M., Herrmann, E., Marcos, E.: A Model Driven Approach for the Alignment of Business and Information Systems Models (2008)
27. Jouault, F., Kurtev, I.: Transforming models with ATL (ATLAS Transformation Language). In: Bruel, J.-M. (ed.) MoDELS 2005. LNCS, vol. 3844, pp. 128–138. Springer, Heidelberg (2006)
28. Touzi, J., Benaben, F., Pingaud, H., Lorré, J.P.: A model-driven approach for collaborative service-oriented architecture design. Int. Journal of Prod. Economics 121(1) (2009)

Panel on Green and Sustainable IS

Barbara Pernici

Politecnico di Milano, piazza Leonardo da Vinci 32, 20133 Milano, Italy
barbara.pernici@polimi.it

Abstract. The panel on Green and Sustainable Information Systems has the goal of providing a forum for discussing research issues on this topic within the Information Systems Engineering research community. Information systems, for their pervasive nature in most of human activities that have an IT support, can give a contribution to improve the environmental impact of IT in two main directions: on one side, information systems can provide a support to improve awareness and to control energy efficiency in a variety of areas, such as for instance smart cities and smart buildings, traffic control, and utility management, on the other hand, information systems themselves use computing resources and related facilities in data center and offices, and therefore they have an impact on the environment. While a great emphasis has been given to produce hardware equipment that is energy efficient, only recently the theme of considering energy efficiency in connection to information systems and their design has emerged. The panel will discuss the different types of green and sustainable information systems and will discuss emerging research topics and possible future research developments and goals.

H. Mouratidis and C. Rolland (Eds.): CAiSE 2011, LNCS 6741, p. 681, 2011.

Author Index